DIGITAL CONTROL SYSTEMS
Theory, Hardware, Software

McGraw-Hill Series in Electrical Engineering

Consulting Editor
Stephen W. Director, Carnegie-Mellon University

Networks and Systems
Communications and Information Theory
Control Theory
Electronics and Electronic Circuits
Power and Energy
Electromagnetics
Computer Engineering
Introductory and Survey
Radio, Television, Radar, and Antennas

Previous Consulting Editors

Ronald M. Bracewell, Colin Cherry, James F. Gibbons, Willis W. Harman. Hubert Heffner, Edward W. Herold, John G. Linvill, Simon Ramo, Ronald A. Rohrer, Anthony E. Siegman, Charles Susskind, Frederick E. Terman, John G. Truxal, Ernst Weber, and John R. Whinnery

Control Theory

Consulting Editor
Stephen W. Director, Carnegie-Mellon University

Athans and Falb: *Optimal Controls*
Auslander, Rabins, and Takahashi: *Introducing Systems and Control*
D'Azzo and Houpis: *Feedback Control System Analysis and Synthesis*
D'Azzo and Houpis: *Linear Control System Analysis and Design: Conventional and Modern*
Emanuel and Leff: *Introduction to Feedback Control Systems*
Eveleigh: *Introduction to Control Systems Design*
Houpis and Lamont: *Digital Control Systems: Theory, Hardware, Software*
Meditch: *Stochastic Optimal Linear Estimation and Control*
Raven: *Automatic Control Engineering*
Schultz and Melsa: *State Functions and Linear Control Systems*

DIGITAL CONTROL SYSTEMS

Theory, Hardware, Software

Constantine H. Houpis, Ph.D.
Gary B. Lamont, Ph.D.

Professors of Electrical Engineering
School of Engineering
Air Force Institute of Technology
Wright-Patterson Air Force Base, Ohio

McGraw-Hill Book Company

New York St. Louis San Francisco Auckland Bogotá Hamburg
Johannesburg London Madrid Mexico Montreal New Delhi
Panama Paris São Paulo Singapore Sydney Tokyo Toronto

This book was set in Times Roman by Eta Services Ltd.
The editors were Sanjeev Rao and David A. Damstra;
the production supervisor was Diane Renda.
The drawings were done by Wellington Studios Ltd.
The cover was designed by Scott Chelius.
Halliday Lithograph Corporation was printer and binder.

DIGITAL CONTROL SYSTEMS
Theory, Hardware, Software

1234567890HALHAL8987654

ISBN 0-07-030480-7

Library of Congress Cataloging in Publication Data

Houpis, Constantine H.
Digital control systems—theory, hardware, software.

Bibliography: p.
1. Digital control systems. I. Lamont, Gary B.
II. Title.
TJ223.M53H68 1985 629.8′95 84-9644
ISBN 0-07-030480-7

To our wives,
Mary S. Houpis and Dolores Lamont,
and to our children,
for their encouragement,
support, and understanding,
without which this book
could have never been written.

CONTENTS

PREFACE

With the advent of the small computer and associated microelectronics, there has emerged a need for a fundamental textbook on sampled-data control theory that emphasizes the use of the digital computer as a controller, i.e., a digital control system. It is important that the reader realize that the implementation of a compensator by a digital computer allows design flexibility and system extendability in an efficient and economical manner. Analog components, on the other hand, are rigid in their realization and do not provide flexibility. Moreover, the real-time adaptability of mode changes (i.e., changing the desired compensator characteristics in real time, e.g., process control) due to plant parameter variations, environmental changes, and requirements modifications can not generally be done readily with a simple analog compensator. There are texts on the market that partially meet the need but essentially restrict themselves to theory. What is needed today is a complete text that merges and interrelates the two general areas which are vital to a practicing digital control engineer: sampled-data control theory and computer engineering.

Essential to this type of integrated text are the design values of sampling times that emphasize to the reader the accuracy required in performing digital control analysis and synthesis. Therefore this textbook provides a clear, understandable, logical development and motivated account that spans sampled-data control theory with computer engineering as an integrated entity. The text also provides digital control system background, analysis techniques, synthesis approaches, and implementation considerations. Computer-aided-design (CAD) packages for sampled-data control systems have been used throughout this text to represent contemporary development environments.

The minimum background required for this textbook is a fundamental course in continuous-time control systems (Chapters 1 to 11 of Reference 1). No background in computer or electrical engineering is required. Some higher-order language programming (FORTRAN, BASIC, PASCAL, etc.) would be useful but not necessary in appreciating the software engineering sections.

The authors have tried to exert meticulous care with explanations, diagrams, calculations, tables, and symbols. They have also tried to ensure that the student is made aware that rigor is necessary for advanced control work. The text provides a strong, comprehensive, and illuminating account of those elements of conventional control theory which have relevance in the design and analysis of sampled-data control systems. The variety of different techniques presented contributes to the development of the student's working understanding of what A. T. Fuller has called "the enigmatic control system." To provide a coherent development of the subject, an attempt is made to eschew formal proofs and lemmas with an organization that draws the perceptive student steadily and surely into the demanding theory of multirate multivariable control systems. It is the opinion of the authors that a student who has reached this point is fully equipped to undertake with confidence the challenges presented by more advanced digital control theories. The importance and necessity of making extensive use of comprehensive computer design packages is also emphasized.

A concise but integrated presentation of the fundamentals of computer engineering is set forth in Chapter 2, which provides an introduction to combinational logic, registers, number representation, and programming for use in digital-controller interfaces and process control logic.

The sampling process is discussed in Chapter 1 while the development of ideal impulse sampling is presented in Chapter 3. A discussion of Shannon's sampling theorem is included to help determine the minimal sampling rate. The establishment of linear difference equations to describe the performance of sampled-data control systems is set forth in this chapter. This discussion is followed by the development of the weighting sequence model for sampled-data systems. A technique for continuous-time signal reconstruction from a sampled signal by means of a data-hold device is also given.

The first portion of Chapter 4 introduces the $\mathscr{Z}$ transform (zee transform) as a method for the analysis and design of sampled-data control systems. The correlation between the pole-zero pattern in the s and the z planes is presented with respect to time-response characteristics. $\mathscr{Z}$-transform theorems similar to those for the Laplace transform are included which simplify and facilitate the application of the $\mathscr{Z}$-transformation method of analyzing system performance. The properties and mathematical representations of open-loop and closed-loop sampled-data control systems are developed, and their corresponding block diagram representations are given. This is followed by the inverse $\mathscr{Z}$-transform, $\mathscr{Z}^{-1}$, method which permits the time-response characteristics of the system to be obtained at the sampling instants $t = kT$. In order to determine additional time-response data between sampling instants, the modified $\mathscr{Z}$-transform, $\mathscr{Z}_m$, method and the corresponding theorems are presented. The digital-controller transfer function and its implementation is discussed in this chapter and further amplified in Chapter 10.

The analysis and design techniques of sampled-data control systems that are stressed in this text are based upon the conventional control-theory techniques used for continuous-time control systems. Thus Chapter 5 presents a capsule of these techniques that are covered in chapters 4 to 11 of Reference 1. This chapter reintroduces the reader to modeling of a desired system control ratio. A detailed analysis and design of a disturbance rejection control system, because of its importance in many practical control systems, is discussed in Section 5-7.

Chapter 6 discusses in detail the analysis of the basic (uncompensated) sampled-data control system. The first item of great importance is: "Is the system stable or for what range of gain and/or sampling time can the system have a stable performance?" A stability analysis is presented in the z domain by applying Jury's stability test or in the w or w' plane by applying Routh's stability criterion. The system's steady-state characteristics are treated in the identical fashion as for continuous-time systems. This analysis includes the relationship between system type and the ability of the system to follow polynomial inputs. Further, this chapter ties together the s-, z-, and w'-plane analyses, with respect to time-response characteristics, by means of the root-locus and frequency-response methods. The effect of sampling time T on the correlation of results from the s- and w'-plane analyses with the results from the z-plane analysis is discussed thoroughly. The very important effect of decreasing T on the design accuracy (number of places required) on the system performance and digital implementation is stressed.

In Chapter 7 a sampled-data control-system approximation design technique is developed. The Padé approximation, the Tustin transformation, and the pseudo-continuous-time (PCT) control system are involved in this technique. The technique relies on doing the analysis and design in the s plane when the discrete approximations are valid.

The basic organization of analog-to-digital (A/D) and digital-to-analog (D/A) converters and I/O programming is presented in Chapter 8 in a logical manner which permits the detailed understanding of input-output in a digital control system. General control transducers are also presented, providing insight to device construction and accuracy.

Chapter 9 provides the foundation for statistical analysis of finite word length discussed in Chapter 10. Presented are the fundamentals of continuous and discrete random variables and random processes. Using this statistical background, Chapter 10 focuses on the effects of finite word length on A/D and D/A conversion, binary number implementation in the computer, addition and multiplication operations, pole-zero shifts, and limit cycles due to quantization.

Chapter 11 discusses in detail all three approaches of analyzing and designing a sampled-data cascade-compensated control system: in the direct (DIR) techniques all work is done in the z plane and for the digitization (DIG) technique all work is accomplished in either the s or the w' plane. The degree of accuracy of the system design based upon the approximations of Chapter 7, the interrelation and comparison of the results by all methods, and the effect of T on these results are thoroughly discussed. The PID controller is also discussed. This chapter also presents pertinent components of digital signal processing such as frequency filter design using analog design techniques and a mean-squared error approach. This chapter is interrelated with Chapters 2, 8 to 10, and 12 to 15 with respect to the digital implementation of the controller.

Feedback-compensated systems are discussed in Chapter 12. The first part deals with the tracking problem, i.e., the desire for the system output to follow the system input. The second part of this chapter deals with disturbance rejection, i.e., the desire for the system output to not follow the system input.

The next two chapters (13 and 14) deal again with computer engineering. The tools and methods are discussed in Chapter 13 for defining digital control software requirements, software design, implementation, and testing. Real-time operating principles are presented in Chapter 14 along with various example organizations permitting the student to grasp the critical components for a distributed digital-control-system implementation, a real-time system.

Chapter 15 is an introduction to the state-variable methods of representing a sampled-data control system. It includes the analysis of system performance by state-space representation. The chapter concludes with a design method for minimizing the effect of plant parameter variations on the system output. This chapter provides a foundation for modern digital-control-system studies.

The two important and major features of this text are uniqueness and flexibility of use. The uniqueness features are integration of sampled-data control theory with computer engineering; degree of accuracy needed as T becomes smaller, i.e., its impact on calculations and digital implementation; computer-aided design (CAD) is stressed; text examples are discussed by both points of view, i.e., control theory and digital computer implementation in the appropriate chapters; extensive development of system analysis and design by root-locus techniques; elaborates more than other texts on the interrelationships between the s, z, and w' planes, i.e., time- and frequency-domain correlation (analysis and design); development of the pseudo-continuous-time (PCT) control system design and analysis in s and w' planes is developed; and it emphasizes control law (algorithm) implementation in a digital computer. The text may be used in advanced undergraduate and first-year graduate courses (two quarters or two semesters in length); for a short course (40 lecture hours); as a self-study text; and for a single course restricted to only sampled-data control theory, i.e., Chapters 1, 3 to 7, 11, 12, and 15.

Thus, the authors feel that with the mastery of the text material, the student should be able to analyze and design [single-input–single-output (SISO), uniform rate] sampled-data control systems and implement a digital controller with the use of CAD. It is felt that the goals of achieving an integrated sampled-data theory and computer engineering text have been accomplished.

The authors express their thanks to the students who have used this book and to the faculty who have reviewed it for their helpful comments and corrections. Appreciation is expressed to Dr. R. E. Fontana, Head of the Electrical Engineering Department, Air Force Institute of Technology, for the encouragement he has given. The continual encouragement and review of the text by Dr. T. J. Higgins, Professor Emeritus, University of Wisconsin, has been a very important catalyst in the completion of this text. Special appreciation is expressed to Dr. Donald

McLean, Senior Lecturer, University of Technology, Loughborough, England, formerly a visiting Professor at the Air Force Institute of Technology, who provided a detailed review of the complete book. His perception and insight have contributed extensively to the clarity and rigor of the presentation. Our association with him has been an enlightening and refreshing experience.

Constantine H. Houpis
Gary B. Lamont

CHAPTER

ONE

INTRODUCTION

1-1 INTRODUCTION

These past four decades have witnessed the firm establishment of conventional and modern control theory for continuous-time control systems.[1,99] Not only has this theory revolutionized industrial processes, but it has enabled humanity to initiate the exploration of the universe. Although sampled-data or discrete-time control theory has been evolving for three decades, its application to the development of practical systems has been slow due to inherent theoretical and physical implementation problems.

Engineers and scientists attempt to design control systems to perfection, i.e., to try to achieve ideal system performance. For a practical control system, physical realizability of components, to a great extent, limits to what extent the ideal system performance can be achieved. The advent of the digital computer as a computational device permits more accurate control in general but also constrains the speed of operation. However, this accuracy has proven to be a critical element in the success of modern space exploration. The advent of the microprocessor and minicomputer and their use as a control element has provided the impetus, not only to enhance the theoretical analysis and synthesis techniques for these systems but to enable control-system designers to move closer to their goal of "ideal system performance."

Since the computer plays a very important role in the implementation of a digital control system, a short discussion of its continuing evolution is important. The first computational devices were fingers, stones, and sticks, with the human mind processing part of the desired computational algorithm. Human beings historically and even today desire faster, more accurate, and repeatable calculations. Thus from the abacus evolved mechanical adders and card storage devices. Extended arithmetic operations and large storage devices exist in this century due to electronic developments (tubes, transistors, integrated circuits). This electronic evolution permits relatively faster operations along with less operational energy required. This evolution continues to offer smaller, more accurate, faster, and more economical computers. The microprocessor is an excellent example. The methods for software or computer program generation have also improved in the last three decades. Concepts associated with structure programming and software engineering are important in realizing any algorithm, especially digital control algorithms. Proper use of these methods permits better structures for understanding, faster testing and evaluation, and economical modifications and maintenance. The structure of the data flow and instruction processing in a real-time digital control system influences performance considerably and must be an integral element of any analysis.

In the digital control field, real-time processing is critical to successful implementation. Thus small and fast microprocessor systems can be used in many contemporary control systems if proper hardware and software design and implementation techniques are employed. The purpose of this text is to present an extensive discussion of digital-control-system terminology, sampled-data control-system analysis and synthesis, and practical implementation techniques and considerations from a software and hardware point of view. Various aspects of a general digital control system are discussed in this text from various levels of observation (theory, hardware, software). The three simplest control-system configurations or architectures are shown in Fig. 1-1. Figure 1-1*a* is an open-loop control system representing many industrial process structures. The other two closed-loop configurations represent the most commonly used control systems where the performance specifications are more restrictive. An introductory background in classical continuous control theory is a necessary prerequisite for a full understanding of the control-theory portion of this text. However, important extensions to digital control theory from the continuous theory are explained.

1-2 DIGITAL-CONTROL-SYSTEM MODELING

A digital-control-system model can be viewed from many different levels, i.e., control law (algorithm), computer program, conversion between analog and digital signal domains, and system performance. One of the most important aspects leading to the understanding of digital control systems is the sampling process level, which is introduced in this section. The associated system terminology, which is very important in understanding digital control concepts, is also presented.

1-2A A Sampling Process[1,2]

In continuous control systems, all system variables are continuous signals as represented by Fig. 1-2*a*. That is, whether the system is linear or nonlinear, all

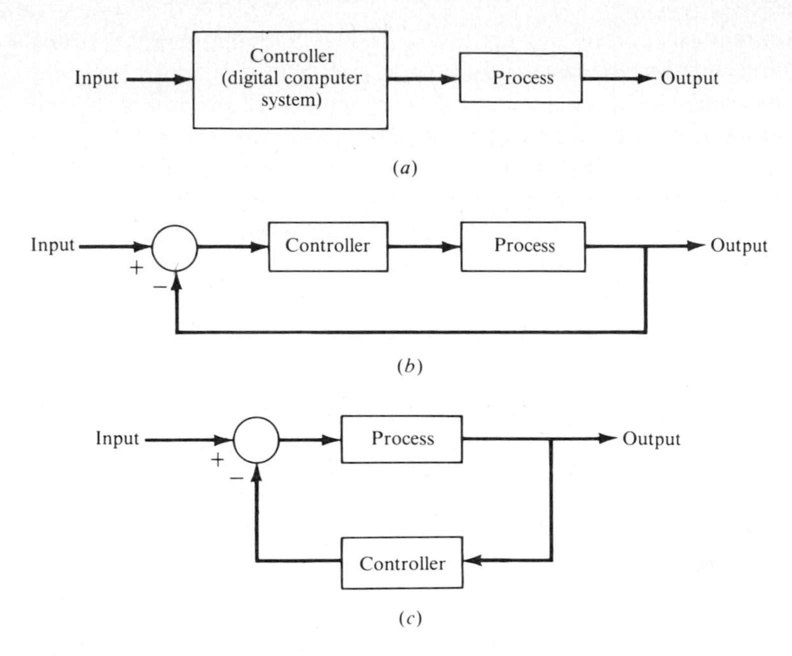

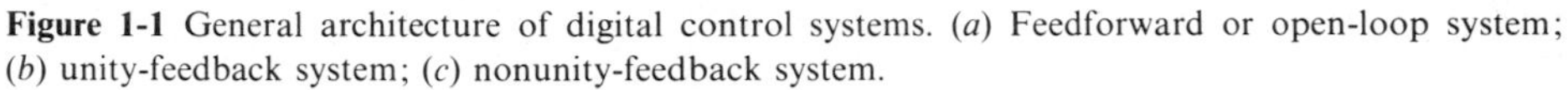

Figure 1-1 General architecture of digital control systems. (*a*) Feedforward or open-loop system; (*b*) unity-feedback system; (*c*) nonunity-feedback system.

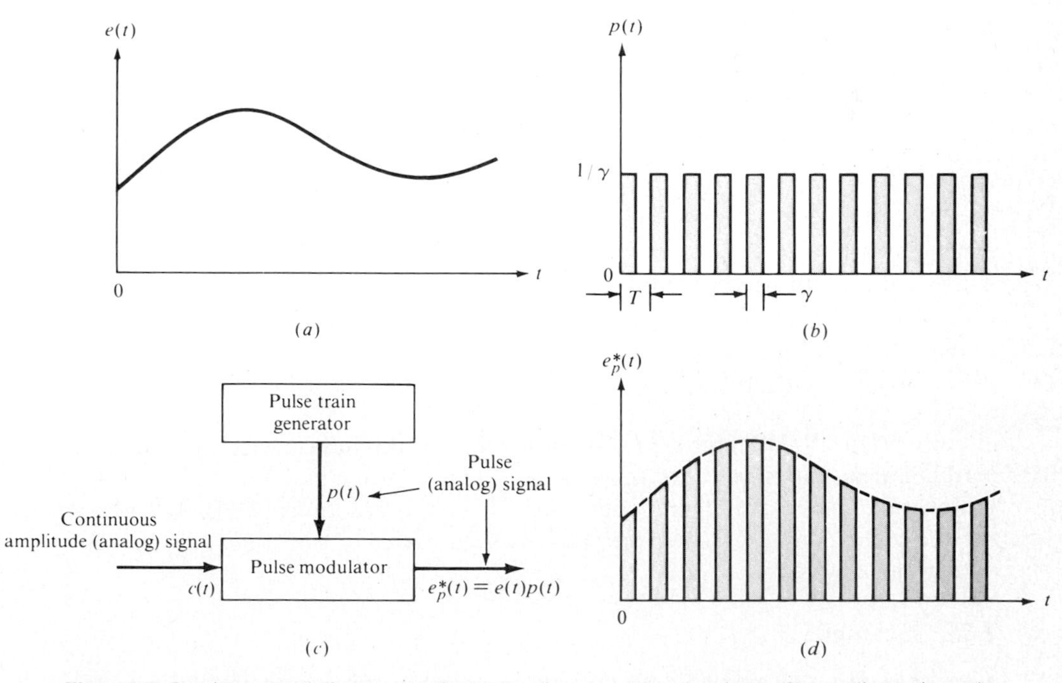

Figure 1-2 Continuous and discrete signals. (*a*) Continuous analog signal $e(t)$; (*b*) sampling pulse train $p(t)$; (*c*) sampling device; (*d*) sampled function.

variables are continuously present and are therefore known at all times. (This text deals only with linear or linearized systems.) Another category of control systems is one in which one signal (or more) is sampled in the manner shown in Fig. 1-2*c* so that it appears as a pulse train of varying amplitudes, as shown in Fig. 1-2*d*. That is, the pulse train of Fig. 1-2*b* is modulated with the continuous-time signal of Fig. 1-2*a* to yield the sampled function of Fig. 1-2*d*. Such sampling may be an inherent characteristic of the system; i.e., a radar tracking system supplies information on an aerospace vehicle's position at discrete periods of time. This information is therefore available at a succession of time intervals as data levels.

1-2B System Terminology

It should be noted that in the control literature the terms digital systems, discrete-data systems, digital control systems, sampled-data systems, and discrete-time systems have been and are being used interchangeably. In the early development of this technology, an analog system not containing a digital device in which some of the signals were sampled (pulse-, amplitude-modulated) was referred to as a *sampled-data system*. With the advent of the digital computer, the term *discrete-time system* denoted a system in which all its signals were in a digital coded form (digitized). Most practical systems today are of a hybrid nature, i.e., contain both analog and digital components. In this text the term *sampled-data control system* is used to describe a system that contains at least one sampled signal. Digital computers are available for performing the computations necessary in a complex control system of this type. Since a digital computer must operate with discrete numbers, it is necessary to first convert a sampled signal to a digital coded form through the use of an analog-to-digital (A/D) converter. A digital-to-analog (D/A) device transforms the computer's digitized control value for input to the continuous plant. Systems in which the digital computer is utilized as a control device is referred to as a *digital control system*. If the digital computer interfaces directly with the plant (or process), the system is referred to as a *direct digital control* (DDC) system.

In practice, the output of the pulse modulator of Fig. 1-2*c* is generally fed into a data-hold device (a device that converts a discrete signal into a continuous signal), as shown in Fig. 1-3*a*. The simplified representation of the sampling and hold devices is shown in Fig. 1-3*b* where the "ideal sampler" represents the unit impulse train of Fig. 1-3*c*, and the output of the sampler is the amplitude-varying impulse train $e^*(t)$ of Fig. 1-3*d*. The utilization of the hold device simplifies the mathematical analysis of the sampled signal. This in turn makes the analysis of the sampled-data control system model easier. Thus the structures of Fig. 1-1, where the controller contains the sampler and the hold device of Fig. 1-3*a*, are sampled-data control systems.

1-2C Examples

Two examples of sampled-data control systems are presented to illustrate the sampling process.

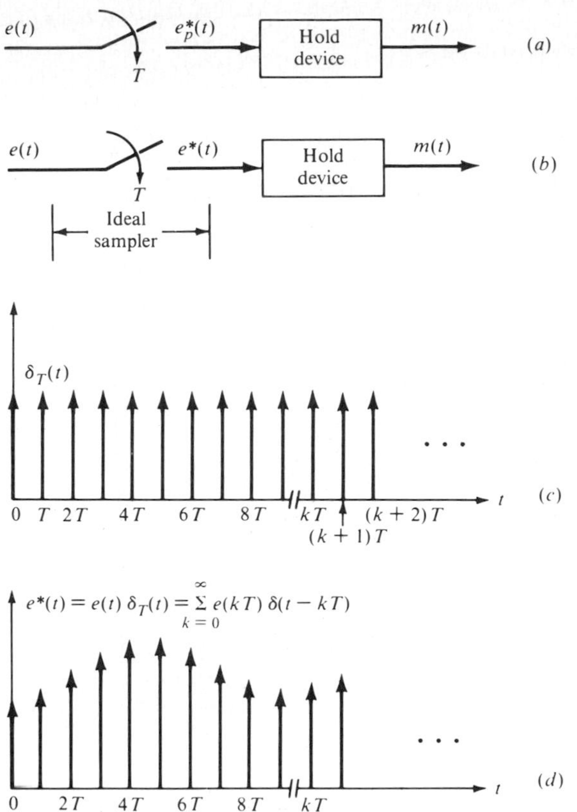

Figure 1-3 Sampling process. (*a*) Pulse-sampling and data-hold devices; (*b*) simplified representation of (*a*); (*c*) ideal sampler representation: (*d*) output of ideal sampler.

The first sampled-data control system to exist is that of a human being walking, which can be illustrated in a modern setting of driving a vehicle. The second example, also in a modern setting, is of a guided missile control system. Both examples typify a digital control system.

Example 1-1: Driving a vehicle This example illustrates the beauty of the human being operating as a digital control system (Fig. 1-4). Without this capability the human would not be able to function and for all practical purposes to exist. The main features of this system are:

Continuous and sampled input signals:

Rearview mirror
Speed limit sign (desired speed)
'S' curve sign
Radar trap
Billboard
Desired path
Condition of road

Figure 1-4 Example of a sampled-data control system: driving a vehicle.

Continuous and sampled output signals:
Actual direction of vehicle
Vehicle speed (speedometer)

The sampler: The human eyes sample the input and output signals either individually or in combinations thereof at a variable rate. This information is transmitted to the human brain.

Information processor: The human brain.

The plant or controlled devices:
Hands
Arms
Legs

In this example, the human head is being modeled as an analog-to-digital (A/D) converter, a digital controller, a digital-to-analog (D/A) converter, and as hold devices. A hold device models the function of constructing a continuous signal from a discrete output-level signal. The continuous signal outputs of the "hold devices" are transmitted to the arms and legs of the vehicle driver. The comparison function of the brain provides a signal to the arms and hands to maintain the actual direction of the vehicle in line with the desired path. In a similar fashion, the signal to the arms and legs maintains the desired speed. This example represents a *multiple-input–multiple-output* (MIMO) control system.

Example 1-2: Command guidance interceptor system[1] Another complex system is a command guidance system (Figs. 1-5 and 1-6) which directs the flight of a missile in space in order to intercept a moving enemy target. The defense uses the missile with the objective of intercepting and destroying the bomber before it launches its bombs. A sketch of a generalized command guidance interceptor system is shown in Fig. 1-5. The target-tracking radar is used first for detection and then for tracking the target. It supplies discrete information for determining target range and angle and their rates of change (time derivatives). This information is continuously fed into the computer, which calculates a predicted course for the target. The missile-tracking radar supplies similar discrete information, which is used by the computer to determine its flight path. The computer compares the two flight paths and determines the necessary change in missile flight path to produce a collision course. The necessary flight-path changes are supplied to the radio command link, which transmits this discrete information to the missile. This electrical information containing error corrections in the flight path is used by a control system in the missile. The missile control system converts the error signals to mechanical displacements of the missile airframe control surfaces by means of actuators. This missile responds to the positions of the aerodynamic control surfaces to follow the prescribed flight path, which is intended to produce a collision with the target. Monitoring of the target is continuous so that changes in the missile course can be corrected up to the point of impact. A

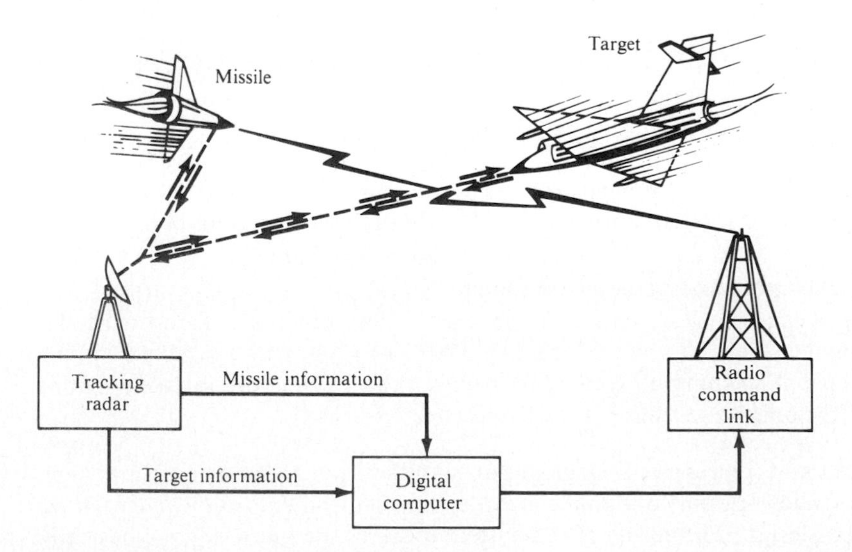

Figure 1-5 Command guidance interceptor system.

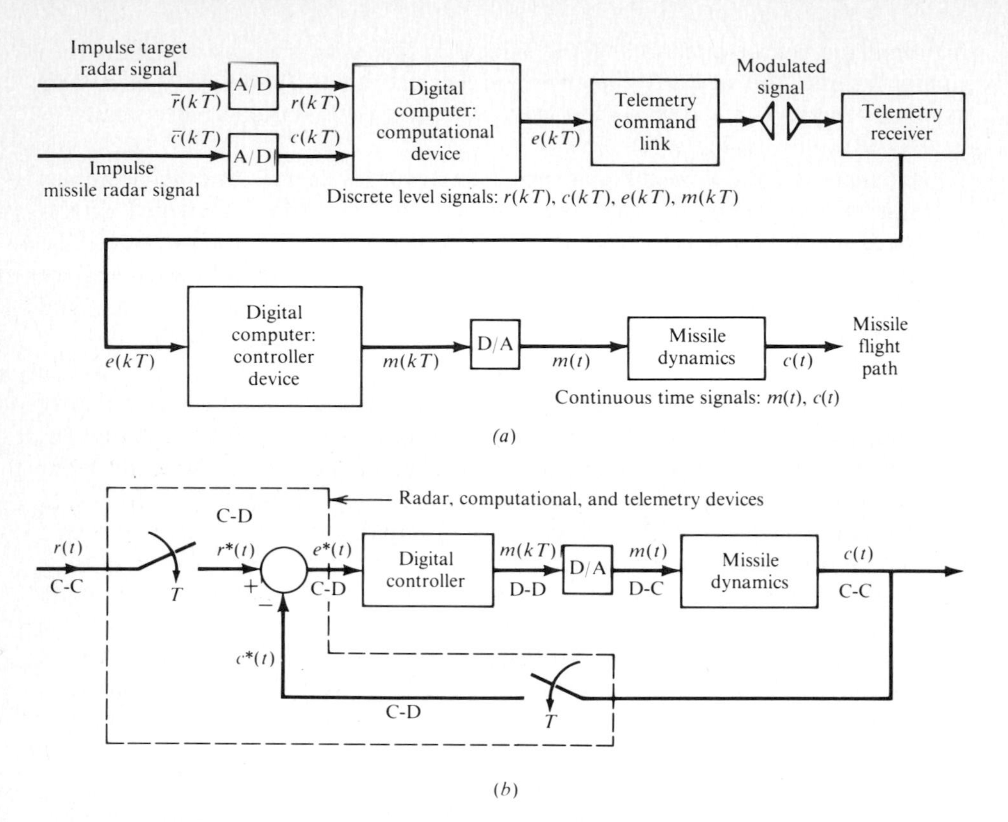

Figure 1-6 A generalized command guidance interceptor system. (*a*) Detailed block diagram representation; (*b*) simplified block diagram representation.

block diagram depicting the functions of this command guidance, neglecting time delays is shown in Fig. 1-6. The signal $e(kT)$ is the computed value of $r(kT) - c(kT)$. The main features of this system are:

Continuous and sampled input signal: Target position: $r(t)$ and $r(kT)$.
Continuous and sampled output signal: Missile position: $c(t)$ and $c(kT)$.
The sampler, hold, and A/D elements: Radar tracker unit.
Information processor: Digital computer.
Digital controller, D/A, and hold elements. These devices are on-board the missile.
Telemetry. The communication link between the ground and missile portions of the control system.

This example represents a single-input–single-output (SISO) control system. Also, it illustrates not only a digital control system but the *interdisciplinary nature* of the system design. The design of this system requires the knowledge of basically five areas: radar, communications, digital signal processing, computer engineering, and control theory. The following sections generalize the previous examples to develop a general-purpose digital-control-system model.

1-2D General Sampled-Data-System Variables

The variables of a sampled-data system can be described in terms of their time and amplitude characteristics. As an example, the general variables of the system shown in Fig. 1-6*b* can be grouped into four general categories: discrete amplitude–discrete time (D-D), discrete amplitude–continuous time (D-C), continuous amplitude–discrete time (C-D), and continuous amplitude–continuous time (C-C). *Note that the first letter refers to the amplitude characteristic and the second letter refers to the time characteristic.* Table 1-1 summarizes these categories.

A general example of the classification of sampled-data control systems is shown in Fig. 1-7. Here the continuous (amplitude and time) input signal $e(t)$ is impulse-sampled, generating a continuous-amplitude–discrete-time signal by definition. The hold circuit of Fig. 1-7 generates a piecewise-continuous step function ("staircase"). Note that the sampled data $e^*(t)$ can be *any* value (an infinite number of values) within some predefined range of amplitude values.

Figure 1-8 presents the various signal classifications in a digital-computer control system. Here again the impulse sampler generates a continuous-amplitude–discrete-time signal. The output $e_D(kT)$ of the A/D quantizes $e^*(t)$ such that it can be only one of a finite number of values within a specific value range. This phenomenon is due to the finite word length of the computer. The digital computer itself manipulates the quantized value (a base 2 number) into another discrete-amplitude–discrete-time signal $f(kT)$. Finally the D/A transforms $f(kT)$ into a discrete-amplitude–continuous-time signal $m(t)$. Note the $m(t)$ in this case as compared to the output of the hold device in Fig. 1-7 has a discrete-amplitude characteristic *because of the quantization of the A/D* which *is not present* in Fig. 1-7. The D/A however does include a hold device and is modeled in Fig. 1-9. The "fictitious" sampler in this *D/A model* is needed to represent the fact that the input value $e^*(t)$ is immediately converted (A/D) and manipulated by the digital computer, resulting in a value at the hold device input during the same sampling instant. That is, there is no computational time delay τ_D in the computer. This instantaneous computation is, of course, impossible. However, if the computational time delay is very much less than the sampling time T, then the assumption

Table 1-1 The nature of sampled-data-system variables

	Time	
Amplitude	Discrete, D	Continuous, C
Discrete, D	D-D	D-C
Continuous, C	C-D	C-C

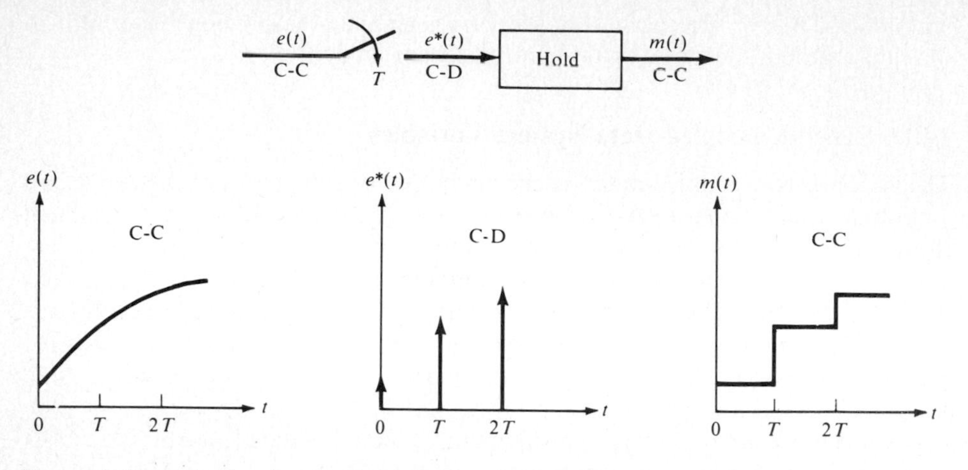

Figure 1-7 Example of analog signal classifications in a sampled-data system (see Table 1-1).

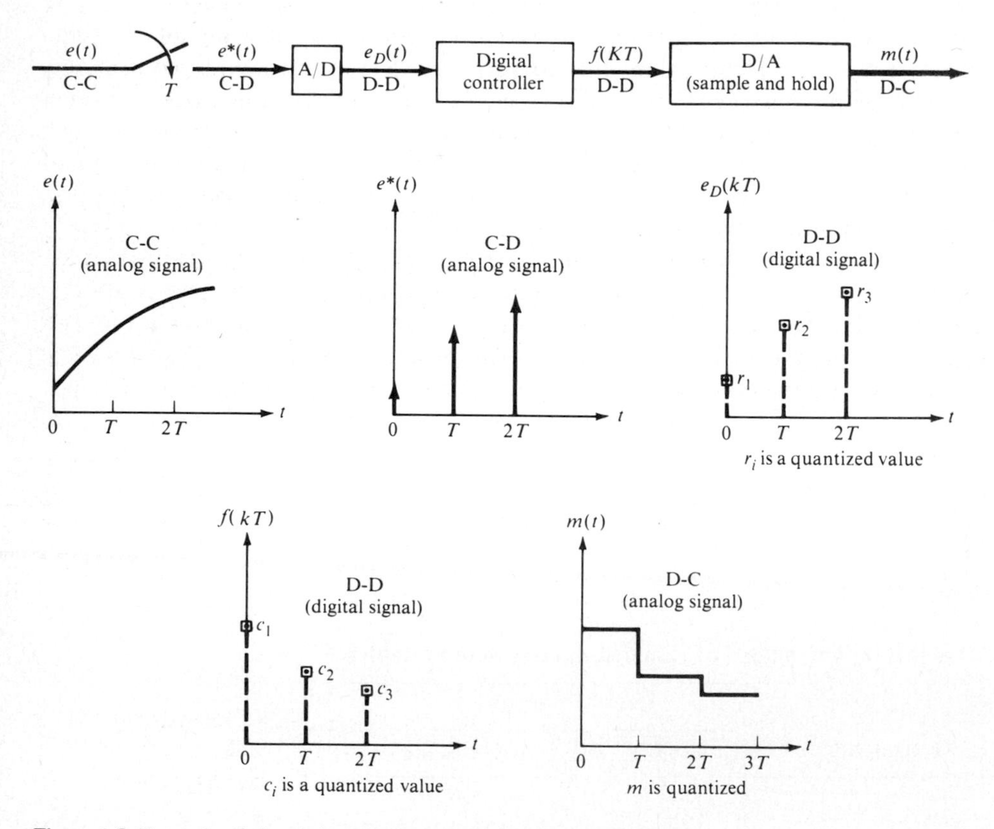

Figure 1-8 Example of signal classifications in a digital control system.

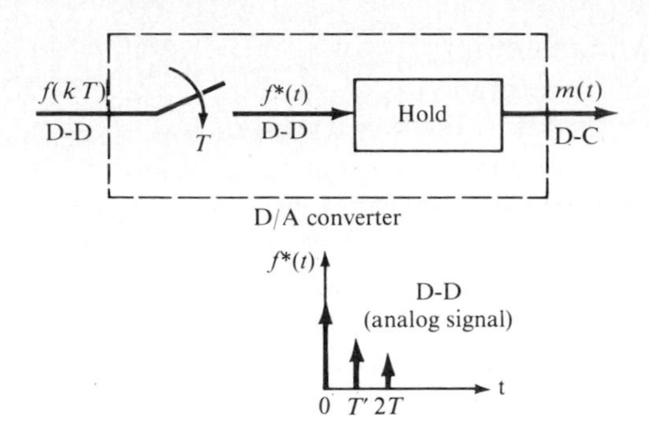

Figure 1-9 D/A converter with a hold device.

that $\tau_D \approx 0$ is appropriate and generates a simplified analytical model. This model is used in the theoretical development of later chapters. However, Chap. 8 discusses in more detail the computational time-delay phenomenon.

1-2E General Modeling

The generation of a plant model for which a controller is realized as a digital information processing device is of primary importance. Generally in an academic environment, the plant model is given and is usually in the form of a linear system with constant coefficients. This type of model is usually easy to analyze with proper techniques. In the real world, most models are nonlinear in nature and are time-varying due to the physics of the system and its environment. With various assumptions (which may not be appropriate) the nonlinear system is linearized and analyzed, and a control law is generated to meet given specifications. Testing is then initiated to validate proper performance and the system model.

Many control problems turn out to be directly related to the use of an incorrect plant model. *The development of the original plant model is probably the most important and probably the most difficult aspect of control engineering.* The application of the various analysis and design techniques usually proceeds in a straightforward manner after the model is obtained.

1-3 WHY USE DIGITAL CONTROL[2]

The choice of designing a continuous-time or a digital control system for a given application may be based upon the knowledge of the following advantages and disadvantages for each approach (this list is not intended to be all-inclusive).

1-3A Advantages

1. *Improved sensitivity*. Permits the use of sensitive control elements with relatively low-energy signals. Thus tremendous amounts of power can be controlled by relatively low-power-level signals. The sampling operation results in very little loading of these instruments.
2. *Digital transducers*. A transducer is a device that takes energy in one form and

transforms it; i.e., it converts energy from one form to another. For example, a dc tachometer (a separately excited dc generator) is a transducer often used in continuous control systems that transforms the input velocity (radians per second) at one level of energy to an output voltage (volts) at another (or identical) level of energy. In other words, a transducer transmits information from one level of energy and signal dimensions to another level of energy having the same or different signal dimensions. A digital transducer is a coupling device that transforms data into some form of a discrete code, usually binary numbers. The advantages of using a digital transducer is the relative immunity of its digital signals to distortion by noise and nonlinearities and its relatively high accuracy and resolution as compared to analog transducers.

3. *Digital coded signals.* The employment of discrete or digital signals provides for the design and development of complex and sophisticated control systems. This phenomenon results from the ability of a digital control system to store discrete information for long time intervals, to process complicated algorithms, and to transmit discrete information with high accuracy. Moreover, some contemporary small digital computers provide very high-speed filter or compensation processing.
4. *System design.* For some control-system applications, better system performance may be achieved by a sampled-data control system design over a continuous-data control system design. For example, one method of counteracting the effect of transport lag, which decreases the degree of stability of the system, is to insert a pure differentiating unit in the forward loop of the control system. For a continuous system, the differentiator, an active device, not only may enhance the existing noise problem but may also be a source of additional noise. In a sampled-data control system the differentiating action can be achieved by a digital computer without increasing the system noise problem.
5. *Telemetry.* Requires only one communication channel to transmit discrete control signals for more than one control system (multiplexing).
6. *Control systems with inherent sampling.* A radar tracking system sends out and receives signals that are in the form of pulses. The radar scanner unit effectively acts as a "sampling switch" that converts both the azimuth and elevation continuous-time data signals into discrete-time data signals.
7. *Digital computers.* Digital processing is playing a very important role in the design and development of modern control systems. The required processing is usually accomplished in economically small digital systems such as *minicomputers* and *microprocessors. These low-cost, low-weight, and low-power computers* provide the required continuous plant *compensation or filtering.* In addition, these machines *can perform sophisticated algorithms* such as high-dimensional state-variable manipulations, Kalman filtering, stochastic system control, and adaptive mode control. The utilization of the computer as a controller has provided flexibility and versatility in the design of control systems. The technology of digital information processor machines continues to improve, and thus future systems will rely on them for improved control-system performance.

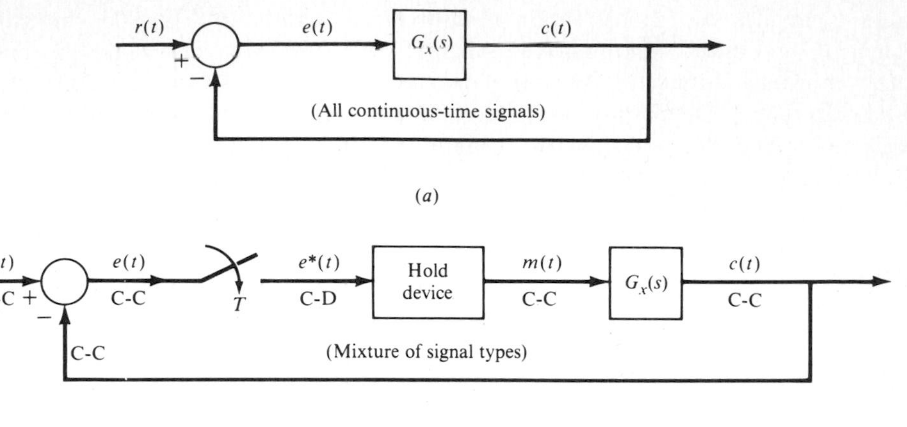

Figure 1-10 Control systems. (*a*) Continuous; (*b*) sampled.

1-3B Disadvantages

1. *System design.* The *mathematical analysis* and design of a sampled-data control system is *sometimes more complex and more tedious* as compared to continuous-data control systems.
2. *System stability.* In general, *converting a given continuous-data control system* (Fig. 1-10*a*) *into a sampled-data control* system (Fig. 1-10*b*) without changing any system parameter, except for the addition of the required hold device, degrades the system stability margin.
3. *Signal information.* The purpose of the hold device in Fig. 1-10*b* is to reconstruct the continuous signal $e(t)$ from the discrete signal $e^*(t)$. The best that can be achieved is the reconstructed signal $m(t)$ that approximates $e(t)$. *Thus there is a loss of signal information.*
4. *Software errors.* The complexity of the control process is in the software-implemented algorithms which may contain errors; i.e., the software is not correct.
5. *Controller dynamic update.* Because the A/D, D/A, and digital computer in reality delay the signal, performance objectives may be difficult to achieve since the theoretical assumes no delay.
6. *Power requirements.* Power in some analog controllers can be extensive.

1-4 DIGITAL-CONTROL-SYSTEMS ARCHITECTURE

The successful development of a digital control system requires an interdisciplinary understanding of numerous fields. This can be shown by considering again Fig. 1-1. The general system can be modeled as shown in Fig. 1-1 by two boxes, process

and controller. In reality, both contain many serial- and parallel-structured elements that must be understood in their entirety for proper system design. There are transducers that interface the discrete (digital) world to the continuous (analog) world. Depending upon the selected level, a set of process variables is selected to monitor and control. In many cases, linear operations hold for only relatively small variable excursions. Thus the modeling activity is critical to achieving the desired controlled performance at the selected level.

The digital computer itself has inherent delays since every interval operation is initiated with a clock signal. How fast one can sample depends upon this clock rate. The accuracy depends upon how many binary digits the computer defines as a word. What is the proper control algorithm given a specific computer organization, or vice versa? The language selected for algorithm implementation may not permit easy understanding and testing! Another consideration is the energy utilized by the controller. The digital computer has many characteristics which make it difficult to meet specifications. To understand and design digital control systems, selected elements of computer science and computer engineering are essential.

Transducers (electric, mechanical, chemical, etc.) must be modeled not only in terms of linear system equations, but also the associated variable accuracy, threshold, noise level, impedance, and other detailed characteristics that impact overall performance must be known. The general transducers used in a digital control system are A/D and D/A converters. These converters permit the data transfer from an analog signal (voltage, current, temperature, speed, etc.) into a discrete signal represented by a binary (base 2) number. Characteristics of individual converters depend upon their organization. Important parameters for digital-control-system consideration are conversion time, accuracy (number of binary digits), and internal error sources. Chapter 8 presents a summary of the general techniques used in converting to and from the continuous domain to the discrete. The inherent delay of a conversion device is generally modeled as a hold device (as mentioned previously).

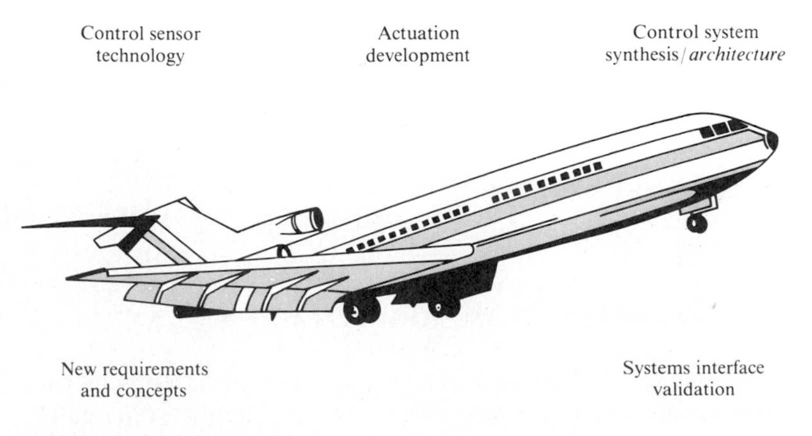

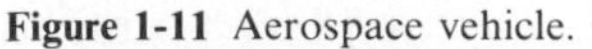

Figure 1-11 Aerospace vehicle.

Figure 1-11 illustrates an integrated avionics system which involves, for a general aerospace vehicle, the five basic areas that are shown. All these areas play an important role in achieving a well-designed aerospace-vehicle control system. This text involves many of these areas: control system analysis and synthesis techniques and computer software and hardware that are used to implement a satisfactory control-system design.

1-5 TECHNIQUES OF CONTROL-SYSTEM ANALYSIS AND SYNTHESIS

Since digital-control-system development is interdisciplinary in nature, the techniques employed cover a wide spectrum. Included are the technique of classical control theory and its extensions, discrete mathematical procedures, as well as computer-related design programs and simulators and computer engineering. All these techniques are used in approaching a satisfactory design.

1-5A Time Domain/Frequency Domain

The analysis of sampled-data control systems relies *heavily on the extension of the complex frequency- and the time-domain methods developed for continuous-time control systems.*[1,21] The major approaches in analyzing sampled-data control systems are listed below.

1. *Complex frequency domain*
   a. s-transform method (Laplace transformation)
   b. Root-locus method
   c. Frequency-response method
   d. $\mathscr{Z}$-transform method and approximations
   e. w'-transform method (bilinear transformation)
   f. State-variable method
2. *Time domain*
   a. Linear difference equation method
   b. Impulse-response method
   c. State-variable method

The $\mathscr{Z}$ and w' transforms and difference equations methods are additional approaches to those used for continuous-time control system analysis and synthesis that are required for designing sampled-data control systems.

The previous methods yield valid results as long as the sampler in the system can be assumed to be an ideal sampler. *By use of an approximation for the hold device, the sampled-data system can be approximated by and analyzed as a continuous-time system. Approximation techniques* provide comparatively easy methods of *converting* the *continuous-time portion* of the sampled-data control system from the continuous domain to the discrete domain.

1-5B Stochastic Analysis

By incorporating probabilistic models of noise into a digital-control-system model, a better understanding of system performance can be achieved. However, this area of study requires extensive understanding of random processes which is generally beyond the scope of this text. But a brief introduction to random processes is presented in order to model the accuracy effects of finite word length, i.e., roundoff and truncation. Further development and applications of stochastic processes as encountered in digital control systems can be found in Refs. 3 and 4.

1-5C Aids

To develop digital control systems in an efficient manner, various aids are suggested that perform accurate and repeatable calculations and permit relatively easy modification of system constants/parameters and software development. Examples of various general aids are simulators, software development systems, and computer-aided-design programs.

1 Simulation To evaluate a digital-control-system design, a realization of all system components can be accomplished. However, this may not be cost-effective due to state-of-the-art design problems as well as delays in implementation. Another possibility then is a computer simulation; either a hybrid analog and digital simulation or a complete digital simulation. In a hybrid simulation an analog computer is employed to represent the continuous plant or process that is to be digitally controlled. In a totally digital simulation, a digital computer is used for both the process and digital controller. In general, this type of simulation is not in real time, i.e., operates at same speed as final realization.

2 Software development systems The importance of a software development facility for the generation and testing of digital-control-system software cannot be overemphasized. A facility of this type permits easy, economical, and efficient generation of programs and associated documentation. This facility must include mass storage (disks, etc.), a fast CRT terminal, and an extensive operating system with utilities. Current programs should be easily modified, stored, retrieved, and executed in a simulation mode using software engineering methods.

3 Computer-aided design and analysis A control engineer must be proficient in the use of available digital-computer programs provided for the programmable calculators and those computer-aided control-system design programs. These programs minimize and expedite the tedious and repetitive calculations involved in the synthesis and analysis of a control-system design. To understand and use a computer-aided analysis and design package, one must first achieve a conceptual understanding of the theory and processes involved in the analysis and synthesis of control systems. It should be noted that for complicated problems these program packages may not provide the required accuracy.

1-5D Design Approach

If the analysis of the basic system of Fig. 1-10*b* reveals that the desired system performance specifications have not been achieved, then a compensator must be inserted into the system. The compensator is designed so that the system achieves the desired performance specifications. Two possible approaches for achieving cascade compensation are shown in Fig. 1-12. Figure 1-12*a* illustrates the utilization of a continuous-data compensator $G_c(s)$ in cascade with the basic plant $G_x(s)$. An alternate method of compensation can be achieved by inserting in cascade an RC network (a passive network) with a hold device or a digital controller (computer), as shown in Fig. 1-12*b*. Since the latter approach is more versatile and comparatively speaking easier to apply, it is the approach covered in this text. Henceforth, the classification, C-C, D-D, D-C, etc., of the system variables are not denoted in the figures except where needed for clarity.

The digital computer not only plays an important role as a control element in a sampled-data control system, but it is of extreme value during the analysis and synthesis of a desired sampled-data control system as a computational device. After a compensator or filter is defined, its implementation in a computer requires a detailed analysis of control law (algorithm) structure, accuracy requirements, computational and delay times, and memory requirements. These real-world considerations and constraints may require a redesign of the compensator.

Note that in Fig. 1-12*b* there are two hold devices. In general, the one on the input to the digital computer is ignored in the control-system dynamic model due to its timing characteristics (a very small "relative" time constant). The timing characteristics of the hold device in the D/A unit must be taken into account in the dynamic model of the system because of its analog output.

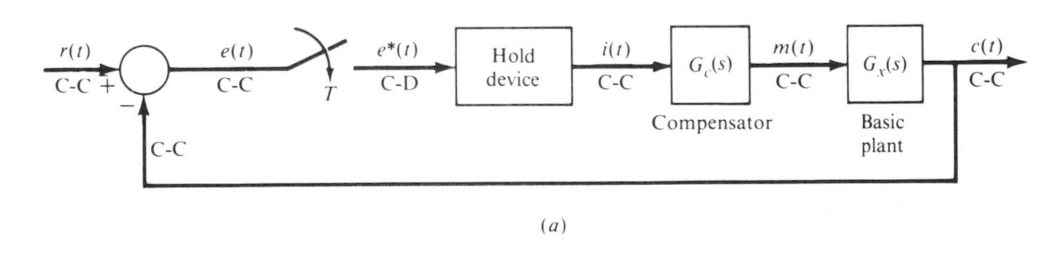

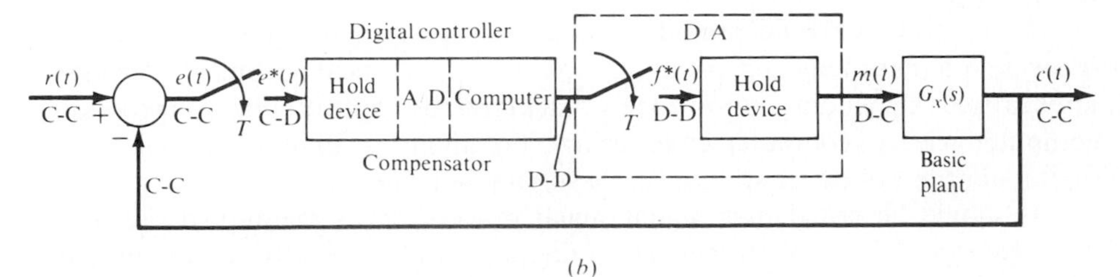

Figure 1-12 Cascade compensation. (*a*) Continuous-data compensator; (*b*) digital controller.

1-6 THE INTERDISCIPLINARY FIELD OF DIGITAL CONTROL

As summarized in this chapter already, many disciplines are involved with developing a successful digital control system. These elementary disciplines include:

Differential and difference equations
Classical control theory
Numerical analysis
Discrete control theory
Computer systems architecture (hardware/software)
Digital integrated circuits
Signal converters
Information structures
Control-algorithm design
Digital signal processing
Software engineering
Test generation

Additional areas providing extended capabilities for analysis and synthesis are:

Matrix algebra[71]
Discrete stochastic processes[3,4]
Modern control theory[3,4,19,70,71,85]
Estimation theory[3,4]

It is not the intent of this text to focus on all elementary disciplines as separate areas but to use appropriate topics within each of them to support the overall theme, digital control systems. Thus some fields such as numerical analysis are only discussed in terms of discrete control algorithms or difference equation approximations. The main areas of concern are discrete control theory, control-algorithm design, and control-system structure at various levels of observation.

Based upon the interdisciplinary nature of digital control systems, the area is difficult to master. The general nature of their design and synthesis requires considerable background in diverse areas of study. A group of individual experts is therefore generally employed to develop an operational digital control system. However, this text should provide not only an overview of the field but also the associated terminology and the ability to realize the critical parameters in a given application. Depending upon one's background (control, signal processing, computer science, computer engineering), the ability to design and synthesize a digital control system at specific levels should be achieved.

It should be noted that digital signal processing, as mentioned previously, provides insight into the processing of digital control signals within a computer. Many of the topics associated with digital signal processing are discussed here because of the large role this field plays in design and synthesis.

It is in the area of compensation of the system or equivalently digital filtering of signals within the system that digital control theory and digital signal processing are essentially equivalent in terms of the design function. In the case of discrete-data control systems (see Fig. 1-12), a digital compensator can be introduced into the forward loop or feedback loop of the system directly connected to the controlled process (plant). In digital signal processing the filter usually operates on the plant output, producing a filtered result for off-line analysis.

A comparison of the design methodology used in the two disciplines shows that in the area of digital-filter design, the type of specifications used to define the desired filter and the method of approximation used to model the filter reflect the primary differences. The specifications for digital control systems are usually given in terms of time constraints. The use of magnitude-squared criteria, as frequently used in digital signal processing for filter design, does not take into account the phase response of the digital filter. In a control system with feedback, the phase response is important to controller design. In each of the two disciplines the representation of the discrete transfer functions of the digital filters are the same. The same considerations of word length in the difference equation coefficients and converters also exist.

In summary, the disciplines of digital signal processing and digital control are equivalent in the area of digital-filter synthesis; the difference possibly lies in the particular methodology used in the design and the application.

1-7 DIGITAL CONTROL DEVELOPMENT

With the decreasing cost of digital hardware, economical digital control implementation is now feasible. Such applications include process control, automatic aircraft stabilization and control, guidance and control of aerospace vehicles, and numeric control of manufacturing machines. An extensive example is state-feedback control systems for which the computer is used to estimate the inaccessible states and to help minimize parameter variations. The development of digital control systems may be illustrated by the following examples which center on digital flight and process control systems.

Example 1-3: Aircraft digital control development Modern technology has brought about some numerous changes in aircraft flight control systems. Initially, flight control systems were purely mechanical, which was ideal for smaller, slow-speed, low-performance aircraft, because they were easy to maintain. However, as the demand for higher-performance airplanes increased, more control-surface force was required, and a hydraulic boost system was added to the mechanical control as shown in Fig. 1-13*a*. This modification still maintains the direct mechanical linkage between the pilot and the control surface. This system is modified to that shown in Fig. 1-13*b*. For those aircraft, the pilot cannot provide the necessary power to directly control the control surface.

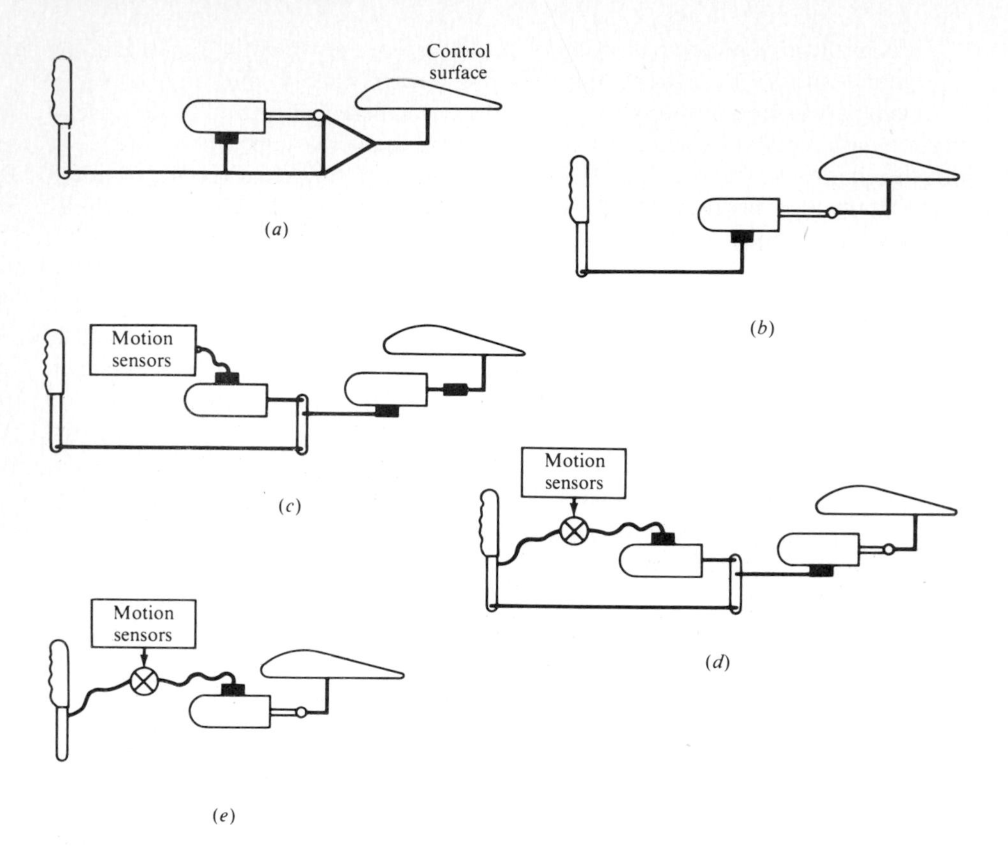

Figure 1-13 The evolution of a fly-by-wire control system. (*a*) Hydraulic boost; (*b*) fully powered controls; (*c*) stability augmentation systems (SAS); (*d*) control augmentation systems (CAS); (*e*) fly-by-wire (FBW). (*Control Systems Development Branch, Flight Dynamics Laboratory, Wright-Patterson Air Force Base, Ohio.*)

Aircraft became much larger, faster, heavier, and harder to control because the increased performance allowed greater variations in flight conditions and also allowed the possibility for more aircraft instability. Although all planes were originally designed to be statically stable, under certain flight conditions large changes in the longitudinal stability began to occur. At this point, the basic airframe could no longer provide all the required flight stability, and the flight control system was called upon to aid in performing this function. A stability augmentation system (SAS) was added to the hydraulic-boosted mechanical regulator system to make the aircraft flyable under all flight configurations (see Fig. 1-13*c*). Motion sensors were used to detect aircraft perturbation and to forward electric signals to an SAS analog computer which, in turn, calculated the proper amount of servo actuator force required to counteract unwanted aircraft responses.

When a higher-authority SAS was required, as with advanced aircraft, both series- and parallel-pitch axis dampers were installed. This so-called command augmentation system (CAS), a tracker system, as shown in Fig. 1-13*d*, allowed greater flexibility in control because the parallel damper could provide full-authority travel without restricting the pilot's stick movements.

The next step in the evolution of flight control systems was the use of a fly-by-wire (FBW) control system shown in Fig. 1-13*e*. In this design, all pilot commands were transmitted to the control-surface actuators through electric wires as shown in Fig. 1-14. Thus all mechanical linkages forward of the servo actuators were removed from the aircraft. The FBW system offered the advantages of reduced weight, improved survivability, and decreased maintenance. Its major disadvantage was the pilot's sense of insecurity with the realization that there was no mechanical backup to a totally electric system. However, the increased survivability was provided by using redundancy throughout the entire flight control system. Individual component reliability was also increased by replacing the older analog circuitry with newer digital hardware. These updated systems were referred to as digital flight control systems (DFCS) and incorporated in the aircraft shown in Fig. 1-15.

The use of airborne digital processors further reduced the cost, weight, and maintenance of modern aircraft. Other advantages associated with newer digital equipment included greater accuracy, increased modification flexibility through the use of software changes, improved in-flight reconfiguration techniques, and more reliable preflight and postflight maintenance testing. The analog computers used in earlier flight control systems had one major disadvantage: They were special-purpose processors built for one application or they were general-purpose computers modified for a particular task. In both

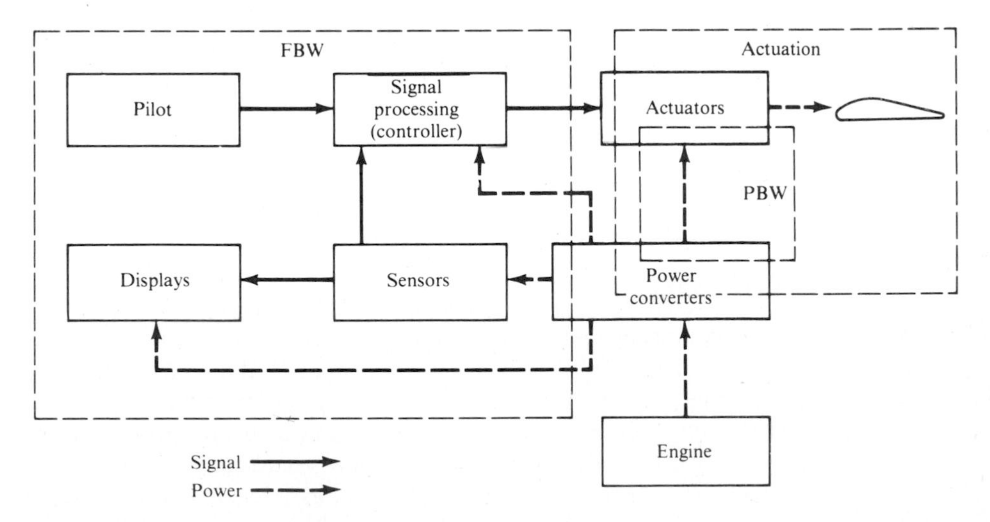

Figure 1-14 Fly-by-wire (FBW) and power-by-wire (PBW) control systems. (*Control Systems Development Branch, Flight Dynamics Laboratory, Wright-Patterson Air Force Base, Ohio.*)

Figure 1-15 Test aircraft exemplifying digital flight control concepts. (*Control Systems Development Branch, Flight Dynamics Laboratory, Wright-Patterson Air Force Base, Ohio.*)

cases, the equipment could not be easily reconfigured to accomplish another purpose. The development of the digital processor has eliminated this major deficiency.

The Digitac aircraft shown in Fig. 1-15 is being used in the development of FBW, fly-by-light (fiber optics), analytical redundancy techniques, and multimode control laws.

Example 1-4: Process control In many modern manufacturing environments, automatic data acquisition and control elements are utilized due to the process accuracy requirements or the environmental severity. The increasing speed of assembly lines and the more sophisticated manufacturing processes as developed over the last century are requiring more measurements and more control of associated parameters. Examples include chemical processes, numerical control, steel processing, automobile manufacturing, and robotics. Most position control systems historically have used a feedback combination of position, integral, and derivative (rate) (PID) signals. Proper analog gains for these signals were selected by tuning the control system empirically. In the following chapters, a more mathematically acceptable approach is discussed for implementation in the discrete domain with extended accuracy.

1-8 GENERAL NATURE OF THE ENGINEERING CONTROL PROBLEM

In general, a control problem can be divided into the following steps:

1. A set of performance specifications is established.
2. As a result of the performance specifications a control problem exists.
3. A set of linear difference and/or differential equations that describe the physical system is formulated.
4. Using the conventional control-theory approach aided by available or specially written computer programs:
   a. The performance of the basic (or original) system is determined by application of one of the available methods of analysis (or a combination of them).
   b. If the performance of the original system does not meet the required specifications, cascade or feedback compensation must be added to improve the response.
5. Or using the modern control-theory approach, the designer specifies an optimal performance index for the system. With the help of computer programs, the design yields the necessary structure to minimize the specified performance index, thus producing an optimal system.
6. Or an alternate modern control-theory approach is the method of entire eigenstructure assignment. First the desired closed-loop eigenvalue spectrum is selected. Then the desired contribution of each mode to each state and output response is selected. The eigenvector spaces are identified, and the eigenvectors are assigned which best meet the selected modal composition of the states and outputs.

Design of the system to obtain the desired performance is the control problem. The necessary basic equipment is then assembled into a system to perform the desired control function. To a varying extent, most systems are nonlinear. In many cases the nonlinearity is small enough to be neglected, or the limits of operation are small enough to allow a linear analysis to be made. In this textbook linear systems or those which can be approximated as linear systems are considered. Because of the relative simplicity and straightforwardness of this approach, the reader can obtain a thorough understanding of linear systems. After mastering the terminology, definitions, and methods of analysis for linear control systems, the engineer will find it easier to undertake a study of nonlinear systems.

A basic system has the minimum amount of equipment necessary to accomplish the control function. The differential and difference equations that describe the physical system are derived, and an analysis of the basic system is made. If the analysis indicates that the desired performance has not been achieved with this basic system, additional equipment must be inserted into the system or new control algorithms employed. Generally this analysis also indicates the characteristics for the additional equipment or algorithms that are necessary to achieve the desired performance. After the system is synthesized to achieve the desired performance, based upon a linear analysis, final adjustments can be made on the actual system to take into account the nonlinearities that were neglected. For digital control systems it is necessary to use good structured programming techniques and to document on a continuing basis all aspects of the software development.

1-9 TEXT OUTLINE

The beginning chapters of this text attempt to develop an understanding of digital-control-system fundamentals. Chapter 2 on computer system architecture presents the basic building blocks of a digital-computer controller. The various computer system elements are discussed as they relate to a controller. They include the memory unit, the arithmetic-logic unit, the input unit, the output unit, and the overall computer task sequencer. With these basic components, a digital controller can be implemented in hardware. The mathematical foundation on which all digital computers are based is the theory of boolean algebra. An overview of boolean logic formulation is discussed also because of its importance in algorithm design as well as in understanding digital-controller interfaces to the continuous environment. Also presented in Chap. 2 is an introduction to data representation as found in a digital-controller algorithm. Selection of proper data representations is very important in ease of system implementation and testing.

Chapters 3 to 5 describe the linear system models, both for the continuous case and for the discrete case. Although it is assumed that the reader has a background in basic continuous control theory including Laplace transforms, the fundamental elements of linear systems modeling is reviewed. Differential and difference equations are employed to represent continuous and discrete linear systems models which are used throughout the text. Using this understanding, the sampling process is structured as a pulse train and in the limit as an impulse-sampling model. Both these models provide insight into the real-world sampling process and the theoretical assumptions used in the following chapters to create a digital controller. To represent a discrete linear difference equation in the frequency domain, a mathematical transform called the $\mathscr{Z}$ transform (zee transform) is developed. Using this frequency- or phasor-domain representation, many of the techniques used in designing continuous controllers with the Laplace transform can be reused in this new environment. Since frequency-domain understanding is very important in analyzing controllers, the relationship between the s domain, z domain, and the Fourier transform is also described. Once the technique of writing the system equations (and, in turn, the $\mathscr{Z}$ transforms) that describe the performance of a dynamic system has been mastered, the ideas of block diagrams and $\mathscr{Z}$ transforms are developed. When sampled-data systems are described in terms of block diagrams and transfer functions, they exhibit basic servo characteristics similar to those for continuous-time systems. Finally, the introductory material reviews the basic aspects of continuous-controller design and analysis methods to provide an integrated foundation on which the study of discrete controllers can be initiated. Single-input–single-output (SISO) systems are used to facilitate an understanding of the synthesis methods.

Various design techniques exist for developing discrete or digital controllers. The basic method presented in this text is based upon the root-locus formulation using the poles and zeros of the transfer function in the z domain. The design approaches relate the s-plane root-locus techniques to the z-plane approaches. Also, the pseudo s plane or w plane (or w' plane) is developed as an alternative approach for digital-controller design. In the w or w' plane, the identical s-plane methods from classical or conventional control theory can be applied. In essence, one can design a continuous controller using s-plane methods and then discretize the resulting differential equation into a difference equation. On the other hand, the entire design process can occur in the z-plane, resulting also in a difference equation representing the controller. In all methods presented, the frequency- and time-domain transformations are also discussed.

As mentioned previously, the real-world implementation of a digital controller requires a conversion from a continuous or analog signal to a digital signal. To accomplish this task, a so-called analog-to-digital (A/D) converter is required. Also, the digital "control signal" generated within the computer must be converted into a continuous signal through a digital-to-analog (D/A) process. Different organizations for performing these operations are presented and analyzed as to their applicability in various control systems. Also, appropriate measurement transducers are described in regard to their physical structure and operation. Proper selection of interface components is critical in meeting performance objectives.

Another aspect of component selection is the real-time digital computer itself. Of primary importance is the word length or binary digit (bit) length of the data to be manipulated (i.e., the solution of the difference equation). To study the impact of finite word length, a statistical or probalistic approach is used. The fundamentals of random variables and random processes are developed in the text and provide the tools for analyzing the impact of finite word length for specific difference equation implementation.

Although the rudiments of digital-computer architecture are presented in Chap. 2, the implementation of a digital controller requires an appreciation of software also. Various chapters describe the essence of programming language levels and the proper techniques for structuring a software program, especially a digital control program in a real-time environment. A limited number of skills from software engineering are presented for proper development of control programs. The embedding of such a program in a commercial real-time executive structure requires a detailed understanding of the executive structure. The text attempts to describe many of the general organizations available and their operation. From this discussion, a unique executive can also be defined to meet specific performance requirements.

The major portion of the text primarily focuses on scalar models for representing controllers. Since many contemporary approaches for continuous-controller design use matrix and vector models, it is appropriate to extend the discrete scalar models to discrete matrix models. Chapter 15 introduces the state-vector modeling approach.

Finally, all the numerous analysis, design, and implementation techniques are integrated together in formal discussions throughout the text. Emphasis is placed upon the trade-offs between them and the associated ramifications.

1-10 SUMMARY

This chapter has introduced the model of a sampled-data control system. Such systems with at least one sampled analog signal and a hold device are used in numerous applications. Contemporary technology permits the use of a digital computer as a controller in many sampled-data systems.

The development of a discrete real-time digital controller requires the understanding of many theoretical and empirical subjects, many of which are presented in this text. Thus it is appropriate to define an individual who is involved in this development as a *systems engineer* or an interdisciplinary engineer. Although it may be difficult for one person to be able to understand and apply all the diverse techniques required in digital-control-system analysis, design, implementation, tests, and integration, the terminology presented permits an individual to communicate with other individuals involved in the development. Such individuals may have expertise in the areas of mechanical engineering, electrical engineering, aeronautical engineering, software engineering, computer systems programming, computer engineering, computer science, etc.

The many and varied methods described in this text result of course in different controller structures and organizations. Some may have performance characteristics very similar in nature; others may be quite different across a wide spectrum. Many may have the same theoretical performance, but as implemented in a digital controller the responses reflect sharp differences.

CHAPTER

TWO

COMPUTER SYSTEM ARCHITECTURE

2-1 INTRODUCTION

This chapter begins the detailed discussion of digital-control-system development. Various concepts from the interdisciplinary fields of computer science and computer engineering are discussed as they relate to digital-control-system development. Such topics include number representations, logical descriptions of algorithm processes, computer arithmetic operations, and computer system hardware and software. To discuss in detail all the abstractions of computer science and computer engineering is of course impossible in a text of this kind. The reader is directed to the many detailed references on particular subjects. The focus in this text is mainly on the computer solution to the difference equation of the form

$$c(kT) = \sum_{i=0}^{m} a_i r[(k-i)T] + \sum_{j=1}^{n} b_j c[(k-j)T]$$

which represents a control law. Performance is directly related to the computer representation of the coefficients a_i and b_j, the processes involved in performing the arithmetic operations, and the data input and output.

2-2 COMPUTERS AND NUMBERS

Many accuracy considerations in digital control systems involve the study of number representations. Moreover, the testing, evaluation, and maintenance of digital control systems rely on an ability to convert between base 10 (the real-world number system) and base 2 (the computer number system). Therefore, the initial subject discussed is an introduction to computer number representation and associated arithmetic operations. Extending from this foundation is the representation of fixed-point signed numbers presented in Sec. 2-5. The symbolic notation may seem cumbersome at first, but the insight provided proves useful in understanding number conversion and representation within the digital control environment.

2-2A Number Systems

In the *decimal* number system, there are 10 digits or *symbols*: 0, 1, 2, 3, 4, 5, 6, 7, 8, and 9. The construction of numbers in this system involves the use of these symbols consecutively: first in the position defined as units, then in the tens position, next in the hundreds position, and so forth. The decimal digit locations are shown symbolically as follows:

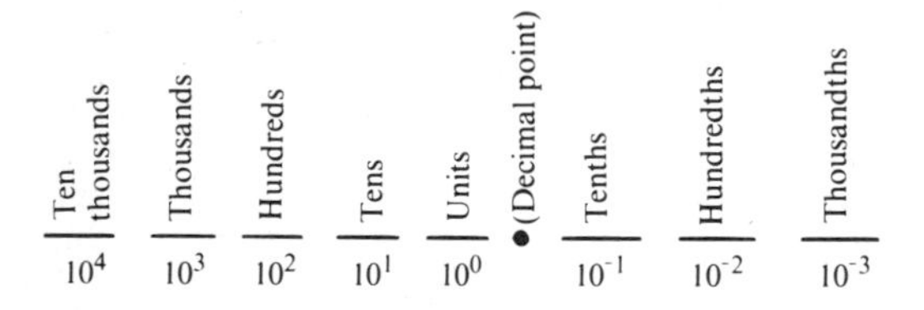

For example, the decimal number $N = 7429.138$ is $7 \times 10^3 + 4 \times 10^2 + 2 \times 10^1 + 9 \times 10^0 + 1 \times 10^{-1} + 3 \times 10^{-2} + 8 \times 10^{-3}$. An identical procedure is used for constructing numbers in other number systems. The base 5 (or *quinary*) number system uses the symbols 0, 1, 2, 3, 4 and employs the same consecutive procedure as the base 10 number system.

In general, a number N in any base can be represented by

$$N_r = b_M \times r^M + \cdots + b_0 \times r^0 + b_{-1} \times r^{-1} + \cdots + b_L \times r^L \tag{2-1}$$

where b_i is the ith digit or symbol of the number system with base, *radix*, r; b_M is defined as the most significant digit (MSD); and b_L is defined as the least significant digit (LSD). Thus $L < 0 \leq M$, $0 \leq b_i < r - 1$, and $i = L, \ldots, -1, 0, 1, \ldots, M$. Equation (2-1) can also be represented by the series sum

$$\begin{aligned} N_r &= \sum_{i=L}^{M} b_i r^i \\ &= b_M r^M + b_{M-1} r^{M-1} + \cdots + b_0 r^0 + b_{-1} r^{-1} + \cdots + b_L r^L \end{aligned} \tag{2-2}$$

or by the format of

$$N_r = b_M b_{M-1} \cdots b_0 \cdot b_{-1} \cdots b_L \tag{2-3}$$

Thus for $r = 10$ (decimal system), Eq. (2-2) yields

$$N_{10} = \sum_{i=L}^{M} b_i 10^i = b_M 10^M + \cdots + b_L 10^L \tag{2-4}$$

where $b_i = 0, 1, 2, \ldots, 9$. For $r = 2$ (binary system), Eq. (2-2) yields

$$N_2 = \sum_{i=L}^{M} b_i 2^i = b_M 2^M + \cdots + b_L 2^L \tag{2-5}$$

where $b_i = 0, 1$. Note that decimal multiplication and summation on the right-hand side of Eq. (2-2) always yields the equivalent base 10 number for the number N_r (that is, $r = 10$)

Example 2-1 Consider 06.50 in the base 10 system. Thus $r = 10$, $M = 1$, $L = -2$, $b_1 = 0$, $b_0 = 6$, $b_{-1} = 5$, and $b_{-2} = 0$. Therefore, from Eq. (2-2),

$$N_{10} = 06.50_{10} = b_1 r^1 + b_0 r^0 + b_{-1} r^{-1} + b_{-2} r^{-2}$$
$$= 0 \times 10^1 + 6 \times 10^0 + 5 \times 10^{-1} + 0 \times 10^{-2}$$

Note that the term-by-term expansion of the series summation of Eq. (2-2) is presented with the highest-order term, $b_M r^M$, first.

The number N_{10} can also be represented by Eq. (2-3) as follows:

$$N_{10} = 06.50_{10} = b_1 b_0 . b_{-1} b_{-2} = 06.50$$

The digit b_i and its associated position weight r^i for this example can be represented by the following format:

Digit	b_1	b_0	b_{-1}	b_2
r^i	10^1	10^0	10^{-1}	10^{-2}

In the binary or base 2 number system, the digit and r^i locations are shown symbolically as follows:

Digit	...	b_4	b_3	b_2	b_1	b_0	b_{-1}	b_{-2}	b_{-3}	...
r^i	...	2^4	2^3	2^2	2^1	2^0	2^{-1}	2^{-2}	2^{-3}	...

Most, if not all, contemporary digital computers implement the binary system at the processor level. Thus it is important to be able to connect the processor base 2 implementation of control law coefficients with the decimal-world coefficient design.

Example 2-2 Consider the base 2 number 110.10_2. Thus $r = 2$, $M = 2$, $L = -2$, $b_2 = 1$, $b_1 = 1$, $b_0 = 0$, $b_{-1} = 1$, and $b_{-2} = 0$. Therefore, from Eq. (2-5),

$$N_2 = 110.10_2 = b_2 2^2 + b_1 2^1 + b_0 2^0 + b_{-1} 2^{-1} + b_{-2} 2^{-2}$$
$$= 1 \times 2^2 + 1 \times 2^1 + 0 \times 2^0 + 1 \times 2^{-1} + 0 \times 2^{-2} = 6.50_{10}$$

which is also represented by the format

Digit	1	1	0	1	0
r^i	2^2	2^1	2^0	2^{-1}	2^{-2}

Table 2-1 Corresponding decimal, binary, and octal numbers

Base 10 number	Base 2 (binary)	Base 8 (octal)	Base 10 number	Base 2 (binary)	Base 8 (octal)
0	0	0	5	101	5
1	1	1	6	110	6
2	10	2	7	111	7
3	11	3	8	1000	10
4	100	4	9	1001	11

The *octal* (or base-8) number system uses the symbols 0, 1, 2, 3, 4, 5, 6, 7. Since the first three integer-digit positions in the binary system represent the numbers 0 to 7, the first three integer digits can be combined to form one octal digit. For example,

$$111_2 = 7_8 \qquad 011_2 = 3_8$$

The octal number system is related to the binary system because $2^3 = 8^1$. Thus one octal digit can be equated to three binary digits. Many digital-computer systems incorporate the octal number representation for direct output for economy and ease of reading.

Table 2-1 gives the binary and octal numbers corresponding to the base 10 numbers 0 to 9.

Another useful system is the *hexadecimal* (or base 16) number system which requires 16 digit symbols. Since there are only 10 symbols available for number representation, 0, 1, 2, 3, 4, 5, 6, 7, 8, 9, six additional symbols are required for the hexadecimal system. It is common to let these additional symbols be the first six letters of the alphabet, A, B, C, D, E, F; thus the base 16 symbols are 0, 1, 2, 3, 4, 5, 6, 7, 8, 9, A, B, C, D, E, F. One hexadecimal digit can be equated to four binary digits because $2^4 = 16$. The usefulness of the hexadecimal system in digital machine output is also its economy and ease of reading as compared to the binary system. Many microprocessor systems use hexadecimal representation. *The reduction in digit number length is the attribute achieved as the radix increases.* Table 2-2 presents some representative number systems with their associated symbolic digits.

Table 2-2 Positional number systems

Number system	Radix (base)	Associated radix symbols
Binary	2	0, 1
Ternary	3	0, 1, 2
Quaternary	4	0, 1, 2, 3
Quinary	5	0, 1, 2, 3, 4
Octal (octonary)	8	0, 1, 2, 3, 4, 5, 6, 7
Decimal (denary)	10	0, 1, 2, 3, 4, 5, 6, 7, 8, 9
Duodecimal	12	0, 1, 2, 3, 4, 5, 6, 7, 8, 9, A, B
Hexadecimal	16	0, 1, 2, 3, 4, 5, 6, 7, 8, 9, A, B, C, D, E, F

2-2B Conversion of Number Systems

It is the intent of this subsection to discuss the basic concepts of conversion from a number system with a specified base to a different number system with a new base. The conversion of the binary into the decimal number system, and vice versa, is used to illustrate the basic conversion fundamentals. Conversions to other bases are accomplished by using similar procedures. Conversion of integer decimal numbers to integer binary numbers is treated first, followed by fractional conversion.

Integer conversion The conversion procedure for integer numbers is different from the procedure for fractional numbers. This procedural separation is due to the existence of negative exponents for fractional values and positive exponents for integer numbers. Integer numbers are considered first. Let the binary number consist of $M + 1$ digits, denoted by b_i, where $i = 0, \ldots, M$. Thus expanding Eq. (2-5) yields

$$N_{10} = b_M 2^M + b_{M-1} 2^{M-1} + \cdots + b_i 2^i + \cdots + b_1 2^1 + b_0 2^0 \tag{2-6}$$

where the terms $b_i 2^i$, when added together using decimal arithmetic, equal N_{10}. The objective is to determine the b_i's (0 or 1) where N_{10} is the given decimal number in the form $b_m 10^m \cdots b_0 10^0$. To do this, Eq. (2-6) is divided by 2 to yield

$$2^{-1} N_{10} = b_M 2^{M-1} + \cdots + b_1 2^0 + b_0 2^{-1} \tag{2-7}$$

Since $b_i = 0, 1$, the last term of Eq. (2-7) due to the negative exponent always has the value of either 0 or 1/2, whereas the other terms remain integer values. That is, if N_{10} is an odd number, then $2^{-1} N_{10}$ yields a number that is the sum of an integer plus 2^{-1}. The fraction 2^{-1} must come from the $b_0 2^{-1}$ term of Eq. (2-7) with $b_0 = 1$. If N_{10} is an even number, then $2^{-1} N_{10}$ is an integer which results in $b_0 2^{-1} = 0$, or $b_0 = 0$, in Eq. (2-7). Considering the remaining integer terms as a new integer decimal number and again dividing by 2 generates in a similar manner the value of $b_1 2^{-1}$. This procedure is repeated until the remaining b_i's are determined. Note that the numerator of the fraction resulting from the division operation defines the value of the binary digit b_i.

Example 2.3 Given $N_{10} = 17$, find N_2.

SOLUTION

	$17 = N_{10}$	
Divide integer by 2:	$8 + 1/2$	$b_0 = 1$
Divide integer by 2:	4	$b_1 = 0$
Divide integer by 2:	2	$b_2 = 0$
Divide integer by 2:	1	$b_3 = 0$
Divide integer by 2:	$0 + 1/2$	$b_4 = 1$

The binary representation of 17_{10} utilizing Eq. (2-3) is $N_2 = b_4 b_3 b_2 b_1 b_0 = 10001_2$.

Equation (2-6) is also applicable for the conversion of an integer binary number to a decimal number. Since all the b_i's are known in this case, the straightforward procedure is to add all the terms of the right-hand side of Eq. (2-5) using decimal addition. An alternative formula for this conversion can be derived by factoring and rearranging the equation as follows for $M = 4$ and $L = 0$:

$$N_{10} = \{[(2b_4 + b_3)2 + b_2]2 + b_1\}2 + b_0 \tag{2-8}$$

Fractional conversion The conversion from a decimal fraction to a binary fraction is similar to the previous procedure except multiplication by 2 is employed instead of using the techniques of division by 2. If N_{10} is a pure decimal fraction, then it can be represented as a binary number with the formula

$$N_{10} = b_{-1}2^{-1} + b_{-2}2^{-2} + \cdots + b_L 2^L \tag{2-9}$$

Note: A fraction in any number system is a fraction in any other number system. To find the b_i's, Eq. (2-9) is multiplied by 2 to yield

$$2N_{10} = b_{-1}2^0 + b_{-2}2^{-1} + \cdots + b_L 2^{L+1} \tag{2-10}$$

The first term on the right-hand side in Eq. (2-10) is an integer, where the remaining terms are still fractional. Thus multiplying N by 2 generates either a 1 or a 0 in the units position corresponding to the value of b_{-1}. That is, if $2N_{10} < 1$, then $b_{-1} = 0$, and if $2N_{10} > 1$, then $b_{-1} = 1$. The determination of b_{-2} proceeds in the same manner used for determining b_{-1}, by operating on the $2^{-1}N_{10}$ equation in the same manner as for N_{10}; that is, find b_{-2} from $2^2 N_{10} - b_1$, and so forth. Use the integer part (a 0 or a 1) to define each binary digit. By applying this procedure iteratively, the remaining bits are found. An example best illustrates the procedure.

Example 2-4 Given $N_{10} = 0.4231$, with $L = -5$, find N_2.

SOLUTION

	$0.4231 = N_{10}$	
Multiply fractional part by 2:	0.8462	$b_{-1} = 0$
Multiply fractional part by 2:	1.6924	$b_{-2} = 1$
Multiply fractional part by 2:	1.3848	$b_{-3} = 1$
Multiply fractional part by 2:	0.7696	$b_{-4} = 0$
Multiply fractional part by 2:	1.5392	$b_{-5} = 1$

The binary number corresponding to 0.4231_{10} using the five significant digits ($L = -5$ for base 2) is $N_2 = 0.01101_2$. Note also that a finite number of fractional decimal digits does not necessarily generate a binary number with a finite number of fractional digits.

As in the integer case, Eq. (2-7) is also applicable for the conversion of a binary fraction to a decimal fraction. Since the b_i's are known, the procedure consists of adding the terms using decimal addition. An equivalent procedure can be used that factors powers of 2^{-1} similar to Eq. (2-8); for example, for $L = -4$,

$$N_{10} = 2^{-1}\{b_{-1} + 2^{-1}[b_{-2} + 2^{-1}(b_{-3} + 2^{-1}b_{-4})]\} \tag{2-11}$$

Example 2-5 Given $N_2 = 0.01101_2$, with $L = -5$, find N_{10}.

SOLUTION Applying Eq. (2-11) for $L = -5$ yields

$$\begin{aligned} N_{10} &= 0 \times 2^{-1} + 1 \times 2^{-2} + 1 \times 2^{-3} + 0 \times 2^{-4} + 1 \times 2^{-5} \\ &= 0.5(0 + 0.5[1 + 0.5(1 + 0.5\{0 + 0.5 \times 1\})]) \\ &= 0.5(0.5[1 + 0.5(1 + 0.25)]) \\ &= 0.5(0.5)[1.625] = 0.40625_{10} \end{aligned}$$

Note in Example 2-4 that $N_{10} = 0.4231_{10}$ is converted to $N_2 = 0.01101_2$. Thus the conversion process in Example 2-4 is good only to one significant position. If in Example 2-4, $L = -6$ (for base 2), then $N_2 = 0.011011_2$ yields $N_{10} = 0.421875_{10}$, which is accurate to two significant digits when converted using the process in Example 2-5. Note $L = -4$ yields $N_{10} = 0.375$.

Note that one decimal digit is equivalent to approximately 3.32 binary digits. To derive this relationship, let d be the number of decimal digits and b be the number of binary digits. The relationship between d and b is then given by

$$10^d = 2^b \qquad \text{or} \qquad 10^{d/b} = 2$$

$$\frac{d}{b} = \log_{10} 2$$

and

$$b = \frac{d}{\log_{10} 2} \sim 3.32d \tag{2-12}$$

In general, if d digits of decimal accuracy are required, then the computer word length (binary) must be the next integer value of $3.32d$. It is left to the reader to check the previous examples. The coefficient implementation approximations should be analyzed in terms of control-system performance; that is, how many binary digits (bits) must be available for control-law implementation in order to achieve the desired accuracy performance? Detailed discussion on this issue is presented in the following chapters.

The inclusion of the octal conversion when converting between decimal and binary can reduce the number of arithmetic operations and make the process more efficient. Of course many calculators and computer programs can do these conversions for you.

Base 8 conversion Octal or base 8 conversion to decimal, or vice versa, is performed in a similar manner as for the previous cases except that (1) r is 8 instead of 2 and (2) multiplication by 8 or 8^{-1} is done instead of multiplication by 2 or 2^{-1}, respectively.

Binary to octal conversion involves 3-bit fields starting at the binary point. Proceeding left or right generates the octal representation. Each group of 3 bits is converted to an octal symbol.

Hexadecimal conversion To convert between base 16 and base 2, the same basic rules of octal conversion are applied except that 4 bits are grouped together instead of 3; that is, $2^4 = 16^1$. Although direct conversion methods between base 16 and base 10 are available, the arithmetic manipulation of the symbols A, B, C, D, E, and F is somewhat difficult.

2-2C Basic Computer Arithmetic

In implementing a digital control algorithm, generally a difference equation, the four arithmetic operations are used. The basic techniques employed in addition, subtraction, multiplication, and division of decimal numbers can be used for the arithmetic operations in any number system: in the computer's case, the base 2 number system. For example, the carry phenomenon occurs identically for addition in any base. Consider the base 2 number system. Any digit in base 2 number is either 0 or 1. Thus, for the addition of 2 bits, the four combinations are:

	Case *a*	Case *b*	Case *c*	Case *d*
Augend digits	0	1	0	1
Addend digits	0	0	1	1
Sum digits	0	1	1	0 + carry

where case *d* also requires a carry of 1 into the left adjacent position. In binary addition, if the respective augend and addend digits are both 1 or both 0 (with no carries), the sum is 0. If the augend bit is different from the respective addend bit (with no carries), the sum is 1. The sum is 0 if the carry into this specific position is 1 and either the augend or the addend bit is 1 (but not both). If the augend and addend digits are both 1, with a carry, then the sum is a 1 with a carry.

Example 2-6 Perform the addition of the given N_2 and N_{10} numbers.

Carry	$1111.0_2 =$	1.0_{10}
Augend	$1101.1_2 =$	13.5_{10}
Addend	$0110.1_2 =$	06.5_{10}
Sum	$10100.0_2 =$	20.0_{10}

Subtraction For binary subtraction, the equivalent rules of decimal subtraction are used. Instead of carries being generated, borrows occur if the subtrahend digit is larger than the minuend digit. In the base 2 number system, the identical technique is used as employed in decimal subtraction. Since the only digits for $r = 2$ are 0 and 1, only four possible configurations of minuend and subtrahend digits exist:

	Case *a*	Case *b*	Case *c*	Case *d*
Minuend digit	1	0	1	0
Subtrahend digit	1	0	0	1
	—	—	—	—
Difference digit	0	0	1	1 + borrow

where case *d* requires a borrow from the left adjacent position. A borrow then occurs only if the minuend bit is 0 and the subtrahend is 1. One should check subtractions by doing the inverse addition process, which requires adding the difference to the subtrahend to obtain the minuend.

Example 2-7 Perform the subtraction of the given N_2 and N_{10} numbers.

Borrows	1	100.0	10.0
Minuend	1	$101.1_2 =$	13.5_{10}
Subtrahend	0	$110.1_2 =$	06.5_{10}
Difference	0	$111.0_2 =$	07.0_{10}

Multiplication In decimal multiplication, each multiplier digit multiplies the multiplicand on a digit-by-digit basis, generating partial sums. In other words, the multiplicand is added to itself the number of times specified by the multiplier digit minus 1. These partial sums are then added together in a proper shifting fashion to generate the product. As indicated, multiplication is essentially a particular combination of additions as it is in the base 10 number system as well. In the base 2 number system the four possible combinations of the 2 binary bits are:

	Case *a*	Case *b*	Case *c*	Case *d*
Multiplicand digit	0	1	0	1
Multiplier digit	0	0	1	1
	—	—	—	—
Product digit	0	0	0	1

Thus the product digit is 1 if and only if the multiplicand and multiplier digits are both 1.

Example 2-8 Perform the multiplication of the given N_2 and N_{10} numbers.

Multiplicand	$1101.1_2 =$	13.5_{10}
Multiplier	$10.1_2 =$	2.5_{10}
Partial product	11011	675
	00000	270
	11011	
Product (sum of the partial products)	$100001.11_2 =$	33.75_{10}

Note that two significant digits are to the right of the radix point due to the fact that the multiplicand and the multiplier each have one significant fractional digit.

Complements As is shown later in this chapter, the employment of complements is very useful in performing binary arithmetic operations in a computer, specifically the representation of negative numbers. The *complement of a digit* is defined as the *positive* difference between the digit and the largest digit $(r - 1)$ of the specified number system. The complement of a digit is denoted as $\bar{b}_i$; that is,

$$\bar{b}_i \equiv (r - 1) - b_i \tag{2-13}$$

For example, $b = 3$ and $r = 10$ generates the complement

$$\bar{b} = (10 - 1) - 3 = 9 - 3 = 6$$

or for $b = 0$ and $b = 1$, respectively, in the binary system $(r = 2)$ generates

$$\bar{b} = 1 - 0 = 1 \qquad \text{and} \qquad \bar{b} = 1 - 1 = 0$$

The complement of a number, N, is defined differently and is a function of the specified word length (number of digits). It is the positive difference between the number r^M and the number N_r. M, a positive integer, is the number of digits used to represent N_r. The complement of a number is denoted as

$$\bar{N} \equiv r^M - N_r \tag{2-14}$$

and is defined as the r's *complement*. Note that $r^M = b_M b_{M-1} \cdots b_0 = 10 \cdots 0$, which has M zeros and $M + 1$ digits. Note also that the subtract operation is executed by using base r arithmetic.

The $(r - 1)$'s *complement* can also be constructed and is defined mathematically as

$$\bar{N}_{-1} = r^M - N_r - r^L \tag{2-15}$$

where $r^L = b_{-1} b_{-2} \cdots b_L = 0 \cdots 1$, which has $|L + 1|$ zeros. L is negative as defined from Eq. (2-1). In these two cases, the word length is $M - 1 - L$ digits. These complements are, respectively, (1) for a decimal system, $\bar{N} = 10^M = N_{10}$ and $\bar{N}_{-1} = 10^M - N_{10} - 10^L$ and (2) for a binary number system, $\bar{N} = 2^M - N_2$ and $\bar{N}_{-1} = 2^M - N_2 - 2^L$. Note that for $L = 0$, $r^L = 1$.

Example 2-9 Given $r = 10$, $M = 6$, and $N_{10} = 12{,}345$, find $\bar{N}$ and $\bar{N}_{-1}$.

Solution Applying Eqs. (2-14) and (2-15) with $L = 0$ yields, respectively,

$$\bar{N} = 100{,}000 - 12{,}345 = 87{,}655_{10}$$

and

$$\bar{N}_{-1} = 100{,}000 - 12{,}345 - 00001 = 87{,}654_{10}$$

It should be noted that the $(r - 1)$'s complement can also be obtained directly from N_r and is

$$\bar{b}_{n-1} \bar{b}_{n-2} \cdots \bar{b}_0 = \bar{N}_{-1}$$

and therefore

$$\bar{N} = \bar{N}_{-1} + r^L$$

Thus the 10's complement = 9's complement + 10^L.

Example 2-10 Given $r = 2$, $M = 3$, and $N_2 = 110$, find $\bar{N}_{-1}$ using Eq. (2-15). Since $2^3 = 1000$ and $2^0 = 1$, then

$$\bar{N} = 1000 - 110 = 010_2$$

$$\bar{N}_{-1} = 1000 - 110 - 001 = 001_2$$

or

$$\bar{N}_{-1} = \text{complement of each digit of } N_2$$

$$= \overline{1}\overline{1}\overline{0} = 001_2$$

Note that the 2's complement = 1's complement + 2^L. It should be realized that the 1's complement of a binary number can be found by complementing, toggling, or switching each bit separately. This phenomenon is important in constructing a digital computer with electronic switching.

The basic reason for generating complements is their use in the subtraction operation. Instead of subtracting B from A, the complement of B is found and added to A; that is, $A - B$ is equivalent to

$$A + \bar{B} = A + (r^M - B) = r^M + (A - B) = A - B$$

The term r^M, as defined previously, has a 1 in the $M + 1$ digit location. Thus for a computer word of M digits, the $M + 1$ digit location is neglected, i.e., not stored. Later on in the chapter, signed number representations are presented which employ this complement operation. Note that number complements play a role in computational accuracy. This topic is covered in Chap. 10.

2-2D Binary Coding of Alphanumeric Information

Information processors use the binary system for internal operations, although human number manipulation is in the decimal system. It is therefore desirable to electronically perform a translation between the user's environment (decimal system) and the internal operations (binary system) of the machine. That is, the decimal data (or information) is encoded into binary information, specified operations performed on the binary data, and the results are then decoded back to the decimal system for the user's evaluation. Since it is desired, in general, that the machine also have the capability of manipulating alphabetic characters, in addition to other special characters (punctuation symbols, etc.), binary coding of such characters is also required. The machine employs a specified binary sequence or number to represent a character and manipulates them according to some specified procedure. Messages to real-time process control system users employ these codes.

The selection of a *binary-coded-decimal* (BCD) number code in which a

Table 2-3 Four-bit BCD code

Decimal symbol	3-4-2-1	Decimal symbol	3-4-2-1
0	0000	5	0101
1	0001	6	0110
2	0010	7	0111
3	0011	8	1000
4	0100	9	1001

decimal digit is represented by 4 bits is based upon the application requirements of the hardware or software mechanization. Some of the possible 4-bit codes are weighted codes. A decimal digit d_{10} is defined as

$$d_{10} = \sum_{i=0}^{3} b_i w_i = b_3 w_3 + b_2 w_2 + b_1 w_1 + b_0 w_0 \tag{2-16}$$

where the summation terms are listed in the manner shown and w_i is the numerical weight assigned to the bit b_i. For example, let $w_i = 2^i$; thus

$$d_{10} = \sum_{i=0}^{3} b_i 2^i = b_3 8 + b_2 4 + b_2 2 + b_0 1 \tag{2-17}$$

This weight code is thus referred to as the 8-4-2-1 BCD code (see Table 2-3).

Cyclic codes *Cyclic* codes or *gray* codes have the property that only 1 bit of the code number (word) changes from one adjacent code word to another. Moreover, the last code word only has to change 1 bit in order to become the first code word. Thus the last code word of the complete cyclic code is adjacent to the first code word. Table 2-4 presents examples of 2- and 4-bit cyclic codes. The 4-bit code has two separate cycle patterns nested or included in the main cycle pattern. Cyclic codes are very useful for encoding cyclic analog data such as the positions of a rotating shaft for positional control. It should be noted that cyclic codes have to have an even number of code words.

Table 2-4 Cyclic codes (gray codes)

Decimal digit symbol	2 bits	4-bits	
1	00	0000	1100
2	01	0001	1101
3	11	0011	1111
4	10	0010	1110
5		0110	1010
6		0111	1011
7		0101	1001
8		0100	1000

Table 2.5 Some standard alphanumeric codes

Character	ASCII 7-bit (octal) codes	Character	ASCII 7-bit (octal) codes
A	101	3	063
B	102	4	064
C	103	5	065
D	104	6	066
0	060	7	067
1	061	8	070
2	062	9	071

The representation of alphabetic characters as well as numbers, referred to as *alphanumeric encoding*, is very desirable in a digital system. The use of these characters is important in the development and use of computer languages and communication between humans and machine and between machine and machine. In addition to the alphanumeric characters, there are special characters that can be employed for arithmetic and logical operations and for editing. The nonalphanumeric symbols on a typewriter are an example of such special characters.

Assuming that there are 36 special characters in addition to the 36 alphanumeric symbols, at least 7 bits ($2^7 > 72$) are required for coding. If the number of special symbols is equal to or less than 28, then only 6 bits are required. Table 2-5 presents some 7-bit ASCII (American standard code for information interchange) codes; if 8 bits are used, the eighth bit is a *parity bit*. The parity scheme is defined as odd or even depending upon the number of "1" bits in the entire 8-bit representation.

In many control systems, especially in process control, many messages are sent from the computer to the system manager for direction of the process. Proper computer programming of all these transmitted messages is very important in achieving reliable system performance over the many days and years of operations.

2-3 COMPUTER LOGIC AND BOOLEAN ALGEBRA

The development of physical hardware and switching systems for the manipulation of binary information requires a logical design procedure. Binary number and alphanumeric code manipulation can be easily defined by the employment of a symbolic or logical design method. Implementation may be in terms of logic circuits or even a computer programming language. In designing the switching operations in a complex process control system, a symbolic design method is necessary. This section discusses the development of the symbolic design technique in terms of logical decisions. Also, the relationship between this logical procedure and digital system operations is briefly discussed.

The development of a design and analysis procedure is initiated with an introduction to logic. Various logic relations are presented along with logic symbols and simple implementations. *Boolean algebra* is presented as a precise method for describing these logical operations. The detailed procedures for the design of large logic circuits can be found in appropriate references.[6,8]

2-3A Logical Reasoning

The most important attributes of a digital system are to store, transfer, and manipulate information. To understand the logical representation of these operations requires an application of logic—the study of the principles of reasoning within a formal structure. The structure used in this application is boolean algebra. Before discussing the specific characteristics of this formal structure, consider first the general aspects of logical reasoning employed daily by an individual.

To discuss logical reasoning, it is necessary to define some terms. The following statements, used typically in everyday communication, are each categorized as a declaration or as an assertion:

The show was fantastic.
All quarterbacks are intelligent.
The race results indicated a new record.
All cars use gas economically.

Of course, these statements (applied to a specific event or class) can be either false or true. The desire to determine the correctness or falseness of a statement thus depends upon an individual's ability to employ logical reasoning.

There are various types of statements or declarations which provide greater insight into digital system operations. Three such types of statements are those which are:

1. Observable
2. Obvious
3. Demonstrable (inferential)

Examples of observable true statements are "all control systems work," "Bob owns a control system," and "politics is an interesting sport." Such statements are assumed to be true based upon experience, not on logical reasoning. Observable statements, like the last example, can be quite controversial. Thus observable statements are not of interest for use in the study of digital systems.

The second type of statement is either obviously true or obviously false. This type of statement is evident as to its correctness or falseness. For example the statement "either Jane has a boy friend or she does not" is obviously true since both alternatives are specified in the given statement. Thus such a type of statement is also not of interest since its correctness or falseness is evident from the statement form. It is essentially a constant.

The third type of statement is of interest because of the employment of correct reasoning or correct inference. For example, if it is true that "all control systems work" and that "Bob owns a control system," then the application of logical inference to the combination of these two statements implies that "Bob owns a working control system." Thus it is demonstrated (or inferred) that Bob's system is operational. Of course, it is assumed that the first two statements are true and that a correct or logical inference is made. This type of reasoning is employed in the design and development of digital systems. It was indicated earlier that a digital-computer system internally processes only the digits 1 and 0, which represent coded information. This coded information can be thought of as a binary (base 2) number. The computer manipulates these 1s and 0s based on a logical process. To understand this process, a 1 or a 0 is used to indicate the trueness or falseness of a particular statement. *The formal structure of boolean algebra is basically a study in symbolic logic, or the study of logical processes.*

2-3B Logic Operations and Truth Tables

Statements must be classified as being either true or false and are constrained in this manner as to be identified as *logical statements*. Statements that are "perhaps true" or "maybe false" are not acceptable. To develop a design and analysis technique for logical statements of the proper form, it is required to develop a set of symbols and relationships (equations). That is, verbal or written statements must be translated into symbolic statements.

A letter may be used to indicate a logical statement. This statement is defined as T or F to indicate that a "1" bit is associated with T and a "0" bit with F. Consider the following problem statement concerning the development of a simplified home entry alarm:

"An alarm signal is to be generated if the front or the back door is open, *provided* that the hallway and living room lights are out and the system switch is on."

Statement	Symbol
The alarm is generated	A
The back door is open	B
The front door is open	E
The hallway light is out	H
The living room light is out	L
The system switch is off	S

For symbolic notation, a statement symbol is assigned to each statement as shown in the above table. Associated with the statement symbols which are also statement variables are the symbols T and F (true and false). Such a variable can be either T or F, depending upon the correctness or falseness of the associated statement.

Returning to the problem, consider the alarm sounding statement, statement A, as the output (dependent variable) of the system and the other statements as input (independent variables) statements. It is desired to derive a functional relationship between A and the input variables. This symbolic relationship, if found, can be used to define a logical process that performs the required information processing.

Before doing so, additional symbols are required to indicate the functional relationship between A and the other variables. Consider the statement

"The front door is NOT opened"

This is defined as a negative statement as compared to

"The front door is opened"

which is considered to be a positive statement. Since the positive statement in this case is identified by E, it is desired to have a symbolic indicator that would change a positive statement into a negative statement. A line (bar) is used over the statement symbol to indicate the negative form. Thus, $\bar{E}$ is called the NEGATION of E. This operation is very useful in the manipulation of statement variables. Other symbolic relations are also required to reflect the OR and AND aspects of the original problem. That is, symbolically it is desired to represent A as a function of E, H, L, and S involving the relationships between these input variables. Consider the AND of the hallway light (H) and the living room light (L). Define the relationship

$$H \cdot L$$

to mean that the combined statement $H \cdot L$ is true if and only if both statements H *and* L are true, T. The dot ($\cdot$) relationship connects the two statements in an AND operation.

Next, for the OR relationship, consider the combined statement from the problem:

"The front OR back door is open"

This combined statement is described symbolically as

$$B + E$$

which means that the combined statement $B + E$ is *true if and only if either B or E is true or both.* The plus ($+$) sign thus symbolically connects the two statements B and E in an OR operation.

Returning to the original objective of generating a functional relationship between system input and output, it is necessary to decompose the given problem statement into the desired relationships by using the AND, OR, and NEGATION symbols. Using these symbols, the statement S (system switch off) is combined with $H \cdot L$ in the proper AND representation as

$$H \cdot L \cdot \bar{S}$$

Thus A is defined by the combined statement expression

$$(B + E) \cdot (H \cdot L \cdot \bar{S})$$

The parentheses are employed to enclose or indicate the various parts of the problem statement. The verb "provided" in the problem statement generates the AND relationship connecting the two main parts of the problem statement. It should be mentioned that the various relationships AND, OR, and NEGATION are sometimes referred to as *connectives.*

A final symbol to be described is the equal (=) sign:

$$A = (B + E) \cdot (H \cdot L \cdot \bar{S})$$

means that A is true if and only if the combined statement on the right-hand side of the "=" is true. Thus a boolean equation representing the logical switching process has been derived. Using this notation, modeling or designing of process control switching operations can be described.

In this section, so far, an attempt has been made to describe the connectives AND, OR, NEGATION, and = and the relationships using these connectives. To obtain a better perspective on these operations, consider connecting two general statement variables (A, B). Since there are only four combinations of T and F for these two variables, a table depicting these values can completely define the logic connective. Such a table for the AND operation is shown as follows for both T, F and 1, 0 encoding:

A	B	$A \cdot B$
F/0	F/0	F/0
F/0	T/1	F/0
T/1	F/0	F/0
T/1	T/1	T/1

This is defined as positive logic since T = "1" and F = "0." Negative logic occurs if, instead, the encoding is T = "0" and F = "1." This table is also known as a *truth table* since it shows all the possible true values of the combined statement ($A \cdot B$). Truth tables for the other operations can also be generated as follows:

A	B	$A + B$	$A = B$	$\bar{A}$	$\bar{B}$
F	F	F	T	T	T
F	T	T	F	T	F
T	F	T	F	F	T
T	T	T	T	F	F

The four connectives are generally referred to as *logical* or *boolean* operations.

The electronic equivalent of the AND function is the AND gate. It is called a gate since, by the previous switch analogy, the input signal is "gated" or allowed to pass only if both switches are closed (true). The logic symbol for an AND gate is

Table 2-6 Standard logic gate symbols

Symbol	Name	Boolean equation
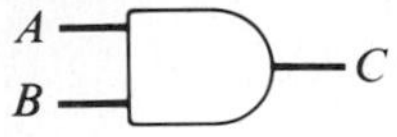	AND	$C = AB$
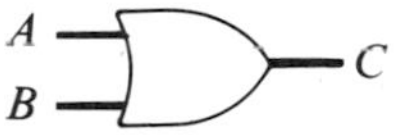	OR	$C = A + B$
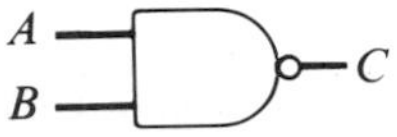	NAND (NOT-AND)	$C = \overline{AB}$
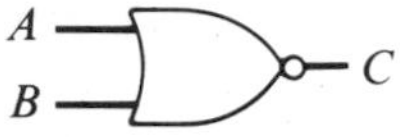	NOR (NOT-OR)	$C = \overline{A + B}$
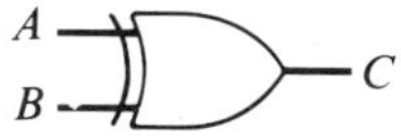	EXCLUSIVE OR	$C = A\bar{B} + \bar{A}B = A \oplus B$

adopted from the logic symbol standard IEEE Std. 91-1973 (ANSI-Y32.14-1973). The standard gate symbols are shown in Table 2-6.

When applying the NOT function in manipulating symbolic statement functions, the following relationships due to DeMorgan are useful:

$$\overline{A + B + C + \cdots + N} = \bar{A} \cdot \bar{B} \cdot \bar{C} \cdots \bar{N} \tag{2-18}$$

$$\overline{A \cdot B \cdot C \cdots N} = \bar{A} + \bar{B} + \bar{C} + \cdots + \bar{N} \tag{2-19}$$

Thus the complement of the AND function (NOT-AND or NAND) as shown in Table 2-6 is the OR of the complements of the inputs. In digital logic, this means that in a NAND gate the result is true if any input is false.

If the EXCLUSIVE OR is complemented, that is $A \oplus B = C$, the EXCLUSIVE NOR (COINCIDENCE or COMPARISON) function is obtained. If A is equal to B, then C is true. This may be most readily seen by complementing (negating) the output in the following truth table for the EXCLUSIVE OR.

EXCLUSIVE OR				EXCLUSIVE NOR/ COINCIDENCE		
A	B	C	$\bar{C}$	A	B	$\bar{C}$
0	0	0	1	0	0	1
1	0	1	0	1	0	0
0	1	1	0	0	1	0
1	1	0	1	1	1	1

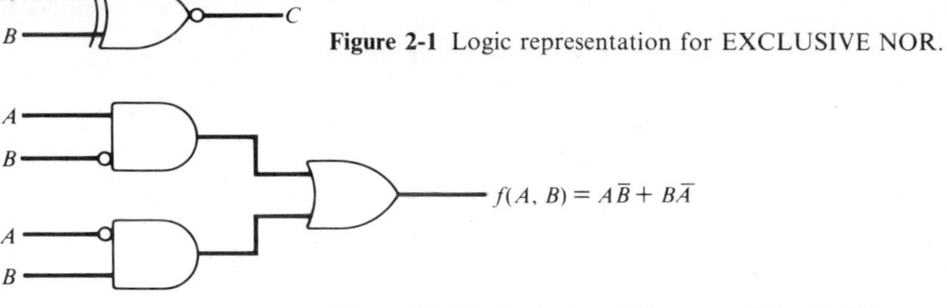

Figure 2-1 Logic representation for EXCLUSIVE NOR.

Figure 2-2 The logic circuit diagram of Eq. (2-20).

That is, the comparator gives a true output when both inputs are equal. This function can be obtained from the EXCLUSIVE OR by adding an inverter to the output as shown in Fig. 2-1.

The logic circuit symbols can now be used to generate *logic circuit diagrams* that represent an appropriate boolean function. Since this function involves boolean variables and the operators $+$, $\cdot$, $-$, then these operations can symbolically be represented.

Example 2-11 Consider the boolean function (EXCLUSIVE OR)

$$f(A, B) = A\bar{B} + B\bar{A} \tag{2-20}$$

whose logic circuit diagram is shown in Fig. 2-2. This logic circuit is defined as a *three-level logic* since the signals have to pass through three levels of gating including the inverters. This same function can also be represented as

$$F(A, B) = (A + B)(\bar{A} + \bar{B}) \tag{2-21}$$

Figure 2-3 is the logic circuit diagram for this presentation. The circuits of Figs. 2-2 and 2-3 require three gates and two inverters.

The same logic symbols can be used for three, four, or more inputs. The boolean operator indicated by the circuit symbol operates on all the inputs. For example if $f(A, B, C) = A + B + C$, then a three-input OR gate is required as illustrated in Fig. 2-4.

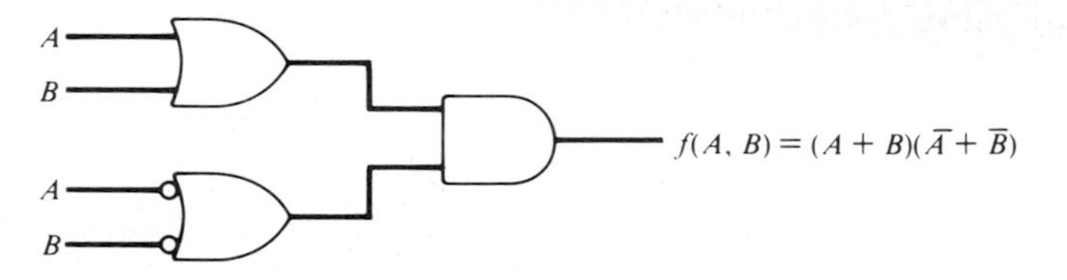

Figure 2-3 The logic circuit diagram of Eq. (2-21).

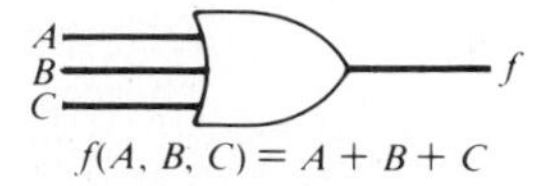

Figure 2-4 Logic circuit diagram for three inputs.

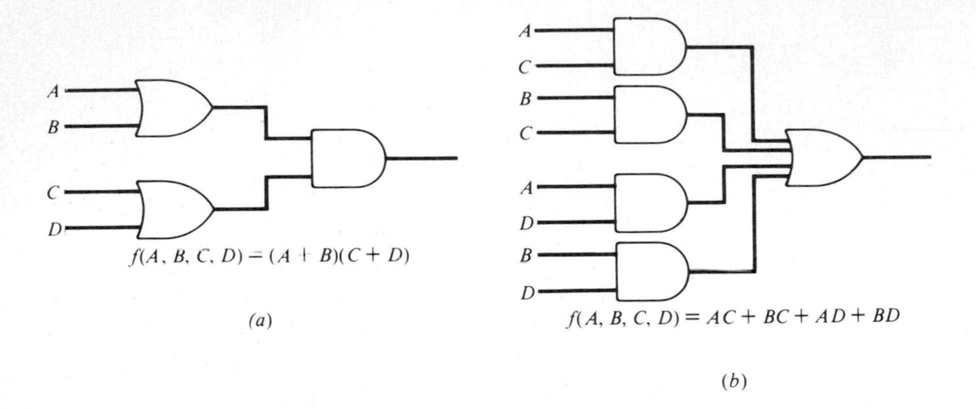

Figure 2-5 Logic circuit diagram for (*a*) Eq. (2-22*a*) and (*b*) Eq. (2-22*b*).

Example 2-12 Consider

$$f(A, B, C, D) = (A + B)(C + D) \tag{2-22a}$$

$$= AC + BC + AD + BD \tag{2-22b}$$

A logic circuit diagram for each of these representations of $f(A, B, C, D)$ is generated (Fig. 2.5). In general, the circuit in Fig. 2-5*a* is preferred since three logic gates are required instead of five gates as in Fig. 2-5*b*. The problem addressed here is the development of a procedure that ensures the minimum number of logic gates without having to compare circuit diagrams in various configurations.[8]

Returning to the alarm example in Sec. 2-3A, the logic diagram corresponding to the derived boolean equation is shown in Fig. 2-6.

2-4 COMPUTER ARCHITECTURE

The five parts of an algorithm or logical process are input, output, calculation, control, and specification of information. In logical processing, all computers also have five basic parts: one for input, one for output, one for calculation, one for

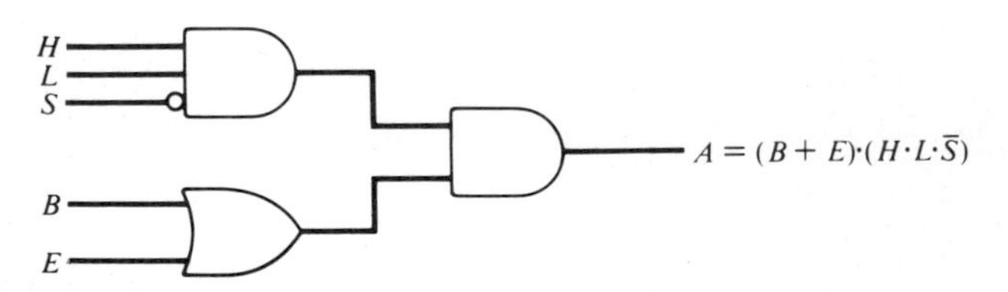

Figure 2-6 Logic circuit diagram of alarm system.

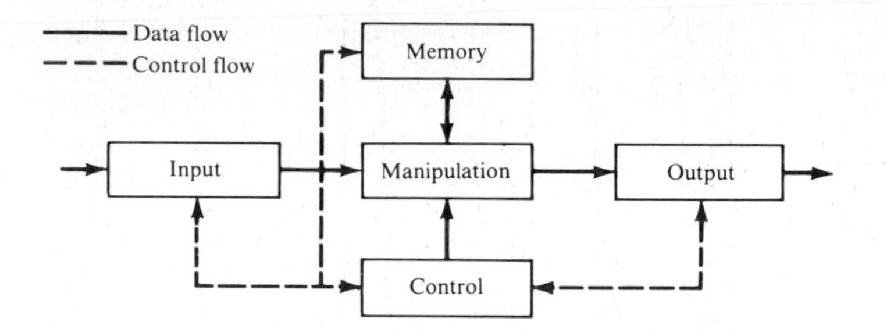

Figure 2-7 Basic functional units of an information processing system.

control, and one for data storage (the memory). The correlation between the first four parts of an algorithm and the first four parts of a computer is obvious. The relation, however, between the information specification part of an algorithm and the memory is not quite so obvious. The information specification part of an algorithm tells what information has to be stored and the form in which it has to be stored. The memory is that part of the computer which stores the information as directed.

While the five basic parts of a computer are common to all computer system hardware organizations, the exact way they are constructed and connected varies. Figure 2-7 depicts a simple computer architecture. This structure emphasizes the fact that the control unit controls directly the flow of data between the other units. The control unit specifies the sequence, origin, and destination of the data. The control unit acts as a buffer between interunit communication. All data passes through the manipulation unit. Note that the combination of the control unit and the manipulation unit is sometimes called the *central processing unit* (CPU).

A basic disadvantage of this structure is that it is generally too slow. Since all information must pass through the manipulation unit, generally only one action can be accomplished at a time; each operation must wait until the last operation has been completed.

Figure 2-8, however, illustrates a typical contemporary configuration. Also, note that information communication is by means of a shared bus (set of wires with minimal logic). This structure generally permits easy connection of additional devices to the system. Many contemporary computer architectures have various buses

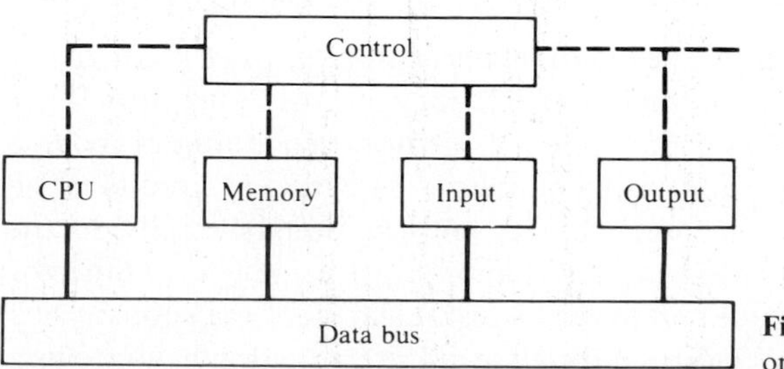

Figure 2-8 A contemporary variation on functional unit interconnections.

for this purpose as well as for increased speed. The memory is the center of the system. Most information is transferred to the memory instead of through the manipulation unit. Now, operations can be overlapped (parallel/concurrent). The control unit can signal a transducer to start reading and signal the memory where to store the incoming reliable data. As soon as the first values are in the memory, the control unit can start the manipulation unit solving a difference equation. Simultaneously, then, the input unit can read data for the next variable values and the output unit can transfer control signals to the plant. The details of this algorithm are rather involved, as is seen later. However, the speed gain is significant, which is usually very important in a control-system implementation. A selection of a particular architecture depends not only on the control laws being implemented but also on practical engineering and management constraints.

2-4A Input-Output Devices

Besides analog-to-digital and digital-to-analog converters, many types of devices are available for input and output of information. For example, terminals and typewriters are often used as input devices. Magnetic tape and "floppy" disks are also used for both inputting data and storing the output data. The number of possible input devices which can be used is limited only to technology and human enterprise. Examples include devices that "read" variable quantities like speed, acceleration, thickness, temperature, humidity, and air pressure; devices that "read" the number of objects in a given region or passing a given point; and devices that "read" printed material. All these input devices can be very useful in a digital control environment depending upon the objective.

Output devices are as numerous and diverse as the input devices. The mechanical printer is essentially a sophisticated typewriter which prints a whole line at a time. A magnetic disk [flexible (floppy) and hard] is one of the fastest of all the output devices available. It is often used for storing results produced by one program and as the data for another program. That is, variable data are brought into the system with a converter interface and stored on the disk. When time is available, these data are transferred to main memory and the CPU for processing. The main aspects influencing a selection depend upon efficiency (speed, size, etc.) vs. economics (cost, energy, etc.).

2-4B Memory and Information Specification

Computer main memories are constructed in many different ways with many different characteristics. The distinctions between the various types are fairly technical and are not discussed here. Examples of memory technologies include magnetic core, magnetic disks, and semiconductor integrated circuits. The technique to locate and retrieve memory binary information is to divide the memory into a series of small words each capable of storing a single number or a few characters and assigning an address to each word. Each word may consist of a string of 1 to 60 bits depending on the specific system architecture. Each word has a unique location called an *address*. Thus the memory addressing schemes play an important role in control-law implementation. The addressing techniques used for information storage and retrieval influence the speed of algorithm execution and, moreover, the ease of implementing a digital control system.

2-4C Manipulation Unit (Arithmetic-Logic Unit)

The manipulation unit is an arithmetic and logic processor. For example, given two binary numbers, it can generate either their sum, difference, product, or quotient. In this sense it does exactly the same thing (and is designed the same way) as an ordinary desk calculator, cash register, accounting machine, or any other machine designed to do arithmetic. It can also do logic or boolean operations on separate bits of information as permitted by the machine design and computer language used.

2-4D Control Unit

One of the most important parts of any computer is the control unit or register-data-transfer sequencer. The control unit automatically activates the units in the correct order for translation of each machine-level instruction. Note that this means the control unit must decode the instruction into a sequence of transfer and manipulation commands or tasks. This unit is a translator of machine language (bits) into data-transfer operations inside the computer. The details of the associated operations are beyond the scope of this text. It is clear that this *control sequence process is in reality a translator from one level of observation to another.*

As implied in the last section, computer system software can be discussed in terms of a hierarchy. The general interface between software and hardware is usually denoted at the machine-language level, i.e., instruction defined in terms of 1s and 0s. As the hierarchy is transcended (see Fig. 2-7), one realizes that numerous translators are required to move from one level of observation to another. In addition to languages, language translator and operating systems, special utility programs for ease of system use are included. Examples of utility programs include editors, file systems, and libraries. All these utility programs are very useful in the economics of digital-control-system development. The following sections discuss two of these aspects. The following chapters present more detailed discussions.

2-4E Operating Systems

Since all the functional units of a digital control system work at different speeds (i.e., a real-time transducer is usually much slower than the manipulation unit), scheduling the work tasks requires considerable effort. Logically, the operating system is a master task-scheduling program that supervises the control unit, indicating the next program to initiate transferring input and output information.

The control unit controls the other four parts of the computer during the execution of each individual instruction, and the operating system sequences or supervises the tasks associated with the digital-control-system program at the higher level.

2-4F Application Programs

In the implementation of a digital control system, some of the many activities include:

Transferring data from the analog world to the digital or discrete world
Computing a difference equation solution and storing appropriate information
Transferring control data from digital to analog plant input
Checking for erroneous data and performing corrective action

In implementing these processes, an individual must understand the program requirements, evaluate alternative algorithm structures, select the optimal approach, use good programming structure and style, and employ software engineering methods. The design and development of computer system hardware involves the application of physical laws and good engineering practice. The control programmer usually has a very limiting set of constraints (especially time). However, a control-law algorithm can exist in many forms. This phenomenon generally prohibits interchanging of existing programs for different computer systems. Also, it may become very difficult to understand, modify, and test someone else's program. Thus the concepts of good programming style, structure programming, and software engineering have evolved.

2-5 SIGNED NUMBER REPRESENTATIONS AND ARITHMETIC PROCESSES

One of the major advantages of a machine that can perform manipulations on binary data is its application to arithmetic operations. This section focuses on the development of a coding process for signed binary numbers, discusses various arithmetic algorithms using these coded numbers, and presents the design and analysis of digital networks that reflect specific arithmetic algorithms.

2-5A Representation of Signed Binary Arithmetic Numbers

The sign of a binary number can be represented by adding a digit b_{M+1} to the left of the most significant digit b_M. That is,

$$N = b_{M+1} + b_M r^M + \cdots + b_L r^L = b_{M+1} + N_2 \tag{2-23}$$

where N_2 is the binary magnitude of the number. A signed binary is defined therefore in the following fashion:

$$\text{If } b_{M+1} = \begin{cases} 0 & \text{the number is positive} \\ 1 & \text{the number is negative} \end{cases}$$

Signed *positive numbers*, $b_{M+1} = 0$, are always represented in the form of *sign + magnitude*. Thus from Eq. (2-23)

$$N = 0 \times 2^{M+1} + N_2 = 0 \times 2^{M+1} + \sum_{i=L}^{M} b_i 2^i \tag{2-24}$$

Signed *negative numbers*, $b_{M+1} = 1$, using Eq. (2-24), may be represented in three different ways:

1. *Sign plus magnitude*:

$$N = 1 \times 2^{M+1} + N_2 = 1 \times 2^{M=1} + \sum_{i=L}^{M} b_i 2^i \tag{2-25}$$

2. *Sign plus 2's complement*:

$$\begin{aligned} N &= 1 \times 2^{M+1} + \bar{N}_2 \qquad \text{(2's complement of } N_2) \\ &= 1 \times 2^{M+1} + \sum_{i=L}^{M} \bar{b}_i 2^i + 2^L \\ &= 1 \times 2^{M+1} + (2^{M+1} - N_2) \end{aligned} \tag{2-26}$$

3. *Sign plus 1's complement*:

$$\begin{aligned} N &= 1 \times 2^{M+1} + \bar{N}_2 \qquad \text{(1's complement of } N_2) \\ &= 1 \times 2^{M+1} + \sum_{i=L}^{M} \bar{b}_i 2^L \\ &= 1 \times 2^{M+1} + (2^{M+1} - N_2 - 2^L) \end{aligned} \tag{2-27}$$

Note: Confusion may arise in the literature when the three classifications above are used as titles for three different mathematical representations. It must be remembered that *positive numbers are always represented mathematically as sign bit* ($b_{M+1} = 0$) + *binary magnitude regardless of the representation specified.*

Example 2-13 Consider $N_{10} = +7$ and -6, converted to a base 2 number N:

Representation	+7	−6
Sign plus magnitude	0111	1110
Sign plus 2's complement	0111	1010
Sign plus 1's complement	0111	1001

Comparison of numbers To determine which of two binary numbers is larger, bit comparison is used. This method first compares the sign bits. If they are different, the number with the positive sign ($b_{M+1} = 0$) is larger. If both numbers are positive, then the bits are compared starting with the most significant bit (MSB).

The number with the first larger bit (i.e., the number with the first "1" bit occurring in the same position as a "0" bit in the other number) is the larger number. If both numbers are negative, the sign plus magnitude approach is employed. The bits are compared in the same order and the larger number is the one with the first smallest digit. For negative numbers in sign plus 2's complement or sign plus 1's complement representation, the same procedure is used except that the larger number is the one with the first largest bit ($1 > 0$).

Fixed-point representation For structural reasons pertaining to arithmetic algorithms, the placement of the radix point is usually fixed before the MSB. The $M + 1 = 0$ bit then represents the sign of the number, the magnitude of the number being a fraction. If the magnitude of the number to be operated on then is greater than or equal to 1, a shift is made such that the number becomes a fraction. If the sign bit is coded in the 2^0 position, then the following formulas define the fixed-point binary number presentation (note that L is a negative number):

$$N = b_0 2^0 + \sum_{i=L}^{-1} b_i 2^i$$

Sign plus magnitude (sign bit plus binary magnitude)

$$N = 0 \times 2^0 + \sum_{i=L}^{-1} b_i 2^i$$

Sign plus 2's complement: positive numbers (sign bit $b_0 = 0$ plus binary magnitude)

$$N = 1 \times 2^0 + \sum_{i=L}^{-1} \bar{b}_i 2^i + 2^L$$

Sign plus 2's complement: negative numbers (sign bit $b_0 = 1$ plus 2's complement)

$$N = 0 \times 2^0 + \sum_{i=L}^{-1} b_i 2^i$$

Sign plus 1's complement: positive numbers (sign bit $b_0 = 0$ plus binary magnitude)

$$N = 1 \times 2^0 + \sum_{i=L}^{-1} \bar{b}_i 2^i$$

Sign plus 1's complement: negative numbers (sign bit $b_0 = 1$ plus 1's complement)

Example 2-14 This example presents the fixed-point binary representation of $+.375_{10}$ and $-.375_{10}$ using a 5-bit word.

Representation	$+.375_{10}$	$-.375_{10}$
Sign plus magnitude	00110	10110
Sign plus 2's complement	00110	11010
Sign plus 1's complement	00110	11001

Note from this example that the sign bit can be treated as just another number bit when generating a negative representation. To generate the negative sign plus 1's complement representation of a positive binary number, for example, toggle each bit including the sign bit.

In the various representations, N is now bounded in the following range where b_0 is considered as part of the number:

$$0 \leq N < 2$$

where $0 < N < 1$ when the number is positive, and where $1 \leq N < 2$ when the number is negative.

The placement of the binary point at the indicated position means that the multiplication process does not generate a number greater than or equal to 1. Thus in the multiplication process there is no overflow into the sign-bit location. In the addition, subtraction, and division processes, there can be an overflow. If there is, the programmer or digital system must identify this condition before continuing processing.

The binary adder Consider the addition of two bits x_i and y_i and the associated sum s_i and carry c_{i+1}.

The truth table for adding two 1-bit binary numbers $x_2 = x_i$ and $y_2 = y_i$, with the associated carry c_{i+1}, as shown below, is given in Table 2.7.

Carry		c_{i+1}	c_i
Augend	$x_2 =$		x_i
Augend	$y_2 =$		y_i
Sum	$f_2 =$		f_i

The *boolean expressions* for performing this *binary addition* use the EXCLUSIVE OR as follows: $s_i = x_i \oplus y_i$ where $c_{i+1} = x_i y_i$.

For the addition of 2-bit binary numbers $x_2 = \cdots x_{i+1} x_i x_{i-1} \cdots$ and $y_2 = \cdots y_{i+1} y_i y_{i-1} \cdots$, it is necessary to develop the concept of a *full adder*. The symbol w is used for the resulting generated carries for the full adder as shown below:

Carry		$\cdots w_{i+1}$	w_i	w_{i-1}	$\cdots$
Augend	$x_2 =$	$\cdots x_{i+1}$	x_i	x_{i-1}	$\cdots$
Augend	$y_2 =$	$\cdots y_{i+1}$	y_i	y_{i-1}	$\cdots$
Sum	$f_2 =$	$\cdots f_{i+1}$	f_i	f_{i-1}	$\cdots$

Table 2-7 Truth table for addition of x_i and y_i

x_i	y_i	s_i	c_{i+1}
0	0	0	0
0	1	1	0
1	0	1	0
1	1	0	1

Table 2-8 Truth table for the addition of x_i and y_i and the carry w_i

x_i	y_i	w_i	f_i	w_{i+1}
0	0	0	0	0
0	0	1	1	0
0	1	0	1	0
0	1	1	0	1
1	0	0	1	0
1	0	1	0	1
1	1	0	0	1
1	1	1	1	1

The truth table for various combinations of w_i, x_i, and y_i is presented in Table 2-8. The boolean expressions for performing the binary addition and obtaining f_i and the carry w_{i+1} are, respectively.

$$f_i = \bar{x}_i\bar{y}_i w_i + \bar{x}_i y_i \bar{w}_i + x_i \bar{y}_i \bar{w}_i + x_i y_i w_i$$

$$= x_i \oplus y_i \oplus w_i$$

$$w_{i+1} = x_i y_i + x_i w_i + y_i w_i = x_i y_i \oplus x_i w_i \oplus y_i w_i$$

$$= x_i y_i + x_i w_i \oplus y_i w_i = x_i y_i + (x_i \oplus y_i) w_i$$

2-5B Binary Addition Algorithms

In binary addition, the addition operation is considered by using the various binary number representations. Algorithms are generated based on these operations and the assumptions of a finite register length.

For the sign plus magnitude representation, let X and Y be two arbitrary fractional binary numbers, i.e.,

$$X = x_0 2^0 + \sum_{i=L}^{-1} x_i 2^i = x_0 2^0 + X_2$$

$$Y = y_0 2^0 + \sum_{i=L}^{-1} y_i 2^i = y_0 2^0 + Y_2$$

where X_2 and Y_2 are defined as the base 2 magnitudes of X and Y, respectively. The objective is to determine algorithms which generate the algebraic sum F of two signed binary numbers in sign plus magnitude representation given by the expression

$$F = f_0 2^0 + \sum_{i=L}^{-1} f_i 2^i = f_0 2^0 + F_2$$

where F_2 is the magnitude of F. In this representation, separate algorithms must be generated for f_0 and F_2.

Take the case where $x_0 = y_0$ (numbers of the same sign). Then

$$f_0 = x_0 = y_0$$

and by straight binary addition,

$$F_2 = X_2 + Y_2$$

From the results of the previous subsection,

$$f_i = x_i \oplus y_i \oplus w_i \qquad i = L, \ldots, -1$$

where

$$w_L = 0$$

$$w_{i+1} = x_i y_i \oplus x_i w_i \oplus y_i w_i \qquad i = L, \ldots, -1$$

where w_i represents the carry from the adjacent and least significant bit position due to addition. The $\oplus$ operation is the logical EXCLUSIVE OR and is called the *modulo 2 sum*. Note that in modulo 2 arithmetic the equation for w_{i+1} yields 1 if at least two of x_i, y_i, and w_i are 1s. These relationships can easily be defined symbolically and mechanized with logic gates as shown previously.

In practice, the binary addition procedure consists of first using modulo 2 arithmetic to generate $x_i \oplus y_i$ and w_i, then summing these values. All binary sums are usually stored in a register of the arithmetic unit called the *accumulator register*.

For the sign plus 2's complement representation, the algebraic sum of two numbers is found by straight binary addition of the numbers. The sign bit is treated as part of the number, and any carry beyond the sign bit is ignored. (This is shown in the following examples.)

Consider X, Y, and F to be binary numbers in sign plus 2's complement representation.

$$X = x_0 2 + X^* = x_0 2^0 + \sum_{i+L}^{-1} x_i^* 2^i$$

$$Y = y_0 2^0 + Y^* = y_0 2^0 + \sum_{i=L}^{-1} y_i^* 2^i$$

$$F = X + Y = f_0 2^0 + F^* = f_0 2^0 + \sum_{i=L}^{-1} f_i^* 2^i$$

where the operator $*$ is defined as follows:

$$X^* \triangleq X_2 = \sum_{i=L}^{-1} x_i 2^i \qquad \textit{if X is positive}$$

$$X^* \triangleq \bar{X}_2 \qquad \text{(2's complement)} = 2^0 - X_2$$

$$= \sum_{i=L}^{-1} \bar{x}_i * 2^i + 2^L \qquad \textit{if X is negative}$$

It remains to be shown that straight binary addition of X and Y yields the correct answer. It should be obvious that this is true when both numbers are positive, since they are both in sign plus magnitude form. *Note*: When $x_0 = y_0$, the programmer must assure that $X_2 + Y_2$ does not result in a carry into the 2^0 position since the sign bit is treated as part of the number.

Example 2-15 Given the addition of +0.01001 and +0.00111.

(*a*) Straight addition

$$\begin{array}{ll} X = +0.01001 & \\ \underline{Y = +0.00111} & \\ F = +0.10000 & F_2 = X_2 + Y_2 \end{array}$$

(*b*) Sign plus 2's complement representation addition. (Remember, positive numbers are always in sign plus magnitude, regardless of representation used.)

$$\begin{array}{l} X = 0.01001 = 0 \times 2^0 + X_2 \\ \underline{Y = 0.00111} = 0 \times 2^0 + Y_2 \\ F = 0.10000 \qquad X_2 + Y_2 = F_2 \end{array}$$

(*c*) Addition algorithm (manual technique with carries propagate)

$$\begin{array}{ll} X = 0.01001 & x_0, x_i = x_i^* \\ \underline{Y = 0.00111} & y_0, y_i = y_i^* \\ F = 0.10000 & X_2 + Y_2 = F_2 \end{array}$$

(*d*) Addition algorithm (modulo 2 bit addition)

$$\begin{array}{ll} X = 0.01001 & x_0, x_i = x_i^* \\ \underline{Y = 0.00111} & y_0, y_i = y_i^* \\ \quad 0.01110 & x_i \oplus y_i \\ \underline{\quad 0.11110} & w_i, w_L = 0, w_{i+1} = x_i y_i \oplus x_i w_i \oplus y_i w_i \\ F = 0.10000 & x_i \oplus y_i \oplus w_i \\ F = 0.10000 \text{ or } F = +0.10000 & \end{array}$$

Example 2-16 Given $x_0 = y_0 = 1$.

(*a*) Straight addition

$$\begin{array}{ll} X = -0.01001 & \\ \underline{Y = -0.00111} & \\ F = -0.10000 & F_2 = X_2 + Y_2 \end{array}$$

(*b*) Sign plus 2's complement representation addition

$$\begin{array}{l} X = 1.10111 = 2^0 + X^* = 2^0 + (2^0 - X_2) \\ \underline{Y = 1.11001 = 2^0 + Y^* = 2^0 + (2^0 - Y_2)} \\ (1)1.10000 = 2^1 + 2^0 + [2^0 - (X_2 + Y_2)] \\ \text{Ignore} \qquad\quad = 2^1 + 2^0 + [2^0 - F_2] \\ \qquad\qquad\quad \text{Ignore Signbit 2's complement} \end{array}$$

Thus $F = 1.100000$ (sign plus 2's complement form) or $F = -0.10000$. Note that the 2's complement of 0.10000 is 0.10000.

(*c*) Addition algorithm

1.10111	x_0, x_i^*
1.11001	y_0, y_i^*
0.01110	$x_0 \oplus y_0, x_i^* \oplus y_i^*$
11.11110	$w_i, w_L = 0, w_{i+1} = x_i^* y_i^* \oplus x_i^* w_i^* \oplus y_i^* w_i^*$
Ignore	
1.10000	$f_0 = x_0 \oplus y_0 \oplus w_0, f_i = x_i \oplus y_i \oplus w_i$

Thus $F = 1.10000$ (sign plus 2's complement form) or $F = -0.10000$.

2-5C Floating-Point Number Representation

Floating-point representation of binary numbers by definition "floats" the binary point through the use of an exponent operation. The following general form indicates this capability:

$$N_2 = \text{mantissa} \times 2^{\text{exponent}}$$

where the mantissa is a fixed-point binary number with the sign bit in the 2^0 position. The exponent generally consists of fewer digits than the mantissa and also can be a positive or negative binary number.

The previous signed number representations can be used for the mantissa and the exponent. Other representations are possible. As an example consider $+1101.101_2$ represented as $0.1101101 \times 2^{0100}$. Note that the word length of the computer has a rather strange influence on the accuracy of the mantissa and the range of possible numbers. In many small-word-length machines, two computer words are used to represent a floating-point number structure.

In regard to arithmetic operations, fixed-point operations are executed separately on the mantissa and the exponent after proper shifting (normalization). Various references[7,9] discuss these methods.

In general, owing to floating-point software or hardware delays, many digital control systems are implemented by using fixed-point number structures. Throughout the remainder of this text, fixed-point representation is assumed unless otherwise denoted.

2-6 ALGORITHMS

In an individual's daily life, the solutions to everyday problems involve the use and execution of processes. These processes usually require some initial or a priori stored information. This information is then manipulated according to some procedure. The results of such a manipulation are sometimes referred to as decisions, or approximations, or control outputs. These tasks then involve the concept of information processing. That is, "input information" is transformed through some processing operation, resulting in "output information." These processes can be quite abstract, such as making a value judgment, or may be simple information processes, such as turning on a light switch. Note the extension from the discussion of logical processes in Sec. 2-3.

The information processes considered in this text are restricted to those that can be executed by a digital computer. Processes restricted in this manner are defined as *algorithms* (logical processes). A further restriction is the emphasis on digital-control-system algorithms. For the computer to manipulate the information properly it must be coded in the machine's language. The result of this logical translation process is defined as an *algorithm realization.*

An algorithm must (1) describe the information, (2) describe the required actions or manipulations, and (3) define the sequence or order in which these actions are to be executed. The basic elements of an algorithm are:

Information specification
Input procedure
Manipulations (actions, calculations, etc.)
Output procedure

Information specification encompasses data, control, and status types depending on how it is used. This information may be of the form of a circuit, a voltage, pressure, paper numbers, names, or codes. The input procedure again depends upon the process and its implementation. The input procedure may involve normal movement of files, electron gating, or a control transducer data input. The required manipulation of the information depends upon the specific process. This operation may involve the numerical solution of a differential equation for control or the updating of an instrument data file. The output procedure translates the results into proper output form such as a feedback transfer.

To perform input, calculations, and output, a logical control procedure is required. This logical procedure is accomplished by the use of an orderly sequence structure. The sequence of the algorithm steps can take many, many forms. This phenomenon permits numerous ways of realizing the same logical process. Thus there are many system designs or computer programs that have the same input-output relationship when viewed from a high level of observation.

An algorithm can be pictorially described by a flowchart. This type of diagram depicts the sequence structure of the program in an easily recognized manner. An algorithm that has been realized in computer software (i.e., everything that is not hardware) is called a *computer program.* Of course, there are various levels of computer programming languages. The ease of effectively translating an algorithm into a computer language depends upon the existence of a suitable language.

It will be evident that there are many levels of observation on which algorithms can be described. An algorithm can describe a logic gate design at a low

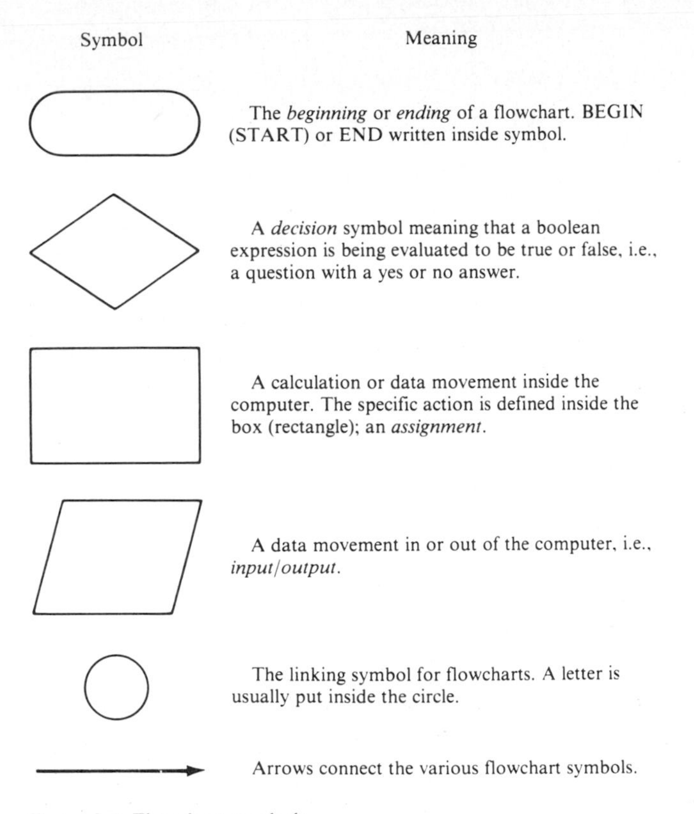

Figure 2-9 Flowchart symbols.

hardware level or a complex chemical plant control process. In essence, all levels of a computer system architecture or digital control process can be described by algorithms. In digital-control-algorithm implementation, many parameters must be considered. They include analog-to-digital conversion time, computation time, sequencing of algorithm tasks when many system states are to be monitored and controlled, and consideration of numerical accuracy.

In order to describe in detail a digital control process, a proper sequence of instructions must be defined. This sequence is in reality a *program control structure* or *flow of instructions*. For example, a correct instruction sequence must be generated so that data is properly inputted, added, multiplied, stored, and outputted. A general way of representing this process is through the use of flowcharts. The general standard flowchart symbols are shown in Fig. 2-9. Considering the general solution to the difference equation, the flowchart of Fig. 2-10 is the starting point for producing a software program. It should be noted that a flowchart can be used at various software levels; i.e., it could depict the algorithm from the control engineer's symbolic point of view or it could represent each instruction at the machine level. Flowcharts can thus be used at various levels of algorithm description depending upon the documentation required and the

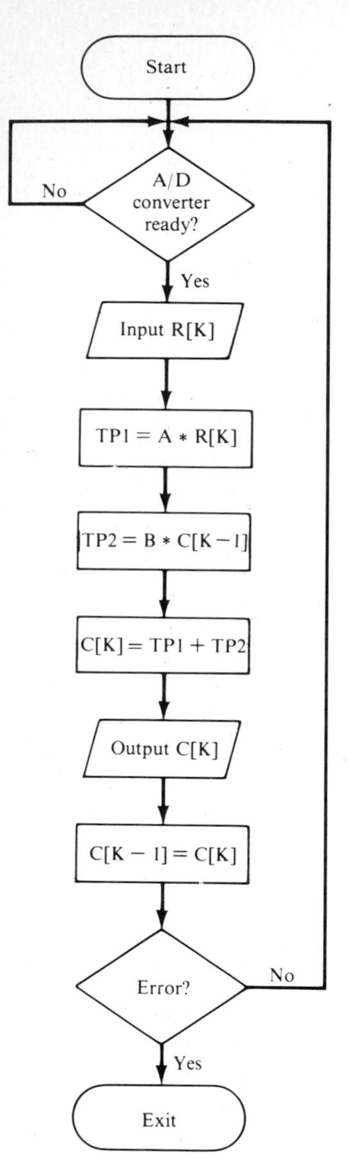

Figure 2-10 Flowchart for solving the difference equation

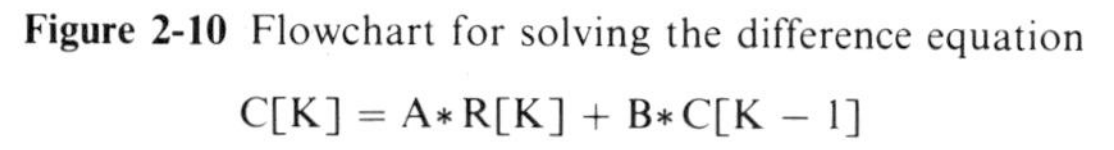

$$C[K] = A*R[K] + B*C[K-1]$$

software level of implementation. Many variations of these symbols exist. The choice is left to a user and experience.

Another way of representing the flow of an algorithm is *pseudo-English* or *pseudocode*. Pseudocode is no more than a very constrained subset of English permitting logical and nonambigious statements. It uses three basic constructs for defining program control flow. They are:

```
        BEGIN
LOOP:       If A/D conversion not complete THEN check again ELSE input R[K]
            Multiply R[K] by A and store as temporary variable TP1
            Multiply C[K - 1] by B and store as temporary variable TP2
            Generate C[K] by adding TP1 to TP2
            Output C[K]
            Update C[K - 1] by transferring the value C[K] to C[K - 1]
            If error condition is not true THEN jump to loop ELSE exit
        END
```

Figure 2-11 Pseudocode for solving difference equation C[K] = A∗R[K] + B∗C[K − 1].

Sequence {begin ··· end}
Selection {if [boolean logical expression]† is true, then ···, if false else ···}
Iteration {while [boolean logical expression]† is true, do ···}

In regard to the difference equation solution for digital control, the flowchart of Fig. 2-10 can also be represented in pseudo-English as shown in Fig. 2-11. In both figures, the implicit representation of T is not shown; that is, $C[K] = C[KT]$. This is done since inside the computer the only parameter of interest is the argument of the current sequence, K. T is defined by the sampling-time interval, the interface to the analog world.

Either of the two documentation techniques can be used. The selection depends upon the control engineer's experience and value judgments. It should be noted that either or both can be part of the total software engineering documentation effort as is discussed in Chap. 13. The following sections discuss specific languages for control-law implementation.

2-7 SOFTWARE LANGUAGE HIERARCHY

To direct a machine (the computer), a series of logical commands relating to the general difference equation solution must be logically defined. These sequential machine commands must be part of a language that the machine understands and are therefore written (coded) and stored in the computer's memory as 0s and 1s and are fetched in sequence as instructions to be executed. Of course, the machine is assumed to be able to perform the arithmetic operation defined by the difference equation as well as the storage of initial, transient, and final data. The form of this data storage or data structure is discussed later in this chapter. *The data structure bears a strong relationship to such things as computation time and memory requirements.*

Writing software or software programs does not have to be done with the 0 and 1 (binary)-level machine language. Obviously, mnemonics and labels can be

† The boolean expression must be evaluated to be *true* or *false*. Thus the expression (condition) indicates whether specific operational statements (···) are executed or not.

used to represent machine commands and memory address locations, permitting easier program development. But a translator must be provided that eventually generates machine language.[10] If the mnemonics very closely represent machine-language instructions, the associated language is called *assembly*. This term refers to translating each mnemonic individually to binary and then "assembling" each of the binary statements into the complete machine-language program. The translator is therefore called an *assembler*.

If the mnemonics are considerably more extensive; i.e., relate closely to the form of the human problem statement, a *higher-order language* (HOL) or *compiler language* is defined. As an example of a HOL implementation, consider the alarm system of Sec. 2-3. In this case, the specific boolean equation can be implemented in "software" directly; that is,

```
ALARM=(BACKDOOROPEN.OR.FRONTDOOROPEN).AND.
        (HALLLIGHTOUT.AND.LIVROOMLIGHTOUT.AND.SYSSWON)
```

Here, the variables are assumed to be properly defined as boolean variables in the chosen HOL. Also, the specific value of these variables which are being monitored continuously have to be read in with an additional series of instructions. Details of this procedure are left to Chap. 8. The ease of boolean equation HOL implementation for process control (or the compensator for digital control) does not immediately imply that programming real-time digital systems is difficult. However, the input-output operations can have a visible impact on performance as well as the finite word length of the computer. Chapter 10 focuses on this aspect of design and implementation.

In the digital control system's case, if the difference equation of the form

$$c[kT] = b_1c[(k-1)T] + b_2c[(k-2)T] \\ + a_0r[kT] + a_1r[(k-1)T + a_2r[(k-2)T]$$

can be directly translated into a HOL, i.e., a language closer to the control application with a language statement as follows:

```
C[K] = B[1] * C[K - 1] + B[2] * C[K - 2]
        + A[1] * R[K] + A[2] * R[K - 1] + A[3] * R[K - 2]
```

Note that the index on A has changed due to an addressing restriction of no "zero" indexes. Also, the variables are now capitalized and the multiplication operation is explicit.

A continuous spectrum of programming languages exists. A particular choice depends on the engineer's experience, the languages available in a given software development facility, and the problem constraints such as timing and space. Some particular HOL languages of interest to the control engineer are FORTRAN, BASIC, PASCAL, PL/1, and ADA. Many facilities provide HOL compilers (translators) for these systems. Note that at the machine- or assembly-language level the translators are usually unique to the hardware and associated programs are not transportable as compared to HOL.

Once the software program is in the equivalent of machine language, it must be loaded into the computer memory. A special program called a *loader* puts the machine instructions into memory along with the data addresses. Of course, the instructions can be loaded by hand through the computer's switch panel, but the loader is much more efficient.

2-7A Assembly Language

Assembly language usually has a constrained structure, since it is very close to machine-language organization. The fact that this relationship is so close permits efficient use of hardware resources (memory, CPU) and optimal execution speed. Each assembly-language instruction is usually described by one line of text, and each line has four fields named *label*, *code* (operation), *operand address*, and *comments*. The label field is often blank. If it is not blank, the symbol in that field represents the instruction address and may be used as a synonym for that location throughout the assembly-language program. The code field contains mnemonics for the operation codes, and the operand field contains source and destination address operands (usually in the form of labels). Descriptive remarks are placed in the comments field and separated from the other fields. The comment field is ignored by the assembler but is a very important documentation in terms of understanding the program.

In addition to mnemonics for instruction codes (assembly-language-equivalent machine instruction code), particular assembly languages have other operational instructions. These codes are called pseudoinstructions. An important example of a pseudoinstruction is the ORG instruction which specifies the origin or beginning of the program. In assembly language the symbol END denotes the end.

Appendix F presents the instruction set for the LSI-11 microprocessor as an example of assembly-language mnemonics. This is a relatively simple but extensive instruction set which is used in some chapter examples. This appendix also introduces the architecture of the LSI-11/PDP-11 family.

In analyzing a machine-language program the reader no doubt may have some difficulty reading the binary-coded instructions. The difficulty encountered in such a small procedure may be magnified many times if a large machine-language program is used. Reading and working with binary, hexadecimal, or octal numbers and remembering the meaning of numeric operation codes is difficult. Another problem in the use of machine language is that the use of actual binary addresses makes it inconvenient to modify a program. Inserting a new instruction causes the location of all the succeeding instructions and data constants to change. This requires an address change in each instruction that addresses any of these succeeding instructions or constants.

The major objective of assembly language is to substantially eliminate these difficulties. Assembly language achieves this by allowing the programmer to use symbolic names, called *symbols* or *labels*, for numbers. Since an instruction is composed of several fields, each of which contains a number, each field value may be written as a symbol. The number that a symbol represents may be an operation code, an address, a displacement, a register number, or any other number used in the composition of an instruction.

The value of a symbol is defined to be an actual binary number, after translation of course. Every symbol used in an assembly-language procedure must eventually have a value. But, the symbols or labels are defined within the assembly-language procedure by the assembly-language programmer.

The fundamental unit of any assembly language is the symbolic instruction called the *statement*, consisting generally of the four fields mentioned previously. In general, three classes of statements can be defined. They are:

Machine command statements
Assembler command statements (pseudoinstructions)
Data statements

To show how specific terms can be used to form complete statements, a simple structure is presented. The notation

(name of term)

is employed to indicate the different parts of a statement and the general function it serves. For example, the notation

(blank)

serves to indicate a sequence of spaces (delimiters) that separates other terms. The notation

(command)

is used to indicate a mnemonic symbol corresponding to one of the possible machine-language commands. Using these ideas, the "sentence" structure of the various types of statements can be studied as follows.

Command statements A command statement or instruction is translated by the assembler machine-language instruction. The structure of these statements depends on the type of operation that they represent. For the simple assembler, each command statement can have the following general structure:

(label): (blank)(command)(blank)(address information)(blank)(comment)

Not every statement contains all these parts. Note that the colon is also being used as a delimiter for labels.

In general, most machine command instructions reference data storage locations with the address information. These locations may be in main memory, and the associated instruction is defined as a *memory reference instruction* (MRI). Other instructions of this type reference CPU registers with the address information. These instructions are called *register reference instructions* (RRI).

The term

(address information)

is present if and only if the term

(command)

is the mnemonic for a machine-language memory reference instruction. If the instruction is a *direct* MRI, then this term is either an octal number representing the absolute address of a memory location or the label used to identify some memory location as mentioned previously. If the instruction is an *indirect* MRI, then the complete address structure has the form

1 (blank) (address information)

where 1 is used to indicate that the address information involves an indirect memory reference. The term (address information) is thus an octal number or label (symbol) indicating the memory location in which the address information is stored. *Note the difference between a numerical address and the data (value) stored in that address.*

The term

(label)

is present if an identifying label or symbol is attached to the statement so that it can refer to the statement in some other command statement. The colon following (label) is included as a signal to the assembler that the term (label) is to be used to identify the memory location that contains the indicated statement. Labels, which may be omitted in any statement, cannot duplicate any command mnemonics or assembler pseudoinstructions (ORG, END).

The term

(comment)

consists of a semicolon (;) followed by any sequence of symbols. The semicolon indicates to the assembler that a comment follows. A comment can be appended to any statement, or it can stand by itself. Comments are provided for documentation. Comments are ignored by the assembler program during the assembly process.

Data statements Data statements are very closely related to command statements. They are used to enter specific values of data into memory for use by the program. Statements of this type have the general structure

(label): (blank)(data)(blank)(comment)

The terms (label) and (comment) have the same form as they do in command statements, and they may or may not be present.

The term

(data)

can take on a variety of forms. It can be a binary, octal, or decimal number, for example. The specific type defined is associated with a symbol. Octal numbers are usually assumed by default.

Assembler command statements For simplicity, the two *pseudocommand* formats ORG and END are presented here. The statements are defined as follows.

The origin statement is represented as

ORG (blank)(octal address)(blank)(comment)

This pseudocommand or pseudoinstruction statement indicates to the assembler program that the machine-language code following this pseudocommand is to be placed in the memory location indicated by the term (octal address). This then is the first program statement, the starting address. All subsequent machine-language instructions are placed in successive memory locations until another ORG pseudocommand is encountered. If ORG is not specified at the beginning of the program, the assembler usually assumes a value of 0. The START command is an equivalent mnemonic.

The END command must be the last statement in the assembler-language program, and it indicates to the assembler that the end of the program has been reached.

The example shown in Fig. 2-12 reflects an assembly-language program for the simple controller of Figs. 2-10 and 2-11, using LSI-11 instructions. The LSI-11 assembler uses dot commands for origin (.=) and end (.END). Other assembler pseudoinstructions are listed in Appendix F. The %R symbol refers to a specific hardware register R. Note that this program is not completely documented.

2-7B Program Loaders

All computer instructions are stored in memory in binary-encoded form. In principle, it is possible to manually load a program or sequence of instructions into memory from the console or front panel of the processor by setting each bit in a word and loading one word at a time. However, from a practical standpoint, this method is extremely tedious. Even a short program of some 15 to 20 instructions takes several minutes, including error checks to assure that the proper bit settings have been made for each word. The initial program loading is therefore performed in two steps. The computer loading instructions are contained in a *bootstrap program* (all binary). The *bootstrap loader* is stored in a hardware ROM. This feature allows the user to load the bootstrap loader simply by pushing a button on the computer's front panel. Usually the bootstrap loader loads in a more extensive loader from mass storage. This loader then loads a program into the computer and executes it. In most computer systems this first program is the operating system. Through the operating system, applications are loaded into main memory and executed. More on operating systems later in Chap. 14.

In its most primitive form, a loader only reads an assembled program in object code from an external storage medium and places it in a proper area of memory.

```
;PROGRAM FOR CALCULATING C(K) = A*R(K) + B*C(K - 1)
INITIAL:                              ;INITIALIZE VALUES TO ZERO
    MOV     #0, %6
    MOV     #0, CK                    ;INITIALIZE C(K) TO ZERO
    MOV     #0, CKM1                  ;INITIALIZE C(K - 1) TO ZERO
FILTER:
;   GET STATUS FROM INPUT PORT (A/D CONVERTER)
    MOV     @ STATADD, %1             ;MOVE A/D STATUS INFORMATION TO
                                      ;REGISTER R1
    CMPB    %1, RDY                   ;CHECK FOR READY
    BEQ     FILTER                    ;IF NOT READY LOOP
;   GET R(K) FROM INPUT PORT
    MOV     @ INADD, %1               ;MOVE INPUT A/D DATA TO R1
;   DETERMINE AR(K)
    MPY     %1, A                     ;ASSUME MULTIPLY INSTRUCTION
                                      ;WITH TRUNCATED RESULT IN
                                      ;REGISTER R6
    MOV     %6, TP1                   ;STORE TEMPORARY VALUE TP1
;   DETERMINE B*C(K - 1)
    MPY     CKM1, B                   ;GENERATE B*C(K - 1)
    MOV     %6, TP2                   ;STORE TEMPORARY VALUE TP2
;   DETERMINE C(K)
    ADD     CK, TP1                   ;GENERATE CK
    ADD     CK, TP2
    MOV     CK, #OUTADD               ;OUTPUT C(K)
    MOV     CK, CKM1                  ;UPDATE C(K - 1)
;   LOOP
    JMP     FILTER:                   ;RETURN TO LOOP-NO ERROR CHECK
                                      DONE
;   I/O ADDRESSES
;   STATADD                           ;I/O ADDRESSES FOR STATUS,
                                      ;INPUT AND OUTPUT DEPEND UPON
                                      ;PARTICULAR INTERFACE HARDWARE
;   INADD
;   OUTADD
A:          .WORD       10            ;VALUE OF A COEFFICIENT
B:          .WORD       01            ;VALUE OF B COEFFICIENT
RDY:        .WORD   174004            ;SPECIFIC VALUE DEPENDS UPON
                                      ;UNIQUE A/D INTERFACE
CK:         .WORD        0            ;STORAGE LOCATION OF CURRENT
                                      ;OUTPUT C(K)
CKM1:       .WORD        0            ;STORAGE LOCATION OF LAST
                                      ;OUTPUT C(K - 1)
TP1:        .WORD        0            ;TEMPORARY STORAGE LOCATION
TP2:        .WORD        0            ;TEMPORARY STORAGE LOCATION
            .END
```

Figure 2-12 LSI-11 program for simple difference equation solution (see Figs. 2-10 and 2-11).

To facilitate this loading process, assemblers partition the assembled program into blocks of words. Into each block is recorded information that controls how the block is to be loaded. Usually the information specifies the starting address of the block and the number of words in the block. The additional information plus the machine-language code defines the *object program.* During loading, the loader examines the control information and places the block in memory at the specified address.

There is a somewhat more versatile type of loader called a *relocating loader*. This type of loader can be used to perform the following tasks in addition to the basic loading operation:

A program can be displaced from its assembled address during loading, and the loader automatically modifies the addresses in the program so that the program executes properly in its displaced position.

Two or more separately assembled programs can be loaded together in non-conflicting areas of computer memory. Memory addresses in one program that refer to labels in another program can be filled in by the loader so that cross references among programs are correct. This is a *linking loader*.

Programs can call on standard algorithms such as sine, cosine, and square root, and the loader automatically searches a library for them.

When relocation is performed by a loader, the result is essentially the same as can be obtained by modifying the origin instruction at the beginning of the program. The advantage of using a relocating loader is that the program need not be reassembled to relocate it, and relocation is considerably simpler than assembly with a new origin.

2-8 OPERATING SYSTEMS FOR CONTROL

In developing or using an existing operating system for digital control implementation, consideration must be given to the "overhead" needed to provide the operational monitoring or sequencing of activities.[67] The primary purposes of a *real-time operating system* are to relieve the programmer of providing programs to:

Input data
Output data
Check for errors
Update constants
Sequence tasks (clocking)
Allocate stage (memory management)
Define interactive communications

The extent to which a commercial operating system provides these capabilities depends upon the hardware architecture and the overhead allowed in a given application. The use of an extensive software development facility can include this type of operating system for incorporation into the final digital-controller software system. For "simple" first- or second-order control equations, an extensive operating system is usually not required since only one input and output are defined. A very simple supervisor or monitor can be incorporated into the code for input-output operations and simple checks on data to ensure system integrity.

On the other hand, digital control systems for large process control plants, such as those found in the chemical industry or in space vehicles, require a complex real-time operating system (*monitor*, *supervisor*, *executive*) due to the numerous individual control and sequencing operations and their interaction. Elaborate error checks are critical to a successful operation.

Note that hardware does fatigue and software is generally not perfect (correct for all situations). Also, updating control-system parameters is sometimes desired to be on-line, i.e., while the control system is executing. Thus provisions for testing and inclusion of the new parameter values must be provided.

Finally, a real-time operating system can provide composite reports and current status information. System operation, then, as directed by the program must reflect human-good interfacing design. This area of human factors can be critical to the success of a project. Operating systems, for example, can provide extensive libraries and error checking for digital control systems. General real-time operating systems are presented in Chap. 14.

2-9 LANGUAGE SELECTION

The selection of a language for implementation of a digital controller depends upon many things, including:

Accuracy requirements
Timing (speed) requirements
Programmer's familiarity
Language standardization within organization
Program development tools

Accuracy requirements not only depend on the machine-language word length but on the HOL compiler's characteristics associated with arithmetic computation. In assembly language, the software engineer can write programs to provide the desired accuracy in floating-point or fixed-point representation. Library routines may also be available. Of course, there is a speed vs. accuracy trade-off. Note that HOLs usually permit a select number of bits for the floating-point mantissa or the fixed-point magnitude value. Thus, before selecting a HOL, determine if the accuracy requirements can be achieved with the compiler's mantissa or fixed-point magnitude accuracy.

Timing is probably one of the most critical dimensions of digital control implementations. Speed is not only determined by internal machine clock rate but by the accuracy mentioned previously and the design of the algorithm itself. *In general, the data transfer and data storage (CPU, memory, input, output) are the most critical operations in determining timing performance.* In other words, the bookkeeping operations must be analyzed in detail for changes leading to the meeting of timing specifications. On the other hand, the algorithms and associated code implementation should be as simple as possible for ease of understanding and ease of maintenance. These two design directions provide for continuing trade-offs.

The familiarity and ability to use a particular language is also important. A software engineer can efficiently and effectively develop and test programs in a familiar language. Programming costs are relatively high compared to hardware costs. The choice of a HOL can make a software engineer more productive not only in lines of code but also in terms of associated documentation and testability. On the other hand, association of HOL instructions with I/O timing diagrams is very difficult without proper tools.

Although software engineers may be able to program in a wide range of languages, a management directive to limit the number of languages for implementation is suggested to keep programming costs low. This approach results in (1) the standardization of software development tools, (2) similar or even possibly identical programs used for different products, and (3) easier management of personnel changes, i.e., personnel transferred from one project to another.

If hardware standardization is not achievable, then a HOL standard with appropriate compilers should be the obvious choice. If execution time is critical, then assembly language is usually selected. Note, however, that critical algorithms can be written in assembly language and those that are not critical can be written in a HOL. The various source programs are linked together after translation into object code. This integrated concept permits the economical use of a HOL, yet provides the strength of assembly language for real-time control. Of course, the software development tools must provide the linking capability. Some languages such as ADA and the C programming language permit bit- and byte-level manipulations as well as the use of HOL-structured programming constraints.

2-10 PROGRAM DEVELOPMENT

Once the algorithm describing the general structure of the digital controller (computation) has been defined, a computer program can be developed in a straightforward logical manner. The general approach consists of a series of steps. The *first step* is to generate the flowchart depicting the control flow of the process, i.e., the digital-controller requirements. Pseudocode can also be used. The *second step* is to define the data (data structures) and associated operations required to carry out the digital control process in the selected programming language. The *third step* in the programming process is to decide how the data and instructions are to be generally organized in memory as representative of the flowchart or pseudocode. This means that labels (memory locations) for each piece of data and each operation in the program must be assigned, usually in sequence. These locations may be logical if programming in a HOL or physical if programming in assembly language. *The generation of the entire program organization is considered as an incremental development process.* The operations called for by the different parts of the digital control algorithm (flowchart/pseudocode) are converted into an instruction sequence. As this development process proceeds, additional locations in memory are assigned for the intermediate data that are generated during the execution of the program. This process continues until the total algorithm has been completely programmed. A complete program may appear to be a very complex structure. However, most *programs are developed by repeatedly applying this iterative technique.*

More specifically the development method is essentially the repeated application of the following iterative steps:

Step 1: Define and understand the problem The software developer (software engineer) develops a clear understanding and definition of the functions that must be performed by the process and the associated data types and data movement. The developer must also have a clear and precise definition of the environment (the context) in which the program operates and any operational constraints. For example, the input and output interfaces must be precisely described (format accuracy, speed).

Step 2: Define functional modules of algorithm The major functional modules are identified and described in terms of a general flowchart or pseudocode. In defining these functional modules, care should be taken to minimize the interconnections between them. These are the major building blocks of the program, and functional modularity demands that the interfaces between them be explicit and as simple as possible.

Step 3: Define required data structures The data structure required by each functional module is defined and organized. Differences between data internal to functional modules and external data (transferred between modules) are analyzed for correct and proper selection and ease of understanding.

Step 4: Define the detail of each functional module The functional modules and data defined in steps 2 and 3 are used to define the detailed process for performing the functions of the program. Again this can be done by representing the detailed algorithms with flowcharts or pseudocode. In defining these detailed algorithms, the software engineer should watch for redefinitions of the functional modules that can reduce the number of modules, simplify the algorithms, and minimize the number of module interconnections (coupling). Some of the functional modules may be simple and small enough so that they can be combined with other functions into a single procedure (module). However, most of the modules are generally implemented as separate procedures, physically as well as functionally (good module cohesion).

The development method should be applied to each separate module. Each major functional module becomes a new, relatively isolated subalgorithm. The method can then be applied to each of these new algorithms. No knowledge of the other modules is required except a definition of the interfaces. This is the iterative aspect of the method. Each iteration produces a new level of increased detail. Iteration is continued until the structure of each of the program procedures is specified in enough detail such that coding of the procedure is straightforward and simple. The flowchart or pseudocode at this last level should be very close to the programming language selected in terms of associated operations and data structures.

In a software realization of a control algorithm, modules can be written for a generalized first- or second-order filter. The appropriate filter or compensator coefficients can be stored in an array and called by using an addressing variable. A main program module can do the bookkeeping and executive operations. Also, I/O modules can be developed for the proper A/D and D/A interface control. Note that the use of general compensator modules permits relatively easy cascade or parallel filter implementation in a digital computer.

The main aspects of proper software development have been presented. However, many value judgments must be made in using this logical approach. Elaborate tools and techniques exist that make the software engineering effort easier and more understandable but they must be available. These tools and techniques are discussed again in Chap. 13, along with the software life-cycle concept.

2-11 SUMMARY

In this chapter, various aspects of the interdisciplinary field of computer science and computer engineering are presented. These include the structure of binary or base 2 numbers, the computer representation of signed binary numbers, the definitions of a logical process, and its representation in boolean algebra symbols and programming. Also, the general organization of a modern computer system consisting of the five basic units (input, output, memory, manipulation, control) is discussed. Various digital-control-system implementation considerations are woven into this overview.

Another objective of this chapter is to introduce the control engineer to the many facets of implementing a digital control law in a computer. In fact, the control engineer has to take on the responsibilities of a software engineer. Such responsibilities include algorithm design, program development, information structure determination, language selection, and test procedures. In essence, the control engineer becomes a computer engineer requiring both hardware and software implementation knowledge. This interdisciplinary area cannot be completely presented in one chapter and not even in one textbook. The digital control engineer should consult other references for total system implementation[29,68,88] and realize that experience plays an important role. In the following chapters involved with the theoretical abstractions of digital control theory, one should continually consider the realistic ramifications of control law implementation in a digital computer.

CHAPTER

THREE

LINEAR SYSTEMS AND THE SAMPLING PROCESS

3-1 INTRODUCTION

The mathematical techniques being used in this text are based upon the consideration of linear time-invariant (LTI) systems. Therefore, unless otherwise stated, only LTI systems are henceforth considered. Chapter 1 introduced the concept of the sampling process. As is seen in this chapter, a sampled continuous signal yields a signal in the form of discrete values. The purpose of this chapter is to discuss the sampling process in more detail and to present some time- and frequency-domain representations of sampled-data systems in terms of linear equations. Another purpose is to present techniques for continuous-signal reconstruction from a sampled signal by means of a data-hold device. In sampled-data control systems in which at least one component operates in continuous time, it is necessary to include a data-hold device to maintain control over the system between sampling instants.

3-2 LINEAR TIME-INVARIANT (LTI) SYSTEM

A system is said to be *linear* if and only if the superposition theorem can be satisfied. This theorem states that "the output response of an LTI system to a number of simultaneously applied inputs is equal to the summation of the system responses when each input is applied individually." For example, consider a single-input–single-output (SISO) system, as shown in Fig. 3-1*a*, whose output is $c(t) = c_1(t)$ due to an input $r(t) = r_1(t)$. This system is said to be linear if for the

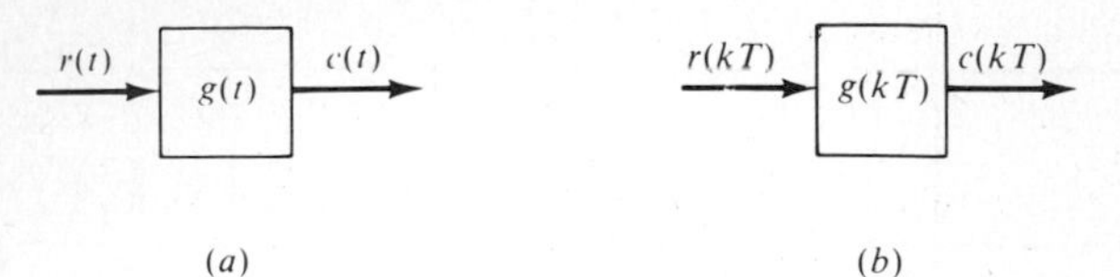

Figure 3-1 SISO system: (*a*) continuous time; (*b*) discrete time.

input $r(t) = a_1 r_1(t) + a_2 r_2(t)$ the output is given by

$$c(t) = a_1 c_1(t) + a_2 c_2(t) \tag{3-1}$$

When a_1 and a_2 are constant real numbers, and c_i is the respective output due to r_i, in terms of function notation, the composite output may be expressed as

$$c = f[r] = f[a_1 r_1 + a_2 r_2] = a_1 f[r_1] + a_2 f[r_2] \tag{3-2}$$

The notation f represents the manner in which the system function g operates on the input signal or transforms the input signal to yield an output signal.

For the system whose input and output quantities represent discrete or sampled quantities, then Eq. (3-2), using the superposition theorem, can be expressed as

$$c(kT) = f[a_1 r_1(kT) + a_2 r_2(kT)] = a_1 f[r_1(kT)] + a_2 f[r_2(kT)] \tag{3-3}$$

The results for a SISO system may be extended to a multiple-input–multiple-output (MIMO) system having m inputs and p outputs. The $m \times 1$ input column vector $\mathbf{r}$ and the $p \times 1$ output column vector $\mathbf{c}$ are expressed, respectively, as

$$\mathbf{r} = \begin{bmatrix} r_1 \\ r_2 \\ \vdots \\ r_m \end{bmatrix} \qquad \mathbf{c} = \begin{bmatrix} c_1 \\ c_2 \\ \vdots \\ c_p \end{bmatrix}$$

where r_i and c_j are the elements of the respective vectors. Therefore, for either a continuous- or a discrete-time system to be linear, the relationship

$$\mathbf{c} = f[a_1 \mathbf{r}_1 + a_2 \mathbf{r}_2] = a_1 f[\mathbf{r}_1] + a_2 f[\mathbf{r}_2] \tag{3-4}$$

where a_1 and a_2 are constant real numbers and $\mathbf{r}_1$ and $\mathbf{r}_2$ are any two input vectors, must be satisfied. If this relationship is not satisfied, the system is said to be *nonlinear*.

Example 3-1 The following first-order differential equation describes a system having the input $r(t)$:

$$c(t) = r(t) + \alpha \tag{3-5}$$

To determine if the system is linear, proceed as follows:

SOLUTION Consider $r(t) = a_1 r_1(t) + a_2 r_2(t)$.

Step 1. For $r_1(t)$,

$$c_1 = r_1 + \alpha \tag{3-6}$$

Step 2. For $r_2(t)$,

$$c_2 = r_2 + \alpha \tag{3-7}$$

Step 3. For $r = a_1 r_1 + a_2 r_2$,

$$c = a_1 r_1 + a_2 r_2 + \alpha \tag{3-8}$$

Step 4. Substituting Eqs. (3-6) and (3-7) into Eq. (3-1) yields

$$c = a_1 r_1 + a_2 r_2 + (a_1 + a_2)\alpha \tag{3-9}$$

Step 5. As Eqs. (3-8) and (3-9) are not identical, Eq. (3-5) is not linear.

The system is nonlinear since Eq. (3-4) is not satisfied.

Example 3-2 An approximation for numerical differentiation [see Eq. (3-62)] of $c(t) = \dot{r}(t)$ is given by the discrete-time equation

$$c(kT) = \frac{1}{T}\{r(kT) - r[(k-1)T]\} \tag{3-10}$$

where T is the constant sampling time of $r(t)$ and $k = 0, 1, 2, \ldots$. Is Eq. (3-10) a linear equation representing a linear system?

SOLUTION

Step 1. For $r_1(kT)$,

$$c_1(kT) = \frac{1}{T}\{r_1(kT) - r_1[(k-1)T]\} \tag{3-11}$$

Step 2. For $r_2(kT)$,

$$c_2(kT) = \frac{1}{T}\{r_2(kT) - r_2[(k-1)T]\} \tag{3-12}$$

Step 3. For $r(kT) = a_1 r_1(kT) + a_2 r_2(kT)$,

$$c(kT) = \frac{a_1}{T}\{r_1(kT) - r_1[(k-1)T]\} + \frac{a_2}{T}\{r_2(kT) - r_2[(k-1)T]\} \tag{3-13}$$

Step 4. Substituting Eqs. (3-11) and (3-12) into the discrete version of Eq. (3-1) yields

$$c(kT) = a_1c_1(kT) + a_2c_2(kT)$$

$$= \frac{a_1}{T}\{r_1(kT) - r_1[(k-1)T]\} + \frac{a_2}{T}\{r_2(kT) - r_2[(k-1)T]\} \qquad (3\text{-}14)$$

Step 5 As Eqs. (3-13) and (3-14) are identical, Eq. (3-10) is linear.

Example 3-3 To determine if $c(kT) = r^2(kT)$ is a linear equation proceed as follows.

SOLUTION

Step 1. For $r_1(kT)$,

$$c_1(kT) = r_1^2(kT) \qquad (3\text{-}15)$$

Step 2. For $r_2(kT)$,

$$c_2(kT) = r_2^2(kT) \qquad (3\text{-}16)$$

Step 3. For $r(kT) = a_1r_1(kT) + a_2r_2(kT)$,

$$c(kT) = a_1^2r_1^2(kT) + 2a_1a_2r_1(kT)r_2(kT) + a_2^2r_2^2(kT) \qquad (3\text{-}17)$$

Step 4. Using the discrete version of Eq. (3-1) and substituting Eqs. (3-15) and (3-16) into it yields

$$c(kT) = a_1c_1(kT) + a_2c_2(kT) = a_1r_1^2(kT) + a_2r_2^2(kT) \qquad (3\text{-}18)$$

Step 5. As Eqs. (3-17) and (3-18) are not identical, the given equation $c(kT) = r^2(kT)$ is not linear.

A system is said to be *time-invariant* when the *coefficients* of the differential or difference equation relating the system's output to its input do not depend upon time. In general, if a specified input is applied to a given system at any time t_i or t_j, and if the output responses to each application of this input are the same, the system is said to be time-invariant. The concept of time invariance can be illustrated by using the notation of a function with a time delay of τ seconds and a block diagram as shown in Fig. 3-2. The delay in path 1 acts on both the implicit and explicit time variables in the equation for $c(\cdot)$. The delay in path 2 acts only on the implicit variable in the equation for $c(\cdot)$. Thus if the output quantities of path 1 are equal to the corresponding outputs of path 2, the system is said to be time-invariant, i.e.,

$$\mathbf{c}(t-\tau) = \mathbf{f}[r(t-\tau)] \qquad \textit{for a continuous-time system} \qquad (3\text{-}19)$$

$$\mathbf{c}[(k-p)T] = \mathbf{f}\{r[(k-p)T]\} \qquad \textit{for a discrete-time system} \qquad (3\text{-}20)$$

where $\tau = pT$ and $p = 1, 2, 3, \ldots$.

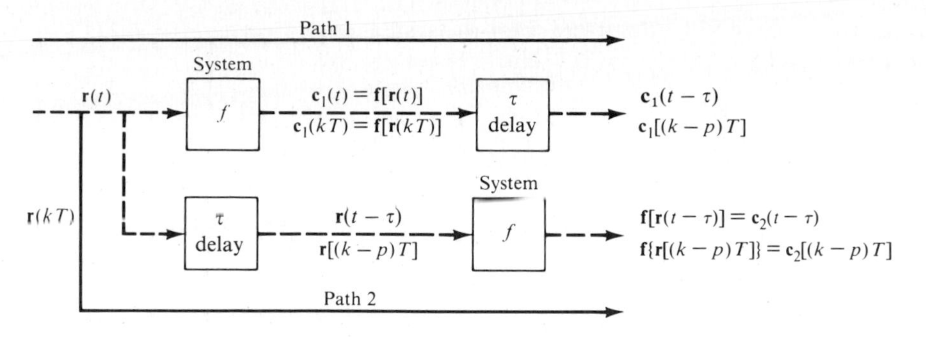

Figure 3-2 Time-invariance concept.

Example 3-4 To determine if the system described by

$$\dot{c}(t) + c(t) = tr(t) \tag{3-21}$$

is time-invariant, proceed as follows.

SOLUTION

Path 1:

$$\dot{c}_1(t) + c_1(t) = tr \rightarrow \dot{c}_1(t-\tau) + c_1(t-\tau) = (t-\tau)r(t-\tau)$$

$$= tr(t-\tau) - \tau r(t-\tau) \tag{3-22}$$

Path 2:

$$\dot{c}_2(t-\tau) + c_2(t-\tau) = tr(t-\tau)$$

$$= tr(t-\tau) - \tau r(t-\tau) + \tau r(t-\tau) \tag{3-23}$$

As Eqs. (3-22) and (3-23) are not identical, $c_1(t-\tau) \neq c_2(t-\tau)$. Therefore the system is not time-invariant.

Example 3-5 The solution for determining if the discrete-time system described by $c(kT) = kTr(kT)$ is time-invariant is as follows.

SOLUTION

Path 1:

$$c_1(kT) = kTr(kT) \rightarrow c_1[(k-p)T] = [(k-p)T]r[(k-p)T] \tag{3-24}$$

Path 2:

$$c_2[(k-p)T] = kTr[(k-p)T]$$

$$= [(k-p)T]r[(k-p)T] + pTr[(k-p)T] \tag{3-25}$$

As Eqs. (3-24) and (3-25) are not identical,

$$c_1[(k-p)T] \neq c_2[(k-p)T]$$

Therefore the system is not time-invariant.

Example 3-6 Repeat Example 3-5 for the system described by $c[(k+1)T] - c(kT) = r(kT)$. As both paths 1 and 2 result in

$$c[(k+1-p)T] - c[(k-p)T] = r[(k-p)T]$$

the system is said to be time-invariant.

From these examples it can be stated that a continuous (discrete) system is time-invariant if the variable $t(kT)$ does not explicitly appear in the system's differential (difference) equation, i.e., as a coefficient. All system models in this text are assumed LTI owing to the general techniques used in control-system analysis and synthesis. Also most systems can be approximated over some range of operation as LTI.

3-3 SAMPLING PROCESS (FREQUENCY-DOMAIN ANALYSIS)

As stated in Chap. 1, the sampling process can be considered as a modulation process in which the particular pulse train $p(t)$ of Fig. 3-3a is multiplied by a continuous function $e(t)$, as shown in Fig. 3-3b, to produce the sampled function

$$e_p^*(t) \equiv p(t)e(t) \tag{3-26}$$

as shown in Fig. 3-3c where T is the sampling period. It may be noted in Fig. 3-3a that as γ approaches zero, the process approaches impulse sampling; that is, $p(t)$ approaches an impulse train, each impulse being of unit strength. Also, $e(kT) \approx e(kT+\gamma)$ as shown in Fig. 3-3c.

To effectively reconstruct $e(t)$ from $e_p^*(t)$, it is necessary that the frequency content of the pulse signal contain all the frequency information initially in $e(t)$. One way to functionally determine this frequency characteristic is to perform a Fourier series analysis[1] of $p(t)$. The physical phenomenon of pulse sampling produces an output signal $e_p^*(t)$, which contains periodic components. To determine these components, it is first necessary that $p(t)$ be represented by the following trigonometric or Fourier series:[92]

$$\begin{aligned} p(t) &= f_0 + \sum_{m=0}^{\infty} F_m \cos(m\omega_s t + \theta_m) \\ &= f_0 + \sum_{m=0}^{\infty} F_m \frac{\epsilon^{j(m\omega_s t + \theta_m)} + \epsilon^{-j(m\omega_s t + \theta_m)}}{2} \end{aligned} \tag{3-27a}$$

where $m = 0, 1, 2, \ldots, \infty$. This equation can also be recast in the form

$$p(t) = \sum_{m=-\infty}^{\infty} \mathbf{P}(jm\omega_s)\epsilon^{jm\omega_s t} \tag{3-27b}$$

where now $m = 0, \pm 1, \pm 2, \ldots, \pm\infty$ and $\mathbf{P}(jm\omega_s)$ are defined as the Fourier coefficients. Equation (3-27b) is referred to as the complex Fourier series expansion of $p(t)$ (see Appendix A). Equations (3-27a) and (3-27b) are equivalent for a

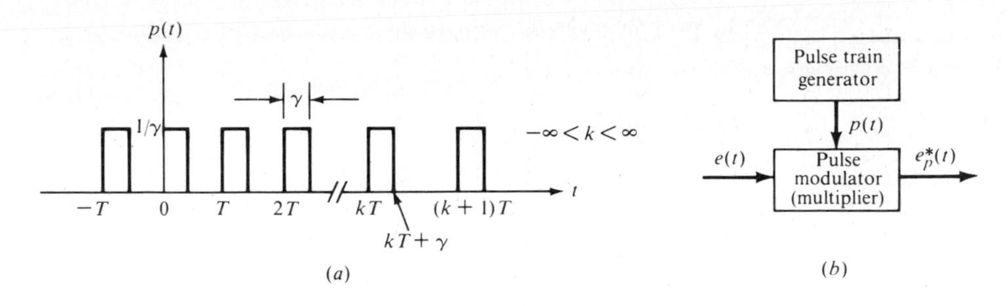

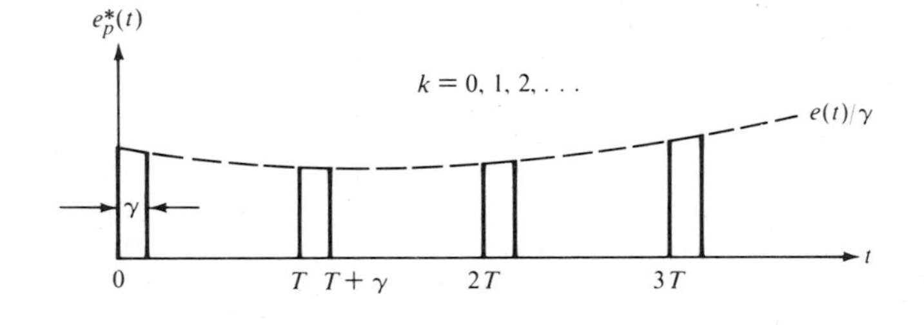

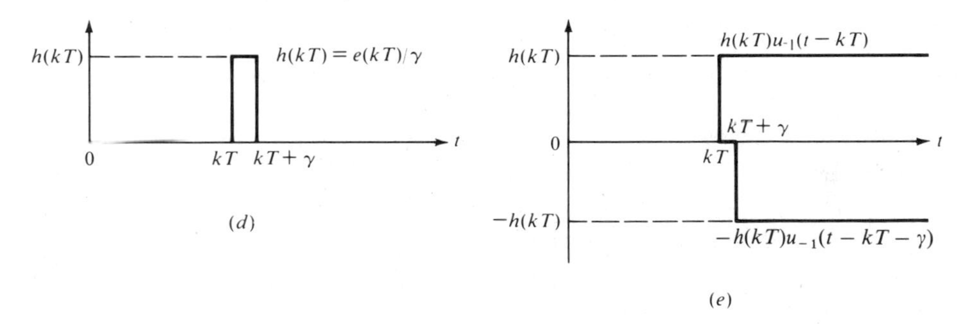

Figure 3-3 Sampling process ($0 < \gamma < T$). (*a*) Pulse train; (*b*) sampling device; (*c*) pulse signal $e_p^*(t)$; (*d*) a pulse of height $h(kT)$ and strength $e(kT)$ of $e_p^*(t)$: (*e*) mathematical equivalency of $h(kT)$.

periodic function. Since the sampling frequency is $\omega_s = 2\pi/T$, and

$$p(t) = \begin{cases} \dfrac{1}{\gamma} & \text{for } kT \le t < kT + \gamma \\ 0 & \text{for } kT + \gamma \le t < (k+1)T \end{cases}$$

then the Fourier coefficients $\mathbf{P}(jm\omega_s)$ only have to be evaluated over the time period $0 \le t < T$ and are

$$\mathbf{P}(jm\omega_s) = \frac{1}{T}\int_0^{\gamma} p(t)\epsilon^{-jm\omega_s t}\,dt = \frac{1}{T}\frac{\sin(m\omega_s\gamma/2)}{m\omega_s\gamma/2}\epsilon^{-jm\omega_s\gamma/2} \tag{3-28}$$

The Fourier transform $\mathbf{P}(j\omega)$ of $p(t)$ is defined as:

$$\mathbf{P}(j\omega) \equiv \int_{-\infty}^{+\infty} p(t)\epsilon^{-j\omega t}\, dt \tag{3-29}$$

Substituting from Eq. (3-27b) into Eq. (3-29) yields

$$\mathbf{P}(j\omega) - \int_{-\infty}^{+\infty} \sum_{m=-\infty}^{+\infty} \mathbf{P}(jm\omega_s)\epsilon^{j(m\omega_s-\omega)t}\, dt$$

$$= \sum_{m=-\infty}^{+\infty} \mathbf{P}(jm\omega_s) \int_{-\infty}^{+\infty} \epsilon^{-j(-m\omega_s+\omega)t}\, dt = \sum_{m=-\infty}^{\infty} \mathbf{P}(jm\omega_s)\delta(\omega - m\omega_s) \tag{3-30}$$

where $\delta(\omega - m\omega_s)$ denotes an impulse of unit strength occurring at $\omega = m\omega_s$. Hence, in the frequency domain, $\mathbf{P}(j\omega)$ is a train of impulses, the members being of strength $\mathbf{P}(jm\omega_s)$ and occurring at $\omega = m\omega_s, m = 0, +1, +2, \ldots$. Thus substituting from Eq. (3-28) into Eq. (3-30) yields the Fourier transform

$$\mathbf{P}(j\omega) = \frac{1}{T} \sum_{m=-\infty}^{\infty} \frac{\sin(m\omega_s\gamma/2)}{m\omega_s\gamma/2} \epsilon^{-(jm\omega_s\gamma/2)}\delta(\omega - m\omega_s) \tag{3-31}$$

To obtain $\mathbf{E}_p^*(j\omega)$ for $e_p^*(t) = p(t)e(t)$ to determine the periodic frequency components of $e_p^*(t)$ requires the application of the frequency-domain convolution theorem. Note that this is the counterpart of the time-domain convolution to the multiplication of two functions in the frequency domain (see Appendix B). Thus applying the frequency-domain convolution theorem to Eq. (3-26) yields

$$\mathbf{E}_p^*(j\omega) = \int_{-\infty}^{+\infty} \mathbf{P}(j\omega')\mathbf{E}[j(\omega - \omega')]\, d\omega'$$

$$= \int_{m-\infty}^{+\infty} \frac{1}{T}\left[\sum_{m-\infty}^{+\infty} \frac{\sin(m\omega_s\gamma/2)}{m\omega_s\gamma/2} \epsilon^{-(jm\omega_s\gamma/2)}\delta(\omega' - m\omega_s)\right] \mathbf{E}[j(\omega - \omega')]\, d\omega'$$

$$= \frac{1}{T} \sum_{m=-\infty}^{+\infty} \frac{\sin(m\omega_s\gamma/2)}{m\omega_s\gamma/2} \epsilon^{-(jm\omega_s\gamma/2)} \int_{-\infty}^{+\infty} \delta(\omega' - m\omega_s)\mathbf{E}[j(\omega - \omega')]\, d\omega'$$

The integral portion only has a nonzero value when $\omega' = m\omega_s$, that is,

$$\int_{-\infty}^{\infty} \delta(\omega' - m\omega_s)\mathbf{E}[j(\omega - \omega')]\, d\omega' = \mathbf{E}[j(\omega - m\omega_s)]$$

Thus

$$\mathbf{E}_p^*(j\omega) = \frac{1}{T} \sum_{m=-\infty}^{+\infty} \left[\frac{\sin(m\omega_s\gamma/2)}{m\omega_s\gamma/2}\right] \epsilon^{-(jm\omega_s\gamma/2)}\mathbf{E}[j(\omega - m\omega_s)] \tag{3-32}$$

A plot of $|\mathbf{E}(j\omega)|$ vs. ω and of $|\mathbf{P}(j\omega)|$ vs. ω are shown in Fig. 3-4a and b, respectively.

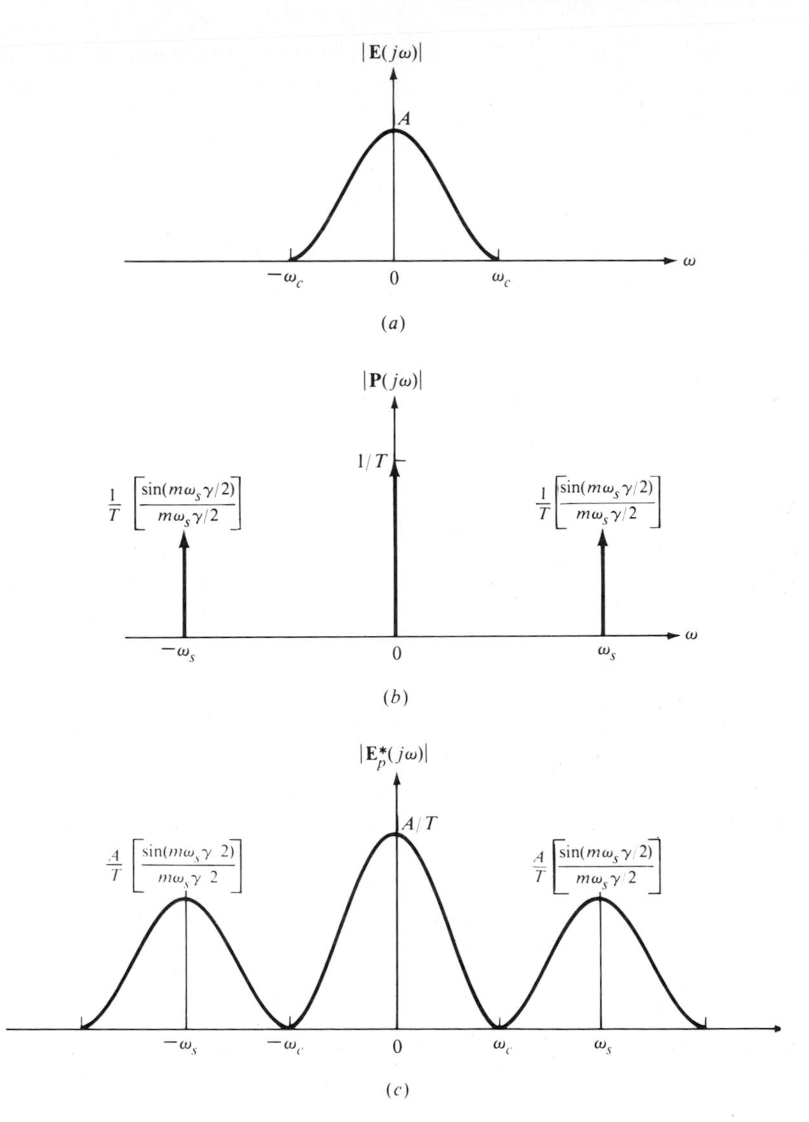

Figure 3-4 Frequency spectra for finite-pulse-width sampling frequency spectrum for $\omega_s = 2\omega_c$. (*a*) Band-limited frequency spectrum of a continuous function $e(t)$; (*b*) frequency spectrum for the pulse train $p(t)$; (*c*) frequency spectrum for a pulse-sampled function $e^*(t)$.

Note that the bracketed term in Eq. (3-32) as $\gamma \to T$ has the value

$$\lim_{\gamma \to T} \frac{\sin (m\pi\gamma/T)}{m\pi\gamma/T} = \frac{\sin m\pi}{m\pi} = 0 \qquad \text{for } m \neq 0$$

and all the folded components of $\mathbf{E}(j\omega)$ vanish. Only the fundamental component of $\mathbf{E}_p^*(j\omega)$ remains since $[|\sin m\pi|/m\pi]_{m=0} = 1$. In this case, the modulator of Fig. 3-3 is a *continuous multiplier of $1/T$ for all time.*

Assuming that the amplitude spectrum of $e(t)$ is band-limited and of maximum value A, as shown in Fig. 3-4a, then the plot of $|\mathbf{E}_p^*(j\omega)|$ vs. ω is shown in Fig. 3-4c, where ω_s is selected to equal $2\omega_c$. It is seen that the sampling process produces periodic components. The fundamental spectrum, $-\omega_c < \omega < \omega_c$, is similar in shape to that of the continuous function. It also produces a succession of complementary or folded spectra which are shifted periodically by a frequency separation $m\omega_s$ ($m = \pm 1, \pm 2, \ldots$). If the sampling frequency ω_s is sufficiently high, there is essentially no overlap between the fundamental and complementary spectra as shown in Fig. 3-4c. If ω_s is relatively low, overlap or frequency folding occurs. This aspect is discussed more fully in the next section.

3-4 IDEAL SAMPLER

To simplify the mathematical analysis of sampled-data control systems, it is necessary to develop the concept of an ideal sampler (impulse sampling). If the duration γ of the sampling pulse is much less than the sampling time T and much smaller than the smallest time constant of $e(t)$, the output of the pulse modulator, as shown in Fig. 3-3c, can be approximated as an ideal impulse train. Assuming the sampling duration is very small, that is, $\gamma \ll T$, so that the height of $e_p^*(t)$ is essentially constant, then each pulse of Fig. 3-3c may be represented as the difference between two unit-step functions (see Fig. 3-3d and e). Assuming a finite-pulse-width sampler, then the pulse signal of Fig. 3-3c may be mathematically represented as:

$$e_p^*(t) \approx \frac{1}{\gamma} \sum_{k=0}^{\infty} e(kT)[u_{-1}(t - kT) - u_{u-1}(t - kT - \gamma)] \tag{3-33}$$

where

$$e_p^*(t) = \begin{cases} \dfrac{e(kT)}{\gamma} & \text{for } kT \le t < kT + \gamma \\ 0 & \text{for } kT + \gamma \le t < (k+1)T \end{cases}$$

and $k = 0, 1, 2, 3, \ldots$. The expression for the ideal impulse function may be derived by taking the Laplace transform of Eq. (3-33), which yields

$$E_p^*(s) \approx \sum_{k=0}^{\infty} e(kT) \frac{1 - \epsilon^{-\gamma s}}{\gamma s} \epsilon^{-kTs} \tag{3-34}$$

or by the Fourier transform approach, which is done later. Assuming the sampling duration γ is very small results in the following approximation:

$$1 - \epsilon^{-\gamma s} = 1 - \left[1 - \gamma s + \frac{(\gamma s)^2}{2!} - \frac{(\gamma s)^3}{3!} + \cdots\right] \approx \gamma s$$

where s is the Laplace transform operator and γ is appropriately small. Utilizing this approximation in Eq. (3-34) yields

$$E_p^*(s) \approx \sum_{k=0}^{\infty} e(kT)\epsilon^{-kTs} \tag{3-35}$$

Taking the inverse of the Laplace transform of Eq. (3-35) yields

$$e_p^*(t) \approx \sum_{k=0}^{\infty} e(kT)\delta(t - kT) \equiv e(t)\delta_T(t) \tag{3-36}$$

where

$$\delta_T(t) \equiv \sum_{k=0}^{\infty} \delta(t - kT) \tag{3-37}$$

is defined as a unit impulse train as shown in Fig. 3-5a and $\delta(t - kT)$ *represents an impulse of unit area occurring at time* $t = kT$, *and* $e(t) = 0$ *for* $t < 0$. Therefore, from Eq. (3-36), the following can be written:

$$e_p^*(t) \approx e(t)\delta_T(t) \equiv e^*(t) \equiv \sum_{k=0}^{\infty} e(kT)\delta(t - kT) \tag{3-38}$$

where $e^*(t)$ represents an impulse train. Thus Eq. (3-38) reveals that the pulse modulator of the sampling device of Fig. 3-3*b* can be modeled by an impulse modulator which is represented by the symbolic diagram of Fig. 3-5*b*.

The one-sided Laplace transform of the output of the ideal sampler, Eq. (3-38), is

$$E^*(s) = \mathscr{L}[e^*(t)] = \sum_{k=0}^{\infty} e(kT)\epsilon^{-kTs} \tag{3-39}$$

This equation represents the transform in the s domain of any impulse-train sampled signal.

Another approach to determining $E^*(s)$ is to utilize the convolution integral (see Appendix B).

$$E^*(s) = E(s) * \Delta_T(s) = \int_{-\infty}^{\infty} E(s - s')\,\Delta_T(s')\,ds' \tag{3-40}$$

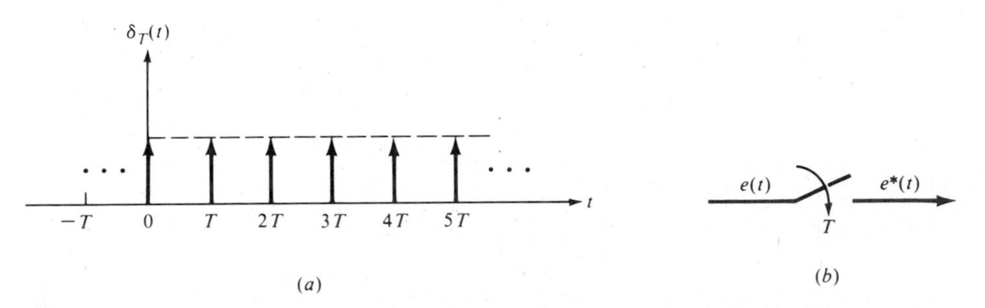

Figure 3-5 Ideal sampler. (*a*) Unit impulse train $\delta_T(t)$; (*b*) symbolic diagram of impulse modulator.

Since $e^*(t) = e(t)\delta_T(T)$, where

$$\delta_T(t) = \lim_{\gamma=0} p(t) \tag{3-41}$$

substituting Eq. (3-27*b*) into Eq. (3-41) yields

$$\delta_T(t) = \sum_{m=-\infty}^{\infty} \frac{1}{T} \epsilon^{jm\omega_s t} \tag{3-42}$$

Taking the Laplace transform of Eq. (3-42) and using the complex translation theorem[1] results in

$$\Delta_T(s) = \int_0^{\infty} \sum_{m=-\infty}^{\infty} \frac{1}{T} \epsilon^{jm\omega_s t} \epsilon^{-st} dt$$

$$= \sum_{m=-\infty}^{+\infty} \frac{1}{T} \int_0^{\infty} \epsilon^{-(s-jm\omega_s)t} dt = \sum_{m=-\infty}^{\infty} \frac{1}{T} \delta(s - jm\omega_s) \tag{3-43}$$

But from Eq. (3-40),

$$E^*(s) = \int_{-\infty}^{\infty} E(s - s') \sum_{m=-\infty}^{+\infty} \frac{1}{T} \delta(s' - jm\omega_s)\, ds' \tag{3-44}$$

Since the convolution of a function, $f(\cdot)$, with a unit impulse function, $\delta(\cdot)$, yields the function $f(\cdot)$ itself, that is,

$$\int_{-\infty}^{\infty} E(s - s')\delta(s' - jm\omega_s)\, ds' = E(s - s')|_{s' = jm\omega_s}$$

then Eq. (3-44) yields

$$E^*(s) = \sum_{m=-\infty}^{\infty} \frac{1}{T} E(s - jm\omega_s) \tag{3-45}$$

This expression for $E^*(s)$ permits an analysis of the frequency spectrum of the ideal sampler output.

The relationship between the pulse and impulse functions may be illustrated as follows: For an impulse function the Fourier series representation is

$$e^*(t) \equiv \lim_{\gamma \to 0} e_p^*(t) = \lim_{\gamma \to 0} [p(t)e(t)] = \lim_{\gamma \to 0} \left[e(t) \sum_{m=-\infty}^{\infty} \mathbf{P}(jm\omega_s)\epsilon^{jm\omega_s t} \right]$$

Thus substituting Eq. (3-28) into this equation yields

$$e^*(t) \equiv \lim_{\gamma \to 0} e_p^*(t) = \frac{e(t)}{T} \lim_{\gamma \to 0} \left[\sum_{m=-\infty}^{\infty} \frac{\sin(m\omega_s\gamma/2)}{m\omega_s\gamma/2} \epsilon^{-jm\gamma/2} \right] \epsilon^{jm\omega_s t}$$

Applying L'Hopitâl's theorem to this equation yields

$$e^*(t) \equiv \lim_{\gamma \to 0} e_p^* = \sum_{m=-\infty}^{+\infty} \frac{e(t)}{T} \epsilon^{jm\omega_s t} \qquad 0 \le t < \infty \tag{3-46}$$

where m is an index variable relating the multiples of frequency (ω_s). Next, note that the sampling process model Eq. (3-38) yields

$$e^*(t) = \sum_{k=0}^{\infty} e(kT)\delta(t - kT) \qquad 0 \le t < \infty \tag{3-47}$$

where k is an index variable relating to multiples of time T. Assuming $e(t) \equiv 1$ and the period $0 \le t < T$ ($k = 0$), then the output of the impulse sampler during this period must be an impulse; thus from Eq. (3-46)

$$e^*(t) = \sum_{m=-\infty}^{+\infty} \frac{\epsilon^{jm\omega_s t}}{T} \qquad 0 \le t < T \tag{3-48}$$

Now considering Eq. (3-47) for the first period, that is, $k = 0$ yields with Eq. (3-48)

$$e^*(t) = \delta(t - kT)|_{k=0} = \delta(t) = \sum_{m=-\infty}^{+\infty} \frac{\epsilon^{jm\omega_s t}}{T} \qquad 0 \le t < T \tag{3-49}$$

Thus Eqs. (3-46) and (3-47) are equivalent if one considers each time interval for which $e^*(t)$ is being represented.

Equations (3-46) to (3-49) relate the pulse signal to the impulse signal in the time domain. The corresponding relationship in the frequency domain is as follows:

$$\mathbf{E}^*(j\omega) \equiv \lim_{\gamma \to 0} \mathbf{E}_p^*(j\omega) \tag{3-50}$$

Then substituting from Eq. (3-32) into this equation yields

$$\begin{aligned} \mathbf{E}^*(j\omega) &= \frac{1}{T} \sum_{m=-\infty}^{+\infty} \left[\lim_{\gamma \to 0} \frac{\sin(m\omega_s\gamma/2)}{m\omega_s\gamma/2} \right] \mathbf{E}[j(\omega - m\omega_s)] \\ &\equiv \frac{1}{T} \sum_{m=-\infty}^{+\infty} \mathbf{E}[j(\omega - m\omega_s)] = [\text{Eq. (3-45)}]_{s=j\omega} \\ &= \frac{\mathbf{E}(j\omega)}{T} + \frac{\mathbf{E}[j(\omega \pm \omega_s)]}{T} + \frac{\mathbf{E}[j(\omega \pm 2\omega_s)]}{T} + \cdots \end{aligned} \tag{3-51}$$

Note that the distinction between the pulse and impulse functions is the value of the term $\sin(m\omega_s\gamma/2)/(m\omega_s\gamma/2)$. For the pulse it is a variable quantity which is a function of γ and reduces the amplitude of the folded $\mathbf{E}(j\omega)$'s as seen in Fig. 3-4. For the impulse it is a constant of unit value, or in other words, from Eq. (3-31),

$$[\mathbf{P}(j\omega)]_{\text{impulse}} = \frac{1}{T} \sum_{m=-\infty}^{+\infty} \delta(\omega - \omega_s) = \lim_{\gamma \to 0} [\mathbf{P}(j\omega)]_{\text{pulse}}$$

Using Eq. (3-51), the frequency spectra of the continuous and impulse-sampled functions are obtained as shown in the examples of Fig. 3-6. It is seen from Eq. (3-51) that the sampling process *produces a fundamental spectrum similar in shape to that of the continuous function, with an amplitude 1/T of the continuous*

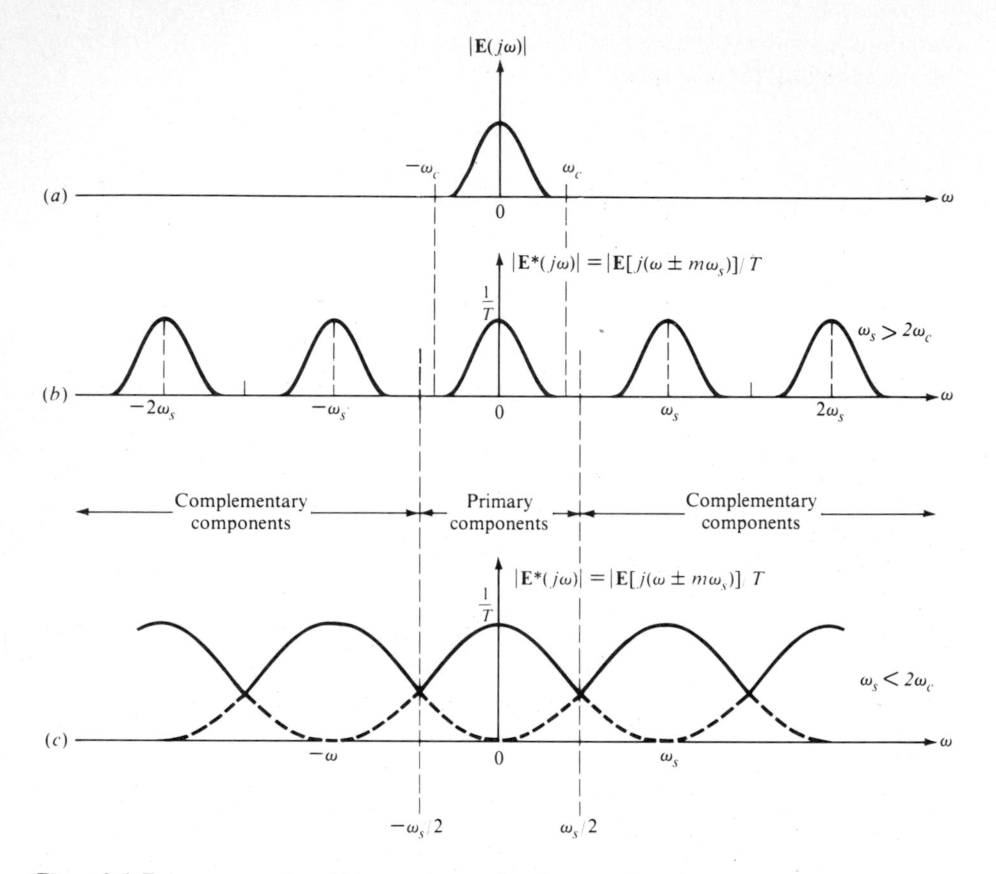

Figure 3-6 Frequency spectra of (*a*) a continuous function $e(t)$; (*b*) an impulse-sampled function $e^*(t)$, where $\omega_s > 2\omega_c$; (*c*) an impulse-sampled function $e^*(t)$, where $\omega_s < 2\omega_c$.

function. It also produces a succession of complementary or folded spectra which are shifted periodically by a frequency separation $m\omega_s$ ($m = \pm 1, \pm 2, \ldots$) and have the same amplitude as the fundamental spectrum if the sampling frequency ω_s is sufficiently high; that is, if $\omega_s \geq 2\omega_c$, where ω_c is the highest frequency component of $\mathbf{E}(j\omega)$, then there is essentially no overlap between the fundamental (primary) and complementary frequency spectra as shown in Fig. 3-6*b*. If the sampling frequency π/T is too low, that is, $\omega_c > \omega_s/2$, then the fundamental and complementary frequency spectra overlap one another, as shown in Fig. 3-6*c*, and aliasing† occurs (see Sec. 7-2). Note that

$$f_s = \frac{1}{T} \qquad \text{Hz} \tag{3-52}$$

† The sampling process is unable to distinguish between two sinusoids whose respective radian frequencies have a sum or difference of an integral multiple of $2/T$. Effectively, all radian frequencies are folded (or aliased) into the interval $0 \leq \omega < \omega_s/2$.

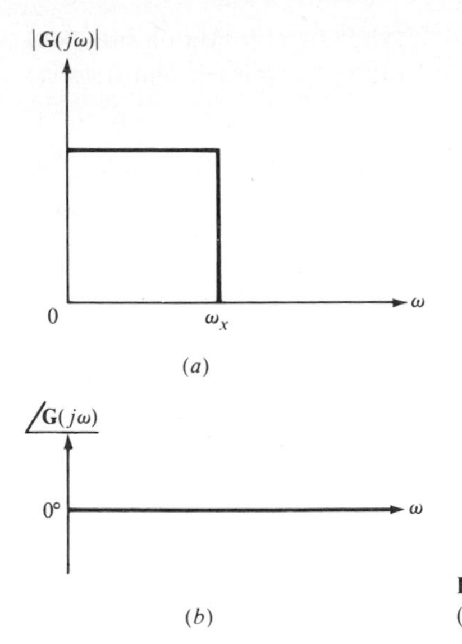

Figure 3-7 Ideal low-pass-filter frequency transfer function. (*a*) Magnitude characteristic; (*b*) phase characteristic.

and

$$\omega_s = 2\pi f_s \qquad \text{rad/s} \tag{3-53}$$

For the situation represented by Fig. 3-6*b*, an *ideal low-pass filter*, having the magnitude and phase characteristics shown in Fig. 3-7, *can essentially extract the spectrum of the continuous input signal by attenuating the spurious higher-frequency spectra*. Of course, a band-limited signal is assumed here.

The forward-frequency transfer function **G**$(j\omega)$ *of a control system generally has a low-pass characteristic*, so the system responds with a greater degree of accuracy to a continuous signal. In other words, the low-pass characteristics of the system attenuates the complementary components. Obviously, in the situation of Fig. 3-6*c*, it is difficult to reconstruct $e(t)$ from $e^*(t)$ because it is difficult to perform the necessary low-pass filtering. Thus this situation must be avoided; that is, ω_s must be sufficiently high to minimize overlapping of the primary and the complementary components of the input signal.

3-5 SHANNON'S SAMPLING THEOREM[12]

Shannon's sampling theorem states that "a function of time $e(t)$ which contains no frequency components greater than ω_c rad/s (band-limited) can be reconstructed by the values of $e(t)$ at any set of sampling points that are spaced apart by

$$T < \frac{\pi}{\omega_c} \qquad \text{s}$$

This theorem gives the maximum value of T permissible in order to be able to reconstruct the original sampled signal and avoid the overlapping as illustrated in

Fig. 3-6*b*. In other words, there exists an upper bound that specifies that a sufficient number of *impulse* samples be taken. These sampled values completely characterize $e(t)$. Therefore, the following sampling-frequency criterion is used:

$$\omega_s > 2\omega_c \tag{3-54}$$

(see Fig. 3-6*b*). The selection of a sampling rate must also consider noise, data quantization, system resonant frequencies, and design technique (as discussed in Chaps. 2 and 8). These effects usually dictate a nominal value of sampling frequency at least eight times greater than ω_c depending upon the specific value of the above parameters.

3-6 UNIFORM-RATE SAMPLING

Figure 3-8 illustrates open-loop systems having one or more samplers. The impulse sampling schedule for each of the samplers may be one of the following:

Uniform-rate sampling
Multirate sampling
Skip sampling
Variable sampling
Random sampling

Figure 3-9 illustrates these types of sampling rates. This text deals only with uniform-rate sampling since most control systems only require this sampling schedule. This type of scheduling assumes that the sampling switch model instantly closes and opens periodically; that is, the switch opens and closes at discrete periodic instants

$$t = 0,\ T,\ 2T,\ 3T,\ \ldots,\ kT,\ \ldots$$

For systems having more than one sampler, it is assumed that all samplers have the same uniform-rate sampling schedule, and they operate synchronously. This

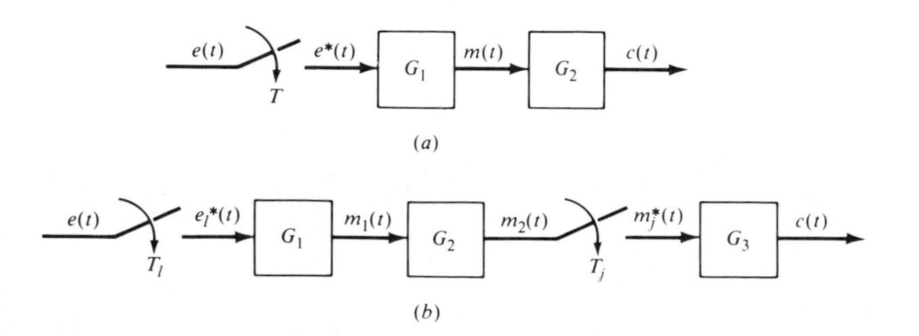

Figure 3-8 An open loop system with (*a*) one sampler; (*b*) two samplers.

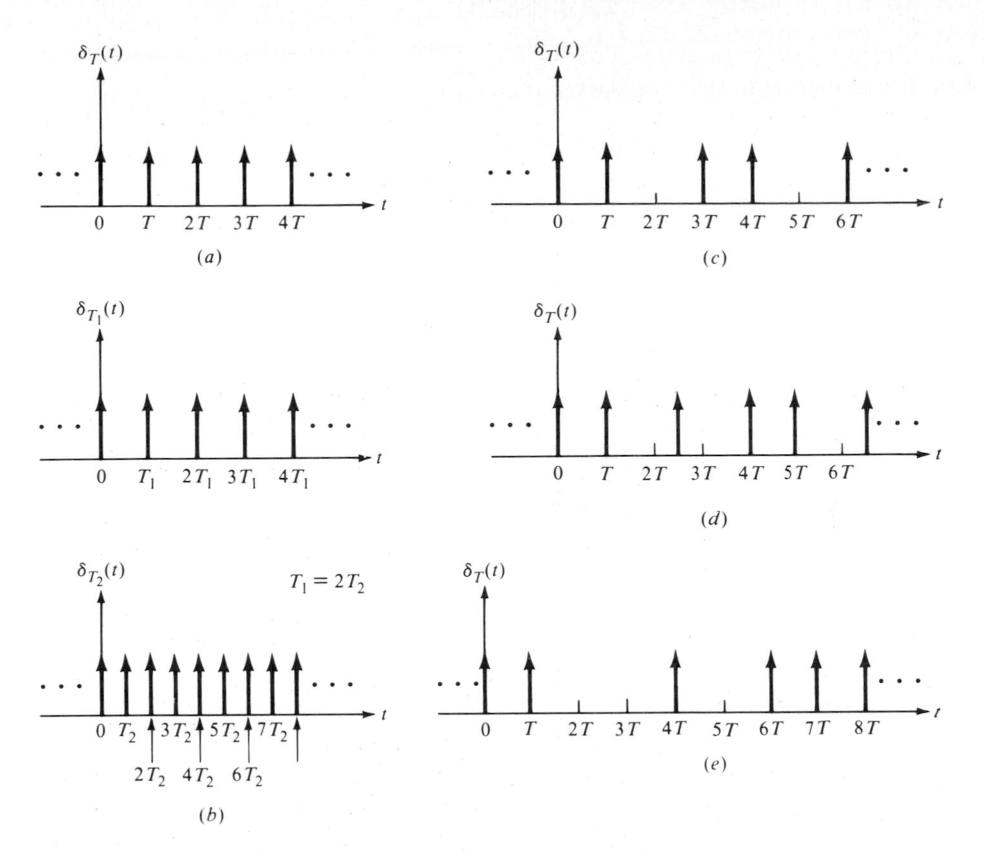

Figure 3-9 Various sampler schedules. (*a*) Uniform rate; (*b*) multirate; (*c*) skip; (*d*) variable; (*e*) random.

constraint simplifies the analysis of a sampled-data control system. Thus for the system of Fig. 3-8*b* ($T_l = T_j$), then

$$e_l^*(t) = \sum_{k=0}^{\infty} e_l(kT) \qquad \text{and} \qquad m_j^*(t) = \sum_{k=0}^{\infty} m_j(kT)$$

3-7 GENERATION AND SOLUTION OF LINEAR DIFFERENCE EQUATIONS: SYSTEM MODEL

Sampled-data control systems may be modeled by linear difference equations since some or all of the linear system variables are discrete quantities. Thus to model digitally a continuous-time or sampled-data system performance, it is necessary to discuss methods of discretizing continuous variables. The approach of this section focuses on the use of a difference operator as an approximation to a differential or integral operator.

Given the set of discrete values $c(NT)$, $c[(N-1)T]$, ..., $c(0)$, where $N = 1, 2, 3, \ldots$, one defines the *first-backward difference* ∇ as

$$\nabla c(NT) = c(NT) - c[(N-1)T] \tag{3-55}$$

the *second-backward difference* ∇^2 as

$$\begin{aligned}\nabla^2 c(NT) &= \nabla[\nabla c(NT)] = \nabla c(NT) - \nabla c[(N-1)T] \\ &= c(NT) - c[(N-1)T] - \{c[(N-1)T] - c[(N-2)T]\} \\ &= c(NT) - 2c[(N-1)T] + c[(N-2)T]\end{aligned} \tag{3-56}$$

and the rth-*backward difference* ∇^r as

$$\nabla^r c(NT) = \nabla^{r-1} c(NT) - \nabla^{r-1} c[(N-1)T] \tag{3-57}$$

where $r = 1, 2, 3, \ldots$. The representation of a sampled-data control system by difference equations permits its solution on a digital computer.

Example 3-7 Evaluate

$$c(t) = \int_0^t e(\tau)\, d\tau \tag{3-58}$$

by means of a numerical integration technique.[13] The integral of Eq. (3-58) is approximated (see Fig. 3-10) by

$$c(NT) = \int_0^{NT} e(\tau)\, d\tau \approx \sum_{k=0}^{N-1} e(kT)\, \Delta t = \sum_{k=0}^{N-1} Te(kT) \tag{3-59}$$

where $k = 0, 1, 2, \ldots$, $\Delta t = T$, and N is a constant integer greater than zero. $e(kT)$ represents a constant function for the time interval $kT \le \tau < (k+1)T$. This equation represents the definition of the integration process and is called

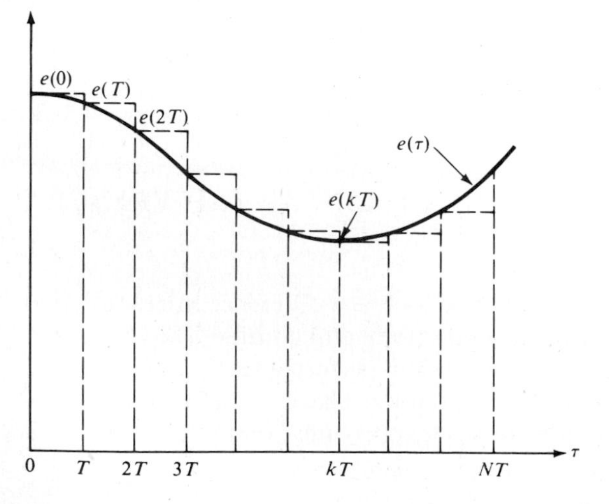

Figure 3-10 Numerical integration technique applied to

$$c(t) = \int_0^t e(\tau)\, d\tau$$

the *first-difference approximation* (*rectangular integration*). To obtain $c(NT)$ by using the form

$$\sum_{k=0}^{N-1} Te(kT)$$

requires the retention of all the N values of $e(kt)$ and then performing the summation. This is a *nonrecursive* formulation since $c(NT)$ is given explicitly in terms of $e(kT)$.

If the formula for $c(NT)$ explicitly contains past values of c, that is, $c[(N-1)T]$, $c[(N-2)T]$, ..., as well as values of $e(kT)$ then this formulation is defined as being *recursive*. The amount of storage space required when a digital computer is used to evaluate Eq. (3-59) may be substantially reduced by first evaluating

$$c[(N-1)T] = \int_0^{(N-1)T} e(\tau)\,d\tau \approx \sum_{k=0}^{N-2} Te(kT) \tag{3-60}$$

and subtracting it from Eq. (3-59) to obtain the first-backward difference equation

$$c(NT) - c[(N-1)T] = Te[(N-1)T] = \nabla c(NT) \tag{3-61}$$

Use of this recursive equation to evaluate the present value of $c(NT)$ requires the retention of only the immediate past sampled value $e[(N-1)T]$ and the immediate past value of the integral, $c[(N-1)T]$.

To illustrate the savings of the storage space requirement, first note from Eq. (3-58) that

$$c(0) = \int_0^0 e(\tau)\,d\tau = 0$$

Next, evaluate $c(NT)$ from Eq. (3-61) as follows:

$k = 0$ $\qquad c(T) = Te(0) + c(0) = Te(0)$

$k = 1$ $\qquad c(2T) = Te(T) + c(T)$

$$= T[e(T) + e(0)]$$

$k = 2$ $\qquad c(3T) = Te(2T) + c(2T)$

$$= T[e(2T) + e(T) + e(0)]$$

... $\qquad$ ... $\qquad$ ...

$k = N-1$ $\qquad c(NT) = Te[(N-1)T] + c[(N-1)T]$

$$= T\sum_{k=0}^{N-1} e(kT)$$

Analyzing the above iteration reveals that the present value of $c(NT)$ is readily obtained by adding its immediate previous value $c[(N-1)T]$ to the immediate previous value $e[(N-1)T]$ multiplied by T.

The first-backward difference $\nabla c(kT)$ may also be obtained by applying the same procedure to $c(kT) = \int_0^{kT} e(\tau)\,d\tau$ to obtain

$$c(kT) - c[(k-1)T] = Te[(k-1)T] = \nabla c(kT) \tag{3-62}$$

This concept is very important in realizing a digital control law or equation which will prove to be a linear difference equation. The specific form implemented can have advantages in terms of memory space and speed of execution. This is discussed in more detail in later chapters.

Example 3-8 Obtain the discretization of the differentiation process. Differentiation is defined as

$$\dot{c}(t) \equiv Dc(t) \equiv \frac{dc(t)}{dt} \equiv \lim_{T \to 0} \frac{c(kT) - c[(k-1)T]}{T} \tag{3-63}$$

where $c(t)$ has the appropriate continuity characteristics and D is the derivative operator. Equation (3-63) may be approximated by the first-backward difference

$$Dc(kT) = \left.\frac{dc(t)}{dt}\right|_{t=kT} \approx \frac{c(kT) - c[(k-1)T]}{T} = \frac{1}{T}\nabla c(kT) \tag{3-64}$$

Also, the second derivative of a function is defined as

$$D \cdot Dc(t) \equiv D^2 c(t) \equiv \frac{d^2 c(t)}{dt^2} \equiv \lim_{T \to 0} \frac{Dc(kT) - Dc[(k-1)T]}{T} \tag{3-65}$$

which may be approximated by

$$D^2 c(kT) = \left.\frac{d^2 c(t)}{dt^2}\right|_{t=kT} \approx \frac{Dc(kT) - Dc[(k-1)T]}{T} \tag{3-66}$$

The term $Dc(kT)$ in Eq. (3-66) is given by Eq. (3-64). The term $Dc[(k-1)T]$ is expressed in the format of Eq. (3-64) by replacing k by $k-1$ to yield

$$\frac{1}{T}\nabla c[(k-1)T] = \frac{c[(k-1)T] - c[(k-2)T]}{T} \tag{3-67}$$

Substituting Eqs. (3-64) and (3-67) into Eq. (3-66) yields the second-backward difference

$$\begin{aligned} D^2 c(kT) &\approx \frac{1}{T^2}\{\nabla c(kT) - \nabla c[(k-1)T]\} \\ &\approx \frac{1}{T^2}\{c(kT) - 2c[(k-1)T] + c[(k-2)T]\} = \frac{1}{T^2}\nabla^2 c(kT) \end{aligned} \tag{3-68}$$

for the second derivative. The rth derivative is approximated by the following rth-backward equation:

$$D^r c(kT) \approx \frac{1}{T^r}\{\nabla^{r-1}c(kT) - \nabla^{r-1}c[(k-1)T]\} = \frac{1}{T^r}\nabla^r c(kT) \qquad (3\text{-}69)$$

where $r = 1, 2, 3, \ldots$ and $k = 0, 1, 2, \ldots$. For a given value of r, Eq. (3-69) can be expanded in a similar manner as is done for the second derivative to obtain the expanded representation of $D^r c(kT)$.

Example 3-9 The transfer function of a second-order linear time-invariant (LTI) continuous system is

$$G(s) = \frac{C(s)}{E(s)} = \frac{10(s+2)}{s^2+s+1} = \frac{10(s+2)}{s+0.5 \pm j0.866} \qquad (3\text{-}70)$$

(*a*) It is desired to determine the time response $c(t)$ at discrete points ($t = kT$) for a given input $e(t)$ by the use of difference equations. First, rearrange Eq. (3-70) and then take the inverse Laplace transformation as follows:

$$(s^2+s+1)C(s) = 10(s+2)E(s)$$

$$(D^2+D+1)c(t) = 10(D+2)e(t) \qquad (3\text{-}71)$$

Using Eqs. (3-64) and (3-68), Eq. (3-71) is approximated by the following equation:

$$\frac{1}{T^2}\{c(kT) - 2c[(k-1)T] + c[(k-2)T]\} + \frac{1}{T}\{c(kT) - c[(k-1)T]\} + c(kT) = \frac{10}{T}\{e(kT) - e[(k-1)T]\} + 20e(kT)$$

This equation can be rearranged as follows to solve for $c(kT)$:

$$\frac{T^2+T+1}{T^2}c(kT) + \frac{-T-2}{T^2}c[(k-1)T] + \frac{1}{T^2}c[(k-2)T] = \frac{20T+10}{T}e(kT) - \frac{10}{T}e[(k-1)T]$$

Letting $A = T^2 + T + 1$ and solving for $c(kT)$ yields

$$c(kT) = \frac{1}{A}\{(20T^2+10T)e(kT) - (10T)e[(k-1)T] + (T+2)c[(k-1)T] - c[(k-2)T]\}$$

This last equation is of the (recursive) form

$$c(kT) = q_0 e(kT) + q_1 e[(k-1)T] - d_1 c[(k-1)T] - d_2 c[(k-2)T] \qquad (3\text{-}72)$$

where

$$q_0 = \frac{20T^2 + 10T}{A} \qquad q_1 = \frac{-10T}{A} \qquad c(t) = 0 \text{ for } t \leq 0$$

$$d_1 = -\frac{T+2}{A} \qquad d_2 = \frac{1}{A}$$

For $T = 0.1_{10}$ s and assuming a 10-bit word length, the coefficients in base 2 (sign plus magnitude) are

$$q_0 = +1.0810811_{10} = +1.000101001_2$$

$$q_1 = -0.9009009_{10} = -0.111001101_2$$

$$d_1 = -1.8918919_{10} = -1.111001000_2$$

$$d_2 = +0.9009009_{10} = +0.111001101_2$$

Note that the base 2 number for q_0, q_1, d_1, and d_2 are approximations to the base 10 numbers (8 digits) due to finite word length. An exact base 10 representation needs 27 bits (see Chap. 2). Thus, the value of $c(kT)$ does not represent the true value of $c(t)$ at the sampling instants. Also, if this transfer function $G(s)$ is to be used as a controller, further analysis of accuracy requirements and word length must be accomplished. (This topic is discussed in Chap. 10.)

Equation (3-72) represents the approximate solution of Eq. (3-71). Also, Eq. (3-72) is a linear second-order difference equation that permits the evaluation at discrete points of the system's time response $c(t)$ to a general input $e(t)$. Equation (3-72) may be programmed to yield values of $c(t)$ at the computation times kT. An accurate evaluation of $c(t)$ requires that the sampling time T satisfy the previously discussed constraints. The maximum permissible value of T is determined by the system dynamics as illustrated in part *b*.

(*b*) Given that $e(t) = u_{-1}(t)$, for $t > 0$,† it is desired to determine the appropriate value of T for accurately evaluating values of $c(t)$ at $t = kT$. The dominant poles of Eq. (3-70) are $p_{1,2} = \sigma \pm j\omega_d = -0.5 \pm j0.866$ which results in an underdamped stable response as shown in Fig. 3-11. From the values of $t_p = \pi/\omega_d \approx 3.63$ s and $T_s = 4/|\sigma| = 8$ s, it can be concluded that $T < t_p = \pi/\omega_d$. In general, for accurate values of $c(kT)$, $T \ll t_p$.

For this example, using the value of $T = 0.1$ s results in the values shown in Table 3-1 obtained by use of Eq. (3-72). The values obtained by directly solving $c(t) = \mathscr{L}^{-1}[C(s)]_{t=kT}$ are also given in this table for comparison purposes.

† See definition of a unit-step forcing function.

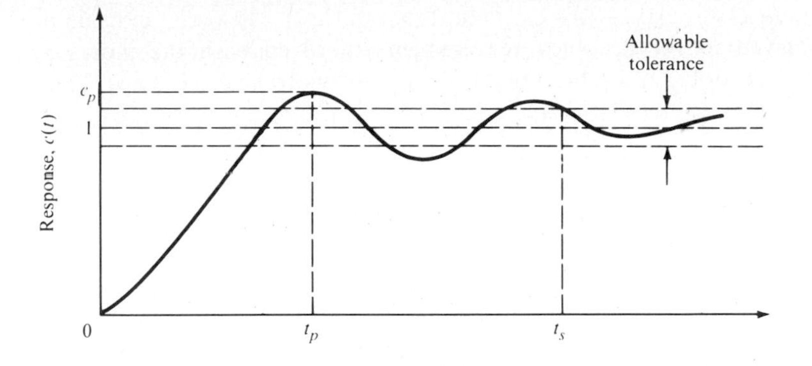

Figure 3-11 Typical underdamped response to a unit-step function. (*From Ref. 1.*)

Table 3-1 Values of $c(kT)$: Example 3-9

k	Eq. (3-72)	$c(t) = \mathscr{L}^{-1}[C(s)]$
0	0	0
1	1.081081	1.046709
2	2.225469	2.174026
3	3.41658	3.36356
4	4.63906	4.59808
5	5.87888	5.86157
6	7.12288	7.13926
...	...	...

Note that owing to finite word length, decreasing T increases the computation time required and increases the roundoff and truncation errors by use of Eq. (3-72). Thus the final selection of T is based upon a trade-off when all factors are considered. If it is not small enough, aliasing or frequency folding problems will exist.

The general linear difference equation of an LTI nth-order SISO system is

$$c(kT) = q_0 e(kT) + q_1 e[(k-1)T] + \cdots + q_w e[(k-w)T] - d_1 c[(k-1)T] - d_2 c[(k-2)T] - \cdots - d_n c[(k-n)T] \quad (3\text{-}73)$$

where $w \leq n$. This inequality is required to avoid an anticipatory model which generally cannot be solved in real time. Equation (3-73) may be derived in the same manner as employed in Example 3-9. This equation can characterize the approximation of a continuous system or may characterize exactly a discrete system.

Equation (3-73) can also be solved on a digital computer. The output $c(kT)$ is generated at discrete instants of time kT as a linear function of the past values of the output $c[(k-1)T]$, $c[(k-2)T]$, ... and the input $e(t)$. However, consideration must be given to the accuracy requirements involved with the indicated multiplications, additions, and subtractions due to finite word length (Chaps. 2 and 10).

3-8 WEIGHTING SEQUENCE (SAMPLED-DATA SYSTEM)

The purpose of this section is to extend the concept of system weighting or transfer function for a continuous system[1] to a discrete system and to present another method of modeling a continuous-time system. Referring to Fig. 3-12*a*, if $e(t) = \delta(t)$, a unit impulse at $t = 0$, then the impulse response is $c(t) = g(t)$, where $g(t) = \mathscr{L}^{-1}[G(s)]$ is the system weighting function. Thus, for an input $e(t) = 0$ for $t < 0$, the output response is given by the convolution integral

$$c(t) = \int_0^t e(t-\tau)g(\tau)\,d\tau = \int_0^t e(\tau)g(t-\tau)\,d\tau \tag{3-74}$$

This integral may be evaluated by approximating $g(t)$ by a piecewise-constant (P.C.) function. Consider the impulse response $g(t)$ shown in Fig. 3-13. Also shown is its approximation by a P.C. function. The values of g_k are chosen so that the area under the $g(t)$ and the P.C. curves, for $kT \le t < (k+1)T$, are essentially equal. Each P.C. portion represents a pulse function of height g_k and of width T s, as shown in Fig. 3-14*a*. This pulse function can be mathematically decomposed into two step functions, one delayed from the other by T s, as shown in Fig. 3-14*b*. Thus the P.C. curve of Fig. 3-13 can be mathematically expressed as

$$g(t) \approx \sum_{k=0}^{M-1} g_k[u_{-1}(t-kT) - u_{-1}(t-kT-T)] \tag{3-75}$$

where t_x is the value of time for which the value of $c(t)$ is to be determined,

$$M = \frac{t_x}{T} = 1, 2, 3, \ldots$$

$$g(t) \approx 0 \qquad \text{for } t \ge MT$$

$$u_{-1}(\alpha) = \begin{cases} 1 & \text{for } \alpha \ge 0 \\ 0 & \text{for } \alpha < 0 \end{cases} \qquad \text{where } \alpha \equiv t - kT$$

Equation (3-75) is called a *pulse weighting sequence*. Using Eq. (3-74), the value $c(t)$ at $t = t_x = MT$ is approximately given by inserting Eq. (3-75) into this equation. Thus

$$c(MT) \approx \int_0^{MT} e(MT-\tau) \sum_{k=0}^{M-1} g_k[u_{-1}(\tau-kT) - u_1(\tau-kT-T)]\,d\tau$$

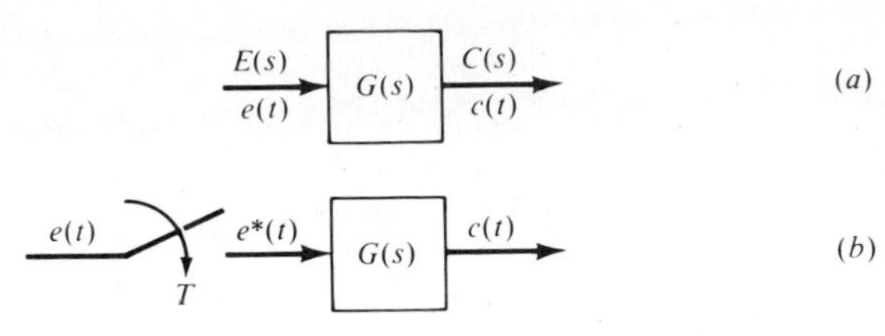

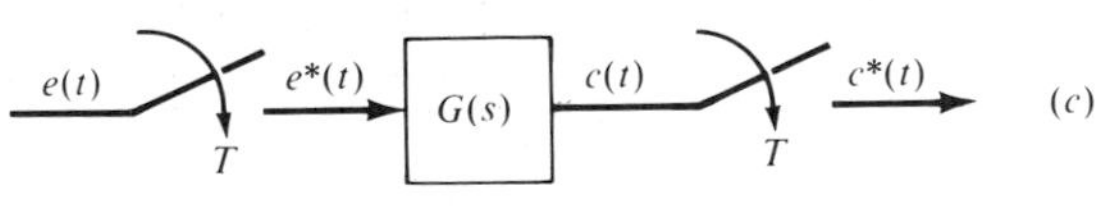

Figure 3-12 A linear system with: (*a*) all continuous signals; (*b*) sampled input; (*c*) sampled input and output.

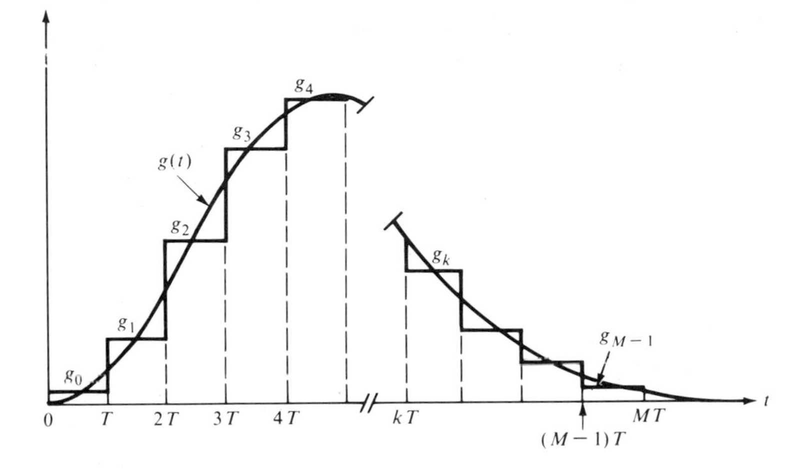

Figure 3-13 An impulse response $g(t)$ of a system and its piecewise-sampled approximation.

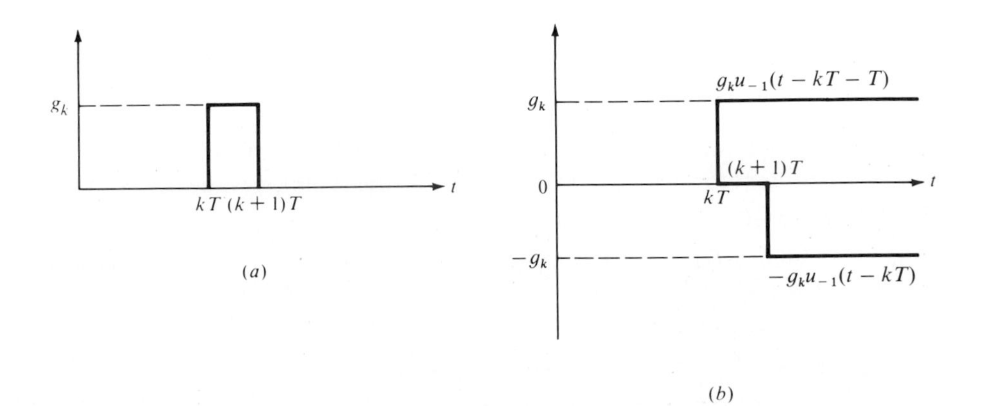

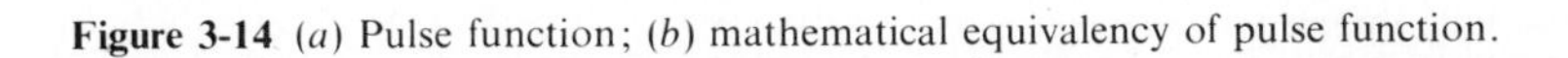
Figure 3-14 (*a*) Pulse function; (*b*) mathematical equivalency of pulse function.

which may be expressed as

$$c(MT) \approx \sum_{k=0}^{M-1} g_k \int_{kT}^{kT+T} e(MT-\tau)\,d\tau \tag{3-76}$$

A change of variable is made to obtain a simpler form. Let $\beta = MT - \tau$. Then $d\beta = -d\tau$, and the upper limit becomes

$$\tau = MT - \beta = kT + T \rightarrow \beta = MT - kT - T = (M - k - 1)T$$

The lower limit is

$$\tau = MT - \beta = kT \rightarrow \beta = MT - kT = (M - k)T$$

The integral portion of Eq. (3-76) becomes

$$\int_{MT-kT}^{MT-kT-T} -e(\beta)\,d\beta$$

Thus Eq. (3-76) becomes

$$c(MT) \approx \sum_{k=0}^{M-1} g_k \left[\int_{MT-kT-T}^{MT-kT} e(\beta)\,d\beta \right] \tag{3-77}$$

For T small, the bracketed term represents the area under the curve of $e(\beta)$ vs. β for $MT - kT - T < \beta < MT - kT$, and this area is approximated by rectangular integration, yielding

$$Te[(M-k)T] \tag{3-78}$$

Substituting from Eq. (3-78) into Eq. (3-77) yields

$$c(MT) \approx T \sum_{k=0}^{M-1} g_k e[(M-k)T] \tag{3-79}$$

which is another method of approximating a continuous-time system by a difference equation. If the word length is assumed to be infinite, then choosing a smaller value of "sampling time," $T > 0$, results in a more accurate approximation.

Equation (3-79) can be considered as yielding the value of $c(t)|_{t=MT}$ of Fig. 3-15. In this figure the zero-order-hold device holds the sampled value of $e(kT)$ for one period; that is,

$$m(t) = e(kT) \qquad \text{for } kT \leq t < (k+1)T \tag{3-80}$$

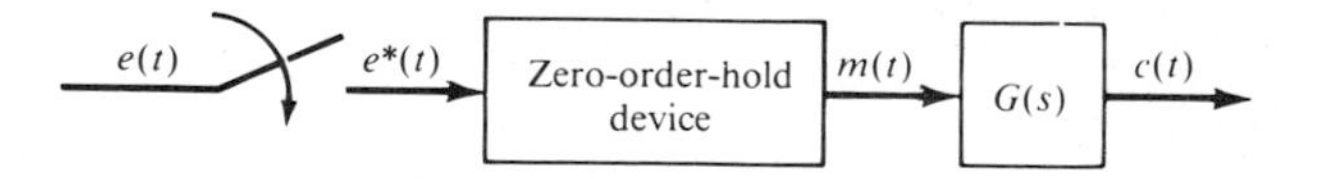

Figure 3-15 An open-loop sampled-data system.

(Zero-order holds are discussed in more detail in Secs. 3-9 and 8-8.) Applying Eq. (3-74) to the open-loop system of Fig. 3-15 yields

$$c(t)|_{t=MT} = \int_0^{MT} e(MT-\tau)g(\tau)\,d\tau = \sum_{k=0}^{M-1} \int_{kT}^{(k+1)T} e(MT-\tau)g(\tau)\,d\tau \tag{3-81}$$

Based on Eq. (3-80), Eq. (3-81) can be rewritten as

$$c(MT) = \sum_{k=0}^{M-1} e(MT-kT) \int_{kT}^{(k+1)T} g(\tau)\,d\tau \tag{3-82}$$

Next by use of rectangular integration, that is,

$$\int_{kT}^{(k+1)T} g(\tau)\,d\tau \approx Tg(kT)$$

Eq. (3-82) becomes

$$c(MT) = T\sum_{k=0}^{M-1} g(kT)e[(M-k)T] \tag{3-83}$$

which is identical to Eq. (3-79) where $g_k = g(kT)$. Thus the open-loop sampled-data system, utilizing a zero-order-hold device as shown in Fig. 3-15, is an equivalent discrete simulation of the continuous-time open-loop system of Fig. 3-12*a*.

Example 3-10 Given $G(s) = 1/(s+1)$, $E(s) = 1/s$,

$$c(t) = \mathscr{L}^{-1}[G(s)E(s)] = 1 - \epsilon^{-t} \qquad \text{and} \qquad c(t)|_{t=0.5} = 0.39347$$

Use Eq. (3-79) to obtain the value of $c(t)$ at $t = 0.5$. For this overdamped system, then select $T \ll T_s = 4/|\sigma| = 4$. For a unit-step input, $e[(M-k)T] = 1$. Since $g(t) = \mathscr{L}^{-1}[G(s)] = \epsilon^{-t}$, then $g(kT) = \epsilon^{-kT}$. The value of 0.1 is chosen to satisfy $T \ll T_s$ and so that $\nabla c(kT)$ is small in between sampling intervals; that is, $\nabla c(kT) = c(kT) - c(kT-T) \approx T\dot{c}(t)|_{t=kT} = Te^{-kT}$ is small. An approximate method (trapezoidal integration) for evaluating g_k, as illustrated by this example, is

$$g_k \approx \frac{g(kT) + g(kT+T)}{2}$$

and values of g_k are given in Table 3-2. Thus, using Eq. (3-79), where $e[(M-k)T] = 1$, yields

$$c(5T) = c(0.5) = 0.1\sum_{k=0}^{4} g_k = 0.39379$$

which is close to the exact value of 0.39347.

Table 3-2 Values of g_k for Example 3-10

k	g_k	k	g_k
0	0.95242	3	0.70557
1	0.86178	4	0.63843
2	0.77978		

Consider the case where a unit impulse $\delta(t)$ is applied to the system of Fig. 3-12c (sampled input and output). The system's continuous-time response is $c(t) = g(t)$. For the case where the two samplers are synchronized to sample at $t = kT$, the sampled output, based upon Eq. (3-47), is described by

$$c^*(t) = g^*(t)$$

$$= \sum_{k=0}^{\infty} g(kT)\delta(t - kT) = g(0)\delta(t) + g(T)\delta(t - T) + g(2T)\delta(t - 2T) + \cdots \tag{3-84}$$

This sequence is defined as the *impulse weighting sequence* of the system. The values of

$$g(kT) = c(kT) = c(t)|_{t=kT} = \mathscr{L}^{-1}[G(s)]|_{t=kT}$$

for the system impulse response shown in Fig. 3-13 are present in Fig. 3-16 where $g(0) = 0$. Note that Eq. (3-84) can describe the output of a real or fictitious ideal impulse sampler placed at the system output as shown in Fig. 3-17. The sequence $g(kT)$ is the counterpart of the impulse response $g(t)$ for a continuous-data system. The Laplace transform of the impulse weighting sequence, Eq. (3-84), yields the *impulse transfer function*, i.e., the *impulse transform* of $G(s)$,

$$G^*(s) = \sum_{k=0}^{\infty} g(kT)\epsilon^{-kTs} \tag{3-85}$$

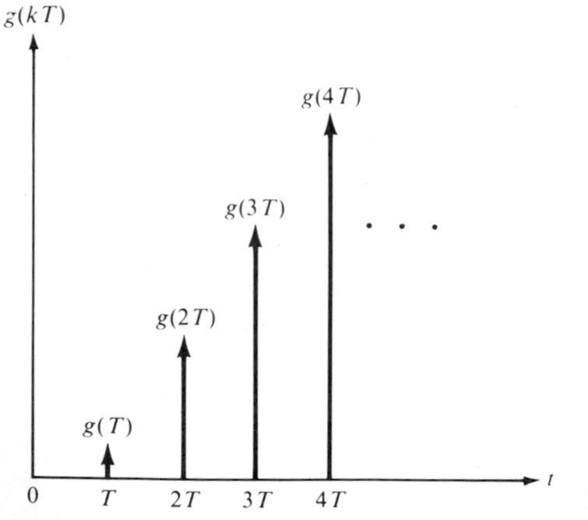

Figure 3-16 Impulse weighting sequence.

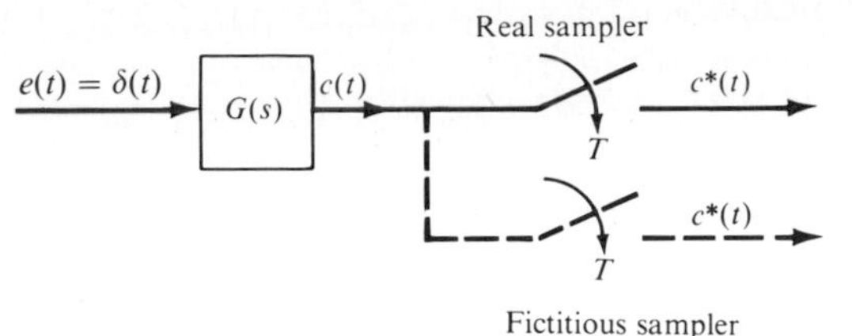

Figure 3-17 Sampled output response.

The impulse-response method may be used as a means of representing a sampled-data system and of analyzing its system response. As an introduction to this method consider the system of Fig. 3-12*b* that has the impulse response of Fig. 3-13 for $e(t) = \delta(t)$. Now consider the unit impulse $\delta(t - T)$, which represents a delay of one period; it is applied to the system at time $t = T$. This delayed impulse input results in the impulse response shown in Fig. 3-18. This response, when compared to the response in Fig. 3-13, is translated to the right by one "sampling" period due to the delay of one period of the input signal. Next consider the sampled-data system of Fig. 3-12*c* whose input $e(t)$ is the one shown in Fig. 3-19*a* and whose $g(t)$ is given by Fig. 3-13. The output of the ideal sampler, $e^*(t) = e(t)\delta_T(t)$, is a train of impulses, as shown in Fig. 3-19*b*, and with the use of Eq. (3-37) can be expressed as

$$e^*(t) = e(t) \sum_{k=0}^{\infty} \delta(t - kT) = \sum_{k=0}^{\infty} e(kT)\delta(t - kT) \tag{3-86}$$

where $e(t) = 0$ for $t < 0$. The Laplace transform of $e^*(t)$ is

$$E^*(s) = \mathscr{L}[e^*(t)] = \sum_{k=0}^{\infty} e(kT)\epsilon^{-kTs} \tag{3-87}$$

Using the superposition theorem and the system's impulse weighting sequence, the output $c(t)$ due to any arbitrary input $e^*(t)$ may be expressed for any

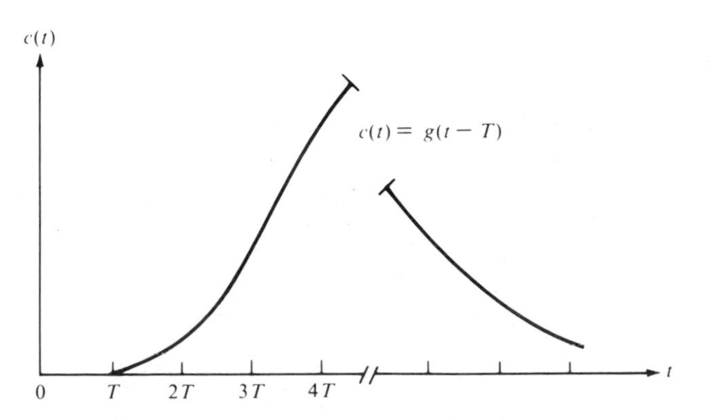

Figure 3-18 An impulse response of a system due to the unit impulse $\delta(t - T)$.

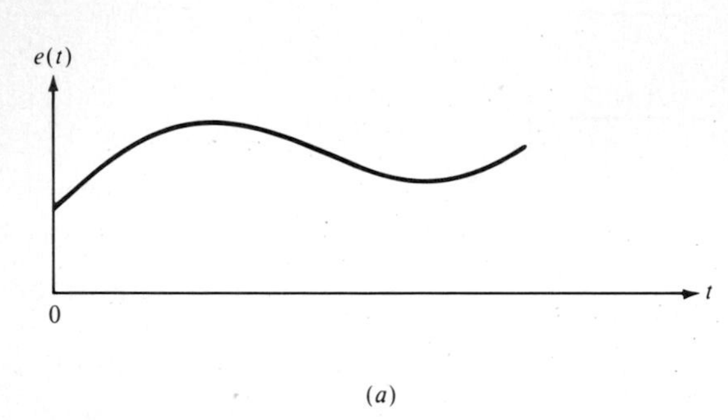

(a)

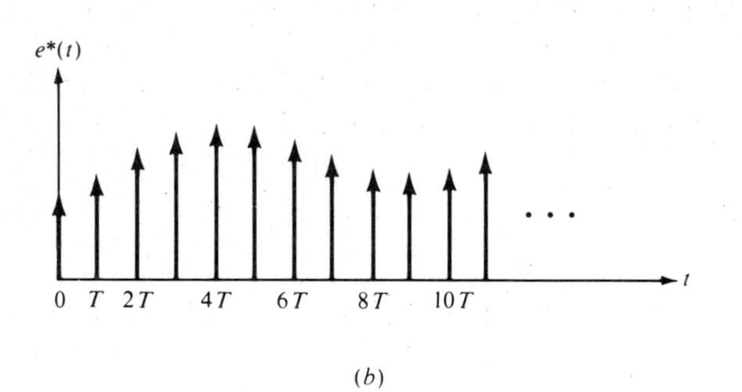

(b)

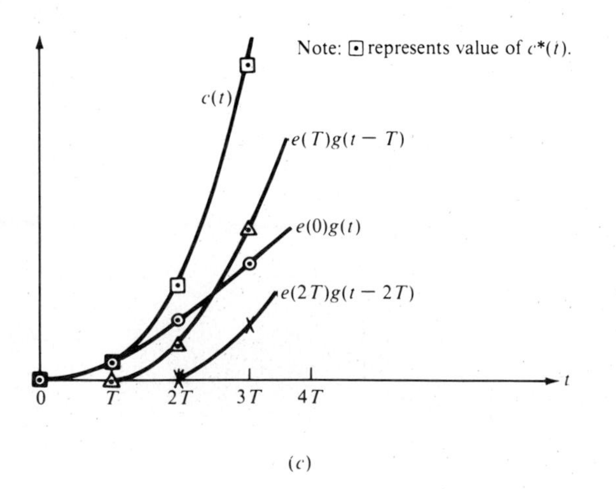

(c)

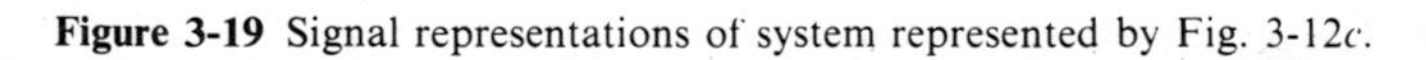

Figure 3-19 Signal representations of system represented by Fig. 3-12*c*.

time period $0 \le t \le MT$ as follows:

$$c(t) = \sum \text{(of all impulse responses)} = \sum_{k=0}^{M} e(kT)g(t - kT) \tag{3-88}$$

where $k = 0, 1, 2, \ldots, M$. Applying this equation to the example illustrated in Fig. 3-19 and having the inputs $e(0)$, $e(T)$, and $e(2T)$ yields the continuous output

$$c(t) = e(0)g(t) + e(T)g(t - T) + e(2T)g(t - 2T) + \cdots \tag{3-89}$$

The sampled output $c^*(t)$ at $t = 2T$ $(M = 2)$ is obtained by substituting $t = 2T$ into Eq. (3-89). Thus

$$c(2T) = e(0)g(2T) + e(T)g(T) + e(2T)g(0) \tag{3-90}$$

Therefore the value of $c(MT)$ is given by

$$c(MT) = \sum \text{(of the sampled values of all the impulse responses)}$$

or

$$c(MT) = e(0)g(MT) + e(T)g[(M - 1)T] + \cdots + e(MT)g(0)$$

$$= \sum_{k=0}^{M} e(kT)g[(M - k)T] \tag{3-91}$$

The similarity between this equation and the convolution integral of Eq. (3-74) for continuous-data systems is evident. Therefore, Eq. (3-91) is called the *convolution summation*. Equations (3-79) and (3-91) are equivalent as the pulse approaches an impulse.

3-9 DATA CONVERSION PROCESSES

The sampled signal $e^*(t)$, as shown in Fig. 3-6, not only contains the frequency components of the continuous signal $e(t)$ but high-frequency complementary (or folded) components resulting from the sampling process. The low-pass filtering characteristics of the system components to which $e^*(t)$ is applied are, in general, not sufficient to completely minimize the effects of high-frequency components. In most sampled-data control systems these high-frequency components must be removed (or filtered out of the system) in order to be able to accurately reconstruct a continuous control signal from the sampled data. This must be accomplished because these complementary components may adversely affect the system component characteristics (model). The effectual minimization of these undesirable complementary components can be accomplished by the use of a "smoothing" or "holding" device, i.e., a filter. This device *converts* the impulse-input signal $e^*(t)$, either by means of extrapolation or interpolation of the input impulses, into a continuous signal $m(t)$, whose form approximates the form of the input signal $e(t)$.

The most widely used holding device is the *zero-order hold* (*ZOH*) as discussed in Sec. 8-8. A simple ZOH device is shown in Fig. 3-20*a* along with a sampling

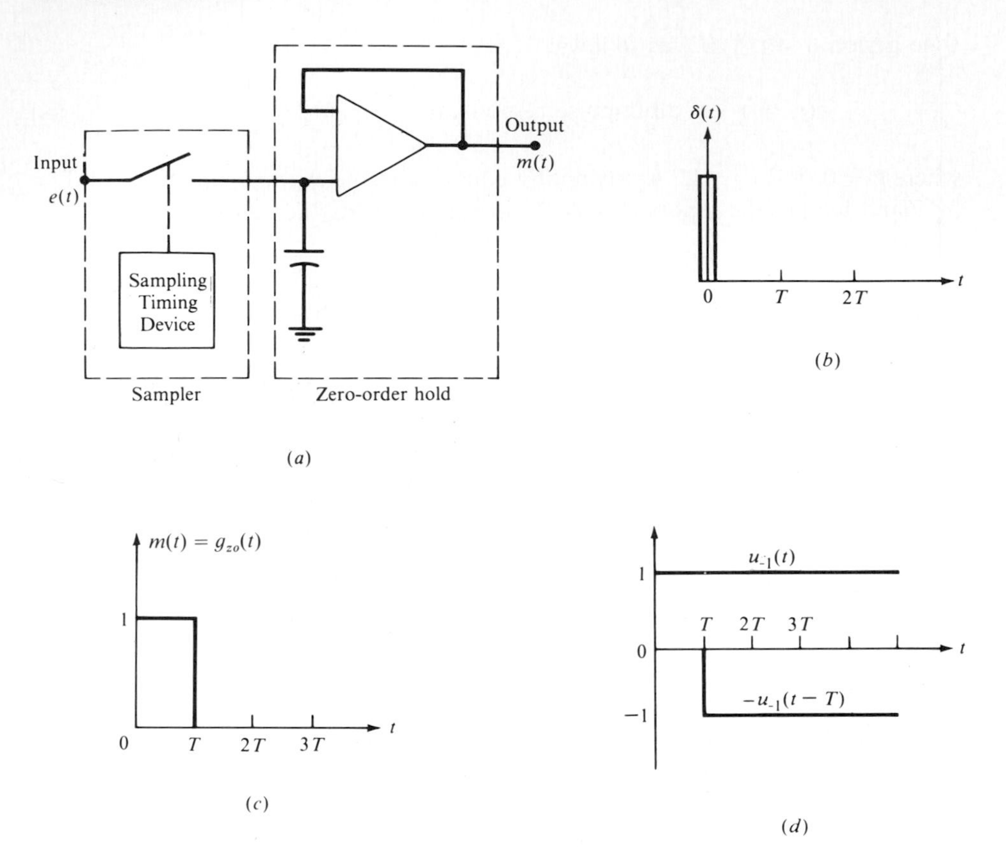

Figure 3-20 Zero-order hold and sampler operation. (*a*) ZOH and sampler circuit; (*b*) unit impulse input; (*c*) ZOH output response $m(t)$; (*d*) decomposition of $m(t)$.

circuit. To model the ZOH device, it is assumed that the capacitor is instantly charged to the impulse voltage $e(kT)$ at the sampling instant $t = kT$. Applying the single unit impulse input $e(0) = \delta(t)$ of Fig. 3-20*b* to the ZOH device yields the output shown in Fig. 3-20*c*. That is, during the sampling interval T in which the sampling switch is open, the capacitor holds the charge until the next sampling instant. Therefore, the ZOH seems to convert an input impulse into an approximate rectangular wave. By proper design of the ZOH unit and selection of the sampling time, the approximation of $e(t)$ can be made very good. The response of the ZOH device to the unit impulse input, as shown in Fig. 3-20*c*, is equivalent to the difference of two unit-step functions (see Fig. 3-14) as expressed by

$$g_{zo}(t) = u_{-1}(t) - u_{-1}(t - T) \tag{3-92}$$

This equation defines the weighting function $g_{zo}(t)$ of the ZOH unit.

Consider that the signal $e(t)$ being sampled at a uniform rate is the one shown in Fig. 3-21*a*. The sampled values $e(kT)$ are applied to a ZOH device. The resulting

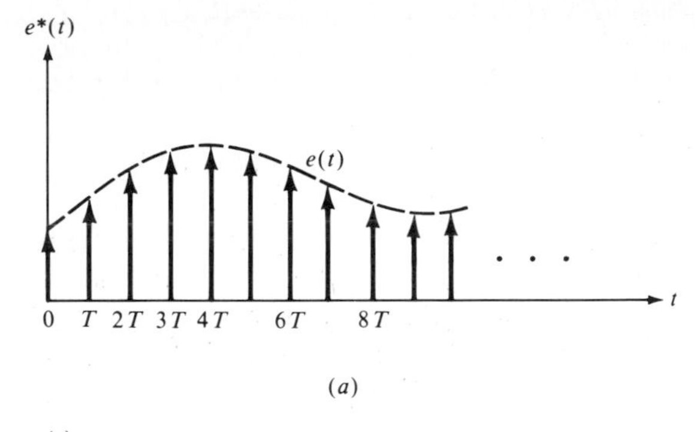

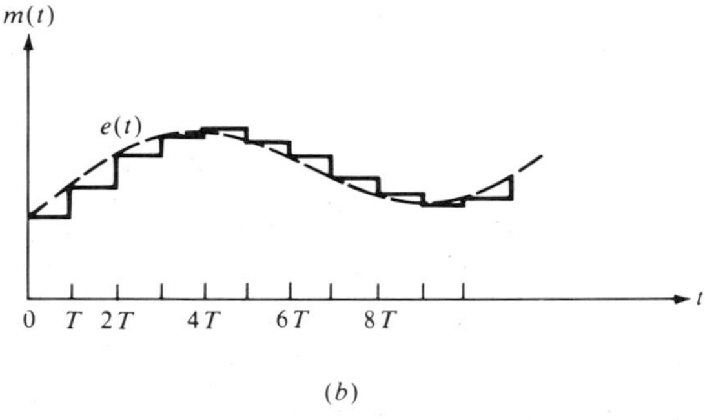

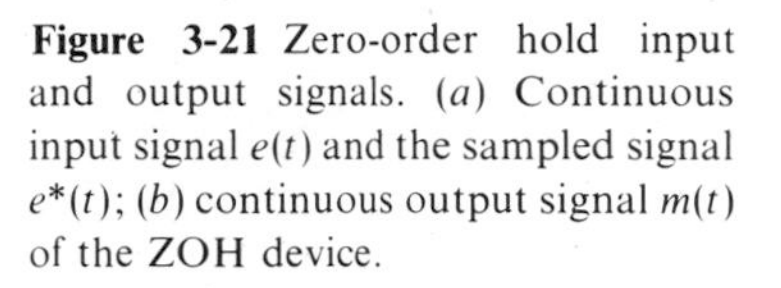
Figure 3-21 Zero-order hold input and output signals. (*a*) Continuous input signal $e(t)$ and the sampled signal $e^*(t)$; (*b*) continuous output signal $m(t)$ of the ZOH device.

holding device output is shown in Fig. 3-21*b*. From this figure it is noted that the output signal $m(t)$ is a step approximation to the actual continuous signal $e(t)$. This approximation, based upon an impulse input, may be improved by decreasing the sampling time T.

Figure 3-21 represents one example of converting an impulse signal into a continuous signal by means of a ZOH device. In this case, since $e^*(t)$ represents sampled values of $e(t)$, it is referred to, therefore, as a *reconstruction process*. Another example is the conversion of the discrete output of a digital computer to a continuous signal. This conversion, a *construction process*, can also be modeled by a ZOH device.

The transfer function representing the ZOH device is obtained by taking the Laplace transform of Eq. (3-92); thus

$$G_{zo}(s) = \frac{1 - \epsilon^{-Ts}}{s} = G_z^*(s)G_o(s) \tag{3-93}$$

where $G_o(s) = 1/s$ and

$$G_z^*(s) = 1 - \epsilon^{-Ts} \tag{3-94}$$

To obtain the frequency-response characteristic of the ZOH device, let $s = j\omega$ in Eq. (3-93). This yields

$$\mathbf{G}_{zo}(j\omega) = \frac{1 - \epsilon^{-j\omega T}}{j\omega} \tag{3-95}$$

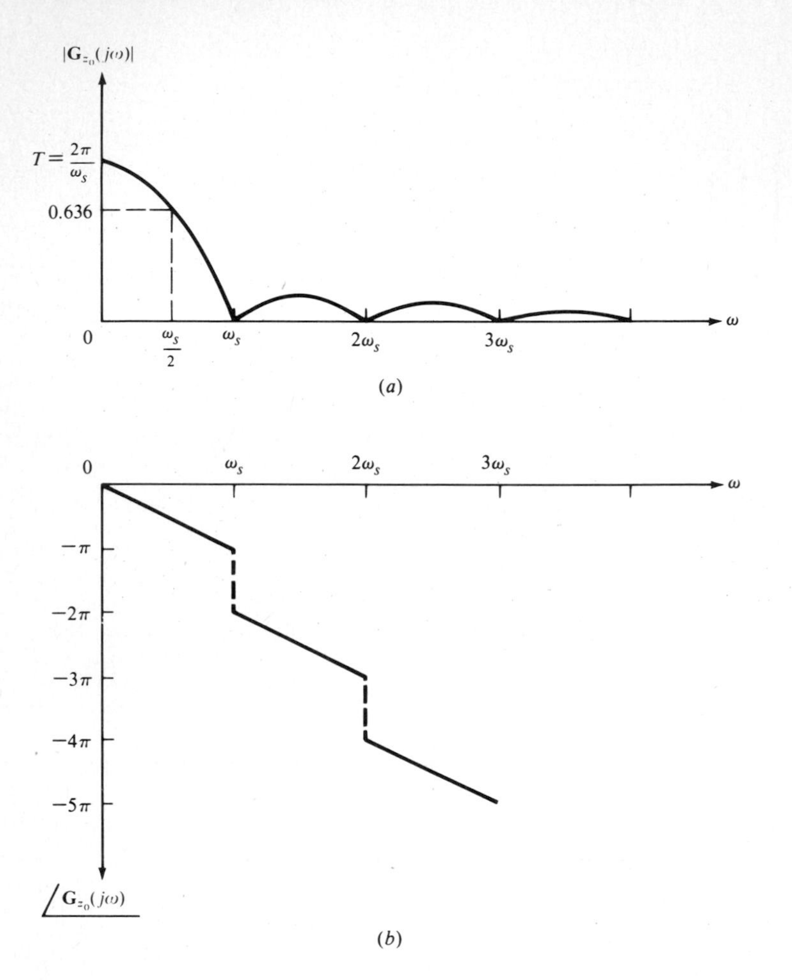

Figure 3-22 ZOH device. (*a*) Gain and (*b*) phase characteristics.

Rearranging this equation results in

$$\mathbf{G}_{zo}(j\omega) = \frac{2\epsilon^{-j\omega T/2}(\epsilon^{j\omega T/2} - \epsilon^{-j\omega T/2})}{j2\omega} = \frac{2\epsilon^{-j\omega T/2} \sin(\omega T/2)}{\omega} \tag{3-96}$$

Since $T = 2\pi/\omega_s$ s and $\omega_s = 2\pi f_s$ rad/s, Eq. (3-96) can be rewritten as

$$\mathbf{G}_{zo}(j\omega) = \frac{2\pi}{\omega_s} \frac{\sin \pi(\omega/\omega_s)}{\pi(\omega/\omega_s)} \epsilon^{-j\pi(\omega/\omega_s)} = |\mathbf{G}_{zo}(j\omega)| \underline{/ -[(\pi\omega/\omega_s) + m\pi]} \tag{3-97}$$

where $m = 0, 1, 2, \ldots$ and the expression involving the sine term provides an angular contribution of $m\pi$. The plots $|\mathbf{G}_{zo}(j\omega)|$ vs. ω and $\underline{/\mathbf{G}_{zo}(j\omega)}$ vs. ω are shown in Fig. 3-22*b*. Figure 3-22*a* reveals the low-pass filter characteristic of the ZOH device. However, the lag characteristic, of this device (see Fig. 3-22*b*) degrades the degree of system stability. Also from Fig. 3-22*a* it is seen that the hold device does not have the ideal filter characteristic as depicted by the dashed lines. The ideal filter characteristic may be approached by decreasing the sampling time T. Decreasing T increases the bandwidth but decreases the magnitude of the frequency spectrum as depicted in Fig. 3-22*a*. Although the primary component approaches the ideal filter characteristic the bandwidth increases. This increased bandwidth degrades the ZOH unit's capability to minimize the unwanted higher (noise) frequencies.

It should be noted that the term "zero-order" refers to the capability of the device to pass without distortion a constant, e.g., a zero-order polynomial. A *first-order hold device* then by definition passes without distortion a first-order polynomial signal. In effect, the first-order hold is a signal (data) extrapolator using the first-difference equation. The nth-order hold devices then pass without distorting nth-order polynomials. In general, only zero-order hold devices are implemented in hardware. Higher-order hold algorithms are implemented in computer software since it is most difficult to physically implement higher-order hold devices.

3-10 SUMMARY

This chapter presents the concept of a linear time-invariant system. The sampling process associated with an LTI system is analyzed by using the pulse-function representation of the sampled quantity. This analysis is extended to the ideal impulse-function representation of the sampled quantity. The effective use of the concept of an ideal sampler is based upon the adherence to Shannon's sampling theorem. Linear difference equations are introduced in this chapter and are used to model a continuous-time or a sampled-data control system based on the approximation of differentiation. The application of the convolution summation results in the derivation of the impulse weighting sequence. (This derivation is based upon the approximation of integration.) An important aspect of a sampled-data control system is the data conversion process which is modeled by a zero-order hold device.

CHAPTER

FOUR

DISCRETE SYSTEMS MODELING

4-1 INTRODUCTION

This chapter introduces the $\mathscr{Z}$ transform as a method for the design and analysis of sampled-data control systems.[97] The techniques employed are similar to those used in continuous control system analysis and design utilizing the Laplace transformation. To enhance this analysis (1) a correlation is made between the pole-zero pattern in the s and the z planes with respect to time-response characteristics, and (2) the properties and mathematical representations of open- and closed-loop sampled-data control systems are developed. Once the analysis and design of the sampled-data system in the z domain is completed, it is then necessary to apply the inverse $\mathscr{Z}$-transform method to obtain the time-response characteristics of the system at the sampling instants $t = kT$. To secure additional response data in between sampling instants, $kT < t < (k + 1)T$, the modified $\mathscr{Z}$-transform method is presented. The digital transfer function and its implementation is also discussed in this chapter. Once the concepts of this chapter are thoroughly understood and sufficient problems have been worked by hand, readers are then urged to use a computer-aided control system program to expedite their analysis and design of a sampled-data system[14,15] (see Appendix E).

4-2 DEFINITION AND DETERMINATION OF THE $\mathscr{Z}$ TRANSFORM

The one-sided Laplace transform of a sampled function $e^*(t)$ has been derived in Chap. 3 [see Eq. (3-39)] and is repeated below.

$$E^*(s) = \mathscr{L}[e^*(t)] = \sum_{k=0}^{\infty} e(kT)\epsilon^{-kTs} \tag{4-1}$$

where $e^*(t) = 0$ for $t < 0$ (one-sided) and $k = 0, 1, 2, \ldots$. Since Eq. (4-1) shows that the function $E^*(s)$ contains the factor ϵ^{-Ts}, then to utilize the algebraic manipulation techniques that are used in the Laplace domain a change of variable must be made. Let

$$z \equiv \epsilon^{Ts} \tag{4-2a}$$

$$s = \frac{1}{T} \ln z \tag{4-2b}$$

Hence

$$E(z) = E^*(s)|_{z=\epsilon^{Ts}} \equiv \sum_{k=0}^{\infty} e(kT)z^{-k} \tag{4-3}$$

which is defined as the $\mathscr{Z}$ transform (read as "zee" transform) $E(z)$ of the function $e^*(t)$. Thus the negative power series

$$E(z) \equiv \mathscr{Z}[e^*(t)] = E^*(s)|_{z=\epsilon^{Ts}} \tag{4-4}$$

is called the *one-sided $\mathscr{Z}$ transform* of $e^*(t)$, a power series in z^{-1} where z can be real or imaginary. Referring to Fig. 4-1, if for a given magnitude of z, say:

1. $|z| > r$, then Eq. (4-4) has a finite value; i.e., the series converges. This value of z, by definition, lies in the convergence region of the z plane.
2. $|z| < r$, the series of Eq. (4-3) becomes unbounded (negative or positive infinity), i.e., undefined. Thus Eq. (4-3) exists only if $|z| > r$. If $|z|$ does not satisfy this restriction, if $|z| < r$, one simply cannot write Eq. (4-3). This value of z therefore lies in the divergence region which is contained within the circle of radius r.
3. $|z| = r$, on the boundary the infinite series may or may not converge depending upon the sequence $e(kT)$.

Thus, in the z plane, the magnitude has a significance equivalent to that of σ in determining the convergence region of the Laplace transform of an s-plane

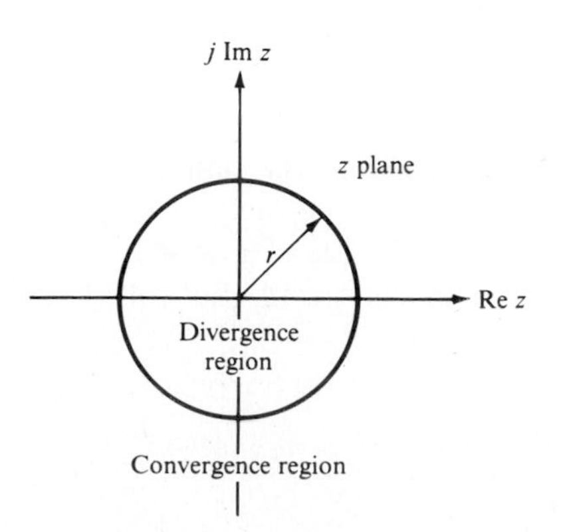

Figure 4-1 Power-series regions of divergence and convergence.

function (see Appendix D).† Depending on the sequence values of $e(kT)$, the one-sided $\mathscr{Z}$ transform may exist everywhere ($|z| \geq r = 0$), or in some infinite region ($|z| > r > 0$), or nowhere ($|z| > r = \infty$). Fortunately, *for most practical applications, the one-sided $\mathscr{Z}$-transform power series has a closed-form representation in its region of convergence.* This representation is given by the rational function

$$E(z) = \frac{K(z^w + c_{w-1}z^{w-1} + \cdots + c_1 z + c_0)}{z^n + d_{n-1}z^{n-1} + \cdots + d_1 z + d_0} \tag{4-5a}$$

which is a ratio of polynomials. Dividing the numerator and denominator by z^n yields a rational function in powers of z^{-1}; that is,

$$E(z) = \frac{K(z^{-n+w} + \cdots + c_1 z^{-n+1} + c_0 z^{-n})}{1 + d_{n-1}z^{-1} + \cdots + d_1 z^{-n+1} + d_0 z^{-n}} \tag{4-5b}$$

As shown later, the nature of the system time response is more readily evident from the factored form of the $\mathscr{Z}$ transform. Thus factoring the polynomials of Eq. (4-5*a*) yields

$$E(z) = \frac{KN(z)}{D(z)} = \frac{K(z - z_1)(z - z_2) \cdots (z - z_w)}{(z - p_1)(z - p_2) \cdots (z - p_n)} \tag{4-6}$$

where p_j and z_i are referred to as the poles and zeros, respectively, of $E(z)$. The values $p_1, p_2, \ldots, p_n$ in the finite plane that make the denominator $D(z)$ equal to zero are called *zeros* of the denominator. These values of z, which may be either real or complex, also make $|E(z)|$ infinite, and so they are also called *poles* of $E(z)$. Therefore, the values $p_1, p_2, \ldots, p_n$ are referred to as zeros of the denominator or poles of the complete function in the finite plane; i.e., there are n poles of $E(z)$. A zero of $N(z)$ is defined as a value of z that results in a zero value for $E(z = z_i)$; that is, $E(z_i) = 0$. The power-series representation for Eq. (4-6) converges for all $z \neq p_j$. Although a particular $\mathscr{Z}$-transform power-series representation does not converge within the r circle, the concept of analytical continuation is employed (see Appendix D). After the closed-form representation is generated, the concept of analytical continuation guarantees that the closed-form representation is analytical within the r circle except at $z = p_j$. The same concept is used in showing that the $E(s)$ is analytical for the entire s plane except for the poles.‡ The location of these poles and zeros in the z plane determines the time-response characteristics of $e(kT)$. In a similar manner, this correlation is equivalent to the correlation between the location of the poles and zeros of $C(s)$ in the s plane and the time-response characteristics of $c(t)$. The symbols X and O are used to denote the plot of poles and zeros, respectively, in the z plane as shown in the example of Fig. 4-2.

The following two examples illustrate the application of Eq. (4-4) in generating a $\mathscr{Z}$ transform.

† The subject of convergence of infinite series is considered in most advanced calculus texts.[16]

‡ Singularities other than poles (such as essential singularities and branch points) are not being considered and do not need to be considered in this textbook.

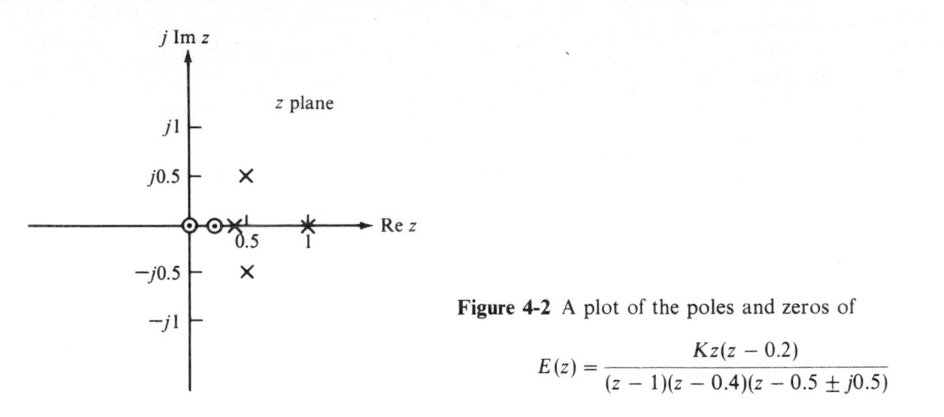

Figure 4-2 A plot of the poles and zeros of

$$E(z) = \frac{Kz(z - 0.2)}{(z - 1)(z - 0.4)(z - 0.5 \pm j0.5)}$$

Example 4-1 Given an impulse sampling of $e(t)$,

$$\begin{aligned} e^*(t) &= \sum_{k=0}^{\infty} e(kT)\delta(t - kT) = e(0)\delta(t) + e(T)\delta(t - T) \\ &\quad + e(2T)\delta(t - 2T) + \cdots + e(iT)\delta(t - iT) + \cdots \end{aligned} \tag{4-7}$$

where the mathematical summation represents the sampled sequence of numbers $e(0)$, $e(T)$, $e(2T)$, ... Taking the Laplace transform of Eq. (4-7) yields

$$\begin{aligned} E^*(s) &= \mathscr{L}[e^*(t)] = \sum_{k=0}^{\infty} e(kT)\epsilon^{-kTs} \\ &= e(0) + e(T)\epsilon^{-Ts} + e(2T)\epsilon^{-2Ts} + \cdots + e(iT)\epsilon^{-iTs} + \cdots \end{aligned} \tag{4-8}$$

The $\mathscr{Z}$ transform of Eq. (4-8) is [using Eq. (4-2a)]

$$\begin{aligned} E(z) &= e(0) + e(T)z^{-1} + e(2T)z^{-2} + \cdots + e(iT)z^{-i} + \cdots \\ &= \sum_{k=0}^{\infty} e(kT)z^{-k} \end{aligned} \tag{4-9}$$

Analyzing Eqs. (4-7) to (4-9) reveals that $z^{-1} = \epsilon^{-Ts}$ represents a delay of one sampling period, i.e., a delay of T seconds. Thus $z^{-i} = \epsilon^{-iTs}$ represents a delay of i periods. In other words, the quantity ϵ^{-Ts} in Eq. (4-8) represents a delay operator in the s plane. This corresponds to a delay of the impulse quantity of $e(T)$ in the time domain of T seconds (one sampling period) as illustrated in Fig. 4-3a. In the z domain the corresponding delay operator is represented by z^{-1}. Note that if Eq. (4-9) is multiplied by z^{-1} it yields

$$E_a(z) = z^{-1}E(z) = e(0)z^{-1} + e(T)z^{-2} + \cdots$$

Thus multiplying $E(z)$ by z^{-1} delays the response by one period. Likewise, if $E(z)$ is multiplied by $z = \epsilon^{Ts}$, it advances the response by one period as illustrated in Fig. 4-3b. Thus $z = \epsilon^{Ts}$ represents an advance operator of T seconds.

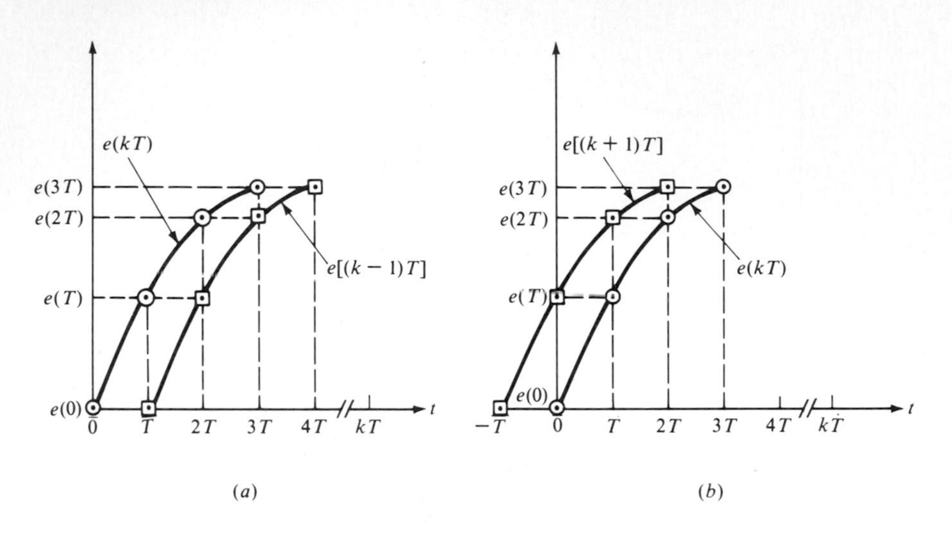

Figure 4-3 Delay or advance of a discrete function, $e(kT)$: (a) delay, $e[(k-1)T]$, (b) advance, $e[(k+1)T]$.

Example 4-2 Given $e^*(t) = e(t)\delta_T(t)$, where $e(t) = u_{-1}(t)$, determine $E(z) = \mathscr{Z}[e^*(t)]$.

SOLUTION

$$e^*(t) = \sum_{k=0}^{\infty} \delta(t - kT) = \delta(t) + \delta(t - T) + \delta(t - 2T) + \cdots$$

$$E^*(s) = 1 + \epsilon^{-Ts} + \epsilon^{-2Ts} + \epsilon^{-3Ts} + \cdots$$

$$E(z) = E^*(s)|_{z=\varepsilon^{Ts}} = 1 + z^{-1} + z^{-2} + z^{-3} + \cdots \qquad (4\text{-}10)$$

Since the expression for $E(z)$, given by Eq. (4-10), is an infinite series, it is referred to as the *open-form* expression of $E(z)$. The *closed form* of the infinite series given by Eq. (4-10) is

$$E(z) = \frac{z}{z-1} = \frac{1}{1-z^{-1}} \qquad \text{for } |z| > 1 \text{ (region of convergence)} \qquad (4\text{-}11)$$

It should be noted that the sequence of $e^*(t)$ and its $\mathscr{Z}$ transform $E(z)$ do not uniquely define the original time function $e(t)$. This is discussed further in a later section.

The closed form of the $\mathscr{Z}$ transform, being of a similar mathematical structure as the Laplace transform, permits the utilization of the equivalent s-plane algebraic manipulations for system analysis. Therefore, the $\mathscr{Z}$ transform should be applicable to the solution of linear difference equations in much the same manner as the Laplace transform is applied to solve ordinary differential equations.

A more general and expeditious approach for evaluating the closed-form $\mathscr{Z}$

transform of any signal $e(t)$, whose Laplace transform is $E(s)$ and whose sampled function is given by $e^*(t)$, is given by the following definition:†

$$E(z) = \sum \left[\text{residues of } \frac{E(\alpha)}{1 - \epsilon^{-T(s-\alpha)}} \text{ evaluated } \textit{only at the poles of } E(\alpha) \right]_{z=\epsilon^{Ts}} + \beta$$

$$\equiv \hat{E}(z) + \beta \tag{4-12}$$

where

$$\beta \equiv \lim_{s \to \infty} sE(s) - \lim_{z \to \infty} \hat{E}(z) \tag{4-13}$$

and $E(s)$ is a rational function. The value of β determined by Eq. (4-13) ensures that the initial values $e(0)$ of $E(s)$ and $E(z)$ are identical. For the case where $E(s)$ contains only simple poles, then the number of residues of Eq. (4-12) equals the number of poles n of $E(s)$. When $E(s)$ has a pole α_i of multiplicity m_j then the number of residues of Eq. (4-12) is given by $n - (m_j - 1)$. In general, when $E(s)$ contains l poles having multiplicities, then the number of residues n' of Eq. (4-12) is given by

$$n' = n - \sum_{j=1}^{l} m_j + l$$

where $j = 1, 2, \ldots, l$. In other words, the pole of multiplicity m_j is counted as a single pole for the purpose of evaluating Eq. (4-12). If the n poles of $E(s)$ are simple (nonrepeated), then Eq. (4-12) can be simplified to

$$E(z) = \sum_{i=1}^{n} \left\{ \left[\frac{N(\alpha)}{dD(\alpha)/d\alpha} \right]_{\alpha=\alpha_i} \frac{1}{1 - \epsilon^{\alpha_i T} z^{-1}} \right\} + \beta = \hat{E}_s(z) + \beta \tag{4-14}$$

The bracketed term, which is evaluated at $\alpha = \alpha_i$, is by definition the residue of the simple pole α_i. In Eq. (4-14), $E(\alpha)$ is $E(s) = N(s)/D(s)$ with s replaced by α; α_i are the nonrepeated poles of $E(\alpha)$; and $E(\alpha) = N(\alpha)/D(\alpha)$. Note that $N(\alpha)$ represents the numerator factors (or polynomial) and $D(\alpha)$ represents the denominator factors (or polynomial) of $E(\alpha)$. Also, note that if $N(s)$ contains the factor $(1 - \epsilon^{-sT})^p = (1 - z^{-1})^p$, where p is some integer greater than zero, it must not be included in $N(\alpha)$, since it is already expressed in the z domain. This factor, with ϵ^{-sT} replaced by z^{-1}, must multiply the summation terms of Eq. (4-14). To show how this evaluation is accomplished, the following examples are presented.

Example 4-3 Repeat Example 4-2 by utilizing Eq. (4-14).

SOLUTION Since $E(s) = 1/s$, then $\alpha_i = \alpha_1 = 0$, $N(\alpha) = 1$, $D(\alpha) = \alpha$, and $dD(\alpha)/d\alpha = 1$. Applying Eqs. (4-13) and (4-14) yields $\beta = 0$ and

$$E(z) = \frac{1}{1 - z^{-1}} = \hat{E}(z)$$

which agrees with Eq. (4-11).

† Those readers having a complex-variable theory background are referred to Ref. 2 for the derivation of Eq. (4-12). Note that in this definition s is replaced by α.

Example 4-4 Given $e(t) = 1 - \epsilon^{-2t}$, determine $E(z) = \mathscr{Z}[e^*(t)]$, using Eq. (4-14).

SOLUTION Since $E(s) = \mathscr{L}[e(t)] = 2/[s(s+2)]$ then $N(\alpha) = 2$, $dD(\alpha)/d\alpha = 2\alpha + 2$, $n = 2$, $\alpha_1 = 0$, and $\alpha_2 = -2$. Applying Eqs. (4-13) and (4-14) yields

$$\hat{E}(z) = E(z) = \frac{2}{2}\frac{1}{1 - z^{-1}} + \frac{2}{-2}\frac{1}{1 - \epsilon^{-2T}z^{-1}}$$

$$= \frac{z^{-1} - \epsilon^{-2T}z^{-1}}{(1 - z^{-1})(1 - \epsilon^{-2T}z^{-1})} = \frac{(1 - \epsilon^{-2T})z}{(z-1)(z - \epsilon^{-2T})}$$

For transfer functions that have a pole α_i of multiplicity m the residue for this pole, by definition, is given by the following equation:[17]

$$\hat{E}_{\alpha_i} = \text{residue of pole } \alpha_i = \left.\frac{d^{m-1}[W(\alpha)/(m-1)!]}{d\alpha^{m-1}}\right|_{\alpha=\alpha_i} \tag{4-15}$$

where

$$W(\alpha) = \frac{(\alpha - \alpha_i)^m E(\alpha)}{1 - \epsilon^{-T(s-\alpha)}} \tag{4-16}$$

The following example illustrates the application of Eq. (4-15) in generating $\mathscr{Z}$ transforms.

Example 4-5 Case of repeated poles Given

$$E(s) = \frac{10(s+2)}{s(s+1)^2}$$

determine $E(z)$.

SOLUTION The poles of $E(\alpha)$ are $\alpha_1 = 0$, and $\alpha_2 = -1$ with a multiplicity of $m = 2$. Thus $n' = 3 - 2 + 1 = 2$. The residue for the simple pole at $\alpha_1 = 0$ can be found in the same manner as in Example 4-4. Thus, the residue for $\alpha_1 = 0$, using Eq. (4-14), is

$$\hat{E}_s(z) = \left[\frac{N(\alpha)}{dD(\alpha)/d\alpha}\right]_{\alpha=\alpha_1}\frac{1}{1 - z^{-1}} = \frac{20}{1 - z^{-1}}$$

Applying Eq. (4-15) to this example yields

$$W(\alpha) = \frac{10(\alpha + 2)}{\alpha(1 - \epsilon^{-T(s-\alpha)})}$$

$$E_{\alpha_2}(z) = \frac{dW(\alpha)}{d\alpha}$$

$$= \frac{10(\alpha - \alpha\epsilon^{-T(s-\alpha)}) - 10(\alpha + 2)(1 - \epsilon^{-T(s-\alpha)} - \alpha T\epsilon^{-T(s-\alpha)})}{(\alpha - \alpha\epsilon^{-T(s-\alpha)})^2}\Bigg|_{\alpha=-1}$$

$$= \frac{10(-2 + 2\epsilon^{-T}z^{-1} - T\epsilon^{-T}z^{-1})}{(\epsilon^{-T}z^{-1} - 1)^2}$$

$$\hat{E}(z) = \hat{E}_s(z) + \hat{E}_{\alpha_2}(z) = \frac{20}{1 - z^{-1}} + \frac{10(-2 + 2\epsilon^{-T}z^{-1} - T\epsilon^{-T}z^{-1})}{(\epsilon^{-T}z^{-1} - 1)^2}$$

Therefore applying Eqs. (4-12) and (4-13) yields $\beta = 0$ and $E(z) = \hat{E}(z)$. A table of $\mathscr{Z}$ transforms of functions that appear often in equations describing sampled-data systems are available in many handbooks and texts.[2,18,19] An abbreviated table of transforms is given in Table 4-1.

The availability of a table like Table 4-1 permits the use of the following expansion procedure as another means of obtaining $E(z)$.

Step 1 Perform a partial-fraction expansion on $E(s)$ to obtain

$$E(s) = \frac{A_0}{s + \alpha_0} + \frac{A_1}{s + \alpha_1} + \cdots + \frac{A_n}{s - \alpha_n}$$

Step 2 Thus

$$E(z) = \mathscr{Z}[E(s)] = \mathscr{Z}\left[\frac{A_0}{s + \alpha_0}\right] + \mathscr{Z}\left[\frac{A_1}{s + \alpha_1}\right] + \cdots + \mathscr{Z}\left[\frac{A_n}{s + \alpha_n}\right]$$

The $\mathscr{Z}$ transform of each term on the right-hand side of this equation is obtained from Table 4-1. Note that it is rather common to use the shortened notation of $\mathscr{Z}[e(t)]$ or $\mathscr{Z}[E(s)]$ instead of the proper form of $\mathscr{Z}[e^*(t)] = \mathscr{Z}[e(t)|_{t=kT}]$ or $\mathscr{Z}[E^*(s)]$ when specifying the $\mathscr{Z}$ transform.

Example 4-6 Given

$$E(s) = \frac{5}{s(s + 1)(s + 5)}$$

determine $E(z)$.

SOLUTION

(1)
$$E(s) = \frac{1}{s} - \frac{1.25}{s + 1} + \frac{0.25}{s + 5}$$

(2)
$$E(z) = \frac{z}{z - 1} - \frac{1.25z}{z - \epsilon^{-T}} + \frac{0.25z}{z - \epsilon^{-5T}}$$

Table 4-1 Table of transforms

	Time function	Laplace transform	$\mathscr{Z}$ transform of $e^*(t)$	Modified $\mathscr{Z}$ transform
	$e(t)$	$E(s)$	$E(z)$	$E(z, m)$
	$e(kT)$		$k \geq 0$	$k \geq 1$
1	$e(0)\delta(t)$	$e(0)$	$e(0)$	0
	$e(0)\delta(0)$			
2	(Transport lag) $e(t)\delta(t - lT)$, l is any positive l integer	$\epsilon^{-lTs}E(s)$	$z^{-l}E(z)$	$z^{-l-1+m}E(z, m)$
	$e(kT)\delta[t - (k + l)T]$			
3	$u_{-1}(t)$	$\frac{1}{s}$	$\frac{z}{z - 1}$	$\frac{1}{z - 1}$
	$\delta(t - kT)$			
4	t^{m-1}, m is any positive integer	$\frac{(m - 1)!}{s^m}$	$\lim_{b \to 0} \left[(-1)^{m-1} \frac{\partial^{m-1} \frac{z}{z - \epsilon^{-bT}}}{\partial b^{m-1}} \right]$	$\lim_{b \to 0} \left[(-1)^{m-1} \frac{\partial^{m-1} \frac{\epsilon^{-bmT}}{z - \epsilon^{-bT}}}{\partial b^{m-1}} \right]$
	$(kT)^{m-1}$			
5	ϵ^{-at}	$\frac{1}{s + a}$	$\frac{z}{z - \epsilon^{-aT}} = \frac{z}{z - d}$	$\frac{\epsilon^{-amT}}{z - \epsilon^{-aT}}$
	$\epsilon^{-akT} = d^k$			
6	$\frac{\epsilon^{-bt} - \epsilon^{-at}}{a - b}$	$\frac{1}{(s + a)(s + b)}$	$\frac{1}{a - b}\left(\frac{z}{z - \epsilon^{-bT}} - \frac{z}{z - \epsilon^{-aT}}\right)$	$\frac{1}{a - b}\left(\frac{\epsilon^{-bmT}}{z - \epsilon^{-bT}} - \frac{\epsilon^{-amT}}{z - \epsilon^{-aT}}\right)$
	$\frac{\epsilon^{-bkT} - \epsilon^{-akT}}{a - b}$			

7	$\frac{1}{a}[u_{-1}(t) - \epsilon^{-at}]$ $\frac{1}{a}(1 - \epsilon^{-akT})$	$\frac{1}{s(s+a)}$	$\frac{1}{a}\frac{(1-\epsilon^{-aT})z}{(z-1)(z-\epsilon^{-aT})}$	$\frac{1}{a}\left(\frac{1}{z-1} - \frac{\epsilon^{-amT}}{z-\epsilon^{-aT}}\right)$
8	$\frac{1}{a}\left(t - \frac{1-\epsilon^{-at}}{a}\right)$ $\frac{1}{a}\left(kT - \frac{1-\epsilon^{-akT}}{a}\right)$	$\frac{1}{s^2(s+a)}$	$\frac{1}{a}\left[\frac{Tz}{(z-1)^2} - \frac{(1-\epsilon^{-aT})z}{a(z-1)(z-\epsilon^{-aT})}\right]$	$\frac{1}{a}\left[\frac{T}{(z-1)^2} + \frac{amT-1}{a(z-1)} + \frac{\epsilon^{-amT}}{a(z-\epsilon^{-aT})}\right]$
9	$\sin at$ $\sin akT$	$\frac{a}{s^2+a^2}$	$\frac{z \sin aT}{z^2 - 2z(\cos aT) + 1}$	$\frac{z\sin(amT) + \sin(1-m)aT}{z^2 - 2z(\cos aT) + 1}$
10	$\cos at$ $\cos akT$	$\frac{s}{s^2+a^2}$	$\frac{z(z-\cos aT)}{z^2 - 2z(\cos aT) + 1}$	$\frac{z(\cos amT) - \cos(1-m)aT}{z^2 - 2z(\cos aT) + 1}$
11	$\frac{1}{b}\epsilon^{-at}\sin bt$ $\frac{1}{b}\epsilon^{-akT}\sin bkT$	$\frac{1}{(s+a)^2+b^2}$	$\frac{1}{b}\frac{z\epsilon^{-aT}\sin bT}{z^2 - 2z\epsilon^{-aT}(\cos bT) + \epsilon^{-2aT}}$	$\frac{1}{b}\frac{\epsilon^{-amT}[z(\sin bmT) + \epsilon^{-aT}\sin(1-m)bT]}{z^2 - 2z\epsilon^{-aT}(\cos bT) + \epsilon^{-2aT}}$
12	$\epsilon^{-at}\cos bt$ $\epsilon^{-akT}\cos bkT$	$\frac{s+a}{(s+a)^2+b^2}$	$\frac{z^2 - z\epsilon^{-aT}\cos bT}{z^2 - 2z\epsilon^{-aT}(\cos bT) + \epsilon^{-2aT}}$	$\frac{\epsilon^{-amT}[z(\cos bmT) - \epsilon^{-aT}\sin(1-m)bT]}{z^2 - 2z\epsilon^{-aT}(\cos bT) + \epsilon^{-2aT}}$
13	$t\epsilon^{-at}$ $kT\epsilon^{-akT}$	$\frac{1}{(s+a)^2}$	$\frac{Tz\epsilon^{-aT}}{(z-\epsilon^{-aT})^2}$	$\frac{T\epsilon^{-amT}[\epsilon^{-aT} + m(z-\epsilon^{-aT})]}{(z-\epsilon^{-aT})^2}$

Although a $\mathscr{Z}$ transform can be obtained directly from the time-domain function by means of Eq. (4-4), the understanding of z-plane pole and zero location characteristics can be understood better with a knowledge of s-plane pole and zero location characteristics.[1] The correlation between the time domain and the z domain, by means of the s plane, can thus help reveal the discrete system performance characteristics.

4-3 MAPPING BETWEEN s AND z DOMAINS

For a continuous control system, a designer can synthesize a desired z-plane control ratio, i.e., a desired system transfer function.[1] The synthesis is based upon the correlation between the pole-zero placement in the s plane and the desired response in the time domain. This phenomenon can also be used to establish a correlation between the pole-zero location in the z plane and the time-domain characteristics. The z-plane correlation is done by mapping the s plane into the z plane by means of $z = \epsilon^{Ts}$.

Section 3-4 states that the frequency range of the signal $\mathbf{E}^*(j\omega)$, with $\omega_s > 2\omega_c$, can be divided into primary and complementary strips. That is, Eq. (3-44) reveals that an important property of the output of an ideal sampler is that $E^*(s)$ is a periodic function with a period $j\omega_s$. This can also be shown by substituting in Eq. (4-1) $s + jl\omega_s$ for s, where l is an integer. Thus

$$E^*(s + jl\omega_s) = \sum_{k=0}^{\infty} e(kT)\epsilon^{-kT(s+jl\omega_s)} = \sum_{k=0}^{\infty} e(kT)\epsilon^{-kTs}\epsilon^{-jkl\omega_s T} \tag{4-17}$$

Since $\omega_s = 2\pi/T$ then $\epsilon^{-jkl\omega_s T} = \epsilon^{-j2\pi kl} = 1$ which simplifies Eq. (4-17) to

$$E^*(s + jl\omega_s) = \sum_{k=0}^{\infty} e(kT)\epsilon^{-kTs} = E^*(s) \tag{4-18}$$

Therefore, another important property is: If $E(s)$ has a pole (zero) at $s = s_1$, then $E^*(s)$ has poles (zeros) at $s = s_1 \pm jl\omega_s$, where $l = 0, \pm 1, \pm 2, \pm 3, \ldots, \pm\infty$. Equation (4-18) means that, given any pole (or zero) $s = s_1$ in the s plane, the sampled function $E^*(s)$ has the same value at all periodic frequency points $s_1 \pm jl\omega_s$. Figure 4-4 represents a periodic function $E^*(s)$, where $E(s)$ has a pole at $s = s_1$. Since it is assumed that the low-pass analog filtering characteristic of the continuous plant and the ZOH device attenuate the responses due to the poles in the complementary strips, only the poles in the primary strip generally need be considered for analysis and synthesis of a desired control ratio. Note that the primary pole (zero) s_1 and its associated complementary poles (zeros) map onto the same point in the z plane! The following discussion attempts to clarify this phenomenon.

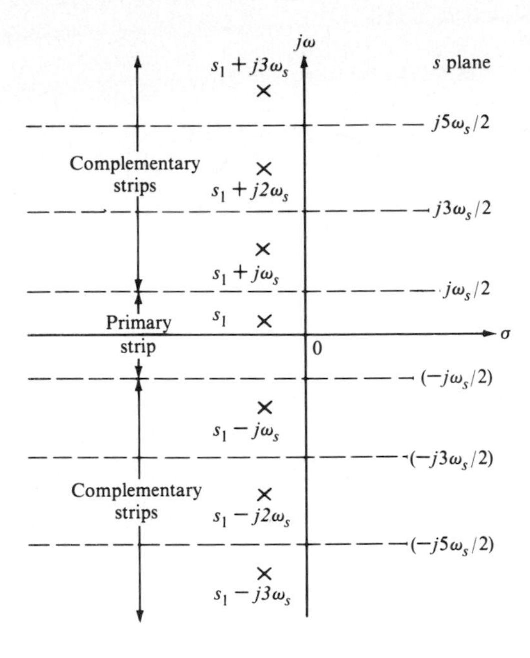

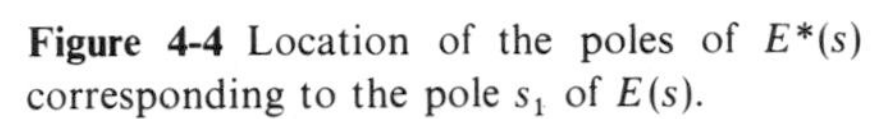

Figure 4-4 Location of the poles of $E^*(s)$ corresponding to the pole s_1 of $E(s)$.

4-3A Mapping of the Primary Strip

The boundary *abcdea* of the primary strip in the s plane of Fig. 4-5a is mapped into the z plane, utilizing

$$\mathbf{z} = \epsilon^{Ts} = \epsilon^{(\sigma \pm j\omega)T} = \epsilon^{\sigma T}\epsilon^{\pm j\omega T} = \epsilon^{\sigma T}\underline{/\pm\omega T} \tag{4-19}$$

Table 4-2 presents the values of $\mathbf{z}$ at various points on the boundary of the contour in the z plane. The superscript minus denotes a magnitude just slightly less than that given within the parentheses. Note that the value of $\mathbf{z}$ varies in the following manner as the contour *abcdea* is traversed counterclockwise in the z plane:

Interval $[a, b)$:

$$\mathbf{z} = 1\underline{/\omega T}, \qquad 0 \le \underline{/\omega T} \le (180°)^-$$

Interval $[b, c)$:

$$\mathbf{z} = |\mathbf{z}|\underline{/(180°)^-}, \qquad 1 \ge |\mathbf{z}| > 0$$

Interval $[c, d]$:

$$\mathbf{z} = |\mathbf{z}|\underline{/\pm\omega T}, \qquad 0^+ \ge |\mathbf{z}| \ge 0,\ (180°)^- \ge \underline{/\pm\omega T} \ge -(180°)^-$$

Interval $(d, e]$:

$$\mathbf{z} = |\mathbf{z}|\underline{/-(180°)^-}, \qquad 1 \ge |\mathbf{z}| > 0$$

Interval (e, a):

$$\mathbf{z} = 1\underline{/-\omega T}, \qquad -(180°)^- < \underline{/-\omega T} < 0°$$

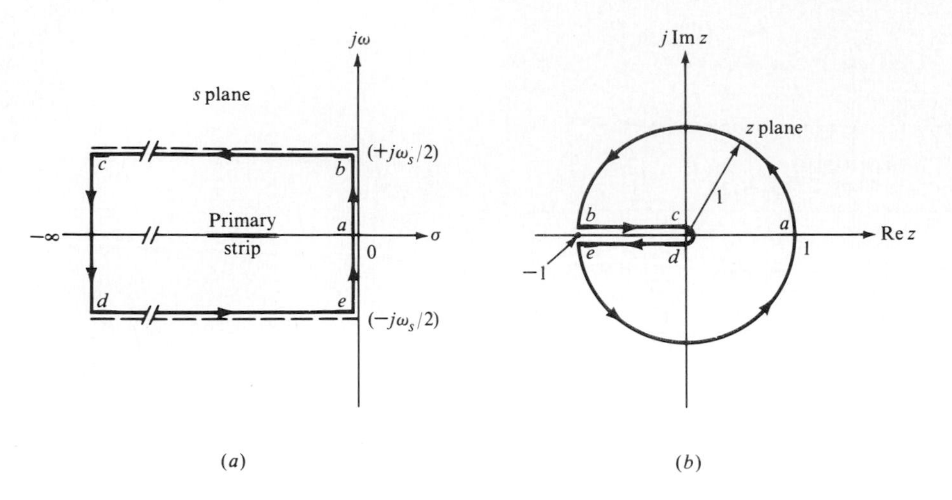

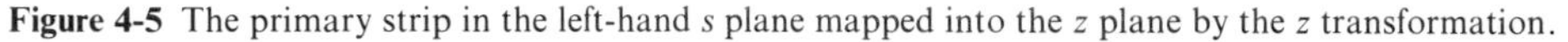
Figure 4-5 The primary strip in the left-hand s plane mapped into the z plane by the z transformation.

Table 4-2 Mapping of s-plane points into the z plane

Point	$s = \sigma \pm j\omega$	$\mathbf{z} = \epsilon^{\sigma T} \angle \pm \omega T$
a	$0 + j0$	$1\angle 0°$
b	$0 + j(\omega_s/2)^-$	$1\angle (180°)^-$
c	$-\infty + j(\omega_s/2)^-$	$0\angle (180°)^-$
d	$-\infty - j(\omega_s/2)^-$	$0\angle -(180°)^-$
e	$0 - j(\omega_s/2)^-$	$1\angle -(180°)^-$

where the bracket denotes starting or ending (closed endpoint) "at" and the parenthesis denotes starting or ending (open endpoint) "close" to a point of the interval under consideration. In the limit, the boundary bc approaches the value $\omega_s/2$ and the boundary de approaches the value $-\omega_s/2$, from the directions shown in Fig. 4-5a. The primary strip in the left-half s plane then maps onto a circle with a "two-sided rounded slot" in it in the z plane centered at the origin as shown in Fig. 4-5b. The above boundaries in the z plane, in the limit, approach the 0 to $-1 + j0$ portion of the negative real z axis. See Fig. 4-5b. For *convenience of expression*, the slotted figure in the limit is *usually* referred to as a *unit circle* (UC).

4-3B Mapping of the Constant Frequency Loci

The mapping of the constant frequency loci (see Fig. 4-6), $\omega = \omega_1 < \omega_s/2$, in the s plane into the z plane is expressed by $\mathbf{z} = \epsilon^{\sigma T} \angle \omega_1 T$. Thus any constant frequency loci in the s plane, a horizontal line, is mapped into a radial line in the z plane emanating from the origin where its angle is given by $\angle \mathbf{z} = \omega_1 T$ rad.

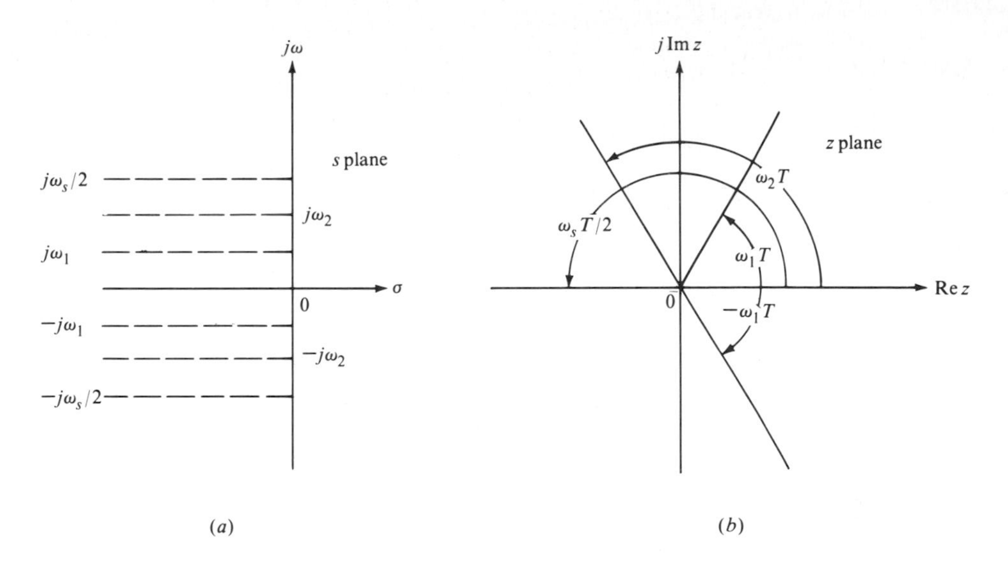

Figure 4-6 Mapping of the constant frequency loci from s plane into the z plane.

4-3C Mapping of the Constant Damping-Coefficient Loci

The mapping of a constant s-plane damping-coefficient loci (see Fig. 4-7), $\sigma = \sigma_1$, of a second-order denominator factor into the z plane is expressed by $\mathbf{z} = \epsilon^{\sigma_1 T} \underline{/\omega T}$. The line segments of the s-plane loci from $-j\pi/T$ to $j\pi/T$ are mapped onto the circle boundaries in the z plane as shown in Fig. 4-7. Thus repeated boundary mappings of these line segments occur as ω goes from $-\infty$ to ∞. Therefore, any constant damping-coefficient locus (a vertical line in the s plane) is mapped onto a circle of radius $|\mathbf{z}| = \epsilon^{\sigma_1 T}$ centered at the origin of the z plane. Since $\sigma_1 < 0$, then this circle lies within the unit circle, whereas, for $\sigma_2 > 0$, the corresponding circle lies outside the unit circle. The value σ_1 (or σ_2) of Fig. 4-7a is correlated to its corresponding value $|\mathbf{z}|$ in the z plane as shown in Fig. 4-7c. This correlation is between the $j\omega$ axis in the s plane and the unit circle in the z plane.

4-3D Mapping of the Constant Damping-Ratio Loci

A point on the damping-ratio (ζ) line, in the second quadrant (see Fig. 4-8), for an underdamped second-order denominator factor can be expressed as

$$s = \sigma + j\omega_d = -\zeta\omega_n + j\omega_n\sqrt{1-\zeta^2} \tag{4-20}$$

Since

$$\cot \eta = \frac{|\sigma|}{\omega_d} = \frac{\zeta\omega_n}{\omega_d} = \frac{\zeta}{\sqrt{1-\zeta^2}}$$

then

$$\zeta\omega_n = \omega_d \cot \eta \tag{4-21}$$

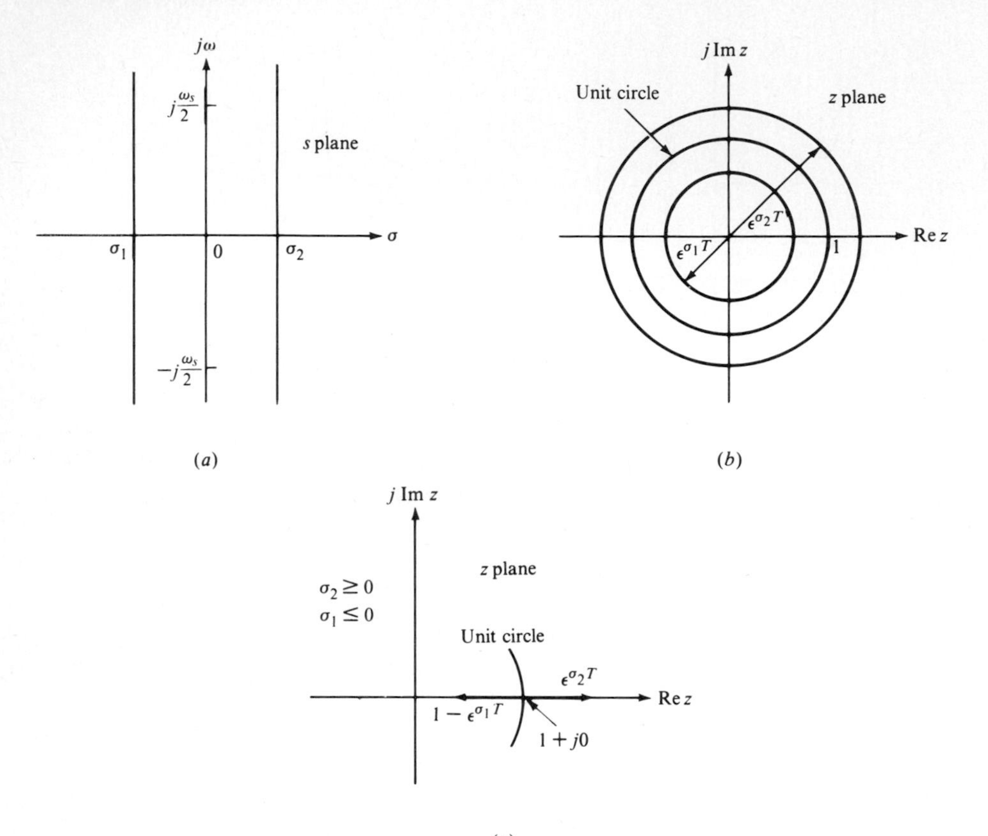

Figure 4-7 Mapping of the constant damping-coefficient loci from the s plane into the z plane.

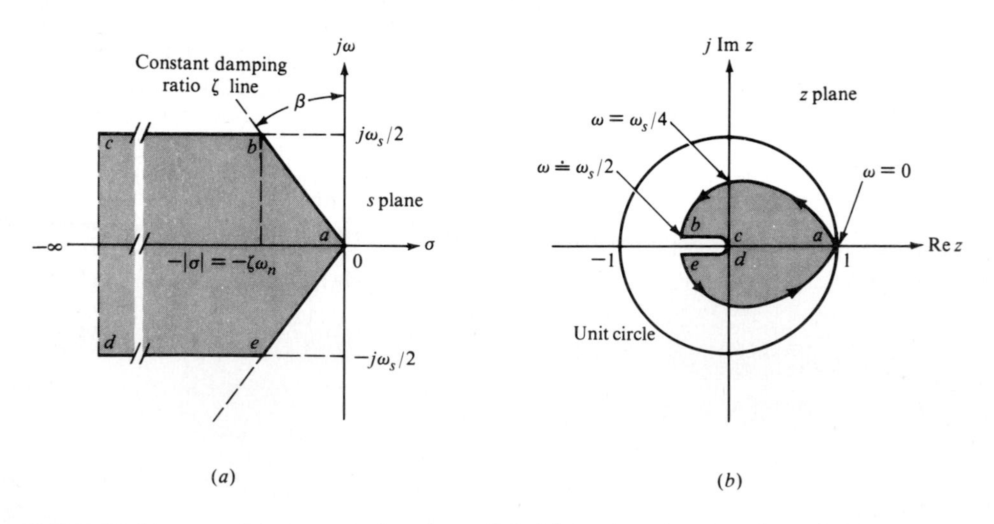

Figure 4-8 Mapping of the constant damping-ratio loci from the s plane into the z plane.

Substituting Eq. (4-21) into Eq. (4-20) yields

$$s = -\omega_d \cot \eta + j\omega_d \tag{4-22}$$

And in turn substituting Eq. (4-22) into ϵ^{st} yields

$$\mathbf{z} = \epsilon^{-(\omega_d \cot \eta + j\omega_d)T} \tag{4-23}$$

Since $T = 2\pi/\omega_s$ then Eq. (4-23) can be expressed as

$$\mathbf{z} = \epsilon^{-(2\pi\omega_d \cot \eta/\omega_s)} \underline{/2\pi\omega_d/\omega_s} = |\mathbf{z}|\underline{/\mathbf{z}} \tag{4-24}$$

Equation (4-24) reveals that for a given constant value of $\zeta = \zeta_1$ (that is, $\cot \eta =$ constant) $\mathbf{z}$ is a function of ω_d only. Therefore, the constant damping-ratio locus in the s plane maps into the z plane as a logarithmic spiral (except for $\eta = 0°$ and $90°$). The portion of the ζ line between $\omega_d = 0$ and $\omega_d = \omega_s/2$ in the s plane corresponds to one-half revolution of the logarithmic spiral in the z plane. The shaded area of the s plane shown in Fig. 4-8*a* corresponds to the shaded area in the z plane shown in Fig. 4-8*b*. By use of Eq. (4-24), plots of constant ζ loci and constant ω_d loci are obtained as shown in Fig. 4-9.

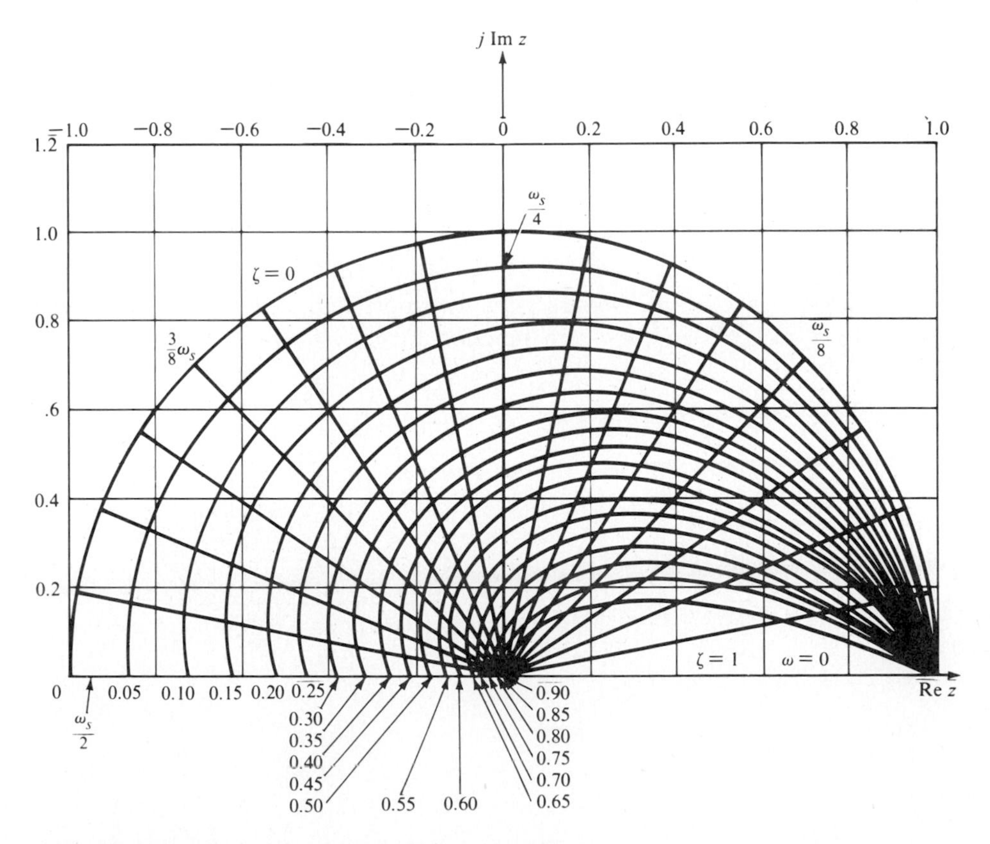

Figure 4-9 Plots of loci of constant ζ and loci of constant ω_d.

Note that the *negative real roots* ($\zeta = 1$, $\omega_d = 0$, and $s_1 = -|\sigma|$) in the s plane map onto corresponding points on the *positive* real axis in the z plane that lie within the UC ($\mathbf{z} = \epsilon^{-|\sigma|T}\underline{/0^\circ}$). Also note that complex roots in the primary strip of the left-half s plane map onto corresponding points inside the UC of the z plane excluding the positive real axis. Note, further, that the frequency of oscillation resulting from complex roots in the first and fourth quadrants of the UC in the z plane is smaller than that resulting from those located in the second and third quadrants.

It can be shown that each complementary strip in the left-half s plane maps onto the UC in the z plane. Further, it can be shown in a similar manner, that the strips in the right-half s plane each map onto the *entire* z plane, excluding the area covered by the UC. In particular, the real axis of the s plane maps onto the nonnegative real axis in the z plane.

Example 4-7 Given

$$E(s) = \frac{s + a}{(s + a)^2 + (b)^2} = \frac{s - \sigma}{(s - \sigma)^2 + \omega_d^2}$$

where $p_{1,2} = -a \pm jb = \sigma \pm j\omega_d$ and the plot of the zero and poles of $E(s)$, in the s plane, are shown in Fig. 4-10*a*. Determine $E(z)$ and plot its zeros and poles in the z plane when $\omega_d = \omega_s/2$.

SOLUTION From Table 4-1, where $b = \omega_d = \pi/T$,

$$E(z) = \frac{z^2 - z\epsilon^{-aT}\cos bT}{z^2 - 2z\epsilon^{-aT}\cos bT + \epsilon^{-2aT}} = \frac{z}{z + \epsilon^{-aT}}$$

Thus, as this example illustrates, a two-pole and one-zero function $E(s)$ maps into a one-pole and one-zero function $E(z)$ as shown in Fig. 4-10*b* when

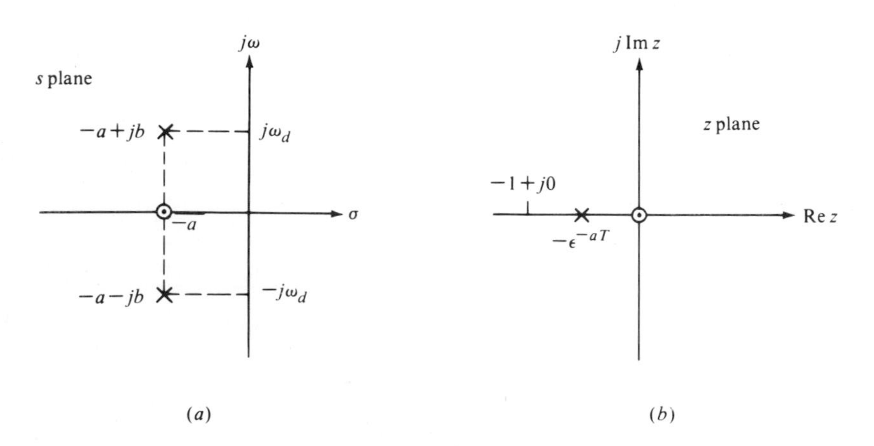

Figure 4-10 Mapping of an $E(s)$ function into the z-plane.

$b = \omega_d = \omega_s/2$. Note that this reduction in order is due to the unique position of the parameter a in the numerator and denominator of $E(s)$.

4-4 $\mathscr{Z}$-TRANSFORM THEOREMS

Theorems similar to those for the Laplace transform exist for the $\mathscr{Z}$ transform. These transforms simplify and facilitate the application of the $\mathscr{Z}$-transformation method of analyzing system performance. The most important theorems are presented here without proof.[2]

Theorem 1: Addition and subtraction Given that $e_a(t)$ and $e_b(t)$ possess Laplace transforms, and that both $E_a(z) = \mathscr{Z}[e_a^*(t)]$ and $E_b(z) = \mathscr{Z}[e_b^*(t)]$ exist, then

$$\mathscr{Z}[e_a^*(t) + e_b^*(t)] = E_a(z) + E_b(z) \tag{4-25}$$

Theorem 2: Multiplication by a constant Given that $E(z) = \mathscr{Z}[e^*(t)]$ exists and A = constant, then

$$\mathscr{Z}[Ae^*(t)] = A\mathscr{Z}[e^*(t)] = AE(z) \tag{4-26}$$

Theorem 3: Real translation (shifting) Given that $E(z) = \mathscr{Z}[e^*(t)]$ exists, then shifting to the right (delay) gives

$$\mathscr{Z}[e^*(t - pT)] = z^{-p}E(z) \tag{4-27}$$

and shifting to the left (advance) gives

$$\mathscr{Z}[e^*(t + pT)] = z^pE(z) - \sum_{i=0}^{p-1} e(iT)z^{p-i} \tag{4-28}$$

where, in Eq. (4-27), z^{-p} represents a delay of p periods, and in Eq. (4-28), z^p represents an advance of p periods. The integer $p \geq 0$.

Theorem 4: Complex translation Given that $E(z) = \mathscr{Z}[e^*(t)]$ exists and A is a constant, then

$$\begin{aligned}\mathscr{Z}[e_A^*(t)] &= \mathscr{Z}[E^*(s + A)] \\ &= [E(z)]_{\text{with } z \text{ replaced by } z\epsilon^{\pm AT}} \\ &= E(z\epsilon^{\pm AT}) = E_A(z)\end{aligned} \tag{4-29}$$

where $e_A(t) = \epsilon^{\mp AT}e(t)$.

Theorem 5: Initial-value theorem Given that both

$$E(z) = \mathscr{Z}[e^*(t)] = \sum_{k=0}^{\infty} e(kT)z^{-k}$$

and the $\lim_{z \to \infty} E(z)$ exist, then

$$e(0) = \lim_{\substack{t \to 0 \\ k \to 0}} e^*(t) = \lim_{z \to \infty} E(z) \tag{4-30}$$

Note, this theorem is valid *only* for the one-sided $\mathscr{Z}$ transform, that is, $k \geq 0$ $(n \geq w)$.

Theorem 6: Final-value theorem Given that $E(z) = \mathscr{Z}[e^*(t)]$ exists and that the function $(1 - z^{-1})E(z)$ does not have any poles outside or on the UC in the z plane, then

$$\lim_{t \to \infty} e^*(t) = \lim_{k \to \infty} e(kT) = \lim_{z \to 1} [(1 - z^{-1})E(z)] \tag{4-31}$$

can be applied to determine the final value of $e^*(t)$.

Note that if $E(z)$ has poles on the UC, excluding $z = 1$, or has poles outside the UC (excluding the positive real axis), there is no unique final value of $e^*(t)$. However, if the function $(1 - z^{-1})E(z)$ has poles at $z = 1$, this theorem gives the correct final value of $e^*(\infty)$ $(= \infty$ for this case). This correctly describes the behavior of $e^*(t)$ as $t \to \infty$.

It should be realized that:

1. If the poles are outside the UC and none are on the positive real axis for $|z| > 1$, then $e^*(\infty)$ contains unbounded oscillations $[\max e(\infty) = \pm\infty]$. Therefore an unstable response $e^*(\infty)$ has no unique solution and the final-value theorem cannot be applied.
2. If the poles are outside the UC and all are on the positive real axis, then $|e^*(\infty)| = \infty$. Thus $e^*(\infty)$ has a unique solution, and again the final-value theorem cannot be applied.
3. If more than one pole at $z = 1$ exists for $E(z)$ and the other poles are within the UC, then $|e^*(\infty)| = \infty$ (a unique solution), where these poles, at $z = 1$, can be due to forcing functions of the form t^l $(l = 1, 2, 3, \ldots)$.

Note that if only one pole exists at $z = 1$ (that is, $l = 0$) then $|e^*(\infty)| =$ constant (a unique solution).

Theorem 7: Real convolution Given that $E_1(z) = \mathscr{Z}[e_1^*(t)]$ and $E_2(z) = \mathscr{Z}[e_2^*(t)]$, then

$$\begin{aligned} E_1(z)E_2(z) &= \mathscr{Z}\left[\sum_{i=0}^{k} e_1(iT)e_2(kT - iT)\right] \\ &= \mathscr{Z}\left[\sum_{i=0}^{k} e_1(kT - iT)e_2(iT)\right] \end{aligned} \tag{4-32}$$

Note that

$$e_1(kT)e_2(kT) \neq \mathscr{Z}^{-1}[E_1(z)E_2(z)] = \sum_{i=0}^{k} e_1(iT)e_2(kT - iT)$$

where $\mathscr{Z}^{-1}$ represents the inverse z transform (see Sec. 4-5).

Theorem 8: Differentiation Given that $E(A, z) = \mathscr{Z}[e^*(A, t)]$, where A is a constant or an independent variable, and $f(A, t) = \partial^m e(A, t)/(\partial A^m)$, then

$$F(A, z) = \mathscr{Z}[f^*(A, t)] = \mathscr{Z}\left\{\frac{\partial^m[e^*(A, t)]}{\partial A^m}\right\}$$

$$= (-1)^m \frac{\partial^m E(A, z)}{\partial A^m} \qquad \text{for } m > 0 \quad (4\text{-}33)$$

For certain types of functions $f(A, t)$, this theorem provides an easy method for obtaining the $\mathscr{Z}$ transform of the mth partial derivative of $e(A, t)$, with respect to A. For example,

$$f(A, t) = t^m \epsilon^{-At} = (-1)^m \frac{\partial^m \epsilon^{-At}}{\partial A^m}$$

where $e(A, t) = \epsilon^{-At}$.

Example 4-8 The $\mathscr{Z}$ transform of a unit-step function is

$$E(z) = \mathscr{Z}[u^*_{-1}(t)] = \frac{z}{z - 1}$$

The $\mathscr{Z}$ transform of a unit-step function delayed by $p = 2$ sampling periods is obtained by applying Theorem 3 as follows:

$$\mathscr{Z}[u^*_{-1}(t - 2T)] = z^{-2}E(z) = \frac{z^{-1}}{z - 1}$$

Example 4-9 A sampled-data open-loop system is characterized, with zero initial conditions, by the linear difference equation

$$c(kT) = -2c[(k - 1)T] + c[(k - 2)T] + e(kT) - 0.5e[(k - 2)T]$$

which is of the form of Eq. (3-71). Theorem 3 is utilized as follows to determine the $\mathscr{Z}$-transfer function $G(z)$ of the system's forward loop.

$$c(z) = \mathscr{Z}[c(kT)] = -2z^{-1}C(z) + z^{-2}C(z) + E(z) - 0.5z^{-2}E(z)$$

$$(1 + 2z^{-1} - z^{-2})C(z) = (1 - 0.5z^{-2})E(z)$$

$$G(z) = \frac{C(z)}{E(z)} = \frac{1 - 0.5z^{-2}}{1 + 2z^{-1} - z^{-2}}$$

Example 4-10 The closed-loop $\mathscr{Z}$-transfer function of a sampled-data system is

$$\frac{C(z)}{R(z)} = \frac{2(1 + 0.2z^{-1})}{1 - 0.4z^{-1} + 0.2z^{-2}}$$

To obtain the difference equation for this system, the transfer function is manipulated as follows:

$$(1 - 0.4z^{-1} + 0.2z^{-2})C(z) = (2 + 0.4z^{-1})R(z)$$

Utilizing Theorem 3 and taking the $\mathscr{Z}^{-1}$ of this equation results in

$$c(kT) - 0.4c[(k-1)T] + 0.2c[(k-2)T] = 2r(kT) + 0.4r[(k-1)T]$$

yielding

$$c(kT) = 2r(kT) + 0.4r[(k-1)T] + 0.4c[(k-1)T] - 0.2c[(k-2)T]$$

which is of the form of Eq. (3-71).

Example 4-11 Utilizing Theorem 4, the $\mathscr{Z}$ transform of the function $\epsilon^{-At} \cos \omega t$ is obtained as follows. Considering only the function $e(t) = \cos \omega t$, then

$$E(z) = \mathscr{Z}[e^*(t)] = \frac{z(z - \cos \omega T)}{z^2 - 2z (\cos \omega T) + 1} \tag{4-34}$$

The $\mathscr{Z}$ transform of the desired function, $e_A(t) = \epsilon^{-At} \cos \omega t$, is obtained by replacing z in Eq. (4-34) by $z\epsilon^{+At}$ to yield

$$\mathscr{Z}[e_A^*(t)] = \frac{z\epsilon^{+AT}(z\epsilon^{+AT} - \cos \omega T)}{z^2\epsilon^{+2AT} - 2z\epsilon^{+AT} (\cos \omega T) + 1}$$

$$= \frac{z^2 - z\epsilon^{-AT} \cos \omega T}{z^2 - 2z\epsilon^{-AT} (\cos \omega T) + \epsilon^{-2AT}}$$

Example 4-12 The initial value of the function given by Eq. (4-34) is evaluated by utilizing Theorem 5 as follows:

$$\lim_{t \to 0} e^*(t) = \lim_{z \to \infty} E(z) = 1$$

Example 4-13 The final value of the function

$$E(z) = \frac{2.4z}{(z - 0.5)(z + 0.2)(z - 1)}$$

is evaluated by applying Theorem 6 as follows:

$$\lim_{t \to \infty} e^*(t) = \lim_{z \to 1} [(1 - z^{-1})E(z)] = \lim_{z \to 1} \frac{2.4}{(z - 0.5)(z + 0.2)} = 4$$

Example 4-14 Theorem 8 is illustrated as follows. Given $f(A, t) = t^2\epsilon^{-At}$, where $e(A, t) = \epsilon^{-At}$ and $f(A, t) = \partial^2[e(A, t)]/\partial A^2$ and $E(A, z) = \mathscr{Z}[e^*(A, t)] = z/(z - \epsilon^{-At})$,

$$F(A, z) = \frac{(-1)^2\, \partial^2 E(A, z)}{\partial A^2} = \frac{T^2\epsilon^{-AT}z(z + \epsilon^{-AT})}{(z - \epsilon^{-AT})^3}$$

4-5 THE INVERSE $\mathscr{Z}$ TRANSFORM, $\mathscr{Z}^{-1}$

Once the analysis and design of a sampled-data control system is performed in the z domain, it is often desired to determine the system's time response. Remember that $E(z)$ represents the sampled values $e^*(t)$ of $e(t)$ at the sampling instants kT. Thus the inversion from the z domain to the time domain yields $e^*(t)$ or the values $e(kT)$. Since the inversion process yields values of $e(t)$ only at the sampling instants, then the function of $e(t)$ that possesses these values at $t = kT$ is not unique, as illustrated by Fig. 4-11. In other words, $E(z)$ does not contain any information about the values of $e(t)$ in between the sampling instants.† This is, by definition, a limitation of the $\mathscr{Z}$-transform method. Tables of $\mathscr{Z}$ transforms are available that present $\mathscr{Z}$ transforms for common $e(t)$ functions and their corresponding $e(kT)$ function (see Table 4-1). When tables are not available or are not complete enough, the following two methods are both available for performing the inversion process.

† However, if possible, proper selection of T can generate a unique $e(t)$.

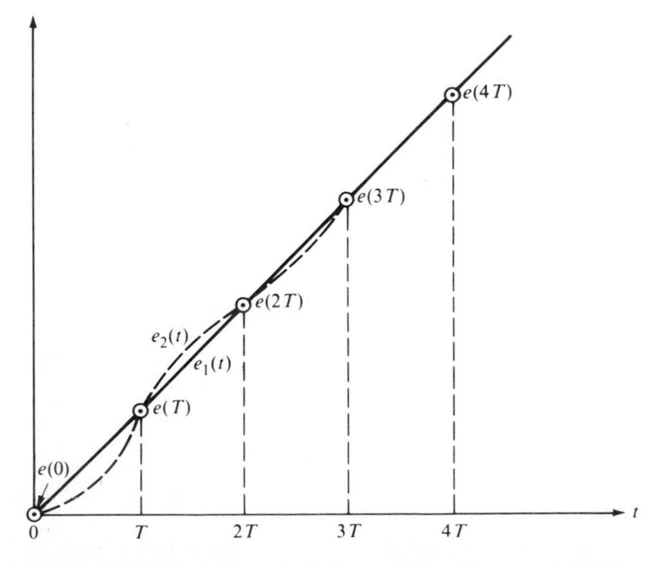

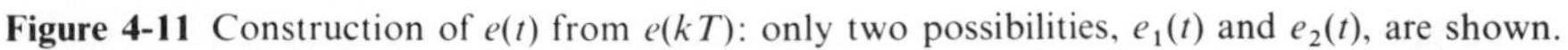
Figure 4-11 Construction of $e(t)$ from $e(kT)$: only two possibilities, $e_1(t)$ and $e_2(t)$, are shown.

4-5A Partial-Fraction Method

Assume that the $E(z)$ function for which the inverse $\mathscr{Z}$ transform is desired is not of any of the simple forms contained in a table of $\mathscr{Z}$ transforms. Thus the partial-fraction expansion method[1] is used to expand $E(z)$ into the sum of functions which exist in the available tables. This is a popular method for obtaining $\mathscr{Z}^{-1}$ because of its "simplicity," and it results in a closed-form solution for $e^*(t)$. Note that this method requires factoring of the denominator of $E(z)$ and also that $E(z)$ be a proper fraction, that is, $n > w$.

Consider $E(z)$ of the form

$$E(z) = \frac{K(z^w + c_{w-1}z^{w-1} + \cdots + c_1 z + c_0)}{z^n + d_{n-1}z^{n-1} + \cdots + d_1 z + d_0} = \frac{KN(z)}{D(z)} \tag{4-35}$$

where $c_0 \neq 0$. Note that $E(z)$ has w zeros and n poles. If $w \geq n$ then before expanding $E(z)$ into a sum of functions, it is necessary to first divide $N(z)$ by $D(z)$ until the order of the remainder polynomial is at least one degree less than $D(z)$. For the case where $w \geq n$ and $w - n = r$, the resulting division process is

$$E(z) = K\,[(g_r z^r + g_{r-1}z^{r-1} + \cdots + g_1 z + g_0) + E'(z)] \tag{4-36}$$

where

$$E'(z) = \frac{k_{n-1}z^{n-1} + \cdots + k_1 z + k_0}{z^n + d_{n-1}z^{n-1} + \cdots + d_1 z + d_0} = \frac{N'(z)}{D(z)} \tag{4-37}$$

Note for the case $w < n$ that $E(z) = KE'(z)$. Thus for either case the partial-fraction expansion method is applied to $E'(z)$.

A basic first-order form of functions of z that appear in Table 4-1 is

$$\mathscr{Z}[e^*(t)] = \frac{A_i z}{z - p_i} \tag{4-38}$$

where $e(kT) = A_i \exp(-\mathbf{b}_i kT)$, and $\mathbf{p}_i = \exp(-\mathbf{b}_i kT)$. If $\mathbf{p}_i$ is a negative real number or a complex number, then $\mathbf{b}_i$ is a complex number (see Example 4-15). Note that all entries in the table have a zero at the origin. The expansion method used in this text is based upon achieving the form of Eq. (4-38). To obtain this form the following procedure is presented.

Case 1: $E'(z)$ has one or more zeros at the origin ($k_0 = 0$, $c_0 \neq 0$), that is, roots of $N'(z)$

Step 1. Obtain $E'(z)/z$. (4-39)
Step 2. Perform partial-fraction expansion on Eq. (4-39).
Step 3. Multiply the resulting equation of Step 2 by z.
Step 4. Obtain $e^*(t)'$ by taking the $\mathscr{Z}^{-1}$ of the resulting form.
Step 5. $e^*(t) = Ke^*(t)'$ for $n > w$.

The first three steps ensure that each function in the expansion has a zero at the origin.

Case 2: $E'(z)$ has no zeros at the origin ($k_0 \neq 0$, $c_0 \neq 0$)

Step 1. Expand $E'(z)$.
Step 2. Multiply the resulting equation of Step 1 by z to obtain $E_d'(z) = zE'(z)$. This guarantees that each function in the expansion has a zero at the origin.
Step 3. Obtain $e_d^*(t)'$ by taking the $\mathscr{Z}^{-1}$ of the equation resulting from Step 2.
Step 4. Since

$$e_d^*(t)' = \mathscr{Z}^{-1}[E_d'(z)] = \mathscr{Z}^{-1}[zE'(z)] \tag{4-40}$$

then by use of Theorem 3, Eq. (4-27),

$$e^*(t)' = \mathscr{Z}^{-1}[E'(z)] = \mathscr{Z}^{-1}[z^{-1}E_d'(z)]$$

$$= \sum_{k=1}^{\infty} e_d'[(k-1)T]\delta(t-kT) \tag{4-41}$$

where $k \geq 1$. Again $e_d^*(t)'$ may be obtained by use of the tables.
Step 5. $e^*(t) = Ke^*(t)'$ for $n > w$. For the situation where $w \geq n$,

$$e^*(t) = K\{\mathscr{Z}^{-1}[g_r z^r + g_{r-1}z^{r-1} + \cdots + g_1 z + g_0] + e^*(t)'\} \tag{4-42}$$

Example 4-15 Obtain $\mathscr{Z}^{-1}$ for

$$E(z) = \frac{10(z^3 - z^2 + 3z - 1)}{(z-1)(z^2 - z + 1)}$$

To obtain $E'(z)$, since $w = n$, it is necessary to first divide $N(z)$ by $D(z)$ to yield

$$E(z) = 10\left[1 + \frac{z(z+1)}{(z-1)(z^2 - z + 1)}\right] \tag{4-43}$$

Since $E'(z)$ has one zero at the origin, the Case 1 procedure must be used.

Step 1

$$\frac{E'(z)}{z} = \frac{z+1}{(z-1)(z^2 - z + 1)} = \frac{z+1}{(z-1)(z-\mathbf{a}_1)(z-\mathbf{a}_2)}$$

where, for this example, $\mathbf{a}_1 = 0.5 + j0.5\sqrt{3} = |\mathbf{a}_1|\underline{/\pi/3} = \exp(\mathbf{b}_1 T) = \epsilon^0\epsilon^{j\pi/3}$ and $\mathbf{a}_2 = 0.5 - j0.5\sqrt{3} = |\mathbf{a}_2|\underline{/-\pi/3} = \exp(\mathbf{b}_2 T) = \epsilon^0\epsilon^{-j\pi/3}$, where $|\mathbf{a}_1| = |\mathbf{a}_2| = 1$, $\mathbf{b}_1 = 0 + j\pi/3$, and $\mathbf{b}_2 = 0 - j\pi/3$. It is left to the reader to determine $\mathbf{b}_1$ and $\mathbf{b}_2$ in terms of T.
Step 2

$$\frac{E'(z)}{z} = \frac{2}{z-1} - \frac{1}{z - \epsilon^{j\pi/3}} - \frac{1}{z - \epsilon^{-j\pi/3}} \tag{4-44}$$

Step 3

$$E'(z) = \frac{2z}{z-1} - \frac{z}{z-\epsilon^{j\pi/3}} - \frac{z}{z-\epsilon^{-j\pi/3}} \tag{4-45a}$$

$$E'(z) = 2\left\{\frac{z}{z-1} - \left[\frac{z(z-\cos\omega T)}{z^2-(2\cos\omega T)z+1}\right]_{\omega T=\pi/3}\right\} \tag{4-45b}$$

Step 4. By using tables, Eq. (4-45*a*) yields

$$e^*(t)' = \sum_{k=0}^{\infty} (2 - \epsilon^{j(\pi/3)k} + \epsilon^{-j(\pi/3)k})\delta(t-kT)$$

$$= \sum_{k=0}^{\infty} (2 - 2\cos k\omega T)\delta(t-kT) \tag{4-46}$$

or

$$e(kT)' = 2 - 2\cos k\omega T \qquad \text{for } k \geq 0 \tag{4-47}$$

If Eq. (4-45*b*) is used, then from the tables the last expression on the right-hand side of Eq. (4-46), or the expression given by Eq. (4-47) is obtained directly.

Step 5. Thus, where $\omega T = \pi/3$,

$$e^*(t) = 10\mathscr{Z}^{-1}[1] + 10e^*(t)'$$

$$= 10\delta(t) + \sum_{k=0}^{\infty} [20 - 20(\cos k\omega T)]\delta(t-kT)$$

or

$$e(kT) = 10\delta(t) + 20 - 20\cos k\omega T \qquad \text{for } k \geq 0 \tag{4-48}$$

Applying the initial-value theorem to Eqs. (4-43) and (4-48) results in $e(0) = 10$. This is a necessary, but not a sufficient, condition on the accuracy of the determination of $\mathscr{Z}^{-1}[E(z)]$.

Example 4-16 Obtain $\mathscr{Z}^{-1}$ for

$$E(z) = \frac{-10(11z^2 - 15z + 6)}{(z-2)(z-1)^2} = -10E'(z)$$

where $n > w$. The Case 2 procedure must be used since there are no zeros at the origin. Note that pole $z_1 = 1$ has a multiplicity of $m = 2$. Assume $T = 1$ s.

Step 1

$$E'(z) = \frac{A_{12}}{(z-1)^2} + \frac{A_{11}}{z-1} + \frac{A_2}{z-2}$$

For nonrepeated poles, the constants A_i, in this example $A_i = A_2$, can be obtained in the usual manner. For the A_i's of repeated poles, the following equation is applied.[1]

$$A_{i(m-q)} = \frac{1}{q!}\frac{d^q}{dz^q}\left[(z - z_i)^m \frac{N'(z)}{D(z)}\right]_{z=p_i} \tag{4-49}$$

where $q = 0, 1, \ldots, m - 1$. Since in this example $q = 0, 1$ then

$$A_{12} = \left.\frac{11z^2 - 15z + 6}{z - 2}\right|_{\substack{z=p_i=1\\q=0}} = -2$$

$$A_{11} = \frac{1}{1!}\left(\frac{d}{dz}\left[\frac{11z^2 - 15z + 6}{z - 2}\right]\right)_{\substack{z=p_1=1\\q=1}}$$

$$= \left.\frac{(z - 2)(22z - 15) - (11z^2 - 15z + 6)}{(z - 2)^2}\right|_{p_1=1}$$

$$= -9$$

which results in

$$E'(z) = \frac{2}{(z - 1)^2} - \frac{9}{z - 1} + \frac{20}{z - 2}$$

Step 2

$$E_d'(z) = zE'(z) = -\frac{2z}{(z - 1)^2} - \frac{9z}{z - 1} + \frac{20z}{z - 2}$$

Step 3

$$e_d'(kT) = -2k - 9 + 20\epsilon^{0.693k} \qquad \text{for } k \geq 0$$

Note that $\epsilon^{-b_iT} = 2$, which yields $b_iT = -0.693$ and, by use of entry 4 in Table 4-1,

$$\mathscr{Z}^{-1}\left[\frac{2z}{(z - 1)^2}\right] = \frac{1}{T}\mathscr{Z}^{-1}\left[\frac{2zT}{(z - 1)^2}\right] = \frac{2}{T}(kT) = 2k$$

Step 4. To obtain $e'(kT)$, which is $e_d'kT$ delayed by one period, replace k on the right-hand side of the equation for $e_d'(kT)$ by $k - 1$ which yields

$$e'(kT) = -2(k - 1) - 9 + 20\epsilon^{0.693(k-1)}$$

Step 5

$$e(kT) = Ke'(kT) = 20(k - 1) + 90 - 200\epsilon^{0.693(k-1)}$$

or

$$e^*(t) = \sum_{k=1}^{\infty} \{e_d[(k-1)T]\delta(t-kT)\}$$

$$= \sum_{k=1}^{\infty} e(kT)\delta(t-kT)$$

where now $k \geq 1$ and $e(kT) = 0$ for $k < 1$, since from Eq. (4-1), $e(t) = 0$ for $t < 0$. As a check on the solution for $e(kT)$ the initial-value theorem yields

$$e(0) = \lim_{k \to 0} e(kT) = \lim_{z \to \infty} E(z) = 0$$

The final-value theorem cannot be applied since one pole lies outside and one on the UC. The initial-value theorem as applied to $e(kT)$, in order to obtain its value at $t = T(k = 1)$, is as follows:

$$e(T) = \lim_{k \to 1} e(kT) = \lim_{z \to \infty} zE(z) = -110$$

Note that since $e(kT)$ holds for $k \geq 1$, then in applying the initial-value theorem to $e(kT)$ the limit on k is 1, and in applying it to $E(z)$ requires that $E(z)$ be advanced by one period which represents the condition of $t \to T$.

For the case where $c_0 = 0$ and $w \geq n$ in Eq. (4-35), it is first necessary, in order to preserve a zero at the origin, to obtain

$$\frac{E(z)}{z} = K \frac{N(z)/z}{D(z)} \tag{4-50}$$

then proceed in the following manner.

Case 3: $E(z)/z$ **is in proper form** [$n = w$ **in Eq. (4-35)**]

Step 1. Perform partial-fraction expansion on Eq. (4-50).
Step 2. Multiply the resulting equation of Step 1 by z.
Step 3. Obtain $e^*(t)$ by taking the $\mathscr{Z}^{-1}$ of the resulting form of Step 2.

Case 4. $E(z)/z$ **is not in proper form** ($w > n$)

Step 1. Obtain the format of Eq. (4-36) by dividing $N(z)/z$ by $D(z)$ in Eq. (4-50).
Step 2. Perform the partial-fraction expansion on the remainder, $E'(z)$.
Step 3. Multiply both sides of the expanded form of Eq. (4-50), as obtained in Steps 1 and 2, by z.
Step 4. Obtain $e^*(t)$ by taking the $\mathscr{Z}^{-1}$ of the resulting form of Step 3.

These four cases illustrate the simplicity of the application of the partial-fraction expansion method. This method, as stated previously, yields a closed-form solution for $e^*(t)$.

4-5B Power-Series Method (Direct Division Method)

The easiest method of obtaining the desired values $e(kT)$ is by dividing $N(z)$ by $D(z)$ to obtain a power series in terms of powers of z^{-1} [see Eq. (4-9)]. This method results in an open form for $E(z)$ and in turn for $e^*(t)$ [see Eq. (4-7)], in contrast to the closed form for $e^*(t)$ that is obtained by the partial-fraction expansion method for physically realizable system $n \geq w$. Henceforth, all $E(z)$'s that are to be considered will satisfy the condition $n \geq w$ unless otherwise stated. Dividing $N(z)$ by the $D(z)$ of Eq. (4-35) yields

$$E(z) = K(B_{w-n}z^{w-n} + B_{w-n-1}z^{w-n-1} + \cdots) \tag{4-51}$$

Comparing the term in parentheses of Eq. (4-51) with Eq. (4-3), it is seen that

$$KB_{w-n} = e[(n-w)T], \qquad KB_{w-n-1} = e[(n+1-w)T], \qquad \ldots$$

which are the values of $e(t)$ at the sampling instants $kT = (n-w)T$, $(n+1-w)T, \ldots$.

Note: Eq. (4-51) is also valid for $w > n$ $(k < 0)$.

Example 4-17 Repeat Example 4-16 by using the power-series method and compare the results. [Note that it is immaterial whether the coefficient of the highest power of z in $N(z)$ is unity or not.]

$$N(z) \rightarrow 11z^2 - 15z + 6 \qquad \Big| \; z^3 - 4z^2 + 5z - 2 \leftarrow D(z)$$

$$\underline{11z^2 - 44z + 55 - 22z^{-1}} \qquad \underbrace{11z^{-1} + 29z^{-2} + 67z^{-3} + 145z^{-4} + \cdots}_{\text{Quotient}}$$

$$29z - 49 + 22z^{-1}$$

$$\underline{29z - 116 + 145z^{-1} - 58z^{-2}}$$

$$67 - 123z^{-1} + 58z^{-2}$$

$$67 - 268z^{-1} \ldots$$

The coefficients in $N'(z)$ and $D(z)$, for Example 4-17, are such that the values of $e(kT)$ obtained by both methods, with $T = 1$, are in close agreement as shown in Table 4-3. In general, this is not the case; the latter method yields less accurate

Table 4-3

	0	T	$2T$	$3T$	$4T$
Example 4-16, $k \geq 1$	0	−110	−289.94	−669	−1448
Example 4-17, $n - w = 1$	0	−110	−290.00	−670	−1450

values of $e(kT)$, especially for large values of k, as is illustrated in the next example. Of course, with these values of $e(kT)$ it is possible to write an expression for $e^*(t)$, an impulse train.

Example 4-18 Determine $e(kT)$ for

$$E(z) = \frac{0.51z^2 + 0.121z}{z^3 + 0.11z^2 + 0.2z + 0.5} \qquad \text{for } k = 0, 1, 2, 3$$

No roundoff (exact division as in Example 4-17):

$$\begin{array}{rr|l}
0.51z^2 + 0.121z & & z^3 + 0.11z^2 + 0.2z + 0.5 \\
\underline{0.51z^2 + 0.0561z + 0.102} & \underline{+\, 0.255z^{-1}} & \overline{0.51z^{-1} + 0.0649z^{-2} - 0.109139z^{-3} - \cdots} \\
0.0649z - 0.102 & -\, 0.255z^{-1} & \\
\underline{0.0649z + 0.007139 + \cdots} & & \\
-0.109139 - \cdots & &
\end{array}$$

Roundoff (to three digits):

$$\begin{array}{r|l}
0.51z^2 + 0.121z & z^3 + 0.11z^2 + 0.2z + 0.5 \\
\underline{0.51z^2 + 0.056z + 0.102 + 0.255z^{-1}} & \overline{0.51z^{-1} + 0.065z^{-2} - 0.109z^{-3} - \cdots} \\
0.065z - 0.102 - 0.255z^{-1} & \\
\underline{0.065z + 0.007 + \cdots} & \\
-\, 0.109 - \cdots &
\end{array}$$

This example illustrates that roundoff must be made if the calculations are not to become cumbersome. The error due to roundoff propagates as the divison process is continued. Thus, for large values of k, the values of $e(kT)$ obtained from the open form of $E(z)$ resulting from the power-series method becomes less accurate as compared to the values obtained from the closed form of $e^*(t)$ obtained by the method of Sec. 4-5A. In the generation of $e(kT)$ from $E(z)$ in digital computers, the effect of finite computer word length has to be considered as reflected in the above analysis. Thus computer-aided-design programs for this mapping must implement accurate and efficient algorithms. The inversion formula method[2] is another means for obtaining $e(kT)$, with which $e^*(t)$ can be constructed, but it is not presented in this text as most students do not have the required knowledge of complex-variable contour integration theory.

4-6 LIMITATIONS

Like any method of analysis, there are limitations to the use of the $\mathscr{Z}$-transform method. In spite of these limitations, however, this method is a very useful and convenient analytical tool because of its "simplicity." Nevertheless, the following limitations must be kept in mind when applying and interpreting the results of the $\mathscr{Z}$-transform method.

1. The use of the concept of an ideal sampler in the analysis of a discrete-data system is based upon the model that the strengths (area) of the impulses in the impulse train, $e^*(t)$, are equal to the corresponding values of the input signal $e(t)$ at the sampling instants kT. For this mathematical model to be "close" to correct, it is necessary for the sampling duration (or pulse width) γ to be very small by comparison with the smallest time constant of the $E(s)$ of concern. Note that it is also very much less than the sampling time T.
2. In Sec. 4-5 it is stated that the inverse $\mathscr{Z}$ transform, $\mathscr{Z}^{-1}[E(z)]$, may not yield a unique $e(t)$ function. The *modified* $\mathscr{Z}$-transform method, $\mathscr{Z}_m$ (see Sec. 4-9), removes this limitation.
3. The impulse responses of rational $E(s)$ functions (*exclusive of a hold device*) that do not have at least two more poles than zeros experience a jump behavior at $t = 0$ [at $t = kT$ for $e^*(t)$]. For these functions

$$\lim_{t \to kT+} e(t) = L_+ \neq \lim_{t \to kT-} e(t) = L_- \tag{4-52}$$

that is, the value of $e(kT)$ as t approaches the value of kT from the low side is not identical to the value obtained when approaching kT from the high side. In other words, a discontinuity occurs in the value of $e(kT)$ at $t = kT$. Fortunately, the $E(z)$ encountered in the analysis of most in-practice control systems have at least two more poles than zeros ($n \geq w + 2$), and this limitation is not of concern for such systems.

4-7 $\mathscr{Z}$ TRANSFORM OF SYSTEM EQUATIONS

The previous sections have dealt only with the properties and mathematical representation of discrete signals. This section uses these properties and mathematical representations to develop the properties and mathematical representation of open-loop and closed-loop sampled-data control systems, i.e., the linear model. A further objective of this section is to express these models in terms of the $\mathscr{Z}$ transform and to develop the appropriate control ratios. The manipulation of transfer-function blocks within a sampled-data control system diagram is presented as an important technique of system analysis.

4-7A Open-Loop Hybrid Sampled-Data Control System

The continuous-time output of the open-loop hybrid sampled-data control system of Fig. 4-12*a* is

$$C(s) = G(s)E^*(s) \tag{4-53}$$

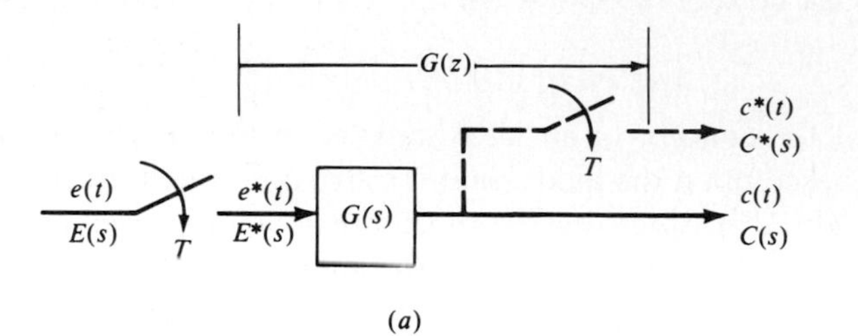

(a)

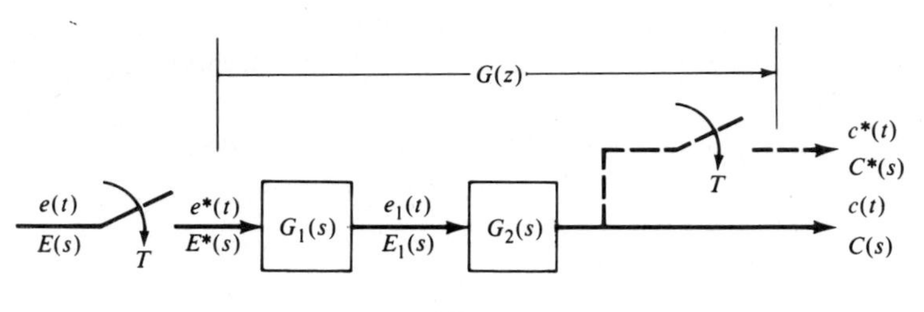

(b)

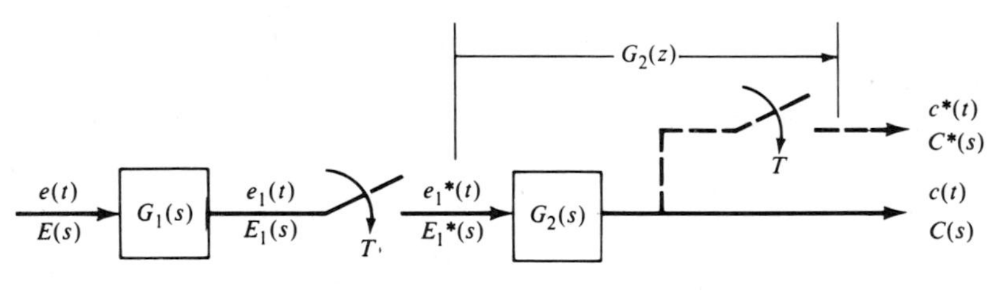

(c)

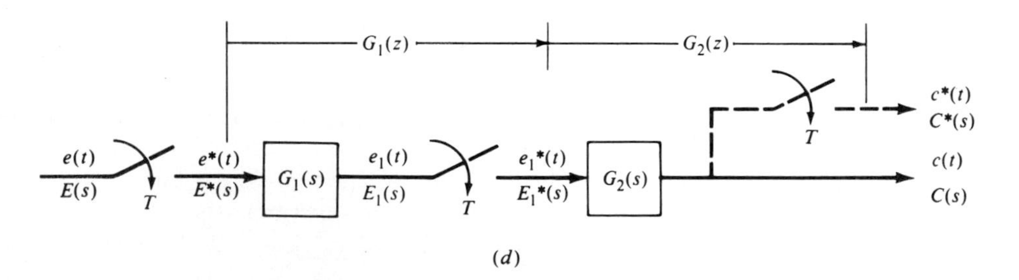

(d)

Figure 4-12 Open-loop sampled-data control systems.

where $G(s)$, the *forward transfer function*, incorporates the ZOH device and the basic plant characteristics. Henceforth, it is assumed that the output of a sampler that is driving a continuous-time plant must contain a ZOH device between the sampler and the plant. That is, realistically, the continuous plant should not be driven by impulses.

To obtain an expression for $C(z)$, it is first necessary to obtain the $C^*(s)$ and $G^*(s)$ functions. The approach used to obtain these functions involves the impulse-response method. This modeling method uses impulses as input signals into a block transfer function and evaluates the output at the sampled time. This technique is essentially the exact $\mathscr{Z}$-transform operation. The presentation here, however, should help you to understand in greater detail the manipulation of transfer-function blocks. The linear system concepts of Chap. 3 are utilized in the development of the impulse-response method. Assume that a unit impulse input $e(t) = \delta(t)$ is applied to $G(s)$. Since $c(t) = g(t)$, then the output $c^*(t)$ of the fictitious sampler in Fig. 4-12*a* is of the same format as Eq. (3-83); that is,

$$c^*(t) = g^*(t) = \sum_{k=0}^{\infty} g(kT)\delta(t - kT) \tag{4-54}$$

where $k = 0, 1, 2, \ldots$ and $g(kT)$ is defined in Chap. 3 as the impulse *weighting sequence* of the system.

Now assume that an arbitrary input function $e(t)$ is applied to the system at $t = 0$. The sampled input to G is then simply the sequence $e(kT)$. Since the system is linear, then, at $t = kT$, the output sampled function $c(kT)$ is equal to the sum of the effects of all sampled quantities of the output signal occurring at, and prior to, $t = kT$, as shown in Fig. 4-13. That is,

$$\begin{aligned} c(kT) &= \sum \{\text{effects of all samples: } e(kT), e[(k-1)T], \ldots, e(0)\} \\ &= e(0)g(kT) + e(T)g[(k-1)T] + \cdots + e[(k-1)T]g(T) + e(kT)g(0) \end{aligned} \tag{4-55}$$

Multiplying both sides of Eq. (4-55) by ϵ^{-kTs} and taking the summation from $k = 0$ to $k = \infty$ yields

$$\begin{aligned} C^*(s) &= \sum_{k=0}^{\infty} c(kT)\epsilon^{-kTs} \\ &= \sum_{k=0}^{\infty} e(0)g(kT)\epsilon^{-kTs} + \sum_{k=0}^{\infty} e(T)g[(k-1)T]\epsilon^{-kTs} \\ &\quad + \sum_{k=0}^{\infty} e(2T)g[(k-2)T]\epsilon^{-kTs} + \cdots + \sum_{k=0}^{\infty} e[(k-1)T]g(T)\epsilon^{-kTs} \\ &\quad + \sum_{k=0}^{\infty} e(kT)g(0)\epsilon^{-kTs} \end{aligned} \tag{4-56}$$

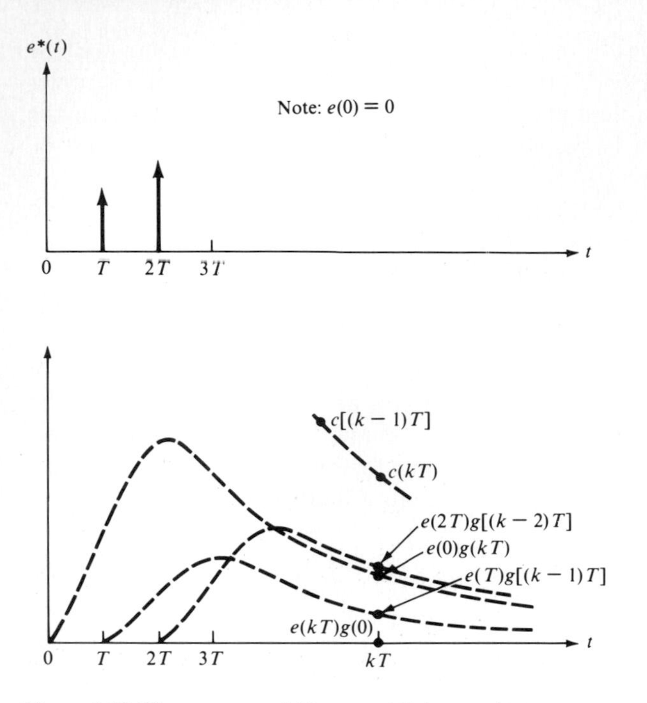

Figure 4-13 The response $c(kT)$ at $t = kT$ due to $e^*(t)$.

Substituting $k' = k - 1$, $k'' = k - 2, \ldots$ into Eq. (4-56) yields

$$\sum_{k=0}^{\infty} c(kT)\epsilon^{-kTs} = \sum_{k=0}^{\infty} e(0)g(kT)\epsilon^{-kTs} + \left[\sum_{k'=-1}^{\infty} e(T)g(k'T)\epsilon^{-k'Ts}\right]\epsilon^{-Ts}$$
$$+ \left[\sum_{k''=-2}^{\infty} e(2T)g(k''T)\epsilon^{-k''Ts}\right]\epsilon^{-2Ts} + \cdots \tag{4-57}$$

Since $e(t) = 0$ for $t < 0$, then $g(t)|_{t<0} = 0$. Thus, for a forcing function $e(T)$, an impulse, delayed by one period, applied to $G(s)$ yields the output response

$$e(T)g(k'T) = e(T)g[(k-1)T]$$

where $g(k'T) = 0$ for $k' < 0$. Likewise, for the forcing function $e(2T)$, the output response is

$$e(2T)g(k''T) = e(2T)g[(k-2)T]$$

where $g(k''T) = 0$ for $k'' < 0$; etc. Therefore, Eq. (4-57) may be written as

$$\sum_{k=0}^{\infty} c(kT)\epsilon^{-kTs} = e(0)\left[\sum_{k=0}^{\infty} g(kT)\epsilon^{-kTs}\right] + e(T)\epsilon^{-Ts}\left[\sum_{k=0}^{\infty} g(k'T)\epsilon^{-k'Ts}\right]$$
$$+ e(2T)\epsilon^{-2Ts}\left[\sum_{k''=0}^{\infty} g(k''T)\epsilon^{-k''Ts}\right] + \cdots \tag{4-58}$$

Noting that all dummy variables k, k', k'', etc., in Eq. (4-58) have a running index of 0, 1, 2, 3, . . ., then this equation may now be expressed with *one common dummy* variable. This common variable is chosen as k; thus

$$\begin{aligned}\sum_{k=0}^{\infty} c(kT)\epsilon^{-kTs} &= e(0)\left[\sum_{k=0}^{\infty} g(kT)\epsilon^{-kTs}\right] + e(T)\epsilon^{-Ts}\left[\sum_{k=0}^{\infty} g(kT)\epsilon^{-kTs}\right] \\ &\quad + e(2T)\epsilon^{-2Ts}\left[\sum_{k=0}^{\infty} g(kT)\epsilon^{-kTs}\right] + \cdots \\ &= [e(0) + e(T)\epsilon^{-Ts} + e(2T)\epsilon^{-2Ts} + \cdots]\left[\sum_{k=0}^{\infty} g(kT)\epsilon^{-kTs}\right] \\ &= \left[\sum_{k=0}^{\infty} e(kT)\epsilon^{-kTs}\right]\left[\sum_{k=0}^{\infty} g(kT)\epsilon^{-kTs}\right] \end{aligned} \tag{4-59}$$

Comparing the three summation terms of Eq. (4-59) with Eq. (4-1) yields the desired functions

$$C^*(s) = E^*(s)G^*(s) \tag{4-60}$$

$$C(z) = E(z)G(z) \tag{4-61}$$

where

$$C^*(s) = \sum_{k=0}^{\infty} c(kT)\epsilon^{-kTs}$$

$$E^*(s) = \sum_{k=0}^{\infty} e(kT)\epsilon^{-kTs}$$

and

$$G^*(s) = \sum_{k=0}^{\infty} g(kT)\epsilon^{-kTs}$$

The function

$$G(z) = G^*(s)|_{z=\varepsilon^{Ts}} = \sum_{k=0}^{\infty} g(kT)z^{-k} \tag{4-62}$$

is defined as the *forward z-transfer function* of the linear system. This function relates the input $E(z)$ to the system output $C(z)$ [see Eq. (4-61)] in the same manner as for all continuous-data open-loop systems [$C(s) = E(s)G(s)$]; thus $G(z) = C(z)/E(z)$. For the system of Fig. 4-12*a*, Eq. (4-61) yields information on $c(t)$ only at the sampled instants $t = kT$. This is not a critical problem if the output response $c(t)$ is overdamped. For the case where the system is lightly damped, the $\mathscr{Z}$-transform analysis may yield misleading results with respect to system performance in between sampling instants. In such cases the modified $\mathscr{Z}$ transform should be used.

The open-loop hybrid sampled-data control system of Fig. 4-12*b* can be considered to be identical to the one in Fig. 4-12*a* where $G(s) = G_1(s)G_2(s)$. Thus the forward z-transfer function and the output $\mathscr{Z}$ transform for this system are, respectively,

$$G(z) = \mathscr{Z}[G^*(s)] = \mathscr{Z}[G_1G_2^*(s)] = G_1G_2(z) \tag{4-63}$$

$$C(z) = G_1G_2(z)E(z) \tag{4-64}$$

where $G^*(s) = G_1G_2^*(s)$ and, in general,

$$G_1G_2(z) \neq G_1(z)G_2(z) = \mathscr{Z}[G_1^*(s)]\mathscr{Z}[G_2^*(s)]$$

Note that the notation of $G_1G_2^*(s)$ and $G_1G_2(z)$, where the variables s and z are not specifically indicated with G_1, implies that $G_1G_2^*(s) = [G_1(s)G_2(s)]^*$ and $G_1G_2(z) = \mathscr{Z}[G_1G_2^*(s)]$, respectively. This notation is used to indicate the impulse transform of p transfer functions in cascade. Thus, for p blocks in cascade, with no samplers in between them, $C(z)$ is given by Eq. (4-61), where

$$G(s) = G_1(s)G_2(s) \cdots G_p(s) \tag{4-65}$$

and

$$G(z) = G_1G_2 \cdots G_{p-1}G_p(z) \tag{4-66}$$

For the control system of Fig. 4-12c, the following equations apply:

$$E_1^*(s) = G_1E^*(s) \tag{4-67}$$

$$C^*(s) = G_2^*(s)E_1^*(s) \tag{4-68}$$

Substituting Eq. (4-67) into Eq. (4-68) yields

$$C^*(s) = G_2^*(s)G_1E^*(s) \tag{4-69}$$

or

$$C(z) = G_2(z)G_1E(z) \tag{4-70}$$

where $E_1(z) = G_1E(z)$ and $C(z) = G_2(z)E_1(z)$. Thus for the open-loop sampled-data control system of Fig. 4-12c, it is impossible to define a $G(z)$-transfer function that relates its output to its input; that is, $G(z) = C(z)/E(z)$. This situation occurs when the actuating signal $e(t)$ is not sampled. The z-domain representations of the open-loop systems of Fig. 4-12 are shown in Fig. 4-14.

4-7B Open-Loop Discrete-Input-Data Control System

The open-loop discrete-input-data control system of Fig. 4-12d, in which all external block signals are sampled, may be considered as being equivalent to putting in cascade two systems. Each system has the configuration of Fig. 4-12a (or Fig. 4-14a) where the one fictitious sampler is now an actual sampler. Thus, based upon the analysis of the system of Fig. 4-14a, the following relationship may be written for the system of Fig. 4-14c.

$$C(z) = G_2(z)E_1(z) \tag{4-71}$$

$$E_1(z) = G_1(z)E(z) \tag{4-72}$$

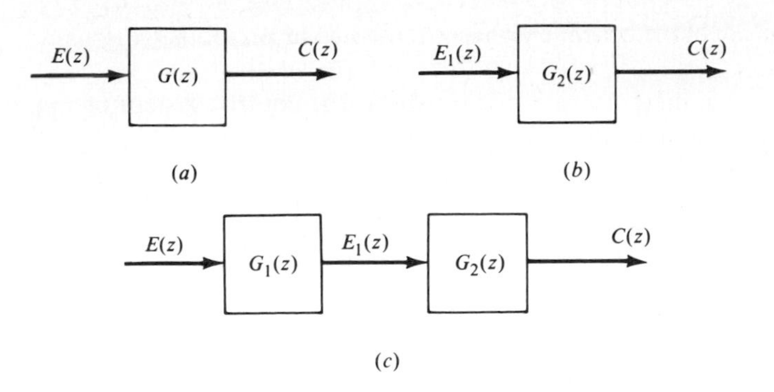

Figure 4-14 z-Domain representations of the open-loop systems of Fig. 4-12.
(*a*) $C(z) = G(z)E(z)$ (Fig. 4-12*a*, *b*, and *d*)
(*b*) $C(z) = G_2(z)E_1(z)$ (Fig. 4-12*c*)
(*c*) $C(z) = G_2(z)E_1(z) = G_1(z)G_2(z)E(z)$ (Fig. 4-12*d*)

Substituting Eq. (4-72) into Eq. (4-71) yields

$$C(z) = G_1(z)G_2(z)E(z) = G(z)E(z) \tag{4-73}$$

where

$$G(z) = \frac{C(z)}{E(z)} = G_1(z)G_2(z) \tag{4-74}$$

Therefore, the forward z-transfer function of a linear system consisting of two subsystems in cascade separated by a sampler is the product of the z-transfer functions of each of the subsystems. The above may be extended to include p subsystems in cascade, each separated from the other by a sampler; that is,

$$G(z) = G_1(z)G_2(z) \cdots G_p(z) \tag{4-75}$$

$$C(z) = G(z)E(z) \tag{4-76}$$

Thus the open-loop system of Fig. 4-14*c* can be represented by the open-loop system of Fig. 4-14*a*.

4-7C Closed-Loop Sampled-Data Control System

Obtaining the $\mathscr{Z}$ transform of an open-loop system, for the configurations shown in Fig. 4-12, is relatively easy. For closed-loop systems, the manipulations required to obtain the control ratio of the system are more involved. The object of this section is, therefore, to present a method for obtaining the relationship between the output and input variables of the closed-loop system. There are two common approaches for obtaining the closed-loop $\mathscr{Z}$ transforms of digital control systems which contain feedback and multiple numbers of synchronized samplers: by algebraic manipulation of block transfer functions or by the signal-flow graph method.[2] This text deals with the former approach, although the latter method may be more applicable for complex systems with equivalent manipulations of these transfer functions. This selection emphasizes the use of previously discussed techniques.

Applying the algebraic method used previously to the control system of Fig. 4-15a yields the following equations:

$$C(s) = G_2(s)G_3(s)E_1^*(s) = G_{2,3}(s)E_1^*(s) \tag{4-77}$$

$$E_1(s) = G_1(s)E^*(s) \tag{4-78}$$

$$E(s) = R(s) - B(s) = R(s) - H(s)C(s) \tag{4-79}$$

The object is to *manipulate these equations to eliminate the internal system variables and to obtain a relationship only between the system output and input variables.* Substituting from Eq. (4-77) into Eq. (4-79) yields

$$E(s) = R(s) - G_{2,3}(s)H(s)E_1^*(s) \tag{4-80}$$

Taking the impulse transform of both sides of Eqs. (4-77), (4-78), and (4-80) results in, respectively,

$$C^*(s) = G_{2,3}^*(s)E_1^*(s) \tag{4-81}$$

$$E_1^*(s) = G_1^*(s)E^*(s) \tag{4-82}$$

$$E^*(s) = R^*(s) - G_{2,3}H^*(s)E_1^*(s) \tag{4-83}$$

Substituting from Eq. (4-82) into Eqs. (4-81) and (4-83) gives

$$C^*(s) = G_1^*(s)G_{2,3}^*(s)E^*(s) \tag{4-84}$$

and

$$E^*(s) = R^*(s) - G_1^*(s)G_{2,3}H^*(s)E^*(s)$$

which yields

$$E^*(s) = \frac{R^*(s)}{1 + G_1^*(s)G_{2,3}H^*(s)} \tag{4-85}$$

This last equation is substituted into Eq. (4-84) to yield the impulse transform of the output in terms of the input $R^*(s)$.

$$C^*(s) = \frac{G_1^*(s)G_{2,3}^*(s)}{1 + G_1^*(s)G_{2,3}H^*(s)} R^*(s)$$

$$C(z) = \frac{G_1(z)G_{2,3}(z)}{1 + G_1(z)G_{2,3}H(z)} R(z) \tag{4-86}$$

Thus the z-transfer function of the *overall transfer function* (i.e., hereafter termed the *control ratio*) is

$$\frac{C(z)}{R(z)} = \frac{G_1(z)G_{2,3}(z)}{1 + G_1(z)G_{2,3}H(z)} \tag{4-87}$$

For the system of Fig. 4-15a, and for all other cases where the input forcing

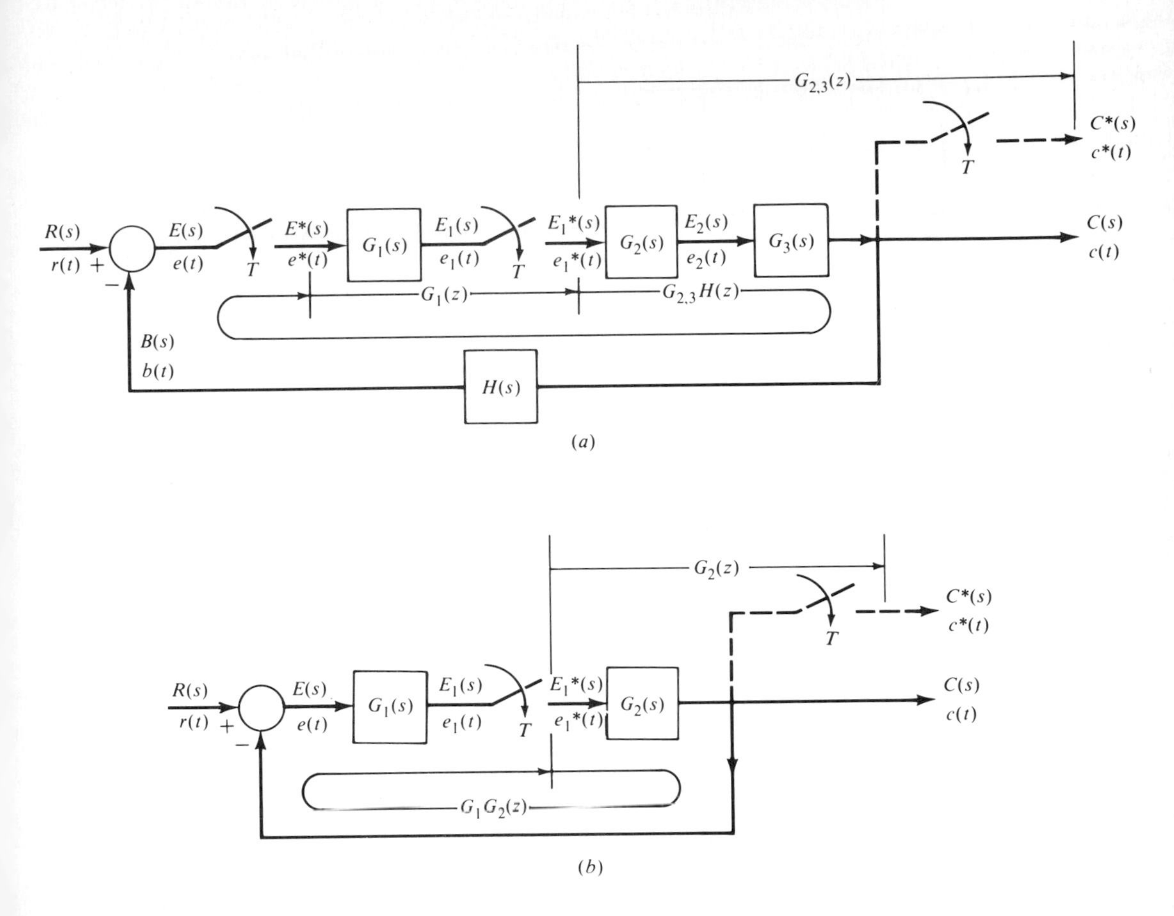

Figure 4-15 Sampled-data control systems.

function *is sampled*, the control ratio $C(z)/R(z)$ may be manipulated to obtain the z-domain block diagram representation of the form of Fig. 4-16. The control ratio for the system represented by Fig. 4-16 is

$$\frac{C(z)}{R(z)} = \frac{G_0(z)}{1 + G_0(z)H_0(z)} \tag{4-88}$$

By letting $G_0(z) = G_1(z)G_{2,3}(z)$ for Fig. 4-15a, the equivalent $H_0(z)$ of Fig. 4-16 may be obtained by equating Eq. (4-88) to Eq. (4-87).

The algebraic method is now applied to the control system of Fig. 4-15b. From this figure the following equations are obtained:

$$C(s) = G_2(s)E_1^*(s) \tag{4-89}$$

$$E_1(s) = G_1(s)E(s) \tag{4-90}$$

$$E(s) = R(s) - C(s) \tag{4-91}$$

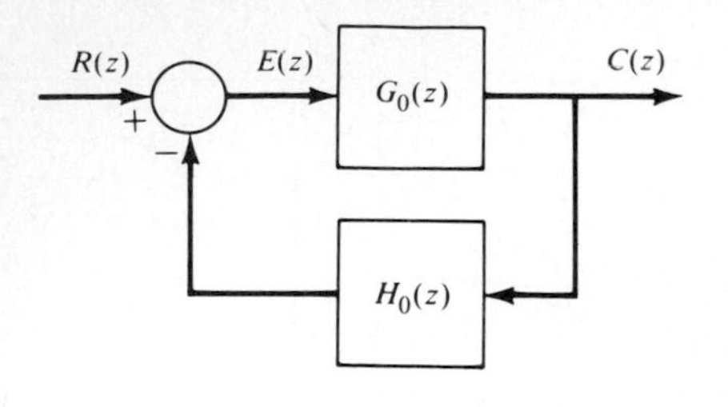

Figure 4-16 The z-domain representation of Fig. 4-15a.

As in the previous example, *these equations are manipulated to eliminate the internal system variables and to obtain a relationship between the input and output variables only*. First substituting from Eq. (4-89) into Eq. (4-91) and then substituting the resulting equation into Eq. (4-90) yields

$$E_1(s) = G_1(s)R(s) - G_1(s)C(s) = G_1(s)R(s) - G_1(s)G_2(s)E_1^*(s) \tag{4-92}$$

Taking the impulse transform of both sides of Eqs. (4-89) and (4-92) results in, respectively,

$$C^*(s) = G_2^*(s)E_1^*(s) \tag{4-93}$$

$$E_1^*(s) = G_1R^*(s) - G_1G_2^*(s)E_1^*(s) \tag{4-94}$$

Equation (4-94) is manipulated to obtain

$$E_1^*(s) = \frac{G_1R^*(s)}{1 + G_1G_2^*(s)}$$

which is substituted into Eq. (4-93) to yield

$$C^*(s) = \frac{G_2^*(s)G_1R^*(s)}{1 + G_1G_2^*(s)} \tag{4-95}$$

or

$$C(z) = \frac{G_2(z)G_1R(z)}{1 + G_1G_2(z)} \tag{4-96}$$

Note that

$$G_1R(z) = \mathscr{Z}[G_1(s)R(s)] \neq G_1(z)R(z)$$

due to the location of the sampler in Fig. 4-15b. Thus it is impossible to obtain the control ratio $C(z)/R(z)$ and a corresponding equivalent z-domain block diagram representation for this system. In comparing both systems of Fig. 4-15, it is seen that if the actuating signal $E(s)$ of a system is sampled, then the control ratio $C(z)/R(z)$ may be obtained, if desired.

4-8 DIGITAL-COMPUTER TRANSFER FUNCTION

When the analysis of the basic sampled-data closed-loop control system of Fig. 4-17a reveals that the desired system performance specifications cannot be met with

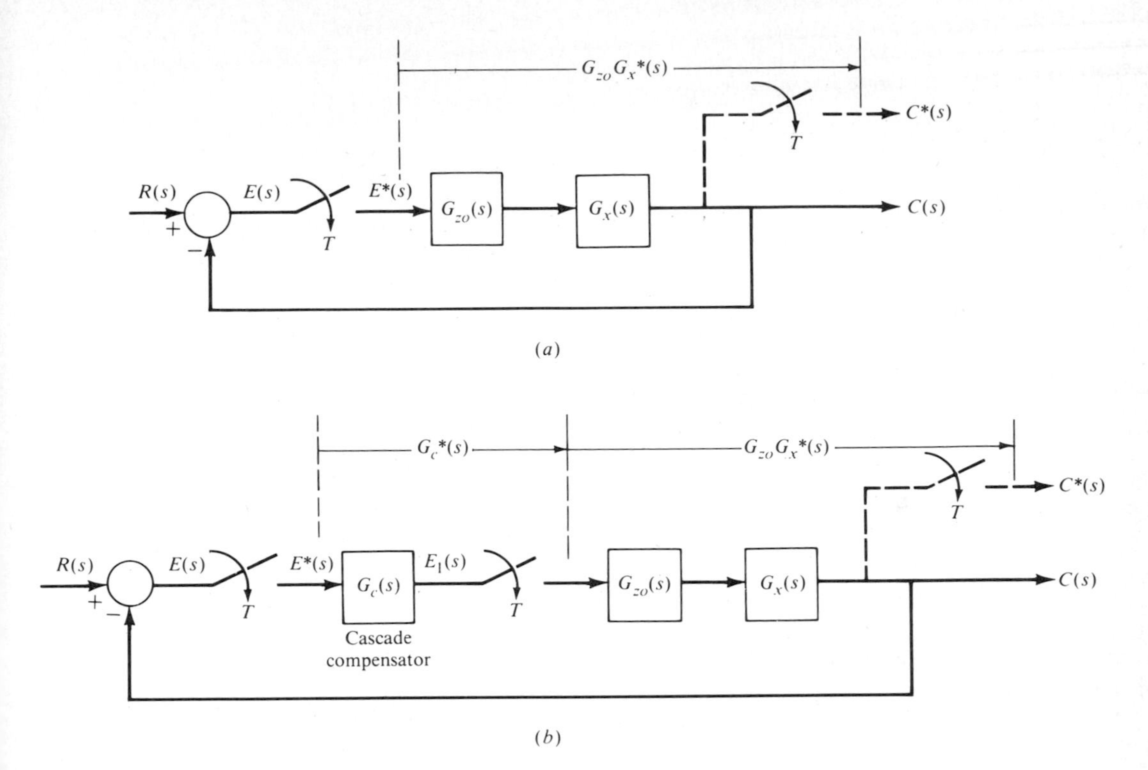

Figure 4-17 (*a*) Basic and (*b*) cascade compensated sampled-data control systems.

only gain and/or sampling-time adjustments, additional elements must then be added to the system. One method of achieving the desired system performance is by inserting a cascade compensator in the forward loop of the control system as shown in Fig. 4-17*b*. Many control systems utilize cascade compensation because of the easily understood cascade design procedures and the ease with which the procedures are applied.[1]

The z-transfer function of a cascade compensator with input $e^*(t)$ and output $e_1^*(t)$ is of the form

$$G_c(z) = \frac{E_1(z)}{E(z)} = \frac{c_w z^w + c_{w-1} z^{w-1} + \cdots + c_1 z + c_0}{z^n + d_{n-1} z^{n-1} + \cdots + d_1 z + d_0} \tag{4-97}$$

where $n \geq w$. Dividing the numerator and denominator of this transfer function by z^n, for the case of $n = w$, Eq. (4-97) is rewritten as

$$G_c(z) = \frac{E_1(z)}{E(z)} = \frac{c_w + c_{w-1} z^{-1} + \cdots + c_1 z^{-(n-1)} + c_0 z^{-n}}{1 + d_{n-1} z^{-1} + \cdots + d_1 z^{-(n-1)} + d_0 z^{-n}} \tag{4-98}$$

The input-output relationship of the compensator may be expressed as a difference equation (a linear recursion equation) by taking the inverse z transform of

$$(1 + d_{n-1}z^{-1} + \cdots + d_1 z^{-(n-1)} + d_0 z^{-n})E_1(z)$$
$$= (c_w + c_{w-1}z^{-1} + \cdots + c_1 z^{-(n-1)} + c_0 z^{-n})E(z) \quad (4\text{-}99)$$

with the use of the real translation theorem, Eq. (4-27). Thus

$$\begin{aligned} e_1(kT) = c_w e(kT) &+ c_{w-1}e[(k-1)T] + \cdots + c_1 e[(k-w+1)T] \\ &+ c_0 e[(k-w)T] - d_{n-1}e_1[(k-1)T] - \cdots \\ &- d_1 e_1[(k-n+1)T] - d_0 e_1[(k-n)T] \end{aligned} \quad (4\text{-}100)$$

Since Eq. (4-100) can model at a relatively high level the input-output sequence relationship of a digital computer, Eq. (4-98) can model the transfer function of a digital computer when it is utilized as a cascade compensator (controller). A computer used in this manner is said to operate in *real time*, and its transfer function can be referred to as the *digital transfer function*. Being able to model the real-time control computer operation by a $\mathscr{Z}$-transfer function permits a $\mathscr{Z}$-transform analysis of the sampled-data closed-loop control system. Of course, certain assumptions and constraints such as computational delays and accuracy as discussed previously must always be considered.[95]

The following PASCAL program implements one version for simulating the difference equation of Eq. (4-100). There are variations of this program that can solve the original equation. The important aspect of this discussion is the accuracy of the various forms due to the finite word length of the digital controller (computer). This topic of numerical accuracy vs. form is expanded in Chap. 10.

```
PROCEDURE DIFFEQ
    [This procedure calculates the value of a system described by a difference equation of
    the form of Eq. (4-100). The values of the coefficients, C(K) and D(K) are assumed
    defined in a higher-level procedure. Also, the procedures for data input,
    READCONTINPUT, and data output, WRITECONTOUTPUT, are assumed to be
    assembly language programs that are "linked" to the overall PASCAL program.]
CONST
    N =       ;    {ORDER OF EQUATION PLUS 1}
    MAXCOUNT =       ;    {MAXIMUM NUMBER OF SAMPLES)
TYPE
    VECTOR = ARRAY [1 ... MAXCOUNT] OF REALS;
    COEFFICIENTS = ARRAY [1 ... N] OF REALS;
VAR
    E1: VECTOR;                  {DIFFERENCE EQUATION OUTPUT}
    E: VECTOR;                   {DIFFERENCE EQUATION INPUT}
    C: COEFFICIENTS;             {NUMERATOR COEFFICIENTS}
    D: COEFFICIENTS;             {DENOMINATOR COEFFICIENTS}
    SAMPLE, OUTPUT: REAL;        {CURRENT SAMPLE INPUT AND OUTPUT}
    K,I,J: INTEGERS;             {INDEXING VARIABLES}
```

```
BEGIN
   K = 1;
   WHILE K ≤ MAXCOUNT DO;      {INITIALIZE INPUT AND OUTPUT)
      BEGIN
         E1(K) = 0.;
         E(K) = 0.;
         K = K + 1;            {INCREMENT INDEX}
   END; {WHILE INITIALIZE}
   K = 1;                      {INITIALIZE INDEX}
   WHILE K ≤ MAXCOUNT DO
      BEGIN
         READCONTINPUT (SAMPLE);      {PROCEDURE FOR SAMPLED
                                 ;                     INPUT}
         E(K) = SAMPLE;     {PUT SAMPLE IN INPUT VECTOR}
         I = 1;     {INITIALIZE INDEX}
         WHILE I ≤ N DO
            BEGIN
               E1(K) = C(N - I)E(K - I) - D(N - I)E1(K - 1) + E1(K)
               I = I + 1
         END;     {WHILE - DIFF EQ LOOP}
         E1(K) = C(N)E(K) + E1(K);     {ADD FIRST NUMERATOR TERM}
         CONTROL = E1(K);
         WRITECONTOUTPUT
                       (CONTROL);     {PROCEDURE FOR
                                 ;            CONTROLLER OUTPUT}
         K = K + 1
   END;     {WHILE - COMPLETE PROCESS}
   END;     {DIFFEQ}
```

4-9 THE MODIFIED $\mathscr{Z}$-TRANSFORM METHOD

The modified $\mathscr{Z}$-transform method alleviates limitation 2 of Sec. 4-6 by permitting the calculation of values of $e(t)$ in the interval $(k-1)T \leq t < kT$. This is accomplished by letting

$$t = (k - \rho)T \tag{4-101}$$

and

$$\rho = 1 - m \tag{4-102}$$

where $0 \leq \rho \leq 1$ and $1 \geq m \geq 0$. Thus when $m = 1$, $\rho = 0$ and $t = kT$ and there is no delay; when $m = 0$, $\rho = 1$ and $t = (k-1)T$ and there is a delay of one full period. For values of m in between, the signal $e^*(t)$ is delayed by a fractional period. This mathematical manipulation is best illustrated by Fig. 4-18. By use of a fictitious time-delay (transport lag) operator $\epsilon^{-\rho Ts}$, sufficient values in between the sampling instants can be determined for any continuous-time signal $e(t)$ in a sampled-data system that accurately describes this signal. In other words, this method provides values of $e(t)$ that exist *during* the time interval $(k-1)T \leq t \leq kT$ by adjusting the value of ρ (or m).

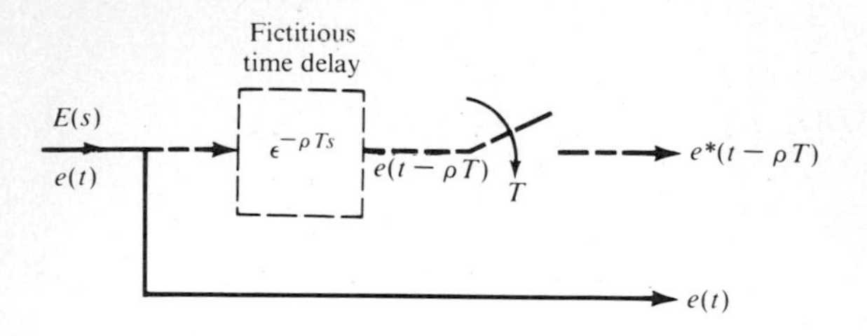

Figure 4-18 Fictitious time-delay representation.

A method for evaluating $e^*(t - \rho T)$ requires the use of the modified $\mathscr{Z}$ transform, which is given by

$$E(z, m) \equiv \mathscr{Z}_m[e^*(t)] \equiv \sum_{k=0}^{\infty} e[(k - \rho)T]z^{-k}$$

$$= \sum_{k=0}^{\infty} e[(k + m - 1)T]z^{-k} \qquad (4\text{-}103)$$

This equation can be considered as the $\mathscr{Z}$ transform of $e[(k + m)T]$ delayed by one period. Thus

$$E(z, m) = z^{-1} \sum_{k=0}^{\infty} e[(k + m)T]z^{-k}$$

$$= z^{-1}\{e(mT) + e[(1 + m)T]z^{-1} + e[(2 + m)T]z^{-2} + \cdots\}$$

$$= e(mT)z^{-1} + e[(1 + m)T]z^{-2} + e[(2 + m)T]z^{-3} + \cdots \qquad (4\text{-}104)$$

The methods for determining $\mathscr{Z}_m$ and $\mathscr{Z}_m^{-1}$, which are similar to those for determining $\mathscr{Z}$ and $\mathscr{Z}^{-1}$, are presented in the following subsections. Another method for evaluating $e^*(t - \rho T)$ in the time domain is presented in Sec. 15-7.

4-9A Obtaining the Modified $\mathscr{Z}$ Transform, $\mathscr{Z}_m$

The manner of obtaining $\mathscr{Z}_m$ is presented in this subsection without proof.[2,20] The $\mathscr{Z}_m$ of $E^*(s)$ is given by

$$E(z, m) = \mathscr{Z}_m[e^*(t - \rho T)]$$

$$= \left\{ z^{-1} \sum \left[\text{residues of } \frac{E(\alpha)\epsilon^{mT\alpha}}{1 - \epsilon^{-T(s-\alpha)}} \right.\right.$$

$$\left.\left. \text{evaluated } \textit{only at the poles of } E(\alpha) \right]_{z=\varepsilon^{Ts}} \right\} + \beta$$

$$= \hat{E}(z, m) + \beta \qquad (4\text{-}105)$$

The manner of evaluating the residues is the same as that discussed in Sec. 4-2, where now

$$\beta = \lim_{s \to \infty} sE(s) - \lim_{\substack{z \to \infty \\ m=0}} z\hat{E}(z, m) \tag{4-106}$$

Note: For multiple poles

$$W(\alpha) = \frac{(\alpha - \alpha_i)^p E(\alpha)\epsilon^{mT\alpha}}{1 - \epsilon^{-T(s-\alpha)}}$$

The term ϵ^{mT} is the delay factor which permits the evaluation of $e^*(t)$ in between the sampling instants.

Example 4-19 Determine $E(z, m)$ for

$$E(s) = \frac{2}{(s+1)(s+2)} = \frac{N(s)}{D(s)} \tag{4-107}$$

where $\alpha_1 = -1$, $\alpha_2 = -2$, $N(\alpha) = 2$, $D(\alpha) = \alpha^2 + 3\alpha + 2$, and $dD(\alpha)/d\alpha = 2\alpha + 3$. Thus

$$E(z, m) = z^{-1} \sum_{i=1}^{n=2} \left\{ \left[\frac{N(\alpha)}{dD(\alpha)/d\alpha} \right]_{\alpha=\alpha_i} \frac{mT\alpha_i}{(1 - \epsilon^{\alpha_i T} z^{-1}} \right\} + \beta$$

$$= \frac{KN(z, m)}{D(z)}$$

$$= z^{-1} \left(\frac{2\epsilon^{-mT}}{1 - \epsilon^{-T} z^{-1}} - \frac{2\epsilon^{-2mT}}{1 - \epsilon^{-2T} z^{-1}} \right) \tag{4-108}$$

where $\beta = 0$.

Equation (4-108) illustrates the use of the following two relationships:

$$\lim_{m \to 0} E(z, m) = z^{-1} E(z) \tag{4-109}$$

and

$$\lim_{m \to 1} E(z, m) = E(z) \tag{4-110}$$

Equation (4-110) is valid as long as the condition of Eq. (4-52) is not satisfied; i.e., the condition $L_+ = L_-$ must exist for Eq. (4-110) to be valid.

For Example 4-19,

$$\lim_{m \to 0} E(z, m) = 2z^{-1} \left(\frac{z}{z - \epsilon^{-T}} - \frac{z}{z - \epsilon^{-2T}} \right) = z^{-1} E(z) \tag{4-111}$$

$$\lim_{m \to 1} E(z, m) = 2 \left(\frac{\epsilon^{-T}}{z - \epsilon^{-T}} - \frac{\epsilon^{-2T}}{z - \epsilon^{-2T}} \right)$$

$$= 2 \frac{(\epsilon^{-T} - \epsilon^{-2T})z}{(z - \epsilon^{-T})(z - \epsilon^{-2T})} = E(z) \tag{4-112}$$

By referring to entry 6 in Table 4-1, it is seen that $E(z, m)$ is, for $m = 0$, the function $E(z)$ delayed by one period. For $m = 1$, $E(z, m)$ reverts to $E(z)$.

In Sec. 4-7, for the system of Fig. 4-12*a*, the relationship $C(z) = G(z)E(z)$ is derived. In a similar manner, if $c^*(t - \rho T)$ is desired for the system of Fig. 4-12*a*, it can be shown that

$$C(z, m) = \mathscr{Z}_m[c^*(t - \rho T)] = G(z, m)E(z)$$

4-9B Evaluation of $e[(k - \rho)T]$

The manner of obtaining values of $e(t)$ in between sampling instants is illustrated by means of an example. Assuming that $T = 1$ s, determine the values of $e[(k - \rho)T]$ for Eq. (4-107) at $t = T, 4T/3$, and $5T/3$. Equation (4-108) is rearranged so that the power-series method may be utilized as follows:

$$E(z, m) = \frac{(2\epsilon^{-mT} - 2\epsilon^{-2mT})z + (0.73576\epsilon^{-2mT} - 0.27068\epsilon^{-mT})}{z^2 - 0.50322z + 0.04979}$$

Applying the power-series method to this equation yields

$$\begin{aligned} E(z, m) &= (2\epsilon^{-mT} - 2\epsilon^{-2mT})z^{-1} + (0.73576\epsilon^{-mT} - 0.27068\epsilon^{-2mT})z^{-2} + \cdots \\ &= [(2 - 2\epsilon^{-mT})\epsilon^{-mT}z^{-1} + (0.73576 - 0.27068\epsilon^{-mT})\epsilon^{-mT}z^{-2} + \cdots]_{T=1} \end{aligned} \tag{4-113}$$

As stated previously $z^{-k} = \epsilon^{-skT}$ represents a delay of k periods. The term ϵ^{-mT} represents a delay of a fractional period. Thus the terms in Eq. (4-113),

$$\epsilon^{-mT}z^{-1} = \epsilon^{-(m+s)T}, \qquad \epsilon^{-mT}z^{-2} = \epsilon^{-(m+2s)T}, \qquad \ldots$$

denote the effect of this fractional delay. This phenomenon allows the quantity $e[(k - \rho)T]$ to be delayed by $(1 - m)T$ seconds. Therefore, *the values of $e(t)$ during the first sampling period are given by the coefficient of the z^{-1} term in Eq.* (4-113), *during the second sampling period by the coefficient of the z^{-2} term, etc*. Note that for the $E(s)$ function of Eq. (4-107), $e(0) = \lim_{s \to \infty} sE(s) = 0$. To verify that Eq. (4-113) yields this value, let $m = 0$ (a delay of one full period; that is, $\rho = 1$) in the coefficient of z^{-1}. That is, with $T = 1$, evaluating at $t = kT = 1$ for the first sampling period (see Fig. 4-19):

$$k = 1: \qquad e(0^+) = [2\epsilon^{-m} - 2\epsilon^{-2m}]_{m=0} = 0$$

which represents the value of the function $e^*(t)$ at $t = 0$ delayed by one period and for

$$k = 1, m = 1: \qquad e(T^-) = 2\epsilon^{-1} - 2\epsilon^{-2} = 0.46508$$

Evaluating at $t = kT - \rho T = 2T - \rho T$ for the second sampling period, $k = 2$,

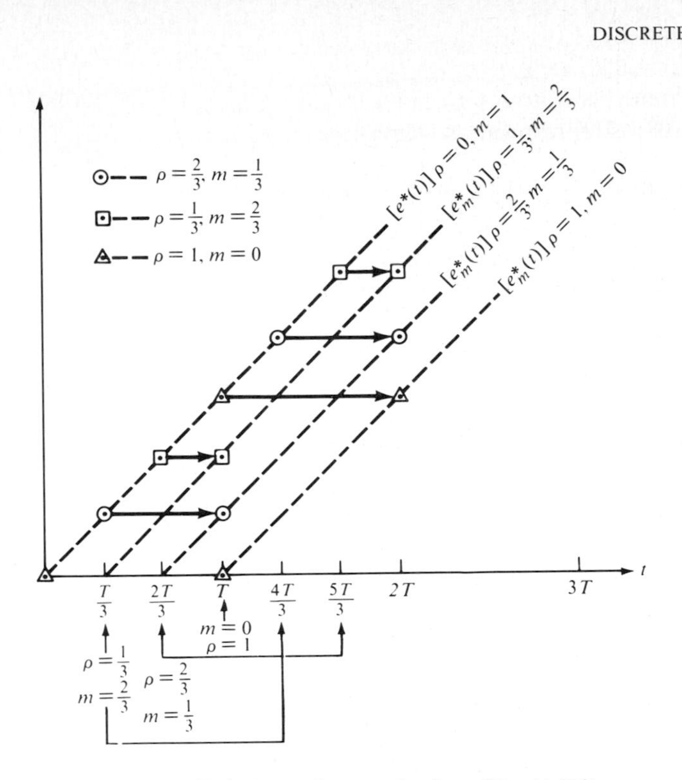

Figure 4-19 Modified $\mathscr{Z}$-transform evaluation of Eq. (4-113).

yields

$$e(T^+) = 0.73576 - 0.27068$$
$$= 0.46508 \qquad \text{for } m = 0$$
$$e\left(\frac{4T}{3}\right) = [0.73576\epsilon^{-m} - 0.27068\epsilon^{-2m}]_{m=1/3}$$
$$= 0.38972$$
$$e\left(\frac{5T}{3}\right) = [0.73576\epsilon^{-m} - 0.27068\epsilon^{-2m}]_{m=2/3}$$
$$= 0.3064$$

An alternate method of evaluating $e[(k + m)T]$ is by use of the following definition,[20] which requires the denominator of $E(z, m)$ to be in factored form:

$$e(kT, m) = \sum [\text{residues of } z^{k-1}E(z, m) \text{ at the poles of } E(z, m)]$$
$$= \mathscr{Z}_m^{-1}[E(z, m)] \qquad \text{(4-114)}$$

The manner of obtaining the residues is the same as for Eq. (4-12). Note that Eq. (4-14) is not applicable to $E(z, m)$.

Example 4-20 Determine $e(kT, m)$ for Eq. (4-108), using Eq. (4-114). To use Eq. (4-114), Eq. (4-108) is rearranged as follows:

$$E(z, m) = 2\frac{(\epsilon^{-mT} - \epsilon^{-2mT})z + (\epsilon^{-2mT}\epsilon^{-T} - \epsilon^{-mT}\epsilon^{-2T})}{(z - \epsilon^{-T})(z - \epsilon^{-2T})}$$

Since $E(z, m)$ has simple poles, the partial-fraction expansion of $z^{k-1}E(z, m)$ is

$$z^{k-1}E(z, m) = \frac{2A_1}{z - \epsilon^{-T}} + \frac{2A_2}{z - \epsilon^{-2T}}$$

where

$$2A_1 = (z - \epsilon^{-T})[z^{k-1}E(z, m)]_{z=\varepsilon^{-T}} = 2\epsilon^{-(k+m-1)T} = 2\epsilon^{-(k-\rho)T}$$

$$2A_2 = (z - \epsilon^{-2T})[z^{k-1}E(z, m)]_{z=\varepsilon^{-2T}} = -2\epsilon^{-2(k+m-1)T} = -2\epsilon^{-2(k-\rho)T}$$

Thus, applying Eq. (4-114) yields, for $k \geq 1$,

$$e(kT, m) = 2A_1 + 2A_2 = 2(\epsilon^{-(k+m-1)T} - \epsilon^{-2(k+m-1)T})$$

or

$$e[(k - \rho)T] = 2(\epsilon^{-(k-\rho)T} - \epsilon^{-2(k-\rho)T}) \tag{4-115}$$

This closed-form solution, as for the inverse $\mathscr{Z}$ transform, yields more accurate values for $e(kT, m) = e[(k - \rho)T]$. That is, the roundoff error does not propagate as it does for the power-series method. It should be noted that the "simple" factored form of Eq. (4-108) permits going to Table 4-1 directly to obtain $e[(k - \rho)T]$; that is, in obtaining $\mathscr{Z}_m^{-1}$, replace k by $k - \rho$ in the left-hand column.

This technique, although useful in reconstruction of a continuous signal, is usually not employed in a digital computer control mechanization owing to the processing time required. However, the technique can be useful in simulation and analysis.

4-9C Modified $\mathscr{Z}$-Transform Theorems

1: Real translation theorem Same as for $\mathscr{Z}$ transform:

$$\mathscr{Z}_m[e^*(t \pm pT)] = z^{+p}E(z, m)$$

2: Initial-value theorem

$$\lim_{t\to 0+} e(t) = \lim_{\substack{m=0\\k\to 1}} e[(k - \delta)T] = \lim_{\substack{m=0\\z\to\infty}} zE(z, m)$$

3: Final-value theorem

$$e(t)^*_{ss} = \lim_{\substack{0\leq m\leq 1\\k\to\infty}} e[(k + m)T] = \lim_{\substack{0\leq m\leq 1\\z\to 1}} \left[\frac{z - 1}{z} E(z, m)\right]$$

4-10 SUMMARY

This chapter presents the definition and evaluation of the one-sided $\mathscr{Z}$ transform. The correlation between the s domain and the z domain with respect to the time-domain characteristics is discussed. $\mathscr{Z}$-transform theorems, the inverse $\mathscr{Z}$ transform, and the limitations of the $\mathscr{Z}$ transform are also discussed in depth. The modeling of a sampled-data system in the z domain is presented first for both open-loop and closed-loop systems. An important aspect of a sampled-data control system is the use of a digital computer to implement a z-domain controller transfer function, i.e., a difference equation. To obtain discrete data in between sampling instants, the modified $\mathscr{Z}$ transform is used. Computer programs are readily available to expedite the analysis of sampled-data system in the z domain (see Appendix E).[14,15]

CHAPTER

FIVE

CONTINUOUS-TIME CONTROL SYSTEM RESPONSE CHARACTERISTICS[1]

5-1 INTRODUCTION

Owing to the correlation between the s and z planes, the highly developed analysis and design techniques for continuous-data control systems form the basis for the analysis and design of sampled-data control systems. Thus a knowledge of the figures of merit of the conventional control-time response for a simple second-order system is required. These figures of merit are used in designing an nth-order system so that its time-response characteristics are essentially those of a simple second-order system. If the analysis of the basic system reveals that the desired specifications cannot be satisfied, then one of two common methods (cascade or feedback compensation) may be used to try to achieve these desired specifications.

The performance of a control system, based upon a step input, generally falls into one of the following four categories.

1. A given system is stable and its transient response is satisfactory, but its steady-state error is too large. Therefore, the gain must be increased to reduce the steady-state error without appreciably reducing the system stability.
2. A given system is stable, but its transient response is unsatisfactory.
3. A given system is stable, but both its transient response and its steady-state response are unsatisfactory.
4. A given system is unstable for all values of gain.

Generally all four categories require the utilization of additional system components to achieve the desired system performance.

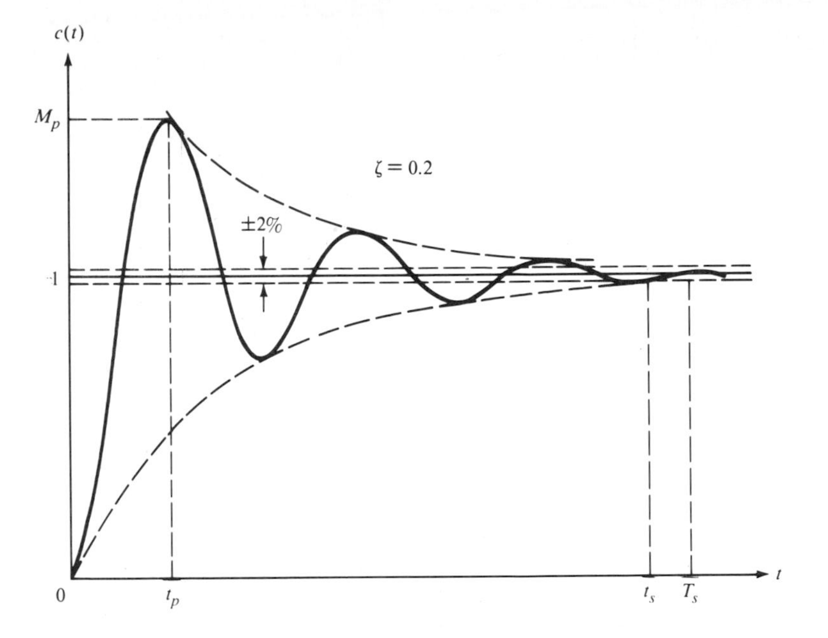

Figure 5-1 Simple second-order system time response.

Note that the part of the control system on which control is exerted is usually called the *plant* or *basic* system. Generally, the basic system is fixed or unalterable except possibly for gain adjustment to improve the system performance. The additional equipment that is inserted into the system, in order to obtain the desired system performance, is referred to as the *controller* or *compensator*. Compensation of a system by the introduction of compensator poles and zeros is thus used to improve the system performance. However, each additional compensator pole increases the number of roots of the closed-loop characteristic equation. If an underdamped response of the form shown in Fig. 5-1 is desired, the system gain can be adjusted so that there is only one pair of dominant complex poles. This requires that any other pole be far to the left or near a zero so that the magnitude of the transient term due to that pole is small and therefore has a negligible effect on the total time response. The effect of compensator poles and zeros on the system's performance [figures of merit (see Sec. 5-2)] can be evaluated rapidly by use of computer-aided-design techniques (see Appendix E).[14,15]

Three common basic types of compensators[1] that are used are shown in Fig. 5-2. They are called *passive compensators* since they consist of only resistors and capacitors in conjunction with an amplifier. The transfer function of the *lag compensator*, shown in Fig. 5-2*a*, is

$$G_c(s) = \frac{A(1 + Ts)}{1 + \alpha Ts} = \frac{A}{\alpha}\frac{s + 1/T}{s + 1/\alpha T} \tag{5-1}$$

where $1 < \alpha < 10$ (nominal for passive networks; for digital implementation it can

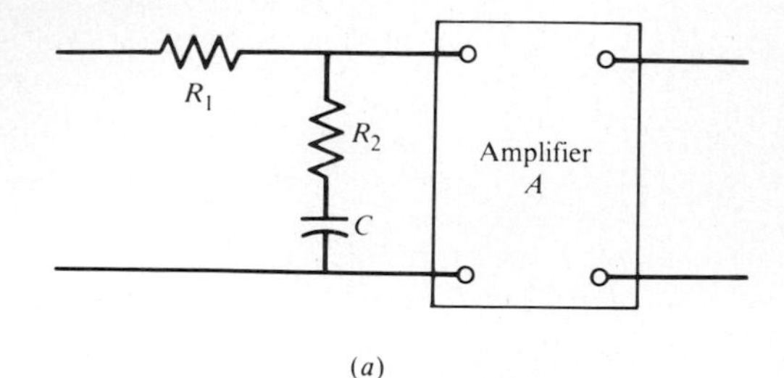

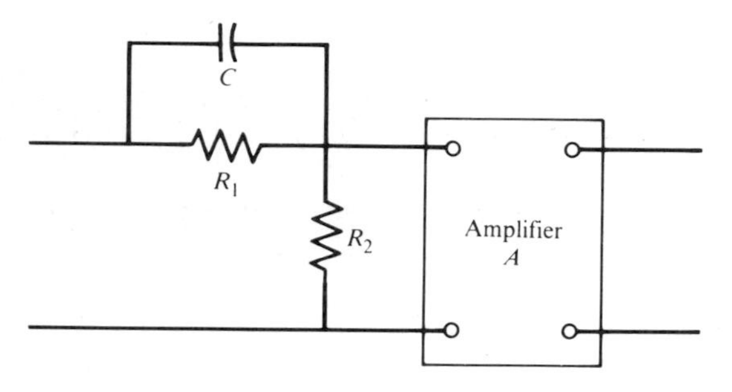

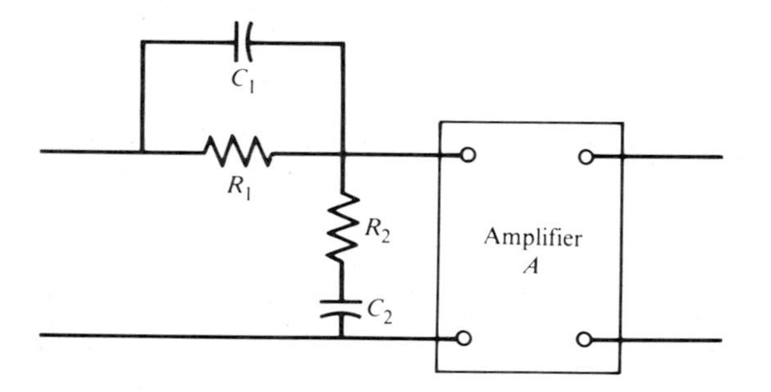

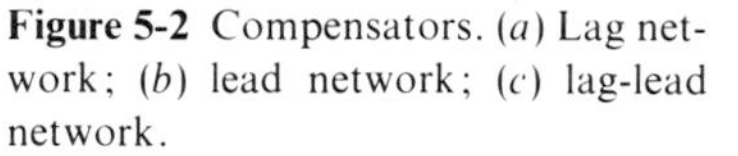
Figure 5-2 Compensators. (*a*) Lag network; (*b*) lead network; (*c*) lag-lead network.

be higher). It is referred to as a *lag* network, since for a sinusoidal input the output phase always lags relative to the input phase. This type of compensator, a low-pass filter, can be used for systems whose performance falls into category 1, where the steady-state error can be reduced approximately by a factor of α. The transfer function of the *lead compensator*, shown in Fig. 5-2*b*, is

$$G_c(s) = \frac{A\alpha(1 + Ts)}{1 + \alpha Ts} = A\,\frac{s + 1/T}{s + 1/\alpha T} \tag{5-2}$$

where $1 > \alpha \geq 0.1$ (nominal for passive networks; for digital implementation it can be lower). It is referred to as a *lead* network, since, for a sinusoidal input the output phase always leads relative to the input phase. The lead network, a high-pass filter, can be used for systems whose performance falls into category 2 or 4.

The improvement in system performance that can be obtained by using each network separately can be achieved by a single network, see Fig. 5-2*c*, which is referred to as a *lag-lead network*. An improvement to system performance for a system that falls into category 3 can be achieved by using a lag-lead network whose transfer function is

$$G_c(s) = A \frac{(1 + T_1 s)(1 + T_2 s)}{(1 + \alpha T_1 s)[1 + (T_2/\alpha)s]}$$

$$= A \frac{(s + 1/T_1)(s + 1/T_2)}{(s + 1/\alpha T_1)(s + \alpha/T_2)} \tag{5-3}$$

where $\alpha > 1$ and $T_1 > T_2$. The fraction $(1 + T_1 s)/(1 + \alpha T_1 s)$ represents the lag compensator, and the fraction $(1 + T_2 s)/[1 + (T_2/\alpha)s]$ represents the lead compensator.

A compensator that is generally used in the process control industry is the *PID controller* of Fig. 5-3. This controller combines the characteristics of the lag and the lead compensators to provide proportional, integral, and derivative control action for improving a system's performance. Its transfer function is

$$G_c(s) = K_p + K_d s + \frac{K_i}{s} \tag{5-4}$$

The analog PID controller of Fig. 5-3 is not of a practical form that can be readily implemented due to the ideal derivative operation. A practical form of this controller is to replace $K_d s$ by a lead network of the form of Eq. (5-2) which is an approximation to the ideal derivative process. Different discrete approximations to Eq. (5-4) are discussed in Chap. 11.

Another procedure for designing a cascade compensator is the Guilleman-Truxal method. This method, which is discussed in Sec. 5-5, is commonly called a *pole-zero* placement technique. In other words, some desired control ratio is synthesized for an *n*th-order system that yields the desired values for the figures of merit.

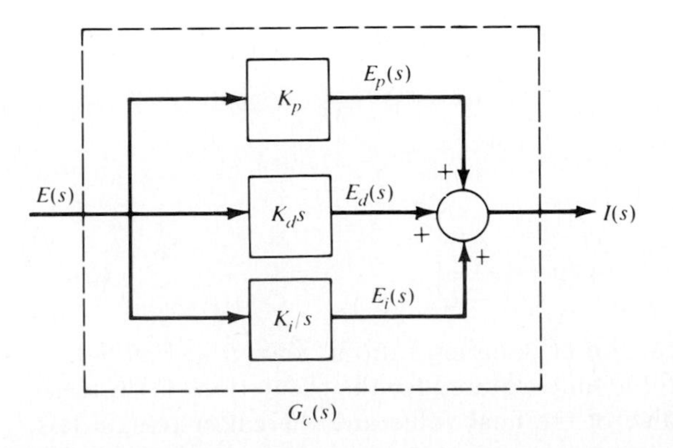

Figure 5-3 A PID controller.

5-2 SIMPLE SECOND-ORDER SYSTEM RESPONSE CHARACTERISTICS

The response to a unit-step-function input is usually used as a means of evaluating the response of a system. As an illustrative example, consider the simple second-order system described by the differential equation

$$\frac{\ddot{c}(t)}{\omega_n^2} + \frac{2\zeta}{\omega_n}\dot{c}(t) + c(t) = r(t) \tag{5-5}$$

This is defined as a *simple* second-order equation because there are no derivatives of $r(t)$ on the right side of the equation. The response to a *unit-step input*, subject to zero initial conditions, is

$$c(t) = 1 - \frac{\epsilon^{-\zeta\omega_n t}}{\sqrt{1-\zeta^2}} \sin[(\omega_n\sqrt{1-\zeta^2})t + \cos^{-1}\zeta] \tag{5-6}$$

where $1 > \zeta > 0$. For the underdamped case, the system oscillates around the final value as shown in Fig. 5-1 for $\zeta = 0.2$. The oscillations decrease with time, and the system response approaches its final value. The *peak overshoot* M_p for the underdamped system is defined as the maximum value of the first overshoot. The time at which the peak overshoot occurs, t_p, can be found by differentiating Eq. (5-6) with respect to time and setting this derivative equal to zero:

$$\dot{c}(t) = \frac{\zeta\omega_n\epsilon^{\zeta\omega_n t}}{\sqrt{1-\zeta^2}} \sin[(\omega_n\sqrt{1-\zeta^2})t + \cos^{-1}\zeta] - \omega_n\epsilon^{-\zeta\omega_n t}\cos[(\omega_n\sqrt{1-\zeta^2})t + \cos^{-1}\zeta] = 0 \tag{5-7}$$

This derivative is zero at $(\omega_n\sqrt{1-\zeta^2})t = 0, \pi, 2\pi, \ldots$. The peak overshoot occurs at the first value after $t = 0$, provided there are zero initial conditions. Therefore the *peak time* is given by

$$t_p = \frac{\pi}{\omega_d} = \frac{\pi}{\omega_n\sqrt{1-\zeta^2}} \tag{5-8}$$

Inserting Eq. (5-8) into Eq. (5-6) gives the *peak overshoot* as

$$M_p = \frac{c_p}{r} = 1 + \exp\frac{-\zeta\pi}{\sqrt{1-\zeta^2}} = 1 + M_o \tag{5-9}$$

where

$$M_o = \frac{c_p - c_{ss}}{c_{ss}} = \exp\frac{-\zeta\pi}{\sqrt{1-\zeta^2}} \tag{5-10}$$

is the per unit overshoot as a function of damping ratio as shown in Fig. 5-4.

The *settling time* is defined as the time required for the oscillations to decrease to within a specified absolute value of the final value and thereafter remain less

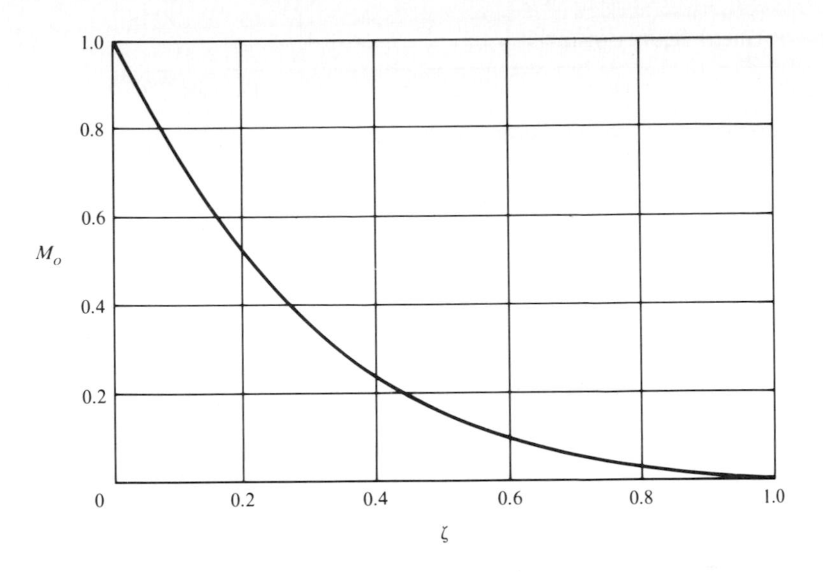

Figure 5-4 M_σ vs. ζ for a simple second-order equation.

than this magnitude. For second-order systems the value of the transient at any time is equal to or less than the exponential $\epsilon^{-\zeta\omega_n t}$. A 2 percent error criterion (error = $r - c$) is often used to determine settling time. Based on the envelope of the transient, the 2 percent criterion yields a settling time T_s, which is approximately given by

$$T_s = \frac{4}{\zeta\omega_n} \tag{5-11}$$

The actual value of the settling time for $c(t)$ is denoted by t_s (see Fig. 5-1).

The simple second-order system response characteristics to a unit-step input are summarized in Table 5-1. These characteristics (M_p, t_p, T_s, and t_s) are referred to as the *conventional control system figures of merit*. The unit-step function is often used as the standard input for which a system is designed to satisfy prescribed values of the figures of merit.

Another important figure of merit is the *static error coefficient*, which is defined only for a stable *unity-feedback system* whose forward transfer function is in the following mathematical form:

$$\begin{aligned} G(s) &= \frac{K_m(1 + b_1 s + b_2 s^2 + \cdots + b_w s^w)}{s^m(1 + a_1 s + a_2 s^2 + \cdots + a_v s^v)} \\ &= \frac{K_m(1 + T_1 s)(1 + T_2 s)\cdots}{s^m(1 + T_a s)(1 + T_b s)\cdots} \end{aligned} \tag{5-12}$$

where $n = m + v$ and $m = -1, 0, 1, 2, \ldots$. A Type -1 system is a velocity control system where the load characteristics on the motor's shaft are described by J, B, and K (inertia, viscous damping, and elastance).[1,21]

Table 5-1 Conventional figures of merit

Symbol	Name	Description	Formula
M_p	Peak overshoot	The amplitude of the system response the first time it overshoots the final value	$M_p = 1 + \exp\dfrac{-\zeta\pi}{\sqrt{1-\zeta^2}}$
t_p	Peak time	The time after $t = 0$ when the peak overshoot is reached	$t_p = \dfrac{\pi}{\omega_n\sqrt{1-\zeta^2}}$
t_s	Settling peak	The time required for the response to reach and remain within some specified percentage of the final value	t_s of $y(t)$
T_s			$T_s = \dfrac{4}{\zeta\omega_n}$ of envelope of $c(t)$ (for $\pm 2\%$ of final value)
N (not shown in figure)	Number of oscillations up to settling time	The number of oscillations that the system output undergoes between $t = 0$ and $t = T_s$	$N = \dfrac{\sqrt{1-\zeta^2}}{\pi\zeta}$ (for $\pm 2\%$ of final value)

When the constant term in each numerator and denominator factor of $G(s)$ is unity, as in Eq. (5-12), the gain constant term K_m appearing in the numerator of $G(s)$ is called the *static error coefficient*. Note that when $m = 0$ then $G(s)$ is referred to as a *Type 0 system*, when $m = 1$ it is referred to as a *Type 1 system*, etc. Therefore, for systems which are Type 0, 1 and 2, the gain constant is referred to as the step (position), ramp (velocity), and parabolic (acceleration) error coefficient, respectively. *The derivations of the error coefficients are independent of the system type. They apply to any system type and are defined for specific forms of the input, i.e., for a step, ramp, or parabolic input. These error coefficients are useful only for stable unity-feedback systems.* The results are summarized in Tables 5-2 and 5-3. The reader is urged to refer to some basic conventional control-theory text for a more detailed review of system types and error coefficients.[1]

Table 5-2 Definitions of steady-state error coefficients for stable unity-feedback systems

Error coefficient	Definition of error coefficient	Value of error coefficient	Form of input signal $r(t)$
Step	$\dfrac{c(t)_{ss}}{e(t)_{ss}}$	$\lim_{s\to 0} G(s)$	$R_0 u_{-1}(t)$
Ramp	$\dfrac{\dot{c}(t)_{ss}}{e(t)_{ss}}$	$\lim_{s\to 0} sG(s)$	$R_1 t u_{-1}(t)$
Parabolic	$\dfrac{\ddot{c}(t)_{ss}}{e(t)_{ss}}$	$\lim_{s\to 0} s^2 G(s)$	$\dfrac{R_2 t^2}{2} u_{-1}(t)$

Table 5-3 Steady-state error coefficients for stable systems

Systems type	Step error coefficient K_p	Ramp error coefficient K_v	Parabolic error coefficient K_a
0	K_0	0	0
1	∞	K_1	0
2	∞	∞	K_2

5-3 HIGHER-ORDER SYSTEM RESPONSE CHARACTERISTICS

The last section analyzed the time response of a simple second-order system. Its response is characterized by the figures of merit which are given by Eqs. (5-8) to (5-11). Since most systems are of higher-order, it is now appropriate to analyze their response characteristics. Before so doing, it is necessary to stipulate that:

1. An underdamped response is desired.
2. It is desired that an *n*th-order system act "effectively" like a second-order system.
3. The complex-conjugate poles of $C(s)/R(s)$ closest to the imaginary axis of the s plane shall be defined as the *dominant poles* of the control ratio.

5-3A Time-Response Characteristics of a Third-Order All-Pole Plant

First consider a third-order closed-loop control system whose control ratio is given by

$$\frac{C(s)}{R(s)} = \frac{K}{(s - p_1)(s - p_2)(s - p_3)} \tag{5-13}$$

where $K > 0$ and the poles are situated in the s plane as shown in Fig. 5-5*a*. [Figure 5-5*b* shows the corresponding locations of the poles of Eq. (5-13) in the z plane, i.e., the mapping of p_1, p_2, and p_3 into the z plane.] The poles (or roots of the system's characteristic equation) are represented by $p_{1,2} = \sigma_D \pm j\omega_D$ and $p_3 = \sigma_3$, where the roots $p_{1,2}$ are the dominant roots. (These roots lie closest to the imaginary axes.) For a step input,

$$c(t) = A_0 + A_D\epsilon^{\sigma_D t}(\sin \omega_D t + \phi) + A_3\epsilon^{\sigma_3 t} \tag{5-14}$$

As an approximate rule, if $|\sigma_3| > 6|\sigma_D|$ then $p_{1,2}$ are said to be *truly dominant* roots since p_3 has little effect on $c(t)$. For example, $A_3\epsilon^{\sigma_3 t}$ dies out very fast as compared to $A_0\epsilon^{\sigma_D t}(\sin \omega_D t + \phi)$. If $|\sigma_3| < 6|\sigma_D|$, then all roots are said to be dominant. An exception to this rule occurs when a root(s), which by definition falls into the dominant category, lies very close to a zero(s) of $C(s)$. If this situation occurs, this (these) root(s) is (are) said to be nondominant. For the case where $p_{1,2}$ are truly dominant, the time response can be approximated by

$$c(t) \approx A_0 + A_D\epsilon^{\sigma_D t} \sin(\omega_D t + \phi) \tag{5-15}$$

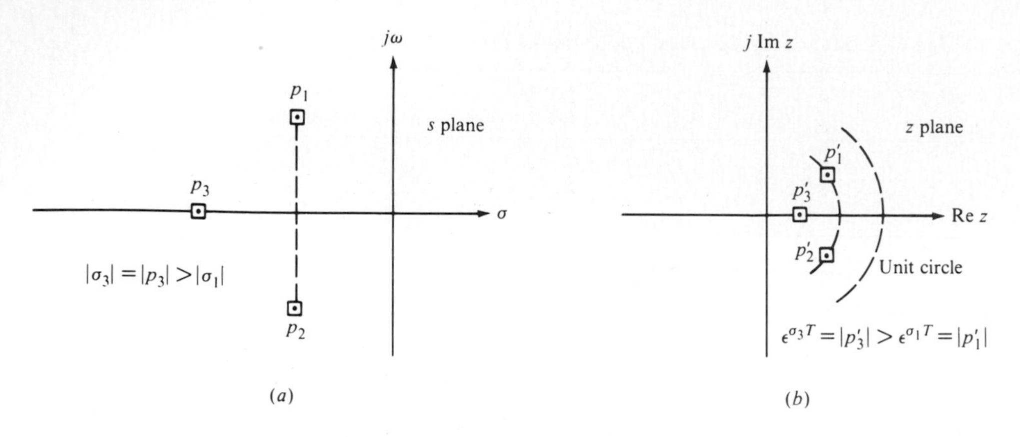

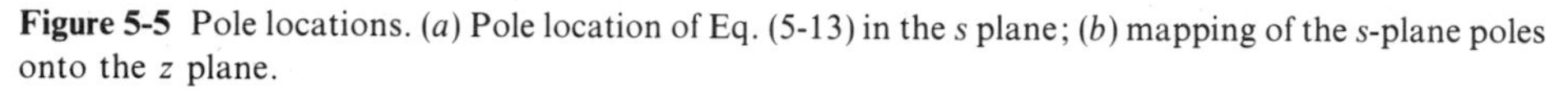

Figure 5-5 Pole locations. (*a*) Pole location of Eq. (5-13) in the *s* plane; (*b*) mapping of the *s*-plane poles onto the *z* plane.

5-3B Time-Response Characteristics of a Sixth-Order Plant

Next, consider a sixth-order closed-loop control system whose control ratio is

$$\frac{C(s)}{R(s)} = \frac{K(s - z_1)}{(s - p_1)(s - p_2)(s - p_3)(s - p_4)(s - p_5)(s - p_6)} \tag{5-16}$$

where $K > 0$. The transform of the output, $C(s)$, for a unit-step input is

$$C(s) = \frac{1}{s}\,\frac{K(s - z_1)}{(s - p_1)(s - p_2)(s - p_3)(s - p_4)(s - p_5)(s - p_6)} \tag{5-17}$$

which by means of the partial-fraction expansion method is put into the form of

$$C(s) = \frac{A_0}{s} + \frac{A_1}{s - p_1} + \cdots + \frac{A_6}{s - p_6} \tag{5-18}$$

The inverse Laplace transform of $C(s)$ is

$$c(t) = A_0 + A_1\epsilon^{p_1 t} + \cdots + A_6\epsilon^{p_6 t} \tag{5-19}$$

The poles (input function and control ratio poles) of Eq. (5-17) are plotted in Fig. 5-6*a*. Since Eq. (5-17) has all simple poles (first-order poles), then the coefficients $A_1, \ldots, A_6$ of Eq. (5-18) can be determined from Fig. 5-6*a* by using the equation

$$A_j = K\,\frac{\text{product of directed distances from each zero of } C(s) \text{ to the pole } p_j}{\text{product of directed distances from all other poles to the pole } p_j} \tag{5-20}$$

Thus Eq. (5-20) shows that the closer a pole p_j comes to a zero, the smaller becomes the value of A_j. For the configuration shown in Fig. 5-6*a* where p_4, p_5, and p_6 are

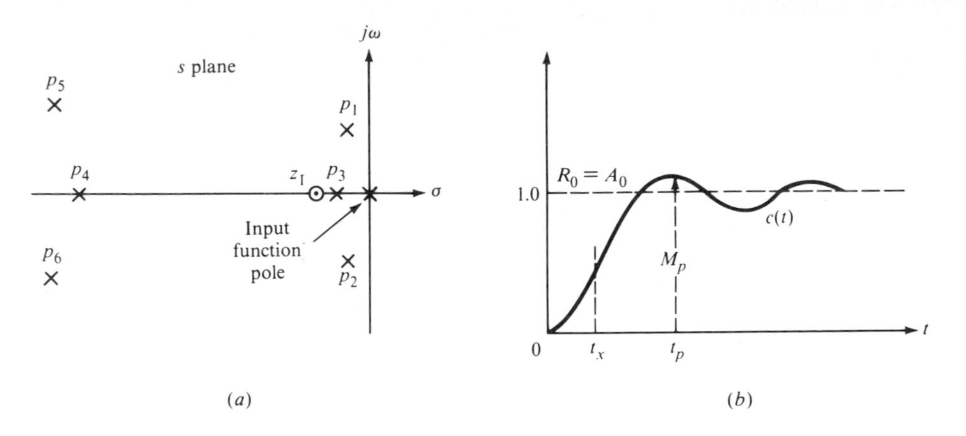

Figure 5-6 Pole and zero locations of Eq. (5-17) and the corresponding time response.

nondominant poles, then A_4, A_5, and A_6 are negligible, compared to A_1 and A_2. If p_3 lies close enough to z_1 so that A_3 is also negligible, then Eq. (5-19) can be rewritten as

$$c(t) \approx A_0 + A_D \epsilon^{\sigma_D t} \sin(\omega_d t + \phi) \tag{5-21}$$

The time response of Eq. (5-21) is shown in Fig. 5-6*b*.

The approximate equations (5-15) and (5-21) have the mathematical format of the output response of a simple second-order system given by Eq. (5-6). Thus, under the condition of a truly complex-conjugate dominant pair of roots, an n- (>3) order system is said to act like a simple second-order system, or, is said to be an *effective simple second-order system*. Under this condition, fairly accurate values of M_p, t_p, and T_s, using Eqs. (5-8) to (5-11), may be obtained for the output responses of the respective systems. These values are determined by using the values of ζ and ω_n obtained from σ_D and ω_D.

5-3C Correlation between Frequency and Time Domain

For a simple *second-order system*, the exact form of the frequency response, $M(\omega)$ vs. ω, and $\alpha(\omega)$ vs. ω, depends upon the system damping ratio in the same way that the time response does, where

$$M(\omega) = \left|\frac{\mathbf{C}(j\omega)}{\mathbf{R}(j\omega)}\right| = \left|\frac{\mathbf{G}(j\omega)}{1 + \mathbf{G}(j\omega)}\right| \quad \text{and} \quad \alpha(\omega) = \angle\frac{\mathbf{C}(j\omega)}{\mathbf{R}(j\omega)} \tag{5-22}$$

The frequency-response curves of a typical second-order system would look like those in Fig. 5-7, which are drawn for a damping ratio of about 0.4. Note that in this case $M(\omega)$ has a peak which is greater than unity. This peak occurs at a frequency of ω_m. $M(\omega)$ always has a peak value greater than unity if the damping

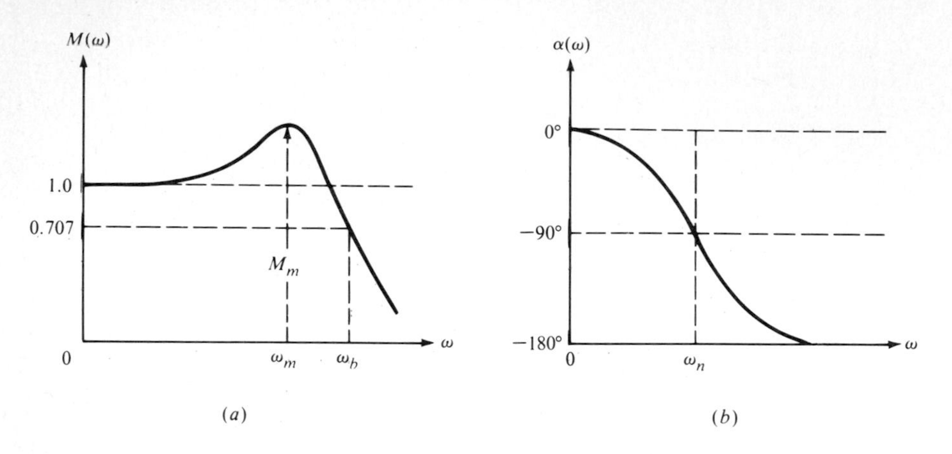

Figure 5-7 Frequency response of a simple second-order system for $\zeta \approx 0.4$.

ratio is less than 0.707. It can be shown that[1]

$$\omega_m = \omega_n\sqrt{1 - 2\zeta^2} \tag{5-23}$$

and

$$M_m = \frac{1}{2\zeta\sqrt{1 - \zeta^2}} \tag{5-24}$$

Recall that the damped natural frequency of a simple second-order system subject to a step input is given by

$$\omega_d = \omega_n\sqrt{1 - \zeta^2} \tag{5-25}$$

Also, recall that if the steady-state response to a unit-step input is unity, the value of the peak overshoot M_p of a simple second-order underdamped system is given by

$$M_p = 1 + \exp\frac{-\zeta\pi}{\sqrt{1 - \zeta^2}} \tag{5-26}$$

If Eqs. (5-25) and (5-26), which give information about the time-domain response, are compared with Eqs. (5-23) and (5.24), the following conclusions can be drawn concerning the correlation between the time and frequency responses:

1. For a given value of ζ, the higher the value of ω_n the faster the system responds: Eq. (5-23) demonstrates that a larger ω_m also means a faster responding system.
2. Both M_m and M_p are functions of ζ only, and the smaller ζ becomes the larger become M_m and M_p. There is a close correspondence between the values of M_p and M_m for $\zeta > 0.4$ (e.g., for $\zeta = 0.6$, $M_p = 1.09$ and $M_m = 1.04$). For $\zeta < 0.4$, the correspondence is qualitative only (e.g., if $\zeta = 0$, $M_p = 2$ but $M_m = \infty$). The acceptable range of M_m is usually $1.0 < M_m < 1.4$.
3. By examining the Nyquist plot of $\mathbf{G}(j\omega)$, it can be seen that the closer this plot

comes to the point $-1 + j0$, the larger M_m becomes and thus the larger M_p becomes. Since it is usually desirable to have the system output follow the system input as closely as possible, it may seem that it would be very desirable for $M(\omega)$ to equal unity at all frequencies from zero to infinity. Not only is this impossible, but it is not even desirable from a practical standpoint. In almost all cases, the system input consists of not only the reference input, which the system must follow, but also the extraneous disturbance inputs.

Moreover, disturbances may enter the system at points other than the input. Examples of such disturbances include the vibration of an electric motor or noise in electronic power supplies. Often these disturbances exist in a frequency band which is higher than the frequency range of the input signals. In such a case it makes good sense to design the system so that all frequencies above some minimum are severely attenuated.

The *bandwidth* of a control system is defined as the frequency band between $\omega = 0$ and some maximum frequency ω_b. It gives a precise indication of the range of frequencies to which the system responds. Of course, the system still exhibits some response to frequencies above the maximum frequency in the band. Since control systems do not generally have what could be described as "sharp" cutoff characteristics, the frequency that is considered to be the upper limit of the operating band may not be the same in all texts. In most cases, ω_b, *the half-power point* shown in Fig. 5-7, is often considered as the upper-frequency bandwidth because it represents a good general value where large-frequency attenuation begins.

Following the previous discussion, it may seem logical to design the system so that for a given ζ, ω_m is as low as possible so that the entire range of input frequencies can still be accommodated. This is only partially true. While setting ω_m as low as possible does result in the system rejecting the maximum amount of disturbing inputs, it also results in a lower value of undamped natural frequency ω_n, and a slower responding system which may be undesirable. Therefore, some compromise between noise rejection and speed of response may be necessary in the selection of ω_m.

5-3D Correlation of the Poles and Zeros of the Control Ratio with Frequency and Time Responses

Whenever the closed-loop control ratio $\mathbf{M}(j\omega) = \mathbf{C}(j\omega)/R(j\omega)$ has the characteristic form shown in Fig. 5-7, the system may be approximated as a second-order system. This usually implies that the poles, other than the dominant complex pair, are either far to the left of the dominant complex poles or are close to zeros. When these conditions are not satisfied, the frequency response may have other shapes. This is illustrated by considering the following three control ratios:

$$\frac{C(s)}{R(s)} = \frac{1}{s^2 + s + 1} \tag{5-27}$$

$$\frac{C(s)}{R(s)} = \frac{0.313(s + 0.8)}{(s + 0.25)(s^2 + 0.3s + 1)} \tag{5-28}$$

$$\frac{C(s)}{R(s)} = \frac{4}{(s^2 + s + 1)(s^2 + 0.4s + 4)} \tag{5-29}$$

The pole-zero diagram, the frequency response, and the time response to a step input for each of these equations are shown in Fig. 5-8. From Fig. 5-8*a*, which represents Eq. (5-27), the following characteristics are noted:

1. The control ratio has only two complex poles, which are dominant, and no zeros.
2. The frequency-response curve has the following characteristics:
   a. A single peak, $M_m = 1.157$
   b. $1.0 < M < M_m$ in the frequency range $0 < \omega < 1.0$

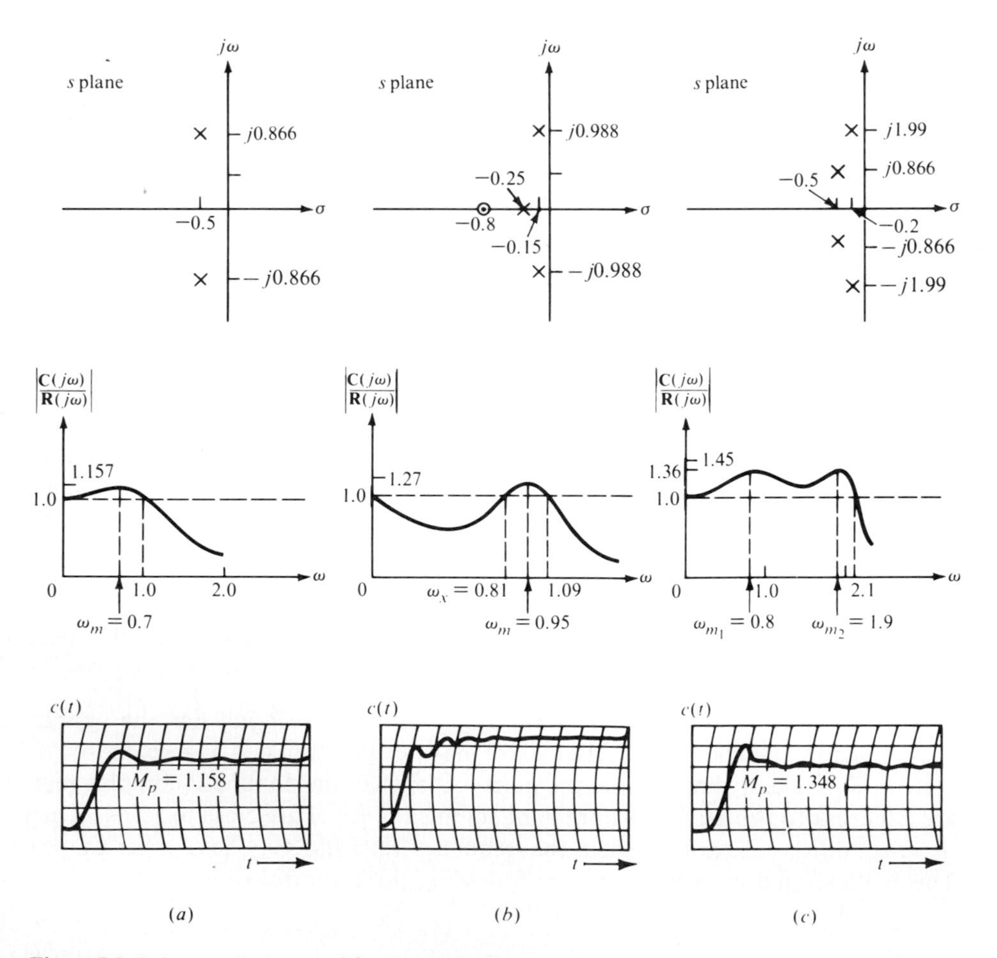

Figure 5-8 Pole-zero diagram and frequency and time responses.

3. The time response has the typical second-order underdamped response. The first maximum of $c(t)$ due to the oscillatory term is greater than $c(t)_{ss}$. The $c(t)$ response after this maximum oscillates around the value of $c(t)_{ss}$.

From Fig. 5-8*b*, for Eq. (5-28), the following characteristics are noted:

1. The control ratio has two complex poles and one real pole, all dominant, and one real zero.
2. The frequency-response curve has the following characteristics:
   a. A single peak, $M_m = 1.27$.
   b. $M < 1.0$ in the frequency range $0 < \omega < \omega_x$.
   c. The peak M_m occurs at $\omega_m = 0.95 > \omega_x$.
3. The time response does not have the conventional waveform. The first maximum of $c(t)$ due to the oscillatory term is less than $c(t)_{ss}$.

From Fig. 5-8*c*, for Eq. (5-29), the following characteristics are noted:

1. The control ratio has four complex poles, all dominant, and no zeros.
2. The frequency-response curve has the following characteristics:
   a. There are two peaks, $M_{m1} = 1.36$ and $M_{m2} = 1.45$.
   b. $1.0 < M < 1.45$ in the frequency range of $0 < \omega < 2.1$.
3. The time response does not have the simple second-order waveform. The first minimum of $c(t)$ for $t > t_p$ in the oscillation is greater than $c(t)_{ss}$. $c(t)$ does not oscillate about a value of $c(t)_{ss}$.

Another example is the system represented by

$$M(s) = \frac{K}{(s^2 + 2\zeta\omega_n s + \omega_n^2)(s - p_3)}$$

$$= \frac{K}{(s + \sigma_D + j\omega_D)(s + \sigma_D - j\omega_D)(s - p_3)} \tag{5-30}$$

where the real pole p_3 and the real parts of the complex poles are equal (see Fig. 5-9*c*). The frequency response is shown in Fig. 5-10*a*. The magnitude at ω_m is less than unity. The corresponding time response to a step input is monotonic; i.e., there is no overshoot. This may be considered as a critically damped response. Other examples are shown in Fig. 5-9. The frequency response corresponding to Fig. 5-9*d* is shown in Fig. 5-10*b*.

The examples discussed in this section show that the time-response waveform is closely related to the system's frequency response. In other words, the system's time-response waveform may be predicted from the shape of the frequency-response plot. Thus, as illustrated in this section, the frequency-response plot may be utilized as a guide in determining (or predicting) time-response characteristics.

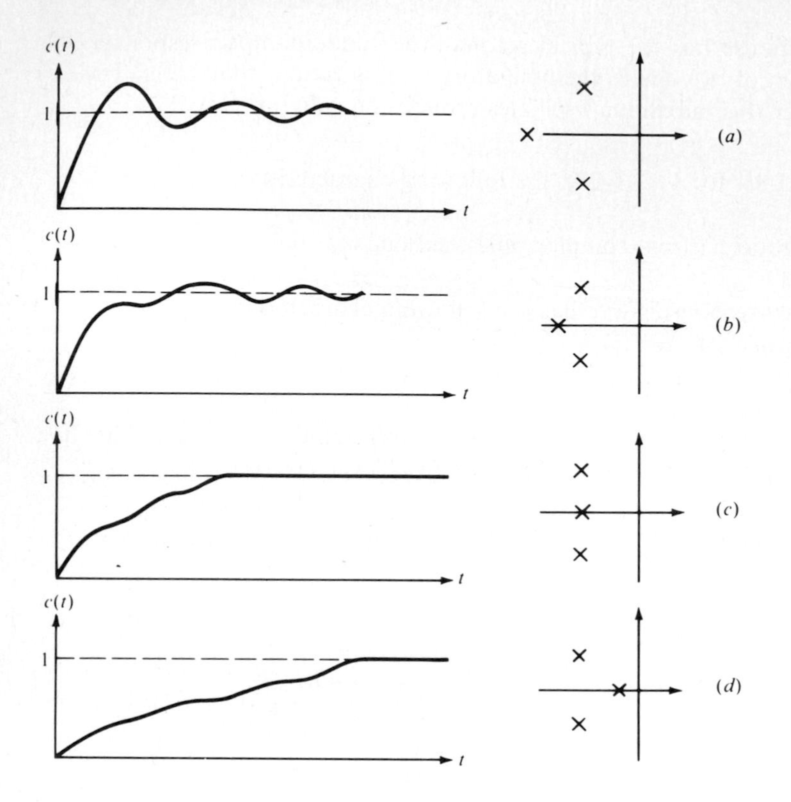

Figure 5-9 Pole-zero diagram and corresponding time responses.

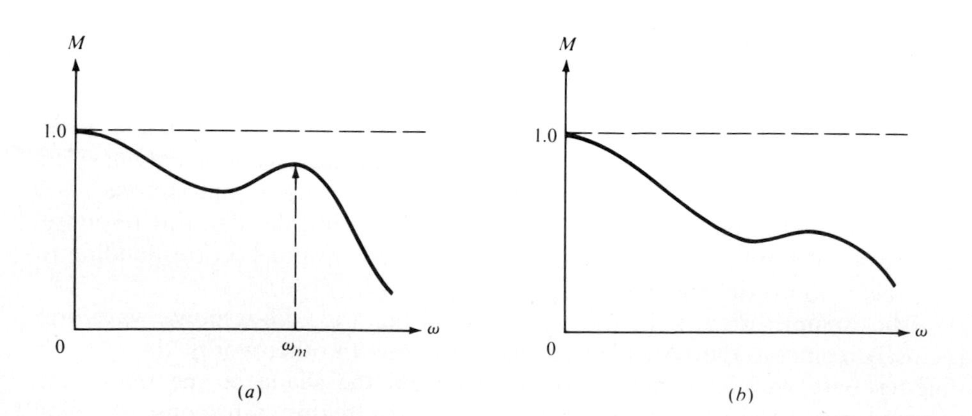

Figure 5-10 Frequency responses.

5-3E Effect of a Third Real Dominant Root

The time response to a unit-step input of a control system, whose control ratio is given by Eq. (5-30), is

$$c(t) = A_0 + A_D\epsilon^{\sigma_D t}(\sin \omega_D t + \phi) + A_3\epsilon^{\sigma_3 t} \tag{5-31}$$

For an effective second-order system the desired pole pattern of this third-order system is shown in Fig. 5-9*a*. The mode due to the real pole p_3 has the form $A_3\epsilon^{\sigma_3 t}$, where A_3 is always negative [see Eq. (5-20)]. When p_3 is also dominant the overshoot M_p is reduced, and settling time t_s may be either increased or decreased. This effect is typical of adding another dominant real pole. The magnitude A_3 depends on the location of p_3 relative to the complex poles. The farther to the left the pole p_3 is located, the smaller is the magnitude of A_3. Hence its effect on the total response is smaller.

The presence of a real zero in addition to the real pole when the pole is not "sufficiently" close to the zero further modifies the transient response. The time response to a unit-step input of a control ratio that has the pole-zero pattern shown in Fig. 5-11*a* still has the form given in Eq. (5-31). (The corresponding pole-zero pattern in the z plane is shown in Fig. 5-11*b*.) However, the sign of A_3 depends on the relative locations of the real pole and the real zero. An analysis of Eq. (5-20) reveals that (1) if the zero is to the left of p_3, A_3 is negative, (2) if the zero is to the right of p_3, A_3 is positive, and (3) the magnitude of A_3 is proportional to the distance from p_3 to z_1. Compared with the response for Eq. (5-30), shown in Fig. 5-9*a*, when p_3 is in the dominant region, then:

1. If z_1 is to the left of the real pole p_3, the response is qualitatively the same as that for a system with only complex poles. However, the peak overshoot is smaller.

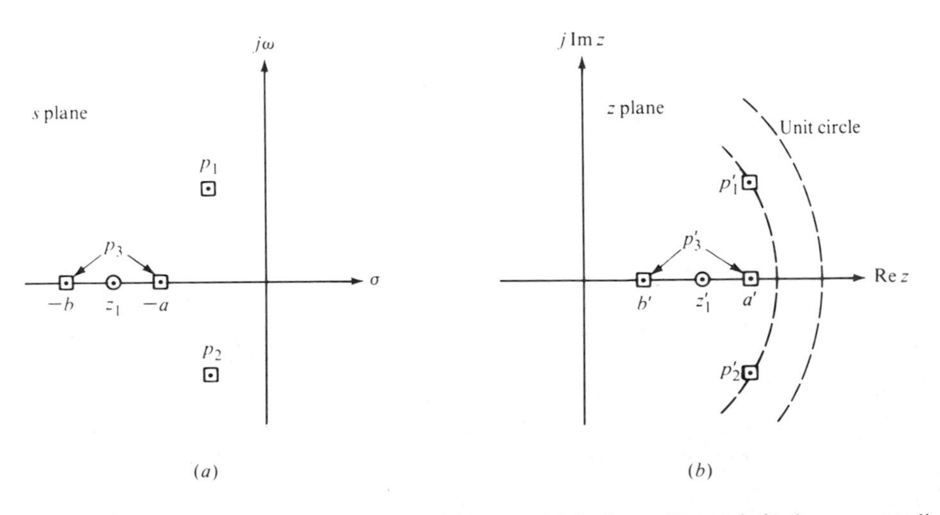

Figure 5-11 Pole-zero pattern for a third-order system (*a*) in the *s* plane and (*b*) the corresponding pattern in the *z* plane.

2. If z_1 is to the right of the real pole p_3, the peak overshoot is greater than that for a system with only complex poles.

Many control systems (for $n > 2$) can be approximated by one having the following characteristics: (1) two complex poles, (2) two complex poles and one real pole, and (3) two complex poles, one real pole, and one real zero. For case 1, Eqs. (5-8) to (5-11) yield accurate values for M_p, T_p, and T_s, and consequently they accurately represent the time response. For cases 2 and 3 these equations yield "approximate" values for the figures of merit.

5-4 CASCADE-COMPENSATOR DESIGN PROCEDURES

This section presents concise design procedures for the three basic compensator types of Fig. 5-2. The general forms of the cascade compensators are shown in Fig. 5-12*a*. The reader is referred to a basic control-theory text[1] for a more detailed discussion of these procedures.

Lag compensator The following procedure is used to design the passive lag cascade compensator. First, the pole $p_c = -1/\alpha T$ and the zero $z_c = -1/T$ of the compensator, Eq. (5-1), are placed very close together. (See Fig. 5-13*a*.) This means that most of the original root locus remains practically unchanged. If the angle contributed by the compensator *at the original closed-loop dominant root* p_1 is less than 5°, the new locus is displaced only slightly. This 5° figure is only a guide and should not be applied arbitrarily. The new closed-loop pole p_1' is, therefore,

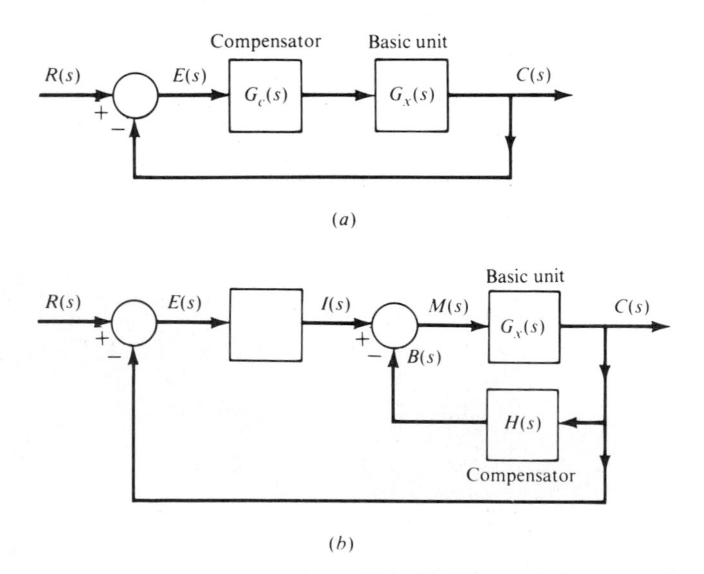

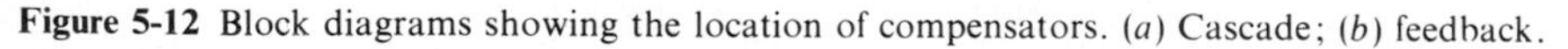
Figure 5-12 Block diagrams showing the location of compensators. (*a*) Cascade; (*b*) feedback.

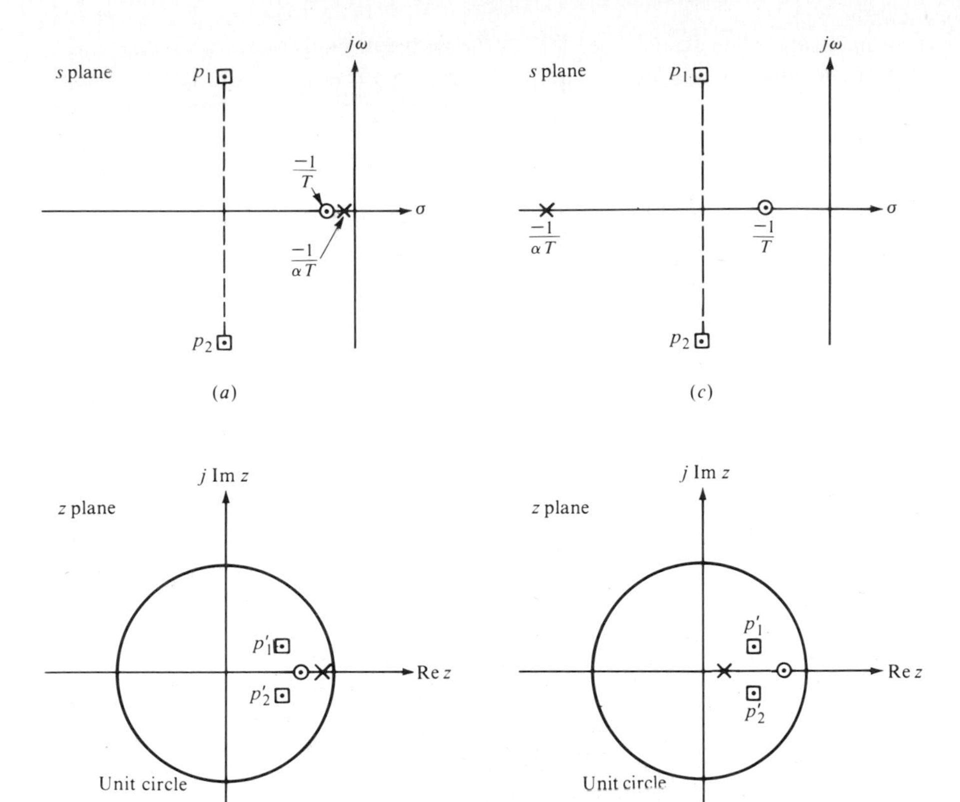

Figure 5-13 Pole-zero location for (*a*) a lag compensator in the *s* plane, (*b*) their corresponding locations in the *z* plane, (*c*) a lead compensator in the *s* plane, and (*d*) their corresponding locations in the *z* plane.

essentially unchanged from the uncompensated value p_1. This satisfies the restriction that the transient response must not change appreciably. As a result, the values $p'_1 + 1/\alpha T$ and $p'_1 + 1/T$ are almost equal, and the values of the static loop sensitivity before and after the compensator is inserted are approximately equal. Also, the values of the static error coefficient (Refs. 1, 21) K_m before and after now differ only by a factor α so that $K'_m \approx \alpha K_m$. The gain required to produce the new root p'_1 therefore increases approximately by the factor α, which is the ratio of the compensator zero and pole. Summarizing, the necessary conditions on the compensator are that (1) the pole and zero must be close together and (2) the zero-pole ratio α must approximately equal the desired increase in gain. These requirements can be achieved by placing the compensator pole and zero very close to the origin.

Lead compensator The design concept for the lead compensator, Eq. (5-2), is based on making α sufficiently small so that its pole $-1/\alpha T$ is far to the left and has a small effect on the important part of the root locus. Near the compensator zero $-1/T$, the net angle of the compensator is due predominantly to the zero. (See Fig. 5-13*c*.) The best location of the zero must be determined by trial and error. However, it is often found that the gain of the compensated system is increased. The maximum increases in gain and in the real part ($\zeta\omega_n$) of the dominant root of the characteristic equation do not coincide. The compensator zero location must then be determined for the desired optimum performance. It can be shown that the loop sensitivity is proportional to the ratio of $|s + 1/\alpha T|$ to $|s + 1/T|$. Therefore, as α decreases, the loop sensitivity increases. The minimum value of α is limited by the size of the parameters needed in the network to obtain the minimum input impedance required. Note also from Eq. (5-2) that a small α required a large value of additional gain A from the amplifier.

The following guidelines are used to apply cascade lead compensators to a Type 1 or higher system:

1. If the zero $z_c = -1/T$ of the lead compensator is superimposed and cancels the largest real pole (excluding the pole at zero) of the original transfer function, a good improvement in the transient response is obtained.
2. If a larger gain is desired than that obtained by guideline 1, several trials should be made with the zero of the compensator moved to the left or right. The location of the zero that results in the desired gain and roots is selected.

For a Type 0 system it is often found that a better time response and a larger gain can be obtained by placing the compensator zero so that it cancels or is close to the second largest real pole of the original transfer function.

An additional method for the design of an appropriate lead compensator G_c placed in cascade with a basic transfer function G_x, as shown in Fig. 5-12*a*, is illustrated in Fig. 5-14. This figure shows the original root locus of a control system. For a specified damping ratio ζ the dominant root of the uncompensated system is p_1. Also shown is p_2, which is the desired root of the system's characteristic equation. Selection of p_2 as a desired root is based on the performance required for the system. The design problem is to select a lead compensator that results in p_2 being a root. The first step is to find the sum of the angles at the point p_2 due to the poles and zeros of the original system. This angle is $180° + \phi$. For p_2 to be on the new root locus, it is necessary for the lead compensator to contribute an angle $\phi_c = -\phi$ at this point. The total angle p_2 is then 180°, and it is a point on the new root locus. A simple lead compensator represented by Eq. (5-2), with its zero to the right of its pole, can be used to provide the angle ϕ_c at the point p_2.

Actually, there are many possible locations of the compensator pole and zero that will produce the necessary angle ϕ_c at the point p_2. Cancellation of poles of the original open-loop transfer function by means of compensator zeros may simplify the root locus and thereby reduce the complexity of the problem. The compensator

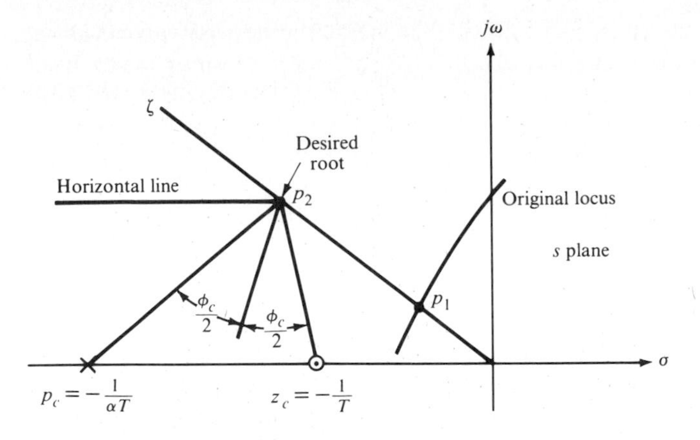

Figure 5-14 Graphical construction for locating the pole and zero of a simple lead compensator.

zero is simply placed over a real pole. Then the compensator pole is placed farther to the left at a location which makes p_2 a point on the new root locus. The pole to be canceled depends on the system type. For a Type 1 system the largest real pole (excluding the pole at zero) should be canceled. For a Type 0 system the second largest pole should be canceled.

The changes introduced by the use of the cascade compensators can be summarized as follows:

1. Lag compensator:
   a. Results in a large increase in gain K_m (by a factor almost equal to α), which means a much smaller steady-state error.
   b. Decreases ω_n and therefore has the disadvantage of producing a small increase in the settling time.
2. Lead compensator:
   a. Results in a moderate increase in gain K_m, thereby improving steady-state accuracy.
   b. Results in a large increase in ω_n and therefore reduces the settling time considerably.
   c. The transfer function of the lead compensator, using the passive network of Fig. 5-2*b*, contains the gain α, which is less than unity. Therefore the additional gain A, which must be added, is larger than the increase in K_m for the system.

Figure 5-13*a* and *c* illustrates the typical locations of the compensators' pole and zero in the *s* plane with respect to the dominant closed-loop poles of the uncompensated system. Also, Fig. 5-13*b* and *d* shows the relative mapping of the *s*-plane compensator pole and zero onto the *z* plane, respectively.

The Guilleman-Truxal method is another approach for designing G_c. It is commonly called a *pole-zero* placement technique. That is, some desired control ratio is synthesized for an nth-order system that yields the desired values for the figures of merit. The next section is devoted to this method since many basic textbooks do not discuss it extensively. The reader is referred to the many available texts[1,21] for details on other cascade- and feedback-compensator methods (see Fig. 5-12b and Secs. 5-6 and 11-6).

5-5 SYNTHESIZING A DESIRED CONTROL RATIO FOR A CONTINUOUS-DATA CONTROL SYSTEM WITH A UNIT-STEP INPUT

A good approach toward understanding the concepts involved in synthesizing a desired control ratio for any given plant $G_x(s)$ is by means of two examples. The first example illustrates the technique of trying to achieve the desired specifications with only the basic system, that is, of trying to achieve the desired values of the figures of merit without the use of additional hardware. The second example uses the Guilleman-Truxal method of trying to achieve the desired performance specifications when they cannot be achieved by the basic system only. Both examples involve the pole-zero placement technique.

Example 5-1 Consider a unity-feedback system where

$$G_x(s) = \frac{K_x}{s(s^2 + 4.2s + 14.4)}$$

$$= \frac{K_x}{s(s + 2.1 + j3.1607)(s + 2.1 - j3.1607)} \tag{5-32}$$

and $K_1 = K_x/14.4$. For this system it is specified that the following figures of merit be satisfied:

$$1 < M_p \leq 1.123 \qquad t_s \leq 3 \text{ s} \qquad e(t)_{ss} = 0$$

$$t_p \leq 1.6 \text{ s} \qquad K_1 \geq 1.5 \text{ s}^{-1}$$

These values are used in Eqs. (5-5) to (5-9) to determine the effective second-order system model. These equations yield the following information:

$$M_p = 1.123 = 1 + \exp \frac{-\zeta\pi}{\sqrt{1 - \zeta^2}} \rightarrow \zeta_D = 0.555$$

$$t_p = \frac{\pi}{\omega_d} \qquad \rightarrow \omega_d = 1.9635$$

which yield $\omega_n = 2.36$, $\sigma_D = -1.31$, and $T_s = 4/|\sigma_D| = 3.05$. The desired control ratio $C(s)/R(s)$ is obtained from this data information; i.e., the dominant complex-conjugate poles are $p_{1,2} = -1.31 \pm j1.9635$. To determine whether these poles are achievable for this system it is necessasy to obtain a

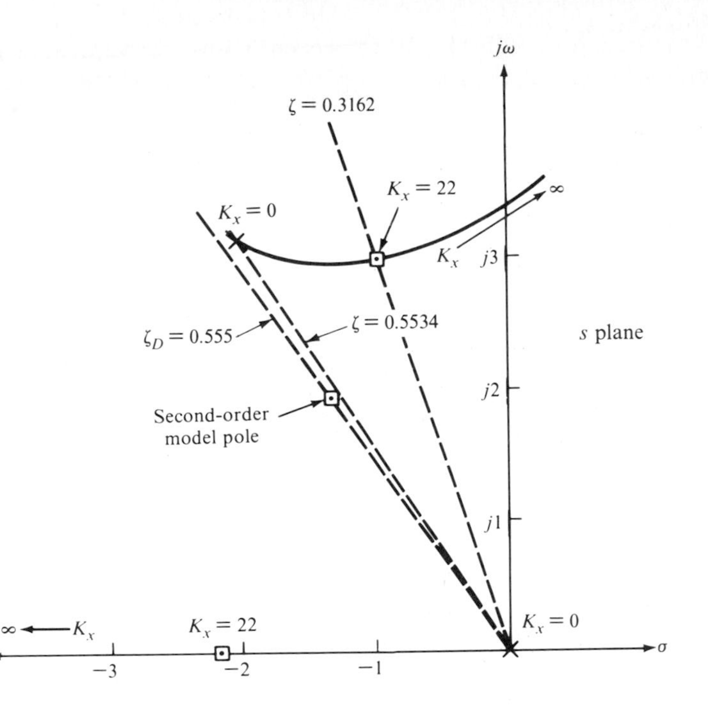

Figure 5-15 Root locus for Eq. (5-32).

root-locus plot. From $G_x(s) = -1$, the root-locus plot of Fig. 5-15 is obtained. An analysis of this plot reveals that (1) the maximum damping ratio achievable is 0.5534, which is less than the desired value of 0.555, and (2) for low values of K_x, the real root is more dominant than the complex roots.

As a consequence of this analysis, the only possibility of achieving the desired specifications is to adjust the gain to obtain the pole pattern of Fig. 5-9*a* and to satisfy the value of $K_1 \geq 1.5$. This allows a damping ratio $\zeta < \zeta_D$, which is offset by the effect of the real root in minimizing the overshoot. Hence, a value of $K_x = 22$ produces complex poles with $\zeta = 0.3162$ and yields the closed-loop transfer function

$$\frac{C(s)}{R(s)} = \frac{22}{(s+1+j3)(s+1-j3)(s+2.2)} \tag{5-33}$$

and $K_1 = 1.528$. The plot of $|\mathbf{C}(j\omega)/\mathbf{R}(j\omega)|$ vs. ω very closely approximates the waveshape of the frequency response for a simple second-order underdamped system as shown in Fig. 5-10*a*. A unit-step input results in the following values:

$$M_p \approx 1.123 \qquad t_p \approx 1.51 \text{ s} \qquad t_s \approx 2.95 \text{ s}$$

Therefore, for this value of K_x the desired specifications have been achieved.

Example 5-2 For the unity-feedback control system whose forward transfer fraction is given by Eq. (5-32), the following performance specifications must be satisfied:

$$1 < M_p \le 1.15 \qquad t_s \le 2 \text{ s} \qquad e(t)_{ss} = 0$$

$$t_p \le 1.6 \text{ s} \qquad K_1 \ge 2.5 \text{ s}^{-1}$$

An analysis of Fig. 5-15 reveals that these specifications cannot be achieved by the basic system $G_x(s)$. An additional unit, a compensator $G_c(s)$, is inserted in cascade with $G_x(s)$ to effect a satisfactory response as shown in Fig. 5-16.

The Guilleman-Truxal method[1] for determining the required $G_c(s)$ that yields the desired specifications first requires the synthesis of a desired control ratio $[C(s)/R(s)]_M$. Inserting the desired values of M_p and t_s into Eqs. (5-8) to (5-11) yields

$$M_p = 1.15 = 1 + \exp \frac{-\zeta\pi}{\sqrt{1-\zeta^2}} \rightarrow \zeta_D = 0.5168$$

$$t_s = 2 \approx T_s = \frac{4}{|\sigma_D|} \qquad \rightarrow |\sigma_D| = \zeta_D \omega_n = 2$$

$$\omega_n = \frac{|\sigma_D|}{\zeta_D} = 3.869$$

$$\omega_d = \omega_n \sqrt{1 - \zeta_D^2} = 3.312$$

Thus based upon $|\sigma_D| = 2$ and $\omega_n \approx 3.87$, the complex-conjugate poles for an "effective" simple second-order response are $p_{1,2} = -2 \pm j3.3$. Since the basic system is not able to satisfy the desired specifications, it is necessary for $G_c(s)$ to be of the form

$$G_c(s) = K_c \frac{s^v + c_{r-1}s^{v-1} + \cdots + c_1 s + c_0}{s^u + a_{u-1}s^{u-1} + \cdots + u_1 s + u_0} \qquad K_c > 0 \tag{5-34}$$

where $u \ge v$ and $u \ge 1$, in order to be able to achieve the desired response. Assume $G_c(s)$ increases the order of the system by 1 over the uncompensated system; thus $n = 4$. Further, assume that $[C(s)/R(s)]_M$ is to have no zeros. Based upon the specifications and the assumptions, the desired control ratio

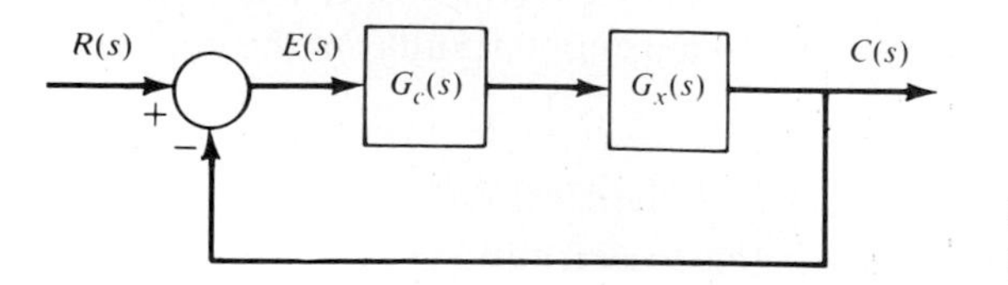

Figure 5-16 A unity-feedback-compensated control system.

must be of the form

$$\left[\frac{C(s)}{R(s)}\right]_M = \frac{K}{(s+2+j3.3)(s+2-j3.3)(s-p_3)(s-p_4)} \tag{5-35}$$

One approach for ensuring an effective simple second-order response is to make the poles p_3 and p_4 nondominant. Thus the initial values of $-p_3$ and $-p_4 = -100$ are selected. With these values, Eq. (5-35) becomes

$$\left[\frac{C(s)}{R(s)}\right]_D = \frac{148{,}900}{s^4 + 204s^3 + 10{,}815s^2 + 42{,}978s + 148{,}900} = \frac{N_D}{D_D} \tag{5-36}$$

where $c(t)_{ss} = \lim_{s \to 0} C(s) = R_0$ yields $K = 148{,}900$. From Fig. 5-12 the actual control ratio is

$$\left[\frac{C(s)}{R(s)}\right]_A = \frac{G_c(s)G_x(s)}{1 + G_c(s)G_x(s)} \tag{5-37}$$

The next step of this method requires that Eq. (5-36) be set equal to Eq. (5-37) and then solved for $G_c(s)$. Thus

$$G_c(s) = \frac{N_D}{(D_D - N_D)G_x(s)} \tag{5-38}$$

Substituting from Eq. (5-32) and inserting the polynomials N_D and D_D into Eq. (5-38) yields

$$G_c(s) = \frac{148{,}900}{K_x} \frac{s^2 + 4.2s + 14.4}{s^3 + 204s^2 + 10{,}815s + 42{,}978}$$

$$= \frac{148{,}900}{K_x} \frac{s^2 + 4.2s + 14.4}{(s + 4.318)(s + 95.91)(s + 103.77)} \tag{5-39}$$

The forward transfer function of the compensated system is

$$G(s) = G_c(s)G_x(s) = \frac{148{,}900}{s(s + 4.318)(s + 95.91)(s + 103.77)} \tag{5-40}$$

where $K_1 = 3.46$. Note that the compensator effectively cancels the poles of $G_x(s)$ and replaces them by two real poles. It also inserts the necessary one real pole. The root locus of the compensated system is shown in Fig. 5-17.

A unit-step input applied to the compensated system results in the following values.

$$M_p \approx 1.148 \qquad t_p \approx 0.95 \text{ s} \qquad t_s \approx 2.0 \text{ s}$$

Therefore the compensator yields values for the figures of merit that satisfy the desired system specifications.

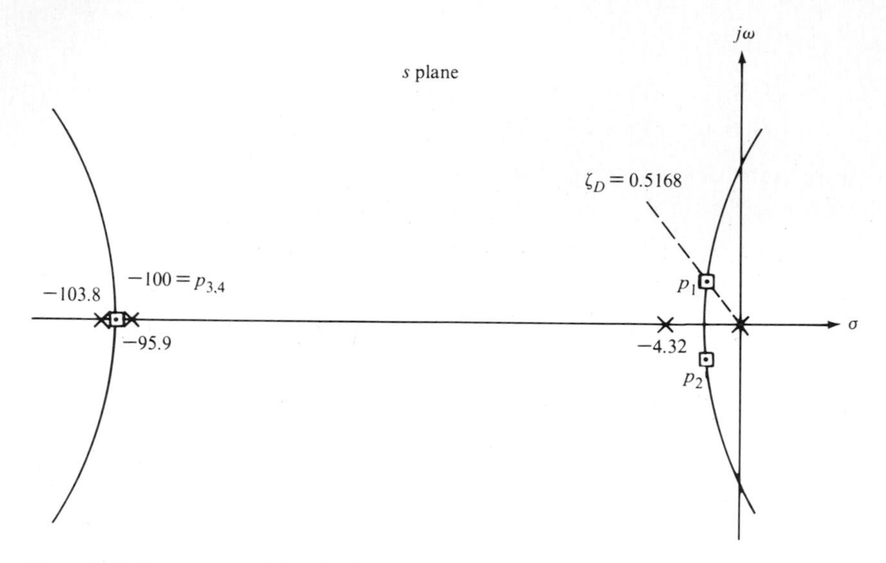

Figure 5-17 Root locus of Eq. (5-40) (not to scale).

5-6 FEEDBACK COMPENSATION

Feedback compensation, as exemplified by Fig. 5.12*b*, employs the three compensators of Fig. 5-2, along with a tachometer whose transfer function is $K_t s$ for position control and K_t for velocity control (where K_t is the tachometer sensitivity). This form of compensation is used when the dominant roots of the control ratio for the basic system lie on the dominant root-locus branches emanating from a pair of open-loop transfer-function complex-conjugate poles. The standard cascade compensators of Fig. 5-2 yield very little improvement in system performance for basic systems that fall into this situation. Feedback compensation is also used for disturbance rejection, as shown in the next section, and used where a faster responding system is desirable over that obtainable by cascade compensation for a given basic system.

The design procedures for feedback compensators are somewhat more involved (may require more trial and error or "fine tuning") and are not reviewed in this text. For review purposes, the reader may refer to a continuous-time control-theory text.[1,21] Examples of feedback compensation for sampled-data systems are presented in Chap. 12 and are based upon the knowledge of these design procedures.

5-7 DISTURBANCE REJECTION[21]

The previous sections of this chapter have dealt with designing a control system whose output follows (tracks) the (desired) input signal. The design approach for

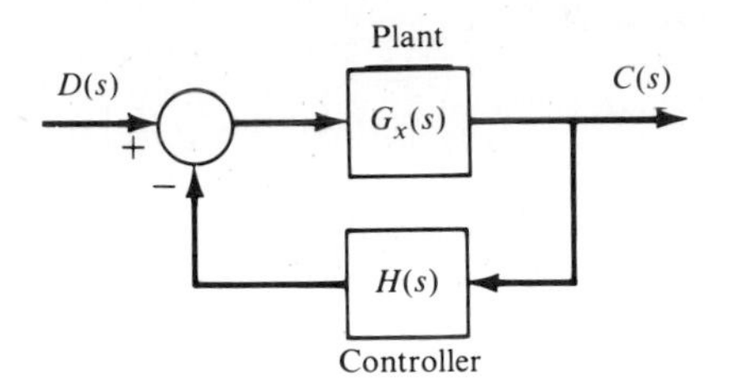

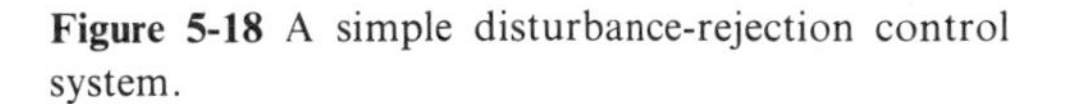
Figure 5-18 A simple disturbance-rejection control system.

the disturbance-rejection problem is in direct contrast to the previous approach. It is based upon the performance specification that the output of a control system not be affected by a system disturbance input $d(t)$. In other words, $c(t)_{ss} = 0$ for $r(t) = d(t) \neq 0$. Consider the control ratio

$$\frac{C(s)}{D(s)} = \frac{K(s^w + c_{w-1}s^{w-1} + \cdots + c_1 s + c_0)}{s^n + \alpha_{n-1}s^{n-1} + \cdots + \alpha_1 s + \alpha_0} \tag{5-41}$$

Thus for a step disturbance $[D(s) = D_0/s]$ and assuming a stable system, then

$$c(t)_{ss} = \lim_{s \to 0} sC(s) = \frac{Kc_0}{\alpha_0} \lim_{s \to 0} sD(s) = \frac{KD_0 c_0}{\alpha_0} \tag{5-42}$$

Therefore, for $c(t)_{ss} = 0$, then $c_0 = 0$ where K, D_0, and α_0 are finite nonzero values. *This requires that the numerator of $C(s)/D(s)$ have at least one zero at the origin.*

For this type of a control problem, not only is it desired that the disturbance have no steady-state effect on the output, but its resulting transient effect must die out as fast as possible with a limit M_D on the maximum value of the output. This added constraint may require the system's characteristic equation to have a pair of dominant complex-conjugate roots in order to try to achieve the smallest t_s possible.

A second-order simple continuous-time *disturbance-rejection-model transfer function* for the system of Fig. 5-18 is

$$\left[\frac{C(s)}{D(s)}\right]_M = \frac{G_x}{1 + G_x H} = \frac{K_x s}{(s+a)^2 + b^2} = \frac{K_x s}{s^2 + 2\zeta\omega_n s + \omega_n^2} \tag{5-43}$$

where $K_x > 0$. For a unit-step disturbance, Eq. (5-43) yields

$$c(t) = \mathscr{L}^{-1}\left[\frac{K_x}{(s+a)^2 + b^2}\right] = \frac{K_x}{b} \epsilon^{-at} \sin bt \tag{5-44}$$

where $a^2 + b^2 = \omega_n^2$, $b = \omega_d$, and $a = |\sigma| = \zeta\omega_n$. Two approaches that may be used to determine appropriate model values for ζ and ω_n, where K_x is assumed fixed, are via the time- or frequency-domain analysis approach.

Time domain In Fig. 5-19a is a plot of the time-varying portion of Eq. (5-44), that is, $\epsilon^{-at} \sin bt$ vs. t. By setting the derivative of $c(t)$ with respect to time equal to

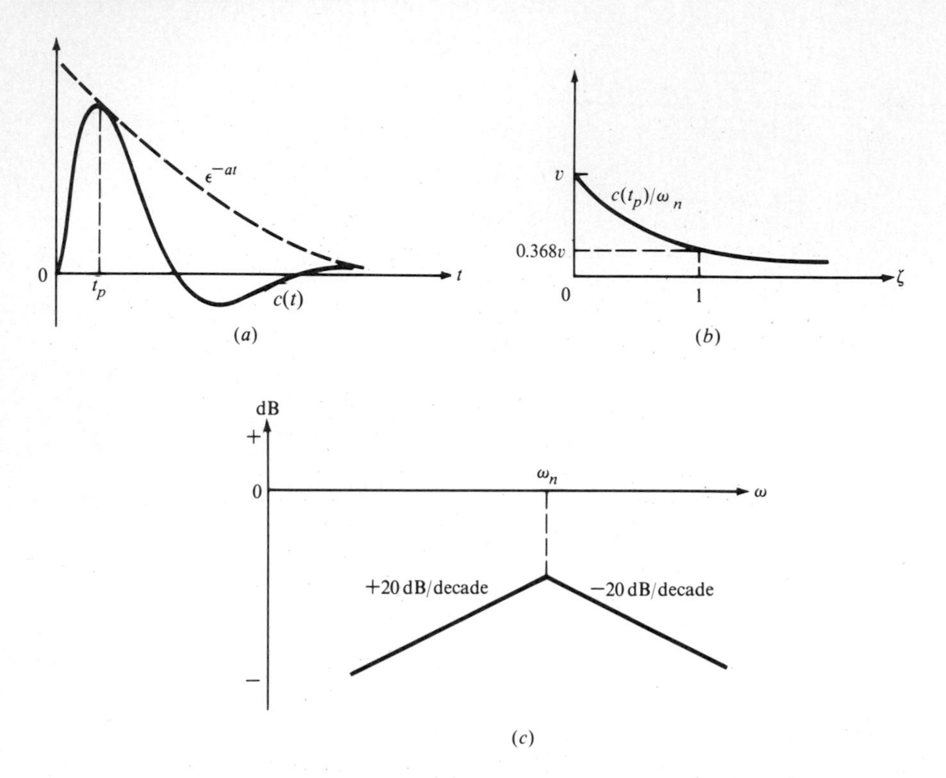

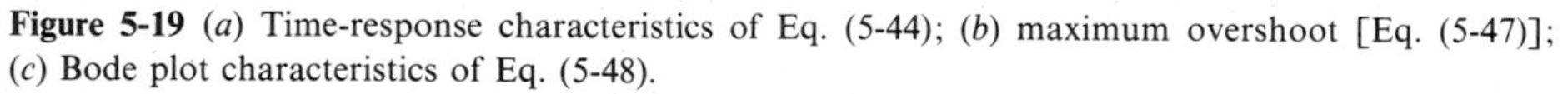

Figure 5-19 (*a*) Time-response characteristics of Eq. (5-44); (*b*) maximum overshoot [Eq. (5-47)]; (*c*) Bode plot characteristics of Eq. (5-48).

zero, it can be shown that the maximum overshoot occurs at

$$t_p = \cos^{-1} \frac{\zeta}{\omega_n \sqrt{1 - \zeta^2}} \tag{5-45}$$

where the maximum value of $c(t)$ is

$$c(t_p) = \frac{K_x}{\omega_n} \exp - \frac{\zeta \cos^{-1} \zeta}{\sqrt{1 - \zeta^2}} \tag{5.46}$$

By letting $K_x = v\omega_n^2$ $(v > 0)$ then Eq. (5-46) is rearranged to

$$\frac{c(t_p)}{\omega_n} = v \exp - \frac{\zeta \cos^{-1} \zeta}{\sqrt{1 - \zeta^2}} \tag{5-47}$$

Equations (5-45) and (5-46) may be used in the following manner:

Step 1. Select a value of ω_n based upon bandwidth considerations.
Step 2. For various values of ζ determine t_p from Eq. (5-45) and $c(t_p)$ from Eq.

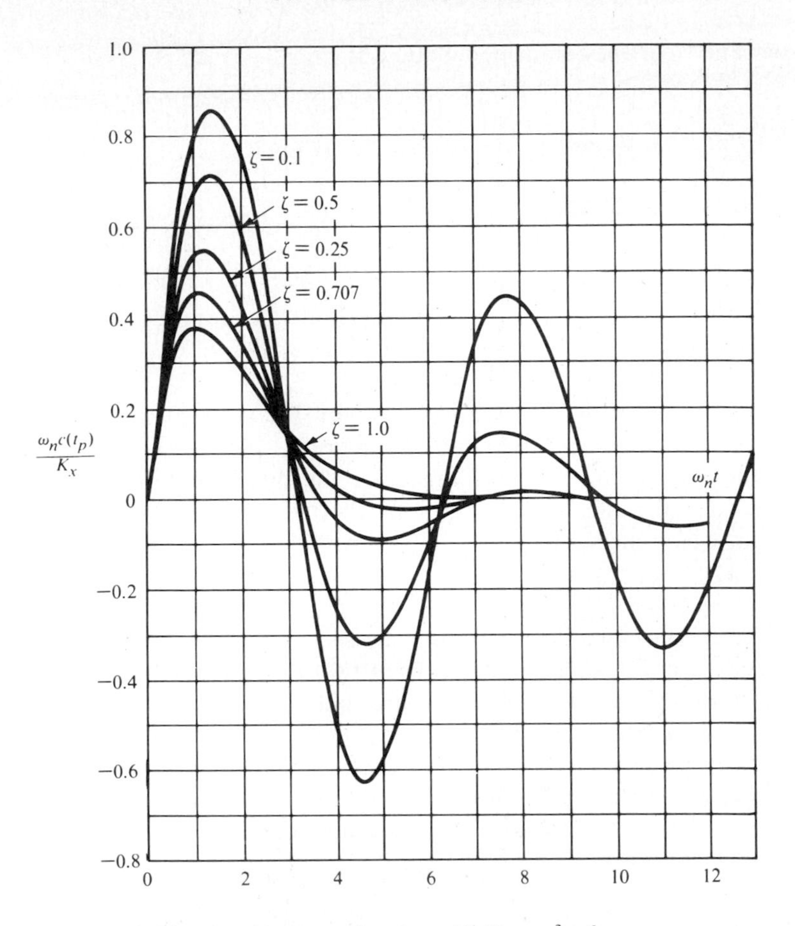

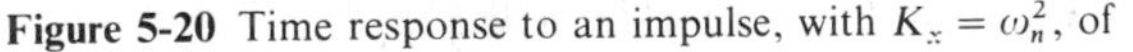

Figure 5-20 Time response to an impulse, with $K_x = \omega_n^2$, of

$$C(s) = \frac{K_x}{s^2 + 2\zeta\omega_n s + K_x}$$

(5-46). An alternate approach is the utilization of Fig. 5-20.[1] This figure may be used to determine ζ for either of the following two cases: $K_x = \omega_n^2$ or $K_x \neq \omega_n^2$. For the former case $c(t_p)/\omega_n$, and for the latter case $\omega_n c(t_p)/K_x$ must be specified.

If there is a value of ζ that satisfies the $c(t_p)$, t_p, and t_s specifications for the value of ω_n chosen, then the simple model of Eq. (5-43) is determined. If the specifications are not met for this first choice of ω_n, a new value of ω_n is chosen and the process is repeated until the specifications are achieved.

An alternate procedure is to use the specified values of t_p and $c(t_p)$ in Eqs. (5-45) and (5-46), and hence by trial and error, in conjunction with Fig. 5-19*b*, compatible values of ζ and ω_n are found.

Frequency domain The frequency transfer function of Eq. (5-43) is

$$[\mathbf{M}(j\omega)]_M = \frac{\mathbf{C}(j\omega)}{\mathbf{R}(j\omega)} = \frac{(K_x/\omega_n^2)(j\omega)}{[1-(\omega/\omega_n)^2] + j(2\zeta\omega/\omega_n)}$$

$$= \frac{K_x}{\omega_n^2}\mathbf{M}'(j\omega) \tag{5-48}$$

A straight-line approximation is used to obtain the Bode plot of Lm $\mathbf{M}'(j\omega)$ vs. ω for various values of ω_n. Figure 5-19*c* shows a typical Bode plot for $\mathbf{M}'(j\omega)$. This plot reveals the following:

1. The height of the plot of Lm $\mathbf{M}(j\omega)$ vs. ω varies with Lm (K_x/ω_n^2).
2. The larger ω_n is, the peak value moves more to the right.
3. If K_x is fixed by the plant, then ω_n is made larger, and the lower becomes the plot of Lm $\mathbf{M}(j\omega)$ vs. ω. This effect is countered by the fact that the breakpoint occurs farther out which means that the plot of Lm $\mathbf{M}(j\omega)$ vs. ω has a larger peak value.

Thus using both the time- and frequency-domain characteristics, a continuous-time model for Eq. (5-43) is determined by trial and error that may meet the specifications. If the specifications are not met with the simple $H(s) = K_H/s$ then a model of the form of Eq. (5-41) is chosen that comes closest to meeting these specifications. This general approach defines a starting point for the design, i.e., determining a "good first choice" of the "desired dominant complex-conjugate poles" with a more complicated $H(s)$, in conjunction, possibly, with a cascade compensator.

Example 5-3 Consider the nonunity-feedback system of Fig. 5-21 that has a disturbance input $D(s)$ and where

$$G_x(s) = \frac{K_G}{s(s+1)} = \frac{1}{s(s+1)} \tag{5-49}$$

and

$$H_x(s) = \frac{K_x}{s+200} = \frac{200}{s+200} \tag{5-50}$$

Design the feedback compensator $H_c(s) = K_cH_c'(s)$ such that the control

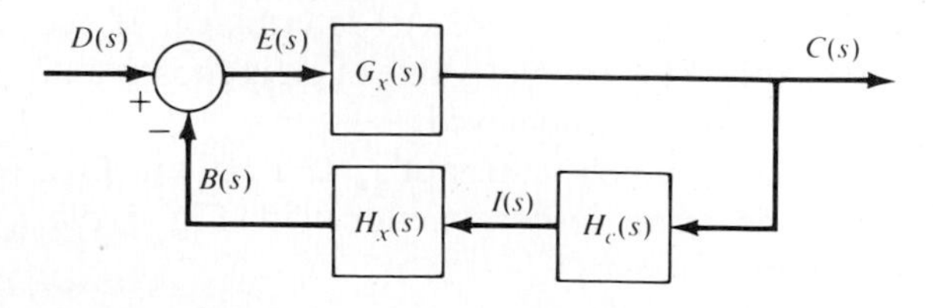

Figure 5-21 A control system with a disturbance input.

system of Fig. 5-21 can satisfy the following specifications for a unit-step disturbance input $d(t)$:

$$c(t_p) \le 0.002 \qquad c(t)_{ss} = 0 \qquad t_p \le 0.20 \text{ s}$$

$$\text{Lm}\,\frac{\mathbf{C}(j\omega)}{\mathbf{D}(j\omega)} \le -54 \text{ dB} = \text{Lm}\, M_D$$

within the bandwidth of

$$0 \le \omega \le 10 \text{ rad/s} = \omega_b$$

In using the frequency-domain approach for designing $H_c(s)$ it is first necessary to present a design technique that utilizes the straight-line Bode plot approximations. For the system's control ratio

$$\mathbf{M}(j\omega) = \frac{\mathbf{C}(j\omega)}{\mathbf{D}(j\omega)} = \frac{\mathbf{G}_x(j\omega)}{1 + \mathbf{G}_x(j\omega)\mathbf{H}(j\omega)} \tag{5-51}$$

where $\mathbf{H}(j\omega) = \mathbf{H}_x(j\omega)\mathbf{H}_c(j\omega)$, consider the cases when

$$|\mathbf{G}_x(j\omega)\mathbf{H}(j\omega)| \ll 1 \qquad \text{and} \qquad |\mathbf{G}_x(j\omega)\mathbf{H}(j\omega)| \gg 1$$

The control ratio can be approximated by

$$\mathbf{M}(j\omega) \approx \begin{cases} \mathbf{G}_x(j\omega) & \text{for } |\mathbf{G}_x(j\omega)\mathbf{H}(j\omega)| \ll 1 \qquad (5\text{-}52) \\[2ex] \dfrac{1}{\mathbf{H}(j\omega)} & \text{for } |\mathbf{G}_x(j\omega)\mathbf{H}(j\omega)| \gg 1 \qquad (5\text{-}53) \end{cases}$$

Still undefined is the condition when $|\mathbf{G}_x(j\omega)\mathbf{H}(j\omega)| \approx 1$, in which case neither Eq. (5-52) nor Eq. (5-53) is applicable. In the approximate procedure this condition is neglected, and Eqs. (5-52) and (5-53) are used when $|\mathbf{G}_x(j\omega)\mathbf{H}(j\omega)| < 1$ and $|\mathbf{G}_x(j\omega)\mathbf{H}(j\omega)| > 1$, respectively. This approximation, along with a root-locus sketch of $G_x(s)H(s) = -1$, allows investigation of the qualitative results to be obtained. After these results are found to be satisfactory, the refinements for an exact solution are introduced.

The design procedure using the approximations of Eqs. (5-52) and (5-53) is as follows:

Step 1. Using the straight-line approximations, plot Lm $\mathbf{G}_x(j\omega)$ vs. ω.

Step 2. Assume an $H_c(s)$ that may, along with $H_x(s)$, yield the desired system performance. For this specified $H(s)$, using the straight-line approximations, plot Lm $1/\mathbf{H}(j\omega)$ vs. ω.

Note:

(*a*) For the condition

$$\text{Lm}\,\mathbf{G}_x(j\omega)\mathbf{H}(j\omega) = \text{Lm}\,\mathbf{G}_x(j\omega) + \text{Lm}\,\mathbf{H}(j\omega) \ge \text{Lm}\,1 = 0$$

assume the equality situation; that is,

$$\text{Lm } \mathbf{G}_x(j\omega) = -\text{Lm } \mathbf{H}(j\omega) = \text{Lm } \frac{1}{\mathbf{H}(j\omega)} \tag{5-54}$$

When the equality situation exists, the intersection of the plots, as specified by Eq. (5-54), represents the boundary between the two regions given by Eqs. (5-52) and (5-53). Thus there must be at least one intersection between the plots in order for both approximations to be valid. For disturbance rejection over the frequency band of $0 \leq \omega \leq \omega_b$, there can only exist one intersection for this example.

(*b*) For the disturbance rejection case, it is desired to have $[\mathbf{C}(j\omega)/\mathbf{D}(j\omega)] < \text{Lm } M_D$ or Eq. (5-53) to hold for $0 \leq \omega \leq \omega_b$ and the specified function $H(s)$ to yield the desired frequency-domain specifications. To satisfy both objectives it is usually necessary for the gain constant of $H(s)$ to be "very high."

(*c*) Plot Lm $\mathbf{G}_x\mathbf{H}$ vs. ω by using the straight-line approximation technique to determine the slope of the plot at the 0-dB crossover point, i.e., slope of Lm $\mathbf{G}_x$ − slope of Lm $1/\mathbf{H}$. For a stable system desire that this slope be greater than −40 dB/decade if the adjacent corner frequencies are not close. Also, desire the loop transmission frequency ω_ϕ (the 0-dB crossover frequency) of Lm $\mathbf{G}_x(j\omega)\mathbf{H}(j\omega)$ be within the allowable specified value in order to attenuate high-frequency loop "noise." Placing zeros of $H_c(s)$ "too close" to the imaginary axis may increase the value of ω_ϕ.

It may be necessary at this stage of the design procedure to modify $H_c(s)$ in order to satisfy the stability requirement and the frequency-domain specifications.

Step 3. Once an $H(s)$ is found that *may* satisfy the frequency-domain specifications, sketch the root locus ($G_xH = -1$) to ascertain (*a*) that at least a conditionally stable system exists and (*b*) that, if a stable system is achievable, it may be possible to satisfy the time-domain specifications. [Section 12-5 discusses the desired locations of the poles of $C(s)$ in order to try to achieve the desired specifications; see also Probs. 5-10 and 5-11.] If the sketch "seems" reasonable, obtain (*a*) the exact root-locus plots from which a value of the static loop sensitivity $K = K_G K_x K_c$ is chosen and (*b*) for the value of K, a plot of $c(t)$ vs. t from which the figures of merit are obtained.

Step 4. If the results of Step 3 do not yield satisfactory results, then based upon these results, select a new $H_c(s)$ and repeat Steps 2 and 3, etc.

It should be noted that, in reality, it may be wise in doing Step 1 to simultaneously sketch the root locus $G_xH = -1$, as discussed in Step 3.

SOLUTION

Step 1. The plot of Lm $\mathbf{G}_x(j\omega)$ vs. ω is shown in Fig. 5-22.

Step 2. Since $G_x(s)$ does not have a zero at the origin and $C(s)/D(s)$ does require one for disturbance rejection, then $H(s)$ must have a pole at the origin. Thus the feedback transfer function is of the form

$$H(s) = H_x(s)H_c(s) = \frac{K_H(s + a_1)\cdots}{s(s + 200)(s + b_1)\cdots} \tag{5-55}$$

where the order of the denominator is equal to or greater than the order of the numerator. As a first trial, let

$$H_c(s) = \frac{K_c}{s} \tag{5-56}$$

where $K_H = K_x K_c = 200K_c$. To determine the value of K_c required to satisfy Lm $\mathbf{C}(j\omega)/\mathbf{D}(j\omega) \le -54$ dB for $0 \le \omega \le 10$ rad/s, initially plot $[\text{Lm } 1/\mathbf{H}_c(j\omega)]_{K_c=1}$. That is, assume $H(s) = H_c(s)$ (see Fig. 5-22) since the corner frequency of $\mathbf{H}(j\omega)$ at $\omega = 200$ lies outside the desired bandwidth. Lower this plot, as shown in the figure, until the plot of Lm $1/\mathbf{H}(j\omega)$ yields the desired value of $\omega_b = 10$ rad/s. For this case Eq. (5-53) satisfies

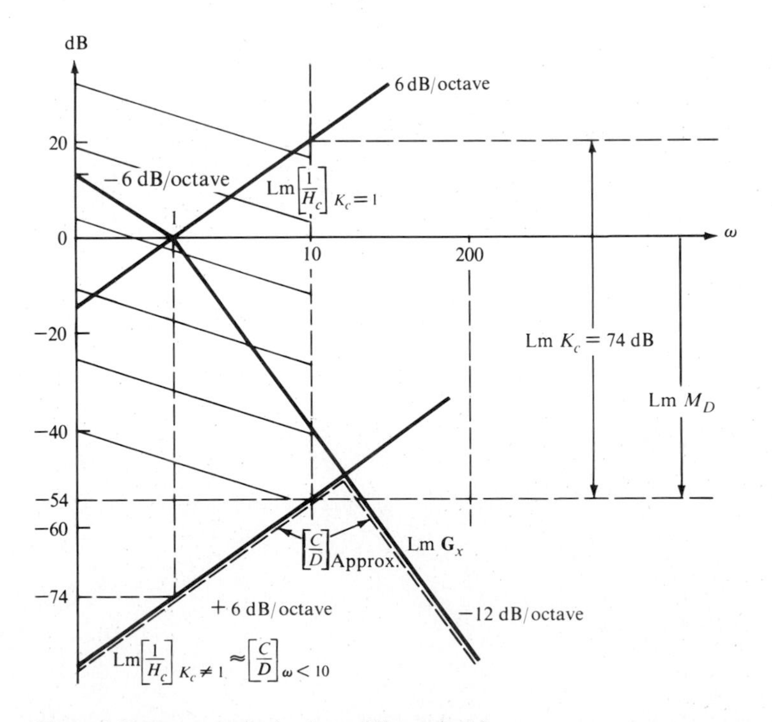

Figure 5-22 Log-magnitude plots of Example 5-3.

the requirement $|\mathbf{M}(j\omega)| \le M_D$ over the desired bandwidth and is below the -54 dB line. The decibels required to lower this plot, that is, Lm $1/K_c = -74$ dB (see Fig. 5-22), yields the required value for K_c. With this value of K_c the resulting first-trial feedback transfer function is

$$H(s) = \frac{K_H}{s(s+200)} \tag{5-57}$$

An analysis of

$$\mathbf{G}_x(j\omega)\mathbf{H}(j\omega) = \frac{K_c}{(j\omega)^2(1+j\omega)\left(1+j\dfrac{\omega}{200}\right)} \tag{5-58}$$

reveals that the plot of Lm $[\mathbf{G}_x(j\omega)\mathbf{H}(j\omega)]$ vs. ω crosses the 0-dB axis with a maximum slope of -40 dB/decade regardless of the value of K_c. This crossover characteristic is indicative of an unstable system.

Step 3. The root-locus sketch for

$$G_x(s)H(s) = \frac{K_H}{s^2(s+1)(s+200)} = -1 \tag{5-59}$$

is shown in Fig. 5-23, which reveals that a completely unstable situation exists for the $H(s)$ of Eq. (5-57).

For the next trial, which is left up to the reader, it is necessary to add at least one zero to Eq. (5-57) in order to try to make the system at least a conditionally stable system while maintaining the desired bandwidth. To achieve the final acceptable design of $H_c(s)$ that results in the system specifications being met, as this example illustrates, requires a trial-and-error procedure.

Example 5-4 Consider the nonunity-feedback system of Fig. 5-21 that has a

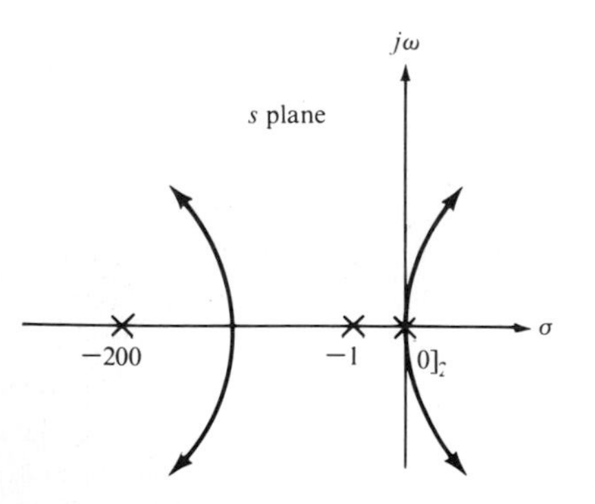

Figure 5-23 A root-locus sketch for Eq. (5-59).

step disturbance input $d(t) = u_{-1}(t)$ and where

$$G_x(s) = \frac{2}{s+2}$$

and

$$H(s) = H_x(s)H_c(s) = \frac{K_H}{s}$$

It is desired to analyze the response $c(t)$ as the gain K_H is increased.
The output response for this system is

$$C(s) = \frac{2}{s^2 + 2s + 2K_H} = \frac{2}{(s - p_1)(s - p_2)} = \frac{A_1}{(s - p_1)} + \frac{A_2}{(s - p_2)} \tag{5-60}$$

where $p_{1,2} = -1 \pm j\sqrt{2K_H - 1}$ (for $K_H > 0.5$)

$$A_1 = \frac{0.1}{j\sqrt{2K_H - 1}}$$

$$A_2 = -A_1$$

Thus

$$c(t) = A_1\epsilon^{p_1 t} + A_2\epsilon^{p_2 t} \tag{5-61}$$

It is seen from

$$\lim_{K_H \to \infty} A_1 = \lim_{K_H \to \infty} A_2 = 0$$

that, for this example, as K_H is made larger the coefficients A_1 and A_2 in Eq. (5-61) become smaller. Thus

$$\lim_{K_H \to \infty} c(t) = 0$$

which is the desired result for disturbance rejection. (Note that the value of K_H is limited by practical considerations.) A similar analysis can be made for a more complicated system (see Prob. 5-11).

The disturbance-rejection design techniques of this section for continuous-time control systems can be used to minimize the disturbance problem for sampled-data control systems. As is shown in Chap. 7, a sampled-data control system may be represented by an equivalent pseudo-continuous-time (PCT) control system. The techniques of this section can then be used to design $H_c(s)$ for the PCT system. When a satisfactory design for $H_c(s)$ is achieved, the corresponding $H_c(z)$, by the use of the Tustin (bilinear) transformation of Chap. 7, is obtained. Because of the limitations (see Chap. 7) of the Tustin transformation, one may wish to limit how far to the left in the s plane the poles of $H_c(s)$ are located. The farther to the left these poles are placed, the larger the value of K_c. This large value of K_c, when used in the sampled-data system, may result in one of two situations. (1) a stable and satisfactory performance (but this large value of K_c may be difficult to implement in the digital controller) or (2) at least one or more poles of the control ratio being on or outside the unit circle, therefore resulting in an unstable performance. For this latter situation it may be possible to achieve a stable and satisfactory performance by merely reducing the gain K_c without altering the location of the poles and zeros of $H_c(z)$.

5-8 SUMMARY

This chapter has presented the standard figures of merit based upon a simple second-order system performance. The qualitative correlation between the frequency response, root locus, and the time domain for second- and higher-order systems is presented as an aid in achieving the desired system performance characteristics. That is, the figures of merit along with this correlation are used as a guidepost for evaluating a system's performance.

CHAPTER

SIX

DISCRETE CONTROL ANALYSIS

6-1 INTRODUCTION

Preceding chapters presented the fundamental concepts, theory, and domain correlation involved in the study of sampled-data systems. These elements are now used to perform an analysis of a basic sampled-data control system. A basic sampled-data control system has the minimum structure (see Sec. 1-8) necessary to achieve the desired controlling action by using only gain adjustment.

This chapter presents methods of analyzing the stability and time-response characteristics of sampled-data systems. There are various approaches that may be used in performing this analysis. These approaches may be divided into two distinct categories: (1) *direct* (DIR) and (2) *digitization* (DIG) or *discrete* digital-control-analysis techniques. The first category involves carrying out the analysis entirely in the z plane. The second category permits the analysis and synthesis of the sampled-data system to be carried out entirely in the w' plane or, by the use of the Padé approximation (see Appendix C), entirely in the s plane. The use of the Padé approximation, along with the information provided by a Fourier analysis of the sampled signal, results in the modeling of a sampled-data control system by a *pseudo-continuous-time* (*PCT*) *control system.* The w'-plane analysis is discussed in Sec. 6-6, whereas the Padé approximation for the ZOH transfer function and the PCT system are discussed in Chap. 7. An overview of the analysis-and-design methods is presented in Fig. 6-1. It is not the intent of this text to discuss all methods in this chart,[3,4,20] only those which solidify the fundamental concepts for the analysis and design of sampled-data control systems. To enhance the analysis and design approaches, a correlation between the frequency and time domain is also presented. Figure 6-2 illustrates, *qualitatively*, the interrelationships between the

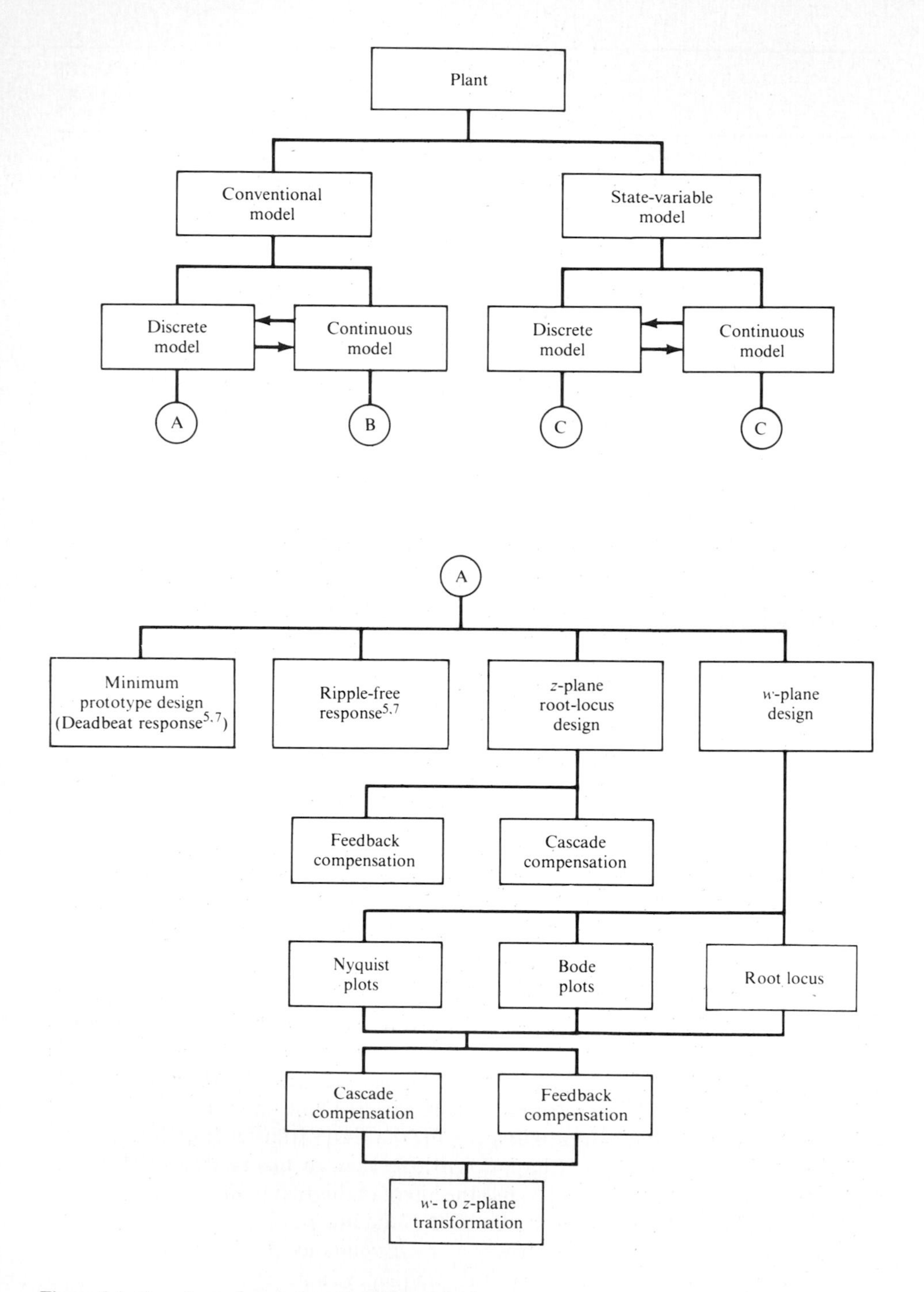

Figure 6-1 Flowchart of analysis-and-design methods.

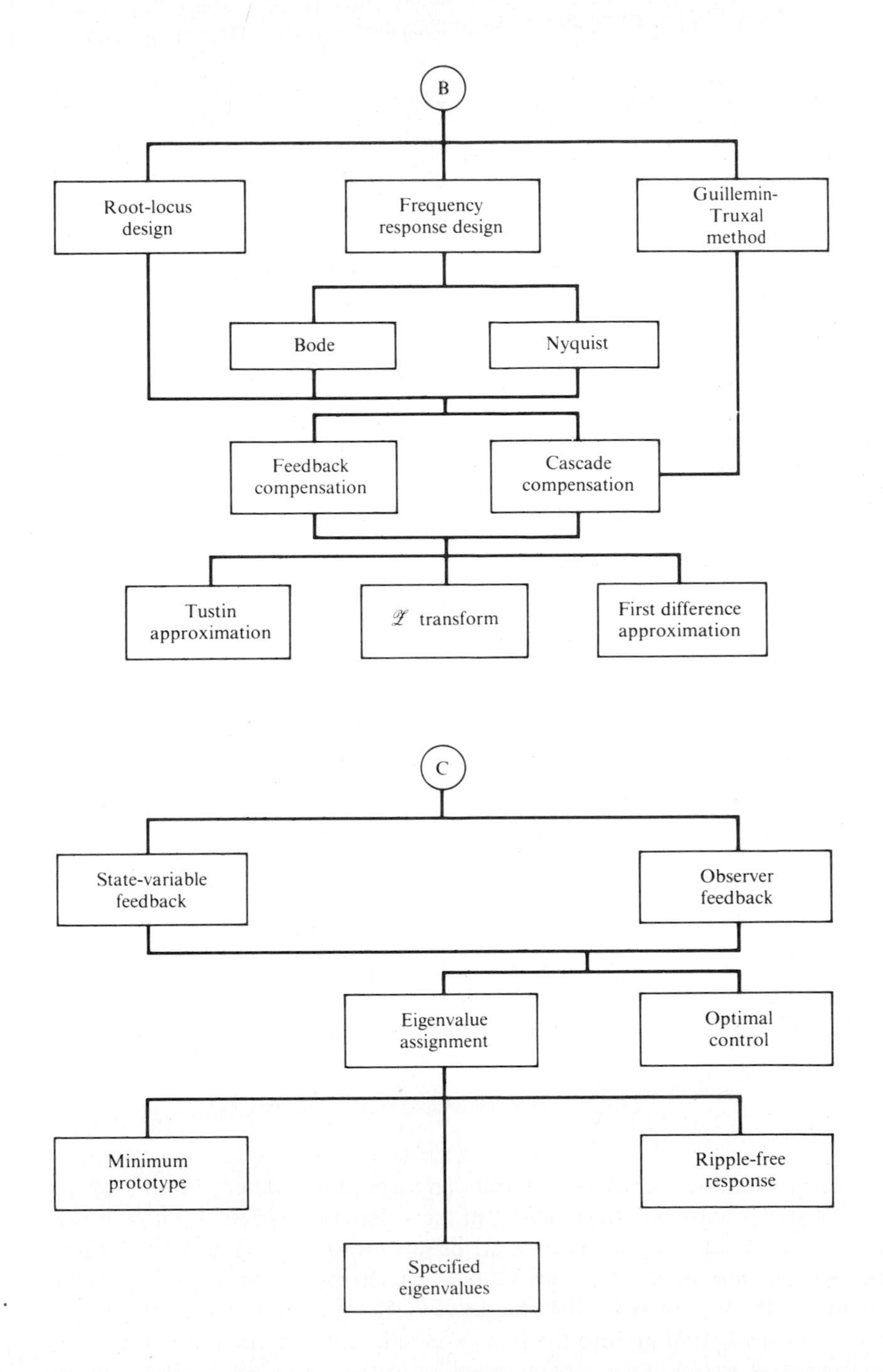
B
Root-locus design
Frequency response design
Guillemin-Truxal method
Bode
Nyquist
Feedback compensation
Cascade compensation
Tustin approximation
$\mathscr{Z}$ transform
First difference approximation
C
State-variable feedback
Observer feedback
Eigenvalue assignment
Optimal control
Minimum prototype
Ripple-free response
Specified eigenvalues

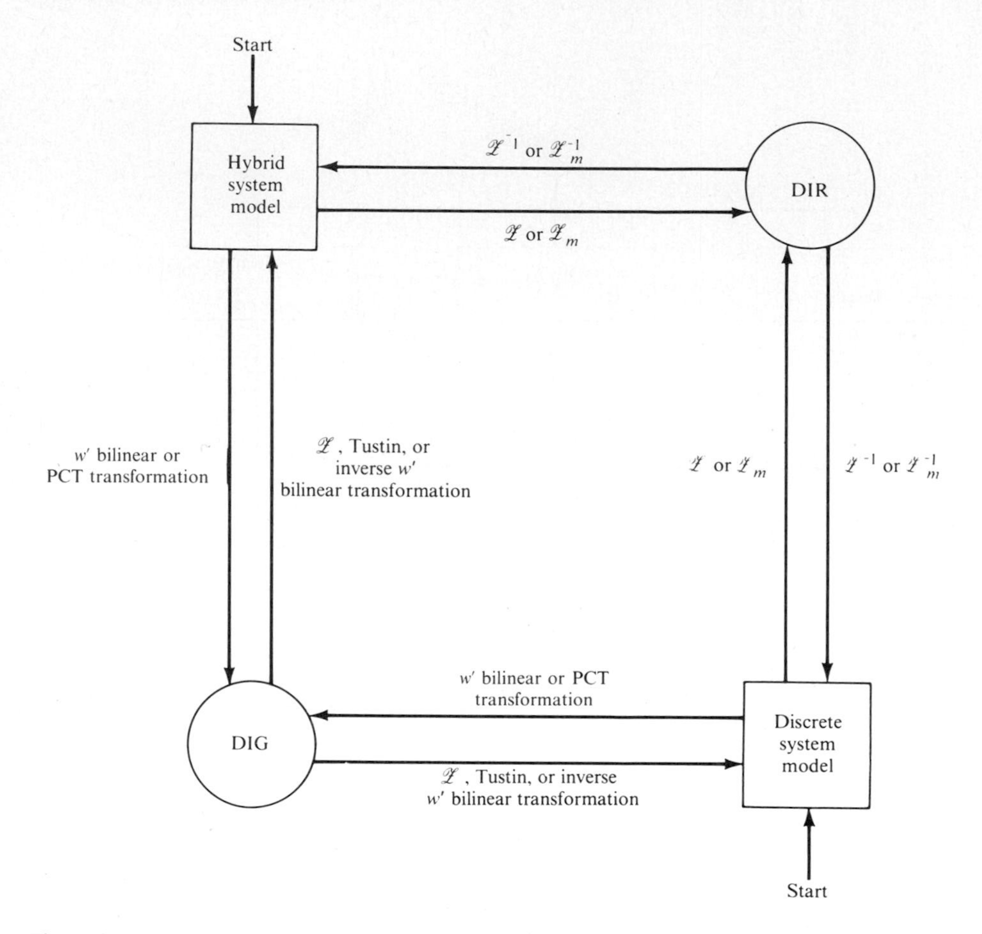

Figure 6-2 Qualitative interrelationships between the direct and digitization categories.

direct and digitization categories. With the help of a digital-computer program,[14,15] it is now possible to perform an analysis and to synthesize an acceptable design with relative ease, using approaches from either category.

6-2 *z*-DOMAIN STABILITY

An approach for determining system stability is through the determination of the roots of the system characteristic equation in the z domain. In Sec. 4-3 it is shown that the primary and complementary strips are transformed into the same (overlapping) portions of the z plane. That is, the strips in the left-half s plane ($\sigma < 0$) map into the region inside the unit circle in the z plane, and the strips in the right-half s plane ($\sigma > 0$) map into the region outside the unit circle (see Fig. 6-3). The imaginary axis $\sigma = 0$ in the s plane maps onto the unit circle in the z plane.

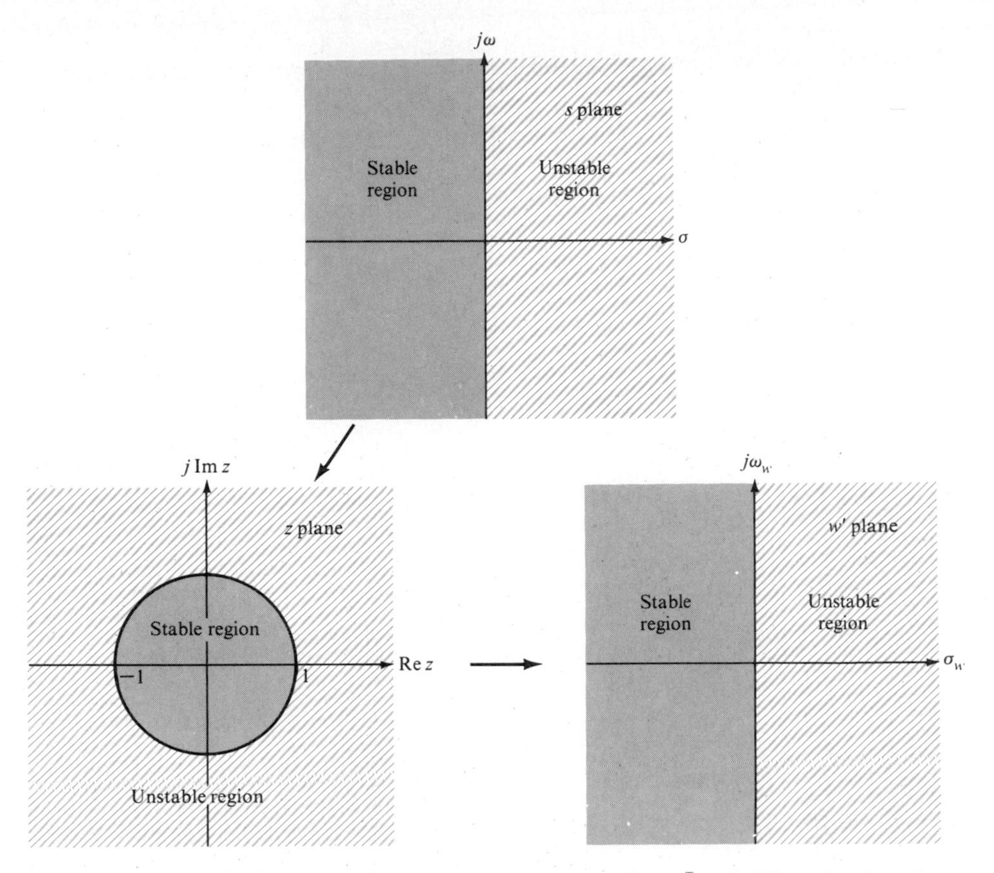

Figure 6-3 The mapping of the s plane into the z plane by means of $z = \epsilon^{Ts}$ and of the z plane into the w plane by means of Eq. (6-60a) or Eq. (6-63).

Therefore, a sampled-data control system is stable if the roots of the z-domain characteristic equation lie inside the unit circle.

The $\mathscr{Z}$ transform of the output-to-input ratio for the open-loop system of Fig. 6-4a is

$$G(z) = \frac{C(z)}{E(z)} = \frac{N(z)}{D(z)} \tag{6-1}$$

and for the closed-loop system of Fig. 6-4b it is given by Eq. (4-88), which is repeated below:

$$\frac{C(z)}{R(z)} = \frac{G_0(z)}{1 + G_0(z)H_0(z)} = \frac{W(z)}{Q(z)} \tag{6-2}$$

The characteristic equation[1] of a sampled-data control system is obtained by setting to zero the denominator of the system transfer function. For example, the

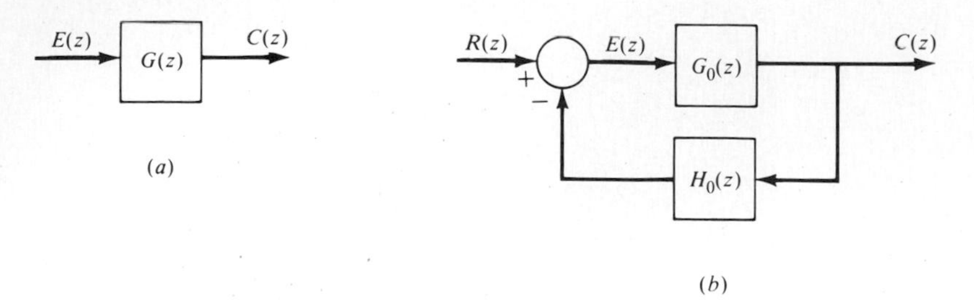

Figure 6-4 z-domain representations of
(*a*) An open-loop system: $C(z) = G(z)E(z)$ (see Fig. 4-12*a*)
(*b*) A closed-loop system: $\dfrac{C(z)}{R(z)} = \dfrac{G_0(z)}{1 + G_0(z)H_0(z)}$ (see Fig. 4-16)

characteristic equations for the open-loop and closed-loop systems represented by Eqs. (6-1) and (6-2) are given, respectively, by

$$D(z) \equiv 0 \tag{6-3}$$

$$Q(z) \equiv 1 + G_0(z)H_0(z) = 0 \tag{6-4a}$$

For the sampled-data system represented by Eq. (4-96), the characteristic equation may be obtained by rearranging this equation to

$$[1 + G_1G_2(z)]C(z) = G_2(z)G_1R(z)$$

Setting the forcing function to zero yields the homogeneous equation

$$[1 + G_1G_2(z)]C(z) = 0$$

which yields the characteristic equation

$$Q(z) = 1 + G_1G_2(z) = 0 \tag{6-4b}$$

In general, for a closed-loop sampled-data control system, the characteristic equation has the form

$$Q(z) \equiv 1 + P(z) = 0 \tag{6-5}$$

As mentioned previously, the nature and location of the roots of the characteristic equations (6-3) and (6-4) determine the stability and the dynamic behavior of the open-loop and closed-loop systems, respectively, in a manner similar to that for continuous-time systems. For a continuous-time system, the root locus for the closed-loop system is based on the characteristic equation $1 + G(s)H(s) = 0$ or, equivalently, $G(s)H(s) = \epsilon^{-j(1+2h)\pi}$, where $h = 0, \pm 1, \pm 2, \ldots$. Similarly a root-locus analysis can be made for a sampled-data system whose closed-loop characteristic equation is given by Eq. (6-5) (see Sec. 6-5).

Example 6-1 Given that the characteristic equation of a sampled-data system is $z^2 - 0.2Az + 0.1A = 0$, determine the range of values of A for stable operation of the system. Consider both positive and negative values of A.

SOLUTION As the roots of the characteristic equation can be real or can be complex conjugates, it is necessary to analyze the expression

$$z_{1,2} = 0.1A \pm \sqrt{0.01A^2 - 0.1A} \tag{6-6}$$

Consider, first, the case that the roots are real and, second, the case that the roots are complex conjugates.

1. *Real roots*
   a. For $A > 0$: To have real roots it is necessary that $0.01A^2 \geq 0.1A$, or $A \geq 10$. This results in at least one real root outside the UC, so the system is unstable.
   b. For $A < 0$: The roots are

   $$z_1 = 0.1A + \sqrt{0.01A^2 - 0.1A} < 1$$

   and

   $$z_2 = 0.1A - \sqrt{0.01A^2 - 0.1A} > -1$$

   Let $A = -|A| = -a$, where $a > 0$. Thus

   $$z_1 = -0.1a + \sqrt{0.01a^2 + 0.1a} < 1$$

   and

   $$z_2 = -0.1a - \sqrt{0.01a^2 + 0.1a} > -1$$

   Assuming the equality condition for the z_1 root (that is, $z_1 = 1$) results in

   $$0.01a^2 + 0.1a = (-1 + 0.1a)^2$$

   This yields $a = -10$, and thus $A = 10$ is not an allowable value. The equality condition for the z_2 root results in $0.01a^2 + 0.1a = (-1 + 0.1a)^2$ which yields $a = 10/3$. Therefore for stable real roots $-10/3 < A < 0$.
2. *Complex roots.* For Eq. (6-6) to have complex roots it is necessary that $A > 0$. Equation (6-6) is rearranged as follows:

   $$z_{1,2} = 0.1A \pm j\sqrt{0.1A - 0.01A^2} = b \pm jc \tag{6-7}$$

   For the complex roots to lie within the UC, Eq. (6-7) indicates that the condition $|z_{1,2}| = \sqrt{b^2 + c^2} \leq 1$ must be satisfied. Using the equality condition, substituting in for b and c, and squaring both sides yields $b^2 + c^2 = 0.01A^2 + 0.1A - 0.01A^2 = 1$; thus $A = 10$. Therefore, the range of values of A for stable operation of the system is $-10/3 < A < 10$.

The manual evaluation method used in Example 6-1 for determining the permissible range of gain A is not practical for systems higher than second-order. Jury's stability test of the next section is a method that can be used for higher-order systems.

6-3 EXTENDED z-DOMAIN STABILITY ANALYSIS: JURY'S STABILITY TEST[20,22]

The response transform $C(z)$ has the general form given by Eq. (6-8), where $R(z)$ is the forcing-function transform

$$C(z) = \frac{W(z)}{Q(z)} R(z)$$

$$= \frac{W(z)}{b_n z^n + b_{n-1} z^{n-1} + b_{n-2} z^{n-2} + \cdots + b_1 z + b_0} R(z) \tag{6-8}$$

Section 4-5 describes the methods used to evaluate the inverse transform $\mathscr{Z}^{-1}[C(z)] = c(kT)$. However, before the inverse transformation can be performed, the denominator polynomial of Eq. (6-8) must be factored. Computer and calculator programs are readily available for obtaining the roots of a polynomial.[14,15,23] Section 4-3 shows that stability of the response $c(kT)$ requires that all roots of $Q(z)$ lie inside the UC. Since it is usually not necessary to find the exact solution when the response is unstable, a simple procedure to determine the existence of roots that lie outside or on the UC is needed. If such roots of $Q(z)$ are found, the system is unstable and must be modified. Jury's stability test, devised by Jury and Blanchard, is a simple method of determining if any roots lie on and/or outside the UC in the z plane without actually solving for the roots of $Q(z)$. It is the counterpart of the Routh criterion for continuous-time systems. It should be noted that the roots of $Q(z)$ are poles of $C(z)$. The characteristic equation is

$$Q(z) \equiv b_n z^n + b_{n-1} z^{n-1} + b_{n-2} z^{n-2} + \cdots + b_1 z + b_0 = 0 \tag{6-9}$$

where all the coefficients are real and $b_n > 0$.

The coefficients of the characteristic equation are arranged in the pattern shown in the first two rows of the following Jury-Blanchard (JB) array. These coefficients are then used to evaluate the rest of the constants to complete the array.

j	Row	z^0	z^1	z^2	$\cdots$	z^{n-j}	$\cdots$	z^{n-1}	z^n
0	$2j+1=1$	b_0	b_1	b_2	$\cdots$	b_{n-j}	$\cdots$	b_{n-1}	b_n
	$2j+2=2$	b_n	b_{n-1}	b_{n-2}	$\cdots$	b_j	$\cdots$	b_1	b_0
1	$2j+1=3$	c_0	c_1	c_2	$\cdots$	c_{n-j}	$\cdots$	c_{n-1}	
	$2j+2=4$	c_{n-1}	c_{n-2}	c_{n-3}	$\cdots$	c_j	$\cdots$	c_0	
2	$2j+1=5$	d_0	d_1	d_2	$\cdots$	d_{n-2}			
	$2j+2=6$	d_{n-2}	d_{n-3}	d_{n-4}	$\cdots$	d_0			
$\vdots$	$\vdots$	$\vdots$	$\vdots$	$\vdots$	$\vdots$				
$n-3$	$2j+1=2n-5$	u_0	u_1	u_2	u_3				
	$2j+2=2n-4$	u_3	u_2	u_1	u_0				
$n-2$	$2j+1=2n-3$	v_0	v_1	v_2					
	$2j+2=2n-2$	v_2	v_1	v_0					

The constants of the $2j + 2$ row consist of the constants of the $2j + 1$ row written in reverse order. The constants c_i, d_i, e_i, and so forth, where $i = 0, 1, \ldots, n - j$, are evaluated as follows:

$$c_i = \begin{vmatrix} b_0 & b_{n-i} \\ b_n & b_i \end{vmatrix}_{j=1}, \quad d_i = \begin{vmatrix} c_0 & c_{n-1-i} \\ c_{n-1} & c_i \end{vmatrix}_{j=2},$$

$$e_i = \begin{vmatrix} d_0 & d_{n-2-i} \\ d_{n-2} & d_i \end{vmatrix}_{j=3}, \ldots, v_i = \begin{vmatrix} v_0 & v_{3-i} \\ v_3 & v_i \end{vmatrix}_{n-j} \tag{6-10}$$

Jury's stability test requires, in order for $Q(z) = 0$ not to have any roots on or outside the UC in the z plane, that the following conditions be both necessary and sufficient:

From Eq. (6-9),

$$Q(z)\Big|_{z=1} > 0 \tag{6-11}$$

and

$$Q(z)\Big|_{z=-1} \begin{cases} > 0 & \text{for } n \text{ even} \\ < 0 & \text{for } n \text{ odd} \end{cases} \tag{6-12}$$

From Eq. (6-10),

$$\left.\begin{array}{l} |b_0| < b_n \\ |c_0| > |c_{n-1}| \\ |d_0| > |d_{n-2}| \\ |e_0| > |e_{n-3}| \\ \cdots\cdots \\ |v_0| > |v_2| \end{array}\right\} n - 1 \text{ constraints} \tag{6-13}$$

The test procedure, for $n \geq 2$, is as follows:

Step 1. Determine if the conditions of Eqs. (6-11) and (6-12) and the $|b_0| < b_n$ constraint are satisfied. If they are not, the instability condition of the system has been determined. If they have been satisfied, proceed with Step 2.

Step 2. Determine the maximum value that j (≥ 0) can have as follows: From $2j + 2 = 2n - 2$ obtain

$$j_{\max} = n - 2 \tag{6-14}$$

If $j_{\max} = 0$ the procedure stops here since enough information is available to determine stability.

Step 3. The maximum number of rows of the array required to determine if the remaining $n - 2$ constraints are satisfied is given by

$$\text{Maximum number of rows} = 2j_{\max} + 1 \geq 0 \tag{6-15}$$

Step 4. Create thc JB array. As each row is completed apply the appropriate constraint. If this constraint is not satisfied, do not proceed with the array creation since the determination of instability has been achieved. Likewise, if all constraints are satisfied, the system is stable. It is only necessary to evaluate the first and last entry of the last required row to determine stability.

Example 6-2 Repeat Example 6-1, where $Q(z) = z^2 - 0.2Az + 0.1A$, using Jury's stability text.

Step 1. $|b_0| < b_2 \rightarrow |0.1A| < 1 \rightarrow |A| < 10$

$$Q(1) = 1 - 0.2A + 0.1A = 1 - 0.1A > 0 \qquad \text{or } A < 10$$

$$Q(-1) = 1 + 0.2A + 0.1A = 1 + 0.3A > 0 \qquad \text{or } A > -10/3$$

Step 2. For $n = 2$,

$$j_{\max} = n - 2 = 0$$

Since $j_{\max} = 0$ based upon Step 1, the range of A for stable operation of the system is $-10/3 < A < 10$ which, naturally, agrees with the results of Example 6-1.

Example 6-3 Given $Q(z) = z^4 - 2z^3 + 1.5z^2 - 0.1z - 0.02$, apply Jury's stability test to determine if $Q(z)$ has any roots on or outside the UC.

SOLUTION For $n = 4$,

Step 1. $|b_0| < b_n \rightarrow |-0.02| < 1$

$$Q(1) = 0.38 > 0 \qquad \text{and} \qquad Q(-1) = 0.58 > 0$$

Step 2. $j_{\max} = 2$
Step 3. Maximum number of rows $= 2j_{\max} + 1 = 5$
Step 4.

	Row	z^0	z^1	z^2	z^3	z^4
0	1	−0.02	−0.1	1.5	−2	1
	2	1	−2	1.5	−0.1	−0.02
1	3	−0.9996	2.002	−1.53	0.14	
	4	0.14	−1.53	2.002	−0.9996	
2	5	0.9796	...	1.2491		

From this array it is seen that the constraint $|c_0| = |-0.9996| > |c_3| = |0.14|$ is satisfied and the constraint $|d_0| = |0.9796| > |d_2| = |1.249|$ is not satisfied. Thus $Q(z)$ has one or more roots on or outside the UC and the system is unstable.

6-4 STEADY-STATE ERROR ANALYSIS FOR STABLE SYSTEMS[1, 21]

The three important characteristics of a control system are (1) stability, (2) steady-state performance, and (3) transient response. The first item of interest in the analysis of a system is its stability characteristics. If there is a range of gain for which a system yields a stable performance, then the next item of interest is the system's steady-state error characteristics; that is, can the system output $c(kT)$ follow a given input $r(kT)$ with zero or a small value of error $[=r(kT) - c(kT)]$. If the first two items are satisfactory, a transient time-response analysis is then made. Sections 6-2 and 6-3 have dealt with the first item. This section discusses the second characteristic, and the remaining sections deal, mainly, with the transient response.

Fortunately the analysis of the steady-state error characteristics of a *unity-feedback sampled-data system* parallels the analysis for a unity-feedback continuous-time stable system based upon system types and upon specific forms of the system input function. For the unity-feedback sampled-data system of Fig. 6-5, the control ratio and the output and error signals expressed in the z domain are, respectively,

$$\frac{C(z)}{R(z)} = \frac{G(z)}{1 + G(z)} \tag{6-16a}$$

$$C(z) = \frac{G(z)}{1 + G(z)} R(z) \tag{6-16b}$$

and

$$E(z) = R(z) - C(z) = \frac{R(z)}{1 + G(z)} \tag{6-17}$$

Thus, *if the system is stable*, the final-value theorem can be applied to Eqs. (6-16*b*) and (6-17) to obtain the steady-state or final value of the output and the error at the sampling instants; that is,

$$c^*(\infty) = \lim_{t \to \infty} c^*(t) = \lim_{z \to 1} \left[\frac{(1 - z^{-1})G(z)}{1 + G(z)} R(z) \right] \tag{6-18}$$

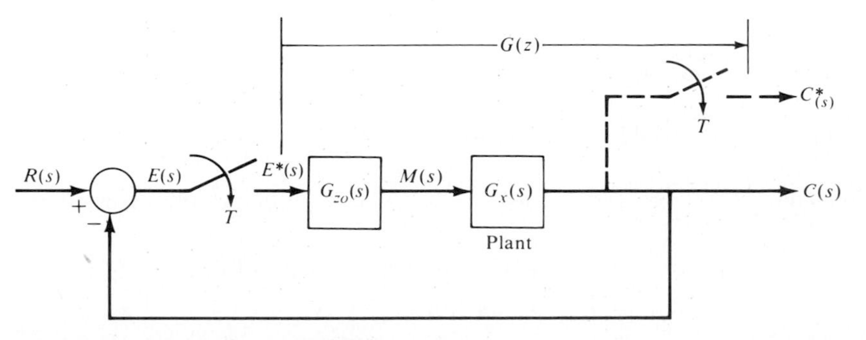

Figure 6-5 A unity-feedback sampled-data system.

$$e^*(\infty) = \lim_{t \to \infty} e^*(t) = \lim_{z \to 1} \frac{(1 - z^{-1})R(z)}{1 + G(z)} \tag{6-19}$$

6-4A Steady-State Error-Coefficient Formulation

The steady-state error coefficients have the same meaning and importance for sampled-data systems as for continuous-time systems; i.e., how well can the system output follow a given type of input forcing function. *The following derivations of the error coefficients are independent of the system type. They apply to any system type and are defined for specific forms of the input. These error coefficients are useful only for stable unity-feedback systems.*

Step input, $R(z) = R_0 z/(z - 1)$ The step error coefficient K_p is defined as

$$K_p \equiv \frac{c^*(\infty)}{e^*(\infty)} \tag{6-20}$$

Substituting from Eqs. (6-18) and (6-19) into Eq. (6-20) yields

$$K_p = \lim_{z \to 1} G(z) \tag{6-21}$$

which applies only for a step input, $r(t) = R_0 u_{-1}(t)$.

Ramp input, $R(z) = R_1 Tz/(z - 1)^2$ The ramp error coefficient K_v is defined as

$$K_v \equiv \frac{\text{steady-state value of derivative of output}}{e^*(\infty)} \tag{6-22}$$

Since Eq. (3-62) represents the derivative of $c(t)$ in the discrete-time domain, use of the translation and final-value theorems [Eqs. (4-27) and (4-31), respectively] permits Eq. (6-22) to be written as

$$K_v = \frac{\lim_{z \to 1} \left[\dfrac{(1 - z^{-1})^2}{T} C(z) \right]}{e^*(\infty)} \tag{6-23}$$

Substituting from Eqs. (6-18) and (6-19) into Eq. (6-23) yields

$$K_v = \frac{1}{T} \lim_{z \to 1} \left[\frac{z - 1}{z} G(z) \right] \qquad \mathrm{s}^{-1} \tag{6-24}$$

which applies only for a ramp input, $r(t) = R_1 t u_{-1}(t)$.

Parabolic input, $R(z) = R_2 T^2 z(z + 1)/2(z - 1)^3$ The parabolic error coefficient K_a is defined as

$$K_a \equiv \frac{\text{steady-state value of second derivative of output}}{e^*(\infty)} \tag{6-25}$$

Since Eq. (3-67) represents the second derivative of the output in the discrete-time domain, use of the translation and final-value theorems permits Eq. (6-25) to be written as

$$K_a = \frac{\lim_{z\to 1}\left[\frac{(1-z^{-1})^3}{T^2}C(z)\right]}{e^*(\infty)} \tag{6-26}$$

Substituting from Eqs. (6-18) and (6-19) into Eq. (6-26) yields

$$K_a = \frac{1}{T^2}\lim_{z\to 1}\left[\frac{(z-1)^2}{z^2}G(z)\right] \quad \mathrm{s}^{-2} \tag{6-27}$$

which applies only for a parabolic input, $r(t) = R_2 t^2 u_{-1}(t)/2$.

6-4B Evaluation of Steady-State Error Coefficients

The forward transfer function of a sampled-data system in the z domain has the general form

$$G(z) = \frac{Kz^d(z-a_1)(z-a_2)\cdots(z-a_i)\cdots}{(z-1)^m(z-b_1)(z-b_2)\cdots(z-b_j)\cdots} \tag{6-28}$$

where a_i and b_j may be real or complex, d and m are positive integers, $m = 0, 1, 2, \ldots$, and *m represents the system type*. Note that the $(z-1)^m$ term in Eq. (6-28) corresponds to the s^m term in the denominator of the forward transfer function of a continuous-time Type m system. Substituting from Eq. (6-28) into Eqs. (6-21), (6-24), and (6-27), respectively, yields the following values of the steady-state error coefficients for the various Type m systems:

$$K_p = \begin{cases} \lim_{z\to 1}\frac{Kz^d(z-a_1)(z-a_2)\cdots}{(z-b_1)(z-b_2)\cdots} = K_0 & \text{Type 0} \quad (6\text{-}29) \\ \infty & \text{Type 1} \quad (6\text{-}30) \\ \infty & \text{Type 2} \quad (6\text{-}31) \end{cases}$$

$$K_v = \begin{cases} \frac{1}{T}\lim_{z\to 1}\frac{Kz^d(z-1)(z-a_1)(z-a_2)\cdots}{(z-b_1)(z-b_2)\cdots} = 0 & \text{Type 0} \quad (6\text{-}32) \\ \frac{1}{T}\lim_{z\to 1}\frac{Kz^d(z-a_1)(z-a_2)\cdots}{(z-b_1)(z-b_2)\cdots} = K_1 & \text{Type 1} \quad (6\text{-}33) \\ \infty & \text{Type 2} \quad (6\text{-}34) \end{cases}$$

$$K_a = \begin{cases} \frac{1}{T^2}\lim_{z\to 1}\frac{Kz^d(z-1)^2(z-a_1)(z-a_2)\cdots}{(z-b_1)(z-b_2)\cdots} = 0 & \text{Type 0} \quad (6\text{-}35) \\ 0 & \text{Type 1} \quad (6\text{-}36) \\ \frac{1}{T^2}\lim_{z\to 1}\frac{Kz^d(z-a_1)(z-a_2)\cdots}{(z-b_1)(z-b_2)\cdots} = K_2 & \text{Type 2} \quad (6\text{-}37) \end{cases}$$

Table 6-1 Steady-state error coefficients for stable systems

System type	Step error coefficient K_p	Ramp error coefficient K_v	Parabolic error coefficient K_a
0	K_0	0	0
1	∞	K_1	0
2	∞	∞	K_2

In applying Eqs. (6-29) to (6-37) it is required that the denominator of $G(z)$ be in factored form to ascertain if it contains any $z - 1$ factor(s). Table 6-1 summarizes the results of Eqs. (6-29) to (6-37).

6-4C Use of Steady-State Error Coefficients

The importance of the steady-state error coefficients is illustrated by means of an example.

Example 6-4 For the system of Fig. 6-5, consider $G(z)$ of the form of Eq. (6-28) with $m = 1$ (a Type 1 system). Determine the value of $e^*(\infty)$ for each of the three standard inputs (step, ramp, and parabolic), assuming that the system is stable.

From the definitions of Eqs. (6-20), (6-22), (6-25) and from Table 6-1, the following results are obtained:

$$e^*(\infty) = \frac{c^*(\infty)}{K_p} = \frac{c^*(\infty)}{\infty} = 0 \tag{6-38}$$

$$e^*(\infty) = \frac{\text{steady-state value of derivative of output}}{K_1} = E_0 \tag{6-39}$$

$$e^*(\infty) = \frac{\text{steady-state value of second derivative of output}}{0} = \infty \tag{6-40}$$

Thus a Type 1 sampled-data stable system (1) can follow a step input with zero steady-state error, (2) can follow a ramp input with a constant error E_0, and (3) cannot follow a parabolic input. Equation (6-39) indicates that the value of E_0 for a given value of R_1 (for the ramp input) may be made smaller by making K_1 larger. This assumes that the desired degree of stability and the desired transient performance is maintained while K_1 is increased in value. A similar analysis can be made for Types 0 and 2 systems (see problems).

Example 6-5 The expression for the output of the control system shown in Fig. 6-6 is

$$C(z) = \frac{G_2(z)G_1R(z)}{1 + G_2G_1H(z)} \tag{6-41}$$

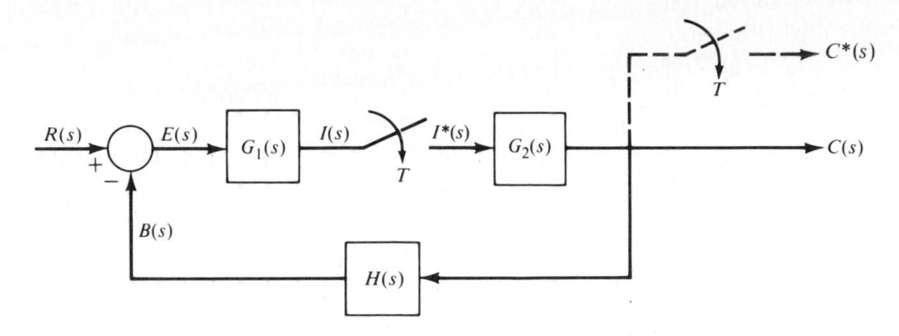

Figure 6-6 A nonunity-feedback sampled-data control system.

For a specified input $r(t)$ (step, ramp, or parabolic), obtain $R(z) = \mathscr{Z}[r(t)]$, as if $r(t)$ is sampled, and then obtain the pseudocontrol ratio $[C(z)/R(z)]_p$; that is, divide both sides of Eq. (6-41) by $R(z)$ to obtain

$$\left[\frac{C(z)}{R(z)}\right]_P = \frac{1}{R(z)} \frac{G_2(z)G_1R(z)}{1 + G_2G_1H(z)} \tag{6-42}$$

By setting Eq. (6-42) equal to Eq. (6-16a), where $G(z) = G_{eq}(z)$, it is then possible to obtain an expression for $G_{eq}(z)$ that represents the forward transfer function of an equivalent unity-feedback system represented by Fig. 6-5. This expression of $G_{eq}(z)$ can then be used to determine the "effective" Type m system that Fig. 6-6 represents and to solve for K_p, K_v, and K_a.

As a specific example, let

$$\left[\frac{C(z)}{R(z)}\right]_P = \frac{K(z^2 + az + b)}{z^3 + cz^2 + dz + e} \tag{6-43}$$

where $r(t) = u_{-1}(t)$, $K = 0.129066$, $a = 0.56726$, $b = -0.386904$, $c = -1.6442$, $d = 1.02099$, and $e = -0.224445$. For these values of the coefficients the system is stable. Setting Eq. (6-43) equal to Eq. (6-16a) and solving for $G_{eq}(z)$ yields

$$G_{eq}(z) = \frac{K(z^2 + az + b)}{z^3 - 1.773266z^2 + 0.947776z - 0.174509} = \frac{N_G(z)}{D_G(z)} \tag{6-44}$$

Applying the final-value theorem to Eq. (6-43) yields $c(\infty) = 1$. Therefore the nonunity-feedback system of Fig. 6-6 effectively acts at least as a Type 1 system. Based upon Eq. (6-28) this implies that $D_G(z)$ contains at least one factor of the form $z - 1$. Dividing $D_G(z)$ by $z - 1$ yields $z^2 - 0.773266z + 0.17451$, which does not contain $z - 1$ as a factor. Thus the nonunity-feedback system is a Type 1 system. Equation (6-44) is rewritten as follows:

$$G_{eq}(z) = \frac{K(z^2 + az + b)}{(z - 1)(z^2 - 0.773266z + 0.17451)}$$

For $T = 0.1$, then,

$$K_1 = \frac{1}{T} \lim_{z \to 1} \left[\frac{z-1}{z} G_{eq}(z) \right] = \frac{K(1 + a + b)}{T(0.401244)} \approx 3.796 \text{ s}^{-1}$$

6-5 ROOT-LOCUS ANALYSIS

A designer can verify whether the design of a control system meets the specifications by determining the actual time response of the controlled variable. By deriving the differential or difference equations for the control system and solving them, an accurate solution of the system's performance can be obtained. If the response does not meet the specifications, it is not easy to determine from this solution just what physical parameters in the system should be changed to improve the response.

A designer wishes to be able to predict a system's performance by an analysis that does not require the actual solution of the differential or difference equations. Also, it is desired that this analysis indicate readily the manner or method by which this system must be adjusted or compensated to produce the desired performance characteristics.

The first thing that a designer wants to know about a given sampled-data system is whether or not it is stable. This can be determined by examining the roots obtained from the characteristic equation $1 + G(z)H(z) = 0$. By applying Jury's stability test to the characteristic equation it is possible to determine whether the system is stable or unstable. Yet this does not satisfy the designer because it does not indicate the degree of stability of the system, i.e., the amount of overshoot and the settling time of the controlled variable. Not only must the system be stable, but the overshoot must be maintained within prescribed limits and transients must die out in a sufficiently short time. The root-locus method, which is used in this text, not only indicates whether a system is stable or unstable but, for a stable system, also shows the degree of stability.

The root-locus method is used to analyze the performance of a sampled-data control system in the same manner as for a continuous-time control system. For either type of system, the *root locus is a plot of the roots of the characteristic equation of the closed-loop system as a function of the gain constant*. This graphical approach yields a clear indication of gain-adjustment effects with relatively small effort compared with other methods. The underlying principle is that the poles of $C(z)/R(z)$ or $C(z)$ (transient-response modes) are related to the zeros and poles of the open-loop transfer function $G(z)H(z)$ and also to the gain. An important advantage of the root-locus method is that the roots of the characteristic equation of the system can be obtained directly; this results in a complete and accurate solution of the transient and steady-state response of the controlled variable. Another important feature is that an approximate control solution can be obtained with a reduction of the required work. As with any other design technique, a person can readily obtain proficiency with the root-locus method. With the help of a computer-aided-design (CAD) package, it is possible to synthesize a compensator, if one is required, with relative ease.

This section presents a detailed summary of the root-locus method. The first subsection details a procedure for obtaining the root locus, the next subsection defines the root-locus construction rules for negative feedback, and the last subsection contains examples of this method. The reader who desires a more in-depth discussion of this method, for review purposes, is referred to Refs. 1 and 21.

6-5A Procedure Outline

The procedure to be followed in applying the root-locus method is outlined in this subsection. This procedure is easier when a CAD program is used to obtain the root locus. Such a program can provide the desired data for the root locus in plotted or tabular form. This procedure, which is a modified version of the one applicable for the continuous-time systems, is summarized as follows:

Step 1. Derive the open-loop transfer function $G(z)H(z)$ of the system.

Step 2. Factorize the numerator and denominator of the transfer function into linear factors of the form $z + a$, where a may be real or complex.

Step 3. Plot the zeros and poles of the open-loop transfer function in the z plane, where $z = \sigma_z + j\omega_z$.

Step 4. The plotted zeros and poles of the open-loop function determine the roots of the characteristic equation of the closed-loop system $[1 + G(z)H(z) = 0]$. By use of the geometrical shortcuts or a digital-computer program, determine the locus that describes the roots of the closed-loop characteristic equation.

Step 5. Calibrate the locus in terms of the loop sensitivity K. If the gain of the open-loop system is predetermined, the location of the exact roots of $1 + G(z)H(z) = 0$ is immediately known. If the location of the roots (or ζ) is specified, the required value of K can be determined.

Step 6. Once the roots have been found in Step 5, the system's time response can be calculated by taking the inverse $\mathscr{Z}$ transform, either manually or by use of a computer program.

Step 7. If the response does not meet the desired specifications, determine the shape that the root locus must have to meet these specifications.

Step 8. Synthesize the network that must be inserted into the system, if other than gain adjustment is required, to make the required modification on the original locus. This process, called *compensation*, is described in Chaps. 7, 11, and 12.

6-5B Summary of Root-Locus Construction Rules for Negative Feedback

The system characteristic equation, Eq. (6-5), is rearranged as follows:

$$P(z) = -1 \tag{6-45}$$

Assume that $P(z)$ represents the open-loop function

$$P(z) = G(z)H(z) = \frac{K(z - z_1) \cdots (z - z_i) \cdots (z - z_w)}{(z - p_1) \cdots (z - p_c) \cdots (z - p_n)} \tag{6-46}$$

where z_i and p_c are the open-loop zeros and poles, respectively, and K *is defined as the static loop sensitivity* (gain constant) when $P(z)$ is expressed in this format. Equation (6-45) falls into the mathematical format for root-locus analysis, that is,

Magnitude condition: $$|P(z)| = 1 \tag{6-47}$$

Angle condition: $$-\beta = \begin{cases} (1 + 2h)180^\circ & \text{for } K > 0 \\ h360^\circ & \text{for } K < 0 \end{cases} \tag{6-48}$$

Thus the construction rules for continuous-time systems, with minor modifications since the plot is in the z plane, are applicable for sampled-data systems and are summarized as follows:

Rule 1 The number of branches of the root locus is equal to the number of poles of the open-loop transfer functions.

Rule 2 For *positive* values of K, the root locus exists on those portions of the real axis for which the sum of the poles and zeros to the right is an odd integer. For *negative* values of K, the root locus exists on those portions of the real axis for which the sum of the poles and zeros to the right is an even integer (including zero).

Rule 3 The root locus starts ($K = 0$) at the open-loop poles and terminates ($K = \pm\infty$) at the open-loop zeros or at infinity.

Rule 4 The angles of the asymptotes of the root locus that end at infinity are determined by

$$\gamma = \frac{(1 + 2h)180^\circ}{[\text{no. of poles of } G(z)H(z)] - [\text{no. of zeros of } G(z)H(z)]} \tag{6-49a}$$

for $K > 0$ and

$$\gamma = \frac{h360^\circ}{[\text{no. of poles of } G(z)H(z)] - [\text{no. of zeros of } G(z)H(z)]} \tag{6-49b}$$

for $K < 0$.

Rule 5 The real-axis intercept of the asymptotes is

$$z_0 = \frac{\sum_{c=1}^{n} \operatorname{Re} p_c - \sum_{i=1}^{w} \operatorname{Re} z_i}{n - w} \tag{6-50}$$

Rule 6 The breakaway point for the locus between two poles on the real axis (or the break-in point for the locus between two zeros on the real axis) can be determined by taking the derivative of the loop sensitivity K with respect to z. Equate this derivative to zero and find the roots of the resulting equation. The root that occurs between the poles (or the zeros) is the breakaway (or break-in) point.

Rule 7 For $K > 0$ the angle of departure from a complex pole is equal to 180° minus the sum of the angles from the other poles plus the sum of the angles from the zeros. Any of these angles may be positive or negative. For $K < 0$ the departure angle is 180° from that obtained for $K > 0$.

For $K > 0$ the angle of approach to a complex zero is equal to the sum of the angles from the poles minus the sum of the angles from the other zeros minus 180°. For $K < 0$ the approach angle is 180° from that obtained for $K > 0$.

Rule 8 The unit circle crossing of the root locus can be determined by setting up the JB array from the closed-loop characteristic equation. Determine the range of values from the root locus that K must have to satisfy the necessary and sufficient conditions for a stable system [see Eqs. (6-10) to (6-12)].

Rule 9 The root loci are symmetrical about the real axis.

Rule 10 The static loop sensitivity calibration K of the root locus can be made by applying the magnitude condition given by Eq. (6-18) as follows:

$$K = \frac{|z - p_1| \cdot |z - p_2| \cdots |z - p_c| \cdots |z - p_n|}{|z - z_1| \cdot |z - z_2| \cdots |z - z_w|} \tag{6-51}$$

Rule 11 The selection of the dominant roots of the characteristic equation is based on the specifications that give the required system performance; i.e., it is possible to evaluate σ, ω_d, and ζ from Eqs. (5-7), (5-8), and (5-10). These values in turn are mapped into the z domain to determine the location of the desired dominant roots in the z plane. The loop sensitivity for these roots is determined by means of the magnitude condition. The remaining roots are then determined to satisfy the same magnitude condition.

A root-locus CAD digital-computer program[14,15] produces an accurate calibrated root locus. This considerably simplifies the work required for the system design. By specifying ζ for the dominant roots or K, use of a computer program can yield all the roots of the characteristic equation.†

It should be remembered that the entire *unbounded* left-half s plane is mapped

† A good engineering design rule as a first estimate for a calculation step size in a CAD program is $T/10$ in order to generate accurate results.

into the UC in the z plane that has a *finite area*. In mapping a given set of poles and zeros in the left-half s plane into the z plane, they migrate to the vicinity of the $1 + j0$ point of the UC as $T \to 0$. Thus in plotting the poles and zeros of $G(z) = \mathscr{Z}[G(s)]$ they approach the $1 + j0$ point as $T \to 0$. For a "small-enough" value of T and an inappropriate plot scale, some or all of these poles and zeros "appear to lie on top of one another." Therefore, *caution* should be exercised in *selecting* the *scale* for the root-locus plot and the *degree of accuracy* that is needed to be maintained for an accurate analysis and design of a sampled-data system.

6-5C Examples

Example 6-6 Example 6-1 is repeated, using the root-locus method. The second-order characteristic equation $z^2 - 0.2Az + 0.1A = 0$ is partitioned[21] to put it into the format of Eq. (6-45) as follows:

$$z^2 = 0.2Az - 0.1A = 0.2A(z - 0.5) = -K(z - 0.5) \tag{6-52}$$

where $K = -0.2A$. Equation (6-52) is rearranged to yield

$$P(z) = \frac{K(z - 0.5)}{z^2} = -1 \tag{6-53}$$

which is of the mathematical format of Eq. (6-45). The poles and zero of Eq. (6-53) are plotted in the z plane as shown in Fig. 6-7. The construction rules applied to this example yield the following information.

Rule 1. Number of branches of the root locus is given by $n = 2$.

Rule 2. For $K > 0$ ($A < 0$) the real-axis locus exists between $z = 0.5$ and $z = -\infty$, and for $K < 0$ ($A > 0$) the real-axis locus exists between $z = 0.5$ and $z = +\infty$.

Rule 3. The root-locus branches start, with $K = 0$, at the poles of $P(z)$. One branch ends at the zero, $z = 0.5$, and one branch ends at infinity for $K = \pm\infty$.

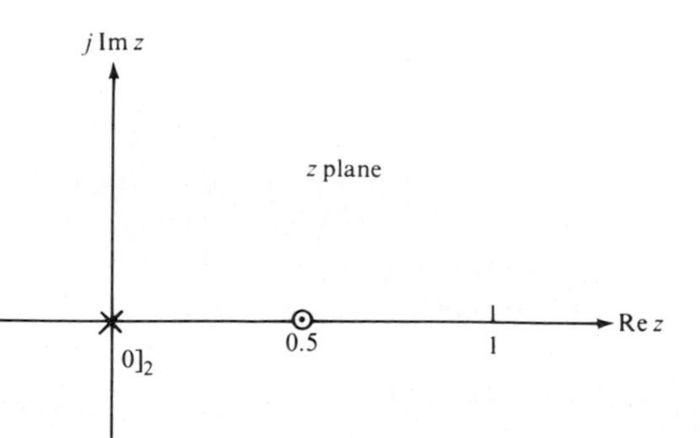

Figure 6-7 Poles and zero of Eq. (6-53).

Rule 4. Asymptotes

For $K > 0$:

$$\gamma = \frac{(1 + 2h)180°}{2 - 1} = (1 + 2h)180° \qquad \text{thus } \gamma = 180°$$

For $K < 0$:

$$\gamma = \frac{h360°}{2 - 1} = h360° \qquad \text{thus } \gamma = 0° \text{ (or } 360°)$$

Rule 5. Not applicable for the asymptotes determined by Rule 4.

Rule 6. For this example there is no breakaway point on the real axis for $K > 0$. For $K < 0$, Eq. (6-53) is rearranged to obtain the function

$$W(z) = \frac{z^2}{-z + 0.5} = K \tag{6-54}$$

Taking the derivative of this function and setting it equal to zero yields the break-in and breakaway points. Thus

$$\frac{dW(z)}{dz} = \frac{z(z - 1)}{(-z + 0.5)^2} = 0$$

which yields $z_{1,2} = 0, 1$. Therefore $z_1 = 0$ is the breakaway point and $z_2 = 1$ is the break-in point.

Rule 7. Not applicable for this example.

Rule 8. The unit circle intersections of the root locus as determined by the Jury stability test of Example 6-2 occur for the static loop sensitivity values of $K = 2/3$ and $K = -2$.

Rule 9. The root locus is symmetrical about the real axis.

Rule 10. The static loop sensitivity calibration of the root locus can be made by evaluating Eq. (6-54) for various values of z that lie on the root locus.

Rule 11. Not applicable for this example.

The root locus for Eq. (6-53) is shown in Fig. 6-8.

Example 6-7 *Given* the unity-feedback sampled-data system shown in Fig. 6-5, where $G_x(s) = K_G/[s(s + 2)]$. The objectives of this example are as follows: (*a*) determine $C(z)/R(z)$ in terms of K_G and T (i.e., the values of K_G and T are unspecified); (*b*) determine the root locus and the maximum value of K_G for a stable response with $T = 0.1$ s; (*c*) determine the steady-state error characteristics with various inputs for this system for those values of K_G and T that yield a stable system response; and (*d*) determine the roots of the characteristic equation for $\zeta = 0.6$, the corresponding time response $c(kT)$, and the figures of merit.

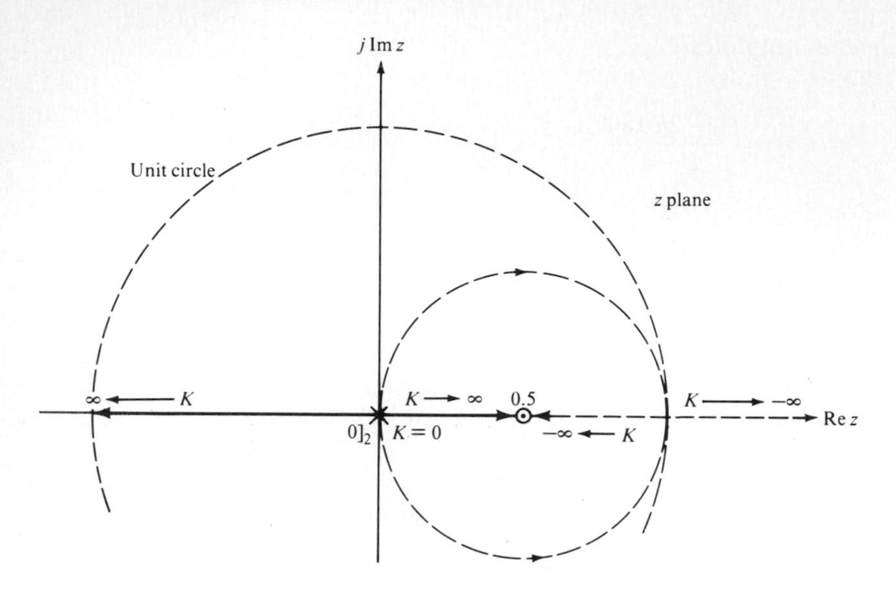

Figure 6-8 Root-locus plot of Example 6-6.

SOLUTION (*a*) The forward transfer function of the open-loop system is

$$G(s) = G_{z0}(s)G_x(s) = \frac{K_G(1 - \epsilon^{-sT})}{s^2(s + 2)} = (1 - \epsilon^{-sT})\frac{K_G}{s^2(s + 2)}$$

$$= G_z^*(s)\frac{K_G}{s^2(s + 2)}$$

Thus, using entry 8 in Table 4-1 yields

$$G(z) = G_z(z)\mathscr{Z}\left[\frac{K_G}{s^2(s + 2)}\right] = (1 - z^{-1})\mathscr{Z}\left[\frac{K_G}{s^2(s + 2)}\right]$$

$$= \frac{K_G[(T + 0.5\epsilon^{-2T} - 0.5)z + (0.5 - 0.5\epsilon^{-2T} - T\epsilon^{-2T})]}{2(z - 1)(z - \epsilon^{-2T})} \qquad (6\text{-}55)$$

Substituting Eq. (6-55) into

$$\frac{C(z)}{R(z)} = \frac{G(z)}{1 + G(z)} = \frac{N(z)}{D(z)}$$

yields

$$\frac{C(z)}{R(z)} = \frac{0.5K_G[(T - 0.5 + 0.5\epsilon^{-2T})z + (0.5 - 0.5\epsilon^{-2T} - T\epsilon^{-2T})]}{z^2 - [(1 + \epsilon^{-2T}) - 0.5K_G(T - 0.5 + 0.5\epsilon^{-2T})]z + \epsilon^{-2T} + 0.5K_G(0.5 - 0.5\epsilon^{-2T} - T\epsilon^{-2T})} \qquad (6.56)$$

(*b*) From Eq. (6-56), the characteristic equation is given by $Q(z) = 1 + G(z) = 0$, which yields

$$G(z) = \frac{K\left(z + \dfrac{0.5 - 0.5\epsilon^{-2T} - T\epsilon^{-2T}}{T + 0.5\epsilon^{-2T} - 0.5}\right)}{(z-1)(z-\epsilon^{-2T})} = -1 \tag{6-57}$$

where $K = 0.5K_G(T + 0.5\epsilon^{-2T} - 0.5)$. For $T = 0.1$ s,

$$G(z) = \frac{K(z + 0.9355)}{(z-1)(z-0.81873)} = -1 \tag{6-58}$$

and $K = 0.004683K_G$. The root locus for Eq. (6-58) is shown in Fig. 6-9. For $T = 0.1$ the maximum value of K for a stable response is $K \approx 0.1938$, which results in $K_{G_{\max}} \approx 41.38$. An analysis of Eq. (6-57) reveals that:

1. The pole at ϵ^{-2T} approaches the UC as $T \to 0$ and approaches the origin as $T \to \infty$.
2. The zero approaches -2 as $T \to 0$ (this can be determined by applying L'Hopital's rule twice) and approaches the origin as $T \to \infty$.
3. Based upon the plot scale chosen it may be difficult to interpret or secure accurate values from the root locus in the vicinity of $1 + j0$ point as $T \to 0$. That is, both poles will "appear" to be superimposed.

As a consequence of items 1 and 2 and considering only the root locus, one may jump to the conclusion that the range of K_G for a stable system decreases as $T \to 0$. This is not the case for this example since K is a function

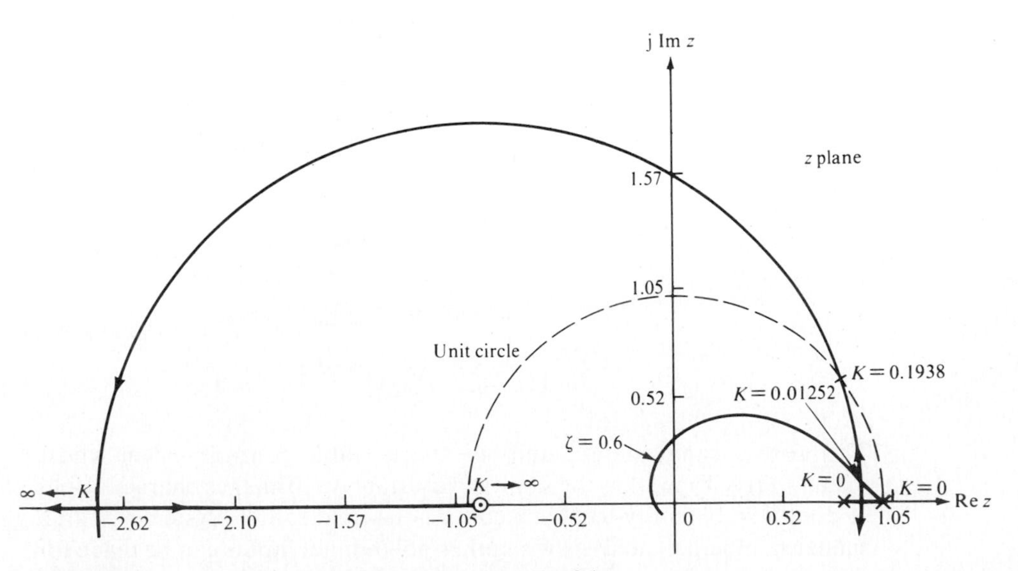

Figure 6-9 A root-locus sketch for Eq. (6-58), where $T = 0.1$ s.

of T, as is shown in the next section. As pointed out in item 3, for an open-loop transfer function having a number of poles and zeros in the vicinity of $z = 1$, it may be difficult to obtain an accurate root-locus plot in the vicinity of $z = 1$ if the plotting area is too large. This accuracy aspect can best be illustrated if the ζ contours of Fig. 4-9 are used graphically to locate a set of dominant complex roots $p_{1,2}$ corresponding to a desired ζ. Trying to determine graphically the values of the roots at the intersection of the desired ζ contours and the dominant root-locus branches is most difficult. Any slight error in the values of $p_{1,2}$ may result in a pair of dominant roots having a larger or smaller value of ζ from the desired value. This problem is also involved even if a computer-aided program is used to locate this intersection, especially if the program is not implemented with the necessary degree of calculation accuracy. Also, the word length (number of binary digits) of the selected digital control processor may not be sufficient to provide the desired damping performance without extended precision. It may be necessary to reduce the plotting area to a small-enough region about $z = 1$ and then to reduce the calculation step size in order to obtain an accurate picture of the range of K for a stable system performance. This aspect of accuracy is amplified at the end of this section.

(*c*) $C(z)$, for a step input $[R(z) = z/(z - 1)]$, is solved from Eq. (6-56). Applying the final-value theorem to $C(z)$ yields

$$c(\infty) = \lim_{z \to 1} [(1 - z^{-1})C(z)] = 1$$

Therefore $e^*(\infty) = 0$ for a stable system.

For other than step inputs one must analyze the steady-state performance characteristics of $E(z)$ for a unity-feedback system. Thus

$$E(z) = \frac{1}{1 + G(z)} R(z) \tag{6-59}$$

Considering the ramp input $R(z) = Tz/(z - 1)^2$, $E(z)$ for this example is

$$E(z) = \left\{ \frac{2(z-1)(z-\epsilon^{-2T})}{2(z-1)(z-\epsilon^{-2T}) + K_G[(T + 0.5\epsilon^{-2T} - 0.5)z + (0.5 - 0.5\epsilon^{-2T} - T\epsilon^{-2T})]} \right\} \frac{Tz}{(z-1)^2}$$

Thus for a stable system

$$e^*(\infty) = \lim_{z \to 1} [(1 - z^{-1})E(z)] = \left. \frac{1}{K} \right|_{K>0} > 0$$

Therefore a sampled-data unity-feedback stable control system whose plant $G_x(s)$ is Type 1 has the same steady-state performance characteristics as does a stable unity-feedback continuous-time control system.[1,21] In a similar manner, an analysis with other polynomial inputs can be made for other Type m plants.

(*d*) It is determined that the roots of the characteristic equation, for $\zeta = 0.6$, are $p_{1,2} = 0.90311 \pm j0.12181$ (where $K = 0.01252$). The output time-response function is

$$c(kT) = 0.01252000r[(k-1)T] + 0.01171246r[(k-2)T] + 1.80621000c[(k-1)T] - 0.83044246c[(k-2)T]$$

and the figures of merit are $M_p \approx 1.113$, $t_p \approx 2.35$ s, $t_s = 3.6$ s, and $K_1 = 1.3368\ \text{s}^{-1}$. Note that $K_G = 2.6735$. It should be noted that the roots can also be determined graphically by the use of the ζ contours of Fig. 4-9 as shown in Fig. 6-9 with limited accuracy.

As discussed previously in this section, Figs. 6-10 and 6-11 illustrate the care that must be exercised in performing a mathematical analysis for a sampled-data

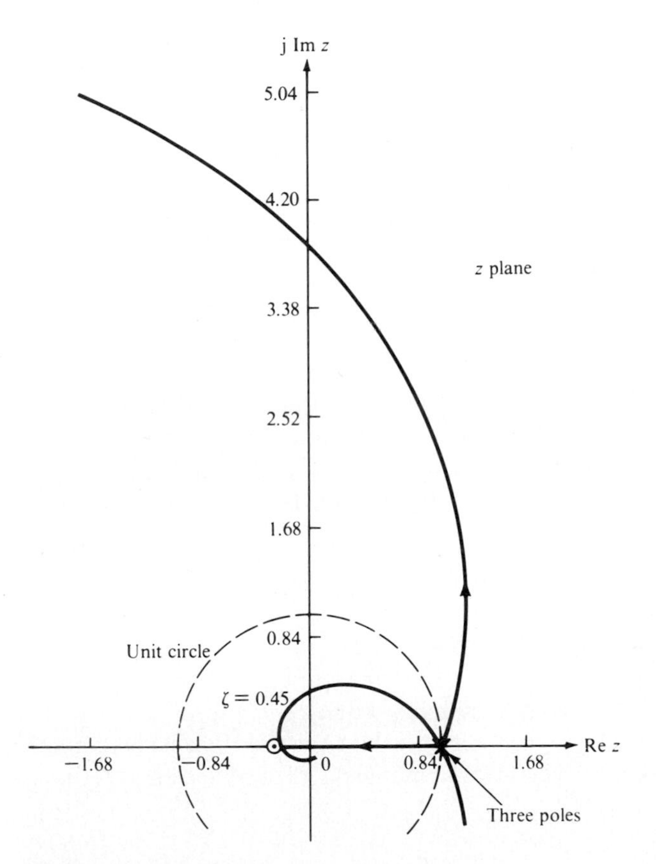

Figure 6-10 Root locus for

$$G(z) = \frac{K_z(z + 0.26395)(z + 3.6767)}{(z - 1)(z - 0.99005)(z - 0.95123)}$$

where $T = 0.01$ s.

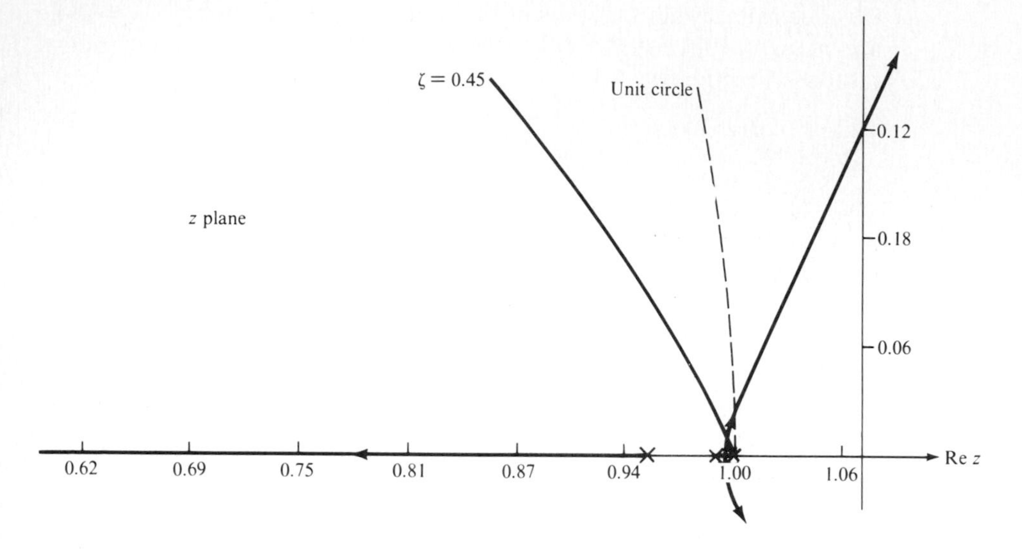

Figure 6-11 Enlargement of the $1 + j0$ area of the root-locus plot of Fig. 6-10 ($T = 0.01$ s).

control system for small values of T. For the value of $T = 0.01$ s the three poles of $G(z)$ [see Eq. (6-86)] in Fig. 6-10 seem to appear to be on top of one another (for the plotting scale used in this figure). Thus, it is most difficult to locate accurately the dominant poles $p_{1,2}$ for $\zeta = 0.45$ in this figure. An error in the graphical interpretation of the values of $p_{1,2}$ can easily put these poles outside the UC or on another ζ contour. The root-locus plot in Fig. 6-11 corresponds to that portion of the root locus in Fig. 6-10 in the vicinity of the $1 + j0$ point. That is, Fig. 6-11 is an enlargement of the area about the $1 + j0$ point of Fig. 6-10. By "blowing up" this region it is possible to plot the root locus accurately in the vicinity of the $1 + j0$ point and determine $p_{1,2}$.

6-6 BILINEAR TRANSFORMATIONS

One manner of extending the Routh stability criterion[1,21] and the s-plane frequency-domain analysis and design of discrete-data control systems is to make use of *bilinear transformations*.[20] The transformations

$$z = \frac{w + 1}{-w + 1} \tag{6-60a}$$

$$w = \frac{z - 1}{z + 1} \tag{6-60b}$$

have mapping properties that provide another useful means of determining system stability. Note that the w domain lacks the desirable property that as the sampling time T approaches zero, w approaches s; that is

$$w\bigg|_{T\to 0} = \lim_{T\to 0}\left[\frac{z-1}{z+1} = \frac{\epsilon^{sT}-1}{\epsilon^{sT}+1} = \frac{sT+(sT)^2/2!+\cdots}{2+sT+(sT)^2/2!+\cdots}\right] = 0 \qquad (6\text{-}61)$$

This situation is overcome by defining

$$w' \equiv \frac{2}{T}w \equiv \frac{2}{T}\frac{sT+(sT)^2/2!+\cdots}{2+sT+(sT)^2/2!+\cdots} \qquad (6\text{-}62)$$

Thus, in the w' plane, the desirable property that $w' \to s$ as $T \to 0$ is achieved. This w'-plane property establishes the conceptual basis for defining a quantity in the w' domain which is analogous to a quantity in the s domain. Substituting $w = Tw'/2$ into Eqs. (6-60a) and (6-60b) yields, respectively, the z- to w'-plane and the w'- to z-plane transformations as follows:

$$z = \frac{Tw'+2}{-Tw'+2} \qquad (6\text{-}63)$$

$$w' = \frac{2}{T}\frac{z-1}{z+1} \qquad (6\text{-}64)$$

Equation (6-64) represents an approximation of $z = \epsilon^{sT}$. *Henceforth, the w' transformation is used throughout the text and the prime designator is omitted.*

6-6A s- and w-Plane Relationship

The relationship between the s and w plane can be found by examination of Eq. (6-63) in terms of $z = \epsilon^{sT}$, that is,

$$w = \frac{2}{T}\frac{z-1}{z+1} = \frac{2}{T}\frac{\epsilon^{sT}-1}{\epsilon^{sT}+1}\frac{\epsilon^{-sT/2}}{\epsilon^{-sT/2}} = \frac{2}{T}\frac{\epsilon^{sT/2}-\epsilon^{-sT/2}}{\epsilon^{sT/2}+\epsilon^{-sT/2}} \qquad (6\text{-}65)$$

which yields

$$w = \frac{2\tanh(sT/2)}{T} \qquad (6\text{-}66)$$

For the s-plane imaginary axis, substitute $s = j\omega_{sp}$ into Eq. (6-66) to obtain

$$w = \sigma_{wp} + j\omega_{wp} = \frac{2\tanh(j\omega_{sp}T/2)}{T} = \frac{j2\tan(\omega_{sp}T/2)}{T} \qquad (6\text{-}67)$$

or

$$\omega_{wp} = \frac{2\tan(\omega_{sp}T/2)}{T} \qquad (6\text{-}68)$$

Thus the imaginary axis in the primary strip of the s plane is mapped onto the entire imaginary axis of the w plane. If $\omega_{sp}T/2$ is small ($\omega_{sp}T/2 \le 0.297$), then $\omega_{wp} \approx \omega_{sp}$.

For the real axis, substitute $s = \sigma_{sp}$ into Eq. (6-66) to obtain

$$w = \frac{2 \tanh (\sigma_{sp} T/2)}{T} = \sigma_{wp} \tag{6-69}$$

By letting $\alpha = \sigma_{sp}T/2$, $\tanh \alpha$ can be expressed in the expanded form

$$\tanh \alpha = \frac{\alpha + \alpha^2/3! + \cdots}{1 + \alpha^2/2! + \cdots} \tag{6-70}$$

For $\alpha^2 \ll 2$, from Eqs. (6-69) and (6-70),

$$\sigma_{sp} \approx \sigma_{wp} \tag{6-71}$$

Thus, when the approximations are valid,

$$w = \sigma_{wp} + j\omega_{wp} = \sigma_{sp} + j\omega_{sp} = s \tag{6-72}$$

If the approximations are not valid, then Eqs. (6-68) and (6-69) must be used to locate the s-plane poles and zeros properly in the w plane. The mapping of the s-plane poles and zeros into the w plane by use of these equations is referred to as *prewarping of the s-plane poles and zeros.* The relationships of Eqs. (6-72) are the basis of a design method (digitization method) that is presented in Chaps. 7, 11, and 12.

Equations (6-67) to (6-69) map the z plane into the w plane with a one-to-one correspondence of points (conformal mapping). Thus the bilinear transformation of Eq. (6-64) maps (1) the interior of the UC in the z plane into the left-half w plane; (2) the remainder of the z plane into the right-half w plane; and (3) the UC onto the $j\omega_w$ axis in the w plane (see Figs. 6-3 and 6-12). Also, note that the s-plane primary strip is mapped into the entire w plane. As the w plane is similar to the s

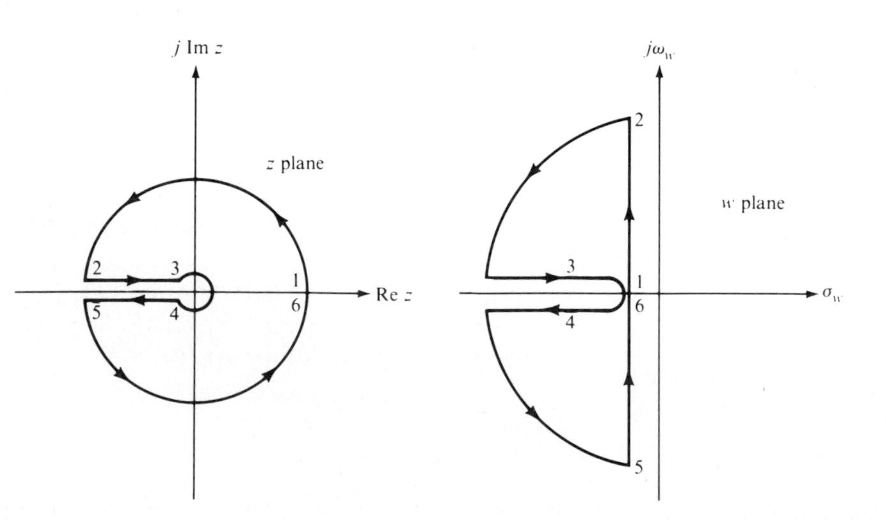

Figure 6-12 Mapping of the unit circle in the z plane into the w plane.

plane, then the s-plane analysis and design methods can be applied to the w plane as "if the w-plane model represents a continuous-time system." Therefore, to perform a z-plane analysis and design (or difference equation representation), the s-plane model is transformed by using the exact z transform, and the resulting z-plane form is transformed to the w plane. The analysis and controller design is now done in the w plane with the resulting compensator transformed back to the z plane where an equivalent difference equation is generated. This difference equation is then implemented on a computer (the digital controller).

A characteristic of a bilinear transformation is that, in general, it transforms an unequal-order transfer function ($n_z \neq w_z$) in the z domain, into one for which the order of its numerator is equal to the order of its denominator ($n_w = w_w$) in the w domain. Further, note that in transforming $G(z) = K_z G'(z)$ to $G(w) = K_w G'(w)$ by means of Eq. (6-64), the value of the gain constant K_w may be positive or negative. The sign of K_w is determined by the coefficients of $G(z)$ which in turn are a function of T. The following simple example illustrates the bilinear transformation characteristics.

Example 6-8 Given

$$G(z) = K_z G'(z) = K_z \frac{z}{(z-1)(z-a)} \tag{6-73}$$

where $K_z > 0$, $n_z = 2$, $w_z = 1$, and $0 \leq a \leq 1$. Determine $G(w)$ and sketch the root locus for $G(w) = -1$. Substituting Eq. (6-63) into Eq. (6-73) yields

$$G(w) = \frac{\dfrac{-K_z}{2(1+a)}\left(w + \dfrac{2}{T}\right)\left(w - \dfrac{2}{T}\right)}{w\left[w + \dfrac{2(1-a)}{T(1+a)}\right]} = K_w G'(w) \tag{6-74}$$

where $K_w = -K_z/[2(1+a)] < 0$,

$$0 \leq \alpha = \frac{2(1-a)}{T(1+a)} \leq \frac{2}{T}$$

and $n_z - w_z = 2 - 1 = 1$ zero at $w = 2/T$. In general, there are $n_z - w_z$ zeros at $w = 2/T$ due to the bilinear transformation. For this example, the zero at $w = -2/T$ is due to the zero of $G(z)$ at the origin. This zero represents a zero in the s plane at $s = -\infty$. Figure 6-13 shows the root-locus sketches for Eqs. (6-73) and (6-74), respectively. For "small" values of T the transformed zero at $w = 2/T$ may be far enough out on the positive real axis so as not to have any appreciable affect on the shape of the root locus in the region where the desired dominant poles are to be located.

6-6B Routh Stability Criterion in w Plane

The substitution of Eq. (6-63) into $C(z)/R(z) = W(z)/Q(z)$ yields $C(w)/R(w) = W(w)/Q(w)$. The Routh stability criterion can now be applied to $Q(w)$ in exactly

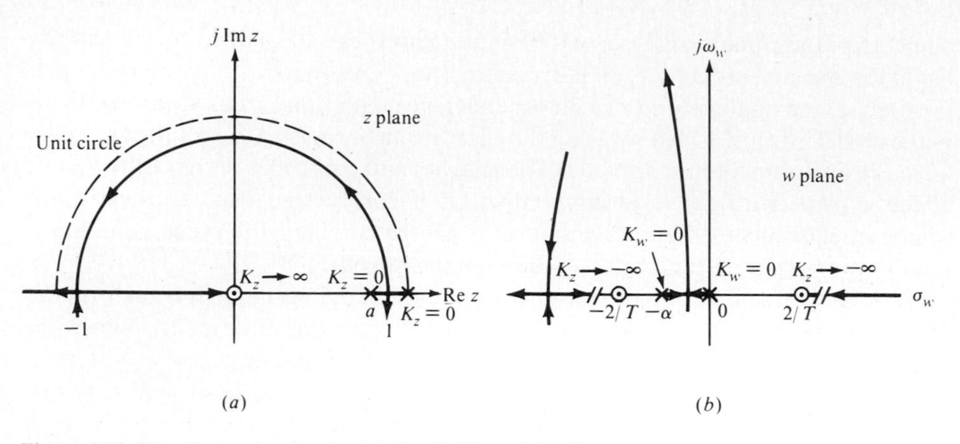

Figure 6-13 Root-locus sketches for (*a*) Eq. (6-73) and (*b*) Eq. (6-74).

the same manner as for continuous-time systems to determine the stability of a sampled-data control system. Three examples illustrate this technique.

Example 6-9 Assume that $Q(z) = z^3 - 4z^2 + 5z - 2 = 0$ represents a characteristic equation and $T = 2$ s. Apply the Routh stability criterion to determine if any of the roots of the characteristic equation lie outside the UC. (Equivalently, if any lie in the right-half w plane.)

$$Q(z) = z^3 - 4z^2 + 5z - 2$$

$$Q_1(w) \equiv Q(z)\Big|_{z=(w+1)/(-w+1)} = \left(\frac{w+1}{-w+1}\right)^3 - 4\left(\frac{w+1}{-w+1}\right)^2 + 5\left(\frac{w+1}{-w+1}\right) - 2$$

$$= \frac{12w^3 - 4w^2}{(-w+1)^3} = \frac{Q(w)}{(-w+1)^3}$$

Since the zeros of $Q(z)$ determine the stability of the system, the zeros of $Q(w)$ determine the stability of the system. The routhian array is

$$\begin{array}{c|cc} w^3 & 12 & 0 \\ w^2 & -4 & 0 \\ w^1 & 0 & 0 \\ w^0 & \cdots & \end{array}$$

An analysis of this array reveals that there is one root in the right-half w plane, due to one change of sign in the first column, and two roots on the $j\omega_w$ axis, due to the zero in the first column. Therefore, the system is unstable; i.e., there is one root of $Q(z)$ outside the UC and two roots on the UC in the z plane $[Q(z) = (z - 2)(z - 1)^2]$.

Example 6-10 Apply the Routh stability criterion to determine, for $T = 2$ s, if the given $Q(z)$ has any roots on or outside the UC.

$$Q(z) = z^3 - 0.8z^2 - 0.03z - 0.17$$

$$Q(w) = \left(\frac{w+1}{-w+1}\right)^3 - 0.8\left(\frac{w+1}{-w+1}\right)^2 - 0.03\left(\frac{w+1}{-w+1}\right) - 0.17$$

$$= \frac{w(1.94w^2 + 3.32w + 2.74)}{(-w+1)^3} = \frac{Q(w)}{(-w+1)^3}$$

$$\begin{array}{c|cc} w^2 & 1.94 & 2.74 \\ w^1 & 3.32 & \\ w^0 & 2.74 & \end{array}$$

The root at $w = 0$ of $Q(w)$ reveals that there is a root of $Q(z)$ at $z = 1$. Thus applying a step input to a sampled-data system having this $D(z)$ results in a ramp output. Consider the polynomial $z^3 - 0.8z^2 - 0.03z - 0.17$ to represent the denominator of $C(z)$ and to contain the factor $z - 1$ due to a step input forcing function. Thus, the remaining factors of the polynomial determine the stability of the system. From the routhian array it is seen that there are no changes in sign in the first column; therefore all roots are in the left-half w plane or within the UC of the z plane; consequently, this system is stable.

Example 6-11 Apply the Routh stability criterion to the system of Example 6-7 to determine the range of values that K_G can have for a stable response and how these values are affected by the sampling time T.

Equation (6-56) is of the form of

$$\frac{C(z)}{R(z)} = \frac{K(z+c)}{z^2 - az + b} \tag{6-75}$$

where

$$K = 0.5K_G(T + 0.5\epsilon^{-2T} - 0.5)$$

$$a = 1 + \epsilon^{-2T} - 0.5K_G(T + 0.5\epsilon^{-2T} - 0.5)$$

$$b = \epsilon^{-2T} + 0.5K_G(0.5 - 0.5\epsilon^{-2T} - T\epsilon^{-2T}) \tag{6-76}$$

$$c = \frac{0.5 - 0.5\epsilon^{-2T} - T\epsilon^{-2T}}{T + 0.5\epsilon^{-2T} - 0.5}$$

Substituting Eq. (6-63) into Eq. (6-75) yields

$$\frac{C(w)}{R(w)} = \frac{K[(1-c)T^2w^2 + 4cTw - 4(1+c)]}{(1+a+b)T^2w^2 + 4(1-b)Tw + 4(1-a+b)} \tag{6-77}$$

The characteristic polynomial of this closed-loop system is given by the denominator of Eq. (6-77); that is,

$$Q(w) = (1 + a + b)T^2w^2 + 4(1 - b)Tw + 4(1 - a + b) \tag{6-78}$$

Applying the Routh criterion to Eq. (6-78) yields the array

$$\begin{array}{c|cc|} w^2 & (1 + a + b)^2T^2 & 4(1 - a + b) \\ w^1 & 4(1 - b)T & \\ w^0 & 4(1 - a + b) & \end{array}$$

Thus for a stable system,

$$1 + a + b > 0 \tag{6-79}$$

$$1 - b > 0 \tag{6-80}$$

$$1 - a + b > 0 \tag{6-81}$$

Substituting for a and b into Eq. (6-79) and solving for K_G yields

$$K_G < \frac{4}{T - (1 - \epsilon^{-2T})/(1 + \epsilon^{-2T})} \tag{6-82}$$

From Eq. (6-80),

$$K_G < \frac{2}{0.5 - (T\epsilon^{-2T})/(1 - \epsilon^{-2T})} \tag{6-83}$$

and from Eq. (6-81)

$$\epsilon^{-2T} < 1 \rightarrow T > 0 \tag{6-84}$$

Table 6-2 indicates the values of K_G obtained for various values of T from Eqs. (6-82) and (6-83). The static loop sensitivity K is given by [see Eq. (6-76)] $K = (T + 0.5\epsilon^{-2T} - 0.5)K_G/2$. A sketch of T vs. the boundary value of K_G for a stable system is shown in Fig. 6-14. Note that $\lim_{T \to 0} K_G = \infty$.

As illustrated by Example 6-11, the bilinear transformation has, in general, the property of converting a transfer function $E(z)$ whose numerator and denominator polynomials are of unequal order into a transfer function $E(w)$ of equal-order numerator and denominator polynomials. Whether this property is manifested in

Table 6-2 Stability requirements for Example 6-10

T, s	Eq. (6-82)	Eq. (6-83)	Value of K_G for stability	K
0.01	$K_G < 3{,}986{,}844$	$K_G < 401.4$	$K_G < 401.4$	0.03987
0.1	$K_G < 12042.7$	$K_G < 41.378$	$K_G < 41.378$	0.2 −
1.0	$K_G < 16.778$	$K_G < 5.8228$	$K_G < 5.8228$	3.3 +
10	$K_G < 4$	$K_G < 4$	$K_G < 4$	38
↓				
∞	0	↓	$K_G \rightarrow 0$	

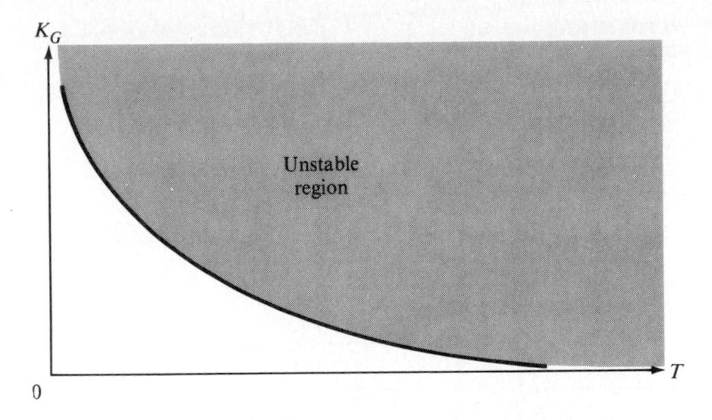

Figure 6-14 A sketch of T vs. K_G depicting the stability and instability regions for the unity-feedback system of Fig. 6-5.

the transformation depends upon the coefficients of the numerator and denominator polynomials of $E(z)$.

Chapters 7 and 11 use the w plane for analyzing and designing sampled-data control systems. The analysis and design procedures are similar to those employed for the s domain assuming s-domain poles and zeros lie in the primary strip and the associated approximations are accurate.

6-7 s-, z-, AND w-PLANE TIME-RESPONSE CHARACTERISTICS CORRELATION

Using the continuous-time plant for the unity-feedback control system of Ref. 1, pp. 357–358,† the effect of the choice of T on the degree of correlation between the s, z, and w planes can be illustrated. This comparative insight is provided by determining the figures of merit and the poles and zeros of $C(\cdot)/R(\cdot)$ for *various values of* T. The basic plant of Fig. 6-5 is

$$G_x(x) = \frac{4.2}{s(s+1)(s+5)} \tag{6-85}$$

Thus for the system of Fig. 6-5 the $\mathscr{Z}$ transform

$$\begin{aligned} G(z) &= K_z G'(z) \\ &= 4.2(1 - z^{-1})\mathscr{Z}\left[\frac{1}{s^2(s+1)(s+5)}\right] \\ &= 4.2\left[\frac{5T - 6z + 6}{25(z-1)} + \frac{(24z - 25\epsilon^{-5T} + \epsilon^{-T})(z-1)}{100(z-\epsilon^{-T})(z-\epsilon^{-5T})}\right] \end{aligned} \tag{6-86}$$

† The value of $K_x = 4.2$ is for dominant closed-loop roots having a $\zeta = 0.45$.

Table 6-3 The gain and the zeros and poles of $G(\cdot)$ for various values of T

Plane	T, s	Gain	Zeros	Poles
Continuous-time system				
s	...	4.2	...	0, −1, −5
Sampled-data system				
z	0.01	6.896×10^{-7}	−0.26395, −3.6767	1, 0.99005, 0.95123
w		1.749×10^{-7}	200, 349.4, −343.4	0, −1, −4.999
z	0.05	8.126×10^{-5}	−0.2483, −3.467	1, 0.9512, 0.7788
w		2.171×10^{-5}	40, 72.43, −66.42	0, −0.9998, −4.974
z	0.1	6.05×10^{-4}	−0.2295, −3.228	1, 0.9048. 0.6065
w		1.697×10^{-4}	20, 37.95, −31.92	0, −0.9992, −4.898
z	1	0.218	−0.04521, −1.315	1, 0.3679, 6.738×10^{-3}
w		2.379×10^{-2}	2, −2.189, 14.71	0, −0.9242, −1.973
z	1.5	0.4863	−0.01915, −0.9741	1, 0.2231, 5.531×10^{-4}
w		-5.047×10^{-3}	1.333, −1.385, −101.6	0, −0.8469, −1.332
z	2	0.8141	−0.009038, −0.7683	1, 0.1353, 4.54×10^{-5}
w		-8.232×10^{-2}	1, −1.018, −7.631	0, −0.7616, −0.9999

and the corresponding w transform, $G(w) = K_w G'(w)$, are obtained for various values of T. The gain (K_z or K_w) and the zeros and poles of $G(\cdot)$ *for each value of* T are given in Table 6-3. The corresponding closed-loop time-response characteristics, for a unit-step forcing function, and the closed-loop poles are given in Table 6-4. The static error coefficients in the w domain are obtained by substituting from Eq. (6-63) into Eqs. (6-21), (6-24), and (6-25). For the example of this section,

$$K_1 = \lim_{w \to 0} \frac{2wG(w)}{2 + Tw} \tag{6-87}$$

By applying the corresponding expression for K_1 for each of the three domains, it is determined that the value of the error coefficient in this example, $K_1 = 0.84\ \text{s}^{-1}$, is unaffected by the transformations (from a continuous-time to sampled-data system and z to w domain). In other words, the steady-state error characteristics of a sampled-data system are determined by the type of plant $G_x(s)$.

As Table 6-4 illustrates, to accurately duplicate the performance of a continuous-time unity-feedback system, having $G_x(s)$ as its plant, by a sampled-data unity-feedback system having the same plant and a ZOH device, the dominant poles and zeros of $C(s)/R(s)$ must satisfy the conditions of $\omega_{sp}T/2 \leq 0.297$ and

Table 6-4 The closed-loop time-response characteristics of $c(\cdot)$† and poles of $C(\cdot)/R(\cdot)$

Plane	T, s	M_p	t_p, s	t_s, s	Poles‡	$\omega_{sp}T/2$ (<0.297)	$\alpha = \sigma_{sp}T/2$	α^2 (≪2)
					Continuous-time system			
s	...	1.202	4.12	9.48	$-0.404 \pm j0.802,\ -5.192$			
					Sampled-data system			
z	0.01	1.205	4.14	9.485	$0.9960 \pm j8.012 \times 10^{-3},\ 0.9494$	0.00401	0.00202	4.08×10^{-6}
w		1.205	4.12	9.490	$-0.4012 \pm j0.8044,\ -5.196$			
z	0.05	1.214	4.1	9.53	$0.9798 \pm j0.03956,\ 0.7704$	0.02005	0.0101	1.02×10^{-4}
w		1.214	4.12	9.550	$-0.3921 \pm j0.8071,\ -5.189$			
z	0.1	1.225	4.1−	9.6−	$0.9595 \pm j0.07788,\ 0.5917$	0.0401	0.0202	4.08×10^{-4}
w		1.225	4.13	9.617	$-0.3808 \pm j0.8101,\ -5.131$			
z	1.0	1.474	4	17	$0.5859 \pm j0.590,\ -0.01516$	0.401	0.202	4.08×10^{-2}
w		1.469	4.34	17.13	$-0.2156 \pm j0.8242,\ -2.062$			
z	1.5	1.614	4.5	34.5+	$0.375 \pm j0.7587,\ -0.01249$	0.601+	0.303+	0.0918
w		1.6412	4.44	24.79	$-0.1535 \pm j0.8204,\ -1.367$			
z	2	1.708	4.0	54	$0.1643 \pm j0.8623,\ -7.328 \times 10^{-3}$	0.802	0.404	0.1632+
w		1.839	4.4	36.1	$-0.1093 \pm j0.8216,\ -1.015$			

† $c(t)_w \approx \mathscr{L}^{-1}[C(w)]$ and $c^*(t) = \mathscr{Z}^{-1}[C(z)]$.

‡ $s_{sp} = \sigma_{sp} \pm j\omega_{sp} = -0.404 \pm j0.802$.

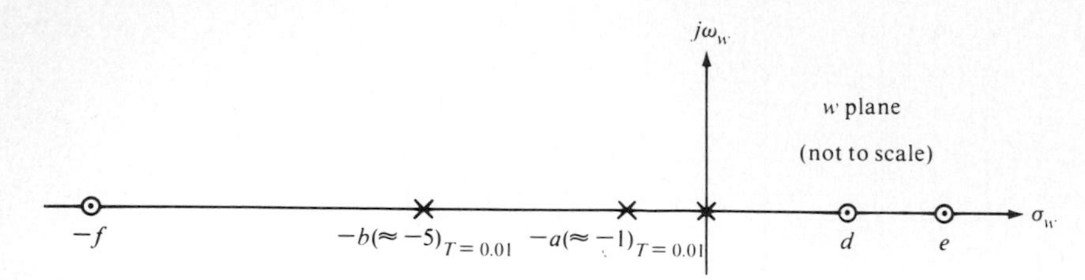

Figure 6-15 The zeros and poles of Table 6-3.

$\alpha^2 \ll 2$. Therefore, the smallest value of T should be chosen as constrained by the processing time of the digital controller.

As pointed out in the last section, a bilinear transformation, in general, results in $G(w)$ having the same number of zeros as poles. Thus the third-order plant $G_x(s)$ of this example results in $G(w)$ having three zeros and three poles. These zeros and poles are listed in Table 6-3 for various values of T and are plotted in Fig. 6-15. As shown in this figure, for $T = 0.01$, the poles of $G(w)$ are located at essentially the same location as the poles of $G_x(s)$. As T is increased, the poles, other than the one at the origin, migrate toward the origin, which explains the divergence of the w-domain time-response characteristics from those for the continuous-time system. Also, Table 6-4 illustrates the performance analysis of a sampled-data system in the z plane. For a given value of T, the performance is essentially the same as obtained by an analysis of this same system in the w plane as long as $\alpha^2 \ll 2$ ($T < 1$ s) for the same value of T. Note that for $T \geq 1$ the values of M_p are much larger than the desired value. By reducing the gain, the value of M_p can be lowered to the desired value. For example, with $T = 1$ s, the gain is reduced from 4.2 to 2.5 to yield $M_p = 1.206$, $t_p = 6$ s, $t_s \approx 13$ s, and $K_1 = 0.494\ \text{s}^{-1}$. Again, note that the bilinear transform (w plane) is only an approximation to ϵ^{sT}. As T becomes relatively larger in a given system representation, the poles and zeros of the w-domain transfer function are changed or warped by means of Eq. (6-65) in the z plane based upon the previous analysis. Note this phenomenon in Table 6-4. The use of the w plane must go hand in hand with a detailed understanding of the transformation phenomenon.

6-8 FREQUENCY RESPONSE

Sections 5-4C and 5-4D discuss the correlation between the frequency and time domain for a continuous-time control system. This correlation applies equally well for sampled-data systems. The values of M_m, ω_m, and ω_b for sampled-data systems are not only indicative of the transient characteristics but also of the frequency bandwidth for noise-rejection properties. The closed-loop frequency response for a sampled-data system is determined by (1) substituting $z = \epsilon^{sT} = {}^{j\omega T}$ into $C(z)/R(z)$ to obtain

$$\frac{\mathbf{C}(j\omega T)}{\mathbf{R}(j\omega T)} = M_z \angle \alpha_z \tag{6-88}$$

where $M_z = |\mathbf{C}(j\omega T)/\mathbf{R}(j\omega T)|$ and α_z is the phase angle; or (2) in the w domain by substituting $w = j\omega_w$ into $C(w)/R(w)$ to obtain

$$\frac{\mathbf{C}(j\omega_w)}{\mathbf{R}(j\omega_w)} = M_w \underline{/\alpha_w} \tag{6-89}$$

This correlation is best illustrated by the means of two examples.

Example 6-12 For the basic plant of Eq. (6-85) determine:

(*a*) M_s vs. ω and α_s vs. ω for the continuous-time system configuration, where $[\mathbf{C}(j\omega)/\mathbf{R}(j\omega)] = M_s \underline{/\alpha_s}$.

(*b*) M_z vs. ω and α_z vs. ω and M_w vs. ω_w and α_w vs. ω_w for the sampled-data configuration where $T = 0.01$ s. Also determine the bandwidth ω_b for all three domains.

SOLUTION (*a*) From Table 6-4 it is noted that the dominant poles for the continuous-time system are $p_{1,2} = -0.404 \pm j0.802$. Thus the value of ω_m is in the neighborhood of $\omega_d = 0.8$. The frequency-response plots of M_s vs. ω and α_s vs. ω are shown in Fig. 6-16. The plot of $\underline{/\alpha_s}$ vs. ω must be a continuous curve, but the plot and scale shown in Fig. 6-15 are due to the plotting routine. Thus the phase angle scale must be interpreted accordingly for each succeeding section of the angle plot. This is true in the remaining angle plots in this text. The figures of merit M_m, ω_m, and ω_b are given in Table 6-5. Notc that $M_m \approx 1.236$ is "close to" the actual value of $M_p = 1.202$.

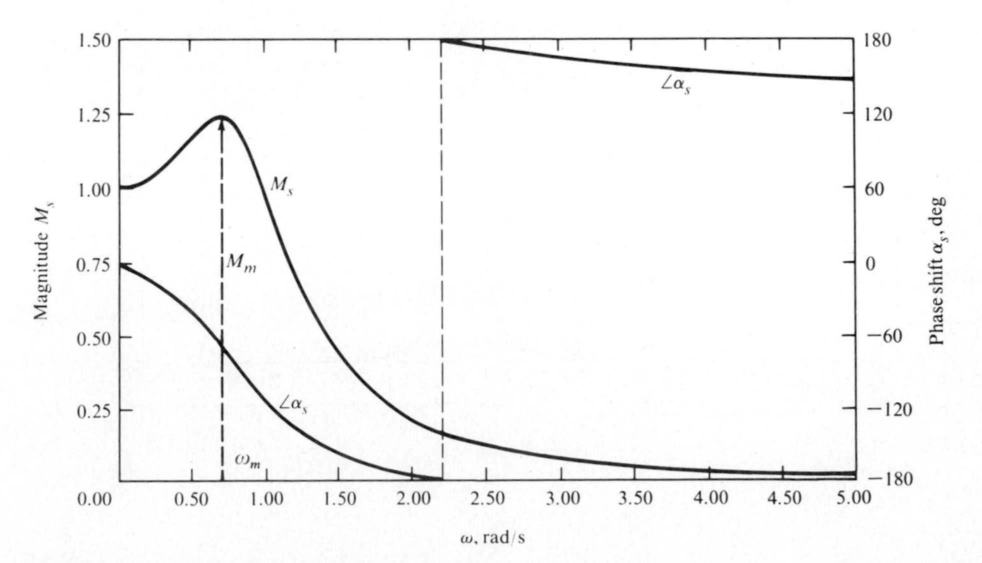

Figure 6-16 M_s vs. ω and α_s vs. ω: Example 6-12.

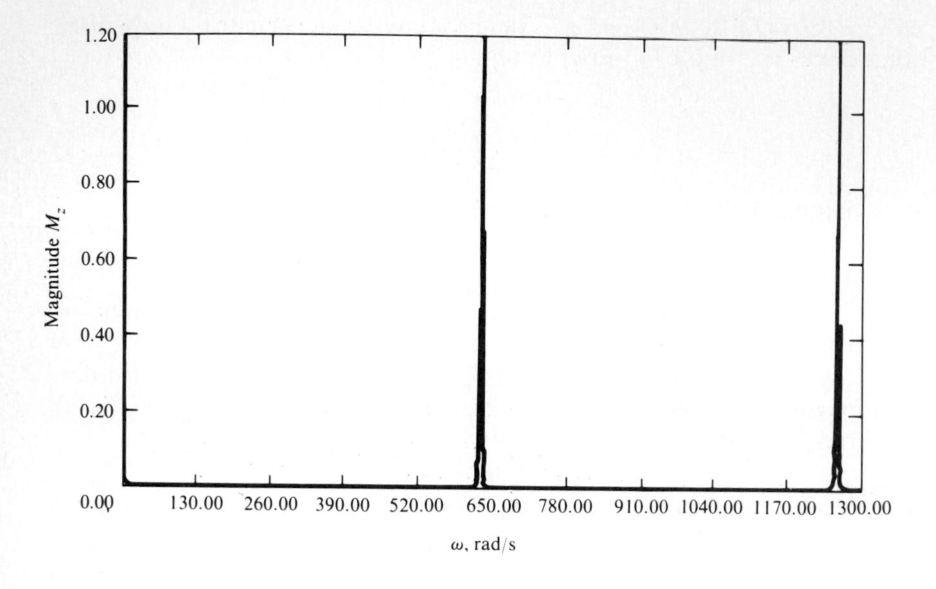

(a)

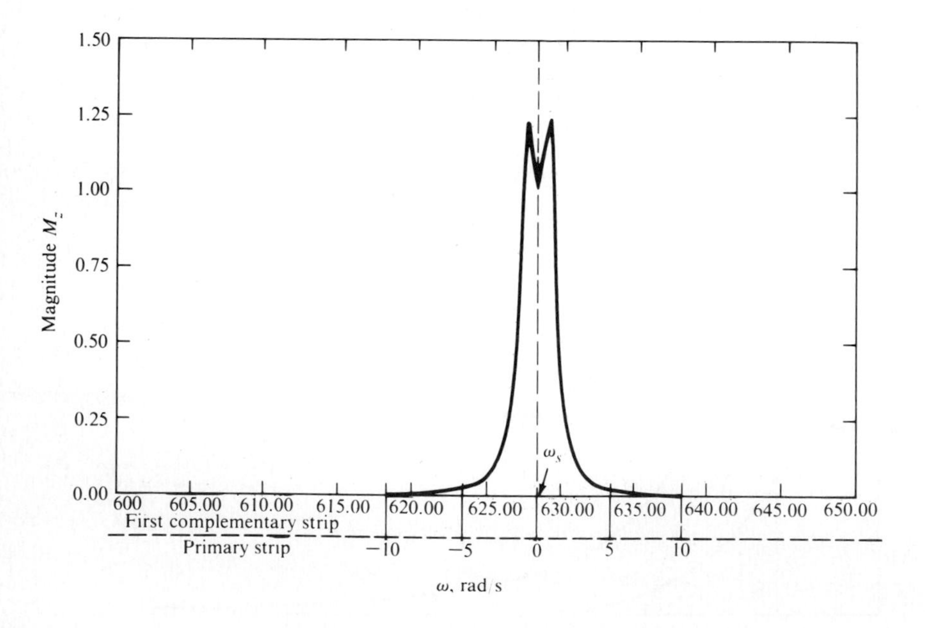

(b)

Figure 6-17 (a), (b) M_z vs. ω, and (c), (d) α_z vs. ω for $T = 0.01$ s: Example 6-12.

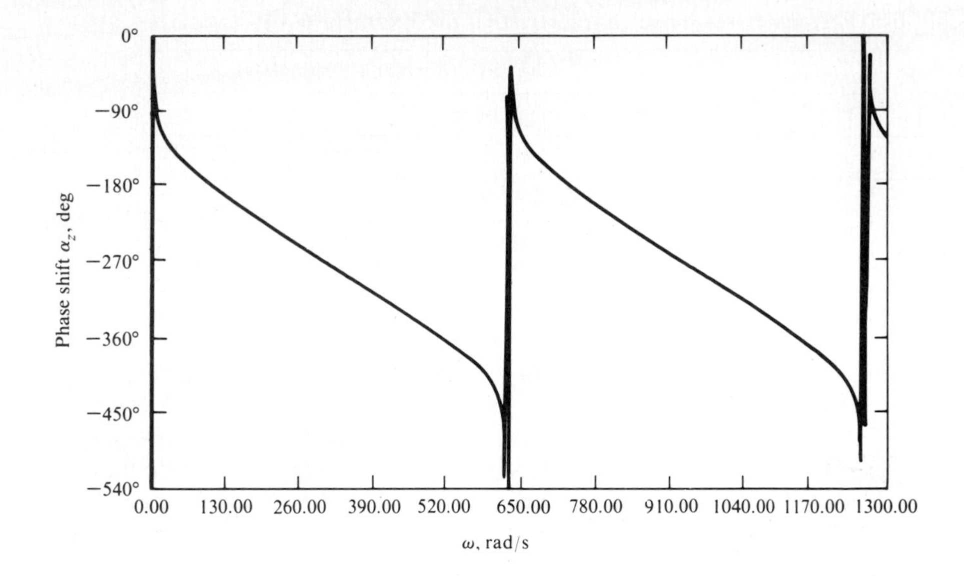
0°
−90°
−180°
−270°
−360°
−450°
−540°
Phase shift α_z, deg
0.00
130.00
260.00
390.00
520.00
650.00
780.00
910.00
1040.00
1170.00
1300.00
ω, rad/s

(c)

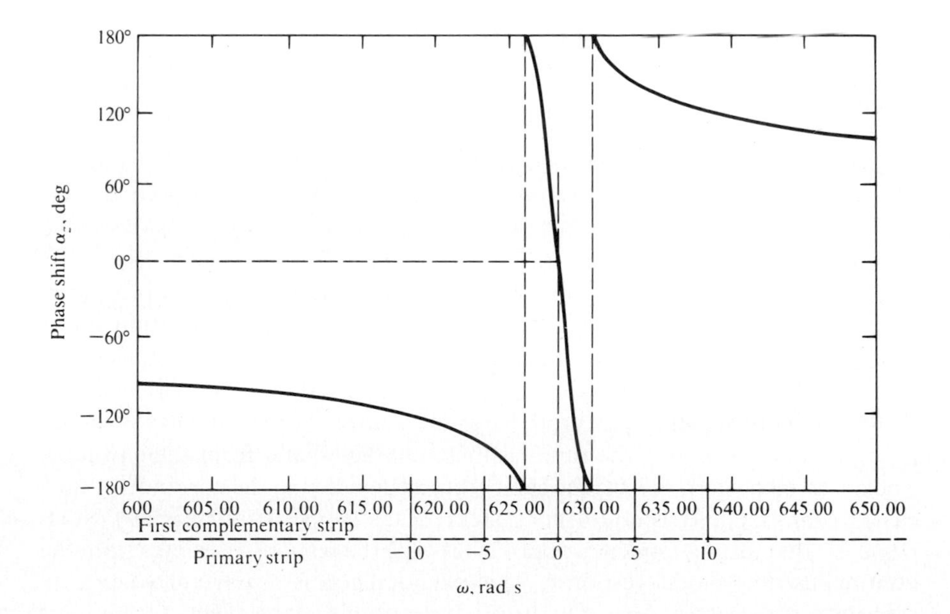
180°
120°
60°
0°
−60°
−120°
−180°
Phase shift α_z, deg
600
605.00
610.00
615.00
620.00
625.00
630.00
635.00
640.00
645.00
650.00
First complementary strip
Primary strip
−10
−5
0
5
10
ω, rad s

(d)

Table 6-5 Frequency-response characteristics for Example 6-12

Domain	M_m	ω_m, rad/s	ω_b, rad/s	$(\alpha)_m$	$(\alpha)_b$
		Continuous-time system			
s	1.236	0.69	1.18 −	−66.7°	−134.2+°
		Sampled-data system ($T = 0.01$)			
z	1.2408	0.691+	1.181 −	−67.04°	−134.9−°
w	1.2405	0.692	1.180+	−67.00°	−134.9°

(*b*) For $T = 0.01$ s, as seen in Sec. 6-7, $s \approx w$, thus $M_s \approx M_z \approx M_w$ and $\alpha_s \approx \alpha_z \approx \alpha_w$ in the primary strip as borne out by the values in Table 6-5. Since the value $T = 0.01$ s satisfies Shannon's sampling theorem, then the frequency-response characteristics of the sampled-data system in the z domain, as depicted in Fig. 6-17, are in conformity with the frequency characteristics of Fig. 3-6*b*. Note that owing to these characteristics and the scale chosen, it is difficult to distinguish between the vertical line representing $\omega = 0$ and M_z vs. ω in the vicinity of $\omega = 0^+$ in Fig. 6-17*a*. The plot of M_z vs. ω in the region of $0 \leq \omega \leq \omega_b$ and the corresponding plots in the complementary strips (see Fig. 6-17*b*) are very close to the plot of M_s vs. ω in Fig. 6-16. That is, for the value of $T = 0.01$ s, the plots of M_s vs. ω and M_z vs. ω, in the vicinity of $0 \leq \omega \leq \omega_b$, lie essentially on top of one another. This is also true for the phase plots of α_z vs. ω and α_s vs. ω. The plots of M_w vs. ω_w and α_w vs. ω_w are essentially the same as those shown in Fig. 6-16 for $T = 0.01$ s.

Example 6-13 Example 6-12 is repeated for $T = 1$ s and $\zeta = 0.45$. The plots of M_z vs. ω and $\angle\alpha_z$ vs. ω for $T = 1$ s are shown in Fig. 6-18, and the plots of M_w vs. ω_w and $\angle\alpha_w$ vs. ω_w are shown in Fig. 6-19. The frequency- and time-response characteristics are summarized in Tables 6-6 and 6-7, respectively. Since T is "large," then the approximation $s \approx w$ is no longer valid, and the correlation between the s and w and z domains is poor. This illustrates the warping effect mentioned previously. The correlation between the w and the z domains is still very good.

The use of frequency plots to portray the control ratio frequency response is helpful in determining the time response characteristics from the envelope. The observed pattern reflects the same form as depicted in the continuous-time case; i.e. for first- and second-order system roll-off (slope characteristics). Note, however, that as the frequency approaches $\omega_s/2$ the response deviates from the continuous-time model response. This phenomenon is important when considering error sources. Also, Fig. 6-18 is an example of the effects of sampling which must be considered in a control system with body-bending modes or other frequency sources above $\omega_s/2$. Thus the frequency plot is very useful in describing this phenomenon.

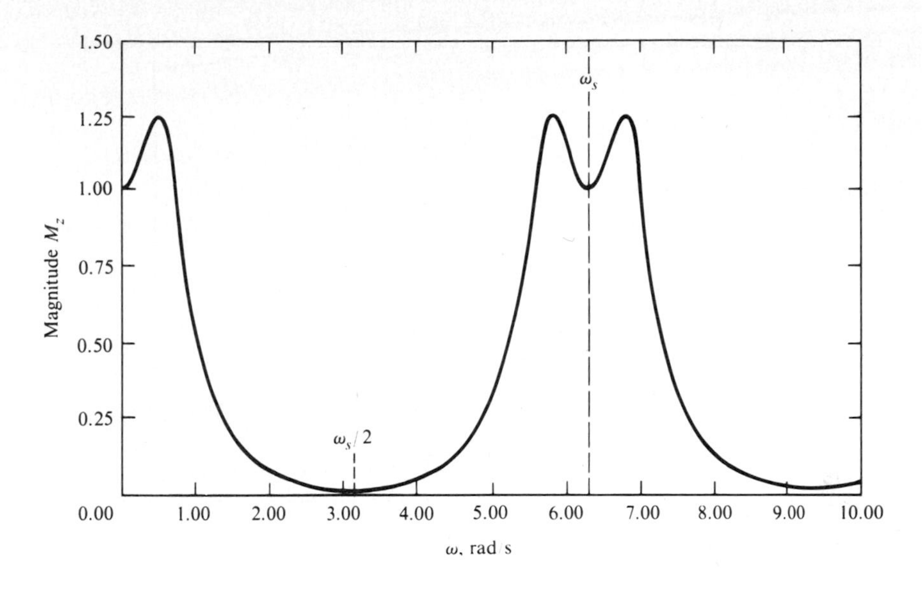

(a)

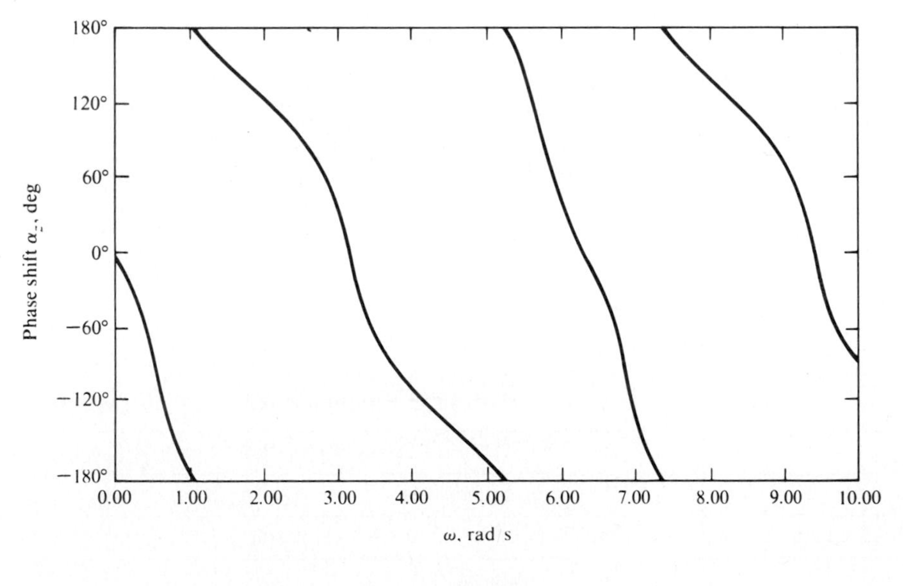

(b)

Figure 6-18 (a) M_z vs. ω and (b) α_z vs. ω for $T = 1$ s.

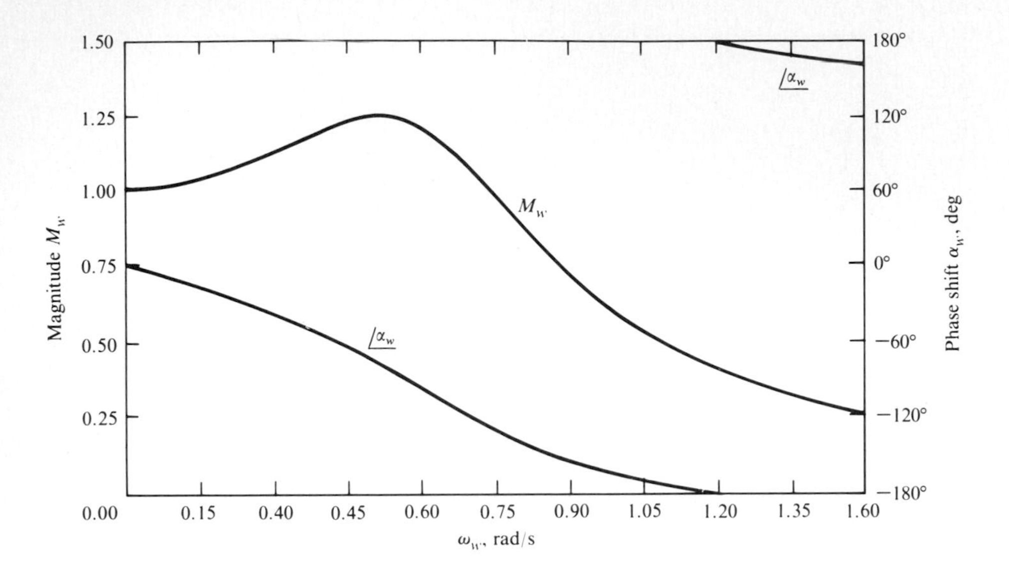

Figure 6-19 M_w vs. ω_w and α_w vs. ω_w for $T = 1$ s.

Table 6-6 Frequency-response characteristics for Example 6-13

Domain	M_m	ω_m, rad/s	ω_b, rad/s	$(\alpha)_m$	$(\alpha)_b$
		Continuous-time system			
s	1.236	0.69	1.18−	−66.7°	−134.2+°
		Sampled-data system			
z	1.252	0.50	0.853	−77.3°	−154°
w	1.250	0.51	0.908	−77.3°	−153.8°

Table 6-7 Transient-response characteristics for Example 6-13

Domain	M_p	t_p, s	t_s, s	K_1, s^{-1}	Poles of $C(\cdot)/R(\cdot)$
			Continuous-time system		
s	1.202	4.12	9.48	0.84	$-0.404 \pm j0.802$, -5.192
			Sampled-data system		
z	1.206	6	13	0.4940	$0.6269 \pm j0.4153$, -9.28×10^{-3}
w	1.2	5.87	13.07	0.4865	$-0.3085 \pm j0.5889$, -2.037

6-9 SUMMARY

In this chapter a stability analysis of a sampled-data system is performed in both the z and w planes. In general, the transfer functions of most plants fall into three categories; they can be identified as Type 0, 1, and 2, with the corresponding definitions of the steady-state error coefficients. These error coefficients are indicative of a unity-feedback stable system's steady-state performance.

As illustrated in this chapter, the root-locus method is as applicable to sampled-data systems as it is to continuous-time systems. When a polynomial is set equal to zero, the equation can be rearranged and put into the mathematical form of a ratio of factored polynomials which is equal to plus or minus one; that is, $N(z)/D(z) = \pm 1$. Once this form has been obtained, the procedure given in this chapter can be used to locate the roots of the polynomial equation.† The root locus permits the analysis of the performance of sampled-data systems and provides a basis for selecting the gain that results in the performance specifications being met. Since the closed-loop poles are obtained explicitly, the form of the time response is directly available. CAD programs[14,15] are available for obtaining $c(kT)$ vs. T for the z-plane analysis or $c(t)$ vs. t for the w-plane analysis. If the performance specifications cannot be met, the root locus can be analyzed to determine the appropriate compensation to yield the desired results. This is covered in Chaps. 11 and 12.

The bilinear transformations are introduced in order to transform from the z domain to the new w domain. The advantage of this transformation is that it allows the continuous-time system analysis and design methods to be applied in the w domain.

A frequency-response analysis in all three domains is made to illustrate the degree of correlation between all these domains. The value of the sampling time is the determining factor in the degree of correlation that exists between all three domains.

† The determination of the roots of a high-order polynomial is a difficult numerical analysis problem especially for equations with coefficients across a wide range of numbers. Special computer programs for each application may have to be developed.

CHAPTER

SEVEN

DISCRETE TRANSFORM ANALYSIS (APPROXIMATIONS)

7-1 INTRODUCTION

In the first six chapters the fundamental concepts and theory involved in the study of sampled-data systems are presented. A correlation by means of the s plane between the time-response characteristics with respect to the z and w planes is made in Chaps. 4 and 6. This correlation is enhanced by the stability analysis in the latter chapter. Chapter 5 presents a review of continuous-time control system response characteristics and one method of synthesizing a desired $[C(s)/R(s)]_M$. The introduction to the study of sampled-data systems and the technique of synthesizing a desired control ratio presented in the preceding chapters now form the basis for the design of sampled-data closed-loop control systems. With the advent of small and inexpensive information processors, a digital controller (or compensator) can be realized that achieves the desired system performance specifications. Using the computer for this purpose requires the preparation of a discrete control law (an algorithm) which is translated into a software program. This software realization usually permits greater flexibility and accuracy as compared to a continuous electronic network controller. Thus, this chapter and Chaps. 11 and 12 stress the use of a digital processor as a digital controller or as a compensator for achieving the desired digital-control-system design, as shown in Fig. 7-1.

Because of the correlation between the s, z, and w planes, the highly developed analysis and design techniques for continuous-time systems can form an approach for the analysis and design of sampled-data control systems. There are two basic approaches for analysis and design of control systems: (1) utilizing the open-loop

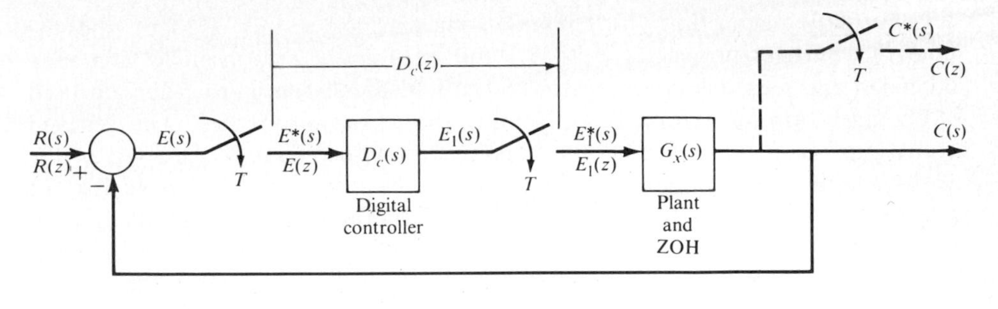

(a)

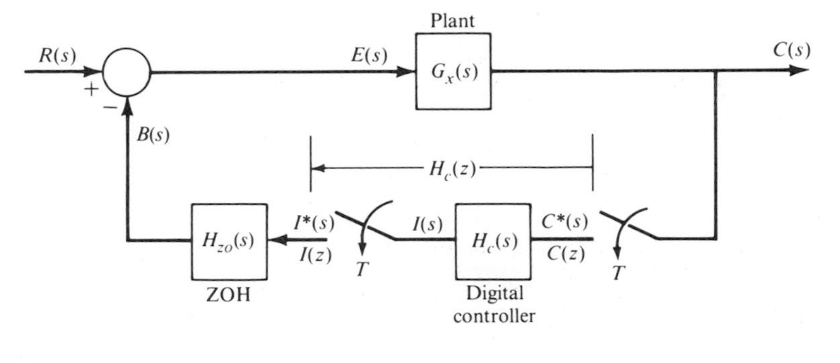

(b)

Figure 7-1 A compensated sampled-data or a digital control system. (a) Cascade and (b) feedback compensation.

transfer function GH or (2) utilizing the closed-loop transfer function C/R to achieve the closed-loop system performance. This latter approach, as stated in Chap. 5, is commonly referred to as the pole-zero placement technique.

The analysis and design of sampled-data control systems may be done entirely in the z plane, referred to as the *direct digital control design* (*DIR*) *technique*, or entirely in the s plane, referred to as the *digitization* or *discrete digital control* (*DIG*) *technique*, which requires the development of the PCT system model. This model also requires the use of the *Padé* (see Appendix C) and *Tustin* transformation approximations. When the analysis and synthesis of the sampled-data system is performed in the w plane by applying the continuous-time system analysis and design techniques, *this approach is also referred to as the DIG technique.*

A controller designed by the DIG technique provides a good base for exhibiting the effects of the sample time parameter of the digitized controller on system performance. The reason for this is that the continuous controller corresponds to the limiting case where the sampling time of $D_c(z)$ is zero. A disadvantage of this method is that $D_c(z)$ may not have all the properties of $D_c(s)$. However, this problem is minimized by the selection of an s- to z- (to w-)

transformation algorithm which maintains the specified properties required of the controller. These properties include stability, impulse response, dc gain, and cascaded and feedback compensator realization. Ideally, an s- to z-domain (and w-domain) transform function maintains the above properties. The selected transform algorithm is basically a mapping technique. Thus the analysis of the various approaches requires an investigation of ranges and domains for the mapping function. The development of the criteria for achieving a good degree of correlation between the w- and z-plane mapping in the use of the DIG technique is presented in Chap. 6. This chapter develops the PCT control system model and the criteria for a good degree of correlation between the s- (or w-) and z-plane mapping. Through the use of either the DIR or the DIG technique, it is hoped that the desired system performance characteristics can be achieved by mere gain adjustment.

7-2 FOLDING OR ALIASING[18]

As indicated in Chap. 3, the selection of the sampling time T determines the width of the primary strip. Thus the sampling time plays a crucial role in specifying a model control ratio in the s domain for a sampled-data system. Also, the value of T determines the type of inputs to which the sampled-data system can accurately respond.

Consider the continuous-time signal

$$e(t) = \sin(\omega_1 t + \theta) \tag{7-1}$$

which is sampled with a sampling time of $\omega_1 = \pi/T$, where $0 \le \omega_1 < \omega_s/2$. Thus letting $t = kT$ yields

$$e(kT) = \sin(k\omega_1 T + \theta) \tag{7-2}$$

Next consider the continuous-time signal

$$q(t) = \sin\left[\left(\omega_1 + \frac{2\pi j}{T}\right)t + \theta\right] \tag{7-3}$$

where $j = 1, 2, 3, \ldots$ and which is sampled with the same sampling time T.

$$\begin{aligned} q(kT) &= \sin\left[\left(\omega_1 + \frac{2\pi j}{T}\right)kT + \theta\right] \\ &= \sin(kT\omega_1 + \theta + 2\pi kj) = \sin(kT\omega_1 + \theta) \end{aligned} \tag{7-4}$$

Comparing Eqs. (7-2) and (7-4) reveals that they are identical; thus it is impossible to differentiate between the two sampled sinusoids whose radian frequencies differ by an integral multiple of $2\pi/T$. Note that Eq. (7-1) lies in the primary strip and Eq. (7-3) lies in the complementary strip. The situation of Eq. (7-3) must be avoided by selecting small-enough T in accordance with the Shannon sampling theorem

$$\omega_s > 2\omega_c \qquad \text{or} \qquad \frac{\pi}{T} > \omega_c$$

In a similar manner consider

$$e(t) = \sin\left[\left(\frac{2\pi j}{T} - \omega_1\right)t + \pi - \theta\right] \tag{7-5}$$

$$e(kT) = \sin(2\pi kj - kT\omega_1 + \pi - \theta) = \sin(-kT\omega_1 + \pi - \theta)$$
$$= \sin(kT\omega_1 + \theta) \tag{7-6}$$

Sampling either Eq. (7-1) or (7-5) results in the identical sequence. Thus it is impossible to differentiate between two sampled sinusoids whose sum of radian frequencies is an integral multiple of $2\pi/T$.

Therefore, the process of uniform sampling is unable to distinguish between two sinusoids whose respective radian frequencies have a sum or difference of an integral multiple of $2\pi/T$. Effectively, all radian frequencies are then *folded* into the interval $0 \leq \omega < \omega_s/2$. The frequency π/T rad/s is commonly referred to as the *folding* or *Nyquist frequency*.

7-3 *s*- TO *z*- (OR *w*-) PLANE TRANSFORMATION METHODS

Several methods have been developed for obtaining the approximate *z*-domain transfer function from the *s*-domain transfer function representation, i.e., an approximation of $z = \epsilon^{sT}$. The transformations from the s to the z domain are accomplished by substituting functions of z for the s^q terms in the *s*-domain transfer function. One of the most popular of these approximation methods is the *Tustin algorithm*.[24,25] The function of z that is substituted for s^q in implementing the Tustin transformation is

$$s^q \approx \left(\frac{2}{T}\frac{1 - z^{-1}}{1 + z^{-1}}\right)^q \tag{7-7}$$

One advantage of the Tustin algorithm is that it is comparatively easy to implement. Also, the accuracy of the response of the Tustin *z*-domain transfer function is good compared with the response of the exact *z*-domain transfer function; however, the accuracy decreases as the frequency increases. Note, for $q = 1$, that Eq. (7-7) is identical to the *w*- to *z*-domain transformation of Eq. (6-64). Thus both equations are henceforth referred to as a Tustin transformation.

Ideally, the *z*-domain transfer function, which is obtained by digitizing an *s*-domain transfer function, maintains all the properties of the *s*-domain transfer function (see Sec. 6-6). This is not practical; however, it is worthwhile to discuss how the transformation exemplifies four particular properties which are of interest to the designer. These four properties are the cascading property, stability, dc gain,

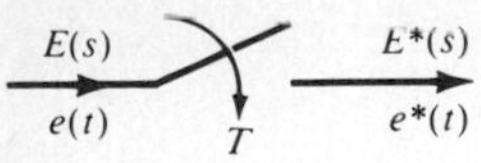

$E(s)$, $e(t)$, T, $E^*(s)$, $e^*(t)$

Figure 7-2 A sampled function.

and the impulse response. A summary of the properties maintained by the Tustin transformation method, as well as others, is shown in Table 7-1.

The Tustin algorithm maintains the cascading property of the s-domain transfer function. The cascading property is maintained when the cascading of two z-domain Tustin transfer functions yields the same result as cascading their s- (or w-) domain counterparts and Tustin-transforming the result to the z domain. This cascading property implies that no ZOH device is involved in the cascading of the s-domain transfer functions. The Tustin transformation of a stable s- (or w-) domain transfer function is stable. The dc gains for the s- (or w-) domain and the z-domain Tustin transfer functions are identical; that is, $F(s)|_{s=0} = F(w)|_{w=0} = F(z)|_{z=1}$. Note that for the exact $\mathscr{Z}$ transfer functions the dc gains are not identical.

Example 7-1 This example is presented to illustrate the properties of the Tustin transformation of a function $e(t)$ that is being sampled as shown in Fig. 7-2. Given $E(s) = E_x(s) = 1/[(s+1)(s+2)]$, *with no ZOH involved*, and $T = 0.1$ s, the exact $\mathscr{Z}$ transform yields

$$E(s) = \mathscr{Z}[e^*(t)] = \frac{0.086107z}{(z - 0.81873)(z - 0.90484)}$$

whereas the Tustin transformation yields

$$[E(z)]_{\mathrm{TU}} = \frac{0.0021645(z+1)^2}{(z - 0.81818)(z - 0.90476)}$$

The initial-value and final-value theorems yield, respectively, for each transform the results shown in the following table.

	$E(s)$	$E(z)$	$[E(z)]_{\mathrm{TU}}$	$\frac{1}{T}[E(z)]_{\mathrm{TU}}$
I.V.	0	0	0.0021645	0.021646
F.V.	0	0	0	0
DC gain	0.5	4.9918	0.499985	4.99985
n	2		2	
w	0	1		

As this example illustrates, the Tustin transformation results in a $\mathscr{Z}$ transform for

Table 7-1 Approximate digitization algorithms

Algorithms	s^{-1}	s^{-2}	s^{-3}	Invariant properties			
				Cascade	Stability	DC gain	Impulse
Tustin	$T_{11} = \frac{T}{2}\frac{1+z^{-1}}{1-z^{-1}}$	$T_{12} = T_{11}^2$	$T_{13} = T_{11}^3$	Yes	Yes	Yes	No
First difference	$T_{21} = \frac{T}{1-z^{-1}}$	$T_{22} = T_{21}^2$	$T_{33} = T_{21}^3$	Yes	Yes	Yes	No
z-form Boxer-Thaler	$T_{31} = \frac{T}{2}\frac{1+z^{-1}}{1-z^{-1}}$	$T_{32} = \frac{T^2}{12}\frac{1+10z^{-1}+z^{-2}}{(1-z^{-1})^2}$	$T_{33} = \frac{T^3}{2}\frac{z^{-1}(1+z^{-1})}{(1-z^{-1})^3}$	No	No	Yes	No
Approximate $\mathscr{Z}$ Transform	$T_{41} = \frac{T}{1-z^{-1}}$	$T_{42} = \frac{T^2 z^{-1}}{(1-z^{-1})^2}$	$T_{43} = \frac{T^3}{2}\frac{z^{-1}(1+z^{-1})}{(1-z^{-1})^3}$	No	Yes	Yes	Yes
Madwed-Truxal	$T_{51} = \frac{T}{2}\frac{1+z^{-1}}{1-z^{-1}}$	$T_{52} = \frac{T^2}{6}\frac{1+4z^{-1}+z^{-2}}{(1-z^{-1})^2}$	$T_{53} = \frac{T^3}{24}\frac{1+11z^{-1}+11z^{-2}+z^{-3}}{(1-z^{-1})^3}$	No	No	Yes	No
Halijak	$T_{61} = \frac{T}{1-z^{-1}}$	$T_{62} = \frac{T^2 z^{-1}}{(1-z^{-1})^2}$	$T_{63} = \frac{T}{2}\frac{1+z^{-1}}{1-z^{-1}} T_{62}$	No	...	...	No

which, in general, $n = w$. Thus, since $n = w$, the initial value of $[E(z)]_{TU}$ is always different from zero. That is,

$$\lim_{z \to \infty} E(z) \neq \lim_{z \to \infty} [E(z)]_{TU}$$

Based upon the dc gain, it is evident that the values of $E(z)$ and $[E(z)]_{TU}$ differ by a factor of $1/T$; that is,

$$E(z) = \mathscr{Z}[e^*(t)] \approx \frac{1}{T}[E(z)]_{TU} \tag{7-8}$$

The reason for this difference is that the Tustin transformation does not take into account the attenuation factor $1/T$ due to the sampling process [see Eq. (3-50)], whereas $E(z) = \mathscr{Z}[e^*(t)]$ does take into account this attenuation factor. This factor must be taken into account, as discussed in Sec. 7-5, when the DIG technique is used for system analysis and design.

The first-difference transformation method consists of substituting the quantity $[T/(1 - z^{-1})]^q$ term in the continuous s-domain transfer function. The term $T/(1 - z^{-1})$ is an approximation for s^{-1}, and it represents s-domain integration through rectangular integration. Owing to the inaccuracy of rectangular integration, the z-domain transfer function resulting from the first-difference transformation has poor accuracy compared with that of the Tustin transformation, i.e., trapezoidal integration. The cascading property, the stability, and the dc gain are preserved by the first-difference transformation.

Table 7-1 lists other methods together with the first-difference method. They are listed on the basis of their properties but are not widely used, being of academic interest only. The Tustin transformation is used in this text for the DIG control design method. Table 7-2 presents a number of polynomials for factors of $(z \pm 1)^q$ to assist in the transformation process. This table may be readily expanded to include higher-order factors or incorporated in a CAD package.[14,15]

7-4 MAPPING APPROXIMATIONS OF $\mathscr{Z}$ TRANSFORM (OR NUMERICAL SOLUTION OF DIFFERENTIAL EQUATIONS)

The purpose of this section is to discuss some of the various approximations to the $\mathscr{Z}$ transform, $z = \epsilon^{Ts}$, in terms of the mapping between the s plane and z plane. These approximation transformations are equivalent, of course, to certain differential approximations (difference equations) as discussed previously.

7-4A First-Backward Difference

Consider the first-backward difference

$$\left.\frac{dx(t)}{dt}\right|_{t=kT} = \nabla^1[x(kT)] = \frac{x(kT)}{T} - \frac{x[(k-1)T]}{T} \tag{7-9}$$

where the $\mathscr{L}$ and $\mathscr{Z}$ transforms of the left- and right-hand sides, with zero ICs, respectively, are

$$[s]X(s) \tag{7-10}$$

and

$$\frac{X(z) - z^{-1}X(z)}{T} = \left[\frac{1 - z^{-1}}{T}\right]X(z) \tag{7-11}$$

Therefore, based upon Eq. (7-9), the equivalent s-plane transformation representation is obtained by equating the bracketed terms of Eqs. (7-10) and (7-11); thus

$$s = \frac{1 - z^{-1}}{T} \tag{7-12}$$

This equation can also be derived by approximating the series $z = \epsilon^{Ts}$. Of course, the implication of this discussion is that there is an equivalence between the concept of replacing derivatives by differences and the s-plane to z-plane mapping relationship.

The exact mapping relationship between the s plane and the z plane implied that the s-plane imaginary axis maps into the unit circle in the z plane with a folding frequency of $\omega_s = 2\pi/T$. Stable roots in the s plane are mapped into the z-plane unit circle.

Returning to the first-difference approximation, z is defined as a function of s, yielding, from Eq. (7-12),

$$z = \frac{T}{1 - sT} \tag{7-13}$$

First, consider the s-plane imaginary axis: $j\omega_{sp}$ mapping, that is,

$$z = \frac{1}{1 - j\omega_{sp}T} \tag{7-14}$$

which can be rewritten as

$$z = \frac{1}{2}\left(1 + \frac{1 + j\omega_{sp}T}{1 - j\omega_{sp}T}\right) = \frac{1}{2}(1 + \epsilon^{j2\tan^{-1}\omega_{sp}T}) \tag{7-15}$$

where the exponent of ϵ is found from

$$\frac{1 + ja}{1 - ja} = \epsilon^{\ln(1 + ja) - \ln(1 - ja)} = \epsilon^{2\tanh^{-1} ja} = \epsilon^{j2\tan^{-1} a}$$

and where

$$\tanh^{-1} x = \frac{1}{2}\ln(1 + x) - \frac{1}{2}\ln(1 - x)$$

$$\tanh^{-1} ja = j\tan^{-1} a$$

$$x = ja$$

Table 7-2 Polynomials of $(z \pm 1)^q$

	z^{12}	z^{11}	z^{10}	z^9	z^8	z^7	z^6	z^5	z^4	z^3	z^2	z^1	z^0
$(z-1)^1$												+1	−1
$(z-1)^1(z+1)^1$											+1	0	−1
$(z-1)^2$											+1	−2	+1
$(z-1)^2(z+1)^1$										+1	−1	−1	+1
$(z-1)^2(z+1)^2$									+1	0	−2	0	+1
$(z-1)^3$										+1	−3	+3	−1
$(z-1)^3(z+1)^1$									+1	−2	0	+2	−1
$(z-1)^3(z+1)^2$								+1	−1	−2	+2	+1	−1
$(z-1)^3(z+1)^3$							+1	0	−3	0	+3	0	−1
$(z-1)^4$									+1	−4	+6	−4	+1
$(z-1)^4(z+1)^1$								+1	−3	+2	+2	−3	+1
$(z-1)^4(z+1)^2$							+1	−2	−1	+4	−1	−2	+1
$(z-1)^4(z+1)^3$						+1	−1	−3	+3	+3	−3	−1	+1
$(z-1)^4(z+1)^4$					+1	0	−4	0	+6	0	−4	0	+1
$(z-1)^5$								+1	−5	+10	−10	+5	−1
$(z-1)^5(z+1)^1$							+1	−4	+5	0	−5	+4	−1
$(z-1)^5(z+1)^2$						+1	−3	+1	+5	−5	−1	+3	−1
$(z-1)^5(z+1)^3$					+1	−2	−2	+6	0	−6	+2	+2	−1
$(z-1)^5(z+1)^4$				+1	−1	−4	+4	+6	−6	−4	+4	+1	−1
$(z-1)^5(z+1)^5$			+1	0	−5	0	+10	0	−10	0	+5	0	−1
$(z-1)^6$							+1	−6	+15	−20	+15	−6	+1
$(z-1)^6(z+z)^1$						+1	−5	+9	−5	−5	+9	−5	+1
$(z-1)^6(z+1)^2$					+1	−4	+4	+4	−10	+4	+4	−4	+1
$(z-1)^6(z+1)^3$				+1	−3	0	+8	−6	−6	+8	0	−3	+1
$(z-1)^6(z+1)^4$			+1	−2	−3	+8	+2	−12	+2	+8	−3	−2	+1
$(z-1)^6(z+1)^5$		+1	−1	−5	+5	+10	−10	−10	+10	+5	−5	−1	+1
$(z-1)^6(z+1)^6$	+1	0	−6	0	+15	0	−20	0	+15	0	−6	0	+1
$(z+1)^1$												+1	+1
$(z+1)^2$											+1	+2	+1
$(z+1)^2(z-1)^1$										+1	+1	−1	−1
$(z+1)^3$										+1	+3	+3	+1

	z^{12}	z^{11}	z^{10}	z^9	z^8	z^7	z^6	z^5	z^4	z^3	z^2	z^1	z^0
$(z+1)^3(z-1)^1$									+1	+2	0	−2	−1
$(z+1)^3(z-1)^2$								+1	+1	−2	−2	+1	+1
$(z+1)^4$									+1	+4	+6	+4	+1
$(z+1)^4(z-1)^1$								+1	+3	+2	−2	−3	−1
$(z+1)^4(z-1)^2$							+1	+2	−1	−4	−1	+2	+1
$(z+1)^4(z-1)^3$						+1	+1	−3	−3	+3	+3	−1	−1
$(z+1)^5$								+1	+5	+10	+10	+5	+1
$(z+1)^5(z-1)^1$							+1	+4	+5	0	−5	−4	−1
$(z+1)^5(z-1)^2$						+1	+3	+1	−5	−5	+1	+3	+1
$(z+1)^5(z-1)^3$					+1	+2	−2	−6	0	+6	+2	−2	−1
$(z+1)^5(z-1)^4$				+1	+1	−4	−4	+6	+6	−4	−4	+1	+1
$(z+1)^6$							+1	+6	+15	+20	+15	+6	+1
$(z+1)^6(z-1)^1$						+1	+5	+9	+5	−5	−9	−5	−1
$(z+1)^6(z-1)^2$					+1	+4	+4	−4	−10	−4	+4	+4	+1
$(z+1)^6(z-1)^3$				+1	+3	0	−8	−6	+6	+8	0	−3	−1
$(z+1)^6(z-1)^4$			+1	+2	−3	−8	+2	+12	+2	−8	−3	+2	+1
$(z+1)^6(z-1)^5$		+1	+1	−5	−5	+10	+10	−10	−10	+5	+5	−1	−1
$(z+1)^7$						+1	+7	+21	+35	+35	+21	+7	+1
$(z+1)^7(z-1)^1$					+1	+6	+14	+14	0	−14	−14	−6	−1
$(z-1)^7$						+1	−7	+21	−35	+35	−21	+7	−1
$(z-1)^7(z+1)^1$					+1	−6	+14	−14	0	+14	−14	+6	−1
$(z+1)^8$					+1	+8	+28	+56	+70	+56	+28	+8	+1
$(z-1)^8$					+1	−8	+28	−56	+70	−56	+28	−8	+1

The mapping relationship defined by Eq. (7-15) indicates that the s-plane imaginary axis corresponds to a z-plane circle whose center is at $z = 1/2$ with a radius of 1/2. Figure 7-3 illustrates this result. It also portrays the mapping of the left-half s plane (stable roots) into the small circle. This is easily verified from Eq. (7-13). The first-difference approximation does map stable s-plane poles (stable continuous controllers) into stable z-plane roots (stable difference equations).

A question that arises is: What is the mapping when T is decreased or, equivalently, when the sampling rate is increased? One would expect the differential approximation to become more precise as $T \to 0$. As T decreases, the relative width of the s-plane primary strip (exact z-plane transformation) increases due to $2\pi/T$. Thus the boundary between the primary and the first complementary

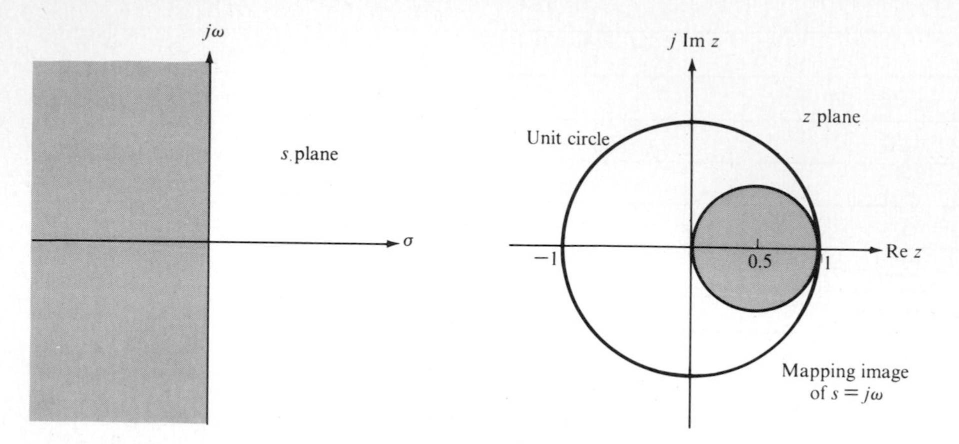

Figure 7-3 First-backward difference s- to z-plane mapping.

strip moves farther and farther away from the poles within the primary strip. Using the exact z transforms, the equivalent z-plane poles move closer and closer to the z-plane positive real axis (0, 1) within the unit circle. Thus, if T is sufficiently small, the s-plane poles fall within the small circle (Fig. 7-3b), reflecting the first-difference approximation. So as $T \to 0$, the first-difference approximation improves. Another way of considering this mapping phenomenon is to reflect on the bandwidth of the controller. As T decreases, the discrete controller bandwidth is mapped into a smaller part of the unit circle, and as $T \to 0$, the bandwidth is concentrated in the vicinity of $z = 1$. Of course, the small circle of Fig. 7-3b and the z-plane unit circle are tangent at $z = 1$. Thus, if T becomes smaller and smaller, and the controller bandwidth and input signal spectrum are located near $z = 1$, the first-difference digital controller accurately approximates the continuous controller.

Returning to Eq. (7-15), it is easily seen that the entire s-plane imaginary axis is mapped once and only once onto the small z-plane circle. Using this first-difference approximation, folding or aliasing problems do not occur. The penalty is a "warping" of the equivalent s-plane poles as shown in Fig. 7-4. This situation is reflected in the relationship between the exact z transformation for $s = j\omega_{sp}$ and the first-difference approximation for $s = j\hat{\omega}_{sp}$, namely, considering Eq. (7-15),

$$z = \epsilon^{j\omega_{sp}T} = \frac{1}{2}[1 + \exp(j2 \tan^{-1} \hat{\omega}_{sp}T)] \tag{7-16}$$

The hat, ^, refers to the first-difference approximation. Thus, a nonlinear relationship or "warping" exists between the two frequencies. To design a controller, using this approximation technique, requires that the original frequency should be modified according to the warping implied by Eq. (7-16). That is, to counter the warping phenomenon, the s-plane poles are frequency-prewarped according to Eq. (7-16). For example, the imaginary term, of a pole in the s plane, which contains the fre-

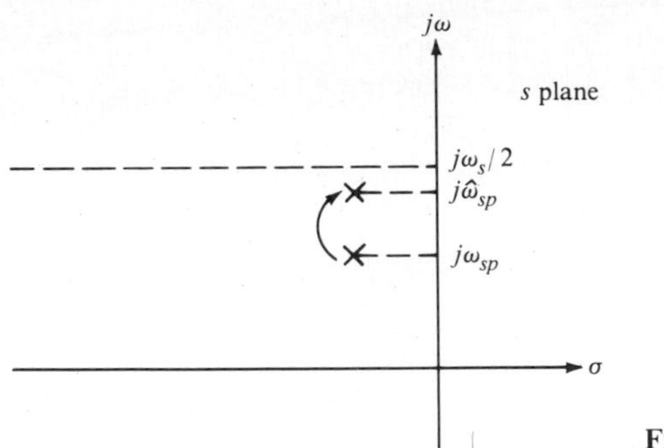

Figure 7-4 Warping effect.

quency information is prewarped by using Eq. (7.16) to generate an $\hat{\omega}_{sp}$ which is used in a new warped s plane. This new pole is then transformed into the z domain by using the first-backward difference approximation of Eq. (7-13). This prewarping technique compensates for the warping of the first-backward difference approach.

Note that for small $\omega_{sp}T$ and using the first two terms of the series approximation of the exponentials of Eq. (7-16) yields

$$1 + j\omega_{sp}T = \frac{1}{2}(1 + 1 + j2\tan^{-1}\omega_{sp}T)$$

or

$$\tan \omega_{sp}T \approx \hat{\omega}_{sp}T \qquad \text{and} \qquad \omega_{sp} \approx \hat{\omega}_{sp}$$

This approximation is good for relatively small $\omega_{sp}T$ (about 17° or less). Thus prewarping is not required for small $\omega_{sp}T$.

In general, the first-backward difference is not used owing to the small values of T that are required in practice. Better approximations are required. Note again that the first-backward difference approach is equivalent to rectangular integration if s^{-1} is considered.

If the first-forward difference approximation is employed, stability problems exist. For example, consider mapping the function

$$\left.\frac{dx(t)}{dt}\right|_{t=kT} \approx \frac{x[(k+1)T] - x(kT)}{T}$$

then analyze stability regions and consider the effects of bandwidth.

A point meriting emphasis in this analysis is that the numerical accuracy of the mapping approach is assumed to be infinite. However, the finite word length of a digital processor requires quantization of the controller poles and zeros, which requires that the z-plane poles and zeros be a certain distance away from the unit circle to prevent oscillatory or unstable conditions. The z-plane poles and zeros can only assume values contained in a finite set of numbers due to the finite word length constraint. The statistical effect of quantization is presented in Chap. 10.

7-4B Tustin Transformation

The Tustin transformation for $q = 1$ is defined as

$$s \equiv \frac{2}{T}\frac{1 - z^{-1}}{1 + z^{-1}} \tag{7-17}$$

which is a bilinear transformation and can be equated to the trapezoidal integration (s^{-1}) method. Also, Eq. (7-17) can be derived by approximating $z = \epsilon^{Ts}$ as a finite series. The operator equation (7-17) is as applicable to matrix equations as it is to scalar differential equations. The following discussion is useful therefore in understanding the mapping result for the vector model also.

To represent functionally the s- to z-plane mapping, Eq. (7-17) is rearranged to yield

$$z = \frac{1 + sT/2}{1 - sT/2} \tag{7-18}$$

Let $s = j\hat{\omega}_{sp}$; thus

$$z = \frac{1 + j\hat{\omega}_{sp}T/2}{1 - j\hat{\omega}_{sp}T/2} \tag{7-19}$$

The exact z transform yields $z = \epsilon^{j\omega_{sp}T}$, where ω_{sp} is an equivalent s-plane frequency. Therefore, using the same approach employed in deriving Eq. (7-16), the following equation is obtained from Eq. (7-19):

$$\epsilon^{j\omega_{sp}T} = \exp\left(j2\tan^{-1}\frac{\hat{\omega}_{sp}T}{2}\right) \tag{7-20}$$

Thus

$$\frac{\omega_{sp}T}{2} = \tan^{-1}\frac{\hat{\omega}_{sp}T}{2} \tag{7-21}$$

or

$$\tan\frac{\omega_{sp}T}{2} = \frac{\hat{\omega}_{sp}T}{2} \tag{7-22}$$

When $\omega_{sp}T/2 < 17°$, or about 0.30 rad, then

$$\omega_{sp} \approx \hat{\omega}_{sp} \tag{7-23}$$

which means that in the frequency domain the Tustin approximation is good for small values of $\omega_{sp}T/2$.

Returning to Eq. (7-20), it is easy to realize that the imaginary axis of the s plane is mapped into the unit circle z plane as shown in Fig. 7-5. The left-half s plane is mapped into the unit circle. The same stability regions exist for the exact $\mathscr{Z}$ transform and the Tustin approximation.

Also, in this approximation, the entire imaginary axis s plane is mapped once and only once onto the unit circle. The Tustin approximation prevents pole and

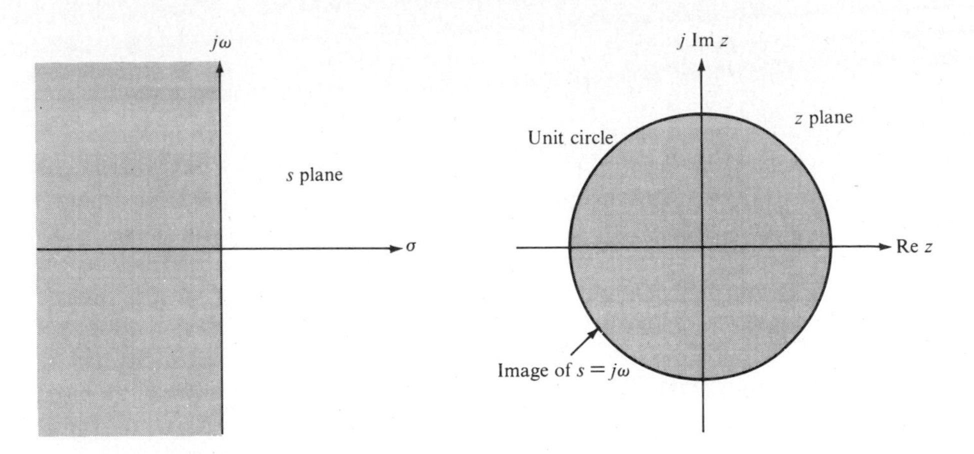

Figure 7-5 Tustin approximation, s- to z-plane mapping.

zero aliasing since the folding phenomenon does not occur with this method. However, there is again a warping penalty. Compensation can be accomplished by using Eq. (7-22), which is depicted in Fig. 7-6. To compensate for the warping, prewarping of ω_{sp} by using Eq. (7-22) generates $\hat{\omega}_{sp}$. The continuous controller is mapped into the z plane by means of Eq. (7-18), using the prewarped frequency $\hat{\omega}_{sp}$. The digital compensator (controller) must be tuned (i.e., its numerical coefficients adjusted) to finalize the design since approximations have been employed. As seen from Fig. 7-6, Eq. (7-23) is a good approximation when $\omega_{sp}T/2$ and $\hat{\omega}_{sp}T/2$ are both less than 0.3 rad.

The prewarping approach for the Tustin approximation takes the imaginary axis s plane and folds it back to $\pi/2$ to $-\pi/2$ as seen from Fig. 7-6. The spectrum of the input must also be taken into consideration when selecting an approximation procedure with or without prewarping. It should be noted that in the previous discussion only the frequency has been prewarped due to the interest in the controller frequency response. The real part of the s-plane pole influences such parameters as rise time, overshoot, and settling time. Thus, consideration of the warping of the real pole component is now analyzed as a fine-tuning approach. Proceeding in the same manner as used in deriving Eq. (7-23), substitute $z = \epsilon^{\sigma_{sp}T}$ and $s = \hat{\sigma}_{sp}$ into Eq. (7-18) to yield

$$\epsilon^{\sigma_{sp}T} = \frac{1 + \hat{\sigma}_{sp}T/2}{1 - \hat{\sigma}_{sp}T/2} \tag{7-24}$$

Replacing $\epsilon^{\sigma_{sp}T}$ by its exponential series and dividing the numerator by the denominator in Eq. (7-24) results in the expression

$$1 + \sigma_{sp}T + \frac{(\sigma_{sp}T)^2}{2} + \cdots = 1 + \frac{\hat{\sigma}_{sp}T}{1 - \hat{\sigma}_{sp}T/2} \tag{7-25}$$

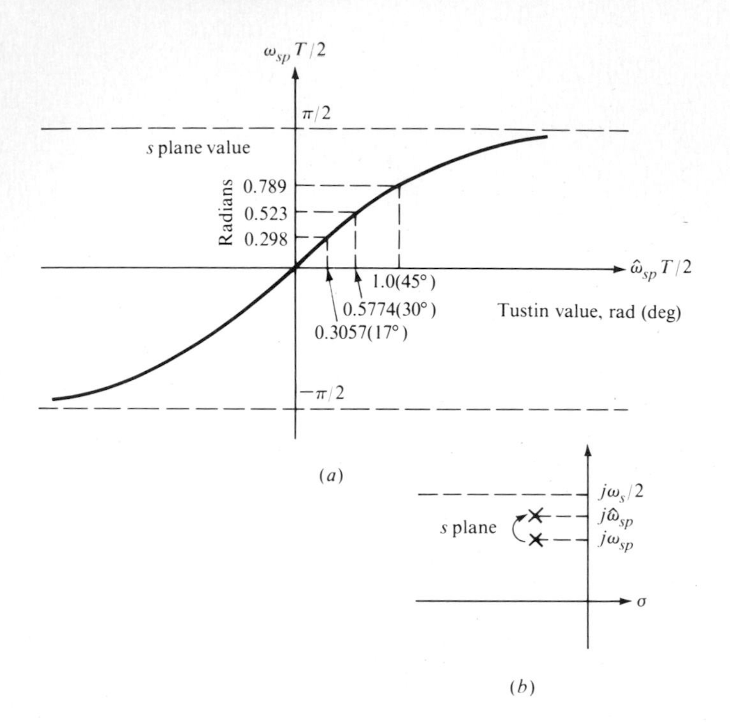

Figure 7-6 Map of $\hat{\omega}_{sp} = 2(\tan \omega_{sp}T/2)/T$. (*a*) Plot of Eq. (7-20) and (*b*) warping effect.

If $|\sigma_{sp}T| \gg (\sigma_{sp}T)^2/2$ (or $1 \gg |\sigma_{sp}T/2|$) and $1 \gg |\hat{\sigma}_{sp}T/2|$, then

$$|\hat{\sigma}_{sp}| \approx |\sigma_{sp}| \ll \frac{2}{T} \tag{7-26}$$

Thus with Eqs. (7-23) and (7-26) satisfied, the Tustin approximation in the *s* domain is good for small magnitudes of the real and imaginary components of the variable *s*. The shaded area in Fig. 7-7 represents the allowable location of the poles and zeros in the *s* plane for a good Tustin approximation. Because of the mapping properties and its ease of use, the Tustin transformation is employed for the DIG technique in this text.

7-5 PSEUDO-CONTINUOUS-TIME (PCT) CONTROL SYSTEM

The DIG method of designing a sampled-data system, in the complex-frequency *s* plane, requires a satisfactory pseudo-continuous-time (PCT) model of the sampled-data system. In other words, for the sampled-data system of Fig. 7-8, the sampler and the ZOH units must be approximated by a linear continuous-time unit $G_A(s)$, as shown in Fig. 7-9*c*. The DIG method requires that the dominant poles and zeros of the PCT model should lie in the shaded area of Fig. 7-7 for a high level

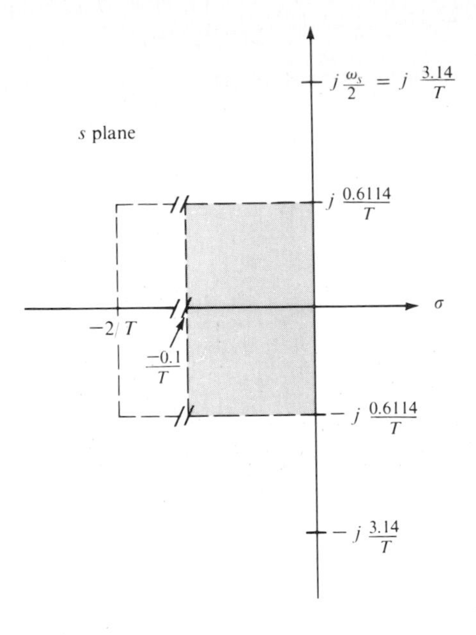

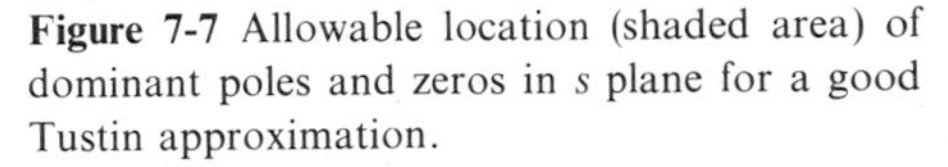

Figure 7-7 Allowable location (shaded area) of dominant poles and zeros in s plane for a good Tustin approximation.

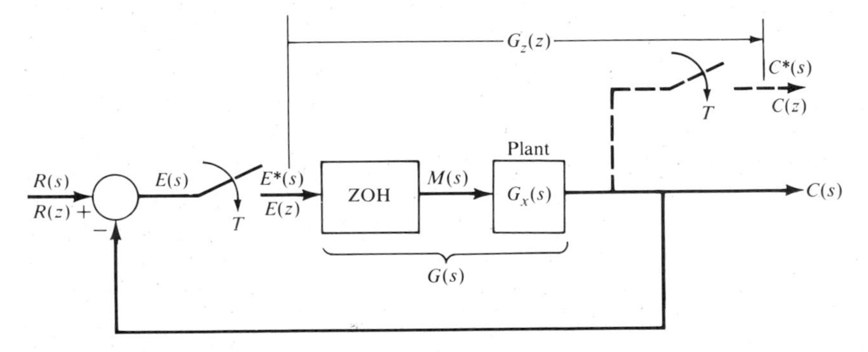

Figure 7-8 The uncompensated sampled-data control system.

of correlation with the sampled-data system. To determine $G_A(s)$, first consider the frequency component of $\mathbf{E}^*(j\omega)$ representing the continuous-time signal $\mathbf{E}(j\omega)$, where all its sidebands are multiplied by $1/T$ (see Fig. 3-6).[26] Because of the low-pass filtering characteristics of a sampled-data system, only the primary component needs to be considered in the analysis of the system. Therefore, the PCT approximation of the sampler of Fig. 7-8*a* is shown in Fig. 7-9*b*.

Using the first-order Padé approximation (see Appendix C), the transfer function of the ZOH device, when the value of T is small enough, is approximated as follows:[89]

$$G_{zo}(s) = \frac{1 - \epsilon^{-Ts}}{s} \approx \frac{2T}{Ts + 2} = G_{pa}(s) \tag{7-27}$$

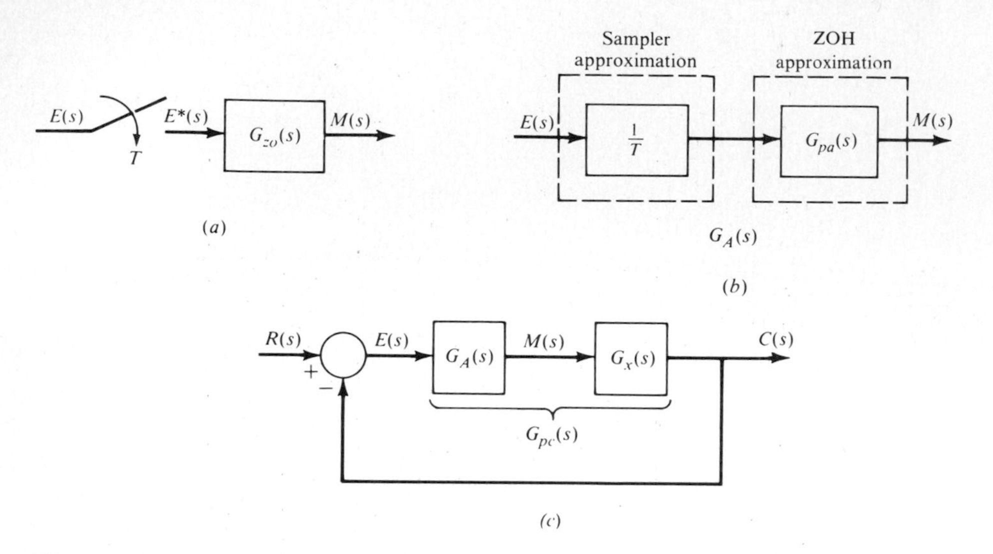

Figure 7-9 (*a*) Sampler and ZOH. (*b*) Approximations of the sampler and ZOH and (*c*) the approximate continuous-time control system equivalent of Fig. 7-8.

Thus the Padé approximation $G_{pa}(s)$ is used to replace $G_{zo}(s)$ as shown in Fig. 7-9*a* and *b*. This approximation is good for $\omega_c \leq \omega_s/10$, whereas the second-order approximation is good for $\omega_c \leq \omega_s/3$ (Ref. 90). Therefore the sampler and ZOH units of a sampled-data system are approximated in the PCT system of Fig. 7-9*c* by the transfer function

$$G_A(s) = \frac{1}{T} G_{pa}(s) = \frac{2}{Ts + 2} \tag{7-28}$$

Since $\lim_{T \to 0} G_A(s) = 1$ then Eq. (7-28) is an accurate PCT representation of the sampler and ZOH units, because it satisfies the requirement that as $T \to 0$ the output of $G_A(s)$ must equal its input. Further note that in the frequency domain (see Fig. 3-6) as $\omega_s \to \infty$ $(T \to 0)$ then the primary strip becomes the entire frequency-spectrum domain which is the representation for the continuous-time system.

Note that in obtaining PCT systems for the sampled-data systems of Fig. 7-1 the factor $1/T$ replaces only the sampler that is sampling the continuous-time signal. This multiplier of $1/T$ is reflected in Eq. (3-45), which states that the fundamental, the frequency of the sampled signal, and all its harmonics are attenuated by $1/T$. *The sampler on the output of the digital controller is replaced by a factor of 1.* To illustrate the effect of the value of T on the validity of the results obtained by the DIG method, consider the sampled-data closed-loop control system of Sec. 6-7. The closed-loop system performance for three values of T and $\zeta = 0.45$ are determined in both the s and z domains, i.e., the DIG and DIR methods,

Table 7-3 Analysis of a PCT system representing a sampled-data control system for $\zeta = 0.45$

Method	T, s	Domain	K_x	M_p	t_p, s	t_s, s
DIR	0.01	z	4.147	1.202	4.16	9.53
DIG		s	4.215	1.206	4.11	9.478
DIR	0.1	z	3.892	1.202	4.2–4.3	9.8+
DIG		s	3.906	1.203	4.33^-	9.90+
DIR	1	z	2.4393	1.199	6	13–14
DIG		s	2.496	1.200	6.18	13.76

respectively. Table 7-3 presents the required value of K_x and time-response characteristics for each value of T. Note that for $T \leq 0.1$ s there is a high level of correlation between the DIG and DIR models. For $T \leq 1$ s there is still a relatively good correlation. (The designer needs to specify, for a given application, what is considered to be "good correlation.")

7-6 THE ANALYSIS OF A BASIC (UNCOMPENSATED) SYSTEM

Figure 7-7 represents a basic or uncompensated sampled-data control system. For the rest of this chapter the plant transfer function

$$G_x(s) = \frac{K_x}{s(s+1)} \qquad \textbf{Case 1} \tag{7-29}$$

is used to illustrate the approaches for improving the performance of a basic system. As mentioned in Chap. 3, the lag characteristic of $G_{zo}(s)$ [see Fig. 3-22*b*, a plot of $\angle \mathbf{G}_{zo}(j\omega)$ vs. ω] reduces the degree of system stability. This degradation is illustrated in some of the examples of this chapter.

7-6A PCT Control-System Model

One approach for designing a sampled-data unity-feedback control system is to first obtain a suitable closed-loop model $[C(s)/R(s)]_M$ for the PCT unity-feedback control system of Fig. 7-9, utilizing the plant of the sampled-data control system. This model is then used as a guide for selecting an acceptable $C(z)/R(z)$. Thus, for the plant of Eq. (7-29),

$$G_{PC}(s) = G_A(s)G_x(s) = \frac{2K_x/T}{s(s+1)(s+2/T)} \tag{7-30}$$

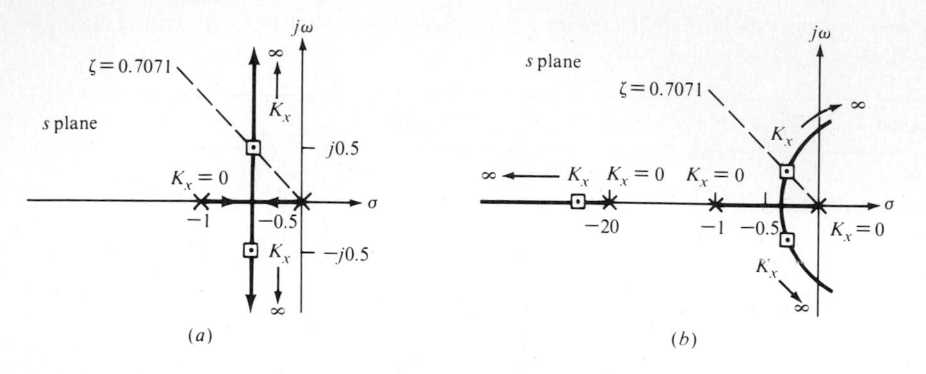

Figure 7-10 Root locus for (*a*) Case 1, Eq. (7-29); (*b*) Case 2, Eq. (7-30). Not to scale.

and for $T = 0.1$ s,

$$\left[\frac{C(s)}{R(s)}\right]_M = \frac{G_{\text{PC}}(s)}{1 + G_{\text{PC}}(s)} = \frac{20K_x}{s^3 + 21s^2 + 20s + 20K_x} \qquad (7\text{-}31)$$

The root locus for $G_{\text{PC}}(s) = -1$ is shown in Fig. 7-10*b*. For comparison purposes the root-locus plot for $G_x(s) = -1$ is shown in Fig. 7-10*a*. These figures illustrate the effect of inserting a lag network in cascade in a feedback control system; i.e., it transforms a completely stable system into a conditionally stable system. Thus for a given value of ζ the values of t_p and t_s (and T_s) are increased. Therefore as stated previously the ZOH unit degrades the degree of system stability.

For the model it is assumed that the desired value of the damping ratio ζ for the dominant roots is 0.7071. Thus for a unit-step input,

$$[C(s)]_M = \frac{9.534}{s(s^3 + 21s^2 + 20s + 9.534)}$$

$$= \frac{9.534}{s(s + 0.4875 + j0.4883)(s + 20.03)} \qquad \textbf{Case 2} \qquad (7\text{-}32)$$

where $K_x = 0.4767$. The real and imaginary parts of the desired roots of Eq. (7-32), for $T = 0.1$ s, lie in the acceptable region of Fig. 7-7 for a good Tustin approximation.

7-6B Sampled-Data Control System

The determination of the time-domain performance of the sampled-data system of Fig. 7-8 may be achieved either by obtaining the exact expression for $C(z)$ or by applying the Tustin transformation to Eq. (7-31) to obtain the approximate expression for $C(z)$. Proceeding with the exact approach first requires the z-transfer function of the forward loop of Fig. 7-8. For the plant transfer function of Eq. (7-29),

$$G_z(z) = \mathscr{Z}\left[\frac{K_x(1-\epsilon^{-sT})}{s^2(s+1)}\right] = (1-z^{-1})\mathscr{Z}\left[\frac{K_x}{s^2(s+1)}\right]$$

$$= \frac{K_x[(T-1+\epsilon^{-T})z + (1 - T\epsilon^{-T} - \epsilon^{-T})]}{z^2 - (1+\epsilon^{-T} + K_x - TK_x - K_x\epsilon^{-T})z + \epsilon^{-T} + K_x - K_x(T+1)\epsilon^{-T}} \tag{7-33}$$

Thus for $T = 0.1$ s and $K_x = 0.4767$,

$$G_z(z) = \frac{0.002306(z+0.9672)}{(z-1)(z-0.9048)} \tag{7-34}$$

or

$$\frac{C(z)}{R(z)} = \frac{0.002306(z+0.9672)}{(z-0.9513 \pm j0.04649)} \qquad \textbf{Case 3} \tag{7-35}$$

The DIG technique requires that the s-domain model control ratio be transformed into a z-domain model. Applying the Tustin transformation to

$$\left[\frac{C(s)}{R(s)}\right]_M = \frac{G_{\text{PC}}(s)}{1+G_{\text{PC}}(s)} = F_M(s) \tag{7-36}$$

yields

$$\frac{[C(z)]_{\text{TU}}}{[R(z)]_{\text{TU}}} = [F(z)]_{\text{TU}} \tag{7-37}$$

This equation is rearranged to

$$[C(z)]_{\text{TU}} = [F(z)]_{\text{TU}}[R(z)]_{\text{TU}} \tag{7-38}$$

As stated in Sec. 7-3,

$$R(z) = \mathscr{Z}[r^*(t)] = \frac{1}{T}[R(z)]_{\text{TU}} \tag{7-39}$$

$$C(z) = \mathscr{Z}[c^*(t)] = \frac{1}{T}[C(z)]_{\text{TU}} \tag{7-40}$$

Substituting from Eqs. (7-39) and (7-40) into Eq. (7-37) yields

$$\frac{C(z)}{R(z)} = [F(z)]_{\text{TU}} \tag{7-41}$$

Substituting from Eq. (7-40) into Eq. (7-38) and rearranging yields

$$C(z) = \frac{1}{T}[F(z)]_{\text{TU}}[R(z)]_{\text{TU}} = \frac{1}{T}[\text{Tustin of } F_M(s)R(s)] \tag{7-42}$$

Table 7-4 Comparison of time responses between $C(z)$ and $[C(z)]_{TU}$ for a unit-step input and $T = 0.1$ s

	$c(kT)$	
k	Case 3 (exact), $C(z)$	Case 4 (Tustin), $[C(z)]_{TU}$
0	0.	0.5672E − 03
2	0.8924E − 02	0.9823E − 02
4	0.3340E − 01	0.3403E − 01
6	0.7024E − 01	0.7064E − 01
8	0.1166	0.1168
10	0.1701	0.1701
12	0.2284	0.2283
14	0.2897	0.2894
16	0.3525	0.3521
18	0.4153	0.4148
20	0.4771	0.4766
22	0.5370	0.5364
24	0.5943	0.5937
26	0.6485	0.6478
28	0.6992	0.6984
30	0.7461	0.7453
32	0.7890	0.7883
34	0.8281	0.8273
36	0.8632	0.8624
38	0.8944	0.8936
40	0.9220	0.9212
42	0.9460	0.9452
44	0.9668	0.9660
46	0.9844	0.9836
48	0.9992	0.9984
50	1.011	1.011
52	1.021	1.020
54	1.029	1.028
56	1.035	1.034
58	1.039	1.038
60	1.042	1.041
61	1.042	1.042
62	1.043	1.043
63	1.043	1.043
64	1.043 } M_p	1.043 } M_p
65	1.043 } M_p	1.043 } M_p
66	1.043	1.043
67	1.043	1.043
68	1.042	1.042
85	1.022	1.022
86	1.021 } t_s	1.021 } t_s
87	1.019 } t_s	1.019 } t_s

Thus based upon Eq. (7-41) the Tustin transformation to Eq. (7-31), with $K_x = 0.4767$, results in a Tustin model of the control ratio

$$\frac{C(z)}{R(z)} = \left[\frac{C(z)}{R(z)}\right]_{\mathrm{TU}}$$

$$= \frac{5.672 \times 10^{-4}(z+1)^3}{(z - 0.9513 \pm j0.04651)(z + 6.252 \times 10^{-4})} \qquad \textbf{Case 4} \qquad (7\text{-}43)$$

Note that the dominant poles of Eqs. (7-43) are essentially the same as those of Eq. (7-35) due to the value of T used which resulted in the dominant roots lying in the good Tustin region of Fig. 7-7. In using the exact $\mathscr{Z}$ transformation the order of the numerator polynomial of $C(z)/R(z)$, Eq. (7-35), is one less than the order of its denominator polynomial. When using the Tustin transformation the order of the numerator polynomial of the resulting $[C(z)/R(z)]_{\mathrm{TU}}$ is in general equal to the order of its corresponding denominator polynomial [see Eq. (7-43)]. Thus $[C(z)]_{\mathrm{TU}}$ results in a value of $c^*(t) \neq 0$ at $t = 0$ which is in error based upon zero initial conditions. Table 7-4 illustrates the effect of this characteristic of the Tustin transformation on the time response due to a unit-step forcing function. The degradation of the time response by use of the Tustin transformation is minimal; i.e., the resulting values of the figures of merit are in close agreement to those obtained by the use of the exact $\mathscr{Z}$ transformation. Therefore, the Tustin transformation is a valid design tool when the dominant zeros and poles of $[C(s)/R(s)]_M$ lie in the acceptable region of Fig. 7-7.

Table 7-5 summarizes the time-response characteristics for a unit-step forcing function of (1) the continuous-time system of Fig. 7-9 for the two cases of $G(s) = G_x(s)$ [with $G_A(s)$ removed] and $G(s) = G_A(s)G_x(s)$ and (2) the sampled-data system of Fig. 7-8 based upon the exact and Tustin expressions for $C(z)$. The ramp error coefficient for Cases 2 to 4 are obtained by applying Eq. (6-24). The table reveals that:

1. In converting a continuous-time system into a sampled-data system the time-response characteristics are degraded.

Table 7-5 Time-response characteristics of the uncompensated system

	M_p	t_p, s	t_s, s	K_1, s^{-1}	Case
Continuous-time system					
$G(s) = G_x(s)$	1.04821	6.3	8.40 to 8.45	0.4767	1
$G_M(s) = G_A(s)G_x(s)$	1.04342	6.48	8.69	0.4767	2
Sampled-data system					
$C(z)$	1.043	6.45	8.65	0.4765	3
$[C(z)]_{\mathrm{TU}}$	1.043	6.45	8.6	0.4765	4

2. The time-response characteristics of the sampled-data system, using the value of gain obtained from the continuous-time model, agrees favorably with those of the continuous-time model. As may be expected, there is some variation in the values obtained when utilizing the exact $C(z)$ and $[C(z)]_{TU}$.

7-7 SUMMARY

Two approaches for analyzing a sampled-data control system are presented: the DIG (digitization) and DIR (direct) techniques. The former requires the use of the Padé approximation and the Tustin transformation for an initial design in the s plane (PCT control-system configuration) or the w transformation for an initial design in the w plane. If the w-plane approximation criteria of Sec. 6-6 and the Tustin approximation criteria of Sec. 7-4B are satisfied, the correlation between all three planes is very good. If the approximations are not valid, the analysis and design of the sampled-data control system should be done by the DIR technique.

CHAPTER

EIGHT

PRINCIPLES OF SIGNAL CONVERSION

8-1 INTRODUCTION

This chapter discusses the general techniques that are employed in the conversion of analog signals for digital processing. The presentation centers on digital-to-analog (D/A) conversion, analog-to-digital (A/D) conversion, signal sample-and-hold techniques, and signal multiplexing methods. Various D/A and A/D conversion techniques are analyzed along with their error sources. Performance measures are shown for numerous conversion system configurations, and techniques for selecting converters are emphasized. Also, various measurement devices are described that relate to digital-control-system transducers. To introduce the topic of signal conversion, timing phenomena exhibited by conversion and computer processing delays in a digital control system are initially presented.

8-2 TIMING CONSIDERATIONS

The development of a digital control law or algorithm usually assumes perfect accuracy and infinitesimal and negligible processing speed. The impact of finite word length of digital processors in a digital control system is discussed in Chap. 10. The purpose of this section is to present the reality of finite delays in converting and processing digital information within a control-system mechanization.

The general control law is usually defined in terms of a discrete control variable (scalar or vector) which is a function (usually linear) of current and past measured variables of the plant and past values of the control. The general form is

$$e_1(kT) = \sum_{i=0}^{n} ae[(k-i)T] + \sum_{j=1}^{m} be_1[(k-j)T] \tag{8-1}$$

This formulation is a difference equation which may be generated from a $\mathscr{Z}$ transfer function representation of the control algorithm as discussed earlier. Note that the equation requires that the control $e_1(kT)$ be generated instantaneously from the input $e(kT)$. In reality, it requires time to convert the continuous $e(t)$ to a discrete variable as well as to perform the indicated arithmetic operations. Also, time is required for the conversion back to the continuous-time domain since $e_1(kT)$ must be transformed to $e_1(t)$ in order to control a continuous plant. Thus, three basic delays are involved in processing the control law (difference equation):

Input (analog-to-digital) delay: $\tau_{A/D}$
Computational delay (difference equation, error checking): τ_C
Output (digital-to-analog) delay: $\tau_{D/A}$

A diagram indicating such a phenomenon is shown in Fig. 8-1. *In order that the theoretical assumption of infinitesimal processing speed be valid in associated control-system realizations, the following relationship must hold*:

$$\tau_{A/D} + \tau_C + \tau_{D/A} \ll T \tag{8-2}$$

That is, the total computer processing time must be much less than the sampling period if the theoretical mathematical controller development is to result in the desired performance. Of course, the sampling time is constrained by the sampling theorem (Nyquist rate) as discussed previously. Thus *T cannot be increased arbitrarily to meet the timing constraint of Eq. (8-2) and vice versa in terms of the sampling theorem.*

Note that considerable time is available between the output of the D/A converter and the taking of the next sample in Fig. 8-1. This time can be used to process as much as possible of the control algorithm [Eq. (8-1)] with existing data already stored in the computer. In other words, the time τ_C should only include the processing of new data, that is, $e(kT)$, in order to minimize τ_C. This approach is discussed in the following chapters where software routines for realizing the difference equation [Eq. (8-1)] are developed. The following sections focus on the values of $\tau_{A/D}$ and $\tau_{D/A}$ as they relate to converter structure and technology.

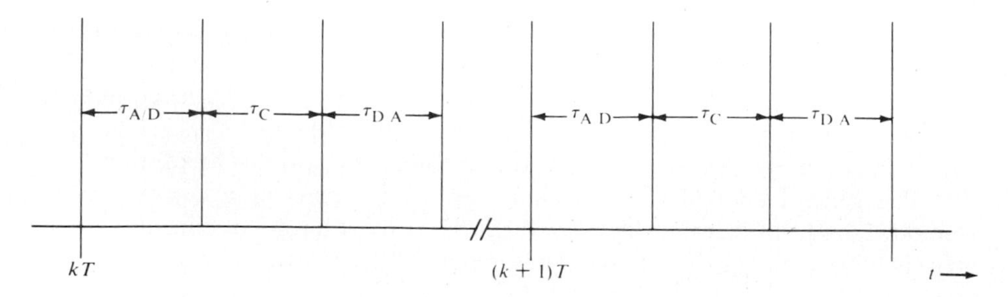

Figure 8-1 Timing delays, $\tau_{A/D}$ = A/D conversion time; τ_C = central processor time; $\tau_{D/A}$ = D/A conversion time; T = sampling period; k defines the kth sample.

8-3 CONVERSION SYSTEMS

Conversion systems can be used to transform analog signals into digital form for digital-computer processing.[96] In some applications, such as digital control systems, digital signals are also transformed to analog signals for controlling the continuous plant. Some analog signal processing is usually employed prior to A/D conversion (an antialiasing filter) or after the D/A conversion (a smoothing filter). Figure 8-2 represents an overall block diagram for a general conversion system. A transducer transforms a physical parameter such as pressure, temperature, velocity, or position into an analog signal (current, voltage, frequency, etc.). Further analog processing is accomplished by using filtering (amplification, spectrum changing, noise depression, etc.). The analog signal is then converted into digital form (binary-coded) for transfer into the digital processor. After appropriate digital processing, digital signals are transferred to a memory device (register, computer memory, etc.) for storage or possible manipulation and then converted to an analog signal for control purposes.

In general, various input and output ports (interfaces) exist on the digital processor (computer) such that numerous channels can be connected. A single port can transfer a number of channels through the use of a multiple sharing device known as a *multiplexer*. Each input channel is connected to the multiplexer output for a specified period of time as controlled by switches (see Sec. 8-9). The circuits which follow the multiplexer are thus time-shared. A sample-and-hold device samples the multiplexer output and holds the voltage level (or equivalent) such that

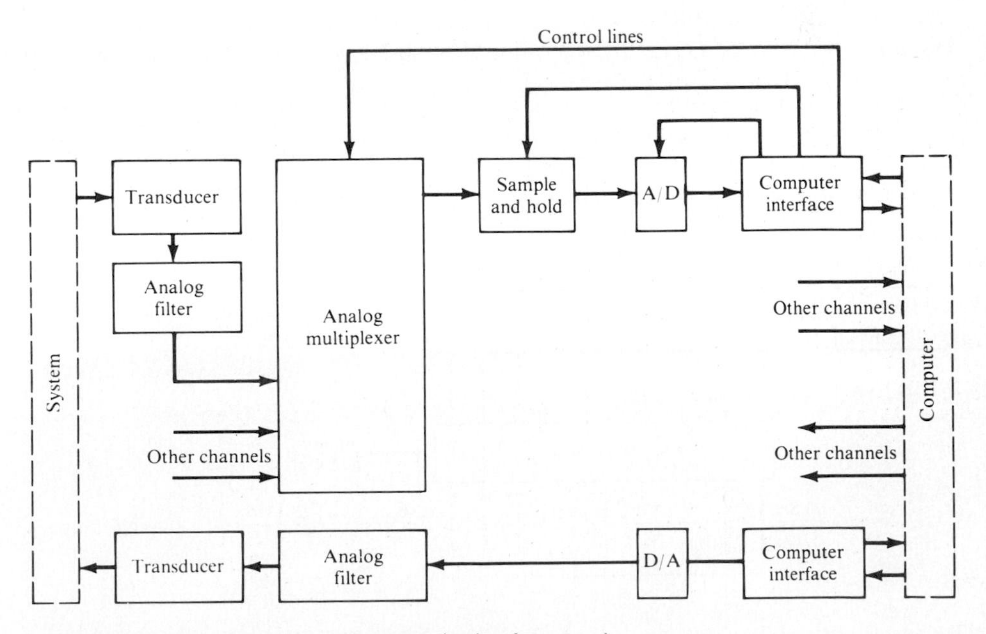

Figure 8-2 General conversion-system organization for control.

the A/D converter has an appropriate period of time to complete its operation. The theoretical reasons for using a sample-and-hold device have been presented previously when discussing the sampling theorem (see Chap. 3) and are further detailed in Sec. 8-8.

The sequencing of the A/D process is controlled by various signals that are generated by the computer interface under software and/or hardware control. The D/A process is usually simpler to control since the output transfer is synchronized to the computer cycle period, and thus data are transferred directly to the output without any interface "handshaking."

8-4 GENERAL DIGITAL-TO-ANALOG (D/A) CONVERSION STRUCTURES

A digital-to-analog (D/A) converter is a device that transforms a q-bit computer word into a continuous analog output. The output may be in units of current, voltage, or some other physical quantity.

The transformation between a binary signal and a continuous analog signal can be accomplished physically through the use of a register, electronic switches, and a passive resistor network as shown in Fig. 8-3. *Each bit* from the *most significant bit* (MSB) to the *least significant bit* (LSB) *of the binary data register is "weighted" by the A/D resistor network in proportion to its binary digit value*, i.e., a relative power of 2. Each switch is controlled by a high or low voltage ("1" or "0") from a register bit. Electronic switches permit a precise reference voltage or ground to be attached to the weighted network, thus achieving a desired degree of output accuracy within this voltage range (reference voltage to ground). The D/A *resolution* is defined as the analog value associated with the LSB and is usually

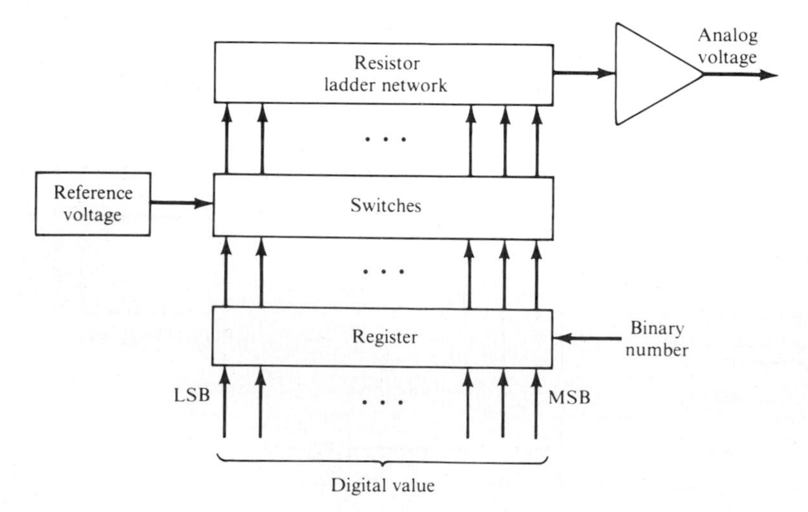

Figure 8-3 Digital-to-analog conversion.

given in volts. For high-speed conversion, the amplifier of Fig. 8-3 is taken out and the current output of the ladder network is used. The amplifier's transfer function and delay are no longer part of the system. However, the current swings of the ladder network can be quite small, and proper electronic devices must be used to transform the current signal into a control signal for the continuous plant.

One method of achieving a binary weighted-current value is shown in Fig. 8-4. Figure 8-4*a* presents a general weighted network while Fig. 8-4*b* defines a transistor-switched circuit. The voltage value of the MSB is $V_{ref}/2$ and for the LSB is $V_{ref}/2^q$ assuming a q-bit register. A series of transistor current sources have their collector currents controlled by emitter resistors. The precision reference voltage V_{ref} is used to bias the bases of all the transistors and permit constant emitter currents. The current-source transistors are switched by the binary register inputs connected through diodes to the emitters. The current flows either through the diode or through the transistor. The weighted currents are summed at the collectors of all the transistors (the input to the operational amplifier).

The weighted-current-source method has the advantage of simplicity and relative high speed. The disadvantage of this general method is the wide range of resistance values required for a high-resolution converter and the resultant effect on both temperature stability and speed. Specifically, matched transistors, usually FETs, and precision resistors are required.

High-resolution D/A converters can be made by using several groups and dividing down the current output of each group. This is illustrated in Fig. 8-5, which shows three groups of current sources with resistive current dividers following groups 2 and 3. If each group has four binary current sources, then the dividers would have to reduce the current outputs of groups 2 and 3 to one-sixteenth of their original value.

Figure 8-5 also shows the method of achieving a bipolar $\pm$ output for a D/A converter. A source with a current equal to the most significant bit (MSB) weight is connected to the output of all the other weighted sources. This offsets (biases) the output of the converter by one-half the full-scale voltage value, thus setting analog zero at one-half digitial full scale. This results in offset binary coding. A discussion on coding is presented in Sec. 8-5.

Another method for D/A conversion is the $R - 2R$ ladder technique. As shown in Fig. 8-6, this method consists of a network of R and $2R$ resistors. The bottoms of the $2R$ shunt resistors are transistor switched between a voltage reference source and ground. The operation of the ladder network is based on the binary division of a current as it moves across the ladder. This can be seen by examination of the junction points between the R resistors in the network. From the right, a resistance R occurs and from the left, a resistance of $2R$. These properties hold for any of the junctions along the ladder. If a $2R$ resistor is switched to the voltage reference source, the source sees a resistance of $2R$ plus $2R$ in parallel with $2R$, or $3R$ total, and a current of $V_{ref}/3R$ flows into the junction. At the junction this current divides equally, with half flowing to the left and half to the right. The right-hand current flows to the next junction where it is again divided in half, and so on, to the right end of the ladder where it becomes part of the total output current. The total

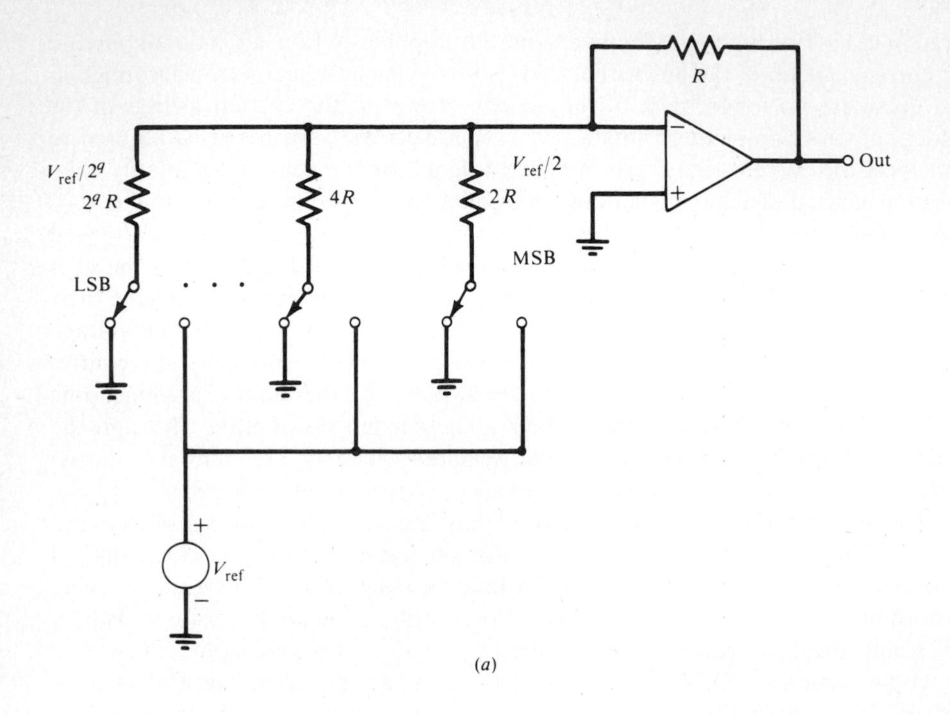

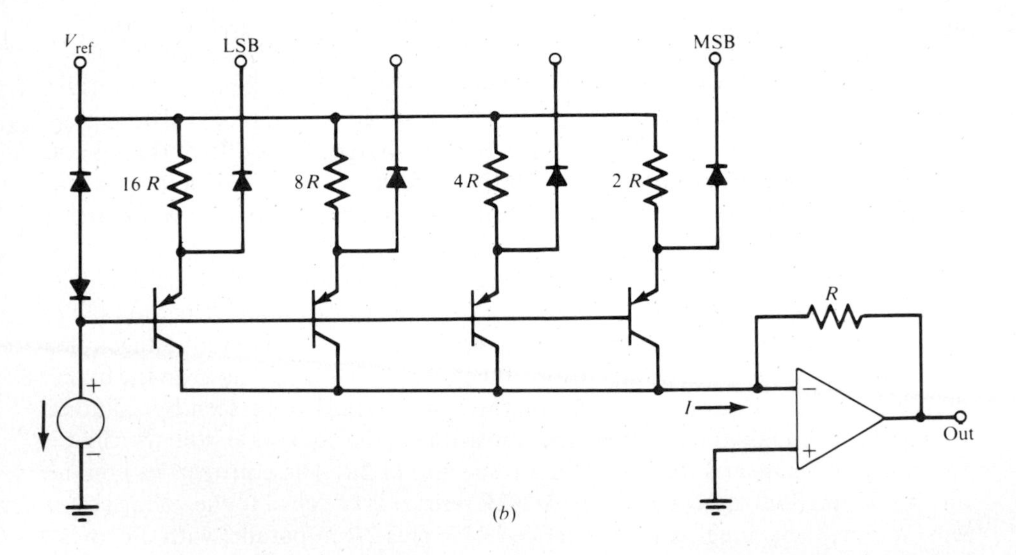

Figure 8-4 Weighted-current-source ladder network using $2^{q-1}R$ resistors. (*a*) General switches; (*b*) transistor switches.

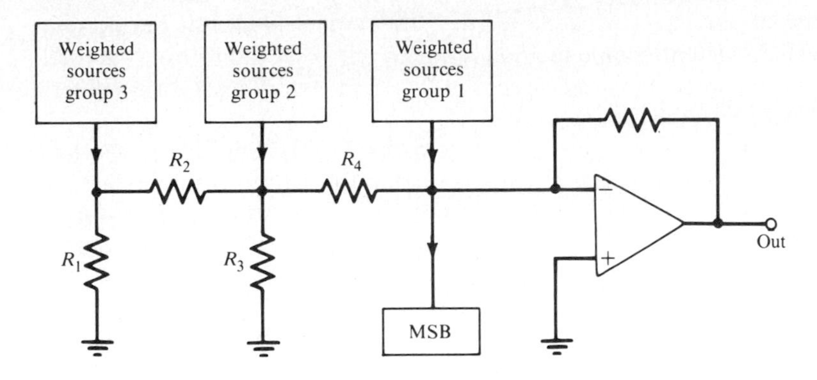

Figure 8-5 Groups of identical binary weighted-current sources to achieve high resolution.

output current is the sum of all the currents from the binary weighted-shunt resistors.

The advantages of the $R - 2R$ method are:

All resistors have values of either R or $2R$, resulting in easy matching and temperature stability.
The output amplifier, if used, always sees a constant-resistance value at its input terminal.
Resistor values can be kept low for high-speed A/D conversion.
Structured for monolithic integrated circuit implementation.

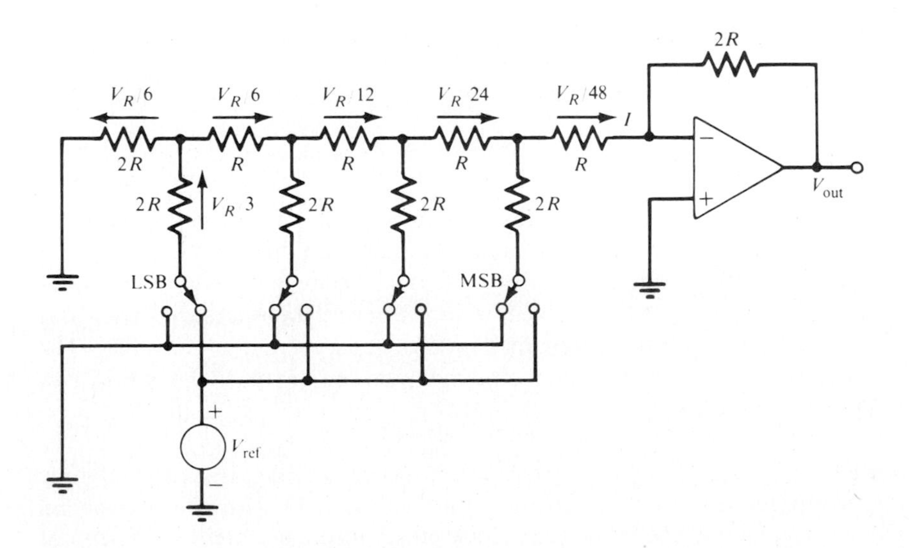

Figure 8-6 $R - 2R$ ladder network.

A disadvantage of the $R - 2R$ ladder is that it requires two resistors per bit whereas the weight-current-source method requires one resistor per bit. It should be noted that implementation variations of the $R - 2R$ network exist due to different integrated circuit production processes.

Many organizations exist for the ladder networks. Multiplying D/A converters exist which generate an output that is proportional to the product of the reference voltage and the digital input. These devices are useful in electronically transforming from one coordinate system to another, for example, in aerospace control systems.

8-5 GENERAL ANALOG-TO-DIGITAL (A/D) CONVERSION STRUCTURES

An analog-to-digtal (A/D) converter transforms an analog signal to a q-bit computer word generally through the use of the comparison operation. The first step of the operation compares an unknown input signal (voltage, current, etc.) with a known value and determines which is larger. The next step of the conversion process can be accomplished by using a variety of techniques and organizations. Organizational parameters that define most types of A/D conversion systems involve the following seven elements:

Basic reference technique
Data transfer (serial-parallel) organization
Information flow structure
Timing structure
Comparator operation
Digital coding of binary information
Scaling of binary information

A brief description of each is presented next.

8-5A Basic Reference Technique

The conversion of an analog variable to a binary word is fundamentally a comparison process. A comparison is made between the analog variable and a precise reference voltage. The distinguishing feature between different techniques is based on the concept of reference generation. That is, the reference varies as a function of either some time-domain or some space-domain variable. The following description of two basic encoding methods encompasses the majority of A/D reference techniques.

Time encoding A precise signal (e.g., ramp) is generated as a function of time which is compared against the analog input. Coincident with the conversion START signal, a gate is opened to pass clock pulses into a counter. At the end of the comparison cycle, the counter contains the number corresponding to the

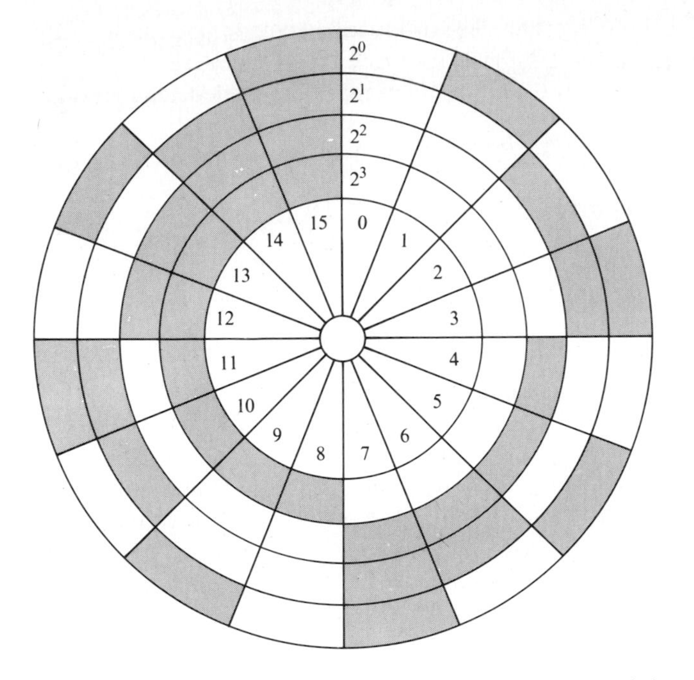

Figure 8-7 Encoder disk. Four tracks represent binary codes consisting of 2^0, 2^1, 2^2, 2^3. Numerals 0–15 represent the possible 16 shaft positions. Note that there is more than one bit change per sector in some cases.

quantized analog quantity. In general, some technique is employed to successively reduce the error between the true value and the encoded value. Specific examples are presented in Sec. 8-6.

Spatial encoding The reference is based on precisely spaced openings on a mask of some type. Physical displacement of a beam, a shaft, etc., as a function of an analog quantity establishes a unique coincidence with a specific position, a space-domain reference (see Fig. 8-7). Reading of such an encoder disk or shaft can be accomplished by electric brushes, magnetic pickups, or optical readers, depending upon disk construction. A binary code corresponding to a sector position represents the digitized result and could be used in a position control system. The resolution of the measurement depends upon the number of information bits encoded on the disk.

8-5B Data-Transfer Organization

The A/D data-transfer organization may be serial, parallel, or a combination. To describe these structures a conversion time step is defined as an interval between comparisons. The number of bits encoded per step can then be defined as follows:

Parallel. This term means that the complete A/D conversion (q bits) is accomplished in one step, implying that q bits are encoded by comparison with 2^q reference levels. That is, all the comparisons are made in parallel.

Serial. This term indicates that each bit in a q-bit register is encoded sequentially one bit at a time.

8-5C Information Flow Structure

The encoding process may also be based on a *feedback* loop or a *feedforward* concept of information flow.

Feedback. This term indicates that the encoding of a given step in the A/D process feeds back to previously used comparison logic to affect subsequent A/D conversion steps. Examples of this structure are found in some serial organizations.

Feedforward. This term means that the results of one step of an encoding "feeds forward" to an as-yet-unused comparison element. Certain advantages in speed occur from this "pipeline" configuration.

8-5D Timing Structure

An A/D organization may require a synchronous or asynchronous clock (see Chap. 2) to sequence the process.

Synchronous. This term indicates that each step of the A/D conversion process is precisely controlled by clock pulses. Therefore the time required to complete the process is predictable within the tolerance of the clock. Most commerical A/D implementations use this type of sequencing control.

Asynchronous. This term means that each step of the A/D process does not depend upon the concurrence of clock pulses for control of signal gating; thus speed advantages may result. It is usually employed in a cascade organization where data flows through all and any components in any asychronous manner.

8-5E Comparator Operation

This structural element involves the modification of either the signal or the reference during the A/D conversion process.

Operation on reference. This term refers to the situation where the reference is altered (i.e., amplitude), the input signal to be encoded is not altered, and the values compared. No input signal transformation occurs during the conversion process.

Operation on signal. This term indicates that the signal is altered during the A/D conversion and the comparison made against a fixed reference. For example, the signal processing may involve amplification, phase shift, attenuation.

8-5F Digital Coding

This term refers to the binary coding (e.g., binary, complemented binary, BCD) internally used in the A/D conversion. The type of coding may have a bearing on speed and accuracy in a given implementation. Both A/D and D/A converters relate analog and digital values by an appropriate digital code. The techniques used are various binary-related codes, the most common of which is *natural binary*.

A q-bit unsigned number, as previously discussed in Chap. 2, is represented as

$$N_{10} = b_M 2^M + b_{M-1} 2^{M-1} + \cdots + b_i 2^i + \cdots + b_1 2^1 + b_0 2^0 \tag{8-3}$$

where, of course, the coefficient b_i assumes the values of "0" or "1" and $q = M + 1$. In an A/D or D/A converter, b_M is the MSB and has a weight of 1/2 of full scale (FS) of the converter; b_{M-1} has a weight of 1/4 FS; and so on down to the last bit b_0, the LSB, which has a weight of $(1/2)^{M+1}$ FS. The resolution of the converter is determined by the number of bits, $q = M + 1$. Remember that the binary code does not necessarily correspond to its decimal equivalent in analog voltage (i.e., scaling problem).

The coding used relates to the set of coefficients of 2^i representing a fractional part of full scale. The binary code 10110 thus represents $(1 \times 1/2) + (0 \times 1/4) + (1 \times 1/8) + (1 \times 1/16) + (0 \times 1/32)$, or 11/16 of full scale of the converter. The full-scale analog value for a voltage converter can be any convenient voltage, but voltages such as 0 to +5, 0 to +10, ±2, ±5, and ±10 are most commonly used. A 12-bit converter, for example, has a resolution of 1 part in 4096. If the full-scale analog voltage is 10 V, then the LSB value is equivalent to 10 V/4096, or approximately 2.4414 mV.

Converters have both *unipolar* and *bipolar* analog ranges and use a number of different binary-related codes. Table 8-1 shows binary coding for a unipolar 8-bit converter with 10 V full scale. Note that the value of 0.039 V is an approximation to 10/256 and that the scale value assumes that the binary point is to the left of the MSB.

Notice that all 1s in the binary code do not correspond to full scale but to $1 - 2^{-q}$ FS. In some converters it is convenient to use reverse binary coding, or

Table 8-1 Binary coding for 8-bit unipolar converters

Scale	+10 V FS (decimal)	Straight binary	Complementary binary
$1 - 2^{-8}$ FS	+9.961	1111 1111	0000 0000
+ 3/4 FS	+7.500	1100 0000	0011 1111
+ 1/2 FS	+5.000	1000 0000	0111 1111
+ 1/4 FS	+2.500	0100 0000	1011 1111
+ 1/8 FS	+1.250	0010 0000	1101 1111
$+ 2^{-8}$ FS	+0.039	0000 0001	1111 1110
0	0.000	0000 0000	1111 1111

Table 8-2 Binary coding for 8-bit bipolar converters

Scale	±5 V FS (decimal)	Offset binary		Sign plus 2s complement	
$1 - 2^{-7}$ FS	+4.92	1111	1111	0111	1111
+ 3/4 FS	+3.75	1110	0000	0110	0000
+ 1/2 FS	+2.50	1100	0000	0100	0000
0	0.00	1000	0000	0000	0000
− 1/2 FS	−2.50	0100	0000	1100	0000
− 3/4 FS	−3.75	0010	0000	1010	0000
$-(1 - 2^{-7})$ FS	−4.92	0000	0001	1000	0001
− FS	−5.00	0000	0000	1000	0000

complementary binary, where the most negative analog value corresponds to a full-scale digital value as shown in Table 8-1. This code is just the binary code with all 1s set to 0s, and vice versa (i.e., negation or 1s complement). For bipolar analog values, the most common codes are *offset binary* and sign plus 2s complement. This is illustrated for an 8-bit converter (7 bits of magnitude) in Table 8-2.

Offset binary is simply a shifted binary code, where 1/2 FS binary corresponds to analog zero. Sign plus 2s complement coding is the same as offset binary except that the MSB (sign bit) is complemented, resulting in a digital code of all 0s corresponding to analog zero (see Chap. 2). Binary-coded decimal (BCD) is also commonly used in converters and is illustrated by the 8-bit coding shown in Table 8-3. For BCD, four binary digits are used to code each decimal digit. This code can also be used for bipolar analog values if a separate sign bit is used. Other codes such as gray code, sign plus magnitude, and sign plus 1s complement are also used (see Chap. 2).

8-5G Scaling

Scaling of the A/D converter is based on the specific application. The external analog value being sampled and the representative discrete value are related by a

Table 8-3 BCD coding for two-digit unipolar converters

Scale	+10 V FS (decimal)	BCD	
+(1 − 1/10) FS	+9.9	1001	1001
+3/4 FS	+7.5	0111	0101
+1/2 FS	+5.0	0101	0000
+1/4 FS	+2.5	0010	0101
+1/10 FS	+0.1	0000	0001
0	0.0	0000	0000

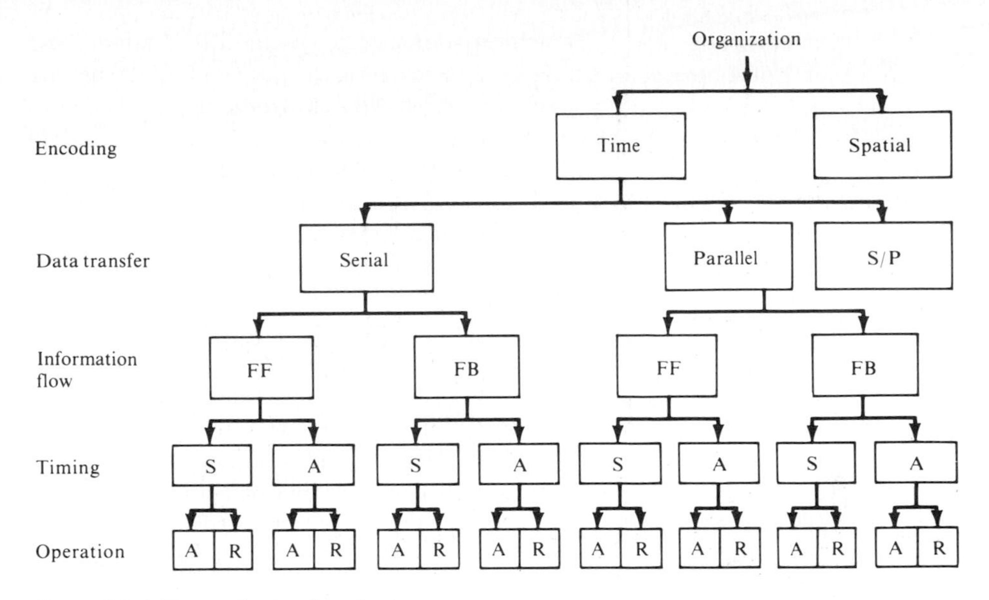

Figure 8-8 A/D organization flowchart.

gain or scale factor as indicated previously. This scaling factor due to shifting impacts the algorithm representation in the computer (i.e., internal gains). *Scaling is critical in meeting performance objectives and cannot be slighted.* General implementation involves shifting the converter binary number so that data overflows or underflows do not occur during arithmetic processing. *Overflow* is when a bit value is shifted off the MSB and lost; *underflow* is when a bit value is shifted off the LSB position and lost. Chapter 10 discusses this problem in more detail.

8-5H Organization Combinations

Figure 8-8 shows a chart that has evolved from consideration of the functions just discussed. As mentioned, many approaches are inherently efficient for only one type of coding. In other cases, economic considerations limit the possible number of variations. The coding parameters are not included in the chart. However, they are significant and must be considered in selecting an A/D converter structure. General A/D conversion systems using aspects of Fig. 8-8 are now presented to provide insight to A/D performance. General performance characteristics are presented first to help understand each individual structure.

8-6 ANALOG-TO-DIGITAL SYSTEMS

Analog-to-digital (A/D) converter structures reflect various organizational design parameters as mentioned in the last section. Each affect overall converter performance, efficiency, and economy in terms of a specific application. The required degree of accuracy, for example, depends upon the frequency of the analog input signal, the conversion time, and associated A/D electronic errors. The conversion delay between the A/D START command and the generation of the desired binary value is different for various A/D structures. The meaning of specific terms and parameters must be understood. Beyond those mentioned in the previous section, the following terms are defined. Following these descriptions, various A/D methods are exhibited for contemporary converter organizations.

Accuracy: exactness of the digital value to the true analog value. Accuracy is constrained by the converter noise, aperture time, *quantization* (1/2 LSB), and the electronic switching process (linearity, gain, temperature, etc.).

Aperture time: the time interval of uncertainty or the "window" in which the analog value occurred. The specific aperture time required in a given application is directly related to the acceptable error between an analog input magnitude within the aperture window and the value of the associated digitized value. This type of error occurs because the input analog signal may change in magnitude during the digitizing process, and thus the digitized value is associated with only one analog value that occurred in the window. In most cases, it is desirable to have the analog value only change a maximum of half the LSB in the window such that the digitized value follows the input accurately. Different conversion techniques reflect various aperture times. A sample-and-hold device can decrease aperture time appreciably since it "holds" an analog signal until the A/D conversion is completed.

Conversion time: time required to transform the analog input into a binary value after a converter START command is given. For some converter organizations, this time is constant; for others, it is a function of the current input amplitude, i.e., the time required to acquire the analog signal in a stepwise fashion. The conversion time includes the aperture time and may be equal to it in some organizations. In these cases, the aperture window is open during the entire conversion time.

Sampling time: T, as defined previously. In reality, T is the time between START commands to the converter. T is usually much, much greater than the conversion time because of the theoretical model (impulse sampling) used in the determination of the digital compensator. If this condition is not met, the physical performance of the digital control system will probably be quite different from the theoretical performance.

The fundamental component of most A/D converters is an analog signal comparator that has a positive logic output of "1" if the analog input signal is greater than an analog reference voltage. If the input value is lower than the reference voltage, the comparator output is defined to be a logical "0." To generate a digitized value of the input, additional logic is added to the comparator circuitry to manipulate the reference voltage and to minimize the error between the input signal (assumed constant) and the reference voltage. The digitized and encoded value of the manipulated reference is the desired value. The technique used in manipulating the reference voltage is the basic difference between A/D implementations.

8-6A Simultaneous or Flash Method (Parallel)

A q-bit simultaneous A/D converter requires $2^q - 1$ comparators, each having a unique precison reference voltage. There are 2^q precision voltage levels required including ground which can be generated with a precision resistive ladder network. Each comparator determines in parallel if the input analog signal is higher than its reference voltage. If the analog signal is higher than the reference signal, the comparator generates an ON logic signal; if not, it generates an OFF logic signal as shown in the example of Fig. 8-9. In this example, the voltage range from V to ground is divided into four ranges that can be coded into 2 bits (that is, $2^2 = 4$). Three comparators are used to provide the four designated levels. Note that the input analog signal is assumed to have a maximum value of V volts for proper binary encoding. Although extremely fast since the delay of a comparator and the associated logic delay define the critical path, many comparators are required for large word length implementations; i.e., for 12 bits of accuracy, 4095 comparators would be required. The simultaneous or flash technique, however, is very useful for

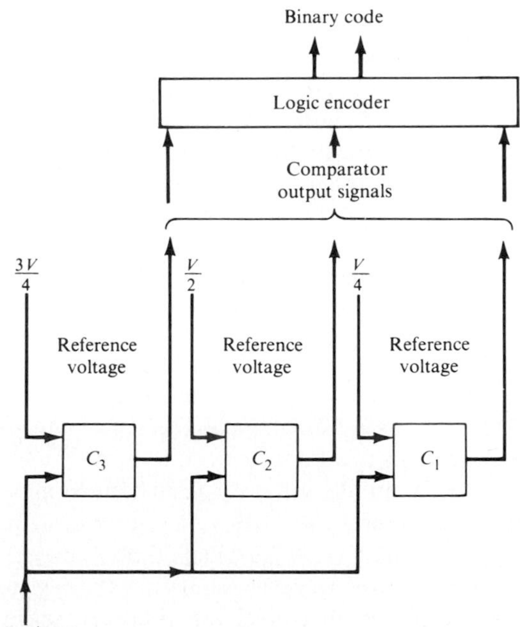

C_1	C_2	C_3	Input voltage	Binary code
Off	Off	Off	0–V/4	00
On	Off	Off	V/4–V/2	01
On	On	Off	V/2–3V/4	10
On	On	On	3V/4–V	11

Figure 8-9 Simultaneous A/D converter.

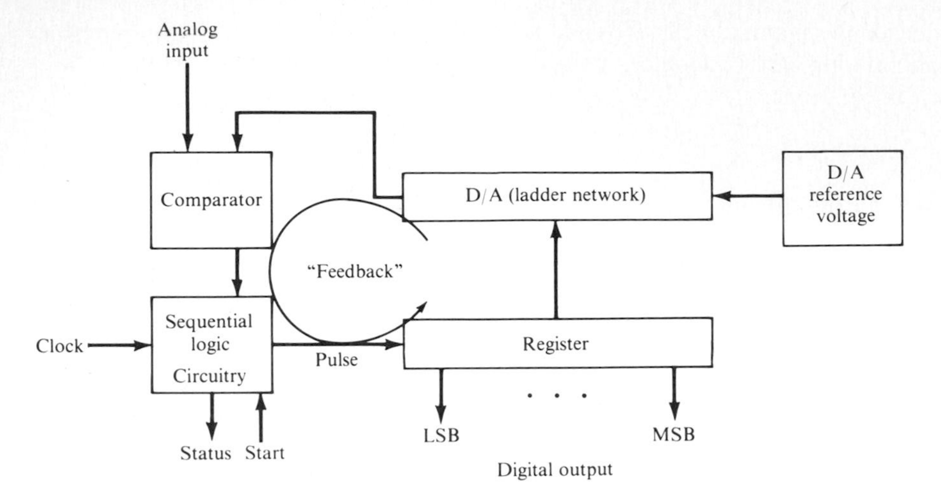

Figure 8-10 A/D converter incorporating a D/A conversion module.

small word length requirements and also in a serial-parallel organization combining the best of both structures.

8-6B Feedback Structures (Serial)

In a closed-loop A/D conversion process, the reference voltage changes as a function of the binary word stored in the associated register. This analog value from a D/A converter is fed back into one comparator along with the input analog signal that is to be converted, as shown in Fig. 8-10. The comparator circuitry generates an error signal if the two input values are different within some tolerance. This error signal then generates a pulse to a counter register and the conversion continues in a serial process. The clocked sequential logic controls the incrementing or decrementing of the register when started and also indicates to the computer when the conversion is complete.

8-6C Counter Method (Serial)

One of the less complex A/D organizations uses a counter as the register in the feedback A/D structure of the last section. This counter is defined to be only an up counter as shown in Fig. 8-11. The counter is initially set at zero and counts up one pulse at a time until the analog signal input and the D/A value are within the predefined error tolerance of the comparator (see Fig. 8-12*a*). Each time a new conversion is to start, the counter must be reinitialized to zero. Also, since the register is incremented by one pulse at a time, the overall conversion time can take a considerable number of counts.

For the counter method, the conversion time is equal to the aperture time. The

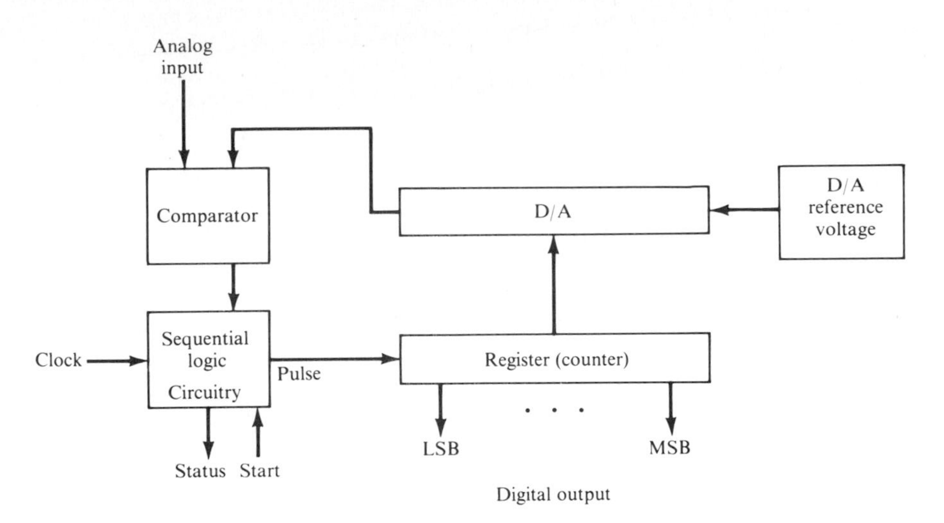

Figure 8-11 Counter-converter structure.

counting continues after activation until the comparator indicates that the difference between the analog input signal and the fed-back analog signal is within a "small" predefined tolerance. During this process, the input analog window (aperture window) is open during the entire conversion.

In digital control systems, the A/D converter is activated with a signal from the computer that is sequenced by the sampling time T, as shown in Fig. 8-12*b*. The maximum conversion time, $\tau_{A/D}$, is of course the time required to count 2^q clock pulses assuming a q-bit conversion. Note that in general, *the aperture time for a counter converter is not constant and depends upon the analog input value*. For a digital control system, one could assume an average time of half the maximum count of the output register. However, in analyzing Eq. (8-2), the maximum value should be used for real-time digital control systems. The counter method requires a relatively small amount of electronics to implement but does have a rather large maximum conversion time.

8-6D Continuous or Tracking Method

Instead of reinitializing the counter converter to zero for each conversion, an up-down counter is provided which "tracks" the input analog voltage. This structure is defined as a continuous converter and is presented in Fig. 8-13. The conversion time for this method can be much faster than the counter method depending upon the input analog signal frequency. An example of the continuous-converter process is shown in Fig. 8-14.

The conversion time is equal to the aperture time in this structure also. If the acceptable error is defined as that caused by the input analog signal changing in magnitude no more than one unit of the LSB during conversion, the aperture time

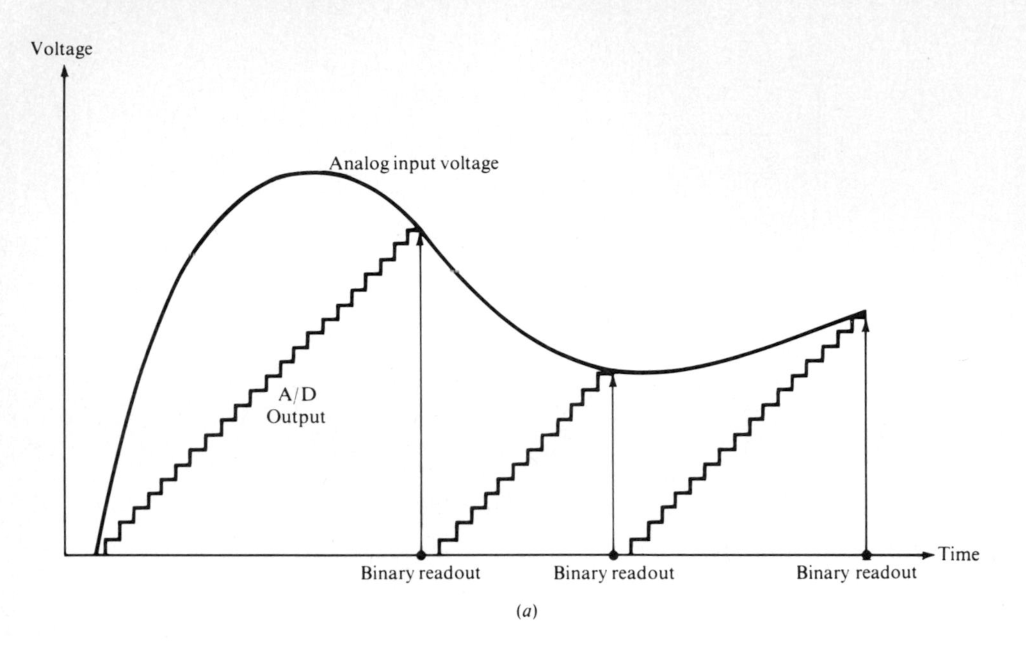

(*a*)

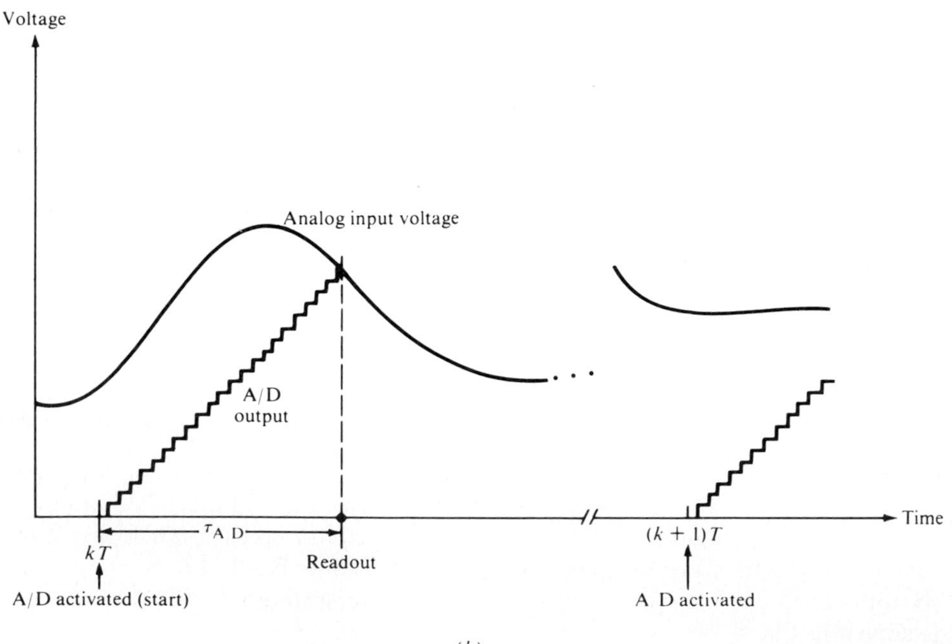

(*b*)

Figure 8-12 (*a*) Counter-converter characteristic; (*b*) Counter-converter characteristic (sampling time T sequencing the A/D converter).

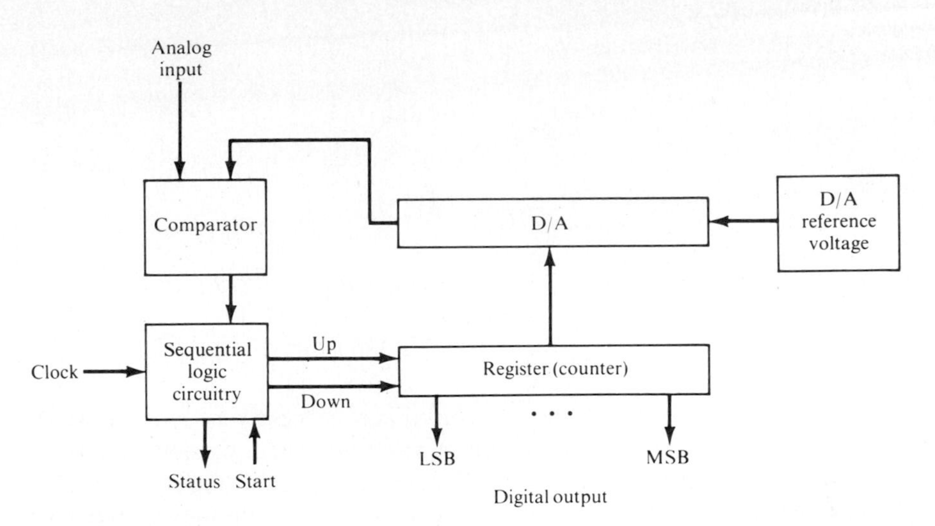

Figure 8-13 Continuous-converter structure.

is the time for the one-step conversion. In essence, the maximum frequency of the input analog signal cannot be greater than the converter speed. The LSB has a value of $r = V_{ref}/2^q$, where the voltage reference is the full-scale A/D voltage. If τ_a is the aperture time, the maximum converter speed is r/τ_a (V/s). This value can be compared to an analog signal with maximum frequency ω and amplitude $V_{ref}/2$. Therefore, the maximum rate of change of the analog signal

$$\omega \frac{V_{ref}}{2} = \frac{r}{\tau_a} \tag{8-4}$$

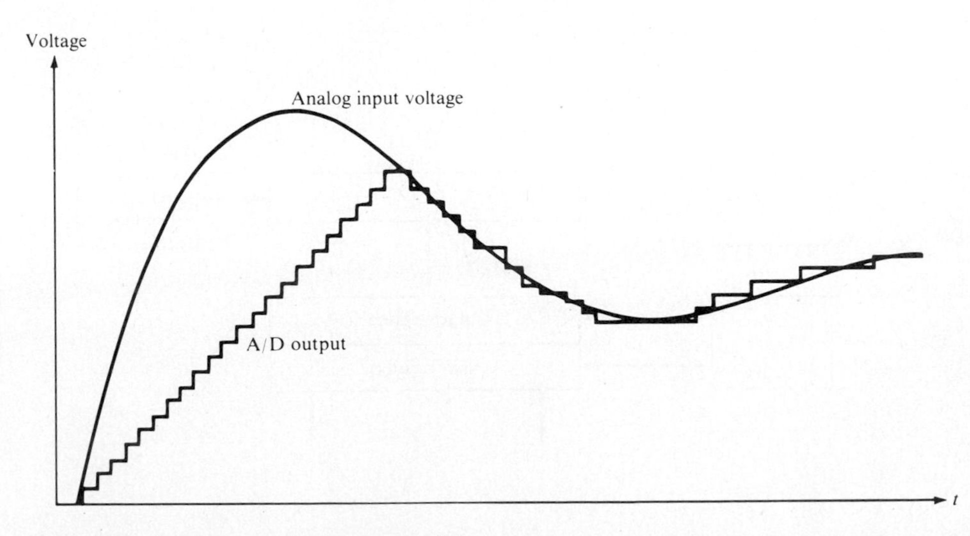

Figure 8-14 Continuous-converter characteristic.

and the maximum input frequency is

$$\omega = \frac{2}{2^q \tau_a} \qquad \text{rad/s}$$

or

$$f = \frac{1}{\pi 2^q \tau_a} \qquad \text{Hz} \tag{8-5}$$

Example 8-1 Let $q = 10$ and $\tau_a = 10^{-6}$, then, using Eq. (8-5),

$$f = \frac{1}{1024 \times 10^{-6} \times \pi} = 310.85 \text{ Hz}$$

Note that this example assumed that the conversion process was continuous. A digital-control-system implementation with noncontinuous conversion (that is, T sampling) would generally have a large A/D aperture time and require a smaller maximum frequency of the input signal if the error is to be minimized (refer to Fig. 8-12*b*). Of course, the A/D subsystem could continually process the analog input with the computer requesting a converted value at the sampling frequency $1/T$. Additional interface logic is required, but with proper synchronization a smaller relative conversion time could result. Also note that a sample-and-hold device is not assumed in the above analysis. Such a device can permit a higher input frequency based upon Shannon's sampling theorem.

8-6E Successive-Approximation Method (Serial)

Instead of modifying the LSB of the output register in the continuous converter, the successive-approximation method starts with the MSB, setting it to 0 or 1, and

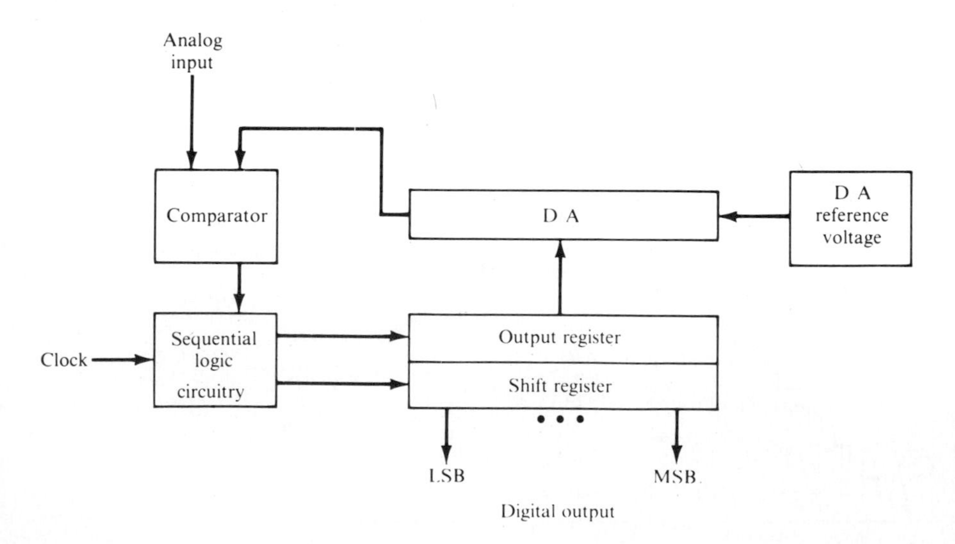

Figure 8-15 Successive-approximation-converter structure.

then doing the same with the next bit, and so forth for q bits. This sequential process of dividing the reference voltage in half at each stage is continued to the LSB. Although this method uses the feedback technique, additional logic is required to set each bit to the proper value as controlled by the output of the comparator. This is accomplished in part by adding a shift register and additional logic as shown in Fig. 8-15. The conversion time, $\tau_{A/D}$, is q clock periods for the q-bit conversion.

Example 8-2 Consider a 4-bit successive-approximation A/D conversion method. Then a specific path in the following tree structure can be traversed in reaching the correct digital value:

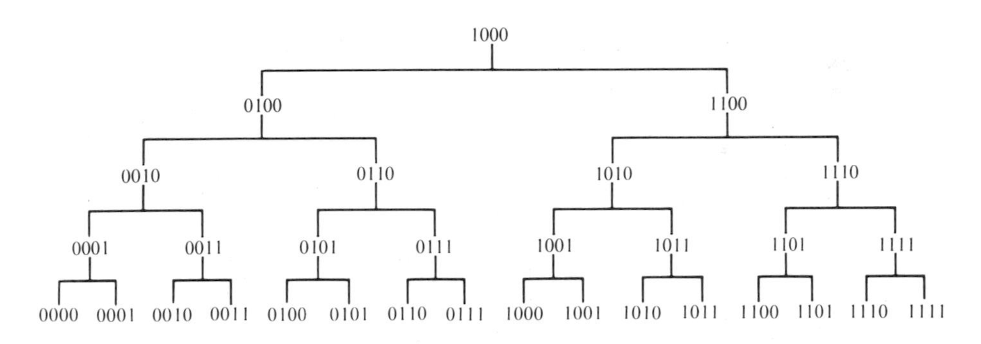

Note that the successive-approximation converter starts with half scale, 1000, and repeatedly halves the scale, resulting in the tree organizations shown above, that is, a binary search. Four iterations are required for the 4-bit word length A/D converter and thus, q steps for a q-bit converter. Sequential logic synchronizes the gating of a shift register as shown in Fig. 8-15 into an output register that eventually contains the digitized value. In this structure, the conversion time is equal to the aperture time, four clock periods, and is always constant as shown in Fig. 8-16 for $q = 4$.

The successive-approximation A/D conversion method is one of the most widely employed based upon reliability, ease of design, relative high speed, and cost considerations. Because the analog input should remain relatively constant during conversion, a sample-and-hold device, ZOH, can be added in front of this type of converter.

As before, the maximum frequency of the input analog signal usually corresponds to a maximum of a $\pm$ LSB change within the aperture window. For an input frequency of 10 kHz, the conversion time (aperture time) assuming the use of a ZOH is found by using Shannon's sampling theorem, that is, $\tau_{A/D} < 1/20{,}000 = 0.00005$ s, since the conversion time includes the entire word. The number of bits to be converted depends upon accuracy requirements. The major error source affecting accuracy is the D/A converter in the feedback loop. Also, this type of converter is highly susceptible to noise.

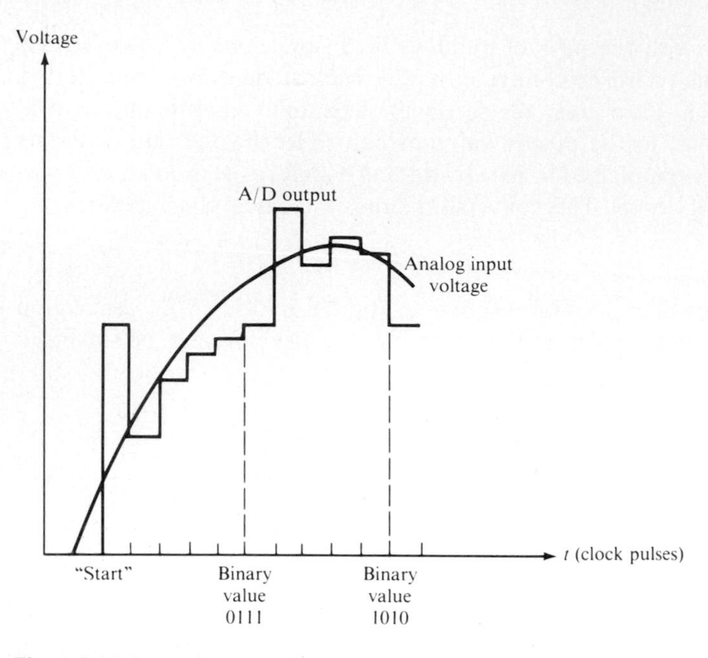

Figure 8-16 Successive-approximation-converter characteristic.

Commercial A/D converters have setup times and may use multiplexers and sample-and-hold devices. These structural aspects increase the conversion time and must be analyzed when selecting an A/D subsystem for digital control applications if the inequality of Eq. (8-2) is to hold.

8-6F Special A/D Conversion Techniques

The A/D conversion techniques described previously are some of the more common methods utilized. Other methods include combinations of these techniques and variations of the feedback implementations. For example, the continuous converter of Sec. 8-6D could be expanded to include an up-down counter that could count in units of 2, 3, or more by using additional comparators (a serial-parallel organization). This structure is sometimes called the *section-counter method.*

The simple up counter of Sec. 8-6C could included a ramp generator in the feedback loop instead of the D/A converter. Here, the ramp is generated through the use of an integrator and continues to increase in magnitude until the comparator error output is zero. During the ramp generation, pulses are also sent to a counter register which stops when the comparator error is zero within the tolerance specified. This organization is usually slower than the successive-approximation method but is cheaper because fewer precision components are required.

A similar method integrates the analog input first and then integrates a reference voltage and counts, comparing results at each step until the comparator error is zero. The integration of the original analog signal provides for good noise rejection assuming zero mean noise (see Chap. 9). This process, called *dual-slope integrating A/D conversion*, is useful in relatively slow speed applications.

Other techniques employ voltage-to-frequency techniques for counting, subranging, or subdividing of voltage ranges and separate A/D converters for each bit. The specific implemented A/D technique to choose for a digital control system depends upon cost, speed, energy, and accuracy requirements along with noise sensitivity. The next section examines the accuracy and speed question. For specific information, performance characteristics can be found in manufacturers' literature.

8-7 MEASURES OF CONVERTER PERFORMANCE

In interpreting manufacturers' data and selecting a converter for a digital control system, the two most important characteristics are probably speed and accuracy. However, it should be noted that the environment (temperature, humidity, vibration, radiation, etc.) must be considered in a converter implementation to ensure satisfactory results. Also, some characteristic values have long-term changes which could affect performance, such as aperture errors.

Accuracy is usually stated as a percent deviation from a given value: in terms of converters, an analog or digital value. On the other hand, various errors are specified as an absolute error such as $\pm$ LSB. For some converter subsystems, an overall error is acceptable. For highly accurate systems, detailed knowledge of each error contributer is required to fine-tune the digital control system. Figure 8-17 presents a composite of various converter error sources.

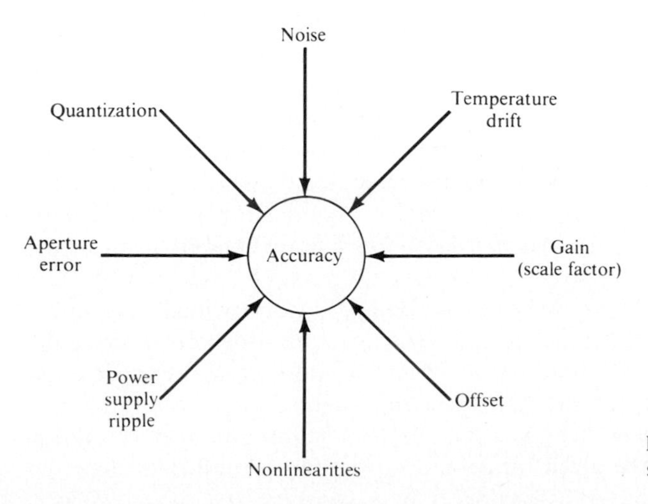

Figure 8-17 Converter error sources.

Absolute D/A accuracy elements include offset, linearity, power supply sensitivity, and scale factor. *Offset* for a D/A is that output value (voltage, current) that corresponds to a zero binary input value. *Linearity* is defined as the deviation from the line drawn from the offset point to the full-scale value on a plot of binary input vs. analog output. Usual values of linearity are within $\pm 1/2$ LSB of the correct value. Also, the error between the physical switching point of the D/A converter and the theoretical binary value can be measured and specified. If an amplifier is used, the gain calibration can also play a critical role in achieving performance requirements. *Power supply sensitivity* corresponds to the percentage change in full-scale output voltage due to a change in the reference voltage. Power supply ripple can, for example, cause oscillations in the D/A output even though the binary input is constant. The D/A scale factor relates the full-scale output value to the maximum binary input value. All these characteristics which may be adjustable are important when implementing a very accurate digital control system.

Some other important D/A parameters include settling time and noise sensitivity. *Settling time* is usually associated with the time required for the analog output to move from a binary input value to an adjacent one within $\pm 1/2$ LSB accuracy. Another definition is the time to settle to $\pm 1/2$ LSB when the input changes from a zero binary value to full-scale input. In both cases, the value of this parameter involves all the error sources. Thus, the maximum conversion rate is generally determined by the settling time. It should be noted that the addition of an operational amplifier adds further delay, although oscillations within the ladder network can be damped.

Noise problems can arise from improper interfacing to a D/A converter with small wire size, no shielding, and mismatch of transmission line parameters. Noise is a critical parameter in meeting digital-control-system performance specifications and must not be slighted.

If more bits are available than needed for a chosen D/A converter, then the accuracy of the D/A converter can be improved by using the next lower-order bit of the computer words to round the D/A input value. This can be done in hardware or software to improve sensitivity if noise levels are relatively low.

A/D error characteristics involve most of the possible D/A errors mentioned previously since most implementations include a D/A converter in the feedback loop. Accuracy terms are given in terms of the digital output rather than analog. Note that a q-bit A/D converter is not really accurate to q bits because of the impact of error sources. These sources include noise, quantization, aperture errors, differential linearity, and the D/A errors. Most errors are defined in terms of the LSB.

The various elements of an A/D system generate noise which may affect accuracy. Most manufacturers indicate that internal noise does not impact the digital output beyond 1/2 LSB. External noise can be modeled and experimental tests will indicate the impact on the A/D switching points.

Quantization error is caused by the fact that an infinite number of analog values are possible within the given range and only a finite number of digitized binary values are transferred to the computer. Thus, the output binary value is accurate to within 1/2 LSB and the associated error is independent of all others.

Aperture error refers to the error inherent in converting an analog value in a finite (nonzero) time to a digital value when the analog value is changing during the aperture time (window). Uncertainty errors of this type are better studied in a given application when most all system parameters are known.

The variation in the size of the analog input which causes the digital output to switch from one binary value to another is called *differential linearity*. In general, this value should be constant; however, it can be considerably different for a small input value and a large input value. Differential linearity is quite good for A/D converters such as the counter converters that always count through all possible binary values. Successive-approximation and selection converters, on the other hand, can have considerable variation in differential linearity. The shorter the word length, the better the differential linearity. In a digital control system, the value of this parameter can be modified on-line by changing offset and the word length through a sophisticated computer interface and associated software.

The speed of an A/D converter depends primarily upon the comparator delay and output register setup time for a parallel converter. For feedback converters, the comparator, sequential logic, and the D/A delays as well as the register setup times dictate the maximum conversion rate. Many commercial A/D converters have their own internal clocks, whereas others can be driven with an external clock.[96]

In selecting a converter system, the literature should only be used as a starting point for the final selection. Power supply characteristics, environmental noise, temperature extremes, and physical size are just a few of the parameters that must be evaluated in a given application. Even though the quantization error is $\pm 1/2$ LSB, the other errors might add to ± 1.0 percent, which may not be acceptable. Each individual source must then be studied for possible error reduction.

8-8 SAMPLE-AND-HOLD OPERATION

When the uncertainty in the aperture window becomes too large, a sample-and-hold device can be used to "freeze" the input before the conversion process is started. To provide this capability, a sample-and-hold device consists of an amplifier that "tracks" an analog voltage and a "hold" circuit that stores a specific value of the input analog voltage when directed by a computer interface. The voltage being held is then the analog input signal to the A/D converter.

The circuit of Fig. 8-18 uses a unity-gain operational amplifier to charge a capacitor during the "track" or "sample" operation as simply presented in Chap. 3. The *acquisition time* of the circuit is that time for the capacitor to acquire the charge associated with the input signal after being started with the *sample* signal from the computer interface. Here the aperture time is associated with the sample-and-hold device. The *hold* signal from the computer interface then permits the capacitor to hold the input value in the feedback loop of the amplifier. The input

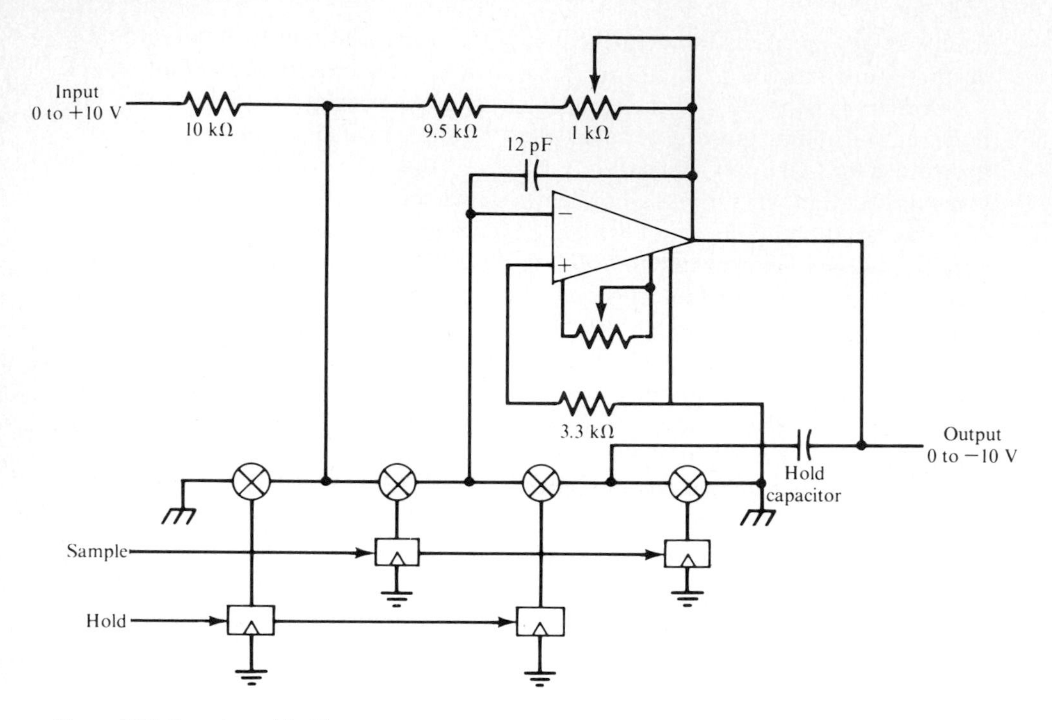

Figure 8-18 Sample-and-hold system.

and feedback resistors are switched to ground in this mode and the amplifier output is fed to the A/D converter.

Many variants of the sample-and-hold organization exist, but they essentially perform the same operation. Various structures have different analog error characteristics as mentioned previously and are therefore useful in a variety of applications.

An example of sample-and-hold performance is shown in Fig. 8-19 for a successive-approximation A/D converter. The aperture time is considerably reduced by the hold device as shown in Fig. 8-19. Note that the conversion delay could have an effect on the real-time digital control model which has assumed impulse sampling.

The sample-and-hold device in conjunction with the A/D converter can be modeled as a ZOH. But, since $\tau_{A/D}$ is assumed to be much less than T, the overall model is just unity gain. The D/A ZOH is still the only one in the standard structure as presented throughout the text. If very accurate design data are required, detailed simulations (hybrid or digital) can be implemented by using realistic sample-and-hold, A/D, and D/A nonlinear mathematical models.

Another hardware method of constructing the analog control signal is through the use of a *first-order-hold* (FOH) device which is also a *polynomial extrapolator*.[89] The FOH uses the last two discrete data points and a derivative

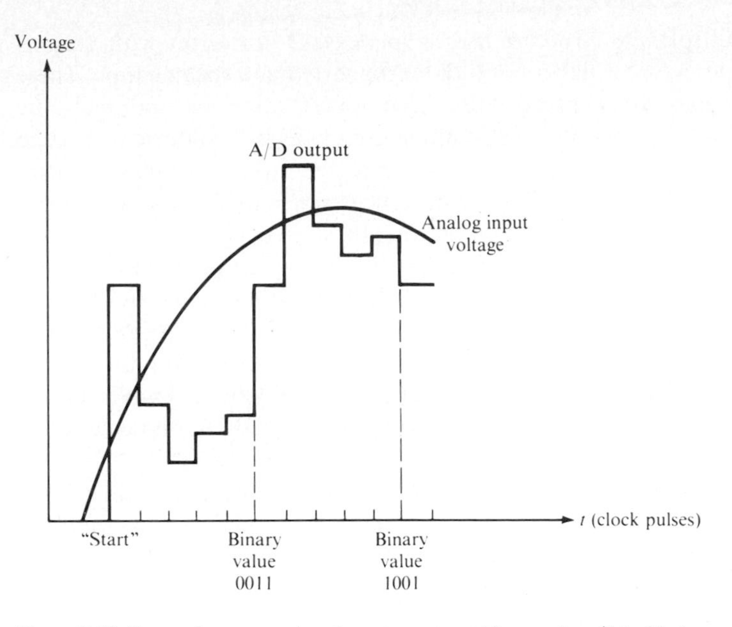

Figure 8-19 Successive-approximation converter with sample-and-hold characteristics.

approximation to construct the analog signal, that is,

$$e_1(t) = e_1(kT) + \beta\left\{\frac{e_1(kT) - e_1[(k-1)T]}{T}\right\}t \tag{8-6}$$

for $0 \leq t < T$ and $\beta = 1$. Note that if $\beta = 0$ the extrapolator is a ZOH device. The ZOH is normally the one used due to its simple construction. Other extrapolator methods such as FOH and the trapezoidal technique require extensive hardware and have accuracy limitations.

Of course, a software program can also provide the output smoothing through digital filters that use, for example, Euler, Adam 5th-order or Runge-Kutta extrapolation methods. But possible delays and undesired frequency-response characteristics can result.

8-9 MULTIPLEXING

With many data input channels of a digital control system, it is sometimes advantageous to share resources of the data acquisition subsystem. Such resources include the A/D converter, buffer register, and digital interface logic. Sharing of these input channel resources is call *multiplexing*, or time-division multiplexing. This technique is useful when the sampling times of each channel are relatively low and other limitations such as space, cost, and energy indicate that such an implementation is efficient.

A common multiplexing structure uses a single A/D converter with various channel switches that define which input is to be converted at a specific time. These semiconductor switches are controlled by a computer interface and real-time software. Switches of this type could also control sample-and-hold devices for each multiplexed channel. Detailed evaluation of such a structure in a digital control environment is required to ensure that performance requirements are being met.

A multiplexed computer interface could also be shared by input channels if proper buffer sizes (1 k^{Ω}, for example) are available. This type of shared resource would be designed to transfer data into memory directly in blocks of information. This type of interface capability is called *direct memory access* (DMA). The intelligence of this interface logic permits the processor to execute in parallel with the conversion process, and thus higher sampling times can be implemented. Similarly, the output D/A converter can be driven with a DMA organization if multioutputs are required.

With the decreasing cost and size of A/D and D/A subsystems and associated digital logic, most digital control systems use dedicated A/D or D/A converters for each channel. In applications where extensive signal conditioning is required (i.e., high voltage, prefiltering), various multiplexing schemes may still be satisfactory. In any case, the input-output software routine plays a critical role along with the hardware in achieving desired timing objectives and therefore must be properly written and tested.

8-10 INTEGRATED COMPUTER CONVERSION INTERFACES

In digital-control-system implementation of the A/D converter subsystem, additional sequencing signals from the computer are required. These signals properly initiate the conversion process and transfer the resulting digital value to a computer processor register. Since the converter is part of the input device, a buffer register is used to store the result of the conversion before transfer to the processor. Specifically, the computer addresses the D/A subsystem by using the I/O address lines, commands a specific input channel to be converted by using a special register, initiates the conversion with control signals, and transfers the data to the processor with an input command. The sampling time T is generally defined by another I/O interface subsystem called the *real-time clock*. A computer program initializes this interface to define the specific sampling period as well as the interval. In many implementations, the real-time clock is on the same board as the A/D and D/A converters.

A general A/D interface organization is depicted in Fig. 8-20. In this case, a two-level multiplexer is shown with 2^5 (32) channels divided into two groups. The specific channel to be converted is defined by a channel-select register which is filled by the computer under program control. The interface operates according to the following sequence:

Step 1. The A/D interface is addressed by the computer processor through the

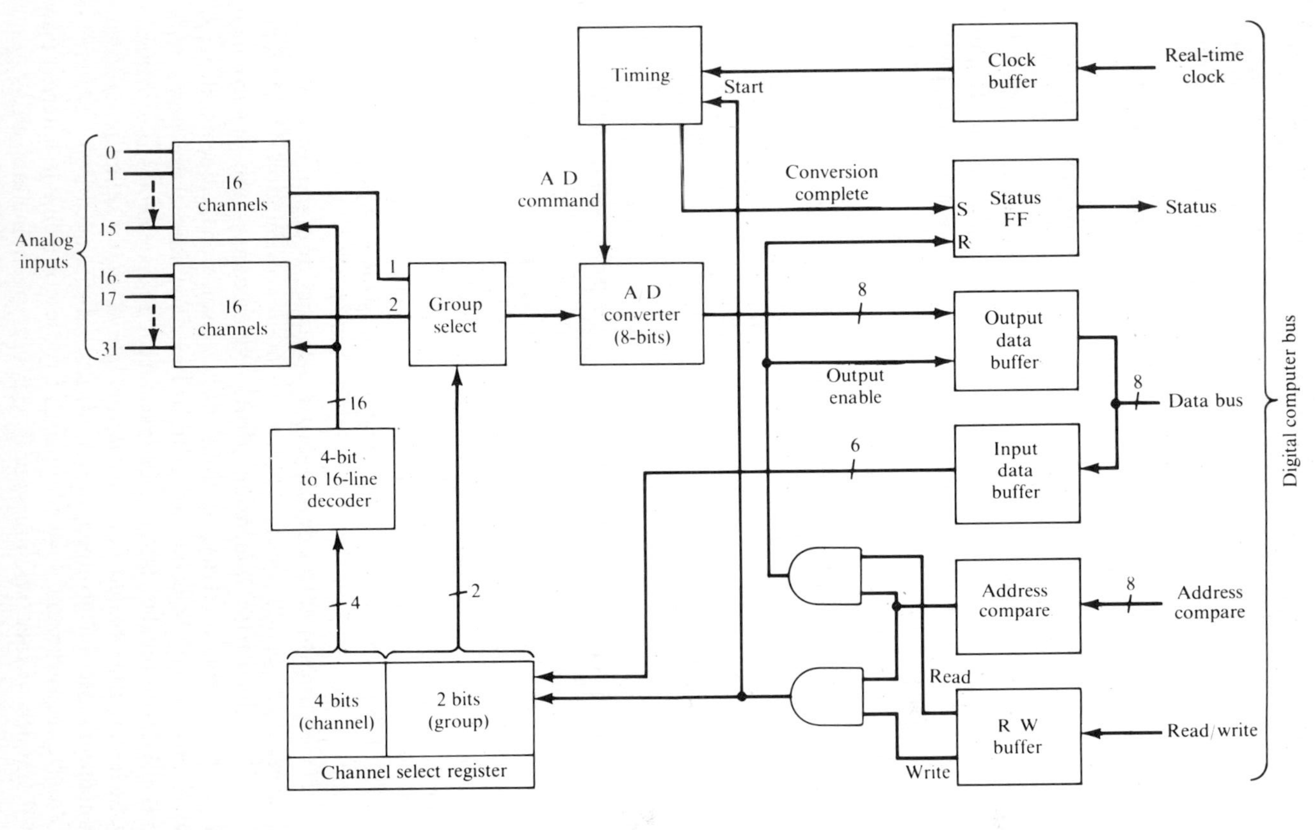

Figure 8-20 Computer A/D conversion interface.

address bus. The A/D channel-select information is concurrently placed on the data lines.

Step 2. The processor generates a WRITE on the Read/Write (R/W) line so that the A/D interface transfers the channel-select information to the channel-select register. After asynchronous delays, the desired analog input is at the input to the A/D converter.

Step 3. The real-time clock interface under control of the processor generates a clock signal initiating the A/D conversion in conjunction with the interface address information and the WRITE on the R/W line.

Step 4. When the conversion is complete, the status flip-flop is set and the digital converted value transferred to the output data buffer.

Step 5. The processor program is in a wait loop continually checking the status before transferring (READ) the converted value to the processor by way of the data bus. This checking requires the interface to be continually addressed. The processor also resets the status flip-flop when enabling the output data buffer. Note that in some systems the status information generates a processor interrupt signal thereby permitting the processor to execute other programs while waiting for the interrupt.

There are many variations of I/O structures in regard to a busing scheme. Each structure must be properly connected and required to achieve the desired performance. Technology is moving toward the integration of all A/D subsystem elements in one dual in-line package (DIP), permitting low-cost and low-power data-acquisition implementation.

8-11 MEASUREMENT IN DIGITAL CONTROL SYSTEMS

The general A/D conversion techniques discussed in the previous sections interface to numerous physical devices. These devices can measure a variety of physical characteristics. Examples include measurements of position, speed, temperature, and pressure. A position may involve vehicle attitude, fluid levels, valve configuration, or the relative place of a physical object. Speed could be associated with object velocity, flow measurements, switching parameters, or even rate of temperature and pressure and humidity changes.

In discussing measurement devices or transducers for digital control systems, general characteristics are desired for performance evaluation. They include accuracy, cost, dead-zone dimension, drift, dynamic response, environmental operation, hysteresis, linearity, quantization (resolution), repeatability, size, sensitivity, and speed. Selection of a transducer includes these items with a specific dependence on the signal to be measured. *Proper choice of a transducer can minimize extensive analog signal conditioning and digital signal processing.*

Transducers can be grouped into two general classifications: passive and active. Passive measurement devices consist of those not using an external power source. Examples of this type use specific physical phenomena such as electromagnetic, photovoltaic, piezoelectric, and thermoelectric. Active transducers require an external power source for measurement. Examples of this classification include devices which use a variable resistance, variable reactance (inductive or capacitive), the Hall effect, or the optoelectric phenomenon. The following sections present some specific devices of digital control transducers. More extensive discussion of transducer operation can be found in the references.[30,31]

8-11A Temperature Measurements

In many process control systems, the product evolves through a series of subprocesses, each of which includes specific parameters. In many cases, a very important subprocess parameter is temperature. Based upon continuing temperature measurements, heating or cooling mechanisms can be controlled to maintain constant temperature or a constant rate of temperature change. Measurement of temperature is generally accomplished by resistance thermometers, thermistors, bridges, thermocouples, or optical pyrometers (radiation measurements). Depending upon the technology used, the temperature units could be associated with the Celsius scale (°C), the Fahrenheit scale (°F), or the Kelvin scale (K).

The principle of resistance thermometers is based upon the electric conductivity in material (metal) at a certain temperature. Examples of appropriate metals are copper, nickel, and platinum. Each metal has to be very pure in order to provide the desired linear relationship. The relationship is usually positive; i.e., as temperature increases so does resistance. A general linear interval is -200 to 1000°C. The current as modeled by $I = V/R$ can be converted to a digital value through the use of an A/D converter. The variation in the resistance R due to temperature can be modeled as a geometric series:

$$R(T_m) = R_0 \sum_{i=0}^{M} a_i T_m^i \tag{8-7}$$

where T_m is expressed in degrees Celsius and R_0 is the value of R at 0°C; thus $a_0 = 1$ by definition. For each particular metal, the values of the a_i's are known and calibrated for a given measurement device. Note that as R increases with constant voltage, the current decreases. For a linear relationship, $a_1 = 1$ and $a_i = 0$, $i = 2, 3, \ldots$. For a nonlinear relationship, $a_i = \text{constant}$, $i = 2, 3, \ldots$.

The resistive measurement of temperature requires the constant current flow through the metal. This flow of electrons itself will heat the metal owing to I^2R causing a measurement inaccuracy. The associated self-heating coefficients are given for specific devices in terms of degrees Celsius per voltage for a given device.

Resistance devices are usually linear over a wide temperature range and are quite accurate. However, they are difficult to use for point measurements because of their relatively large dimensions. Also, a resistance-temperature probe can be an element of a Wheatstone bridge which can produce more accurate results if properly constructed.

Thermistors do permit a finer area measurement of temperature approaching

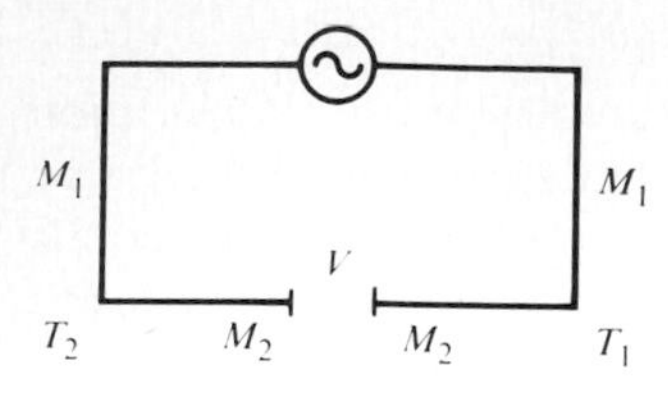

Figure 8-21 Thermocouple-circuit model.

a point measurement. Their resistance value is a decreasing function of temperature. However, the self-heating coefficients are relatively large. The relationship between thermistor resistance and the temperature T_m is given by

$$R(T_m) = R_0^{\alpha(1/T_m - 1/T_0)} \tag{8-8}$$

where T_m is in Kelvins, $T_0 = 273$ K $= 0°$C, and R_0 is the resistance value at 0°C by definition. α is a constant associated with a specific thermistor. Note that in this equation as T_m increases the resistance decreases. Because of the nonlinearity in the equation, the linear range of temperature values measured by a thermistor are very limited. Thermistors then are very sensitive to a small temperature change but are not too accurate over a large range.

Thermocouples are probably the most commonly used temperature measurement device. A thermocouple circuit consists of two different metal electronic conductors connected together as shown in Fig. 8-21. The voltage source creates a voltage drop V, which depends upon the distance across the gap, the two metals, and the two temperatures. If one temperature T_1 is kept constant, the value of T_2 is a function of V, that is, $V(T_1)$. Since this relationship is nonlinear, a plot or table of temperature values is referenced to V for a constant temperature T_1; that is, $V(T_1)$ defines T_2 if certain conditions exist. The conditions include:

The temperature at both ends of the gap are the same.
The pressure at the positions defined by T_1 and T_2 are the same.

The curves for $V(T_1)$ are always referenced to the constant ambient temperature T'. Since T' can take on a very large number of temperatures, a simple linear correction factor is made, referencing to 0°C. The form of the equation is

$$V_{T_0}(T_1) = V_{T_0}(T') + V_{T'}(T_1) \tag{8-9}$$

where the subscript in V refers to the constant-temperature reference. Thus, to use the graph (table), $V_{T'}(T_1)$ must be normalized to 0°C with the correction factor of $V_{T_0}(T')$ to yield $V_{T_0}(T_1)$. For this normalized value of $V_{T_0}(T_1)$ the value of T_2 is found from the table. Note that $V_{T_0}(T_0) = 0$.

The measured value of $V_{T'}(T_1)$ is usually of the order of millivolts. Thus, an isolation amplifier with a large gain is required to interface to commercial A/D converters. The connecting cables have to be calibrated for precise measurements. Analog filtering techniques are also used to measure the constant ambient temperature T'; a resistance measurement could be employed. Finally, the tables can be stored in a computer, and the correction operation for finding $V_{T_0}(T_1)$ executed digitally. With proper thermocouple insulating, accuracies of 0.1 percent can be achieved.

The final temperature-measurement device to be discussed is the optical pyrometer, which in reality measures radiation. The pyrometer is not placed in the medium to be measured but resides outside; thus it cannot be damaged by the temperature. This device is normally used for very-high-temperature measurements. Basically, the device measures either the infrared radiation power emitted or the light transmitted from the source based upon blackbody radiation models. The radiated power is absorbed by a calibrated thermocouple for a very limited band of the spectrum. The other type of pyrometer uses a comparison technique between the measurement source and a calibrated and changeable source. This light-intensity comparison can be measured with a sensitive human eye and the manual control of current in a calibrated light source. Of course, complex optical elements (filters) are required, and the delay caused by human evaluation must not affect control-system performance.

8-11B Pressure Measurements

Pressure is an action of a force per unit area. Such a force can be exerted on some type of sensing device for use in a real-time control system. An example would be the pressure exhibited by a fluid (air, water, etc.). The fluid can be contained in a vessel such that *static* pressure is exerted by the fluid on the vessel walls. Fluid flowing in a pipe exhibits *dynamic* pressure across a cross section of the pipe. A hypostatic gauge could be employed to visually measure such pressures in terms of force per unit area or number of atmospheres. Diaphragms are also used.

Other techniques measure the pressure by deformation of metals through the use of a calibrated gauge (bellows, spiral, etc.). In most cases, the pressure is proportional to the displacement x; that is,

$$\text{Pressure} = kx \tag{8-10}$$

The value of k depends upon the specific linear mechanization. Most devices then use the spring-constant concept reflected in the above equation.

The various pressure-measurement devices can measure absolute, relative, or differential pressures. A specific gauge selection and an associated manufacture depends upon the given environment and the associated accuracy requirements.

Another method for measuring pressure is the linear voltage differential transformer (Fig. 8-22). As indicated in the figure the output voltage V is linearly proportional to the displacement of the core element. The accuracy of this instrument depends upon the associated electronics that amplify the voltage V. To use this device as a pressure measurement, the pressure force is applied to the core external to the transformer through a diaphragm. A linear fixed spring is attached to the other side of the core. Note that the differential transformer could also be used for measuring relative position (linear). The operation of this device is linear over a small range.

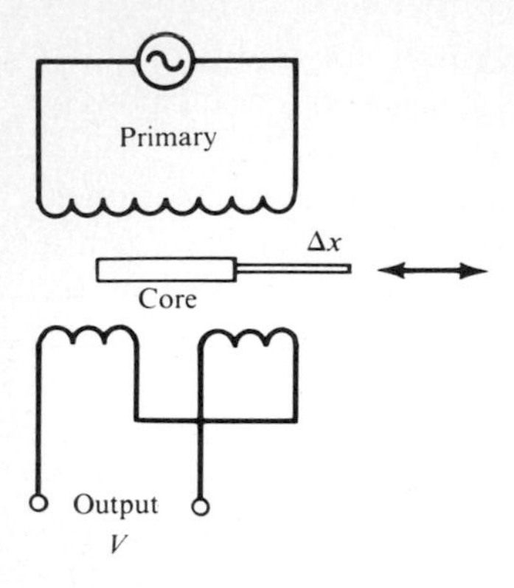

Figure 8-22 Linear voltage differential transformer (LVDT).

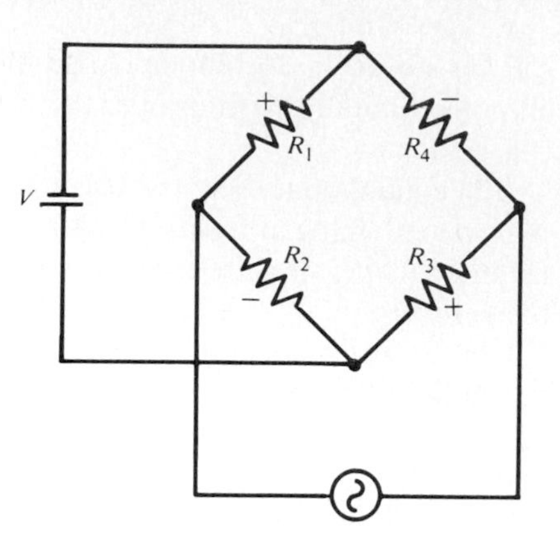

Figure 8-23 Bridge measurement of resistance for linear position or pressure measurement.

A variation in resistance can be made proportional to pressure or linear movement, as depicted in Fig. 8-23, by using a bridge. In essence, this is a variable potentiometer. The dimension of a cavity can be changed under pressure; thus the associated electric capacity of the cavity can be measured through the use of a capacitance bridge, that is, Q (charge)/V, the resulting electric potential being proportional to the capacity (position) where the electric field stays constant. Also piezoelectric pressure transducers can be employed by using a quartz crystal to measure charge variation. These devices exhibit very high resistance and thus because of their small size generate low current measurements. Again filtering and high-gain isolation amplifiers are required. All these devices attempt within certain boundaries and with a specific accuracy to measure relative displacement. The displacement then is proportional to force (pressure or position). Many contemporary devices are quite small, on the order of millimeter dimensions, and require extreme care. To use a mechanical pressure-measuring device in a digital control system, conversion of the pressure information to an electric signal is required. This signal can then be digitized through an A/D convertor.

8-11C Rate Measurement[32]

The specific measurement of rate or speed of movement depends upon the associated range and accuracy requirements. Concern may focus on a very small range of movement around a static equilibrium position or on the speed of a continuously turning shaft. Also, the measurement environment may require special protective packaging of the measurement device. Some examples of rate-measurement devices are tachometers and rate gyros.

A tachometer is basically a voltage generator consisting of a coil properly woven around a turning shaft. The coil rotates in a fixed magnetic field. As the shaft turns, the electric field effect is such that a voltage signal is produced across the fixed coil proportional to the speed of the turning shaft; that is,

$$\text{Speed} = K\omega \tag{8-11}$$

where ω is the rotation speed of the shaft. This device can be used along with special electronics to control the speed of rotation with negative feedback. The rotation signal (velocity) can also be used to damp a step response when the shaft is to be position-controlled. A measurement device is needed here to measure shaft position, i.e., perhaps a precision potentiometer.

A rate gyro is an avionics device used to measure the rate of turn about a specific axis (pitch, yaw, roll) to a very precise value. The airborne or space vehicle can be controlled (stabilized) to a precise attitude with a maximum rate of turn as it moves in a defined coordinate system.

A gyroscope consists of two elements (Fig. 8-24), the gyro element and the gimbals. The gyro element consists of a spinning rotor exhibiting high angular momentum, the rotor drive mechanism, and the rotor support. The gimbal is a support structure which permits a given range of rotational freedom for the spinning rotor (Fig. 8-24). Additional gyro components connect coils for torque or signal generation.

A calibrated torque can be applied to the gimbal for command purposes. The signal generator provides a method for measuring the rotor orientation. The signal generator then produces a signal when an external torque is applied. The gimbal movement is then proportional to ω, the angular velocity, of the spin-axis precession caused by the torque. Physically, the gyro attempts to align the spin axis with the applied torque rotor.

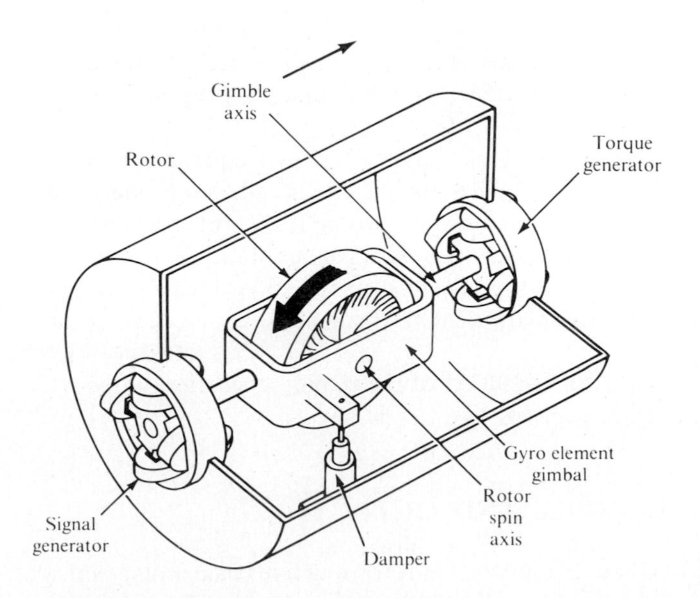

Figure 8-24 Rate gyroscope components.

The basic phenomenon of a gyro that makes it useful is the physical law which states that a rotating device will maintain a fixed orientation (attitude) in space (inertial) unless external torques are applied. Thus the gyro spin axis will tend to remain in a fixed orientation. The external vehicle torques then generate a change in rotor orientation (spin axis) that is picked up by the signal generator.

A gyro with only one gimbal is a single-degree-of-freedom gyro. A two-gimbal system permits the rotor spin axis to have two degrees of rotational freedom, i.e., a two-degree-of-rotational-freedom gyro.

A single-degree-of-freedom gyro functions by summing the applied torques. The rotor precession caused by this torque can be damped by a reaction torque. Such a torque can be achieved by suspending the gyro element in a fluid or by using a mechanical damper. Such damping causes the deflection angle due to precession to be a measure of the integral of the inertial angular rate, i.e., relative change in angular position.

A rate gyro, on the other hand, does not have a damped precession. In this case, the deflective angle is a measure of a constant angular velocity.

Integrating gyros usually have a response delay time of approximately a few milliseconds. Rate gyros have faster response times.

8-11D Position Measurement

Various transducers have already been presented for generating position information. The disk encoder of Fig. 8-7 defines a unique binary number representing a shaft position within a relatively small sector interval. (Another shaft-position device is the 360° wire-wound potentiometer, where the number of wire turns defines the accuracy of the position measurement.) The integrating gyro, as mentioned in Sec. 8-11C, integrates rate to generate position information. The differential transformer (Fig. 8-22) can also be used for measuring position over a wider range. In these methods, the relative position is directly proportional to a voltage value.

One of the newer position devices uses a laser with low-frequency amplitude modulation. The technique employed relies on the measured amplitude of the reflected modulated laser beam. Since the distance to be found must be less than one wavelength of the modulated signal, the reflected amplitude measurement relative to a reference signal is proportional to distance. The device must thus contain circuitry to perform the calculations, converting phase difference to actual distance.

Using any of these devices, the selection of the scaling coefficient after A/D conversion is still an important decision (see Chap. 10).

8-12 PROGRAMMING INPUT AND OUTPUT

The computer structure of Chap. 2 represents hardware as five basic units; control, manipulation (ALU), memory, input, and output. A computer to be useful must interface with the analog world through the input and output (I/O) units. In digital control systems, I/O interfaces to analog and discrete signal transducers are required. Communication through these interfaces is important to the control engineer as they relate to timing, speed of operation, and coding of information. The purpose of this section is to present several different ways of programming an input or output operation based upon hardware capability.

The basic concerns of the control engineer when designing an input or output interface focus on timing, speed, and coding, as mentioned above. Since the analog world is not synchronized to the digital (computer) world, a means of sampling the analog signals must be implemented, resulting in a value for T. The selection of the T value as detailed in the previous chapters is in reality a method of synchronizing the two worlds.

The speed of a digital controller depends upon not only the hardware technology but upon the complete algorithm implementation. Since the analog signal bandwidths in a control environment are relatively low as compared to the internal computer clock frequency, the usual objective is to attempt to keep the computer (processor) busy through proper I/O programming techniques. An analog signal in a control system can be sampled at 100 Hz, for example, but the internal computer clock can be operating at 1 MHz. Thus, 2000 instructions can execute within the sampling interval assuming five clock periods per instruction. To achieve maximum utilization of the processor is quite difficult in a real-time digital control environment because of the continuing interaction between the computer and the analog world.

As mentioned earlier, the coding of information is important in meeting timing objectives. The shifting of data as it is converted through the A/D or D/A converter may require the software to scale the manipulations internal to the processor. The scaling process can take considerable time which could negate the obtaining of system performance requirements.

After briefly presenting I/O interface hardware organization, this section discusses software development for input and output information transfer.

8-12A I/O Hardware Structure

As in any information transfer operation the three types of information involved in I/O structures are data, control, and status information. The data are the information to be transferred. The control information defines the operation such as "input read" or "output write." The status information refers to the state of the I/O interface and device, e.g., "busy," "ready," "error." The hardware interface structure must permit the storage and processing of all this information in order to transfer data. The associated storage registers buffer the high-speed synchronized computer elements from the slower asynchronous analog devices.

Hardware buffer registers (see Fig. 8-25) are used to store data, control (commands), and status information. The data register may be bidirectional, or separate input and output data registers may be implemented. Addressing for separate registers may be required through the assumed general busing organization

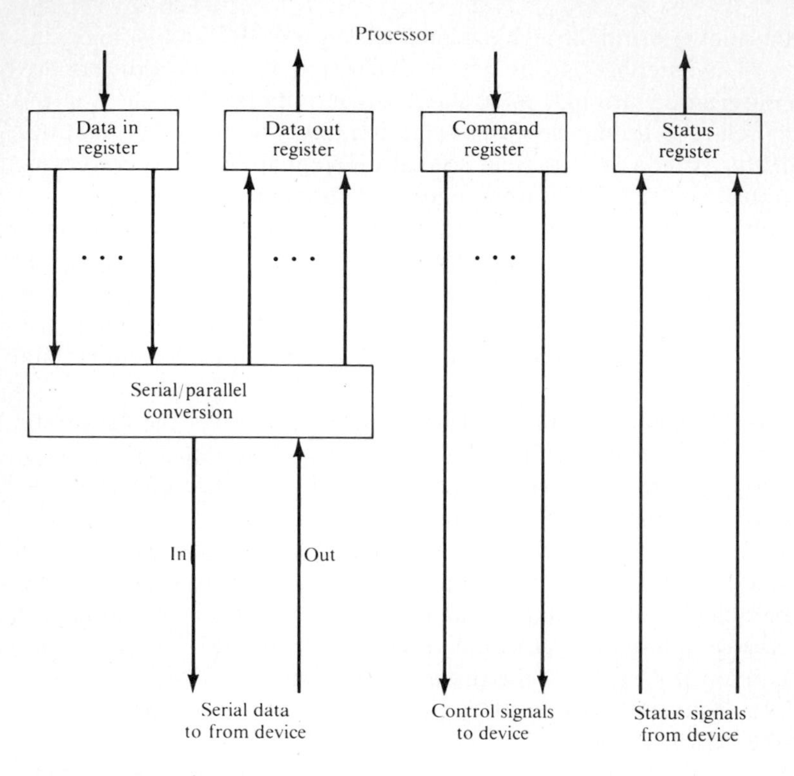

Figure 8-25 I/O buffer organization.

(see Chap. 2). The command register stores command and control information. Each bit or combination of bits controls I/O device functions. Finally, the status register stores bit information relating to I/O device status. In some systems, the command and status registers are combined into a single *command and status register* (CSR). For devices with numerous multiple functions, additional interface registers are required. Each register must be addressed by a unique address through a computer instruction. Variations on the basic hardware interface exist between manufacturers and applications.

8-12B Programmed I/O Mode

In the programmed I/O mode, by definition, the I/O operations are completely controlled by the processor. That is, the processor executes an I/O program that initiates, directs, and terminates the I/O operation. This type of I/O operation is available on virtually every computer system. It is simple to implement, only requiring a few instructions. The primary disadvantage is the loss of overall speed. The processor is essentially slowed to the speed of the I/O device for a single data

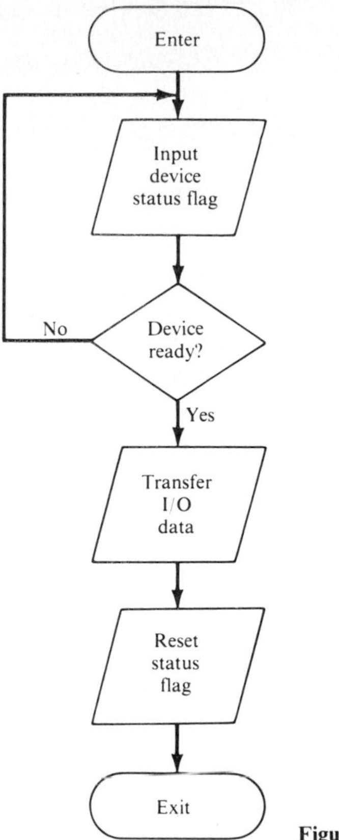

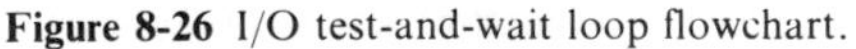
Figure 8-26 I/O test-and-wait loop flowchart.

transfer. Programmed input and output can be implemented by using a *test-and-wait loop* (see Fig. 8-26) consisting of three major operations:

Select (address) the I/O device status register.
Loop (wait) until the device is available (test ready status bit and command device function when ready).
Transfer the information to or from the data register.

The disadvantage of the test-wait loop is the waiting (looping) operation. If the I/O device is not available or ready, the processor sits idle, waiting for the I/O device interface to signal that it is ready. Thus, all the processor speed is negated during the waiting period. Although many I/O devices can be ready at the same time, the processor can only check one I/O device status at a time. The time to check all system devices sequentially may be unacceptable with this approach.

For multiple input devices, some improvement in the system efficiency can be obtained by using the *polled I/O method.* Basically, the processor periodically polls

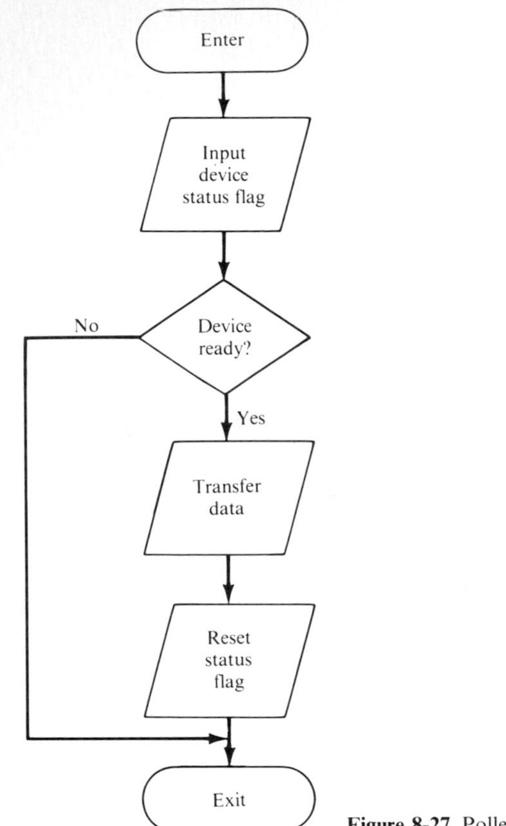

Figure 8-27 Polled I/O flowchart.

or checks all the I/O devices in a predefined sequence (see Fig. 8-27), asking: "Does any device have data to input?" If yes, data is transferred; if no, the processor resumes normal operations. Note that the test-wait loop is eliminated, but the polling operation can still be unacceptably long. Both the test-and-wait or polled method require the processor to check a status register bit to determine if the device is ready to perform the I/O operation.

8-12C Interrupt I/O Devices

In both the test-wait or polled programmed I/O technique, the processor has to check and wait for the availability of I/O devices. In many applications this is impractical because (1) the computer has too many manipulations to perform, and (2) in many digital control applications there are a large number of possible input devices which become ready randomly and must be sampled quickly, before the data changes value. A more advantageous approach can be to transfer data only when there is an I/O device ready. This leads to the concept of I/O interrupts, which allows the processor to respond to stimuli other than those created by its current internal operations. The stimuli come from the I/O device interfaces which interrupt the processor when the I/O devices are ready.

The processor continues its normal execution until some external device needs servicing. The I/O device, consequently, generates a control signal to "interrupt" the currently executing routine (at the end of the current instruction). The interrupt causes the processor to jump to a special routine which is needed to service the interrupt. Once the device has been serviced, normal program execution is resumed. The interrupt, then, can be viewed as a hardware-initiated subroutine jump. The device must directly or indirectly supply the address of the interrupt service subroutine.

There are generally four different types of interrupts that can occur within a computer system:

I/O interrupts. Each I/O device signals to the processor when it is ready for service.
Timer interrupts. An asynchronous counter can be initiated and run independently of the main processor. This real-time clock generates an interrupt when it "times out."
Hardware failure interrupts. Any hardware problem that can be tested can be signaled to the processor by an interrupt.
Program interrupts (traps). Software faults and status indications can generate interrupts if properly implemented.

The priority of these interrupts depends upon the application as defined by the software engineer.

Interrupt structures consist of four general organizations as follows:

Single-level, single-priority: a polled interrupt system. All interrupt requests set a single request flag. The processor must poll the devices to determine the source of the interrupt. An applied ordering or priority is established by the polling order.
Multiple-level, single-priority: a vectored interrupt system. The individual devices set an interrupt flag and then provide a code (the interrupt vector, an address) to identify the requesting device. This structure results in a faster recognition of the interrupt source than in the first type.
Single-level, multiple-priority: a single interrupt flag, with a daisy chain of devices. The device closest to the processor in an electronic sense can inhibit devices farther away.
Multiple-level, multiple-priority: a combination of the two previous methods. The technique usually requires a mask register to inhibit some device levels at any particular time.

The interrupt-servicing procedure consists of five major operations.

Recognize the interrupt and turn off interrupt system.
Save the current processor state.

Determine the source of the interrupt.
Jump to the interrupt routine, turn on interrupt system, and perform the interrupt service.
Restore the processor state and return to (resume) normal operation.

Recognition of the interrupt is usually done at a fixed point in the machine instruction cycle, normally just before the instruction fetch from memory for ease of hardware design. If a multiple-priority interrupt system is in use, the processor must determine if the current interrupt is of sufficiently high enough priority to warrant its recognition. This operation can be done in software or hardware.

The current processor state must be saved so operations can be resumed upon completion of the interrupt service routine. The state may consist of the Program Counter, Processor Status Word, and general register values for those registers which the interrupt service routine uses. The processor state can be stored in memory registers or a register stack in the processor.

Once the interrupt device is identified, the interrupt service routine can be executed, and following restoration of the processor state, the execution of the main program is continued.

In selecting the I/O programming method for a given application, the control engineer must analyze overall system requirements, constraints such as timing and interface information structure, and possible hardware configurations.

8-13 SUMMARY

A variety of physical converters available for real-time digital control systems have been presented. The impact of converter performance characteristics depends upon the specific digital control application. All A/D conversion techniques (the counter, ramp, and continuous converters) generally operate at a considerably faster time per step and produce better differential linearity than the approximation methods such as successive approximation. The readers' understanding of the relationship between the assumed impulse sampling model and the real-world realization of such sampling should have been enhanced through the discussion of the converter speed characteristics and the associated physical constraints.

In general, each digital control application requires unique signal processing and conversion interfaces. Such development relies heavily upon interdisciplinary understanding of the requirements and possible implementations.

The measurement methods presented in reality are only a simplified version of contemporary sophisticated transducers. Detailed understanding of transducer mechanization, associated tolerances, and signal levels are required to develop an integrated digital control system.

CHAPTER

NINE

RANDOM PROCESSES[33, 34]

9-1 INTRODUCTION

In previous chapters, system input signals have been assumed to be deterministic in nature (uniquely determined). This characterization is useful in representing and analyzing discrete-time signals and systems in the time domain and frequency domain by using the $\mathscr{Z}$ transform. However, many control-system input signals are either unknown or random in nature. Such signals cannot be described precisely. This signal representation is generally characterized as *noise* or a *noise process* owing to its unknown nature. Also, the term *stochastic process* is used to characterize noise as a time-dependent random process (discrete or continuous). The purpose of this chapter, however, is not to present a detailed mathematical development but to discuss fundamental probability concepts and random variables and apply them to the characterization of random processes. Chapter 10 then uses these concepts to analyze digital-control-system quantization errors.

The fundamental mathematical basis for the representation of random variables is based upon the theory of probability. The parameters that describe a random variable are usually the mean and variance based upon the long-term behavior of the underlying random process as generated by experiments (i.e., the law of large numbers). The mean and variance can also be mathematically derived from the associated probability distribution. The basic quantity that characterizes the nature of a random variable or a random process is then its probability distribution. From this quantity the various probabilistic parameters such as mean and variance can be generated.

Some additional random signal properties that are useful in analyzing discrete-time random processes as part of a control system are the autocorrelation sequence and the autocovariance sequence. These sequences contain most of the properties of interest. Note that the term *sequence* is employed since a discrete random process can be described by a sequence of numbers each associated with a discrete-time observation. In a discrete system, interest focuses on the system output random process as driven by the noise inputs. However, before discussing the digital-control-system applications in the next chapter, probability theory is introduced and extended to random process descriptions.

9-2 BASIC PROBABILITY

In attempting to analyze an experiment a degree of uncertainty or probability of occurrence is associated with specific events. Examples include the flipping of a coin, the possibility of rain, and the probability of passing an exam. The basic elements of probability theory can be discussed either using abstract mathematics or from an empirical point of view. Although abstract mathematical set theory representations of probability concepts are comprehensive, the empirical approach is more intuitive and thus is preferred. Set theory is still used, however, to support an understanding of basic concepts. The general objective is to be able to analyze random or probabilistic events that may occur in the functioning of a digital control system and determine performance changes.

To develop the probability relationships, specific terminology is required. Initially this is rather simple but is extended in a logical manner to complex models. In determining the empirical probability of a particular event, an *experiment* with a set of specific rules and constraints is performed. An *outcome* of one experiment is defined as the result. The set of all possible outcomes is usually defined as the *sample space* S. For example, in flipping a coin,

$$S = \{\text{heads, tails}\}$$

Note that in this case the set S is discrete and consists of only two outcomes or *elements*. A continuous sample space would be

$$S = \{x \text{ such that } x \in [0, 1]\}$$

or in condensed notation

$$S = \{x | x \in [0, 1]\}$$

Note that x is a continuous scalar variable or element, that is, x can have any value between 0 and 1. The conditional constraints on the sample space variables can be described in any notation that is understandable and consistent. In conducting an experiment, more than one sample space can be defined depending upon the desired information.

Example 9-1 Consider the experiment of tossing an eight-sided die. One sample space (discrete) would be

$$S = \{1, 2, 3, 4, 5, 6, 7, 8\}$$

Another might be

$$S = \{\text{even number, odd number}\}$$

Example 9-2 Consider a digital-controller output through a D/A converter. One experiment consists of observing the output twice to determine if the converter is below saturation (i.e., below the largest value or greater than the smallest value). A sample space of the two observations would be

$$S = \{NN, NR, RN, RR\}$$

where N = nonsaturation and R = saturation.

Another discrete sample space for this example would be

$$S = \{0, 1, 2\}$$

where the elements reflect, respectively, no saturation, one saturated output, and two saturated outputs. Note that the first sample space selected contains more detailed information than the second. *The sample space selected should reflect the level of information desired.*

Continuing the development of pertinent terminology, consider an interest in a combination of outcomes. This combination is called an *event*. An *elementary event* by definition is an event consisting of only one outcome in the sample space. In general, an event is a subset of a sample space.

Example 9-3 From Example 9-2 define an event E in which the number of saturated outputs is at least one during an *observation* (one experiment). Thus

$$E = \{NR, RN, RR\}$$

and E is a collection of elements of S; that is, $E \subset S$. Since each of the elements of E cannot occur simultaneously during one experiment, they are defined as *disjoint*. One additional notation is required to represent an event that contains no outcome: the null set, $\emptyset$.

The probability of an event within a related sample space is associated with a range of numbers from 0 to 1. These numbers then reflect the probability of an event occurring. *If the event always occurs, then the event is defined as certain and the associated probability is 1. If an event never occurs, then by definition the probability of occurrence is 0.* Other events lie somewhere in the range 0 to 1. If the probability is closer to 1, then the likelihood of the event occurring is relatively high, and vice versa. To obtain the numerical value for the probability of any event, the probabilities for each event element (i.e., each elementary event) are added together.

Example 9-4 Consider an unbalanced coin in which heads (H) is likely to occur twice as much as tails (T). If the coin is tossed three times, what is the probability that one or two tails occur?

SOLUTION The selected sample space is

$$S = \{\text{HHH, HHT, HTT, HTH, TTH, THH, THT, TTT}\}$$

To determine the probabilities a weight W, yet unknown, is assigned to H and T. For T it is W and for H it is $2W$. Thus for the three coin tosses, HHH $= 6W$, HHT $= 5W$, HTT $= 4W$, and so forth. Since the total for S must be equal to 1, then

$$6W + 5W + 4W + 5W + 4W + 5W + 4W + 3W = 1$$
$$36W = 1$$

which yields

$$W = 1/36$$

Since

$$E = \{\text{HHT, HTT, HTH, TTH, THH, THT}\}$$

where $E \subset S$, then the probability of the event (one or two tails) is given by

$$\begin{aligned} P(E) &= 5/36 + 4/36 + 5/36 + 4/36 + 5/36 + 4/36 \\ &= 27/36 = 3/4 \end{aligned}$$

Therefore, there is a 75 percent chance of one or two tails! In this example the event E and the event F, where $F \subset S$ consists of only one tail per experiment, are not disjoint.

In general, the *probability* $P(E)$ *of an event* T is

$$0 \leq P(E) \leq 1$$

and $\qquad P(\emptyset) = 0 \qquad$ ($\emptyset$ denotes no event)

and $\qquad P(S) = 1 \qquad$ by definition $\qquad$ (9-1)

The *complement* of the set E is denoted by E^C which contains all the events in S which are not in E. Thus, $P(E^C) = 1 - P(E)$. If E and E' are two events (i.e., subsets of S), then the *union* $E \cup E'$ is the event containing both E and E'. The *intersection* $E \cap E'$ is the event containing outcomes both in E and E'. If the intersection is the empty set $\emptyset$, then the events E and E' are defined as disjoint. Note that for a discrete sequence of events

$$\left(\bigcap_i E_i\right)^C = \bigcup_i E_i^C$$

by DeMorgan's Law (Chap. 2), since "or" can be used to define the union of events and "and" used to indicate the intersection of events.

Consider an experiment that can result in M different but equally likely outcomes. If m of these outcomes correspond to an event E, then $P(E) = m/M$. If equally likely outcomes are not true, then the relative likelihood of each outcome must be known a priori. This may be determined by a large series of recording experiments that are conducted to determine the probability of each outcome. This experimental approach is called the *relative frequency* technique. If an experiment is performed M times and the desired output occurs m of these times, then the probability of an event E is

$$P(E) = \lim_{M \to \infty} \frac{m}{M} \tag{9-2}$$

The law of large numbers is assumed to be applicable here. For example, in flipping a balanced coin, if M is quite large then it can be assumed that $P(E)$ is approaching 0.5 for E, the tossing of a head.

9-3 RANDOM VARIABLE BASICS

Although considerable analysis can be performed with the previous probability concepts, the sample space nomenclature or outcome terminology is usually unique to the given problem, i.e., "heads" and "tails." Thus the obvious extension is the mapping of these specific terms into a more general or symbolic form. Since a numerical representation of the outcome is desired, why not then associate a symbolic variable with each possible outcome that can take on a real number value. For example, associate a "0" with a coin tail and "1" with a head. Any set of real numbers can suffice for this mapping. In general capital letters are used to denote a variable and lowercase letters represent possible real values of the variable. Of course, this variable is defined as a *random variable* since it can assume at any instant a real number value. The specific value depends upon the selected mapping or association within a given experiment.

Example 9-5 Consider again Example 9-4, where $E = \{\text{HHT, HTT, HTH, TTH, THH, THT}\}$. If X, a random variable, is to reflect the occurrence of either one tail or two tails in S, then

$$X = x_1 \text{ or } x_2 \qquad \text{where } x_1 = 1 \text{ tail}, x_2 = 2 \text{ tails}$$

In general, each value of a random variable represents an event. Again, an event is a subset of the sample space. A random variable X can assume only a discrete number of values or it could possibly take on any real value (continuous). A *discrete random variable* can be defined as one whose sample space contains a finite number of possibilities or at least a sequence of events that can be directly associated with the positive integer numbers (a one-to-one map is normally called a *countable infinite sequence*). A *continuous random variable* then requires a sample space with an uncountable infinite number of possible outcomes.

In many experiments requiring the measurement of physical data, the random variables are continuous. Examples are the position of a shaft, the temperature of a furnace, the wind gust on an aircraft, the rotating speed of a motor, and the quantization error of an A/D converter. These situations are encountered in various control-system designs. Examples of discrete random variables are the flipping of a coin, the possible values contained in a computer register, the number of sunshine days per year, and the number of arithmetic overflows in a digital controller.

The intent then is to develop a functional notation using the random variable symbology to graphically represent event probabilities and gain insight to general probabilistic relationships. Again an attempt to generalize provides insight that is very useful in designing and analyzing digital control systems.

One basic desired relationship is the probability that a discrete random variable X is equal to a specific outcome x (real value) within an event. This concept can be presented in the functional form

$$P(X = x) = f(x) \tag{9-3}$$

where $f(x)$ is defined as the *probability distribution function* (PDF) of the discrete random variable X.

Example 9-6 Consider again Example 9-5 with X the occurrence of one or two tails in the sample space S. Then if event E occurs, X is either 1 or 2.

In the new notation. Eq. (9-3) is transformed into the following table where x represents the number of tails occurring in an event:

x	0	1	2	3
$P(X = x)$	1/6	5/12	1/3	1/12

thus

$$P(X = 1 \text{ or } X = 2) = 5/12 + 1/3 = 3/4 = 0.75$$

In general the distribution function for a discrete random variable satisfies

$$f(x) \geq 0$$

and

$$\sum_{x \in s} f(x) = 1 \tag{9-4}$$

One of the more useful functions is the *cumulative distribution function* (CDF). This function is usually denoted by $F(x)$ and defined as

$$F(x) = P(X \leq x) \tag{9-5}$$

where X is a discrete random variable and x is a possible value of X. Then $F(x)$ is the probability that the discrete random variable X is less than or equal to x (an outcome). Note that

$$F(x) \geq 0$$

and

$$F(x) = \sum_{x' \leq x} f(x') \tag{9-6}$$

or, in set notation,

$$P(E) = \sum_{x \in E} f(x)$$

Example 9-7 Using Example 9-5, what is the probability that no more than two tails are tossed within the three-toss experiment, that is,

$$P(X \leq 2) = ?$$

SOLUTION

$$E = \{\text{HHH, HHT, HTH, THH, HTT, TTH, THT}\}$$

$$F(2) = f(0) + f(1) + f(2)$$

$$= 1/6 + 5/12 + 1/3 = 11/12$$

$$P(X \leq 2) = 0.917$$

The cumulative distribution of X is

$$F(x) = \begin{cases} 0 & x \leq 0 \\ 1/6 & 0 < x \leq 1 \\ 7/12 & 1 < x \leq 2 \\ 11/12 & 2 < x \leq 3 \\ 1 & 3 < x \end{cases}$$

$F(x)$ and $f(x)$ can also be plotted as a function of x for easier understanding of the associated probabilities. From Examples 9-5 and 9-6, the diagrams are shown in Figs. 9-1 and 9-2.

In the case of the probability distribution function $f(x)$, the functional values are discrete for discrete values of x. The probability distribution function (PDF), as depicted in Fig. 9.1, shows the expected percentage of zero, one, two, or three tails in three tosses. Thus the PDF indicates the likelihood of the four different

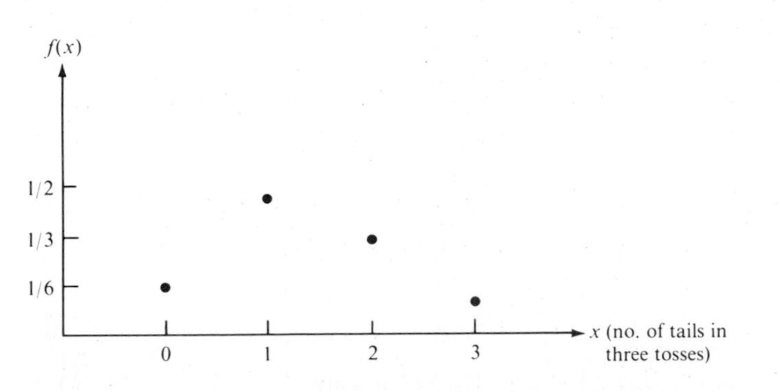

Figure 9-1 Probability distribution function (PDF) example.

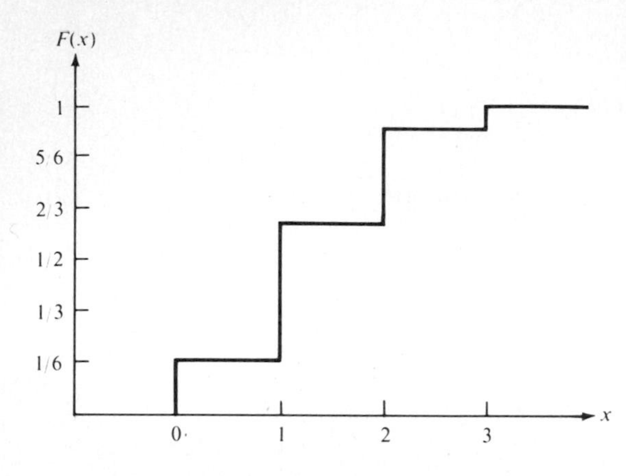

Figure 9-2 Cumulative distribution function (CDF) example.

outcomes. The cumulative distribution function (CDF), shown in Fig. 9.2, portrays the probability or likehood of n or less tails in three tosses.

Associating a physical meaning to these $f(x)$ graphical points (discrete) results in $f(x)$ sometimes being defined as a *probabilistic mass function* (PMF) since the density of $f(x)$ is centered at a discrete number of "mass points."

Continuous random variables also exist and are quite useful in the characterization of control-system signals. Since a continuous random variable X can assume an infinite number of values, the probability of any specific value is zero, i.e.,

$$P(X = x) = 0 \qquad X \text{ a continuous random variable} \tag{9-7}$$

However, the probability that X lies within a range of values can be nonzero; i.e.,

$$\begin{aligned} P(x_1 < X \leq x_2) &= P(x_1 < X < x_2) + P(X = x_2) \\ &= P(x_1 < X < x_2) \end{aligned} \tag{9-8}$$

The inclusion of an endpoint has no contribution to the probability value for the continuous random variable X.

Although the probability distribution of a continuous random variable X cannot be presented in a tabular form, the notation of $f(x)$ is still used, but is now called the *probability density (distribution) function* (PDF) as compared to the probabilistic mass function. Assume that the graph of $f(x)$ is depicted as shown in Fig. 9-3.

The probability that X is within the interval between x_1 and x_2 is

$$P(x_1 < X < x_2) = \int_{x_1}^{x_2} f(x)\, dx$$

Of course, $P(X = x_1) = 0$ since the area $= \int_{x_1}^{x_1} f(x)\, dx = 0$. In general

$$f(x) \geq 0$$

$$\int_{-\infty}^{+\infty} f(x)\, dx = 1 \qquad x \text{ continuous} \tag{9-9}$$

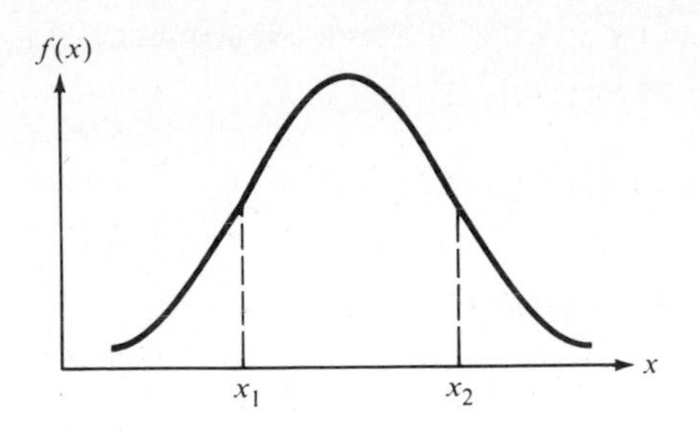

Figure 9-3 Example of continuous random variable PDF $f(x)$.

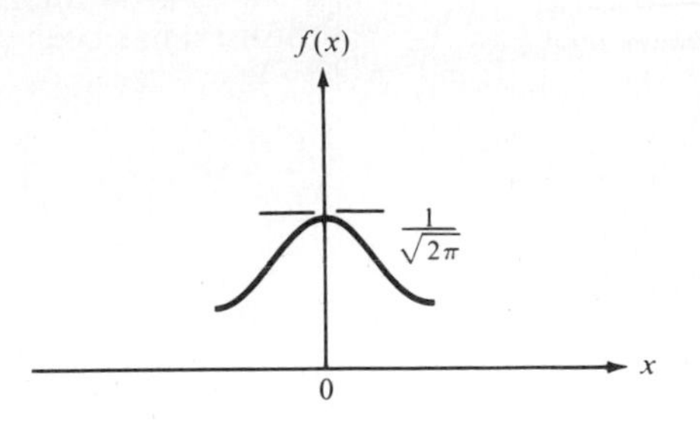

Figure 9-4 Gaussian probability density (distribution) function (PDF).

The determination of $f(x)$ experimentally requires a considerable number of measurements to even approximate the PDF curve. Generally a model of $f(x)$ is assumed based upon theory, insight to a given process, or experimental results. A standard density function is the gaussian or normal density function

$$f(x) = \frac{1}{\sqrt{2\pi}} \epsilon^{-x^2/2^2} \tag{9-10}$$

which results in the bell-shaped curve of Fig. 9-4. This function cannot be integrated in closed form, and therefore tables or approximations are used to determine the desired probabilities. Another standard density function is the uniform density function

$$f(x) = \begin{cases} \text{a constant} & a < x < b \\ 0 & \text{otherwise} \end{cases} \tag{9-11}$$

which is shown in Fig. 9-5.

A cumulative distribution function $F(x)$ of a continuous random variable X can also be defined:

$$F(x) = P(X \le x) = \int_{-\infty}^{x} f(x')\,dx' \tag{9-12}$$

which is very similar to the cumulative distribution function for a discrete random

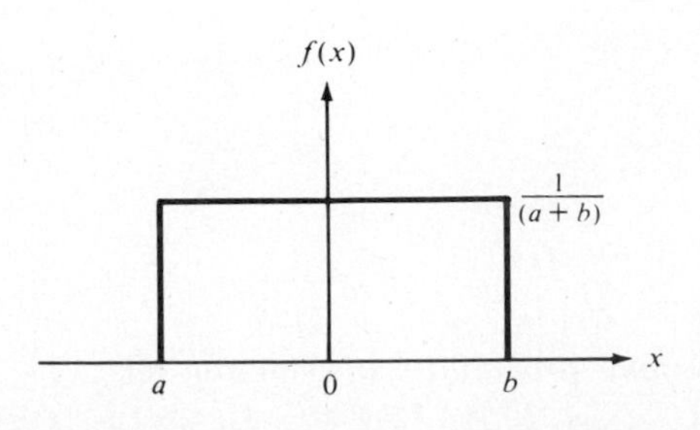

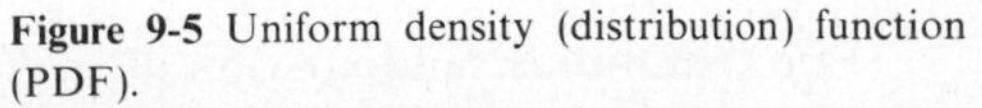

Figure 9-5 Uniform density (distribution) function (PDF).

variable except that the sum has been replaced by an integral. As a consequence of this equation,

$$P(x_1 < X < x_2) = \int_{x_1}^{x_2} f(x)\, dx = F(x_2) - F(x_1) \tag{9-13}$$

$f(x)$ can be found by differentiating Eq. (9-12) to yield

$$\frac{dF(x)}{dx} = f(x) \tag{9-14}$$

if the derivative exists. Note that this operation can be used for discrete random variables if impulses are permitted. Thus a connection between $f(x)$ and $F(x)$ for discrete and continuous random variables has been made.

Example 9-8 Assuming a uniform density function with $a = -1$ and $b = 1$, determine the probability

$$P(0 < X < 1/2)$$

SOLUTION

$$P(0 < X < 0.5) = \int_{0}^{0.5} 1/2\, dx = (1/2)(0.5 - 0.0)$$

$$= 0.25$$

Sometimes two or more random variables are involved in an experiment. One desires to determine the joint probability distribution for their simultaneous occurrence. Notationally then the probability function is $f(x, y)$ for discrete random variables X and Y and

$$f(x, y) = P(X = x, Y = y) \tag{9-15}$$

The *joint probability mass function* $f(x, y)$ for two discrete random variables satisfies

$$f(x, y) \geq 0$$

$$\sum_{x} \sum_{y} f(x, y) = 1$$

and

$$P[(X \leq x)(Y \leq y)] = \sum_{x' \leq x} \sum_{y' \leq y} f(x', y') \tag{9-16}$$

or

$$P(E \cap E') = \sum_{x \in E} \sum_{y \in E'} f(x, y)$$

For continuous random variables, the *joint probability distribution* relationships are

$$f(x, y) \leq 0$$

$$\iint_{-\infty}^{+\infty} f(x, y)\,dx\,dy = 1$$

$$P[(X \leq x)(Y \leq y)] = \int_{-\infty}^{x} \int_{-\infty}^{y} f(x', y')\,dx'\,dy' \tag{9-17}$$

These relationships can easily be extended for any number of random variables.

If the joint probability distribution for two random variables X, Y is known, the individual distributions of the single random variables X and Y can be found. These distributions are called *marginal probability distributions* and are denoted by $g(x)$ and $h(y)$, respectively. For discrete random variables,

$$g(x) = \sum_{y \in E'} f(x, y)$$

$$h(y) = \sum_{x \in E} f(x, y)$$

Similarly, for continuous random variables,

$$g(x) = \int_{-\infty}^{+\infty} f(x, y)\,dy$$

$$h(y) = \int_{-\infty}^{+\infty} f(x, y)\,dx$$

Two random variables are said to be *independent* if their joint distribution may be written as a product of the marginal distributions; that is, for independent X and Y it can be written as

$$f(x, y) = g(x)h(y)$$

or, in set notation,

$$P(E \cap E') = P(E)P(E')$$

These relationships can easily be extended for any number of random variables.

9-4 ESTIMATING RANDOM VARIABLE PARAMETERS[35]

From an empirical point of view, the determination of the average or expected value of a random variable (r.v.) is found by conducting the standard experiment; i.e., determine x for each experiment, add them all together, and divide by the number of experiments. For example, consider the number of heads obtained as a result of tossing a coin. The empirical data for 27 experiments yield 12 heads. Thus the average number of heads is $12/27 = 0.44$. Of course, the expected value of heads is 0.5, assuming a balanced coin. This value of 0.5 is expected after a very large number of experiments or essentially in the limit as the number of experiments increase without bound.

In general, the functional forms of the expected-value operators or *first moments* are

$$E(X) = \begin{cases} \sum_x x^1 f(x) & \text{if } X \text{ is a discrete r.v.} \quad (9\text{-}18) \\ \int_{-\infty}^{+\infty} x^1 f(x)\, dx & \text{if } X \text{ is a continuous r.v.} \quad (9\text{-}19) \end{cases}$$

These equations represent the average value or mean value expected after an unlimited number of experiments. The *nth moment* refers to the exponent of x in Eq. (9-18) or Eq. (9-19). The meaning of the term relates to similar integrals used in introductory physics (mechanics).

Example 9-9 Using Eq. (9-18) in a single coin-tossing experiment, the expected value of heads is

$$E(X) = 0(1/2) + 1(1/2) = 0.5$$

Example 9-10 Determine the expected value of the uniform density function of Example 9-8.

SOLUTION

$$E(X) = \int_{-1}^{+1} \frac{x}{2}\, dx = \left. \frac{x^2}{4} \right|_{-1}^{+1} = \frac{1}{4} - \frac{1}{4} = 0$$

Extending this expected-value operation to determining the expected value of a function $h(X)$, the relationship is

$$E[h(X)] = \begin{cases} \sum_x h(x) f(x) & X \text{ discrete r.v.} \quad (9\text{-}20) \\ \int_{-\infty}^{+\infty} h(x) f(x)\, dx & X \text{ continuous r.v.} \quad (9\text{-}21) \end{cases}$$

If X and Y are two independent random variables, then

$$E(XY) = E(X)E(Y)$$

If they are not independent, then the density function relating to the marginal probability distributions is

$$f(x, y) \neq g(x)h(y)$$

and by definition the double integral cannot be separated.

Another useful concept is the *mean square* value or *second moment* which is defined as

$$E(X^2) = \int_{-\infty}^{+\infty} x^2 f(x)\, dx \tag{9-22}$$

This general form also is proportional to the energy in a given random variable. The structure of the two forms, Eqs. (9-18) and (9-19), can be formalized to the definition of moments; i.e., the nth moment about the origin of X is

$$E(X^n) = \begin{cases} \sum_x x^n f(x) & X \text{ discrete r.v.} \quad 9\text{-}23) \\ \int_{-\infty}^{+\infty} x^n f(x)\, dx & X \text{ continuous r.v.} \quad (9\text{-}24) \end{cases}$$

Letting $m_x = E(X)$, the equation

$$E[(X - m_x)^2] = \begin{cases} \sum_x (x - m_x)^2 f(x) & X \text{ discrete r.v.} \quad (9\text{-}25) \\ \int_{-\infty}^{+\infty} (x - m_x)^2 f(x)\, dx & X \text{ continuous r.v.} \quad (9\text{-}26) \end{cases}$$

defines the *variance* $\sigma^2 = E[(X - m_x)^2]$, or *second central moment*. Note that manipulation of Eq. (9-25) or Eq. (9-26) yields

$$\sigma^2 = E(X^2) - m_x^2 \tag{9-27}$$

where σ is defined as the *standard deviation* statistical measure. The variance by definition is the second moment about the mean. The variance is useful because it contains information about outcomes being close to the expected value or *mean* m_x. The smaller the variance the closer the observed outcomes (values) are to the mean on the average.

Example 9-11 What is the variance of the uniform density function with $a = -1$ and $b = 1$?

SOLUTION Since $m_x = 0$ from Example 9-10,

$$\sigma^2 = E(X^2) - 0$$

$$= \int_{-1}^{+1} \frac{x^2}{2}\, dx = \frac{x^3}{6}\bigg|_{-1}^{+1} = \frac{1}{6} - \left(-\frac{1}{6}\right) = \frac{1}{3}$$

9-5 RANDOM PROCESSES (STOCHASTIC PROCESSES)[34]

So far only random variables with one dimension have been considered. Another dimension can be added, however, and that dimension is time. Random functions are now characterized in terms of both values and time as normally found in digital control systems. The new temporal parameter is associated with the specific time kT that an experimental observation is made. To gain insight into this general definition, consider the flipping of a coin N times and let the random variable X_k be the outcome of the kth toss. Then the specific values for each random variable observed are defined as

$$x(1), x(2), \ldots, x(k), \ldots, x(N)$$

As an example, consider observed values in the form of a sequence, i.e.,

$$01101101110\ldots$$

The number of possible sequences in this example is 2^N. This collection of sequences is defined as the *ensemble* of all possible outcomes.

If the sequence is considered to be infinite ($-\infty < k < +\infty$), it has infinite energy. Since by definition the $\mathscr{Z}$ transform does not exist for this sequence, the use of distribution functions and expected-value operators are appropriate for modeling the random process.

The set of random variables indexed on k, $\{X_k\}$, where $-\infty < k < +\infty$, along with the probabilistic distribution $f(x_k, k)$ for each random variable and all the joint probabilistic distributions define a *discrete random process*. For x_k a specific value, the PDF is defined as

$$f(x_k, k) \equiv P(X_k = x_k) \qquad k = 0, 1, \ldots \tag{9-28}$$

In using this discrete random process definition in a digital control system, a specific sequence of values $x(k)$, $k = 0, 1, \ldots$, is considered to represent one and only one of the ensemble of associated sample sequences. Using the previous definition of distributions, moments can be estimated that are also indexed on k. Thus

$$F(x_k, k) = P(X_k \le x_k)$$

$$= \sum_{x'_k \le x_k} f(x'_k, k) \qquad X_k \text{ discrete r.v.} \tag{9-29}$$

or

$$F(x_k, k) = \int_{-\infty}^{x_k} f(x'_k, k)dx'_k \qquad X_k \text{ continuous r.v.} \tag{9-30}$$

and for X_k a discrete random variable, the *ensemble average* for fixed k is

$$E(X_k) = \sum_{x_k} x_k f(x_k, k) = m_x \tag{9-31}$$

For a continuous random process $X(t)$, the temporal parameter is t. For a specific instant t', $X(t')$ is a random variable. The set of random variables $\{X(t')\}$ along with the PDF $f(x', t)$ for each random variable and all the joint probabilistic distributions define a *continuous random process*. x' is a specific value such that

$$F(x', t) = P\{X(t') \le x'\} \tag{9-32}$$

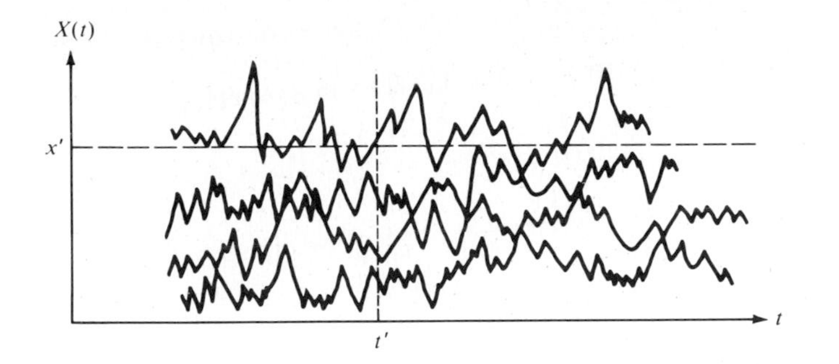

Figure 9-6 Example of continuous random variables for a given random process $X(t)$.

$$f(x', t') = P\{X(t') = x'\} = \frac{\partial F(x', t')}{\partial x} \tag{9-33}$$

The collection of all possible time functions is known as the *ensemble* of all possible outcomes. A specific function of this ensemble is denoted by $x(t)$. Figure 9-6 presents a continuous random process $X(t)$ represented by four example random variables from the set $\{X(t')\}$ associated with four $X(t)$ functions, and the relationships between x' (a given value) and a fixed t, t'. Note that $X(t')$ is a continuous random variable.

Figure 9-7 represents an example PDF $f(x, t')$ for a random variable $X(t')$ that can take on any continuous value. Note that t is fixed at t'.

For a continuous random variable X_k (observed at discrete time kT), the PDF is

$$f(x_k, k) = \frac{\partial F(x_k, k)}{\partial x_k} \tag{9-34}$$

The relationship between two random variables X_k and X of a discrete random process is defined by a *joint probability cumulative distribution function*; i.e.,

$$P(X_k \leq x_k, Y_l \leq y_l) = F(x_k, k, y_l, l) \tag{9-35}$$

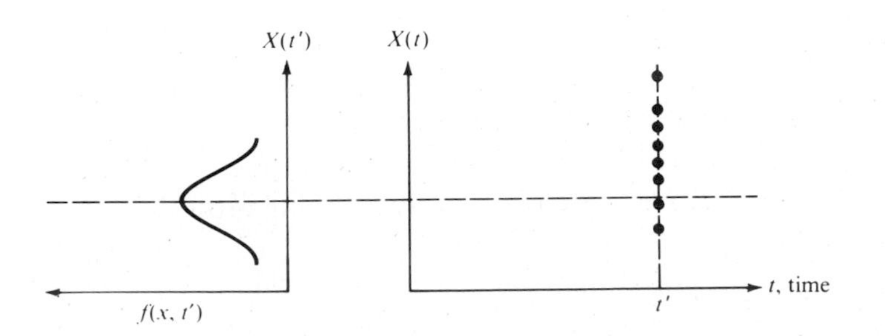

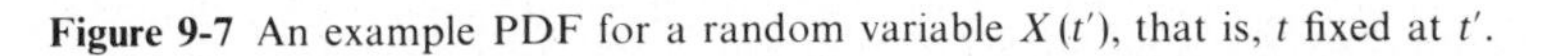

Figure 9-7 An example PDF for a random variable $X(t')$, that is, t fixed at t'.

and the *joint probability density (distribution) function or mass function* for a discrete random process is

$$P(X_k = x_k, Y_l = y_l) = f(x_k, k, y_l, l) \tag{9-36}$$

If

$$F(x_k, k, y_l, l) = F(x_k, k)F(y_l, l)$$

then the random variables are *statistically independent*. Note the difference between linearly independent and statistically independent. This is the case in the coin-tossing experiment, Example 9-11.

To characterize completely a random process, all possible joint probability distributions must be known. For some random processes, all associated probability functions are not a function of a time shift of the origin. Such a random process is said to be *stationary*. The mean and variance of a stationary random process are constants and the *second order or joint distribution* is also independent of the time shift; i.e.,

$$F(x_{k+m}, k+m, y_{l+m}, l+m) = F(x_k, k, y_l, l) \tag{9-37}$$

for random processes $\{X_k\}$ and $\{Y_l\}$. The order of a distribution refers to the number of random variables indexed. Is the joint distribution independent of time for the coin-tossing example? Yes. In a stationary random process, the mean is the same for all random variables.

Other probabilistic functions provide additional information about a random process besides the previous moment structures. For example, the *autocorrelation sequence* (second-order)

$$\begin{aligned} \Phi_{xx}(k, l) &= E(X_k X_l) \\ &= \sum_{x_k} \sum_{x_l} x_k x_l f(x_k, k, x_l, l) \qquad X_k \text{ discrete r.v.} \end{aligned}$$

or

$$\Phi_{xx}(k, l) = \iint_{-\infty}^{+\infty} x_k x_l f(x_k, k, x_l, l)\, dx_k\, dx_l \qquad X_k \text{ continuous r.v.} \tag{9-38}$$

The autocorrelation sequence defines the degree of correlation or relationship between random process values at different times. The *autocovariance sequence*, a two-dimensional sequence, is defined as

$$\Psi_{xx}(k, l) = E[(X_k - m_k)(X_l - m_l)] \tag{9-39}$$

Note that

$$\Psi_{xx}(k, l) = \Phi_{xy}(k, l) - m_k m_l \tag{9-40}$$

for a random process $\{X_k\}$.

The relationship or dependence between two different random processes is the *cross-correlation sequence*

$$\begin{aligned} \Phi_{xy}(k, l) &= E(X_k Y_l) \\ &= \sum_{x_k} \sum_{y_l} x_k y_l f(x_k, k, y_l, l) \qquad X_k \text{ discrete r.v.} \end{aligned} \tag{9-41}$$

or

$$\Phi_{xy}(k, l) = \iint_{-\infty}^{+\infty} x_k y_l f(x_k, k, y_l, l)\, dx_k\, dy_l \qquad X_k \text{ continuous r.v.} \tag{9-42}$$

The *cross-covariance* function is then

$$\begin{aligned}\Psi_{xy}(k, l) &= E[(X_k - m_k)(Y_l - m_l)] \\ &= \Phi_{xy}(k, l) - m_k m_l \end{aligned} \tag{9-43}$$

Returning to the concept of a stationary random process, the first-order statistics (first moments) are independent of time, i.e., the mean and variance. The autocorrelation sequence, which is a second-order statistic, satisfies

$$\Phi_{xx}(k, l) = \Phi_{xx}(k, k + m) = \Phi_{xx}(m) \tag{9-44}$$

This is a one-dimensional sequence now and is only a function of the time difference m.

If a given random process has a constant mean and the autocorrelation sequence is only a one-dimensional sequence, then the process by definition is *stationary in the wide sense*. If all random process joint probability distributions are true time-invariant, then the process is *stationary in the strict sense*.

A final note: If all random variables of a stationary random process are linearly independent, then

$$\Phi_{xx}(m) = \delta(m) \qquad \text{a delta function} \tag{9-45}$$

9-6 RANDOM PROCESS TIME AVERAGES

In applying a random process model to a specific digital-control-system signal, one of many ensemble distributions can be selected. This selection can be based on an intuitive understanding of the underlying process or through empirical measurements. The generation of empirical means, variances, and so forth, can be accomplished by using simple arithmetic on a large number of measurements. For example, a *time average* of a discrete random process $X(k)$ can be defined as

$$\langle X(k) \rangle = \lim_{N \to \infty} \frac{1}{2N + 1} \sum_{l=-N}^{N} x(l) \tag{9-46}$$

Likewise the variance is defined as

$$\langle \{X(k) - \langle X(k) \rangle\}^2 \rangle = \langle X^2(k) \rangle - \langle X(k) \rangle^2 \tag{9-47}$$

where

$$\langle X(k) \rangle^2 = \lim_{N \to \infty} \frac{1}{2N + 1} \sum_{l=-N}^{+N} x^2(l) \tag{9-48}$$

Continuing, the time autocorrelation sequence is

$$\langle X(k)X(k+m)\rangle = \lim_{N\to\infty} \frac{1}{2N+1} \sum_{l=-N}^{+N} x(l)x(l+m) \tag{9-49}$$

In all the time average equations, the limit exists if the random process has a finite mean and is stationary. If these time averages for almost all the possible sample sequences are equal to the same constant, and that constant is the modeled ensemble average, then the underlying random process is said to be *ergodic*. For an ergodic random process,

$$\langle X(k)\rangle = E(X_l) \qquad \text{for all } k, l \tag{9-50}$$

That is, the temporal (time) average is equal to the ensemble average for any k, l. Also,

$$\langle X(k)X(k+m)\rangle = \Phi_{xx}(m) \tag{9-51}$$

Note then that the time averages can be calculated from any random single-sample sequence of a discrete ergodic process. The generation of ensemble averages through time measurements then requires the assumption of an underlying ergodic process.

In reality, N can be quite large, but not infinity. Thus, if the time average values and assumed ensemble average values are relatively equal, then the ergodicity assumptions can be verified. How close the values should be depends upon the degree of confidence desired.

For a continuous random process, the time average moments are defined as

$$\langle X^p(t)\rangle = \lim_{T\to\infty} \int_{-T}^{T} x^p(t)\,dt \qquad p = 1, 2, \ldots \tag{9-52}$$

and if the process is ergodic, then

$$\langle X(t)\rangle = E\{X(t')\} \qquad \text{for any } t' \tag{9-53}$$

that is, $E\{X(t')\} = \langle x(t)\rangle$

and so forth, for the other averages. Figure 9-8 portrays the situation where each $x(t)$ of $X(t)$ has the same PDF.

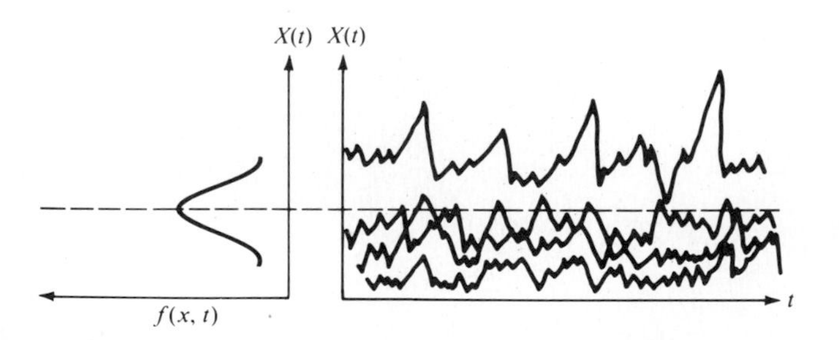

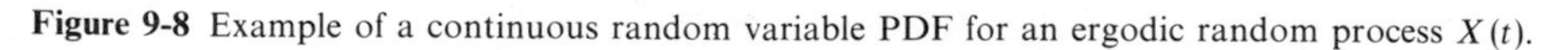

Figure 9-8 Example of a continuous random variable PDF for an ergodic random process $X(t)$.

9-7 LINEAR SYSTEM RESPONSE TO RANDOM SIGNALS

The major objective of this chapter is to develop a means of treating a random signal source in a digital control system. The control system, of course, is modeled as a linear system as discussed previously. Thus a relationship is desired that transforms an input random signal source to a system output response.

Knowing that the input signal source is a deterministic function of time, the linear system output response can be defined by using time-, s-, w-, or z-domain techniques. Now, random process signals can be represented through the use of averages that can also be operated upon by a linear system.

Consider a system $h(k)$ with an input from a wide-sense stationary discrete random process $X(k)$. The specific input $x(k)$ is a sample sequence of $\{X_k\}$. Using the standard time-domain convolution formula for a linear system, the output is

$$y(k) = \sum_{m=-\infty}^{+\infty} h(k-m)x(m) = \sum_{m=-\infty}^{+\infty} h(m)x(k-m) \tag{9-54}$$

Assuming the system is stable, the output is bounded if $x(k)$ is bounded. However, the exact value of $x(k)$ is unknown due to the random process model of the input. Thus, expectation or average operators are used. Since a wide-sense stationary model has been chosen, the mean and autocorrelation averages are employed for both the input and output. Since the input is a random process, then the output is a random process, $\{Y_k\}$. For a fixed k, Eq. (9-54) becomes

$$Y_k = \sum_{m=-\infty}^{+\infty} h(m)X_{k-m} \tag{9-55}$$

where Y_k is a random variable. Using Eq. (9-37) then for all k,

$$\begin{aligned} E(Y_k) &= \sum_{m=-\infty}^{+\infty} h(m)E(X_{k-m}) = \sum_{m=-\infty}^{+\infty} h(m)E(X_k) \\ &= m_x \sum_{m=-\infty}^{+\infty} h(m) \end{aligned} \tag{9-56}$$

Assuming then that the random process is stationary, the linear system output averages can be calculated. Returning to the autocorrelation function relating input to output,

$$\begin{aligned} \Phi_{yy}(k, k+l) &= E(Y_k Y_{k+l}) \\ &= E\left[\sum_{m=-\infty}^{+\infty} \sum_{n=-\infty}^{+\infty} h(m)h(n)x(k-m)x(k+l-n)\right] \\ &= \sum_{m=-\infty}^{+\infty} \sum_{n=-\infty}^{+\infty} h(m)h(n)E(X_{k-m}X_{k+l-n}) \\ &= \sum_{m=-\infty}^{+\infty} \sum_{n=-\infty}^{+\infty} h(m)h(n)\Phi_{xx}(m+l-n) \end{aligned} \tag{9-57}$$

where $\Phi_{yy}(k, k + l) = \Phi_{yy}(l)$ since Φ_{xx} represents a stationary random process. With a change of variable, $i = n - m$, then Eq. (9-57) becomes

$$\Phi_{yy}(l) = \sum_{i=-\infty}^{+\infty} \Phi_{xx}(l - i) \sum_{m=-\infty}^{+\infty} h(m)h(i + m) \tag{9-58}$$

Note that the second term of Eq. (9-58) is the discrete convolution of $h(k)$ with $h(-k)$. Therefore, since all the terms of Eq. (9-58) are finite and the associated $\mathscr{Z}$ transform exists, then

$$\Phi_{yy}(z) = H(z)H(z^{-1})\Phi_{xx}(z) \tag{9-59}$$

Also, from Eq. (9-55) and the definition of the $\mathscr{Z}$ transform,

$$E(Y_k) = H(z)E(X_k)|_{z=\varepsilon^{0T}=1} \tag{9-60}$$

The first- and second-order statistics from Eqs. (9-59) and (9-60) for the output of a linear system are formulated based upon a random input modeled by a wide-sense stationary random process. Of course, other output averages can also be generated by using the derivations and definitions of the previous sections. By assuming that the input process is ergodic, the observed output mean of the system from Eq. (9-50) is

$$\langle y(k) \rangle = E(Y_l) \qquad \text{for all } k, l \tag{9-61}$$

9-8 SUMMARY

The development of random variable theory is based on a fundamental understanding of probability. The mean, variance, and autocorrelation expectation operators are easily transferred from random variable applications to a random process environment. In digital control systems, random signals occur at various levels of observation. Their incorporation into a system model generally depends upon the performance accuracy required. For example, in the next chapter, the effect of finite word length is modeled by a random process. Other areas of interest include the modeling of unknown system disturbances, improper modeling of system coefficients, and extensive pretesting of system performance. Many excellent references [33-38] exist on random processes and should be consulted for a more theoretical understanding.

CHAPTER

TEN

ANALYSIS OF FINITE WORD LENGTH[11,39]

10-1 INTRODUCTION

In implementing digital controllers, compensators, or filters, the effect of the computer's finite word length can be appreciable. The time and frequency responses, for example, may fail to meet the desired requirements after digital realization. Most of the previous development concerning analysis and design of digital controllers essentially assumed infinite accuracy not only of the conversion process but of filter coefficient realization and the associated arithmetic operations. Errors due to finite word length are generated from a number of sources. The continuous analog plant measurements are transformed into discrete binary number representations with a small finite number of digits. The binary coefficients of the digital controller are usually transformations of decimal numbers with relatively large numbers of digits based upon the theoretical design procedures of the previous chapters. The multiplication and addition operations in executing the digital controller also add more error sources which must be analyzed. To determine the effect of each error source, an error model must be developed and justified.

On the other hand, it could always be assumed that a computer with the required word length could be obtained to achieve the desired performance. But what is the relationship between word length and performance to determine the required length? Again, an error model has to be generated.

In a *recursive filter* or controller realization, current and past values of the output and input are stored until the next iteration (execution of the filter). To determine the next controller output, multiplication and addition operations are executed. The multiplication operation continually generates twice as many product bits as the multiplier and multiplicand (assuming they are equal). To retain all product information from the beginning, the number of product bits in the various filter terms increases rapidly. Note that it is normally assumed that all filter coefficients are less than 1 as well as input data in scaled units. The product of two binary numbers then is also always less than 1. Because of finite-word-length limitations and limited computer memory, product results are quantized during each controller iteration. The quantization is generally based upon a truncation operation or a rounding operation. Each of these two approaches has a different error model as is shown later.

The addition operation error is related directly to the overflow or underflow of bits in the finite-word-length representation of sums. The implementer must scale the numerical operators of the controller so that performance degradation does not occur due to overflow, underflow, or loss of bit information.

The purpose of this chapter then is to present the error models for the conversion process, the multiplication quantization, and the filter coefficient quantization. The addition quantization is discussed in terms of proper scaling. Generally, the conversion and multiplication errors are modeled as random "noise" or additional controller inputs. The coefficient error models are represented deterministically in terms of pole-zero migration and associated sensitivity to quantization. Limit-cycle oscillation is also presented as a function of the nonlinear quantization operation.

10-2 CONVERTER ERRORS

As shown in Fig. 10-1, the block diagram model of an analog-to-digital (A/D) converter consists of the ideal sampler, possibly a zero-order hold, a quantizer, and an encoder. Note the use of amplitude-time notation from Chap. 1. The sampler generates discrete values (usually voltage) that can assume any continuous level (value) within the specified input range of the converter. Note that the ZOH holds the sampled input until the conversion is complete. An infinite number of bits are required to represent any value within this range. The sampled values are thus quantized to within the quantization level of the converter. Usually the sampled values can be rounded up or down to the nearest quantization level as defined by the word length. The specific level value is then encoded into a unique binary number within the output range of the finite word length of the converter. Note from Sec. 8-5 that the quantization and encoding processes can execute concurrently in a feedback organization. Also, if the analog input signal is "slow" enough, the ZOH is not required (remember the discussion of Chap. 8).

In any case, the ZOH as part of the computer input conversion process is usually modeled as a "1" (unit gain) since the hold time is much less than the sampling time. Note also that if the input analog signal is above or below the converter range, then "clipping" takes place and additional distortion occurs between the continuous and digital domains.

Returning to the quantizer model, the use of rounding or truncation to

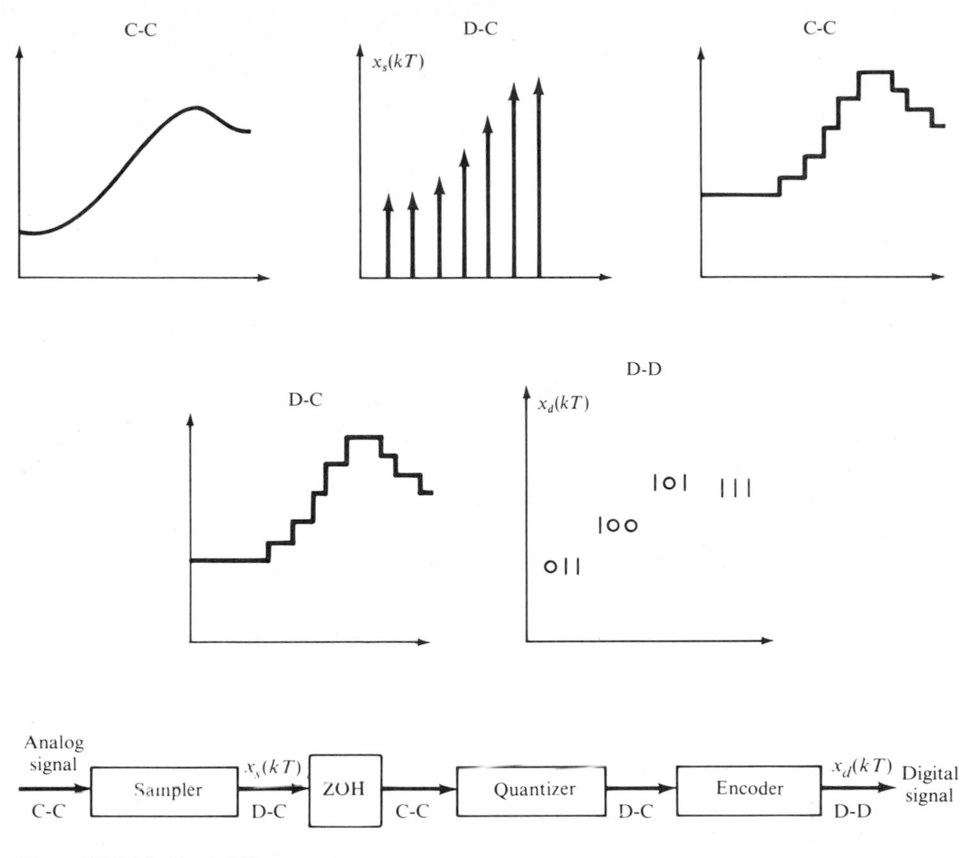

Figure 10-1 Idealized A/D converter.

determine a discrete quantization level is the primary operational characteristic. Figure 10-2 presents the quantizer input-output characteristic for rounding, and Fig. 10-3 depicts the quantizer characteristics for truncation. Note the nonlinear aspects of the quantizer in both cases.

Assuming an M-bit word length, the encoded binary value of the sampled signal and the number of quantization levels are different depending upon the binary number representation used. For example, in sign plus magnitude encoding, $2^M - 1$ quantization levels are available because of positive zero and negative zero defining one quantization level. For sign plus 2s complement, 2^M levels exist and the binary number $1000\cdots$ represents the most negative level. In sign plus magnitude, $1111\cdots$ represents the most negative level. Figure 10-4 shows a rounding quantizer characteristic for 3 bits using sign plus 2s complement encoding.

The difference between the exact sampled signal $x_s(kT)$ and the quantized-encoder signal $x_d(kT)$ is defined as the error $e(kT)$; that is,

$$e(kT) \equiv x_d(kT) - x_s(kT) \tag{10-1}$$

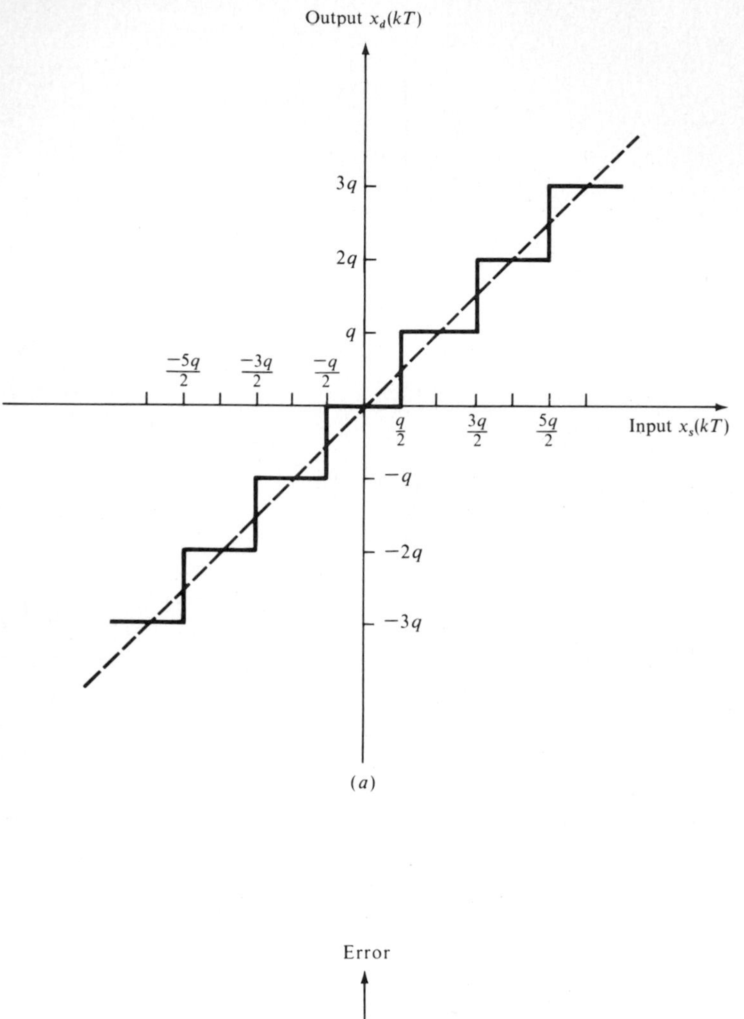

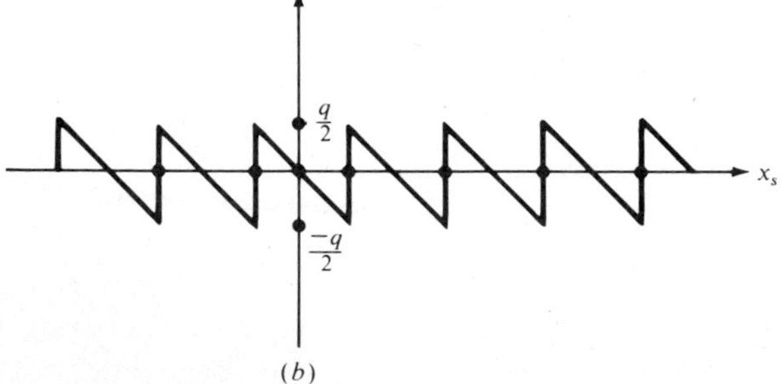

Figure 10-2 (*a*) Rounding input-output relation for a rounding A/D converter assuming quantization levels of nq, $n = 0, +1, \ldots$. (*b*) Rounding error vs. input.

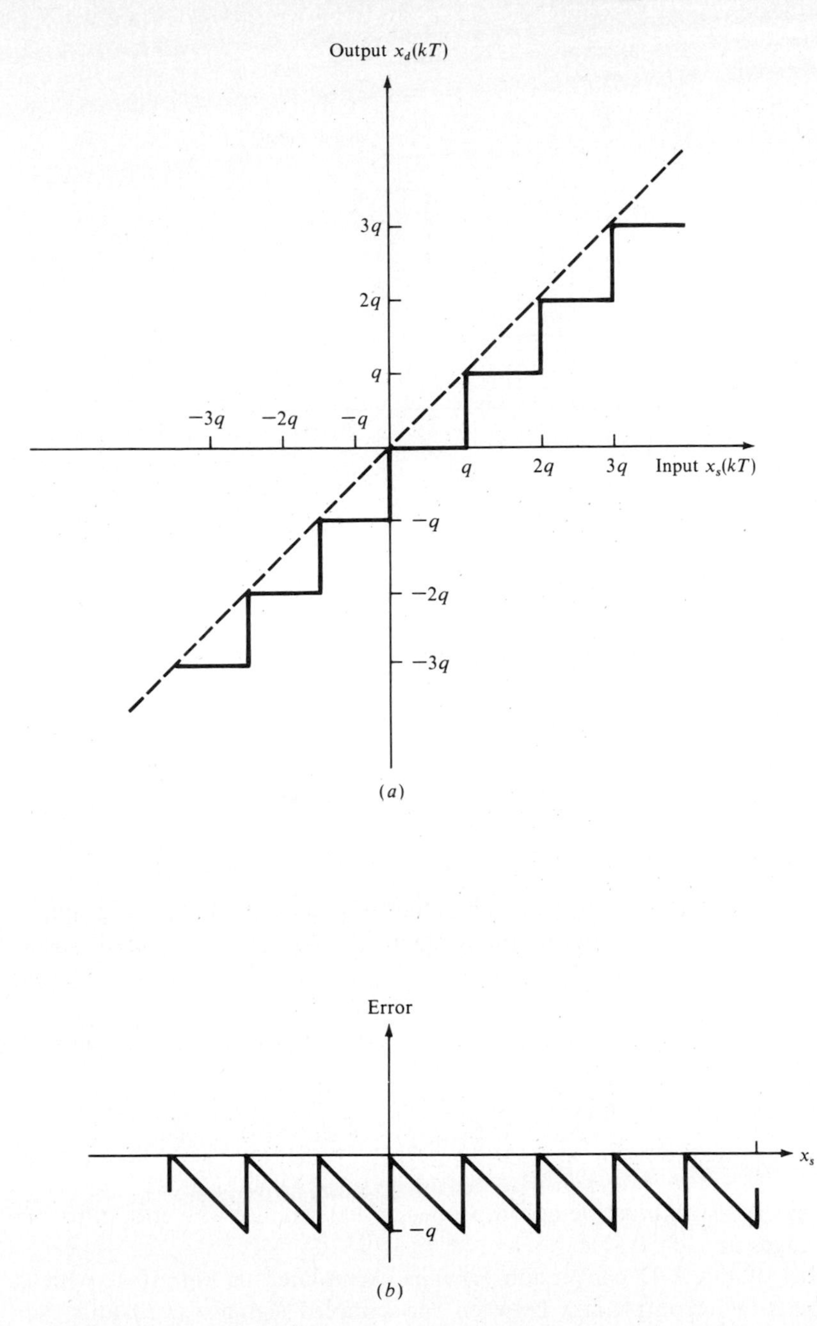

Figure 10-3 (*a*) Truncation (sign plus 2s complement) input-output relation for a truncation A/D converter assuming quantization levels of nq, $n = 0, +1, \ldots$. (*b*) Truncation error vs. input.

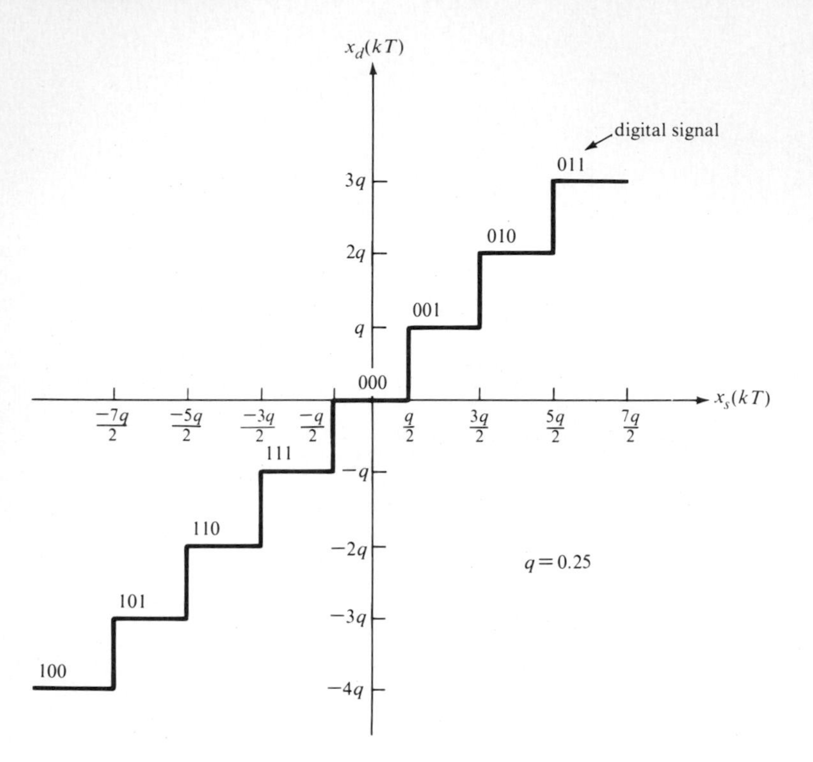

Figure 10-4 Rounding quantizer characters for sign plus 2s complement encoding with 3 bits.

As before $x_s(kT)$ is assumed to be sampled values of a band-limited analog signal with infinite accuracy. For rounding, the range of the A/D converter error for a particular sample is

$$\frac{-q}{2} < e(kT) \leq \frac{q}{2} \qquad \text{with } q = 2^{-M+1} \text{ and } k = 0, 1, \ldots \tag{10-2}$$

The error range for truncation is

$$q < e(kT) \leq 0 \tag{10-3}$$

Some additional analysis must be accomplished to determine if the endpoints of these error ranges are achievable.

The model of the A/D conversion error is exemplified in Fig. 10-1, which reflects a nonlinear error process between the sampled signal $x_s(kT)$ and the digitally encoded signal $x_d(kT)$. Thus, to represent $e(kT)$ as a deterministic function requires diagrams or special symbols. Also, this nonlinear function is difficult to use in the linear system models that have been used previously. Thus, a statistical error model, as shown in Fig. 10-5, in which $e(kT)$ is treated as a noise source can easily be integrated into a linear system model.

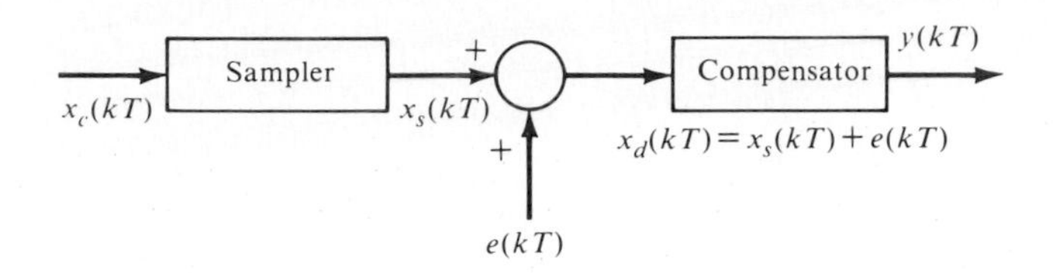

Figure 10-5 Statistical model of quantization-encoding error source $e(kT)$.

To use a statistical error model, certain statistical properties must be assumed. Since the input analog signal is assumed to be sufficiently varying in amplitude such that the error sequence $\{e(kT)\}$ is a sampled sequence of a continuous random process, the error sequence has the following properties:

1. $\{e(kT)\}$ is a sampled sequence of a *stationary random process* (assumed ergodic but not quantized here) with

$$E[e_k] = E[e(kT)] = \text{constant} = m_e \tag{10-4}$$

$$E[e_k e_{k+l}] = \Phi_{ee}(l) \tag{10-5}$$

2. The underlining error random process has a *uniform probability density function* over a finite range. The range depends upon the bounds of the quantization error and is specified in Eqs. (10-2) and (10-3).
3. The error sequence is *linearly independent* or *uncorrelated* with $\{x_s(kT)\}$:

$$E[e(kT)x_s(kT)] = E[e(kT)]E[x_s(kT)] \tag{10-6}$$

4. The error random variables are all linearly independent; that is,

$$E[e_k e_{k+l}] = \Phi_{ee}(l) = \sigma_e^2 \delta(l) \tag{10-7}$$

Thus $\{e_k\}$ is a *white-noise* random process. This white-noise model is used in many practical cases because of its form and associated easy usage. Many real-world systems reflect to some degree this model.

The exact structure of the uniform density function depends upon the type of binary number encoding used. However, for all situations, the rounding-error characteristic ranges over $-q/2$ to $+q/2$ as shown in Fig. 10-6. For sign plus 2s complement encoding with truncation, the error density function is shown in Fig. 10-7. Similarly, the density function for truncation using sign plus magnitude and sign plus 1s complement is shown in Fig. 10-8. The exact forms for each representation are further justified in the next section and in the problems.

The rationale supporting the statistical error model is based upon various experiments and theoretical investigation.[40-50] Assuming this model, the appropriate averaging operations of the last chapter are applied to yield the following results.

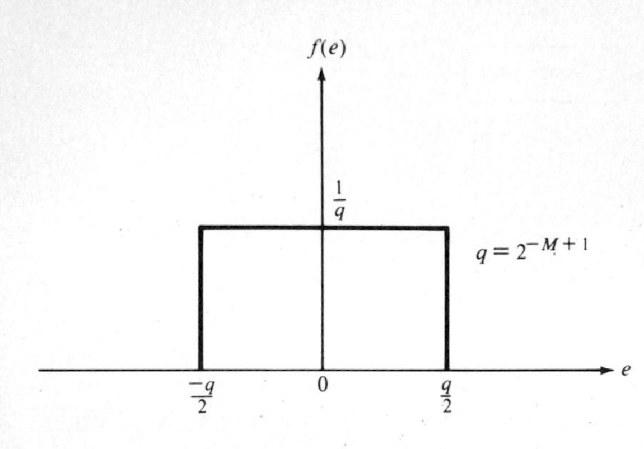

Figure 10-6 Probability density function for rounding.

For rounding (from Fig. 10-6),

$$m_e = 0$$

$$\sigma_e^2 = 2^b \int_{-2^{-b-1}}^{2^{-b-1}} x^2\, dx = \frac{2^{-2b}}{12} = \frac{q^2}{12} \tag{10-8}$$

where $b = M - 1$ and M is the value of computer word length. For truncation, using sign plus 2s complement (from Fig. 10-7),

$$m_e = \frac{-2^{-b}}{2} = \frac{-q}{2} \tag{10-9}$$

$$\sigma_e^2 = \frac{2^{-2b}}{12} = \frac{q^2}{12} \tag{10-10}$$

and

$$E[e_k e_{k+l}] = \Phi_{ee}(kT) = \frac{2^{-2b}\delta(kT)}{12} \qquad \text{for both cases} \tag{10-11}$$

The effect of this A/D quantization on output performance is easily determined by driving the discrete linear system model with the noise source

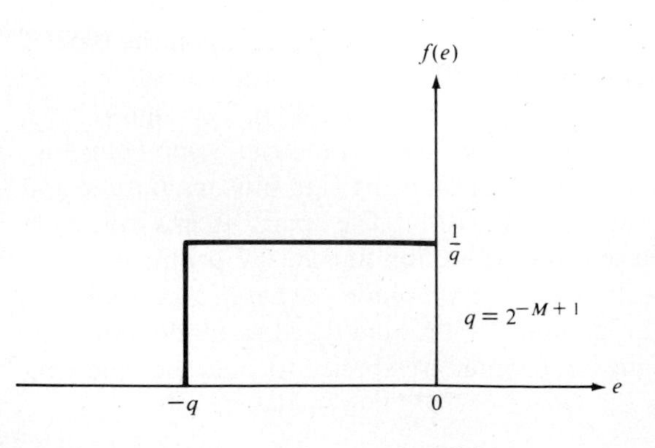

Figure 10-7 Probability density functions for truncation using sign plus 2s complement encoding.

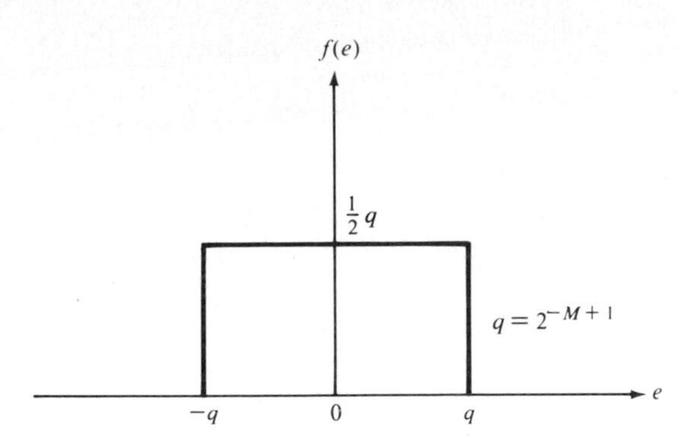

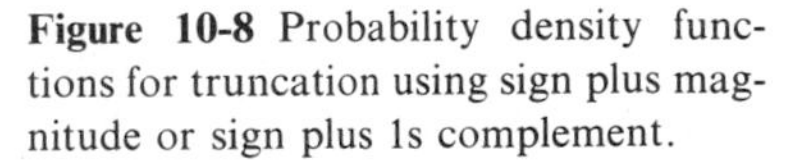

Figure 10-8 Probability density functions for truncation using sign plus magnitude or sign plus 1s complement.

model. From Sec. 9-6, the expected value [Eq. (9-5)] and variance [Eq. (9-52)] of the output y due to A/D rounding are

$$m_y = m_e H(z)|_{z=1} = 0 \tag{10-12}$$

and

$$\sigma_y^2 = \Phi_{yy}(0) = \sum_{l=-\infty}^{+\infty} \Phi_{xx}(0-l) \sum_{k=-\infty}^{\infty} h(kT)h(l+kT)$$

$$= \frac{2^{-2b}}{12} \sum_{k=\infty}^{+\infty} h^2(kT) \tag{10-13}$$

Here the voltage range of the converter is assumed to be 0 to 1. For other ranges the multiplication term $K_{A/D}$ must be included in Eq. (10-13).

Example 10-1 Find the variance of the output whose system transfer function is

$$H(z) = \frac{a}{1 - dz^{-1}} \qquad 0 < d < 1$$

for a 10-bit rounding converter, a gain of 10, with sign plus 2s complement representation and a range of +5 V.

SOLUTION Inverting $H(z)$ to obtain $h(k)$ yields

$$h(k) = ad^k u_{-1}(k) \qquad k > 0$$

$$= 0 \qquad k \le 0$$

Then, using Eq. (10-13),

$$\sigma_y^2 = \frac{2^{-18}(10)^2}{12} \sum_{k=0}^{+\infty} (ad^k)^2 = \frac{2^{-18}(10)^2}{12} a^2 \sum_{k=0}^{+\infty} d^{2k}$$

$$= \frac{2^{-18}a^2}{12} \cdot \frac{10^2}{1-d^2}$$

Example 10-2 In Example 10-1, interpret the value of σ_y^2 for $a = 1$ and $d = 0.5$.

$$\sigma_y^2 = \frac{2^{-18}(10)^2}{12(0.75)} = 4.239 \times 10^{-5}\ \text{V}$$

The overall error performance is relatively small considering that the output variance is quite small and that the mean output is zero in this example. However, the significance of the error depends upon the desired performance accuracy in a given application. The specific relationship (scale factor) between volts and the units of interest is of utmost importance in determining performance.

Example 10-3 Determine the output mean for the model of Example 10-1 with a truncating A/D. Thus $q = 2^{-9}$ and from Eq. (10-9)

$$m_y = m_e H(e^{j0}) = \frac{-10}{2^{10}} \frac{a}{1-b} = \frac{-10}{2^{10}(0.5)}$$

$$\approx -0.0195\ \text{V}$$

To evaluate more extensive forms of Eq. (10-13) involving the summation of $h^2(kT)$, it is easier to use one of *Parseval's* relationships[39] as developed from Eq. (9-54):

$$\sum_{k=-\infty}^{+\infty} h^2(kT) = \frac{1}{2\pi j} \oint_c H(z)H(z^{-1})z^{-1}dz \tag{10-14}$$

$$\sum_{k=-\infty}^{+\infty} |h(kT)|^2 = \frac{1}{2\pi} \int_{-\pi}^{\pi} |H(\epsilon^{j\omega})|^2\, d\omega \tag{10-15}$$

The contour integral of Eq. (10-14) can be evaluated by using the residue theorem; i.e., find the residues of $H(z)H(z^{-1})z^{-1}$ at the poles of $H(z)$. Note that $\sum h^2$ is proportional to energy. For a very narrow "band-limited" frequency-domain model, Eq. (10-15) can be readily evaluated as an approximate constant for determining the summation of $h^2(kT)$. Other techniques[39] also exist for evaluating Eq. (10-14).

Example 10-4 To check the accuracy of the residue approach, reconsider Example 10-2.

SOLUTION Since $H(z) = z(z - 0.5)^{-1}$, to obtain $H(z^{-1})$, replace z by z^{-1} in the expression for $H(z)$; that is,

$$H(z^{-1}) = \frac{1}{1 - 0.5z} = \frac{-2}{z - 2}$$

with the $H(z)$ pole at $z = 0.5$ and the $H(z^{-1})$ pole at $z = 2.0$. The integrand of the contour integral is

$$H(z)H(z^{-1})z^{-1} = \frac{-2}{(z - 0.5)(z - 2)} = \frac{-1/d}{(z - d)(z - 1/d)}$$

Thus, the residue is $1/(1 - d^2)$ where $d = 0.5$ and Eq. (10-12) yields

$$\sigma_y^2 = \frac{\sigma_e^2}{1 - d^2} = \frac{2^{-18}(10)^2}{12(0.75)} = 4.239 \times 10^{-5} \text{ V}$$

Example 10-5 Consider a rounding A/D converter used in a controller of the form

$$H(z) = \frac{1}{1 - 2a(\cos d)z^{-1} - a^2z^{-2}} = \frac{z^2}{z^2 - 2a(\cos d)z - a^2}$$

$$= \frac{z^2}{(z + a\epsilon^{+jd})(z - a\epsilon^{-jd})}$$

whose poles are $p_1 = -a\epsilon^{+jd}$ and $p_2 = a\epsilon^{-jd}$. What is the variance of the output error?

SOLUTION Replacing z by z^{-1} in the expression for $H(z)$ yields

$$H(z^{-1}) = \frac{1}{1 - 2a(\cos d)z - a^2z^2}$$

Applying the residue theorem to the integrand of Eq. (10-14) yields

$$\left.\frac{z}{(z - a\epsilon^{-jd})[1 - 2a(\cos d)z - a^2z^2]}\right|_{z=p_1=-a\varepsilon^{+jd}}$$

$$+ \left.\frac{z}{(z + a\epsilon^{jd})[1 - 2a(\cos d)z - a^2z^2]}\right|_{z=p_2=a\varepsilon^{-jd}}$$

Thus, substituting this result into Eq. (10-13) yields

$$\sigma_y^2 = \sigma_x^2 \frac{1 - a^2}{(1 + a^2)(1 + a^4 - 2a^3 \cos 2d)} \tag{10-16}$$

The use of the residue theorem provides a considerably easier approach for evaluating the "filtering" of an A/D conversion noise source model when $H(z)$ is known. This approach is also useful in determining the effect of multiplication errors due to rounding or truncation involved in the difference equation representation of $H(z)$.

The digital-to-analog (D/A) converter also has an impact on system performance errors. Although it too has a quantization level due to finite word length, the fact that data transferred to it is already in digital form means that no additional error is generated if the D/A word length is the same as the computer's word length. In many cases, this is not true. The computer word length may be generated by arithmetic processing requirements. The value of the D/A word length, however, must reflect the accuracy of the desired overall controller performance.

Note, that the quantization output error for a given $H(z)$ can be increased by faster sampling. For example, let

$$H(z) = \frac{1}{1 - \epsilon^{-dT}z^{-1}} \qquad h(kT) = \epsilon^{-dkT}u_{-1}(kT) \qquad k > 0$$

Assuming a truncation quantization operation with $m_r = q/2$,

$$m_c = m_r H(z)|_{z=1} = \frac{1}{1 - \epsilon^{-dT}}\frac{q}{2} = K\frac{q}{2}$$

For $d = 1$, the following table presents the increased output quantization error gain K due to decreasing sampling time T:

T, s	$\sim K$
1.000	1.6
0.1000	9.5
0.0100	99.5
0.0010	999.5
0.0001	9999.5

Therefore, as T decreases, the statistical output error due to the input truncation error increases. This result is expected since the pole of $H(z)$ is moving closer and closer to the $+1.0$ point in the z domain as $T \to 0$ and the dc gain K of $H(z)$ then increases rapidly.

10-3 ARITHMETIC OPERATIONS

In executing the digital controller in a computer, the arithmetic operations of multiplication and addition occur. *Division can also take place if the denominator coefficient of z^0 in $H(z)$ is not a 1* [see Eq. (4-98)]. Assuming, however, that this coefficient is unity, the forms in this section are used to generate an error model for the first two operations. To develop an error model, truncation and rounding processes associated with the various number representations are presented initially.

For purposes of this discussion, fixed-point representations of binary numbers are assumed along with the binary point being placed to the right of the MSB, the sign bit. In Chap. 2 the definitions for the various number representations are presented. As indicated previously, the effect of truncation or rounding depends upon how negative numbers are represented. First, however, consider positive binary numbers which are represented identically in all these representations. The definition of the error between the exact value $x_d'(kT)$ and the digital value $x_d(kT)$ is $e(kT)$; that is,

$$e(kT) \equiv x_d(kT) - x_d'(kT) \tag{10-17}$$

where x_d' is assumed to be represented by an infinite number of bits and x_d by a finite number of bits, say, M. The truncation operation in Eq. (10-17) discards all but $M - 1$ magnitude bits of x_d' to generate x_d. Then, $e(kT)$ is always less than or equal to zero. If all the bits representing x_d' beyond $M - 1$ magnitude locations are zero, then e is zero. If all the bits are one beyond that point, then e is -2^{-M+1}. Note that the truncated magnitude is always less than the magnitude of x_d' for positive numbers. This is true for all three representations.

Consider the truncation of a sign plus magnitude negative number. Since the magnitude of this negative number is reduced by truncation, the error $e(kT)$ from Eq. (10-17) is positive. If all the bits after M are zero, then the error is zero. If they are all 1s, then the error is 2^{-M+1}. Thus, for sign plus magnitude, the truncation error range is

$$-2^{-M+1} < e(kT) < 2^{-M+1} \tag{10-18}$$

For sign plus 2s complement, truncation to M bits including sign presents a different picture for negative numbers. If all the remaining bits are 0s, then the error is zero. However, if the remaining bits are all 1s, then their deletion results in an increase of magnitude for the truncated negative number because of the 2s complement operation. The maximum value of the error is 2^{-M+1}. Thus for sign plus 2s complement, the error range is

$$-2^{-M+1} < e(kT) \leq 0 \tag{10-19}$$

For sign plus 1s complement, see Prob. 10-1.

For rounding, by definition, $x_d(kT)$ is generated by determining the closest quantization level to the magnitude of $x_d'(kT)$. Thus, $x_d'(kT)$ may be "rounded up" or "rounded down" depending upon the closest quantization level. If $x_s(kT)$ is exactly halfway between two levels, then an arbitrary decision is made. Since the maximum error in either case is one-half the value of the least significant bit $2^{-M+1}/2$, the range of rounding is

$$-2^{-M} < e(kT) \leq 2^{-M} \tag{10-20}$$

Note that Eq. (10-20) assumes that rounding up of $x_d'(kT)$ occurs at a midpoint between two quantization levels. Also, Eq. (10-20) is true for all binary number representations since rounding is only based on magnitude. Specific algorithms are shown in later chapters.

In floating-point arithmetic operations, the generation of errors due to truncation or rounding is very similar to the fixed-point analysis. In fact, the error is relative to the value of the exponent in a multiplicative sense. The error in the mantissa word length is treated identically to the fixed-point rounding and truncation situation. For a rounding operation in a floating-point number with p exponent bits and M mantissa bits including sign, the error range is

$$-2^{-M+p} < e(kT) \leq 2^{-M+p} \tag{10-21}$$

Error ranges due to truncation in floating-point representations are treated in Prob. 10-11.

For multiplication operations, the error ranges can be transformed into a statistical model by generating the appropriate density function along with the appropriate random process properties. As before, the error random process is represented by a uniform density function over the appropriate quantization range depending upon number representation. Also, all the error random processes are assumed to be a linearly independent, stationary ergodic process. Each error sequence is assumed to be linearly independent of the multiplier or multiplicand as well as the other error sources. Thus, the same diagrams (Figs. 10-6 to 10-8) used previously can be employed again to model a random error source. In this case, however, an error source is generated after every multiplication.

Before considering the impact of arithmetic multiplication errors, the controller structure of the realization must be known. That is, the computational path or transfer function between the multiplication operation and the output must be explicitly defined. Various realization forms, in general, generate different transfer functions between multiplication error sources and the controller output. These various realization forms evolve from the following methods:

Direct realization
Cascade realization
Parallel realization

Direct realization is based upon the general controller transfer function

$$H(z) = \frac{N(z)}{D(z)}$$

If $N(z)$ and $D(z)$ are both nth-order polynomials, respectively, of z^{-1}, then $2^n \times 2^n$ possible distinct realizations exist. This number of distinct realizations is due to the different ways of cascading the delay elements z^{-1}, that is, preceding or following a specific multiplier.

A second-order controller with input $X(z)$ and output $Y(z)$ is described by

$$H(z) = \frac{Y(z)}{X(z)} = \frac{a_0 + a_1 z^{-1} + a_2 z^{-2}}{1 + b_1 z^{-1} + b_2 z^{-2}} = \frac{N(z)}{D(z)} \tag{10-22}$$

Figures 10-9 to 10-11 depict three of the direct realization forms with the multiplication error sources e_i shown subscripted for each error source. The error sources can be treated as statistical models via the discussion of Chap. 9. Note that the realization forms in Figs. 10-9 and 10-10 are defined as canonical since by

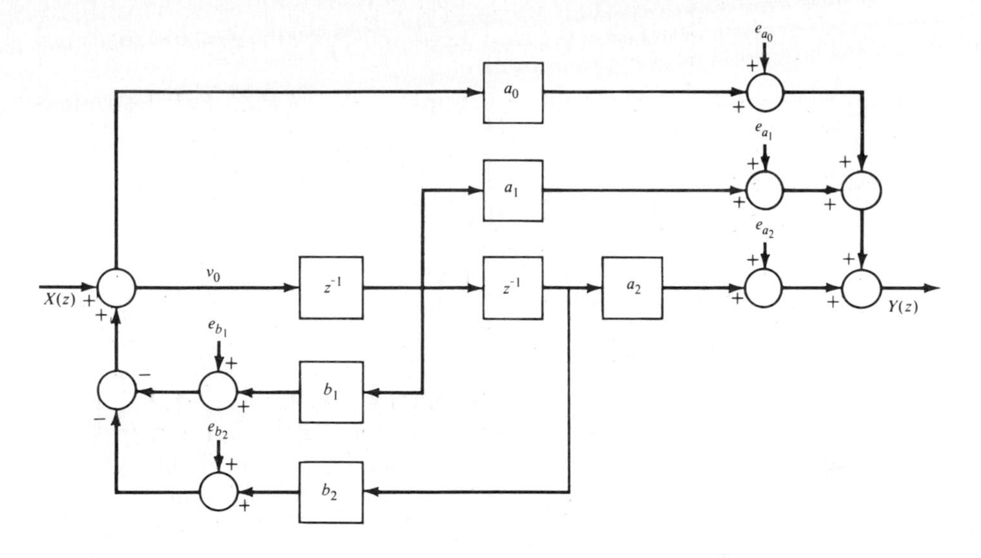

Figure 10-9 Direct canonial realization form of second-order compensator with multiplication error sources noted (1*D*).

$$H(z) = \frac{Y(z)}{X(z)} = \frac{a_0 + a_1 z^{-1} + a_2 z^{-2}}{1 + b_1 z^{-1} + b_2 z^{-2}}.$$

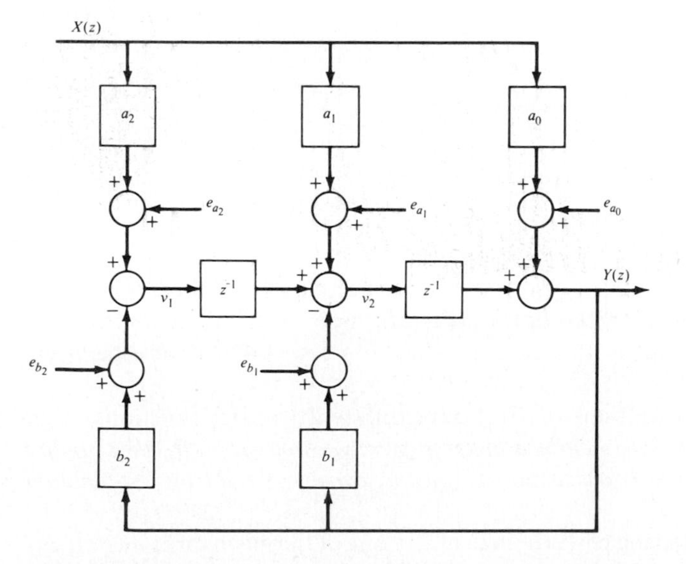

Figure 10-10 Direct canonical implementation forms of the second-order controller with multiplication error sources (2*D*).

definition the number of delays n between input and output is equal to the order of $D(z)$.

Note that the form of Fig. 10-10 is defined[101] as the first direct realization ($1D$) with the factoring

$$H(z) = \frac{a_0}{D(z)} + \frac{a_1 z^{-1}}{D(z)} + \frac{a_2 z^{-1}}{D(z)}$$

The ordered implementation transpose of this equation, that is, the a_2 term first, results in the second direct realization ($2D$) form of Fig. 10-10.

Now, considering the direct implementation of Eq. (10-22) results in the third direct realization ($3D$) of Fig. 10-11. The fourth direct realization ($4D$) is the obvious graphical transpose of the $3D$ form in the figure.

The position of the various multiplying error sources before or after the delay operators can be debated depending upon computer program implementation. Nevertheless, the three different direct implementations of $H(z)$ generate the following transfer function for each error source, i.e., Y/e_i (Chap. 4 developed the manipulative tools):

Error source	Fig. 10-9 ($1D$) implementation	Fig. 10-10 ($2D$) implementation	Fig. 10-11 ($3D$) implementation
e_{a_0}	1	$\frac{1}{D(z)}$	$\frac{1}{D(z)}$
e_{a_1}	1	$\frac{z^{-1}}{D(z)}$	$\frac{1}{D(z)}$
e_{a_2}	1	$\frac{z^{-2}}{D(z)}$	$\frac{1}{D(z)}$
e_{b_1}	$H(z)$	$\frac{z^{-1}}{D(z)}$	$\frac{1}{D(z)}$
e_{b_2}	$H(z)$	$\frac{z^{-2}}{D(z)}$	$\frac{1}{D(z)}$

$D(z) = 1 + b_1 z^{-1} + b_2 z^{-2}$

Other structural implementations also exist[101] for second-order filters where the real and imaginary parts of the complex poles are used explicitly as multipliers. The realized canonical filter in this case is of a cross-coupled structure between z^{-1} elements.

The various implementations of $H(z)$ have different transfer functions for a specific error source. *The selection of a given implementation structure then should be influenced by specific error source models (means, variance) and transfer-function coefficients.*

The relative error-filtering performance of any one of the given three implementations depends upon the pole locations of $D(z)$ and the form of the error source. If the error is considered as a constant, then the dc gain of the filter is the only value

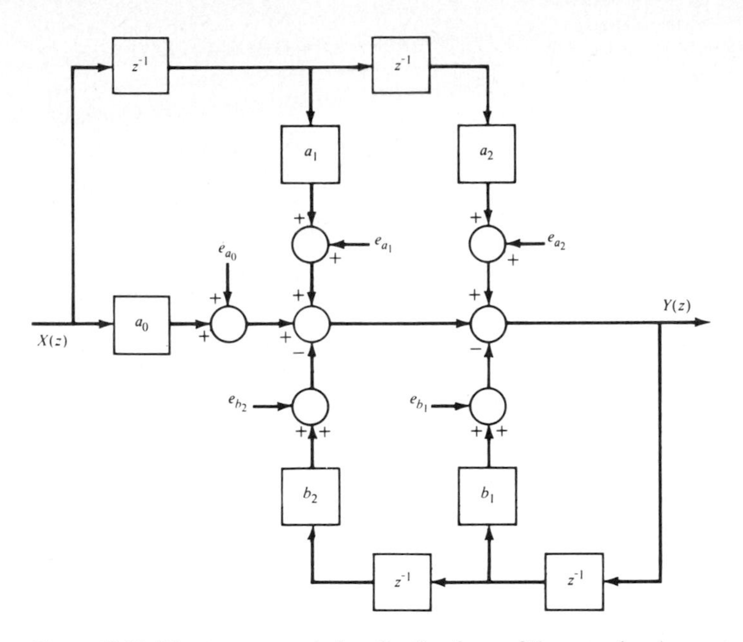

Figure 10-11 Direct noncanonical realization form of the second-order controller with multiplication error sources ($3D$).

of interest. However, if the error contains nonzero frequency components, then the bandwidth of the controllers as related to the poles and zeros helps define a frequency performance interval. Although magnitude may be the major focus of the error analysis, the delay or phase of the error may also affect the transient performance in a given digital control system. Detailed analysis is required in concert with the tools and techniques presented in Chap. 6.

Now, consider the cascading of two first-order controllers and the respective multiplication error source transfer functions. In Fig. 10-12, a second-order

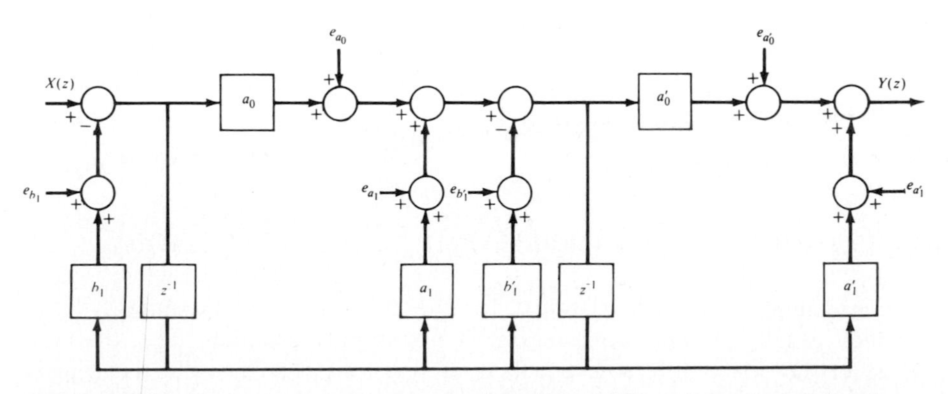

Figure 10-12 Two first-order compensators cascaded with multiplication error sources noted.

controller, having real zeros and poles, is decomposed into two first-order filters as follows:

$$H(z) = \frac{a_0 + a_1 z^{-1}}{1 + b_1 z^{-1}} \frac{a'_0 + a'_1 z^{-1}}{a + b'_1 z^{-1}} = H_1(z)H_2(z) \tag{10-23}$$

The additive error source transfer functions are:

Error source	Fig. 10-12 implementation
e_{a0}	$H_2(z)$
e_{a_1}	$H_2(z)$
e_{b_1}	$H_1(z)H_2(z)$
$e_{a'0}$	1
$e_{a'1}$	1
$e_{b'1}$	$H_2(z)$

From this simple example, it is easily seen from the table that a particular error source's effect on performance depends upon its position in the cascade representation. That is, in general, each succeeding cascade controller filters the previous error sources. Thus the gain of each succeeding filter should generally be less in a given implementation. One way of accomplishing this task is to incorporate poles and zeros that are close together[39] into a subfilter with proper subfilter placement for a cascade compensator.

The same approach is also suggested when implementing second-order controllers in a cascade organization. Since many implementations have complex poles and zeros, this form of structure is normally employed. Chapter 4 on $\mathscr{Z}$ transforms explains specific manipulative techniques for factoring.

The second-order forms can also be realized in a parallel structure (additive). Multiplication error sources then within the specific second-order transfer function directly add to the output.

A final note concerning the assumptions on the error model. The only difference between the truncation and rounding models is the difference in means, the variances are identical. Also, the statistical assumptions on the error models can be inconsistent with reality. For example, in sign plus magnitude truncation, the error is positive for positive signals, and vice versa; i.e., the error is correlated with the signal.

10-4 COMPENSATOR COEFFICIENT REPRESENTATION

In determining the coefficient values for the controller, the theoretical design approach usually assumes considerable if not infinite accuracy. But, the coefficients of the linear difference equation representing the controller have to be implemented by a finite word length. Thus the study of truncation and rounding can again be applied in this situation. In some cases the bits might be drastically changed so that only a shift is required instead of a complete multiply. A shift of course is much faster and thus decreases the time required to generate $y(kT)$. In any implementation then, the desired poles and zeros are shifted due to the finite word length. Since the exact change in the theoretical coefficient is known prior to controller execution, the exact change of the pole or zero is also known. Thus the effect of controller performance can be studied by analyzing the new value of the pole or zero on performance. As stated previously the structure of the filter implementation affects the specific coefficient values. Therefore, if a selected structure yields poor performance due to coefficient realization, then a new structure might yield better results. An important part of the overall analysis of coefficient realization is then the performance sensitivity of coefficient changes due to finite word length. Consideration is given first to pole shifts.

To study the effects of coefficient quantization, a model of the theoretical controller transfer function

$$H(z) = \frac{a_0 + a_1 z^{-1} + \cdots + a_w z^{-w}}{1 + b_1 z^{-1} + \cdots + b_n z^{-n}} = \frac{(1 - z_1 z^{-1}) \cdots (1 - z_w z^{-1})}{(1 - p_1 z^{-1}) \cdots (1 - p_n z^{-1})}$$

must be compared to the implemented transfer function

$$H'(z) = \frac{a'_0 + a'_1 z^{-1} + \cdots + a'_w z^{-w}}{1 + b'_1 z^{-1} + \cdots + b'_n z^{-n}} = \frac{(1 - z'_1 z^{-1}) \cdots (1 - z'_w z^{-1})}{(1 - p'_1 z^{-1}) \cdots (1 - p'_n z^{-1})}$$

Specifically, the changes in the roots of the numerator and denominator polynomials are desired as functions of the polynominal coefficient changes of $H(z)$. The denominator polynominal coefficients are selected for analysis; that is,

$$\mathbf{P}(z, b_1, \ldots, b_n) = 1 + \sum_{j=1}^{n} b_j z^{-j} = \prod_{i=1}^{n} (1 - p_i z^{-1}) \tag{10-24}$$

This equation is defined as the theoretical equation with p_i as a root of $\mathbf{P}(z, b_1, \ldots, b_n) = 0$ [a pole of $H(z)$]. Then

$$\mathbf{P}'(z, b'_1, \ldots, b'_n) = 1 + \sum_{j=1}^{n} b'_j z^{-j} = \prod_{i=1}^{n} (1 - p'_i z^{-1})$$

is defined as the implemented characteristic equation with p'_i as a root [a pole of $H'(z)$]. The change from the theoretical to the implemented coefficients b'_i's is assumed to be quite small such that $b'_j = b_j + \Delta b_j$ and $p'_i = p_i + \Delta p_i$. Thus

$$\begin{aligned}\mathbf{P}'(z, b_1, \ldots, b_n, \Delta b_1, \ldots, \Delta b_n) &= 1 + \sum_{i=1}^{n} (b_i + \Delta b_i) z^{-1} \\ &= \prod_{i=1}^{n} [1 - (p_i + \Delta p_i) z^{-1}]\end{aligned}$$

The original objective is to determine Δp_i as a function of Δb_j which can be described in terms of a Taylor's series expansion of p_i about Δb_j; that is,

$$\Delta z\bigg|_{z=p_i} = \sum_{j=1}^{n} \frac{\partial z}{\partial b_j}\bigg|_{z=p_i} \Delta b_j \qquad i = 1, \ldots, n \tag{10-25}$$

Only single-order poles are assumed in this analysis. For multiple poles a similar form can be generated (see Prob. 10-4). By using the chain rule concept; that is,

$$\frac{\partial \mathbf{P}}{\partial z}\bigg|_{z=p_i} \cdot \frac{\partial z}{\partial b_j}\bigg|_{z=p_i} = \frac{-\partial \mathbf{P}}{\partial b_j}\bigg|_{z=p_i} \quad \text{or} \quad \frac{\partial z}{\partial b_j}\bigg|_{z=p_i} = \frac{-\partial \mathbf{P}/\partial b_j}{\partial \mathbf{P}/\partial z}\bigg|_{z=p_i} \tag{10-26}$$

the value of $\partial z_i/\partial b_j$ can be substituted into Eq. (10-25) to determine Δp_i. To find a solution to Eq. (10-26), consider the derivatives of Eq. (10-24) which are

$$\frac{\partial \mathbf{P}}{\partial z}\bigg|_{z=p_i} = +\sum_{l=1}^{n} p_l z^{-2} \prod_{\substack{k=1 \\ k \neq l}}^{n} (1 - p_k z^{-1})\bigg|_{z=p_i} = p_i^{-1} \prod_{\substack{k=1 \\ k \neq i}}^{n} (1 - p_k p_i^{-1}) \tag{10-27}$$

and

$$\frac{\partial \mathbf{P}}{\partial b_j}\bigg|_{z=p_i} = p_i^{-j} \tag{10-28}$$

Thus, substituting into Eq. (10-25) from Eqs. (10-27) and (10-28) yields

$$\Delta p_i = -\sum_{j=1}^{n} \frac{p_i^{-j+1}}{p_i^{-n+1} \prod_{\substack{k=1 \\ k \neq i}}^{n} (p_i - p_k)} \Delta b_j = -\sum_{j=1}^{n} \frac{p_i^{n-j}}{\prod_{\substack{k=1 \\ k \neq i}}^{n} (p_i - p_k)} \Delta b_j \tag{10-29}$$

Note that in Eq. (10-29), Δp_i is very sensitive to other poles p_j that are nearby. That is, if theoretical poles in a controller are clustered together, truncation or rounding of coefficients can cause major shifts in the realized poles. The same analysis is also applicable to the numerator. The zeros can also be very sensitive to changes in coefficients if the zeros are clustered together. To reduce sensitivity, a design approach can be to change the p_i's or to minimize the Δb_j's.

Returning to Eq. (10-29), Δp_i is in reality a complex number in the z domain. Its angle and magnitude depend upon Δb_j, p_i, and the terms $p_i - p_k$, where $k \neq i$. The assumed form of implementation in this analysis uses the direct form, i.e., the implementation of the nth-order compensator. Considering the cascade or parallel form of implementation generally results in less root sensitivity due to coefficient changes. In both these second-order forms of implementation, the complex-conjugate pair is implemented independent of the others, i.e., its own set of second-order coefficients. Thus these forms are generally selected for implementation over the direct form.

Even within second-order cascade or parallel forms of implementation, different structures can be realized as shown earlier. Each of these structures can be deterministically studied in relationship to finite word length by analyzing all the possible grid locations. The grid is generated by determining all the possible points within the unit circle that can represent a pole (zero) based upon the finite number of coefficient values due to word length. Given a desired theoretical compensator, the "best" implementation is found by using a variety of analysis techniques, including simulation.

10-5 SENSITIVITY COEFFICIENTS

In the determination of Δp_i in Eq. (10-29), the change is relative to the value of p_i. What may be desired is the percentage change in p_i due to a percentage change in b_j, that is, a relationship between $\Delta p_i/|p_i|$ and $\Delta b_j/|b_j|$. With this approach, the relative coefficient influence on a particular polynomial root can be determined. The desired relationship is called the *root sensitivity*; for example,

$$S_j^i \equiv \frac{\Delta p_i/|p_i|}{\Delta b_j/|b_j|} \tag{10-30}$$

Using Eq. (10-29),

$$S_j^i = \frac{-p_i^{n-j}}{\prod_{\substack{k=1 \\ k \neq i}}^{n} (p_i - p_k)} \frac{|b_j|}{|p_i|}$$

In determining the complex value of S_j^i, however, it is sometimes just as easy to use

$$S_j^i = \left. \frac{|b_j|}{|p_i|} \frac{\partial z_i}{\partial b_j} \right|_{z=p_i} = \left. \frac{|b_j|}{|p_i|} \frac{\partial \mathbf{P}/\partial b_j}{\partial \mathbf{P}/\partial z_i} \right|_{z=p_i}$$

as derived from Eqs. (10-25), (10-26), and (10-30).

Example 10-6 Consider the $H(z)$ denominator polynomial to be

$$\begin{aligned} \mathbf{P}(z, b_1, b_2, b_3) &= 1 + 0.3034z^{-1} - 0.8789z^{-2} - 0.4245z^{-3} \\ &= (1 - z^{-1})(1 + 0.6364z^{-1})(1 + 0.6670z^{-1}) \end{aligned}$$

Generate and analyze the sensitivity coefficients. Let $b_1 = 0.3034$, $b_2 = -0.8789$, $b_3 = 0.4245$, $p_1 = 1.0$, $p_2 = -0.6364$, and $p_3 = -0.6670$. From the definition of $\mathbf{P}$, the following derivatives are generated:

$$\begin{aligned} \frac{\partial \mathbf{P}}{\partial z} &= z^{-2}(1 + 0.6364z^{-1})(1 + 0.6670z^{-1}) \\ &\quad - 0.6364z^{-2}(1 - z^{-1})(1 + 0.6670z^{-1}) \\ &\quad - 0.6670z^{-2}(1 - z^{-1})(1 + 0.6364z^{-1}) \end{aligned}$$

$$\frac{\partial \mathbf{P}}{\partial b_1} = z^{-1}$$

$$\frac{\partial \mathbf{P}}{\partial b_2} = z^{-2}$$

$$\frac{\partial \mathbf{P}}{\partial b_3} = z^{-3}$$

For the coefficient b_1, the partial derivatives in Eq. (10-26) are

$$\left.\frac{\partial z}{\partial b_1}\right|_{z=p_1} = \left.\frac{-\partial \mathbf{P}/\partial b_1}{\partial \mathbf{P}/\partial z}\right|_{z=p_1} = \frac{-p_1^{-1}}{p_1^{-4}(p_1 + 0.6364)(p_1 + 0.6670)}$$

$$= \frac{-1}{(1.6364)(1.6670)}$$

$$= -0.3665$$

$$\left.\frac{\partial z}{\partial b_1}\right|_{z=p_2} = \frac{p_2^{-1}}{0.6364 p_2^{-4}(p_2 - 1)(p_2 + 0.6670)}$$

$$= \frac{(0.6364)^2}{(0.6364 - 1)(0.6364 + 0.6670)}$$

$$= -0.8546$$

$$\left.\frac{\partial z}{\partial b_1}\right|_{z=p_3} = \frac{-p_3^{-1}}{0.6670 p_3^{-4}(p_3 - 1)(p_3 + 0.6364)}$$

$$= \frac{(0.6670)^2}{(0.6670 - 1)(1.3134)}$$

$$= -1.0172$$

For b_2, the pole changes from Eq. (10-29) are

$$\left.\frac{\partial z}{\partial b_2}\right|_{z=p_1} = \frac{-1.0}{(1.6364)(1.6670)} = -0.3665$$

$$\left.\frac{\partial z}{\partial b_2}\right|_{z=p_2} = \frac{-0.6364}{-0.4775} = 1.3327$$

$$\left.\frac{\partial z}{\partial b_2}\right|_{z=p_3} = \frac{-0.6670}{-0.4374} = 1.5249$$

and, for b_3, the pole changes are

$$\left.\frac{\partial z}{\partial b_3}\right|_{z=p_1} = \frac{-1}{2.7279} = -0.3665$$

$$\left.\frac{\partial z}{\partial b_3}\right|_{z=p_2} = \frac{1}{-0.4775} = -2.0942$$

$$\left.\frac{\partial z}{\partial b_3}\right|_{z=p_3} = \frac{1}{-0.4374} = -2.2862$$

Thus the sensitivities from Eq. (10-30) are, for b_1,

$$S_1^1 = \frac{(-0.3665)(0.3034)}{1} = -0.1112$$

$$S_1^2 = \frac{(-0.8546)(0.3034)}{0.6364} = -0.4073$$

$$S_1^3 = \frac{(-1.0172)(0.3034)}{0.6670} = -0.4627$$

Thus poles p_2 and p_3 are slightly more sensitive to small changes in b_1 than p_1. For b_2,

$$S_2^1 = (-0.3665)(0.8789) = -0.3221$$

$$S_2^2 = \frac{(1.3327)(0.8789)}{0.6364} = 1.8404$$

$$S_2^3 = \frac{(1.5249)(0.8789)}{0.6670} = 2.0093$$

Here p_1 is not very sensitive to changes in b_2, while the poles p_2 and p_3 are considerably more sensitive. Finally, for b_3,

$$S_2^1 = \frac{(-0.3665)(0.4245)}{1} = -0.1555$$

$$S_3^2 = \frac{(-2.0942)(0.4245)}{0.6364} = -1.3969$$

$$S_3^3 = \frac{(-2.2862)(0.4245)}{0.6670} = -1.4550$$

and p_1 is not as sensitive to changes in b_3. Poles p_2 and p_3 are an order of magnitude more sensitive to b_3 changes. Remember, the sensitivity factors represent the ratio of the percent change in the pole p_i to a percent change in the coefficient b_j. The implementation of the various coefficients in terms of bit accuracy can be analyzed in terms of their sensitivities.

Example 10-7 Consider the following transfer function for a digital controller:

$$H(z) = \frac{Y(z)}{X(z)} = K\frac{z+a}{z+b}$$

What is the sensitivity of the pole or zero to an implementation change in b or a, respectively?

Since $\partial z/\partial a$ and $\partial z/\partial b$ are both 1, the sensitivities are essentially 1 (with $K = 1$) for both the pole and the zero.

Now consider two implementations of this controller, that is,

Set 1: $a = -0.9999$ $b = -0.9998996$

Set 2: $a = -0.999$ $b = -0.9999$

Note that the relative root-locus positions of the pole and zero have been interchanged. A lag network has been changed to a lead network from set 1 to set 2. Although the controller is still stable, the overall system root locus may now reflect unstable performance. Note also that Δa and Δb are relatively quite small; that is,

Set 1: $\Delta a = \pm 0.00005$ $\Delta b = \pm 0.00000005$

Set 2: $\Delta a = \pm 0.0005$ $\Delta b = \pm 0.00005$

Therefore, the associated changes in the zero and pole, respectively, are quite small per the sensitivity relations. But what about the value of the controller output? Considerable differences in performance can occur as shown in the following discussion:

Consider $K = 1$; then the associated difference equation is

$$y(k) = x(k) + ax(k-1) - by(k-1)$$

(Note that T is suppressed as an argument.)

Assume that the controller is driven with a sampled unit step function; that is,

$$x(k) = \begin{cases} 1 & k \geq 0 \\ 0 & k < 0 \end{cases}$$

Thus the form of the iterations is

$$y(0) = 1 + 0 - 0 = 1$$

$$y(1) = 1 + a - b$$

$$y(2) = 1 + a - b(1 + a - b)$$

$$y(3) = 1 + a - b[1 + a - b(1 + a - b)]$$

$$\vdots$$

For set 1 using 8-decimal-digit accuracy ($\sim$26 binary digits), the iterative sequence is

$$y(0) = 1$$

$$y(1) = 1 - 0.9999 + 0.9998996 = 0.9999996$$

$$y(2) = 1 - 0.9999 + 0.9998996(0.9999996) = 0.99999559$$

$$y(3) = 1 - 0.9999 + 0.9998996(0.99999559) = 0.99999518$$

$$y(4) = 1 - 0.9999 + 0.9998996(0.99999518) = 0.99999417$$

and for set 2,

$$y(0) = 1$$

$$y(1) = 1 - 0.999 + 0.9999 = 1.0009$$

$$y(2) = 1 - 0.999 + 0.9999(1.0009) = 1.0017999$$

$$y(3) = 1 - 0.999 + 0.9999(1.0017999) = 1.0026997$$

$$y(4) = 1 - 0.999 + 0.9999(1.0026997) = 1.0035994$$

The trends indicated by the two sets of outputs agree with the final-value theorem; that is,

$$y_i(\infty) = \frac{1 + a}{1 + b}$$

For set 1, $y_1(\infty) = 0.99601594$ (lag network)

For set 2, $y_2(\infty) = 10.000000$ (lead network)

Note also that different values of K affect numerical stability performance since K affects the sensitivity of the zero and the arithmetic truncation process.

The previous comments on truncation imply that coefficient truncation may affect the performance of the control system. That is, when a first trial decision is made on the actual truncation, perform a control system analysis, using the truncated $D_c(z)$, to obtain the system's figures of merit. If these values do not satisfy the desired control system's performance specification, another choice of truncated values is made for $D_c(z)$. This situation becomes more critical as the value of T becomes smaller (that is, $a \to -1$ and $b \to -1$).

10-6 SCALING

In terms of the addition operation, the determination of the computer scaling or scaling constants affect the possibility of overflow or underflow. Although the A/D converter input is usually physically described in units of volts, logically, the units associated with the voltage value can be radians, degrees, feet per second, or temperature, for example, and the controller dynamics can have gains of 10^7 in some applications. Thus *the implementer of a controller must attempt to retain sufficient accuracy within the computer by shifting (scaling) data as they are processed without generating an overflow*. For example, the A/D data may consist of 12 bits, but the computer may have 16 bits of processing capability. Thus the 12 bits can be placed in the most significant position, least significant position, or between them. The choice of position depends upon the selected scaling. Also, input and output data should always be checked to see if they are in range. *For output data, a saturation algorithm should be implemented so that roll-over does not occur.* That is, a high-magnitude positive value should not be changed to a high-magnitude negative value because of overflow into sign bit. A special routine can also check the D/A output prior to output transfer. If the output is greater than the range of the D/A, then a saturation output value is transferred to the D/A.

To determine the scaling constants, the converters and the controller have to be dynamically analyzed. This analysis should determine maximum gains and anticipated maximum values of $y(kT)$ and $x(kT)$ for all k.

A scaling technique for complement number representations uses a scaling multiplier at the front of each cascaded subfilter in the controller. A value of a scaling multiplier, say, K_i, is chosen so that the subfilter inputs are bounded by M. With a proper selection of K_i, the outputs of each subfilter are also bounded by M based upon the condition that overflow does not occur but maximum accuracy is maintained. To find each K_i, consider the convolution summation:

$$y_i(k) = \sum_{l=0}^{+\infty} K_i h_i(l) x(l-k)$$

where y_i is the output of a subfilter with transfer function $H_i(z)$. Thus

$$|y_i(k)| \leq \sum_{l=0}^{+\infty} K_i |h(l)| |x(l-k)|$$

But

$$|x(l-k)| \leq M$$

Therefore

$$|y_i(k)| \leq \sum_{l=0}^{+\infty} |K_i h_i(l)| M$$

But $y_i(k)$ must be less than or equal to M in order to prohibit overflow. Thus

$$\sum_{l=0}^{+\infty} |K_i h_i(l)| < 1$$

or

$$K_i \leq \left[\sum_{l=0}^{+\infty} h_i(l) \right]^{-1}$$

If Eq. (10-31) is satisfied, a subfilter input of magnitude M never generates an output greater than M. Note that different norms[89] can be used, resulting in an equivalent structure of Eq. (10-31). In using any scaling technique that prohibits overflow, scale factors can be chosen that maintain compensator signals at significant bit accuracy permitting the required dynamic signal range.

10-7 SIMULATION

Since no one technique is available for selecting an optimal set of coefficients, another approach might yield better results. That approach is basically a perturbation technique of some kind. The primary ingredient is the initial model that is used for optimization. For example, one can start with a given form of the controller, $H(z)$, and a desired form $H_d(z)$. The error defined as

$$E(z) \equiv [H(z) - H_d(z)]^2 \tag{10-32}$$

is minimized through various gradient techniques based upon the coefficients as parameters (see Sec. 11-8). Of course this can be done without the constraint of finite word length. The desired form H_d can also be defined in terms of a Bode plot. Moreover, each discrete frequency can be weighted with respect to the others, resulting in more emphasis on the control bandwidth.[42]

The use of an analog computer to model (or represent) the continuous plant along with the digital computer which models (or represents) the controller can be used to study the system performance. The digital computer can be programmed to process various word lengths and the performance difference noted for each. With this hybrid computer system considerable tuning and adjustment based upon other constraints can be accomplished, especially scaling. A complete digital simulation can also be of use if it provides the capability to change word length as well as adequately represent the timing difference between the continuous and discrete domains.

10-8 LIMIT-CYCLE PHENOMENON DUE TO QUANTIZATION

The quantization of signals within a digital controller can cause an oscillatory output phenomenon. This effect is due to the nonlinear aspects of the digital controller based upon finite word length and the associated processing. In the statistical analysis of quantization, it is assumed that the noise (quantization model) is not correlated with the signal; i.e., signal levels are much greater in magnitude than quantization step size. However, in a control system, the control objective is to maintain a certain constant output value or follow an input signal, each within a very small tolerance. Since the difference between the desired output and real output is to be kept small, the signal levels being processed by the digital computer can also be small. This negates the assumption employed in the statistical quantization analysis. Thus quantization errors in this situation are highly correlated as well as with the signals being processed. In fact, this correlation can produce a constant output or a continuing oscillation between specific limits, i.e., a limit cycle (see Fig. 10-13). The phenomenon is also defined as a *deadband effect*; i.e., the system output can not be controlled within a certain tolerance (deadband limits).

Example 10-8 As a simple example of the constant limit phenomenon, consider the first-order transfer function

$$H(z) = \frac{Y(z)}{X(z)} = \frac{K}{1 - dz^{-1}} = \frac{1.5}{1 - 0.9z^{-1}} \qquad 0 < d < 1 \tag{10-33}$$

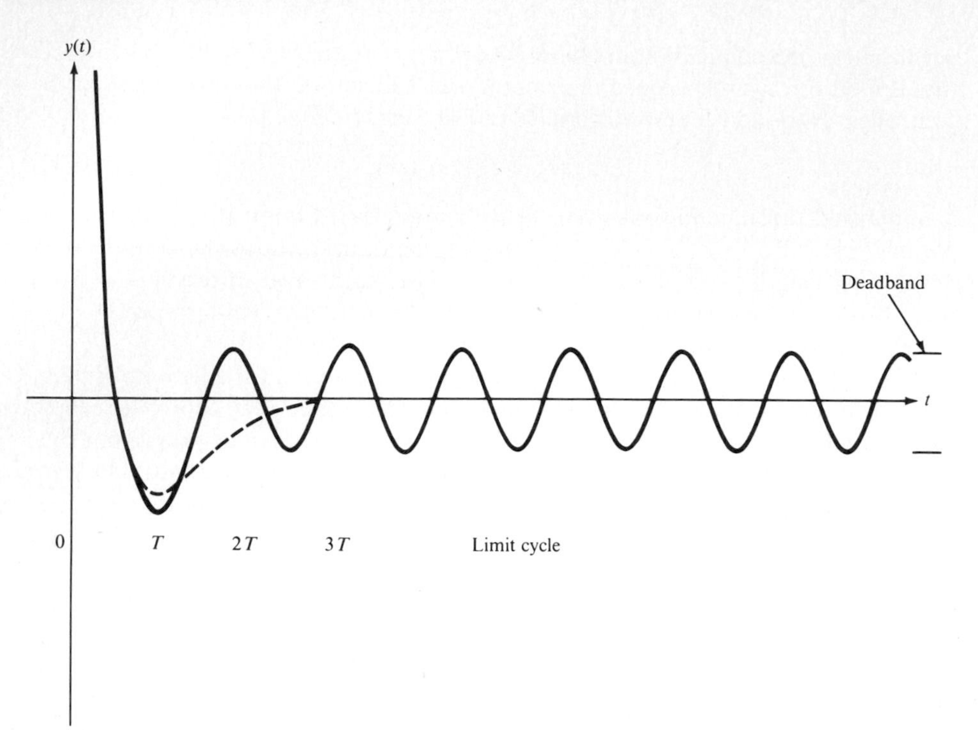

Figure 10-13 Example of limit-cycle phenomenon.

with $x(kT)$ a negative unit impulse at $k = 0$. Assume arithmetic rounding or truncation for one significant decimal digit, that is, $q = 0.1$. Thus

$$y[kT] = 0.9y[(k-1)T] + 1.5x[kT]$$

for $T = 1$ s.

Note that the final value of y, assuming infinite-word-length implementation, is zero. Table 10-1 indicates the values of $y(kT)$ for rounding and truncation, assuming $T = 1$ s. The quantized outputs are also compared to the infinite-word-length solution. Note that for both truncation and rounding, the output reaches a nonzero steady-state value, -0.9 and -0.5, respectively.

To present the general nonlinear aspects of the limit-cycle phenomenon, consider the first-order system

$$H(z) = \frac{Y(z)}{X(z)} = \frac{K}{1 - dz^{-1}} \quad \text{or} \quad h(k) = -K(d)k \tag{10-34}$$

and

$$y(k) = dy(k-1) + Kx(k) \tag{10-35}$$

Table 10-1 Example 10-8, output quantization

	$y(k)$		
k	Infinite word length	Truncated	Rounded
0	−1.5	−1.5	−1.5
1	1.35	−1.4	−1.4
2	1.215	−1.3	−1.3
3	1.0935	−1.2	−1.2
4	0.98415	−1.1	−1.1
5	0.885735	−1.0	−1.0
6	0.7971615	−0.9	−0.9
7	−0.7174454	−0.9	−0.8
8	−0.6457008	−0.9	−0.7
9	−0.5811307	−0.9	−0.6
10	−0.5230177	−0.9	−0.5
11	−0.4707159	−0.9	−0.5
12	−0.4236443	−0.9	−0.5

By definition, assume without loss of generality that the input $x(k)$ is zero after a finite value of k. Thus

$$y(k) = dy(k-1) \qquad k > 0 \tag{10-36}$$

If $Q[\ ]$ is defined as a quantization operator, then

$$y(k) = Q[dy(k-1)] \tag{10-37}$$

If Q is a truncation operator using sign plus 2s complement representation, then Fig. 10-14 represents both Eqs. (10-36) and (10-37). Note that

$$dy(k-1) - q < Q[dy(k-1)] \leq dy(k-1) + q \tag{10-38}$$

From Fig. 10-14, note that for an initial positive value of $y(0)$, say, α, the system output $y(k)$ moves toward the origin due to the truncation operator Q. In Eq. (10-37), then, the value of $y(k)$ becomes smaller and smaller, reaching the origin in finite time, finite k.

Now consider the initial value of $y(0)$, say, $-\alpha$, to be negative. The same phenomenon occurs; that is, $y(k)$ becomes smaller and smaller, but it never reaches the origin due to the intersection of the $1/d$ line with the nonlinear truncation function Q. Note that the 2s complement quantization generates a larger negative number than the original negative number, the error being less than $-q$. At the intersection point, $Q[-d]$ returns the value $y(k) = -\beta$ for all $k > 2$; that is, the output remains at $-\beta$ for all values of $k > 2$, a constant output. This is also reflected in Table 10-1.

Because of sign plus 2s complement representation, the value of y at the intersection point satisfies the following inequality with truncation:

$$by - y < q \qquad \text{where } y = -\beta \tag{10-39}$$

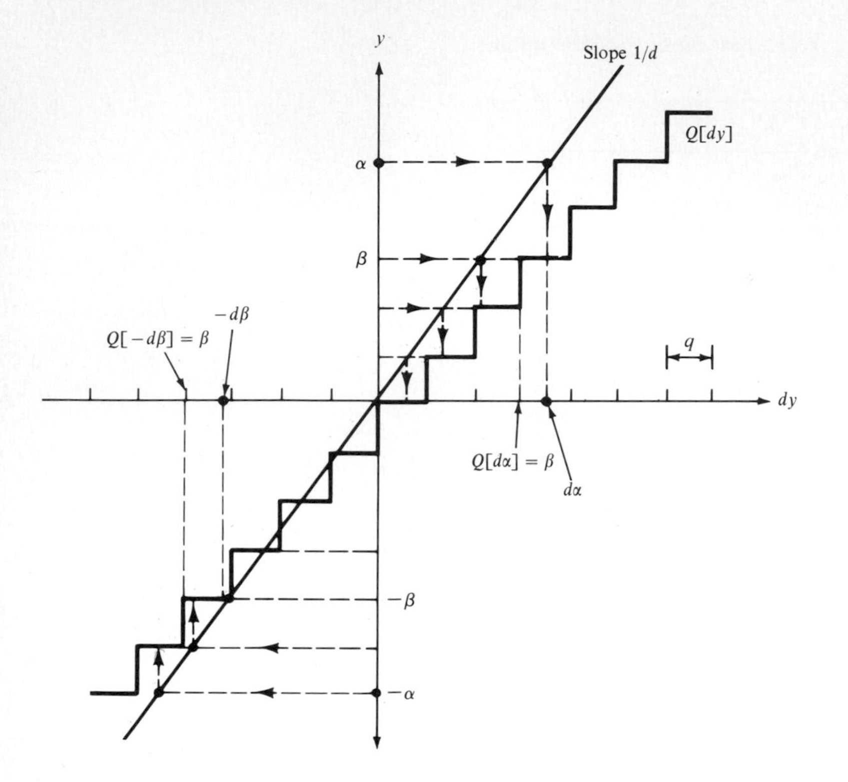

Figure 10-14 Nonzero limit cycle caused by truncation in first-order system (pole = d).

or

$$y < \frac{-q}{1-d} \tag{10-40}$$

For Example 10-8, $y < -q/0.1 = -10q$. Thus, for truncation, $y < -1.0$, and for rounding,

$$y < \frac{-q}{2(1-d)} = -0.5$$

both of which agree very closely with the values of Table 10-1. However, nonlinear phenomena are being approximated in the previous development, and thus the limiting values of Eq. (10-40) are approximations to the real-world results. Also note that the equilibrium point requires that $y = lq$, $l = 0, 1, \ldots$. Thus, from Eq. (10-40),

$$l < \frac{1}{1-d} \qquad 0 < d < 1 \tag{10-41}$$

For Example 10-1, then, $l < 10$ for truncation which is also in close agreement with

Table 10-1; that is, $l = 9$. For rounding,

$$l < \frac{1}{2(1 - d)} = 5 \tag{10-42}$$

which again is in close agreement with Table 10-1; that is, $l = 5$. For a graphical representation of rounding, consider Fig. 10-15 with $y(z) = 1/(1 + dz^{-1})$, $0 < d < 1$. Thus, a limit-cycle oscillation does exist as shown in the figure with amplitude

$$|y| = |\beta| = lq \qquad l = 2 \tag{10-43}$$

which generates the inequality

$$q - dq < \frac{q}{2} \tag{10-44}$$

or

$$l < \frac{1}{2(1 - d)} \qquad d > 0 \tag{10-45}$$

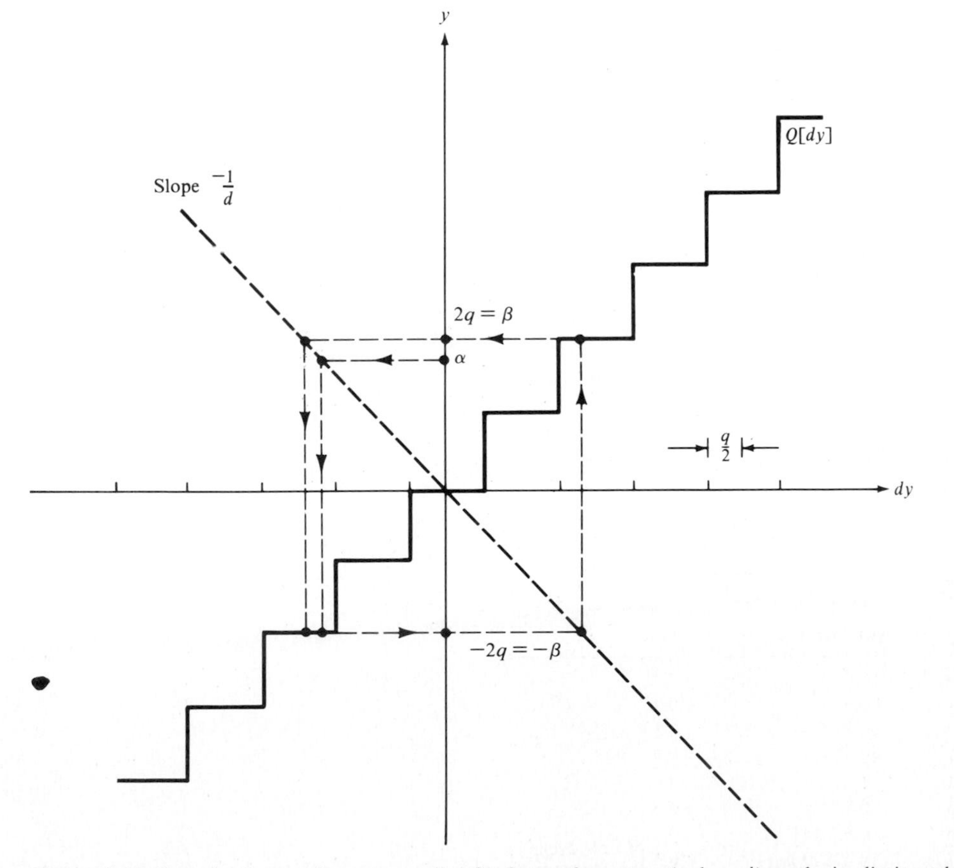

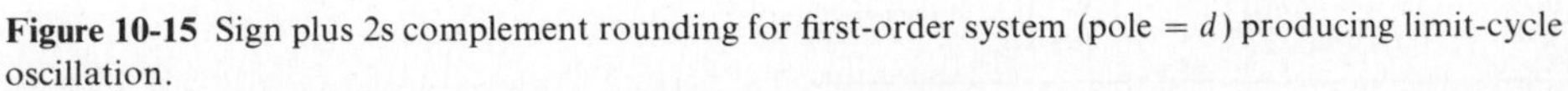
Figure 10-15 Sign plus 2s complement rounding for first-order system (pole = d) producing limit-cycle oscillation.

and

$$|y| < \frac{q}{2(1-d)} \qquad b > 0 \tag{10-46}$$

Note with rounding that

$$by(k-1) - \frac{q}{2} < Q[dy(k-1)] \leq dy(k-1) + \frac{q}{2} \tag{10-47}$$

If a limit-cycle oscillation or a constant output is to exist, a pole of the quantized equation must exist in the unit circle; that is, $z = 1$. For the first-order model, Eq. (10-32), the constant equilibrium output occurs at $z = 1$; that is, the equivalent gain of the quantizer, Q, is equal to $1/d$. For higher-order systems, one determines the equivalent quantizer gain or gains for generating a pole on the unit circle. The resulting output $c(k)$ is a possible limit cycle.

Example 10-9 Consider

$$H(z) = \frac{Y(z)}{X(z)} = \frac{1}{1 + 0.9z^{-1}} \tag{10-48}$$

with $x(k)$ a unit impulse at $k = 0$. Generate the solutions for y, assuming sign plus 2s complement representation with one significant decimal digit. Thus

$$y[kT] = (-0.9)^k \qquad \text{with } T = 1 \text{ s} \tag{10-49}$$

Again, the final value of y is zero, assuming infinite word length. Table 10-2 presents the values of $y(k)$, using rounding; that is,

$$y(k) = -0.9y(k-1) \qquad y(0) = 1$$

which is in concert with Fig. 10-15.

Table 10-2 Example 10-9, output rounding

	$y(k)$		
k	Infinite word length	Rounding	
0	1.0	1.0	
1	−0.9	−0.9	
2	0.81	0.8	
3	−0.729	−0.7	
4	0.6561	0.6	
5	0.57049	−0.5	Limit cycle
6	0.531441	0.5	
7	0.4782969	−0.5	
8	0.43040721	0.5	

To determine the bounds for a second-order controller,

$$H(z) = \frac{1}{1 + \sum_{i=1}^{2} d_i z^{-i}} = \frac{z^2}{(z - p_1)(z - p_2)} \tag{10-50}$$

with fixed-point implementation, a limit-cycle oscillation is assumed with period MT. Thus

$$y(kT) = y(kT + MT) \qquad M = 1, 2, \ldots \tag{10-51}$$

Also, from Eq. (10-50),

$$y(k) = -\sum_{i=1}^{2} d_i y(k - i) \tag{10-52}$$

With quantization

$$y(k) = -\sum_{i=1}^{2} Q[d_i y(k - i)] \tag{10-53}$$

Assuming no overflow and

$$Q[x_i(k)] \equiv x_i(k) + e_i(k)$$

where $e_i(k)$ is the error due to quantization,

$$\begin{aligned} y(k) &= -\sum_{i=1}^{2} d_i y(k - i) + \sum_{i=1}^{2} e_i(k) \\ &= -\sum_{i=1}^{2} d_i y(k - i) + e(k) \end{aligned} \tag{10-54}$$

Here

$$|e(k)| < \sum_{i=1}^{2} |e_i(k)| = q \tag{10-55}$$

which represents a *worst-case analysis*. Thus

$$y(k) = -\sum_{i=1}^{2} d_i y(k + M - i) + e(k + M) \tag{10-56}$$

and $\sum_{i=1}^{2} d_i y(k + M - i)$ is rounded to $y(k)$ if

$$y(k) - q < \sum_{i=1}^{2} Q[d_i y(k + M - i)] \le y(k) + q \tag{10-57}$$

To develop a tighter bound, consider Eq. (10-50) again. Assuming a limit-cycle steady-state output, the error function $e(k)$ is also periodic; that is,

$$e(kT) = e(kT + MT) \tag{10-58}$$

or

$$e(k) = e(k + M)$$

To develop a bound on $y(k)$, the linear convolution operation is also used:

$$y(k) = \sum_{i=0}^{+\infty} e(k-i)h(i) \tag{10-59}$$

where h is the compensator system function. Breaking up the summation intervals based upon the period M yields

$$y(k) = \sum_{i=0}^{+\infty} \sum_{l=0}^{M-1} h(l+iM)e(k-l-iM) \tag{10-60}$$

But, since e is periodic with period MT,

$$e(k-l-iM) = e(k-l)$$

Therefore,

$$y(k) = \sum_{i=0}^{+\infty} \sum_{l=0}^{M-1} h(l+iM)e(k-l) \tag{10-61}$$

But

$$|e(k-l)| \leq q \tag{10-62}$$

Therefore

$$|y(k)| \leq q \sum_{i=0}^{+\infty} \sum_{l=0}^{M-1} h(l+iM) \tag{10-63}$$

To determine $h(l+iM)$, assume distinct real poles inside the unit circle and use the residue theorem.

$$h(l+iM) = \left.\frac{z^{l+iM+1}}{z-p_2}\right|_{z=p_1} + \left.\frac{z^{l+iM+1}}{z-p_1}\right|_{z=p_2} \tag{10-64}$$

and

$$\sum_{i=0}^{+\infty} h(l+iM) = \sum_{i=0}^{\infty} \frac{p_1^{l+iM+1} - p_2^{l+iM+1}}{p_1 - p_2}$$

$$= \frac{1}{p_1-p_2} \frac{p_1^{l+1}}{1-p_1^m} - \frac{p_2^{l+1}}{1-p_2^m} \tag{10-65}$$

Thus, from Eq. (10-63), interchanging the summations and assuming that the infinite series converges yield

$$|y(k)| \leq q \sum_{l=0}^{M-1} \frac{1}{p_1-p_2} \frac{p_1^{l+1}}{1-p_1^M} \frac{p_2^{l+1}}{1-p_2^M} \tag{10-66}$$

To determine a bound on $y(k)$ at this point, a value of M must be chosen. Select $M = 1$ for a constant output and $M = 2$ for a limit-cycle oscillation of π/T.

$$|y(k)| \leq \begin{cases} \dfrac{q}{1+b_1+b_2} & \text{for } M = 1 \quad (10\text{-}67) \\[2ex] \dfrac{q}{1-|b_1|+b_2} & \text{for } M = 2 \quad (10\text{-}68) \end{cases}$$

To generate an absolute bound on $y(k)$, use Eq. (10-59), yielding

$$|y(k)| \leq q \sum_{i=0}^{\infty} |h(i)| \tag{10-69}$$

For the second-order system, then,

$$h(i) = \frac{z^{i+1}}{z - p_2}\bigg|_{z=p_1} + \frac{z^{i+1}}{z - p_1}\bigg|_{z=p_2}$$

$$= \frac{p_1^{i+1} - p_2^{i+1}}{p_1 - p_2} \tag{10-70}$$

and

$$|y(k)| < q \sum_{i=0}^{+\infty} \frac{p_i^{i+1} - p_2^{i+1}}{p_1 - p_2}$$

$$< \frac{q}{1 - |b_1| - b_2} \tag{10-71}$$

which is an absolute bound by definition. For repeated poles, see Prob. 10-23.

10-9 SUMMARY

This chapter presents various techniques for analyzing the effects of finite word length in a digital controller. Consideration is given to the A/D conversion process, the use of finite arithmetic, and the controller coefficient implementation. Error modeling of finite-word-length constraints in the first two cases is done with statistical models. The inherent nonlinearity of the deterministic error sources is overcome with the statistical model. These models require the assumption of various statistical properties which may not be true in a given situation. Thus, other techniques such as simulation can be used as well. Variations in coefficient realization are also studied in terms of pole and zero migrations from theoretical design values. Assuming small variations, a linear relationship is generated for pole and zero migration. Limit-cycle phenomena are also studied as related to the nonlinear quantization function.

Although computer hardware is becoming less expensive for controller implementation, and therefore additional word length bits are relatively inexpensive, the computational time still plays an important role in meeting performance requirements. The longer word length may be more accurate, but the processing time can also be longer.

The study of quantization effects is a continuing effort, and current researchers are attempting to develop better mathematical tasks and automated analysis programs. Current literature generally indicates the diversity of approaches to the finite-word-length problem.

CHAPTER

ELEVEN

CASCADE COMPENSATION

11-1 INTRODUCTION

Chapters 6 and 7 present the DIR and DIG techniques for analyzing the performance of the basic system. Generally, gain adjustment alone is not sufficient to achieve the desired system performance specifications (see Sec. 5-1). Thus, a cascade and/or feedback compensator (controller) can be used to accomplish the design objectives. This chapter discusses the design objective of satisfying the specified values of the conventional control-theory figures of merit by using a cascade compensator. Feedback compensation is discussed in Chap. 12. A more advanced approach (modern control theory) is to develop a digital controller with the design objective of satisfying a performance or optimization criterion.[1]

The analysis of a system's performance may be carried out in either the time and/or frequency domains. If the performance of the basic system is unsatisfactory, then, based on the results of this analysis, a compensator can be developed. Two approaches may be used to design the compensator: the DIR and DIG techniques. *Both these techniques rely heavily on a trial-and-error approach which requires a firm understanding of the fundamentals of compensation.* To facilitate this trial-and-error approach a CAD package is often used. The main advantage of DIR design is that the performance specifications can be met with less stringent requirements on the controller parameters. The disadvantage of the method is that there is not a large amount of knowledge, experience, or engineering tools for effecting a suitable $D_c(z)$ controller.

In the DIG method, once the controller design is achieved in the s or w domain, then one of several transformation methods can be employed to transform the cascade or feedback controller, with or without the s- or w-domain controller gain included, into its equivalent z-domain discrete controller. The reason for not including the gain is discussed in a later section. The advantage of this method is that tried and proven continuous-time domain methods are used for designing a workable $D_c(s)$ or $D_c(w)$ [or $H_c(s)$ or $H_c(w)$] controller.

After $D_c(z)$ [or $H_c(z)$] is included in the system model, as shown in Fig. 11-1, a system analysis is again performed. If the modified system does not meet the desired specifications, the controller is modified accordingly. This process is continued until the desired system specifications are achieved. To illustrate this design process the basic system of Chap. 7 is used for the design of lead, lag, and lag-lead cascade compensators. The design is done in all three domains (s, w, and z domains). For the s-plane analysis the sampled-data control system is transformed to a pseudo-continuous-time (PCT) system.

Feedback-compensator design examples, in Chap. 12, are also presented for two situations: the case where the input is a desired signal and the case where the input is an unwanted disturbance. The compensator transfer function may be determined either by the Guillemin-Truxal method or from the analysis of the open-loop transfer function as illustrated in this and the next chapter. There are many algorithms for expediting the design of a digtal control system. Also, there are a limited number of computer-aided design or interactive techniques (see Appendix E)[14,15] that expedite this modification process.

11-2 DESIGN PROCEDURES

To help the reader visualize the manner in which each of the two techniques is applied, the cascade-compensator design procedures to be followed are first outlined. An important tool for either of the two techniques is a computer-aided design package.[14,15] As the value of T decreases, the use of a computer becomes essential because of accuracy considerations. Also, for accurate root-locus analysis the calculation step size, in the z domain, should be equal to or less than $T/10$. The design procedures are outlined below and discussed in the following sections. This presentation is intended to establish firmly for the reader the manner in which the DIG and DIR techniques are applied to design a compensator: $D_c(\cdot) = (\text{gain})D_c'(\cdot)$.

11-2A DIG (Digitization) Technique

Figure 11-2 represents the trial-and-error design philosophy in applying the DIG technique. If path A does not result in the specifications being met by the sampled-data control system of Fig. 11-1b, then path B is used to try to determine a satisfactory value of K_{zc}. A similar chart may be drawn for the design of the feedback controller of Fig. 11-1d. The design philosophy involves the following considerations:

1. Follow path A if the dominant poles and zeros of $C(\cdot)/R(\cdot)$ lie in the shaded area of Fig. 7-7 (Tustin approximation is good!).

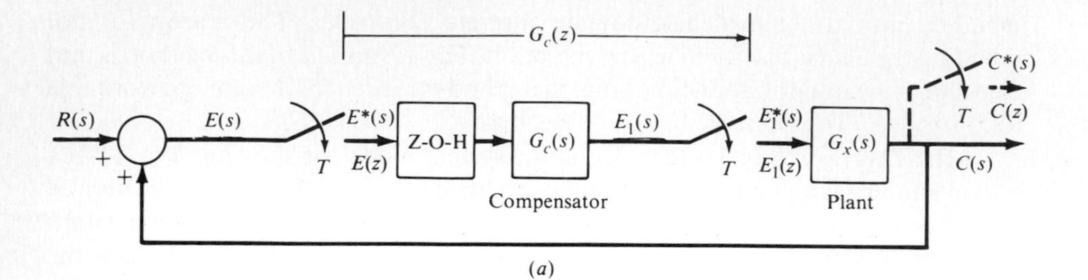

(a)

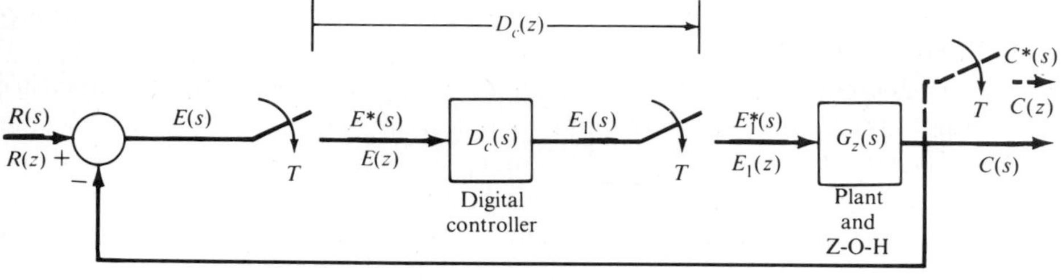

(b)

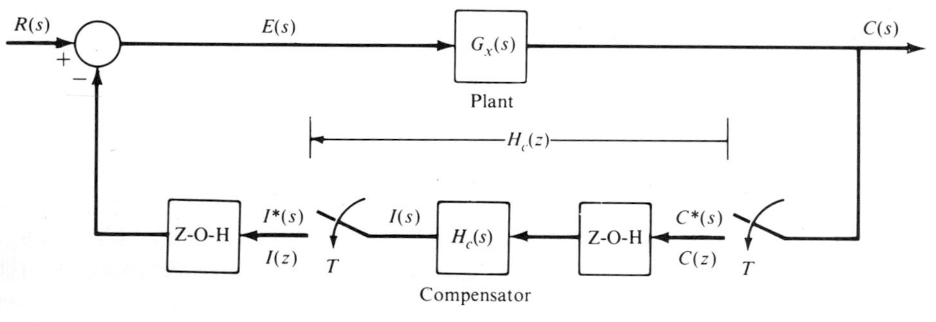

(c)

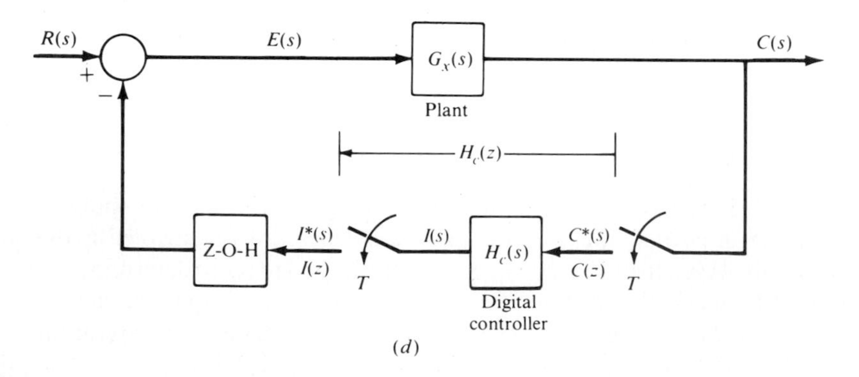

(d)

Figure 11-1 A compensated sampled-data control system. (*a*), (*c*) Cascade and feedback analog compensators, respectively; (*b*), (*d*) cascade and feedback digital controllers, respectively.

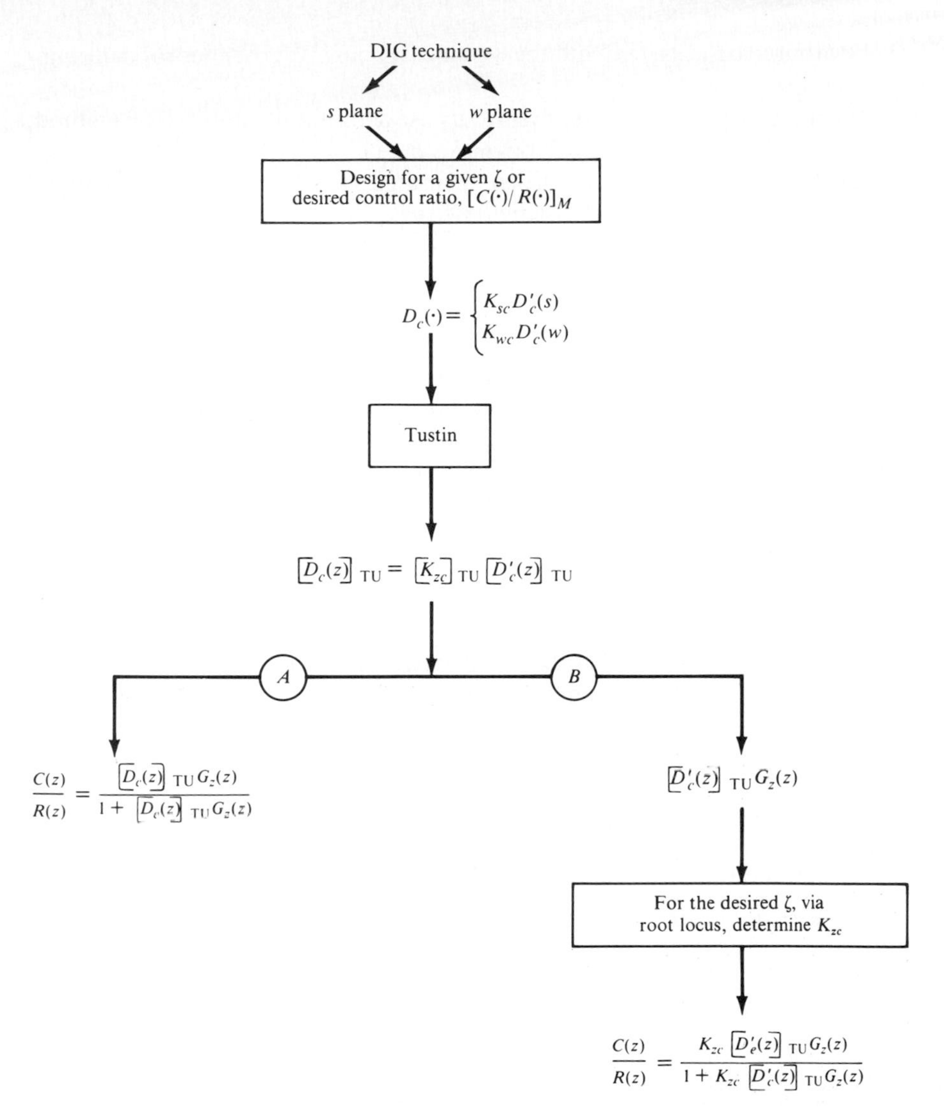

Figure 11-2 DIG design philosophy for Fig. 11-1*a*.

2. Follow path *A* when the degree of warping is deemed not to negatively affect the achievement of the desired design results (see Sec. 11-4A). If the desired results are not achieved, try path *B*.
3. Follow path *B* when severe warping exists.

The DIG design procedure is as follows:

Step 1. Convert the basic sampled-data control system to a PCT control system or transform the basic system into the w plane (try both approaches for determining the best design).

Step 2. By means of a root-locus analysis or by use of the Guillemin-Truxal method, determine $D_c(s) = K_{sc}D'_c(s)$ or $D_c(w) = K_{ws}D'_c(w)$.

Step 3. Obtain the control ratio of the compensated system and the corresponding time response for the desired forcing function. (This step is not necessary if the exact Guillemin-Truxal compensator is used providing there is no warping of the dominant poles and zeros.) If the desired performance results are not achieved, repeat step 2 by selecting a different value of ζ, σ, ω_d, etc., or a different desired control ratio.

Step 4. When an acceptable $D_c(s)$ or $D_c(w)$ has been achieved, transform the compensator, via the Tustin transformation, into the z domain. *Note*: In order for real poles (or zeros) [p_i (or z_i)] that lie on the negative real axis of the s or w plane to lie between 0 and 1 in the z plane, *when utilizing the Tustin transformation* of Eq. (7-17) *requires that* $|p_i| \leq 2/T$ *(or* $|z_i| \leq 2/T$*) be satisfied.* This requirement must be kept in mind when the controller is to be implemented by an equivalent passive analog compensator (RC components). The input to this analog compensator, as shown in Fig. 11-1*a*, must be the output of a ZOH unit.

Step 5. Obtain the z-domain control ratio of the compensated system and the corresponding time response for the desired forcing function. If the desired performance results for the sampled-data control system have been achieved via path A or path B, then the design of the compensator is complete. If not, return to step 2 and repeat the steps with a new compensator design or proceed to the DIR technique.

11-2B DIR (Direct) Technique

The simple lead ($\alpha < 1$) and lag ($\alpha > 1$) compensators in the s domain have the form

$$D_c(s) = \frac{K_{sc}(s - z_s)}{s - p_s} \tag{11-1}$$

where $p_s = z_s/\alpha$. These s-plane zeros and poles are transformed into the z-domain poles and zeros as follows:

$$z_z = \epsilon^{sT}|_{s=z_s} = \epsilon^{z_sT} \tag{11-2}$$

$$p_z = \epsilon^{sT}|_{s=p_s} = \epsilon^{p_sT} = \epsilon^{z_sT/\alpha} \tag{11-3}$$

Thus the corresponding first-order z-domain compensator (digital filter) is

$$D_c(z) = \frac{K_{zc}(z - z_z)}{z - p_z} = \frac{K_{zc}(z - z_z)}{z - z_z/\beta} \tag{11-4}$$

where $p_z = z_z/\beta$.

By taking the natural log of Eqs. (11-2) and (11-3), a relationship between α and β is obtained as follows:

$$z_sT = \ln z_z \tag{11-5}$$

$$\frac{z_s T}{\alpha} = \ln p_z \tag{11-6}$$

Taking the ratio of these equations and rearranging yields $\alpha \ln p_z = \ln z_z$. Thus

$$p_z^{\alpha} = z_z \tag{11-7}$$

and

$$\beta = \frac{z_z}{p_z} = p_z^{\alpha - 1} \tag{11-8}$$

For a lead network, $\alpha < 1$ and p_z is also less than 1. *Therefore, for a lead digital filter $\beta > 1$. For a lag digital filter $\beta < 1$.* (Note that the condition on β is just the opposite of that on α.) The following examples are presented to illustrate the correspondence between the values of α and β.

Example 11-1 Given the lag (la) compensator

$$D_c(s) = \frac{K_{sc}(s + 0.01)}{s + 0.001}$$

where $\alpha = 10$ and $T = 0.1$ s. The corresponding digital filter is

$$D_c(z) = \frac{K_{zc}(z - 0.999)}{z - 0.9999}$$

where $\beta_{la} = (0.9999)^9 = 0.9991 < 1$.

Example 11-2 Given the lead (le) compensator

$$D_c(s) = \frac{K_{sc}(s + 1)}{s + 10}$$

where $\alpha = 0.1$ and $T = 0.1$ s. The corresponding digital filter is

$$D_c(z) = \frac{K_{zc}(z - 0.9048)}{z - 0.36788}$$

where $\beta_{le} = (0.36788)^{-0.9} = 2.4596 > 1$. A nominal value of $\beta_{le} = 10$ is often chosen for a lead digital filter.

Because Eqs. (11-1) and (11-4) are identical in mathematical form, the z-plane compensator design procedures using the DIR techniques are essentially the same as those given in Sec. 5-4 for designing a compensator for a continuous-time system. When analog compensators, with passive components (see Fig. 11-1*a* and *c*) are used, then the corresponding nominal values[1,21] of $1 < \alpha_{la} \leq 10$ and $0.1 \leq \alpha_{le} < 1$ in the z domain, β_{la} and β_{le}, respectively, should be observed.

11-3 GUILLEMIN-TRUXAL COMPENSATION METHOD (DIG)

The Guillemin-Truxal (GT) compensation method is used in this section for improving the system performance of the system of Sec. 7-6 having the plant

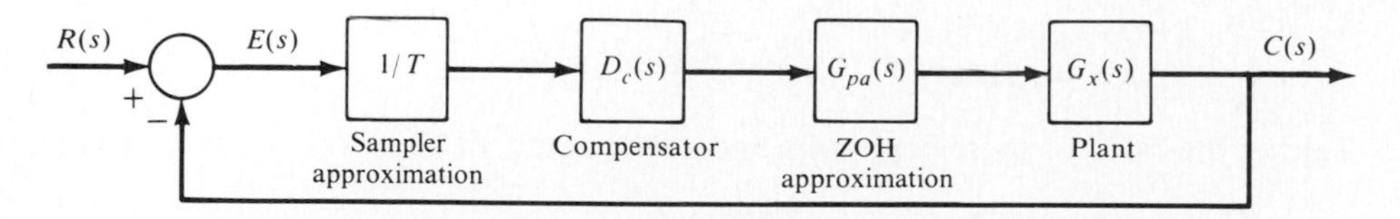

Figure 11-3 The PCT control-system representation of the sampled-data system of Fig. 11-1*b*.

$G_x(s) = K_x/[s(s+1)]$ [see Eq. (7-29)]. Assume that the figures of merit of the basic system are to be improved as follows: t_p and t_s are to be cut by one-half, with some improvement in K_1, while maintaining $M_p \leq 1.10$. Further, assume that the compensator model of Fig. 11-1*b* is constrained to increase the order of the system to 4. Note the order n of the compensated PCT system must be based upon the order n_x of the plant, n_A of $G_A(s)$, and $n_c \geq 1$ for $G_c(s)$. Thus $n = n_x + n_A + n_c$ defines the minimum value of n_c for this system, that is, $n_c \geq n - n_x - n_A = 4 - 2 - 1 = 1$.

Based upon these specifications, the following factors are used to derive the desired control ratio model for the PCT system of Fig. 11-3 that corresponds to the sampled-data system of Fig. 11-1*b*.

1. $|\sigma_{1,2}|$ of the dominant roots is selected to be at least twice that of Eq. (7-32) based upon $T_s = 4/|\sigma_{1,2}|$ s.
2. The dominant roots are selected such that $\zeta = 0.7071$ in order to try to maintain $M_p \leq 1.10$.
3. The s-plane pole-zero combination of $z_1 = -1.4$ and $p_3 = -1.1$ is added to minimize the increase in the overshoot which occurs when transforming from the continuous-time model to the sampled-data model. This selection of values is made to meet the desired performance specifications.

Thus, the following continuous-time control ratio model is achieved:

$$\left[\frac{C(s)}{R(s)}\right]_M = \frac{15.714(s+1.4)}{(s^2+2s+2)(s+1.1)(s+10)} \qquad \textbf{Case 5} \qquad (11\text{-}9)$$

Although the zero-pole combination -1.4 and -1.1 of Eq. (11-9) lies just outside the allowable region of Fig. 7-7, this aspect is overlooked for a first trial design. Applying the Tustin transformation to Eq. (11-9) for $T = 0.1$ s yields

$$\left[\frac{C(z)}{R(z)}\right]_{\mathrm{TU}} = \frac{1.407 \times 10^{-3}(z+1)^3(z-0.8692)}{(z-0.9005 \pm j0.0905)(z-0.8779)(z-0.3333)} = \frac{N(z)}{D(z)} \qquad (11\text{-}10)$$

Applying the GT method, using the exact $\mathscr{Z}$ transform of $G_{zo}(s)G_x(s)$, as illustrated by Example 5-2, yields the following transfer function of the cascade digital compensator of Fig. 11-1*b*:

$$D_c(z) = \frac{N(z)}{[D(z) - N(z)]G_z(z)} = \frac{E_1(z)}{E(z)}$$

$$= \frac{0.6110(z+1)^3(z-0.9048)(z-0.8692)}{(z-0.8567)(z-0.8189)(z-0.3439)(z+0.9672)} \tag{11-11}$$

Note that as a consequence of the use of the GT method and the Tustin approximation, the degree of the numerator of $D_c(z)$ is 1 higher than the degree of the denominator.

A practical approach to achieving a physically realizable $D_c(z)$ is to replace one of the $z + 1$ factors, which appears as a result of the Tustin transformation, by its dc gain factor of $(z + 1)_{z=1} = 2$ in the numerator of $D_c(z)$ to yield

$$D(z) = \frac{K_{zc}(z+1)^2(z-0.9048)(z-0.86921)}{(z-0.8567)(z-0.8189)(z-0.3439)(z+0.9672)} \tag{11-12}$$

where $K_{zc} = 2(0.6110) = 1.2220$. Note that the pole of $D_c(z)$ on the negative real axis does not present a problem since $D_c(z)$ is to be implemented as a digital controller (an algorithm). With this controller, the forward transfer function becomes

$$G(z) = D_c G_z(z) = \frac{0.002306K_{zc}(z+1)^2(z-0.8692)}{(z-1)(z-0.8567)(z-0.8189)(z-0.3439)} = KG'(z) \tag{11-13}$$

where $K = 0.002306K_{zc} = 0.002818$. The root-locus plot of $G(z) = -1$ is shown in Fig. 11-4 (Refs. 14, 15). For this value of K the figures of merit are $M_p = 1.051$,

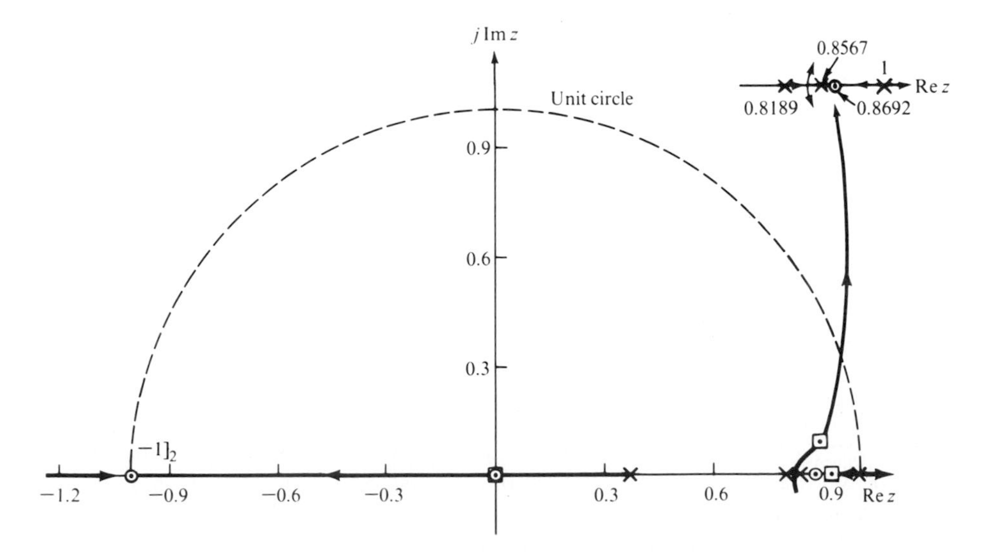

Fig. 11-4 Root-locus plot for Eq. (11-13).

$t_p = 3.2+$ s, and $t_s = 4.4-$ s. Thus with the modifed controller the specifications are essentially met. In general, it may be necessary or desired to do a fine tuning of the gain K_{zc} to achieve or improve on the desired system specifications. For the controller of Eq. (11-12), the following control ratio is obtained:

$$\frac{C(z)}{R(z)} = \frac{2.818 \times 10^{-3}(z+1)^2(z-0.8692)}{(z-0.9056 \pm j0.09455)(z-0.8770)(z-0.3284)} \qquad \textbf{Case 6} \qquad (11\text{-}14)$$

The ramp error coefficient, from Eqs. (6-24) and (11-13), is

$$K_1 = \frac{0.02818(4)(1-0.8692)}{(1-0.8567)(1-0.8189)(1-0.3439)} = 0.86589 \text{ s}^{-1} \qquad (11\text{-}15)$$

Table 11-1 compares the figures of merit obtained from Eqs. (11-9) and (11-14). Thus the desired performance specifications with $D_c(z)$ of Eq. (11-12) are achieved.

As this example illustrates, a physically unrealizable controller may *result when applying the Guillemin-Truxal method to* $[C(z)/R(z)]_{\text{TU}}$. *To maintain the dc gain and achieve a physically realizable controller, one approach is to replace one or more* $z+1$ *numerator factors of* $D_c(z)$ *by their dc gain factor of* 2.

Very often, the output of $D_c(z)$ alternately increases or decreases over a wide range. The Dahlin algorithm[39] includes a further refinement of the design to eliminate these large variations. The variations are caused by poles in the z plane that are located close to the point $z = -1$. For the Dahlin algorithm the value of 1 is substituted for z in the factors whose poles are close to $z = -1$. For example, the factor $z + 0.99$ is replaced by the number 1.99. The result is the elimination of the poles close to the point $z = -1$. Elimination of such high-frequency modes eliminates ringing in the controller output. Experimental tests show that there is no significant degradation of the output signal.[39] In using this design refinement one must ensure that the order of the denominator of the resulting $D_c(z)$ is equal to or higher than the order of its numerator. The damped oscillation of the output of the controller of Eq. (11-12) due to the controller pole at -0.9672 is very small and for most applications can be neglected. Note that the damped oscillation, in this example, is to be expected due to the underdamped response of $c(t)$.

Other approaches to achieving a realizable $D_c(z)$ are:

1. Apply the GT method to Eq. (11-9) to obtain $D_c(s)$; that is,

$$D_c(s) = \frac{1.948(s+1.4)(s+1)(s+20)}{s^3 + 13.3s^2 + 37.6s + 30.03} \qquad (11\text{-}16)$$

   Assuming that the digital controller does not contain a ZOH device, then the $\mathscr{Z}$ transform of Eq. (11-16) cannot be obtained since the order of the numerator is equal to the order of its denominator (see Sec. 4-6). Thus the Tustin approximation can be applied to Eq. (11-16) to yield $D_c(z)$. With this approach a physically realizable digital controller is achieved.
2. Apply the bilinear transformation to Eq. (11-9).

Table 11-1 Figures of merit for a Guillemin-Truxal designed system (DIG)

Compensator	M_p	t_p, s	t_s, s	K_1, s^{-1}	Case
Continuous model	1.0355	3.375	4.275	0.86583	5
$D_c(z)$	1.036	3.3+	4.2+	0.8653	7
Approximate $D_c(z)$	1.051	3.2+	4.4	0.86589	6

The procedure using the bilinear transformation is to first substitute, from Eq. (6-63),

$$z = \frac{Tw + 2}{-Tw + 2} \tag{11-17}$$

into Eq. (7-7), for $q = 1$, to obtain

$$s \approx w \tag{11-18}$$

Equation (11-18) is then substituted into the $[C(s)/R(s)]_M$ to obtain $[C(w)/R(w)]_M$. The GT method is then used to obtain $D_c(w)$. Substituting

$$w = \frac{2(z-1)}{T(z+1)} \tag{11-19}$$

which is the Tustin transformation of Eq. (7-7), into $D_c(w)$ results in a $D_c(z)$ whose degree of its numerator, in general, is equal to the degree of its denominator. Taking $\mathscr{Z}^{-1}[D_c(z)]$ results in a difference equation which may be implemented on a digital computer.

The first approach is illustrated as follows: The Tustin transformation of Eq. (11-16), for $T = 0.1$ s, is

$$D_c(z) = \frac{K_{zc}z(z - 0.9048)(z - 0.8692)}{(z - 0.8567)(z - 0.8189)(z - 0.3439)} \tag{11-20}$$

where $K_{zc} = 2.483$. Thus

$$G(z) = D_c(z)G_z(z)$$

$$= \frac{Kz(z + 0.9672)(z - 0.8692)}{(z - 1)(z - 0.8567)(z - 0.8189)(z - 0.3439)} = KG'(z) \tag{11-21}$$

where $K = 0.002306K_{zc}$, and the corresponding root locus is shown in Fig. 11-5. The control ratio for this compensated system is

$$\frac{C(z)}{R(z)} = \frac{5.726 \times 10^{-3}z(z + 0.9672)(z - 0.8692)}{(z - 0.8999 \pm j0.09039)(z - 0.8781)(z - 0.3359)} \qquad \textbf{Case 7} \tag{11-22}$$

For a unit-step forcing function, Eq. (11-22) results in the following values for the figures of merit: $M_p = 1.035$, $t_p = 3.3$ s, $t_s = 4.2$ s, and $K_1 = 0.8653\ \text{s}^{-1}$. As can

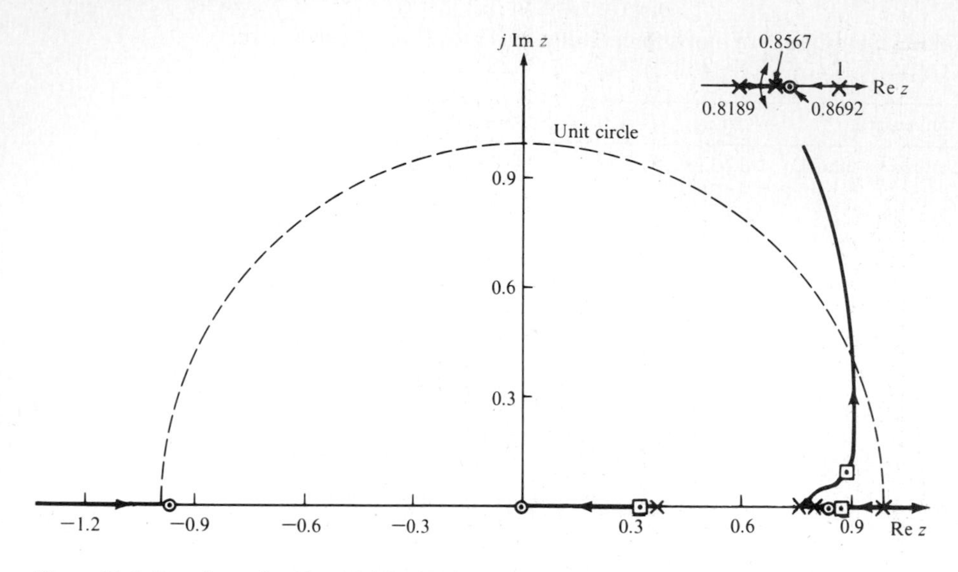

Figure 11-5 Root locus for Eq. (11-21) where $T = 0.1$ s.

be seen from Table 11-1, the Guillemin-Truxal $D_c(z)$ obtained in Cases 6 and 7 result in all specifications being satisfied.

It may be possible to meet the desired specifications by simplifying the controller of Eq. (11-20) by "canceling" the zero at 0.8692 with the pole at 0.8567 since they are "close" to each other. This is left as an exercise for the reader along with applying the second approach to this problem to determine the revised compensator performance. Thus in applying the GT method there is the possibility that the resulting $D_c(z)$ may have a zero and a pole that are very close together. Therefore it may be assumed that they effectively cancel one another, thus simplifying the required controller $D_c(z)$.

11-4 LEAD CASCADE COMPENSATION (DIG AND DIR)

In this section the desired improvement in the performance of the basic sampled-data control system is to be achieved by the use of the conventional s-plane cascade-compensation methods. The first two subsections deal with the DIG technique (s- and w-plane designs), and the remaining subsection presents the DIR technique (z-plane design). *It is assumed in this section that the digital controller does not use a ZOH device* in the A/D model in the input to the controller.

11-4A s-Plane DIG Design

A study of Eq. (7-32) reveals that since $|\sigma_{1,2}| = 0.4875$ (dominant roots) for the basic model system, it is necessary to at least double this magnitude for the real part of the dominant roots, $|\sigma'_{1,2}|$ of the compensated model system (Fig. 11-3), to try to reduce the settling time by one-half. To accomplish this a lead network of the form

$$D_c(s) = K_{sc}\frac{s+a}{s+b} \qquad a < b \tag{11-23}$$

may be inserted in cascade with $G_{PC}(s)$ (see Fig. 7-9). Since the settling time for the basic system is 8.6 s (see Table 7-5), then the desired $t_s \approx T_s$ is 4.3 s. Inserting this value into $T_s = 4/|\sigma'_{1,2}|$ yields

$$|\sigma'_{1,2}| \geq \frac{4}{4.3} = 0.93 \tag{11-24}$$

which is in the region for a good Tustin approximation. Using the cancellation rule for the design of Eq. (11-23), that is, $a = 1$, and assuming $b = 10$ yields, for $\zeta = 0.7071$ in the s domain, the actual value of

$$|\sigma'_{1,2}| = 3.8195 \tag{11-25}$$

This value lies just *outside the shaded region* of Fig. 7-7, for $T = 0.1$ s, but inside the boundary of $\sigma = -2/T = -20$. For this example this does not present a problem, with respect to T_s, since 3.8195 is at least four times greater than the desired value of 0.93. This factor of 4 does not yield $T_s = 4/3.8195$ because $\sigma'_{1,2}$ is not in the shaded region of Fig. 7-7. Since $|\sigma'_{1,2}| > 2|\sigma_{1,2}|$, a definite improvement in T_s is achieved as demonstrated in this example. Thus the compensator

$$D_c(s) = \frac{K_{sc}(s+1)}{s+10} \tag{11-26}$$

for $\zeta = 0.7071$, yields the root locus of Fig. 11-6 and

$$\left[\frac{C(s)}{R(s)}\right]_M = \frac{D_c(s)G_{PC}(s)}{1 + D_c(s)G_{PC}(s)} = \frac{9.534K_{sc}}{s^3 + 30s^2 + 200s + 652.6}$$

$$= \frac{9.534K_{sc}}{(s + 3.8195 \pm j3.8206)(s + 22.36)} \qquad \textbf{Case 8} \tag{11-27}$$

where $K_{sc} = 68.45$ for the continuous-time model. For this system model, $M_p \approx 1.04169$, $t_p \approx 0.875$ s, $t_s \approx 1.15$ s, and $K_1 = 3.263\ \text{s}^{-1}$.

Substituting from Eq. (7-7) into Eq. (11-26), with $T = 0.1$ s, yields the digital compensator or controller form of Fig. 11-1*b*.

$$[D_c(z)]_{TU} = \frac{E_1(z)}{E(z)} = \frac{K_c(21z - 19)}{30z - 10} = \frac{K_{zc}(z - 19/21)}{(z - 10/30)} \tag{11-28}$$

where $K_{sc} = K_c$ and $K_{zc} = 0.7K_c$ and $\beta_{1e} = 2.7143-$. Thus for the system of Fig. 11-1*b*,

$$G(z) = [D_c(z)]_{TU}G_z(z) = \frac{1.6142 \times 10^{-3}K_c(z + 0.9672)}{(z-1)(z - 0.3333)} \tag{11-29}$$

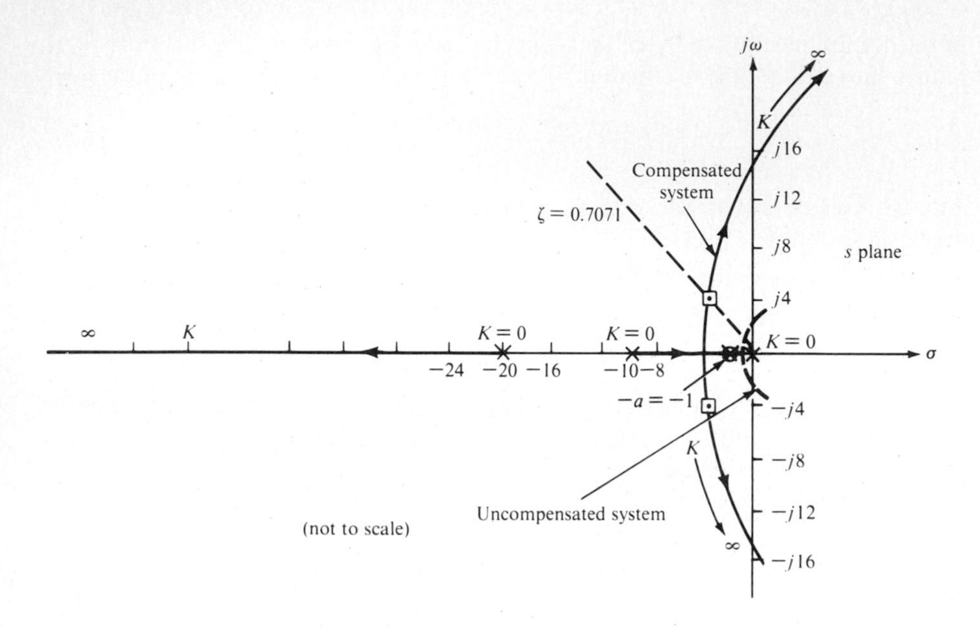

Figure 11-6 Root locus of a compensated system: $D_c(s)G_{PC}(s) = -1$.

The root-locus plot for this design is shown in Fig. 11-7 and, for $\zeta = 0.7071$ in the z domain, the control ratio is

$$\frac{C(z)}{R(z)} = \frac{0.1125(z + 0.9672)}{(z - 0.6104 \pm j0.2638)} \qquad \textbf{Case 9} \qquad (11\text{-}30)$$

where $K_c = 69.694$ is close to the s-plane model value of 68.45 for Eq. (11-27). Thus for this example paths A and B yield essentially the same system performance. This good degree of correlation is due to the fact that the imaginary part of the dominant poles in the s plane lies within the shaded area of Fig. 7-7. Table 11-2 compares the

Table 11-2 Figures of merit for a cascade lead-compensated designed system: s-plane design (DIG)

System	M_p	t_p, s	t_s, s	K_1, s^{-1}	Case
$[C(s)]_M$:					
Uncompensated	1.0482	6.3	8.40–8.45	0.4767	1
Compensated	1.0417	0.875	1.15	3.263	8
$C(z)$:					
Uncompensated	1.043	6.4+	8.6+	0.4765	3
Compensated	1.043	0.8	1+	3.3208	9

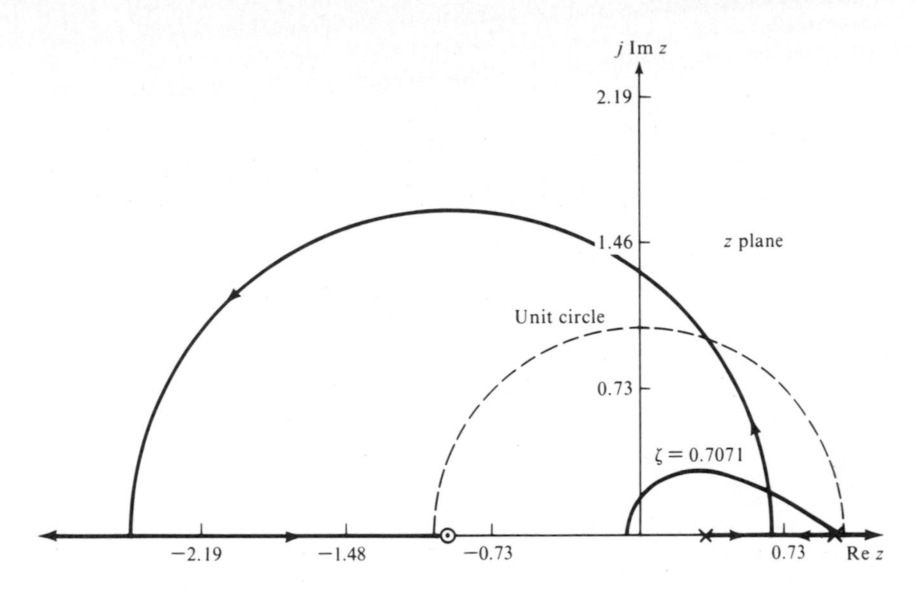

Figure 11-7 Lead compensation: root locus for Eqs. (11-28) and (11-29).

figures of merit of the continuous-time and sampled-data uncompensated and compensated systems. Thus the resulting design of Case 9 achieves the desired design specifications. Taking the $\mathscr{Z}^{-1}$ of Eq. (11-28) yields the difference equation for the resulting compensator as follows:

$$\mathscr{Z}^{-1}[30E_1(z) - 10z^{-1}E_1(z)] = \mathscr{Z}^{-1}[K_c(21 - 19z^{-1})E(z)]$$

$$e_1(kT) = 48.7858e(kT) - 44.13953e[(k-1)T] + 0.333333e_1[(k-1)T]$$

This discrete control law (algorithm) is translated into a software program to be implemented on a digital computer (controller) (see Chap. 10).

11-4B *w*-Plane DIG Design

The w transform of the basic system, for $T = 0.1$ s, is

$$G_z(w) = \frac{-4.163 \times 10^{-5} K_x(w - 20)(w + 1200)}{w(w + 0.9992)} \qquad \textbf{Case 10} \qquad (11\text{-}31)$$

where $K_x = 0.4767$. The basic system time-response characteristics of $C(w)$, for $\zeta = 0.7071$ in the w domain and a unit-step forcing function, are given in Table 11-3.

To achieve the desired improvement a lead compensator $D_c(w)$ of the form of Eq. (11-23) is chosen. Using the cancellation rule,

$$D_c(w) = \frac{K_{wc}(w + 0.9992)}{w + 9.992} \qquad (11\text{-}32)$$

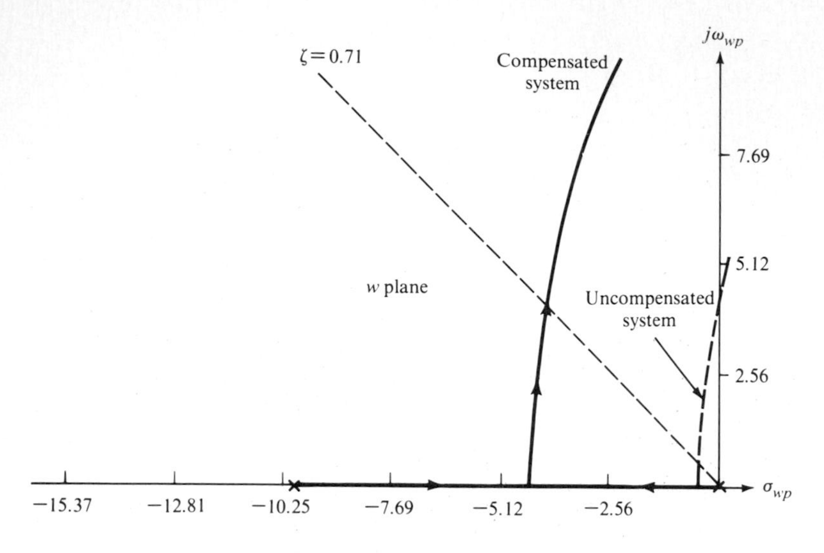

Fig. 11-8 Root loci for Eqs. (11-31) and (11-33).

and

$$G(w) = D_c(w)G_z(w) = \frac{-1.9845 \times 10^{-5} K_{wc}(w - 20)(w + 1200)}{w(w + 9.992)} \tag{11-33}$$

The corresponding root locus is shown in Fig. 11-8 and for $\zeta = 0.7071$,

$$\frac{C(w)}{R(w)} = \frac{-1.438 \times 10^{-3}(w - 20)(w + 1200)}{w + 4.154 \pm j4.155} \qquad \textbf{Case 11} \tag{11-34}$$

The time-response characteristics for a unit-step input are shown in Table 11-3. Since these characteristics are satisfactory,† the Tustin transformation of Eq. (11-32), for $T = 0.1$ s, is

$$D_c(z) = \frac{0.70016 K_{wc}(z - 0.905177)}{z - 0.333689} \tag{11-35}$$

where $K_{zc} = 0.70016 K_{wc}$ and $\beta_{\text{le}} = 2.7126$. Since the zero of Eq. (11-35) is very close to the pole of the basic plant [see Eq. (7-34)], it is assumed that they cancel each other in obtaining

$$G(z) = D_c(z)G_z(z) = \frac{K(z + 0.9672)}{(z - 1)(z - 0.333689)} \tag{11-36}$$

where $K = 0.002306 K_{zc}$ and $K_{zc} = 48.758$. The control ratio, obtained from the root

† Because of the values of T and of the dominant poles and zeros, prewarping is not required (see Secs. 6-6A and 7-4B).

Table 11-3 Figures of merit for a cascade lead-compensated designed system: *w*-plane design

System	M_p	t_p, s	t_s, s	K_1, s^{-1}	Case
$C(w)$:					
Uncompensated, DIG	1.043	6.486	8.69	0.4765	10
Compensated, DIG	1.045	0.796	1.061	3.454	11
$C(z)$:					
Uncompensated	1.043	6.45	8.65	0.4765	3
Compensated, DIG	1.043	0.8	1.05	3.32	9
Compensated, DIR	1.043	0.5	0.7–	4.514	13

locus (essentially the same as in Fig. 11-7), for $\zeta = 0.7071$ in the z domain (path B of Fig. 11-2), is

$$\frac{C(z)}{R(z)} = \frac{0.112436(z + 0.9672)}{z - 0.6106 \pm j0.2638} \qquad \textbf{Case 12} \tag{11-37}$$

and the corresponding time-response characteristics for a unit-step forcing function are given in Table 11-3. Note that the dominant poles of Eqs. (11-27) and (11-34) are close together. The differences are due to the warping effect. Also note that Eq. (11-30) is essentially identical to Eq. (11-37). Figure 11-9 is a plot of the poles of Eqs. (11-27), (11-34), and (11-37), respectively.

11-4C z-Plane DIR Design

The $\mathscr{Z}$ transform of the basic system is given by Eq. (7-34), for $T = 0.1$ s, and is repeated below:

$$G_z(z) = \frac{0.002306(z + 0.9672)}{(z - 1)(z - 0.9048)} \tag{11-38}$$

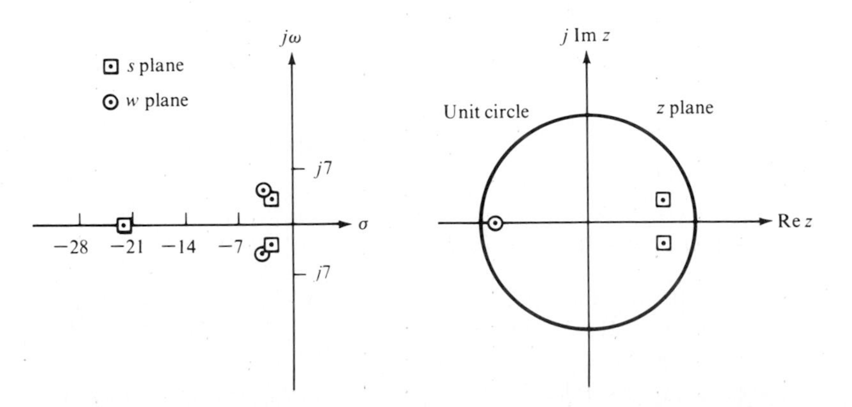

Figure 11-9 Locations of poles of $[C(s)/R(s)]_M$, $C(w)/R(w)$, and $C(z)/R(z)$: Cases 8, 11, and 12.

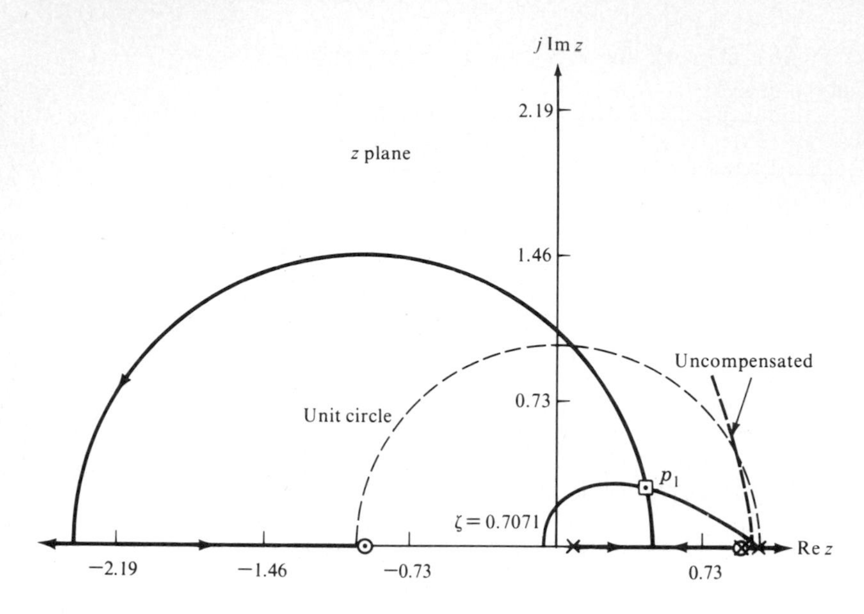

Figure 11-10 Root locus for Eq. (11-40) for $T = 0.1$ s.

The lead compensator of Fig. 11-1*b*, for the basic plant of Eq. (11-30), based upon the cancellation rule and $\beta_{le} = 10$, is

$$D_c(z) = \frac{K_{zc}(z - z_1)}{(z - z_1/\beta_{le})} = \frac{K_{zc}(z - 0.9048)}{z - 0.09048} \tag{11-39}$$

The resulting forward transfer function is

$$G(z) = \frac{K(z + 0.9672)}{(z - 1)(z - 0.09048)} \tag{11-40}$$

and for $\zeta = 0.7071$, based upon the root locus shown in Fig. 11-10, yields

$$\frac{C(z)}{R(z)} = \frac{K(z + 0.9672)}{z - 0.4409 \pm j0.312} \qquad \textbf{Case 13} \tag{11-41}$$

where $K = 0.002306K_{zc} = 0.2087$, and $K_{zc} = 90.5$. The time-response characteristics of Eq. (11-41) are given in Table 11-3.

11-4D Frequency-Response Characteristics

Figures 11-11 and 11-12 present the frequency-response characteristics for both the open-loop and closed-loop systems. Figure 11-11 contains plots of Lm $\mathbf{G}(\cdot)$ vs. ω and $\phi(\omega)$ vs. ω for both the basic and lead compensated systems in the w- and z-domain system representations. As expected, the lead compensator increases the phase margin frequency ω_ϕ of the system as noted in Fig. 7-18*b*, from 0.9+ rad/s to

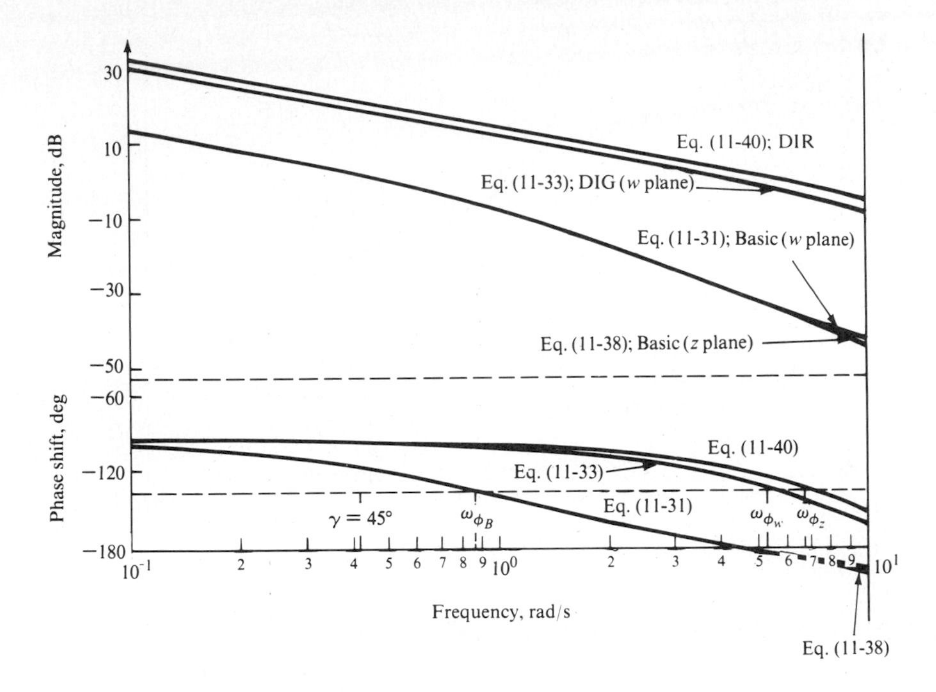

Figure 11-11 Forward transfer-function frequency characteristics: Lm **G**(·) vs. ω and ϕ vs. ω for the basic and lead compensated systems.

5.7 rad/s (w domain) and 7.3 rad/s (z domain). For a given M_m, an increase in the value of ω_ϕ and in turn an increase in the value of ω_m is indicative of a decrease in the values of t_p and t_s. In this example, a decrease in the values of t_p and t_s is indicated by the increase in the value of ω_ϕ and ω_b (since $M_m \leq 1$). Thus the correlation of the frequency and the time domain for continuous-time systems is equally valid for sampled-data systems.

The analysis of the results of the various designs given in Tables 11-2 and 11-3 and Figs. 11-11 and 11-12 reveals the following:

1. The figures of merit for Cases 1, 3, and 10, and Cases 8, 11, and 12 (DIG techniques designs) are in close agreement since the value of T is small enough to satisfy the condition $s = w$ (see Sec. 6-6A). Also, the zero and pole (except for the pole of the lead compensator) of $D_c(w)$ or $D_c(s)$ lie in the allowable region for a good Tustin approximation (see Fig. 7-7) of $D_c(z)$. As pointed out in Sec. 11-4A and noted in Table 11-3, the warping of the lead compensator pole (from 0.368 to 0.333, in the z domain) had a positive effect, in this case, in achieving the desired specifications. Thus when the approximation $s = w$ is valid, the DIG design may be carried out entirely in the s plane, using the conventional s-plane design techniques for continuous-time systems. Once an acceptable design is achieved, the resulting compensator $D_c(s)$ is transformed via the Tustin transformation to $D_c(z)$.
2. The DIR technique, Case 13, gives the best results.

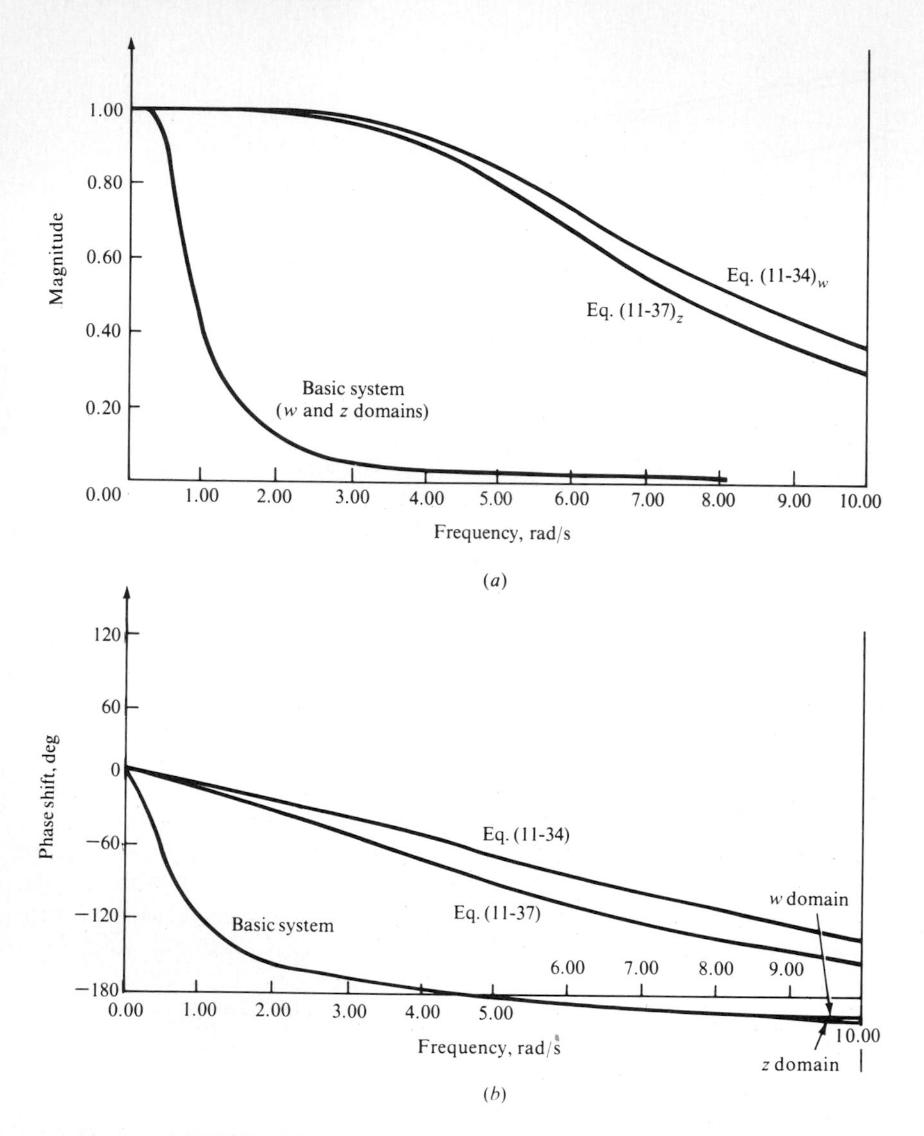

Figure 11-12 Closed-loop frequency-response characteristics. Basic and lead compensated systems: (*a*) $|\mathbf{C}(\cdot)/\mathbf{R}(\cdot)|$ vs. ω and (*b*) α vs. ω.

3. What is not revealed in Tables 11-2 and 11-3 is the effort that may be expended in trying to achieve the desired ζ (0.7071 for this example). As pointed out previously, for small values of T a higher degree of accuracy is required; i.e., a "large" number of significant digits must be maintained in effecting an accurate design in the z domain. Although this effort is minimized by use of a computer-aided design program,[14,15] it still can be further minimized by performing the initial design in the s or w plane where fewer significant digits are required to be

maintained. Once an acceptable $D_c(s)$ or $D_c(w)$ is achieved, then the corresponding $D_c(z)$ is obtained by means of Eq. (7-7) or (11-19). The resulting dominant poles of $C(z)/R(z)$ may have a $\zeta \neq 0.7071$ (for this example) in the z domain. This is due to the approximations involved in going between the s or w and z domains. As a final refinement step in the design, if required, a gain adjustment can be made, maintaining the desired degree of accuracy, until the dominant roots of the root locus in the z plane are located for the desired ζ.

4. For the situation where the value of T is not small enough to use the DIG technique, then without prewarping, the analysis and synthesis of a sampled-data system must be done by the DIR technique.
5. The correlation that exists between the frequency and time domains is a useful adjunct to the analysis and design of a sampled-data control system.

It can be concluded that the design procedure of this and the remaining sections of this chapter are simpler than the Guillemin-Truxal approach. Generally, it can yield a simpler form of $D_c(z)$.

11-5 LAG AND LAG-LEAD COMPENSATION (DIG AND DIR)

The previous section deals extensively with the lead compensation design as the "vehicle" for illustrating the advantages and disadvantages of using the DIG and DIR techniques of analyzing and designing a sampled-data control system. The main purpose of this section is the design of lag and lag-lead compensators. The same basic system and the value of $T = 0.1$ s of the previous section is used for the design of the lag compensator of the form

$$D_c(\cdot) = \frac{A}{\alpha}\frac{(\cdot) + 1/T_1}{(\cdot) + 1/\alpha T_1} \qquad \alpha > 1 \tag{11-42}$$

and of the lag-lead compensator of the form

$$D_c(\cdot) = A\frac{[(\cdot) + (1/T_1)][(\cdot) + (1/T_2)]}{[(\cdot) + (1/\alpha T_1)][(\cdot) + (\alpha/T_2)]} \qquad \alpha > 1 \tag{11-43}$$

where a nominal value of $\alpha = 10$ is chosen and $(\cdot)$ represents either the s or w variable. Note that the portion

$$\frac{(\cdot) + (1/T_1)}{(\cdot) + (1/\alpha T_1)}$$

of Eq. (11-43) corresponds to the lag compensator and the remaining portion corresponds to the lead compensator. The corresponding lag digital filter is of the form of Eq. (11-4), and for the lag-lead digital filter it is

$$D_c(z) = K_{zc}\left[\frac{z - z_{z1}}{z - z_{z_1}/\beta_{\text{le}}}\right]\left[\frac{z - z_{z2}}{z - z_{z_2}/\beta_{\text{la}}}\right] \tag{11-44}$$

11-5A Lag Compensation

When a sizable improvement in the value of the static error coefficient is desired, assuming that the transient characteristics are satisfactory (category 1 of Sec. 5-1), then a lag compensator can be inserted in cascade with the basic plant. Applying the standard design procedure[1] (see Sec. 5-4) for a lag compensator in the w plane for the plant of Eq. (11-31) yields the lag compensator

$$D_c(w) = \frac{K_{wc}(w + 0.01)}{w + 0.001} \tag{11-45}$$

and

$$G(w) = D_c(w)G_z(w) = \frac{K_w(w - 20)(w + 1200)(w + 0.01)}{w(w + 0.9992)(w + 0.001)} \tag{11-46}$$

The root-locus method, for $\zeta = 0.7071$ (see Fig. 11-13), yields

$$\frac{C(w)}{R(w)} = \frac{-1.9845 \times 10^{-5}(w - 20)(w + 1200)(w + 0.01)}{(w + 0.4833 \pm j0.48332)(w + 0.0010195)} \tag{11-47}$$

where $K_{wc} = 1.001$ and $K_w = -1.9845 \times 10^{-5}$. The root-locus branches in the vicinity of the origin, for the scale used in Fig. 11-13, are indistinguishable from the axes.

The corresponding design procedure applied in the z domain (DIR), for $\beta_{la} = 0.9991$, yields for the following transfer function for the digital controller of Fig. 11-1b,

$$D_c(z) = \frac{K_{zc}(z - 0.999)}{z - 0.9999} \tag{11-48}$$

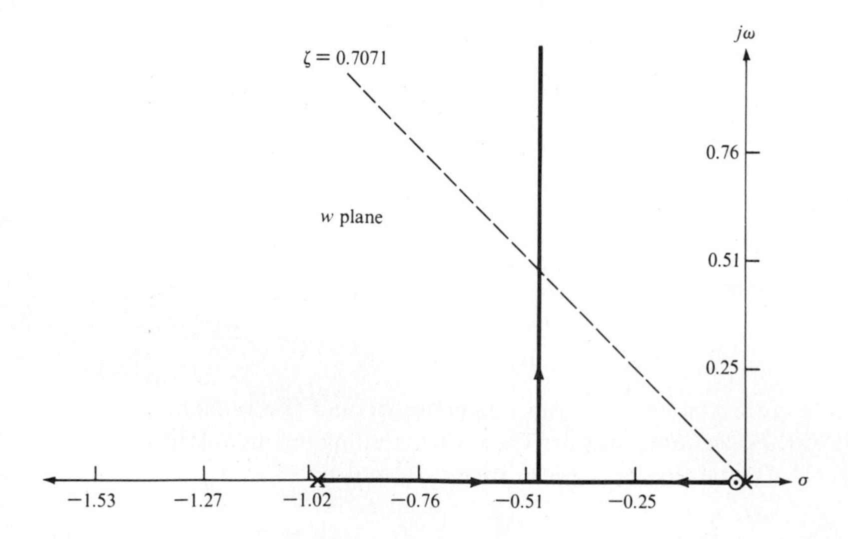

Figure 11-13 Root locus of Eq. (11-46).

where the location of the pole is "close" to 1 and the zero is just to the "left" of it, satisfying the 5° rule. Placing the pole at 1 increases the system type by 1 (see Prob. 11-11). The forward transfer function of the compensated system is

$$G(z) = D_c(z)G_z(z) = \frac{K_z(z + 0.9672)(z - 0.999)}{(z - 1)(z - 0.9048)(z - 0.9999)} \tag{11-49}$$

The root-locus method, for $\zeta = 0.7071$ (see Fig. 11-14), yields

$$\frac{C(z)}{R(z)} = \frac{0.2318 \times 10^{-2}(z + 0.9672)(z - 0.999)}{(z + 0.95175 \pm j0.04622)(z - 0.99898)} \tag{11-50}$$

where $K_{zc} = 1.0876$ and $K_z = 2.318 \times 10^{-3}$. Because of the scale used in Fig. 11-14, the branches of the root locus in the vicinity of $1 + j0$ are indistinguishable from the real axis.

Applying the Tustin transformation, Eq. (11-19), to Eq. (11-45) yields a $D_c(z)$ that has the same pole and zero as Eq. (11-48) but with $K_{zc} = 1.00145$ and $K_z = 2.308 \times 10^{-3}$ (path A of Fig. 11-2). The time-response characteristics obtained by the DIG and DIR techniques are essentially identical as seen in Table 11-4. This table also

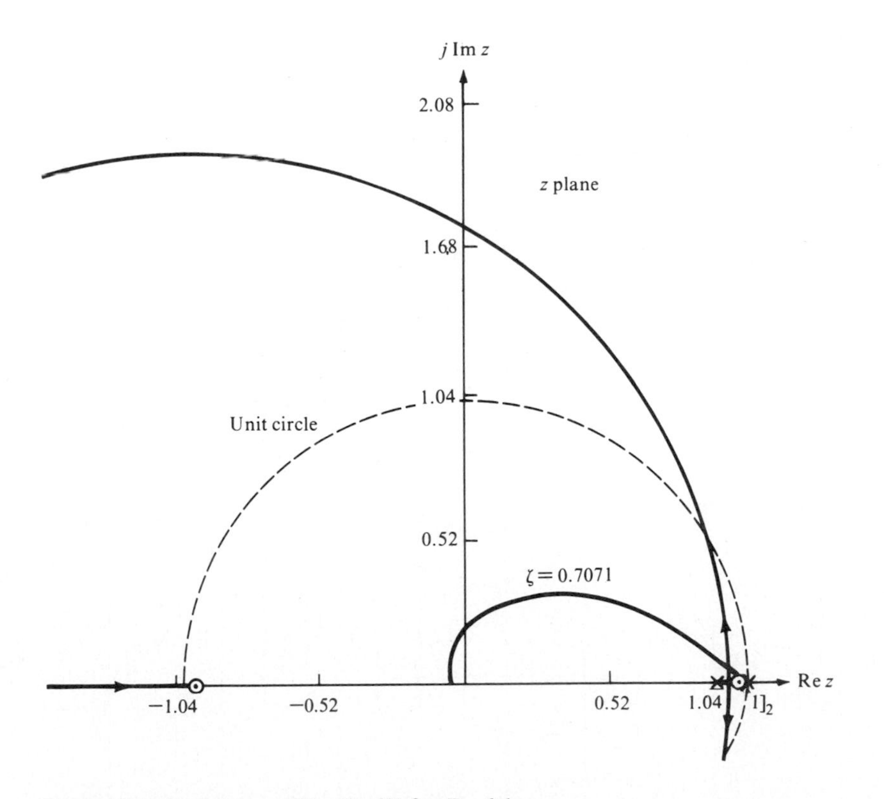

Figure 11-14 Root locus of Eq. (11-49) for $T = 0.1$ s.

Table 11-4 Summary of time- and frequency-response characteristics: basic and compensated systems

Design technique	Equation number	Domain	M_p	M_m	t_p, s	t_s, s	K_1, s^{-1}	ω_m, rad/s	ω_b, rad/s
				Basic system					
	(7-30)	s	1.043	1.00	6.48	8.69	0.4767	...	0.6900
	(11-31)	w	1.043	1.00	6.486	8.69	0.4765	...	0.6905
	(11-38)	z	1.043	1.00+	6.4+	8.6+	0.4767	0+	0.691−
				Lag system					
DIG	(11-46)	w	1.063	1.019	6.54	10.84	4.765	0.090	0.6960
	(11-49)	z	1.063	1.02−	6.4+	10.7	4.78	0.060	0.644+
DIR	(11-49) paths A and B								
				Lead system					
DIG	(11-27)	s	1.042	1.000	0.875	1.15	3.263	...	5.250
	(11-33)	w	1.045	1.001	0.796	1.061	3.454	1.245	6.143
	(11-29) (11-36)	z	1.043	1.000	0.8	1.05	3.320	...	5.68+
DIR	(11-40)		1.043	1.000	0.5	0.7−	4.515	...	8.445
				Lag-lead system					
DIG	(11-51)	w	1.047	1.004	0.797	1.087	34.52	1.220	6.14+
	(11-52) path A	z	1.053	1.003+	0.7	1.05	34.54	1.230	5.970
DIR	(11-52) path B		1.046	1.003−	0.8	1.1−	33.24	0.160	5.7−
	(11-55)		1.045	1.002−	0.5	0.7+	45.1	0.145	8.430

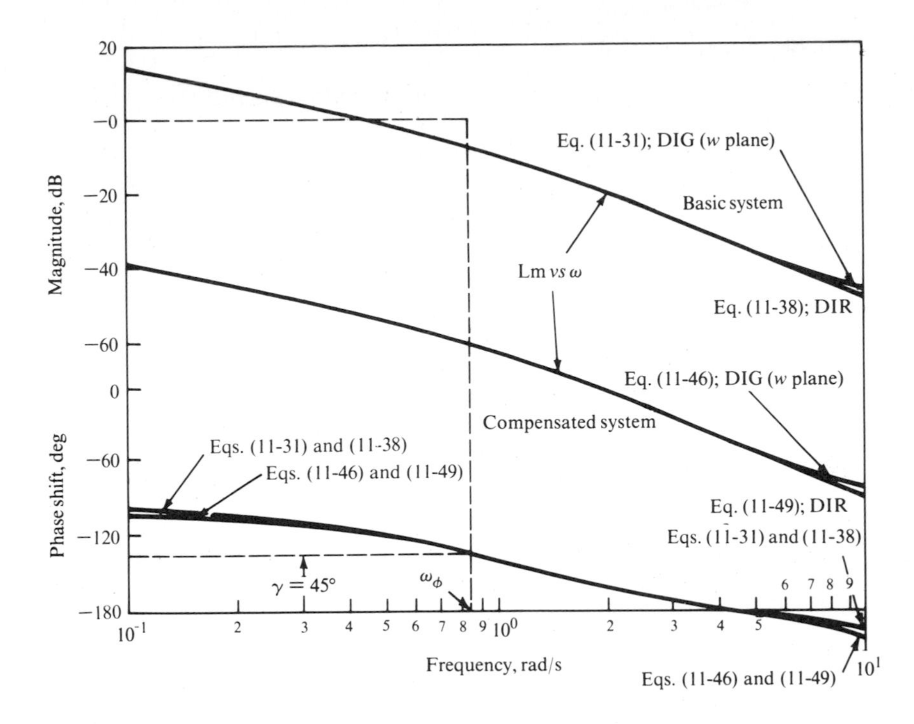

Fig. 11-15 Bode plots of Lm $\mathbf{G}(\cdot)$ vs. ω and ϕ vs. ω for the basic and lag compensated systems (w and z domain).

presents a comparison of the time-response characteristics for the basic, lead, and lag compensated systems.

As expected, an increase in the value of K_1 by use of a lag compensator is achieved at the expense of increasing the values of M_p, t_p, and t_s. Depending upon the pole location of the lead compensator (see Table 11-4), it may be possible to achieve the same improvement in K_1 as by the use of a lag compensator but with a greater improvement in the transient-response characteristics.

The Bode plots of the basic and lag compensated systems for both the w- and z-domain representations are shown in Fig. 11-15. This figure illustrates the effect of using a lag compensator to increase the value of the static error coefficient. That is, assuming that a phase margin of $\gamma = 45°$ is desired, the figure shows that the gain can be increased considerably for a 0-dB crossover at the desired value of ω_ϕ over that which is achievable for the basic system. Table 11-4 contains the values of M_m, ω_m, and ω_b for the various design techniques for both the basic and compensated systems. This table enhances the correlation between the frequency domain, the root-locus analysis, and the time domain. The effect of the implementation of $D_c(z)$ is discussed in Example 10-7. A conditionally stable system is maintained by ensuring truncation or roundoff of the computer-implemented $D_c(z)$ coefficients such that the controller's pole and zero remain inside the UC (see Chap. 10).

11-5B Lag-Lead Compensation

When both a large improvement in the static error coefficient and the transient-response characteristics are desired, then a lag-lead cascade compensator may be used. For the example of this section, the lag and lead portions of Eq. (11-43), for both the DIG and DIR techniques, are chosen to be identical to those used in Secs. 11-4 and 11-5A.

The DIG design yields the lag-lead compensator

$$D_c(w) = \frac{K_{wc}(w + 0.01)(w + 0.9992)}{(w + 0.001)(w + 9.992)} \tag{11-51}$$

which, when transformed into the z domain via Eq. (11-19), becomes

$$D_c(z) = \frac{K_{zc}(z - 0.999)(z - 0.9048)}{(z - 0.9999)(z - 0.33369)} \tag{11-52}$$

and is of the form of Eq. (11-44). The control ratios, for $\zeta = 0.7071$, for both the w and z domain are, respectively,

$$\frac{C(w)}{R(w)} = \frac{D_c(w)G_z(w)}{1 + D_c(w)G_z(w)} = \frac{-1.437 \times 10^{-3}(w + 0.01)(w - 20)(w + 1200)}{(w + 4.1498 \pm j4.1499)(w + 0.010026)} \tag{11-53}$$

where $K_{wc} = 72.43$, and

$$\frac{C(z)}{R(z)} = \frac{D_c(z)G_z(z)}{1 + D_c(z)G_z(z)} = \frac{0.1126(z - 0.999)}{(z - 0.611 \pm j0.2638)(z - 0.998997)} \tag{11-54}$$

where $K_{zc} = 48.829$ (path B of Fig. 11-2). Using the path A approach yields $K_{zc} = 50.735$ and time-response characteristics which are close to those obtained by the path B approach (see Table 11-4).

The DIR design yields the following lag-lead compensator of Fig. 11-1b:

$$D_c(z) = \frac{K_{zc}(z - 0.999)(z - 0.9048)}{(z - 0.9999)(z - 0.09048)} \tag{11-55}$$

Thus the corresponding root locus is shown in Fig. 11-16, and the corresponding control ratio for $\zeta = 0.7071$ is

$$\frac{C(z)}{R(z)} = \frac{0.2085(z - 0.999)}{(z - 0.44149 \pm j0.31196)(z - 0.0100262)} \tag{11-56}$$

Table 11-4 summarizes the transient- and frequency-response characteristics for each of the designs. As is to be expected, by the use of a lag-lead compensator a sizable improvement in the transient-response characteristics is achieved along with a large improvement in the static error coefficient.

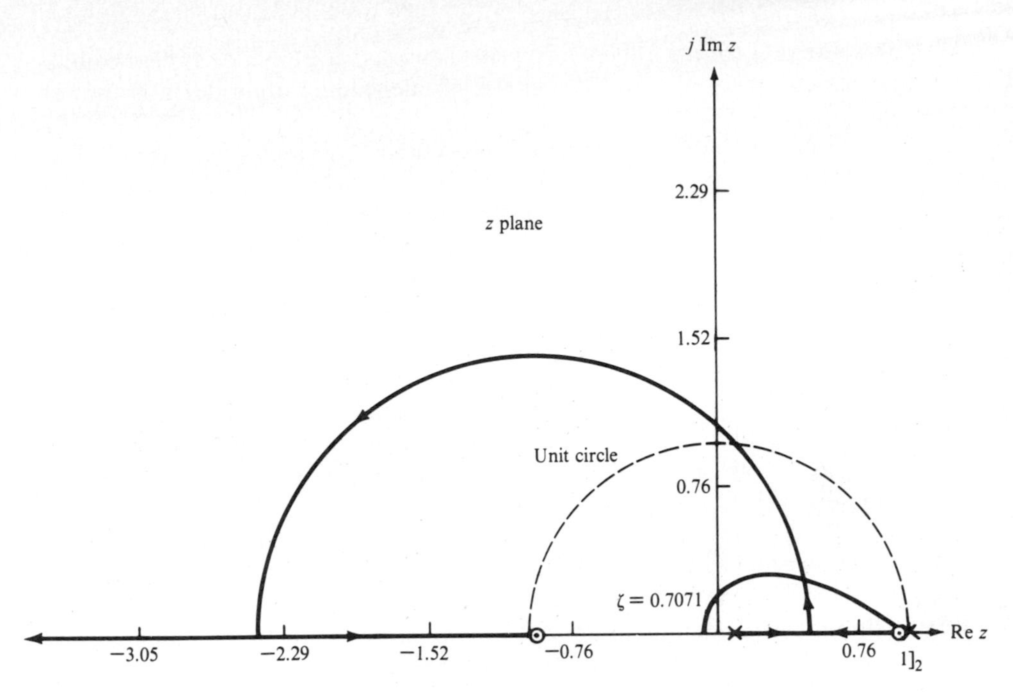

Fig. 11-16 Root locus for Eq. (11-55) where

$$D_c(z)G_z(z) = \frac{K(z + 0.9672)(z - 0.999)}{(z - 1)^2(z - 0.09048)}$$

11-5C PID Controller

The theoretical form of the PID controller is

$$D_{\text{PID}}(s) = K_d s + K_p + \frac{K_i}{s} = \frac{K_d s^2 + K_p s + K_i}{s} \tag{11-57}$$

By use of the approximations of differentiation and integration of Chap. 3, the discrete version of Eq. (11-57) (see Prob. 11-19) is

$$D_{\text{PID}}(z) = \frac{K_{\text{PID}}(z^2 - az + b)}{z(z - 1)} \tag{11-58}$$

which is physically realizable and where $K_{\text{PID}} = K_p + TK_i + (K_d/T)$, $a = (TK_p + 2K_d)/TK_{\text{PID}}$, and $b = K_d/TK_{\text{PID}}$. The approximation for integration can be rectangular, trapezoidal (Tustin, bilinear), or another numerical analysis technique. For differentiation, appropriate approximations are the backward-difference and other Taylor series expansions of ϵ^{sT}. Each approximation generates a different form of Eq. (11-58).

The PID controller is used in the process control industry where the parameters T, K_p, K_i, and K_d are tuned to achieve the desired response characteristics.

Also, in some process control environments, various set points or system equilibrium points are adjusted manually on the job, depending upon demand (power, flow, etc.).

The objective of a PID controller in a process control environment is usually to maintain a particular process set point (temperature, position, etc.) at a given value. Also, this set-point value may be dynamically stepped (manually or automatically) for process control, resulting in a series of step inputs to the PID controller. The proportional component defines the output to be reached depending upon the input set-point value. The integral term, if included, attempts to drive the output to zero steady-state error in conjunction with the negative feedback. Finally, the derivative control permits the PID mechanism to control the rate (damping) at which the plant responds to a set-point step input. All the three gains (K_d, K_p, K_i) determine the dynamic response of the system. Although the set point is based upon a step input as discussed, the input could also include a ramp or parabolic signal as well depending upon the desired system response.

When $D_{\text{PID}}(z)$ is inserted in cascade with a plant, the pole of Eq. (11-58) at $1 + j0$ increases the system type by 1. The two zeros of Eq. (11-58) may be located to counteract the effect of this pole, with respect to the degree of system stability. The zero locations can increase the degree of system stability. The specific tuning of the three gains can be accomplished through a hybrid simulation or an on-site testing. Another approach to determining the gains is the standard z-plane root-locus analysis technique (DIR) as illustrated by Example 11-3.

Example 11-3 Consider the plant of Sec. 7-6B with a PID controller using a root-locus approach. The desired system response is zero steady-state error to a set-point step input. Consideration of a ramp input response should also be studied. The open-loop transfer function of the compensated system is

$$G(z) = G_z(z)D_{\text{PID}}(z) = \frac{K_z K_{\text{PID}}(z^2 - az + b)(z + 0.9672)}{z(z-1)^2(z - 0.9048)} \tag{11-59}$$

where $K_z = 0.002306$ and $z^2 - az + b = (z - z_1)(z - z_2)$. Let $z_1 = 0.9048$ and $z_2 = 0.999$. Note that the portions $(z - 0.999)/(z - 1)$ and $(z - 0.9048)/z$ of the PID controller are comparable to the lag compensator of Eq. (11-48) and the lead compensator of Eq. (11-39), respectively. In general, for the $(z - z_2)/(z - 1)$ portion of the PID controller, Eq. (11-59), the zero z_2 must be close to the pole at 1 to minimize the *lag compensator* effect, i.e., degradation of the system stability or transient-response characteristics. Thus the PID controller when designed in this manner can be viewed as a lag-lead compensator. The resulting compensated forward transfer function is

$$G(z) = \frac{K_z K_{\text{PID}}(z - 0.999)(z + 0.9672)}{z(z-1)^2}$$

Thus for $T = 0.1$ s and $\zeta \approx 0.7071$ the root-locus analysis yields the control ratio

$$\frac{C(z)}{R(z)} = \frac{K(z + 0.9672)(z - 0.999)}{(z - 0.3749 \pm j0.3202)(z - 0.998998)}$$

where $K = 0.2513$, $K_{\text{PID}} = 108.98$, and the system is now Type 2. For $r(t) = u_{-1}(t)$: $M_p \approx 1.042$, $t_p \approx 0.5$ s, and $t_s = 0.6+$ s. Since the system is now a Type 2 system, the static error coefficients are $K_0 = K_1 = \infty$ and $K_2 = 0.04944\ \text{s}^{-2}$, resulting in $e^*(t) = 0$ for step and ramp inputs.

It is seen from Table 11-4 that the values of M_p and t_p of the lag-lead compensation design of Sec. 11-5B are essentially the same as those for the PID design. Locating the pole of the lead portion of the PID controller at the origin and not at 0.09048 results in a smaller value of t_s for the PID design.

11-6 FREQUENCY-DOMAIN COMPENSATION DESIGN USING MEAN-SQUARE ERROR MINIMIZATION

Many approaches to designing a digital or analog compensator are trial and error in nature. Thus, the final performance characteristics resulting from such a method generally do not completely meet the desired system performance specifications. Fine-tuning the design of a system can be accomplished in the frequency domain by attempting to minimize a mean-square-error function. The error function is of the form

$$E = \sum_{i=1}^{N'} |\mathbf{M}(j\omega_i) - \mathbf{M}_d(j\omega_i)|^2 W_M(j\omega_i) + \sum_{i=1}^{N'} [\Theta(j\omega_i) - \Theta_d(j\omega_i)]^2 W_\Theta(j\omega_i) \tag{11-60}$$

where $\mathbf{M}_d(j\omega_i)$ is the desired closed-loop frequency response

$$\mathbf{M}(j\omega) = \frac{\mathbf{C}(j\omega)}{\mathbf{R}(j\omega)} = \frac{\mathbf{G}(j\omega)}{1 + \mathbf{G}(j\omega)\mathbf{H}(j\omega)} \tag{11-61}$$

at a related discrete set of N' frequencies, ω_i, and $\Theta_d(j\omega_i)$ is the associated phase at each discrete frequency. $W_M(j\omega_i)$ and $W_\Theta(j\omega_i)$ are scalar weighting factors with $i = 1, \ldots, N'$. The magnitude of $\mathbf{M}(j\omega_i)$ and the value of $\Theta(j\omega_i)$ come from the initial closed-loop transfer function in polynomial form. For example, consider the initially designed controller to consist of p second-order filters, where each filter is of the form

$$\frac{1 + a_j z^{-1} + b_j z^{-2}}{1 - c_j z^{-1} + d_j z^{-2}} \tag{11-62}$$

The overall transfer function for the p filters in cascade is then

$$H_c(z) = K \prod_{j=1}^{p} \frac{1 + a_j z^{-1} + b_j z^{-2}}{1 + c_j z^{-1} + d_j z^{-2}} \tag{11-63}$$

The stated objective in reality is a constrained minimization problem. Equation (11-60), the error criterion, is to be minimized subject to the constraint of Eq. (11-61). In using Eq. (11-61), the initial values of the coefficients before minimization can be those resulting from the root-locus trial-and-error development (DIR or DIG) or discretization of a system with a continuous controller (DIG).

Note that the reason for using p second-order filters in cascade in Eq. (11-63) is because of the associated relatively low coefficient sensitivity. Other forms may be chosen for Eq. (11-63) such as a $2p$-order compensator (direct-unfactored form) or a partial-fraction (parallel) form. Parallel implementation can be faster and use less memory. The intended mode of implementation should be reflected in the form chosen for the minimization. Chapter 10 emphasizes the coefficient sensitivity considerations as related to implementation form.

The general approach to solving this constrained minimization problem is to use a gradient vector approach. The vector is of dimension $4p + 1$ since Eq. (11-63) has $4p + 1$ unknowns. The classical approach to minimizing Eq. (11-60) consists of symbolically differentiating this equation with respect to the $4p + 1$ parameters and setting the resulting functions (derivatives) equal to zero. This results in $4p + 1$ simultaneous equations which are to be solved for the $4p + 1$ unknowns.

Although $\mathbf{M}(j\omega)$ in Eq. (11-60) is associated with minimizing the overall system frequency-response error, the problem can instead be reformulated with only a desired controller $\mathbf{H}_d(j\omega)$. In this case the general approach is the same, again resulting in the determination of $4p + 1$ parameters. Emphasis is placed on matching a specific controller frequency response. Using this formulation, a more precise set of controller coefficients can be obtained based upon a desired controller frequency response.

Since the magnitude and phase contain transcendental functions of the unknown coefficients, numerical generation (approximation) of the error gradients is selected instead of symbolic differentiation. The numerical gradient is generated for each of the $4p + 1$ parameters. Then the computer algorithm[40] moves along the $(4p + 1)$-dimensional error function curve, Eq. (11-60), in the direction of steepest descent (i.e., negative gradient). In this way a minimum error function value is achieved within the accuracy of the computer employed.

In selecting the weights in Eq. (11-60) specific attention should be given to the various frequency bands and their interrelationships. For example, the control band, the vehicle or equipment body-bending frequencies, and the assumed noise or unknown disturbance bands should be studied.

Phase has been included in the problem formulation owing to the importance of phase delays outside the control band, especially their impact on system stability. Weighting phase in Eq. (11-60), therefore, usually involves the stability criterion.

Solving this constrained minimization problem generally results in a "better" compensator. Such improvement, however, may be very small (incremental). The cost of executing the steepest descent program can be costly, and thus computational analysis of trade-offs should be made.

Note also that the resulting compensator coefficient values can be generated

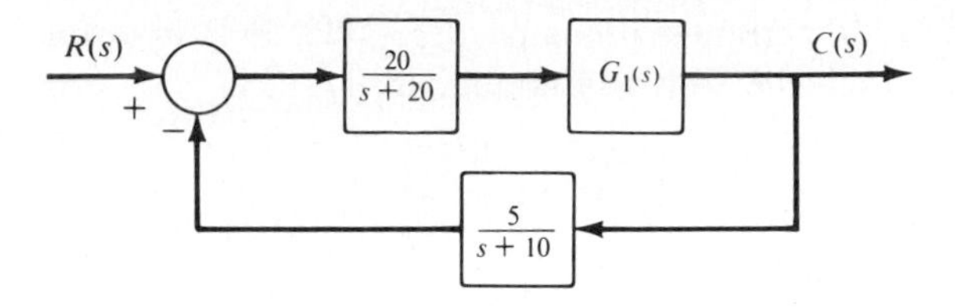

Figure 11-17 Continuous single-axis aircraft control system.

such that they have the same word length as in the final implementation. In this case, the gradient solutions can be generated in a hybrid computer environment.[41] An analog computer can represent the continuous plant which is to be controlled. The new coefficients resulting from the minimization are immediately put into the digital compensator, and immediate performance results relating to the new frequency and time responses can be generated in this real-time design environment. If such results are not completely satisfactory, the weighting and discrete frequencies of Eq. (11-60) can be changed on-line and new coefficients found. Of course, this process is again trial and error but with a fine-tuning capability. Various references (41–44) present good but limited results using the various approaches.

The minimization problem is easier to solve if only the magnitude terms of Eq. (11-60) are to be minimized.[42-44] In this case specific equations can be generated for the $4p + 1$ parameters, resulting in a matrix formulation. The solution to the $4p + 1$ equations results in the optimal parameter values, i.e., the values that minimize the error criterion.

Example 11-4 The minimization technique described previously is applied to the continuous aircraft controller (longitudinal axis pitch)[44] of Fig. 11-17. The corresponding digital-control-system model of Fig. 11-17 is shown in Fig. 11-18 where

$$G_1(s) = \frac{1.45s^3 + 147.6s^2 + 2.37s + 0.83}{s^5 + 20.95s^4 + 16.71s^3 - 46.66s^2 - 0.78s - 0.28}$$

$$G_2(s) = \frac{20}{s + 20}$$

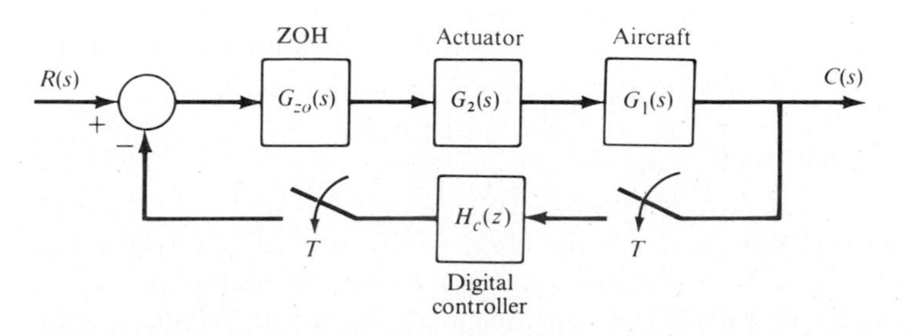

Figure 11-18 Discrete controller for single-axis aircraft system.

Note that $G_1(s)$ has a pole in the right-half plane (unstable). The overall continuous closed-loop system transfer function from Fig. 11-17 is

$$M(s) = \frac{1.45s^4 + 162.1s^3 + 147.8s^2 + 24.53s + 8.3}{s^6 + 30.95s^5 + 226.2s^4 + 127.7s^3 + 270.5s^2 + 3.75s + 1.31}$$

From Fig. 11-18, the closed-loop $\mathscr{Z}$-transfer function is

$$M(z) = \frac{G(z)}{1 + G(z)H_c(z)}$$

where

$$G(z) = \mathscr{Z}[G_{ZOH}(s)G_2(s)G_1(s)]$$

Using $\omega_M(j\omega_i) = 1$ and $\omega_\Theta(j\omega_i) = 0$ and assuming a first order over a first order ($p = 1, b_j = d_j = 0$) for $H_c(z)$ [see Eq. (11-63)] and with $T = 0.1$ s, the minimization routine generates

$$H_c(z) = \frac{0.5113(z - 0.6151)}{z - 0.6065}$$

The continuous and discrete controller frequency responses are shown in Fig. 11-19. Also shown is the frequency response for $\mathbf{C}(j\omega)/\mathbf{R}(j\omega)$ using a Tustin transformation for the original continuous controller, $H_c(s) = 5/(s + 10)$. The variation due to the Tustin transformation is attributed to the associated approximation with this approach.

11-7 DIGITAL FILTERS

In many digital control applications, prefilters are required to filter unwanted information such as noise, body-bending signals, and other external system signals that can incorrectly generate unwanted or disturbance control signals. Various procedures have been developed for continuous filter design. Using these design procedures, the digitization (DIG) approach can be used for mapping from the s plane to the z plane. The continuous low-pass filter approximation techniques such as the Butterworth and Chebyshev methods are often used as the basis for designing a digital filter.[39,42,93,98] The basic form used in the design is the square magnitude of the transfer function

$$|\mathbf{G}|^2 = |\mathbf{G}(j\omega)|^2 = \frac{1}{1 + \xi^2[P_n(j\omega)]^2} \tag{11-64}$$

where $P_n(j\omega)$ is a polynomial in ω, n is the order of the filter, and ξ is a design parameter. Note that the design method uses only frequency magnitude. Since phase is not included, the transients are not explicitly constrained. In this section, the analog design approach is summarized, and the digital design is achieved by

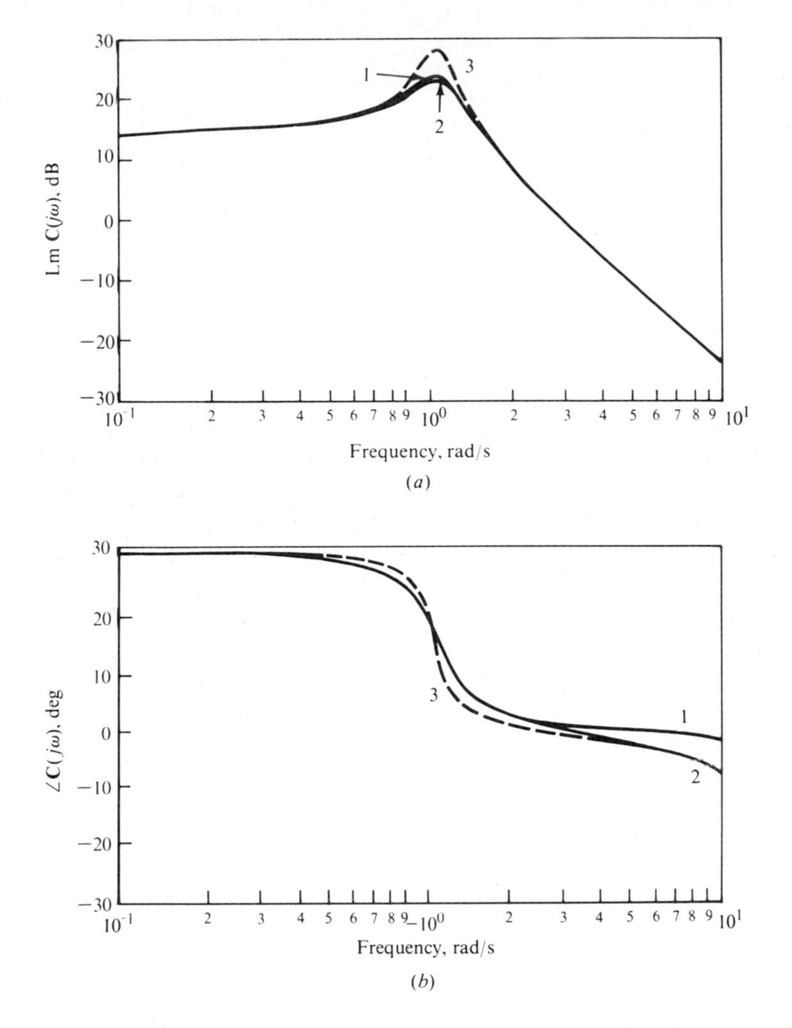

Figure 11-19 Frequency response of $\mathbf{C}(j\omega)$ with a sampling frequency of 10 Hz, $T = 0.1$ s. (*a*) Magnitude; (*b*) Phase. 1 = continuous controller; 2 = frequency-matching controller; 3 = Tustin controller.

using the impulse invariance approach (the exact $\mathscr{Z}$ transform) and the Tustin bilinear approach.

The Butterworth squared function is of the form

$$|\mathbf{G}|^2 = \frac{1}{1 + (j\omega/j\omega_c)^{2n}} \tag{11-65}$$

where $P_n = (j\omega/j\omega_c)$ and $\xi = 1$. The $2n - 1$ derivatives of this nth-order low-pass filter are equal to zero at $\omega = 0$. Therefore, by definition, the frequency response is maximally flat in the so-called passband as shown in Fig. 11-20.

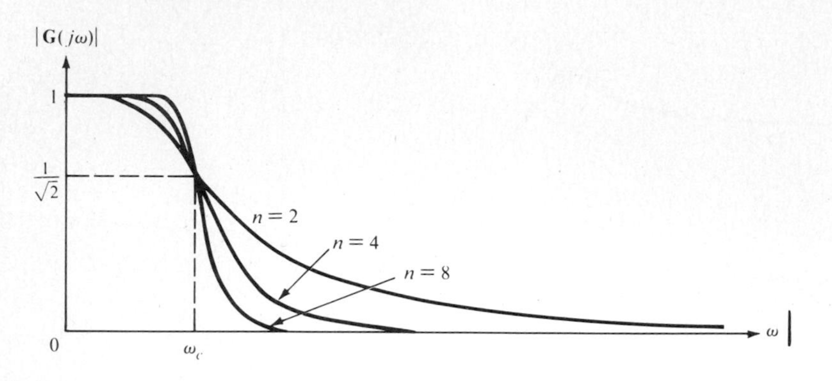

Figure 11-20 General Butterworth frequency response.

With the Butterworth model, the approximation to a perfect low-pass filter is monotonically decreasing in all bands (passband, transitionband, stopband). As n becomes larger, the transitionband becomes smaller. That is, the desired frequency-response specification lies within the general passband- and stopband-selected tolerances depending upon ξ and A (Fig. 11-21).

To determine the poles of the Butterworth filter, note that

$$G(s)G(-s) = \frac{1}{1 + (s/j\omega_c)^{2n}} \tag{11-66}$$

and

$$G(s) = \frac{D_0}{\prod_{k=1}^{n} (s - p_k)} \tag{11-67}$$

where the $G(s)$ poles are

$$p_k = (-1)^{(2k-1)/2n} j\omega_c = \epsilon^{j\pi(2k-1)/2n} j\omega_c \tag{11-68}$$

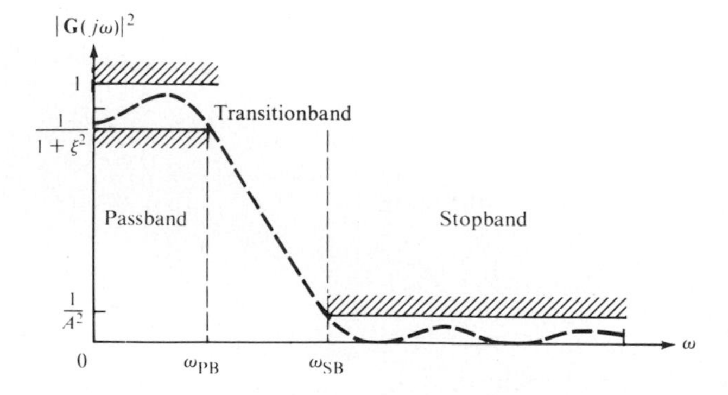

Figure 11-21 General error symbols for low-pass squared-magnitude filter.

$k = 1, \ldots, n$, and D_0 is a normalization constant for steady-state (dc) gain, usually equal to 1 in the passband. The $2n$ poles of Eq. (11-66) are equally spaced, π/n, on a circle of radius ω_c centered at the origin in the s plane. The n poles of $G(s)$ are selected from the left-half plane, i.e., a stable filter.

To determine the value of n, the specific values of the desired passband and the stopband tolerances must be selected (Fig. 11-21) (based upon $|\mathbf{G}(j\omega)|$, that is, $|1 - [1/(1 + \xi^2)^{1/2}]|$ and $1/A$, respectively).

For Butterworth filters

$$n = \frac{\ln r}{\ln L} \tag{11-69}$$

where

$$r \equiv \frac{\xi}{(A^2 + 1)^{1/2}} \qquad \text{and} \qquad L \equiv \frac{\omega_{\mathrm{PB}}}{\omega_{\mathrm{SB}}} \tag{11-70}$$

So, by direct substitutions, the value of n for the Butterworth form can be found.

From the Butterworth model, several properties are apparent:

1. Butterworth filters have a magnitude $\sqrt{1/2}$ at $\omega = \omega_c$.
2. Butterworth filters require the determination of poles only. Zeros of $G(s)$ are all at $s = \infty$.

Using the Tustin transformation in the Butterworth formulation yields, for the z-plane pole locus, another circle as shown in Fig. 11-22. A circle results in the z plane because the bilinear transformation is a conformal mapping. Note that the squared-magnitude function in the z plane has $2n$ zeros at $z = -1$ with the Tustin transformation. The poles of the discrete Butterworth filter are those within the UC and on the Butterworth circle (Fig. 11-22).

The process of generating the z-plane Butterworth filter is to derive the s-plane poles for $G(s)$ and then transform $G(s)$ to $G(z)$ by using the Tustin transform or the exact $\mathscr{Z}$ transform (impulse-invariant). In general, the Tustin transformation results in $[\mathbf{G}(\epsilon^{j\omega T})]_{\mathrm{TU}}$ decreasing faster than that of the exact $\mathscr{Z}$ transformation,

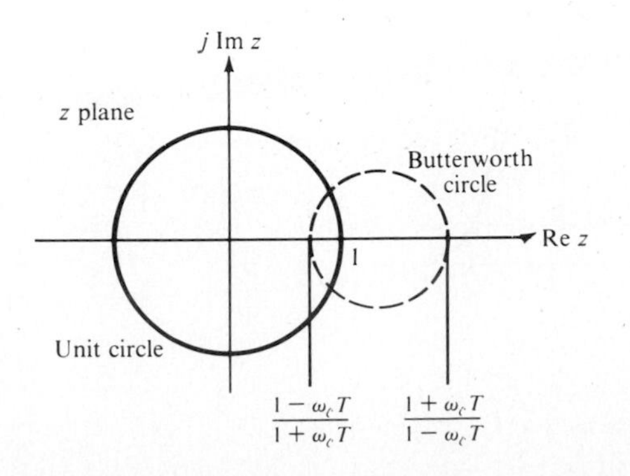

Figure 11-22 Butterworth z-plane circle transformed to z plane by Tustin transformation.

because the Tustin transformation maps the entire s-plane imaginary axis onto the unit circle.

The Chebyshev filter approximation is achieved in a very similar fashion. In the Chebyshev approximation,

$$P_n(j\omega) = \cos\left(n \cos^{-1} \frac{\omega}{\omega_c}\right) \tag{11-71}$$

or

$$P_{n+1}(j\omega) = \frac{2\omega}{\omega_c} P_n(j\omega) - P_{n-1}(j\omega) \tag{11-72}$$

For $\omega/\omega_c > 1$, $\cos^{-1}(\omega/\omega_c)$ is imaginary such that P_n acts like the hyperbolic cosine and therefore increases monotonically in that range. For $\omega/\omega_c < 1$, P_n oscillates. Therefore, $|\mathbf{G}(j\omega)|^2$ oscillates in the frequency range 0 to ω_c and monotonically decreases after $\omega = \omega_c$ (see Fig. 11-23). The value of ω_c has to be calculated and usually lies between ω_{PB} and ω_{SB}. The poles of the characteristic equation lie symmetrically on an ellipse with major-axis radius $\beta\omega_c^2$, where

$$\beta = (1/2)(a^{1/n} + a^{-1/n}) \tag{11-73}$$

and

$$a = \xi^{-1} + (1 + \xi^{-2})^{1/2} \tag{11-74}$$

The minor-axis radius is $\alpha\omega_c$, where

$$\alpha = (1/2)(a^{1/n} - a^{-1/n}) \tag{11-75}$$

The spacing of the poles is the same as for the Butterworth filter, π/n with respect to the imaginary axis. To find n, the following equation can be used:

$$n = \frac{\cosh^{-1}(1/r)}{\ln \dfrac{1 + (1 - L^2)^{1/2}}{L}} \tag{11-76}$$

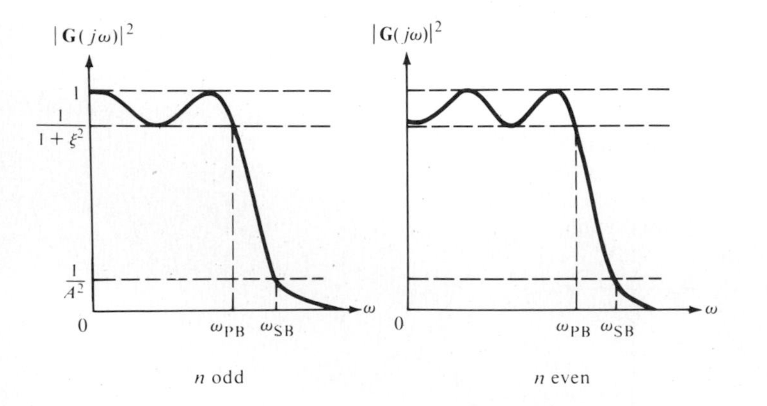

Figure 11-23 Chebyshev approximation to low-pass filter.

where r and L are the same as defined previously for the Butterworth design. To determine the exact Chebyshev filter poles, the equally spaced points are identified. For n odd, a point (pole) is located on the real axis but not on the imaginary axis. For n even, no poles fall on the real axis. CAD programs can easily be developed to generate the analog filters and z-plane equivalents.

From the Chebyshev model, several properties are apparent:

1. The Chebyshev filter magnitude oscillates between zero frequency and ω_c (see Fig. 11-23). The magnitude monotonically decreases beyond ω_c. The passband edge occurs at $\omega/\omega_c = 1$ and the stopband at $\omega = \omega_{SB}$.
2. Chebyshev filter design requires the determination of poles only. Zeros of $G(s)$ are all at $s = \infty$.
3. For a given set of analog filter specifications, a Chebyshev filter meets specifications with a smaller n than that of a Butterworth filter.

With the appropriate selection of P_n in Eq. (11-64) an inverse Chebyshev filter can also be designed.[39,42] This filter magnitude monotonically decreases in the passband and oscillates in the stopband. Other squared-magnitude filters can be designed where P_n is a Jacobi elliptic function. This formulation results in elliptic filters. Elliptic filter magnitudes oscillate both in the passband and stopband. n is usually smaller for elliptic filters than for Butterworth and Chebyshev in meeting performance specifications. In all these cases, equations for n exist and the associated analog filters are easily designed. Low-pass analog filters can also be transformed into high-pass, band-pass, or band-stop filters with the use of simple equations. Transformation to the discrete domain can easily be accomplished by using the Tustin transformation or other techniques, each producing to some degree a good discrete design. Warping considerations are still applicable as well as word length.

One reason for presenting the discussion of continuous filters is their ease of design. Moreover, the designer can study explicitly the impact of filter specifications on n and adjust the specifications if practical to decrease n both in s- and $\mathscr{Z}$-transform implementations.

Example 11-5 Consider the development of an analog prefilter (Butterworth) for the examples of Sec. 11-4. Assume that the control bandwidth (passband) ω_{PB} is zero to one-half the sampling frequency, $10\pi = \omega_s/2$. The "noise" to be filtered is beyond this value of ω_{PB}. Because of the critical amplitudes in the passband, the passband tolerance is selected to be 0.1. Stopband tolerance is selected to be 0.25. Phase also may be important and should be analyzed after the preliminary filter design. To determine n, the value of ω_{SB} is selected to be $2\omega_{PB}$, that is, $\omega_{SB} = 20\pi$.

Thus

$$(1 + \xi^2)^{-1/2} = 0.9 \rightarrow \xi = 0.333$$

$$\frac{1}{A} = 0.25 \rightarrow A = 4$$

and from Eq. (11-70)

$$r = \frac{0.333}{16 + 1} = 0.0196$$

The order of the filter is, from Eq. (11-69),

$$n = \frac{\ln 0.0196}{\ln 0.5} = 5.682$$

which must be rounded off to the nearest integer, which in this case is 6. Inserting the values $\omega_{PB} = 10\pi$ and $n = 6$ into Eq. (11-65), that is,

$$|\mathbf{G}(j\omega_{PB})|^2 = \frac{1}{1 + (\omega_{PB}/\omega_c)^{12}} = (0.9)^2$$

yields $\omega_c = 12\pi$. The transfer function of the resulting filter is

$$G(s) = \frac{D_0}{(s + 9.7603 \pm j36.4136)(s + 26.6553 \pm j26.6553)(s + 36.4136 \pm j9.7603)}$$

where $D_0 = 2.87 \times 10^9$. The filter response (magnitude and phase) is shown in Fig. 11-24. The high gain of this continuous filter is an implementation problem.

Example 11-6 The discrete version of the analog filter from the previous example is to be obtained by use of the Tustin transformation. Since the raw "noisy" signal is to be sampled, a value of the sampling time T_f for the filter

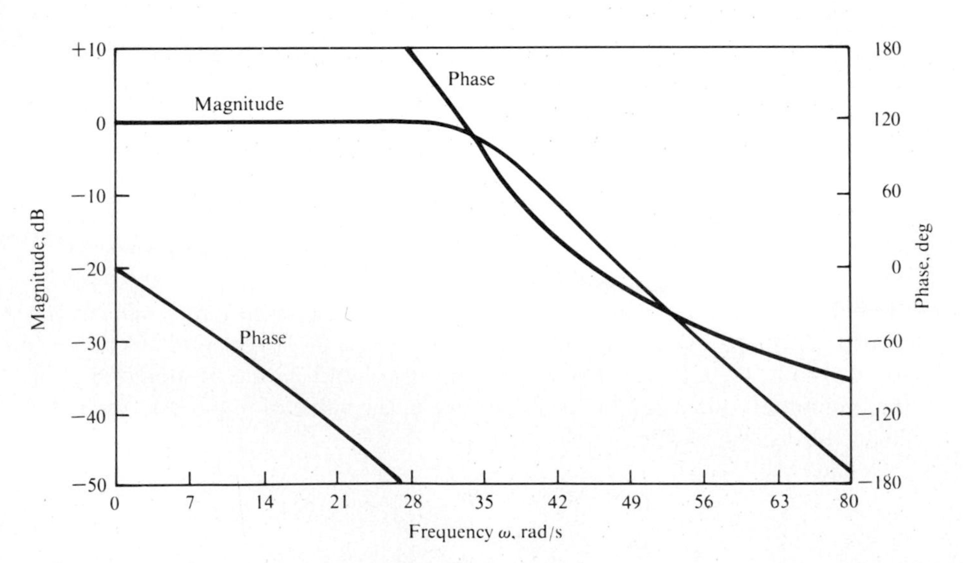

Figure 11-24 Continuous Butterworth frequency response.

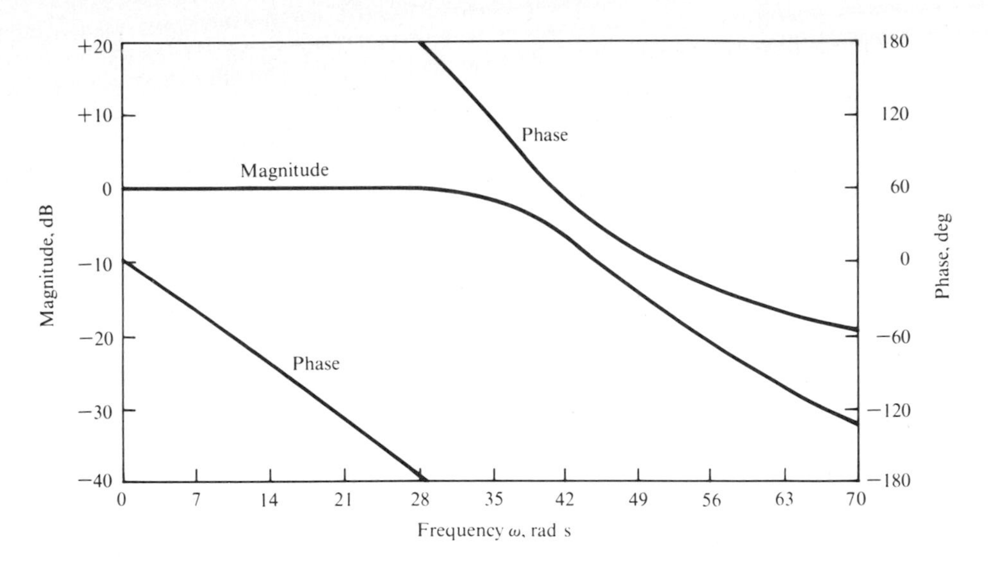

Figure 11-25 Discrete Butterworth frequency response.

must be determined. This value is obviously less than the value selected for the original control problem; that is, $T = 0.1$ s. In this case, assume that the noise band is from 10π to 40π. Therefore, $T_f = 2\pi/80\pi = 0.025$ s. It is also assumed that the digital filter is executed on a separate processor not synchronized with the controller computer. Thus, using the Tustin transformation, $G(s)$ becomes

$$G(z) = \frac{0.001854(z+1)^6}{(z - 0.4120 \pm j0.3529)(z - 0.3648 \pm j0.1144)(z - 0.5306 \pm j0.6209)}$$

with the frequency-domain representation for magnitude and phase as shown in Fig. 11-25. The six zeros of $G(z)$ at $z = -1$ are due to the six zeros of $G(s)$ at $s = \infty$. Note that the discrete Butterworth magnitude falls off more rapidly owing to the Tustin mapping. Also, the phase in the passband ranges between 0 and 180°. In a real-time system the associated delay may cause stability problems.

11-8 SOFTWARE FOR A DIGITAL CONTROLLER

The previous sections of this chapter have stressed the methods for designing a controller (or compensator) to try to achieve the desired control-system performance specifications. In this section the software implementation of $D_c(z)$ is discussed. The basic modules in a digital controller (Fig. 11-26) are data input, data processing (filtering), and data output as called by an executive module. Note that the digital control program in this example provides for two inputs and two

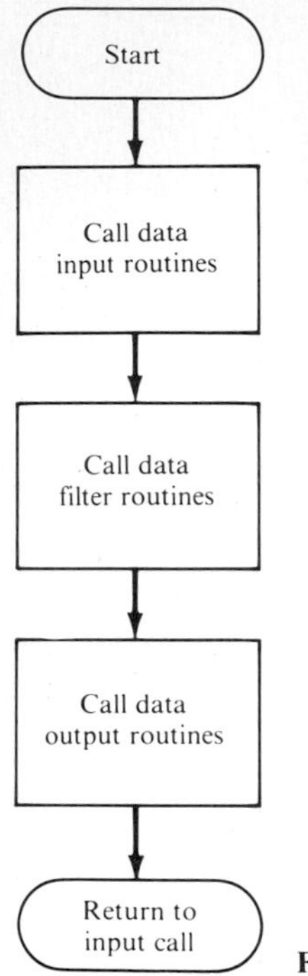

Figure 11-26 General flow diagram for executive program.

outputs; i.e., two separate processes can be controlled with this program. Each filter or compensator is a second-order filter, although in a given application these filters can be executed more than once in cascade to provide higher-order compensation. Of course, the coefficients may have to be changed for each filter execution. An array (vector) of coefficients can be passed to each filter routine to facilitate this capability. For a single-input–single-output compensator, the single-filter implementation is required. On the other hand, both a cascade and a feedback compensator may be implemented by this program for a sampled-data control system that requires both types of compensators to achieve the desired system performance. Figure 11-27 depicts the overall program modular structure.

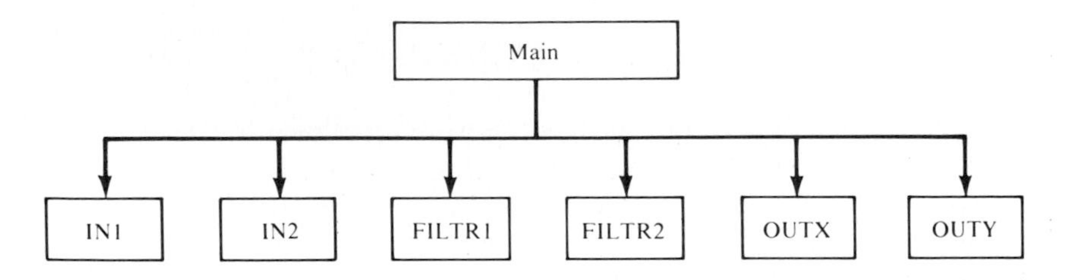

Figure 11-27 Digital-controller program structure chart.

The following list describes in general each operation:

Executive (*MAIN*). The executive routine initializes the A/D conversion subsystem, then enters an infinite loop calling the input, the filter (compensation), and the output routines (Fig. 11-28).

Input (*IN1*, *IN2*). This routine waits for the A/D conversions to be completed,

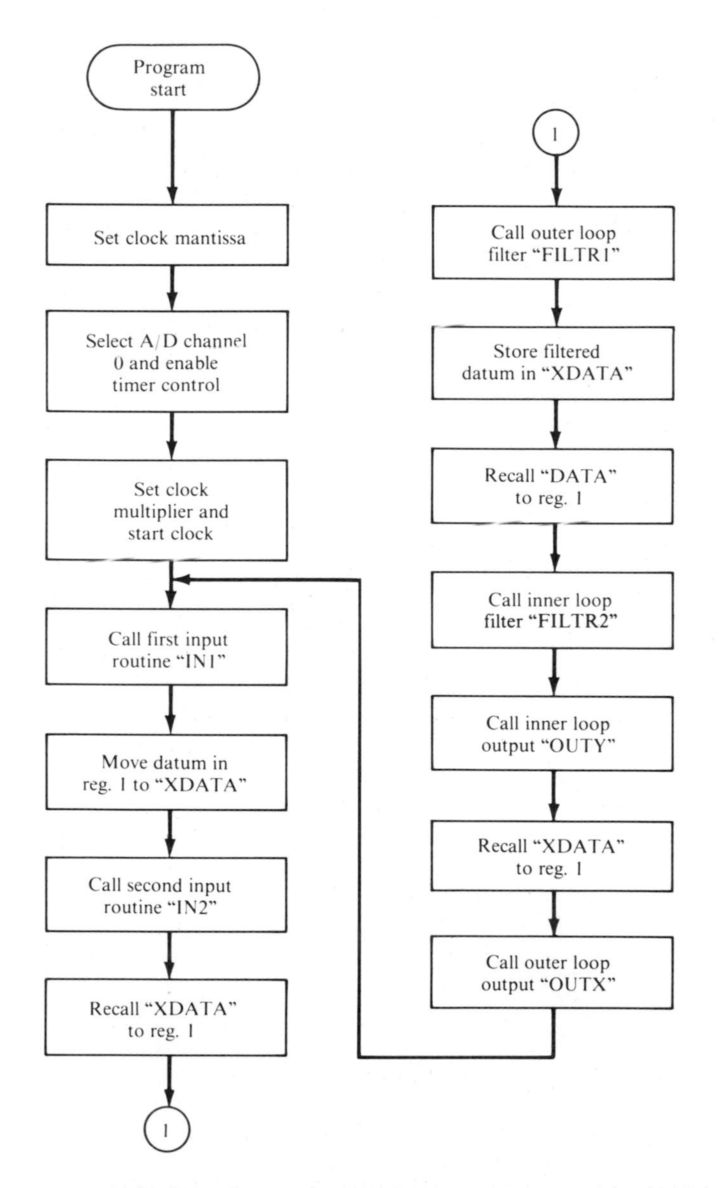

Figure 11-28 Flow diagram for digital plant controller executive MAIN.

then processes the input, checks for overflow and underflow, and left-justifies the data (Figs. 11-29 and 11-30).

Filter (*FILTR1*, *FILTR2*). These routines (Fig. 11-31) allow the user to realize up to a second-order filter by adjusting the coefficients of the difference equation. The coefficient values shown are those of Eq. (11-37). This routine can be called many times to realize a cascade filter.

Output (*OUTX*, *OUTY*). This routine rounds the data, right-justifies the data, then outputs it to the D/A converter for one of two outputs (Figs. 11-32 and 11-33).

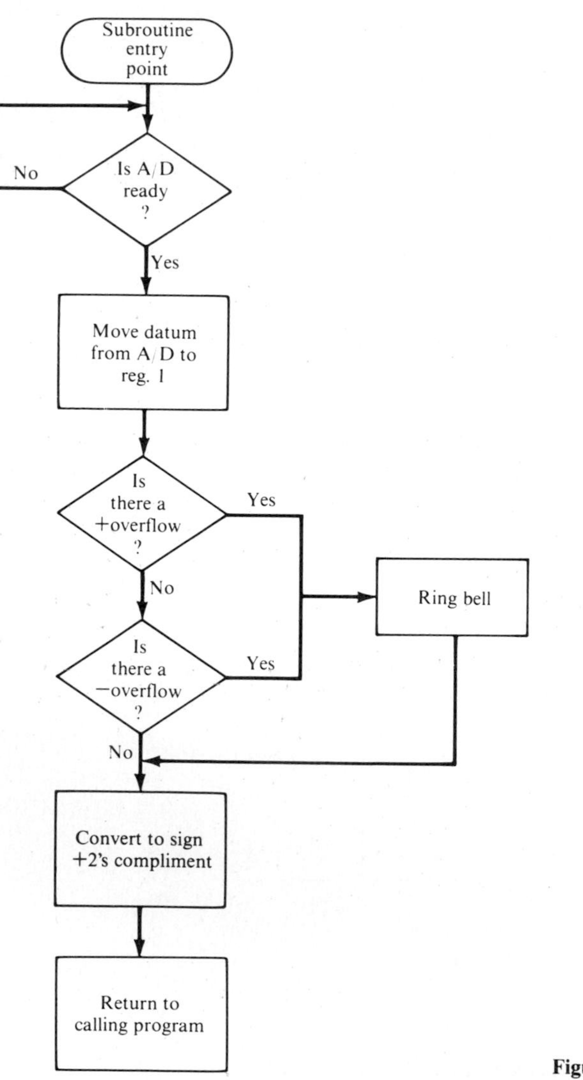

Figure 11-29 IN1 flowchart.

The documentation reflected in the associated PDP-11 relocatable modular program listing (Fig. 11-34) and the flowcharts provide a more detailed discussion of each module. Note the level of documentation and programming style. The following sections discuss each routine.

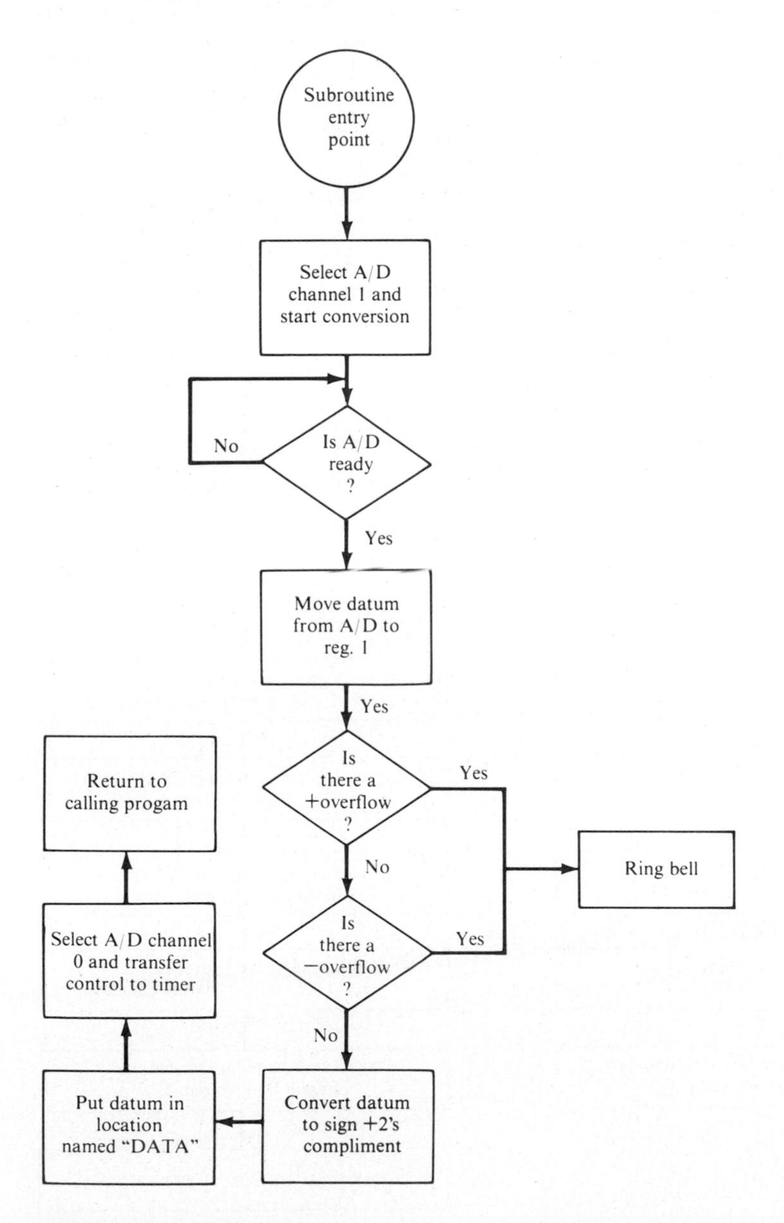

Figure 11-30 IN2 flowchart.

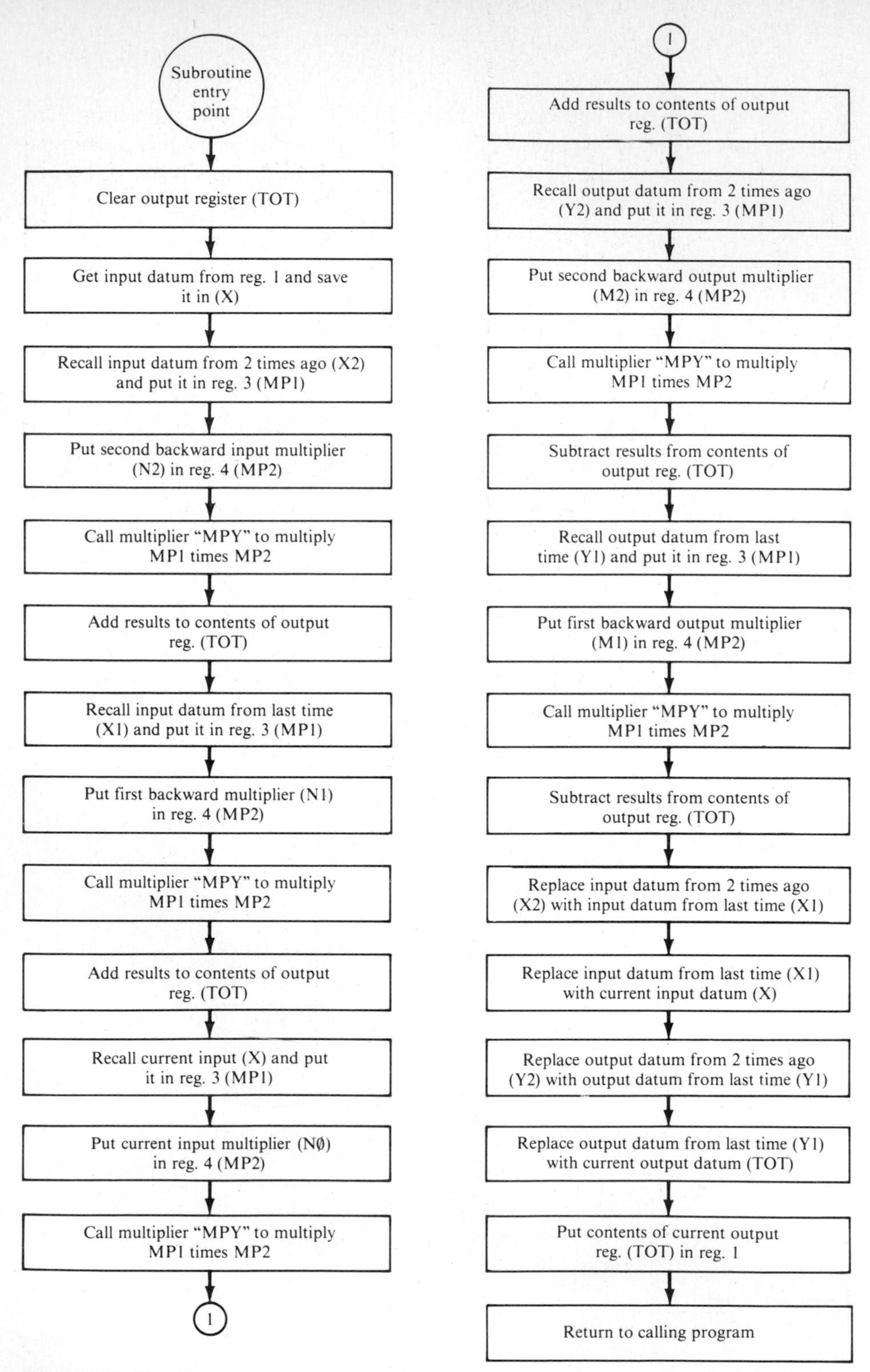

Figure 11-31 Flowchart for filter programs.

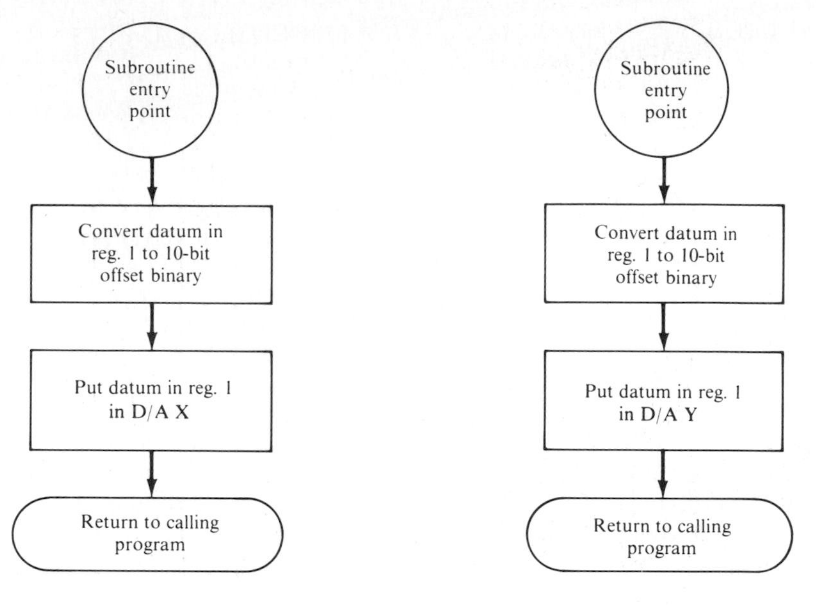

Figure 11-32 OUTX flowchart.

Figure 11-33 OUTY flowchart.

11-8A Program Executive for Digital Control Systems

This program is the main driver for the digital controller. The following programs must be linked with this program to function properly:

Input. These routines obtain the A/D converted input.
Filter. These routines implement the control algorithm.
Output. These routines write the manipulated data out to a buffer for the D/A conversion.

The executive establishes the clock counter at -10, the A/D mode is bipolar, and the input channel is channel 0. The clock is set for repeated intervals, which means it restarts the countdown each time it reaches zero. The clock base frequency is 1 kHz. The clock counter divides 10 into the 1-kHz base clock resulting in a conversion frequency of 100 Hz. For a T of 0.1 s, a counter vaue of -100 is used.

This program contains the following global variables which are known to all program modules:

CONT: temporary storage for value of clock counter
ADMC: temporary storage for A/D mode and channel
CKM: temporary storage for desired clock mode
ADSR: absolute address of A/D status register
CKB: absolute address of clock counter buffer
CKSB: absolute address of clock status buffer
DATA: temporary storage for IN2 data

```
;EXECUTIVE PROGRAM FOR DIGITAL CONTROLLER WITH TWO INPUTS AND
;TWO OUTPUTS

MAIN:   GLOBL   IN1,IN2,OUTX,OUTY,FILTR1,FILTR2,ADSR,CKB,
                CKSB,DATA
        MOV     COUNT,@CKB      ;SET CLOCK COUNTER
        MOV     CKM,@CKSB       ;SET CLOCK AND START
        MOV     ADMC,@ADSR      ;SET MODE AND CHANNEL FOR A/D
START:  JSR     %0,IN1          ;GET INPUT DATA, CH 0
        MOV     %1,XDATA        ;STORE IN1 DATA
        JSR     %0,IN2          ;GET INPUT DATA, CH 1
        MOV     XDATA,%1        ;RESTORE IN1 DATA
        JSR     %0,FILTR1       ;FILTER IN1 DATA
        MOV     %1,XDATA        ;SAVE FILTERED IN1 DATA
        MOV     DATA,%1         ;RESTORE IN2 DATA TO REG 1
        JSR     %0, FILTR2      ;FILTER IN2 DATA
        JSR     %0,OUTY         ;OUTPUT NEW Y DATA FROM REG 1
        MOV     XDATA,%1        ;RESTORE IN1 DATA
        JSR     %0,OUTX         ;OUTPUT NEW X DATA
        BR      START           ;RETURN FOR NEXT INPUT
ADMC:   WORD    00040           ;A/D MODE AND CHANNEL
                                ;BIPOLAR, REAL-TIME CLOCK ENABLE
COUNT:  .WORD   000366          ;2'S COMP OF DESIRED COUNT
                                ;INITIALLY COUNT = -10
CKM:    .WORD   000431          ;DESIRED CLOCK MODE,
                                ;    REPEATED INTERVAL
                                ; MODE.1K HZ CLOCK, CLOCK ENABLE
ADSR:   .WORD   170400          ;A/D STATUS REGISTER ADDRESS
CKB:    .WORD   170406          ;CLOCK PRESET BUFFER ADDRESS
CKSB:   .WORD   170408          ;CLOCK STATUS BUFFER ADDRESS
XDATA:  .WORD   0               ;INTERMEDIATE STORAGE LOCATION
        .END    MAIN

;COMPENSATOR "FILTER" PROGRAM
    GLOBL MPYA,X1,X2,Y1,Y2,X,TOT,NO,N1,N2,M1,M2,FILTR1
    ;ASSIGN REGISTERS
    R1=%1
    MAC1=%2
    MP1=%3
    MP2=%4
FILTR1: CLR TOT             ;CLEAR OUTPUT REGISTER (INITIALIZE)
;       GET R(K) FROM INPUT
        MOV R1,X            ;MOVE INPUT DATA TO X
;       DETERMINE A2*R(K-2) TERM AND ADD TO TOTAL (C(K))
        MOV X2,MP1          ;MOVE R(K-2) TO MULTIPLICAND
        MOV N2,MP2          ;MOVE A2 TO MULTIPLIER
        JSR %0, MPY         ;JUMP TO MULTIPLY ROUTINE
        ADD MAC1,TOT        ;ADD WEIGHTED TERM TOTAL
```

Figure 11-34 List II digital-controller program listing.

```
;       DETERMINE A1*R(K-1) TERM AND ADD TO TOTAL (C(K)).
        MOV X1,MP1          ;MOVE R(K-1) TO MULTIPLICAND
        MOV N1,MP2          ;MOVE A1 TO MULTIPLIER
        JSR %0,MPY          ;JUMP TO MULTIPLY ROUTINE
;       ADD MAC1,TOT        ;ADD WEIGHTED TERM TO TOTAL
;       DETERMINE A0*R(K) TERM AND ADD TO TOTAL (C(K))
        MOV     X,MP1       ;MOVE R(K) TO MULTIPLICAND
        MOV     NO,MP2      ;MOVE AO TO MULTIPLIER
        JSR     %0,MPY      ;JUMP TO MULTIPLY ROUTINE
        ADD     MAC1,TOT    ;ADD WEIGHTED TERM TO TOTAL
;       DETERMINE B2*C(K-2) TERM AND SUBTRACT FROM TOTAL (C(K))
        MOV     Y2,MP1      ;MOVE C(K-2) TO MULTIPLICAND
        MOV     M2,MP2      ;MOVE B2 TO MULTIPLIER
        JSR     %0,MPY      ;JUMP TO MULTIPLY ROUTINE
        SUB     MAC1,TOT    ;SUBTRACT WEIGHTED TERM FROM TOTAL
;       DETERMINE B1*C(K-1) TERM AND SUBTRACT FROM TOTAL (C(K))
        MOV     Y1,MP1      ;MOVE C(K-1) TO MULTIPLICAND
        MOV     M1,MP2      ;MOVE B1 TO MULTIPLIER
        JSR     %0,MPY      ;JUMP TO MULTIPLY ROUTINE
        SUB     MAC1,TOT    ;SUBTRACT WEIGHTED TERM FROM TOTAL
;       REPLACE PAST VALUES WITH PRESENT VALUES FOR NEXT ITERATION
        MOV     X1,X2       ;R(K-1) BECOMES R(K-2)
        MOV     X,X1        ;R(K) BECOMES R(K-1)
        MOV     Y1,Y2       ;C(K - 1) BECOMES C(K-2)
        MOV     TOT,Y1      ;C(K) BECOMES C(K-1)
        MOV     TOT,R1      ;MOVE C(K) TO REGISTER 1 FOR OUTPUT
        RTS     %0          ;RETURN TO CALLING ROUTINE
```

```
; SET UP WORKING (SCRATCH) REGISTERS
X1:     .WORD 0         ;R(K-1)
X2;     .WORD 0         ;R(K-2)
Y1:     .WORD 0         ;C(K-1)
Y2:     .WORD 0         ;C(K-2)
X:      .WORD 0         ;R(K)
TOT:    .WORD 0         ;C(K)--TOTAL
NO:     .WORD 000163    ;A0  (COEFFICIENT OF R(K))
N1:     .WORD 000157    ;A1  (COEFFICIENT OF R(K-1))
N2:     .WORD 0         ;A2  (COEFFICIENT OF R(K-2))
M1:     .WORD 175436    ;B1  (COEFFICIENT OF C(K-1))
M2:     .WORD 000705    ;B2  (COEFFICIENT OF C(K-2))
        .END FILTR1

;INPUT "IN1" PROGRAM
        GLOBL IN1 ADSR
```

```
IN1:    TSTB    @ADSR           ;TEST FOR A/D COMPLETE
        BPL     IN1             ;IF NOT COMPLETE RECHECK
        MOV     @ADBR,%1        ;MOVE A/D OUTPUT TO R1
        CMP     #1777,%1        ;CHECK FOR + OVERFLOW
        BEQ     OVER            ;IF OVERFLOW GO TO OVER
        CMP     #0000,%1        ;CHECK FOR – OVERFLOW
        BNE     OK              ;CONTINUE IF NO OVERFLOW
OVER:   MOV     #7,@PB          ;ECHO BELL TO INDICATE OVERFLOW
OK:     SUB     #1000,%1        ;CONVERT TO SIGN + 2S COMP
        RTS     %0              ;RETURN
ADBR:   .WORD   170402          ;A/D BUFFER REGISTER
PB:     .WORD   177566          ;PRINT BUFFER
        .END    IN1

;INPUT "IN2" PROGRAM
        GLOBL IN2,ADSR,DATA
IN2:    MOV     #000441,@ADSR   ;TEST FOR A/D COMPLETE
INPUT:  TSTB    @ADSR           ;TEST FOR A/D COMPLETE
        BPL     INPUT
        MOV     @ADBR,%1        ;MOVE A/D OUTPUT TO R1
        CMP     #1777,%1        ;CHECK FOR + OVERFLOW
        BEQ     OVER            ;IF OVERFLOW GO TO OVER
        CMP     #0000,%1        ;CONTINUE IF NO OVERFLOW
OVER:   MOV     #7,@PB          ;ECHO BELL TO INDICATE OVERFLOW
OK:     SUB     #1000,%1        ;CONVERT TO SIGN + 2S COMP
        MOV     %1,DATA         ;SAVE DATA
        MOV     #00040,@ADSR    ;REINITIALIZE A/D STATUS
        RTS     %0              ;RETURN
ADBR:   .WORD   170402          ;A/D BUFFER REGISTER
PB:     .WORD   177566          ;PRINT BUFFER
DATA:   .WORD   0
        .END    IN2

;OUTPUT "OUTY" PROGRAM
    GLOBL OUTY
OUTY:   ADD     #1000,%1        ;CONVERT TO OFFSET BINARY
OK:     MOV     %1,@DABR        ;MOVE DATA TO D/A
        RTS     %0              ;RETURN
DABR:   .WORD   170414          ;D/A BUFFER REGISTER
        .END    OUTY

;OUTPUT "OUTX" PROGRAM
    GLOBL OUTX
OUTX:   ADD     #1000,%1        ;CONVERT TO OFFSET BINARY
OK:     MOV     %1,@DABR        ;MOVE DATA TO D/A
        RTS     %0              ;RETURN
DABR:   .WORD   170412          ;D/A BUFFER REGISTER
        .END    OUTX
```

11-8B Data Input

Two signals are assumed to be sampled by the digital controller, each with a magnitude less than 1; thus the input routines do not left-justify the data. The multiplication routine is assumed to be fixed-point multiplication, with a multiplier having 10 binary digits after the binary point. Thus, the input can be represented correctly in fixed point. Sign extension is used to adjust the data to 16-bit accuracy.

The routine IN1 waits (noninterrupt) for the completion of an A/D conversion which is indicated by the real-time clock. The frequency is established by the executive routine at 100 Hz. When the conversion is complete, the input routine obtains the data, manipulates it as discussed above, and returns to the executive. The executive takes care of storing in memory the data that this routine left in the PDP-11 processor's register.

The routine IN2 is called next by the executive. This routine signals an immediate request for an A/D conversion on channel 1; then a wait loop is entered pending completion of the conversion. The data are again massaged as discussed previously. However, to set up for the real-time conversion for channel 0, the A/D status register has to be restored prior to the return to the executive.

Three registers in the AR-11 (the analog conversion system of the PDP-11) control the input of data:

Clock status register (CSR)
Clock buffer/preset register (CKB)
Analog-to-digital status register (ADSR)

Clock status register (CSR) For the purposes of this discussion only, 5 bits are of interest, with the rest set to zero.

Bit 8 (*mode*) is set to 1 for repeated intervals. This means the real-time clock counter is reloaded from the clock preset buffer and the count initiated again.
Bits 1–3 (*sampling rate*) determine the base frequency. This is the rate at which the counter is incremented. Consequently, the conversion rate equals the base frequency divided by the sign plus 2's complement value in the preset buffer. 100_2 sets the base frequency at 1 kHz.
Bit 0 (*clock enable*) starts the counter at the specified rate.

For the AR-11, bit 4 is always 1 so the octal value that is loaded in the clock status register as 000431 is at address 170404. This is accomplished by the executive routine.

Clock buffer/preset register (CKB) The lower 8 bits of the register are used as a divider for the base frequency to extend the number of sampling rates available. This number is the sign plus 2's complement of the number desired. A 100-Hz sampling rate is selected. Since the base frequency is 1 kHz, it is necessary to divide by 10_{10}, where $10_{10} = 00001010_2$ and where the sign plus 2's complement number is

$11110110 = 366_8$. For a divide by 10, the preset buffer at absolute address 170406 is loaded with 000366.

Analog-to-digital status register (ADSR) This register is the only one that has to be changed by the input routines after the real-time clock is started. Since there is one A/D converter with 16 inputs, the ADSR allows an individual to designate the input channel that is desired to be sampled.

Bits 8–11 are used to represent the channel number for the current conversion. Bit 13 is used to designate bipolar or unipolar for the selected input channel. This discussion deals only with bipolar; bit 13 = 0. Bit 1 is set to 1 in the input routine to request conversion IN2. At the completion of a conversion, bit 1 is reset and bit 7 is set indicating the conversion is complete.

When the real-time clock signals a conversion to be initiated, the current values of the ADSR govern the conversion. Consequently the status register has to be reset for channel 0 by the second input routine in sequence.

11-8C Program Filter (Simple Second-Order Filter)

The following equation is implemented by this routine:

$$C(K) = A2*R(K-2) + A1*R(K-1) + AO*R(K) - B2*C(K-2) - B1*C(K-1)$$

where the R(K)'s = past and current inputs
C(K)'s = past and current outputs of this filter

This routine can be used to generate a first order, second order, or simply a gain by setting the coefficients to the appropriate values. The coefficients can be obtained directly from the backward-difference equations for the filter desired. Of course, finite-word-length considerations must be acknowledged. A gain can be implemented by setting the coefficients of the R(K)'s to these values.

11-8D Data Output

The controller requires two output routines. Since the overflow is checked by the filter routine, the output routines only have to adjust the signs for offset binary and output the data to the appropriate D/A buffer register. The absolute address for the X output buffer is 170412, and for the Y output, the address is 170414.

11-8E Important Implementation Notes

When working in bipolar mode, with output -5 to $+5$ V, the D/A hardware in this implementation has an inherent gain of 2.

The coefficients have a specific format in the multiplication routine providing fixed-point multiplication. Consequently, the coefficients can assume decimal values from -31.99902344 to -31.99902344 in steps of 0.00097.

The format consists of 6 binary digits, a binary point, followed by 10 binary digits. The following are example octal numbers and the decimal equivalent:

$$176000 = -1.0$$

$$002000 = +1.0$$

$$001000 = +0.5$$

$$000200 = +0.1$$

In the filter routine, the bulk of the processing takes place and consequently causes the largest processing delay. For a more in-depth understanding of the filter, see the documentation. FILTR1 is used for one input; FILTR2 is the same filter program with different coefficients.

11-9 DEADBEAT RESPONSE[18,20]

The design method for cascade digital controllers emphasized in this text is the root-locus approach, even for a PID structure. Other approaches focus on designing a cascade controller to meet a variety of output characteristics. One example specification is to have the output settle exactly to the input signal in a finite number of sample times. Digital control systems based on this finite settling-time design are sometimes called *deadbeat* systems. This problem is one of a class of regulator problems. The design approach can be defined by using a scalar model or a vector model. A scalar approach is presented here.

As indicated, the objective of a *discrete deadbeat response* system is to match the output to the input signal at each sampling instant. Consider then a general polynomial input; that is,

$$r(kT) = (kT)^p \qquad p \geq 0 \tag{11-77}$$

The system output $c(kT)$ as a function of the discrete system weighting function $m(kT)$ is

$$c(kT) = \sum_{p=0}^{\infty} m(pT)r(kT - pT)$$

or

$$C(z) = M(z)R(z) \tag{11-78}$$

Assuming a cascade controller structure, $D_c(z)$ in the forward loop with a discrete plant transfer function $G_z(z)$, yields for the system function

$$M(z) = \frac{D_c(z)G_z(z)}{1 + D_c(z)G_z(z)} \tag{11-79}$$

or

$$C(z) = \frac{D_c(z)G_z(z)}{1 + D_c(z)G_z(z)k} \mathscr{Z}[(kT)^p] \tag{11-80}$$

The specific objective is to find a $D_c(z)$ that matches the output to the input after a finite (minimum) number of sampling intervals. The derivation[83] yields

$$M(z) = 1 - (1 - z^{-1})^{p+1} \tag{11-81}$$

which results in a $D_c(z)$ of

$$D_c(z) = \frac{1 - (1 - z^{-1})^{p+1}}{(1 - z^{-1})^{p+1} G_z(z)} \tag{11-82}$$

A $D_c(z)$ defined by Eq. (11-82) generates an output $c(kT)$ that is equal to $(kT)^p$ after $p + 1$ sampling intervals. Of course, inputs with order less than p using a $D_c(z)$-type controller also result in a zero steady-state error.

Note that if $G_z(z)$ is in reality a discretized continuous plant, the continuous output $c(t)$ can fluctuate or oscillate appreciably between sampling instants. By not requiring the finite settling time to be minimal, a better overall response can be achieved. This approach uses discrete state-variable modeling techniques and modern control theory.[84]

11-10 SUMMARY

The conventional procedures (see Chap. 5) for applying the cascade compensators are used in the examples of this chapter. These procedures can be used with any system. They involve a certain amount of trial and error and the exercise of judgment based on past experience. The improvements produced by each compensator are shown by application to a specific system. In cases where the conventional design procedures for digital compensators (controllers, filters) do not produce the desired results, then the method of Sec. 11-6 may be used. Other methods of applying compensators are available in the literature.

It is important to emphasize that the conventional compensators cannot be applied indiscriminately. For example, the necessary improvements may not fit the four categories listed in Sec. 5-1, or the use of a conventional compensator may cause a deterioration of other performance characteristics. In such cases, the designer must use ingenuity in locating the compensator poles and zeros to achieve the desired results. Also, the designer, when designing a digital controller, is not restricted to the use of the nominal upper bound of $\alpha = 10$ for a lag compensator and the nominal lower bound of $\alpha = 0.1$ for a lead compensator in the s or w domain. Larger upper and lower bounds, respectively, can be used based upon the flexibility of the computer hardware in terms of word-length accuracy (see Chap. 10). The use of several compensators in cascade and/or in feedback may be required to achieve the desired pole-zero locations for the control ratio.

The PID controller of Chap. 5 can be reformulated as a lead-lag structure as shown in Sec. 11-5C. Results are very similar to the performance resulting from the root-locus design approach. A compensator may be designed by the method of Sec. 11-9 to achieve a deadbeat response with its associated increase in overshoot.

An equally important aspect in the design of a digital controller is its software implementation. The software example of Sec. 11-8 presents considerable documentation for a dual second-order compensator. The reader should realize that this documentation is still only a small part of the overall control system records. Note also that the software programs include in the aggregate many more data-transfer operations than arithmetic operations. In improving controller performance, then, another implementation consideration is the selection of efficient (time and space) data-transfer instructions.

It may occur to the reader that the methods of compensation in this chapter can become laborious because there is some trial and error involved. The use of a CAD control system program[14,15] to solve design problems minimizes the work involved. Using a computer means that an acceptable design of the compensator can be determined in a shorter time. It is still useful to obtain a few values of the solution by a manual analytical or graphical method to ensure that the problem has been properly defined and that the computer program is operating correctly. The examples in this chapter illustrate the accuracy sensitivity of the calculations as the value of T decreases. The reader can validate this fact by trying to duplicate the examples of this chapter on various word-length machines by using CAD techniques.

CHAPTER

TWELVE

FEEDBACK COMPENSATION

12-1 INTRODUCTION

System performance, based upon the system *tracking* or *following* of the desired system input, may be improved by using either cascade or feedback compensation. There are many factors[1,21] that must be considered when making that decision. Some are:

1. The design procedures for a cascade compensator are more direct than those for a feedback compensator. The application of feedback compensators is sometimes more laborious.
2. The environmental conditions in which the feedback control system is to be utilized affect the accuracy and stability of the controlled quantity. This is a serious problem in an airplane or space vehicle, which is subjected to rapid changes in altitude and temperature. It is shown in Sec. 15-8 that control-system performance can be improved by a feedback compensator or by state-variable feedback.
3. When $G(z)$ has a pair of dominant complex poles that result in the dominant poles of $C(z)/R(z)$, the simple first-order cascade compensators discussed in Chaps. 5 and 11 provide minimal improvement to the system's time-response characteristics. For this situation, feedback compensation is more desirable.
4. Besides all these factors, the available components and the designer's experience and preferences influence the choice between a cascade and feedback compensator.

The effectiveness of cascade compensation is illustrated in the preceding chapter. The use of feedback compensation is demonstrated in this chapter, not only for a

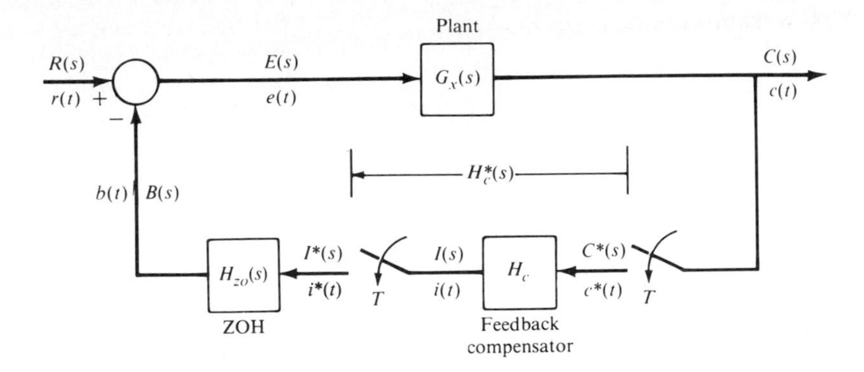

Figure 12-1 Feedback-compensated sampled-data control systems.

control system whose output must track a desired system input but also for a control system whose output should not respond to a system disturbance input.

The feedback-compensation design procedures for continuous-time systems[1,21] may also be applied to sampled-data systems. These procedures are not as straightforward and easy to apply as the design procedures for cascade compensation. Therefore, a design example is employed to best illustrate two possible approaches for designing a feedback compensator (see Fig. 12-1): the Guillemin-Truxal and the root-locus methods. For this example, the transfer function of $G_x(s) = K_x/[s(s+1)]$ used in the examples of Chap. 11 is used again.

12-2 GENERAL ANALYSIS

The following four equations describe completely the control system of Fig. 12-1.

$$E(s) = R(s) - B(s) \tag{12-1}$$

$$C(s) = G_x(s)E(s) = G_x(s)R(s) - G_x(s)B(s) \tag{12-2}$$

$$B(s) = H_{zo}(s)I^*(s) \tag{12-3}$$

$$I^*(s) = H_c^*(s)C^*(s) \tag{12-4}$$

Substituting from Eq. (12-4) into Eq. (12-3) yields

$$B(s) = H_{zo}(s)H_c^*(s)C^*(s)$$

which in turn is substituted into Eq. (12-2) to yield

$$C(s) = G_x(s)R(s) - G_x(s)H_{zo}(s)H_c^*(s)C^*(s) \tag{12-5}$$

The impulse transform of this equation is

$$C^*(s) = G_xR^*(s) - H_{zo}G_x^*(s)H_c^*(s)C^*(s)$$

This equation is manipulated to yield

$$C^*(s) = \frac{G_xR^*(s)}{1 + H_{zo}G_x^*(s)H_c^*(s)} \tag{12-6}$$

and

$$C(z) = \frac{G_xR(z)}{1 + H_{zo}G_x(z)H_c(z)} \tag{12-7}$$

For the given $G_x(s)$ and $R(s) = 1/s$,

$$G_xR(z) = \frac{K_xTz}{(z-1)^2} - \frac{K_x(1-\epsilon^{-T})z}{(z-1)(z-e^{-T})} \tag{12-8}$$

$$H_{zo}G_x(z) = \frac{K_xT}{z-1} - \frac{K_x(1-\epsilon^{-T})}{z-\epsilon^{-T}} \tag{12-9}$$

Let

$$H_c(z) = K_{zc}\frac{N(z)}{D(z)} \tag{12-10}$$

Substituting from Eqs. (12-8) to (12-10) into Eq. (12-7) yields

$$C(z) = \frac{K_x\dfrac{Tz(z-\epsilon^{-T}) - (1-\epsilon^{-T})z(z-1)}{(z-1)^2(z-\epsilon^{-T})}}{1 + K_{zc}K_x\left[\dfrac{T(z-\epsilon^{-T}) - (1-\epsilon^{-T})(z-1)}{(z-1)(z-\epsilon^{-T})}\right]\dfrac{N(z)}{D(z)}} \tag{12-11}$$

$$= \frac{K_x[Tz(z-\epsilon^{-T}) - z(1-\epsilon^{-T})(z-1)]D(z)}{\underbrace{(z-1)}_{\text{Forcing function pole}}\{(z-1)(z-\epsilon^{-T})D(z) + K_{zc}K_x[T(z-\epsilon^{-T}) - (1-\epsilon^{-T})(z-1)]N(z)\}}$$

The characteristic equation

$$(z-1)(z-\epsilon^{-T})D(z) + K_{zc}K_x[T(z-\epsilon^{-T}) - (1-\epsilon^{-T})(z-1)]N(z) = 0 \tag{12-12}$$

is partitioned and rearranged to

$$\frac{K_{zc}K_xN(z)[T(z-\epsilon^{-T}) - (1-\epsilon^{-T})(z-1)]}{(z-1)(z-\epsilon^{-T})D(z)} = \frac{K_{zc}K_x(T-1+\epsilon^{-T})(z-d)N(z)}{(z-1)(z-\epsilon^{-T})D(z)} = -1 \tag{12-13}$$

where $d = (1 - \epsilon^{-T} - T\epsilon^{-T})/(1 - \epsilon^{-T} - T)$.

Equation (12-14) permits a root-locus analysis for the sampled-data control system of Fig. 12-1. The plot of the poles and zeros of Eq. (12-13), assuming that

$$H_c(z) = K_{zc}\frac{z-c}{z-b} \tag{12-14}$$

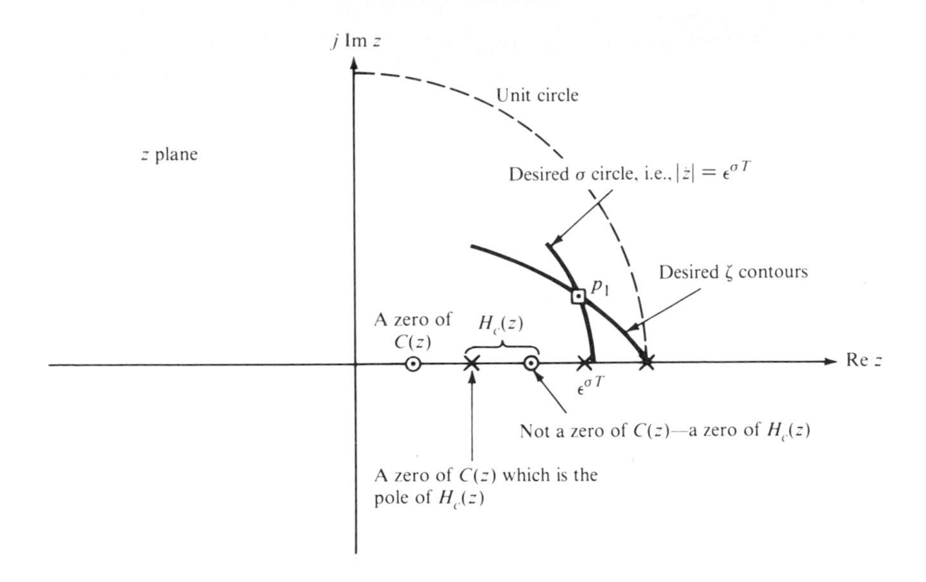

Figure 12-2 A plot of the poles and zeros of Eq. (12-13), where

$$H_c(z) = \frac{K_{zc}(z - c)}{z - b}$$

is shown in Fig. 12-2. For a desired ζ and settling time $T_s \le 4/|\sigma|$, the desired dominant complex pole p_1 is located as shown in this figure. Assume that it is not possible for the basic system to produce the desired dominant poles. By the application of the angle condition it may be possible to locate the zero c and the pole b of H_c that may result in achieving the desired dominant poles. If the simple feedback compensator of Eq. (12-14) is not satisfactory, then the angle condition is again applied but with a more complicated $H_c(z)$ function.

In correlating the root-locus plot with respect to the corresponding time-response characteristics, it is necessary to keep track of which poles and zeros of Eq. (12-13) are zeros of $C(z)$. As mentioned previously a pole of $C(z)$ that lies "very close" to a zero of $C(z)$ may have very little effect on $c^*(t)$.

The feedback-compensation design example of this section is based upon achieving an $M_p \approx 1.1$, $t_s \le 1$ s, and $c(t)_{ss} = 1$, for a unit-step forcing function and $T = 0.1$ s. Solving Eqs. (5-8) and (5-11) for these values of M_p and t_s yields $\zeta \approx 0.59$, $|\sigma| \ge 4$, and $\omega_n \approx 6.7795$. Thus the desired dominant s-plane poles are

$$p_{1,2_s} = \sigma \pm j\omega_d = \sigma \pm j\omega_n\sqrt{1 - \zeta^2} = -4 \pm j5.4738 \qquad (12\text{-}15)$$

which map into the z plane as follows:

$$p_{1,2_z} = \epsilon^{p_{1,2}T} = 0.6703\angle 31.363^\circ = 0.5721 \pm j0.3487 \qquad (12\text{-}16)$$

These desired dominant z-plane poles cannot be achieved by use of only $H_c(z) = K_{zc}$; that is, the desired dominant poles cannot be obtained by only a gain adjustment (see Sec. 7-6).

12-3 DIR TECHNIQUE

The DIR technique is used to design $H_c(z)$ by the two approaches: the Guillemin-Truxal and the root-locus methods.

12-3A Guillemin-Truxal Approach

This approach requires the modeling of a desired control ratio $[C(z)]_M$ that corresponds to Eq. (12-7) and contains the desired dominant complex poles of Eq. (12-16). To synthesize $[C(z)/R(z)]_M$ it is necessary to first determine Eqs. (12-8) and (12-9) for $T = 0.1$ s. Note that $z/(z - 1)$ can be factored from both terms in Eq. (12-8); thus

$$G_x R(z) = \underbrace{\frac{z}{z-1}}_{R(z)} \underbrace{\frac{0.0048374z + 0.004679}{z^2 - 1.9048z + 0.90480} K_x}_{G_y(z)} = R(z)G_y(z) \tag{12-17}$$

From Eq. (12-9) it is seen that

$$G_y(z) = H_{zo}G(z) = \frac{N'}{D'} = \frac{K_x[1 - (1 - \epsilon^{-T})] - K_x[T - (1 - \epsilon^{-T})]}{(z - 1)(z - \epsilon^{-T})}$$

$$= \frac{K_x(0.0048374z + 0.004679)}{z^2 - 1.9048z + 0.9048} = \frac{0.0048374K_x(z + 0.967255)}{(z - 1)(z - 0.9048)} \tag{12-18}$$

Thus based upon Eqs. (12-17) and (12-18), Eq. (12-7) can be manipulated to yield the pseudo control ratio

$$\left[\frac{C(z)}{R(z)}\right]_p = \frac{G_y(z)}{1 + G_y(z)H_c(z)} = \frac{N'}{D' + N'H_c(z)} \tag{12-19}$$

Let

$$\left[\frac{C(z)}{R(z)}\right]_M = \frac{N_M}{D_M} \tag{12-20}$$

represent the synthesized or model control ratio that satisfies the specifications.

Equating (12-19) to (12-20) results in

$$H_c(z) = \frac{G_y(z)D_M - N_M}{G_y(z)N_M} = \frac{N'D_M - D'N_M}{N'N_M} \tag{12-21}$$

Since $G_y(s)$ has a first-order numerator and a second-order denominator, assume that Eq. (12-20) has two zeros and three poles, where one zero is the zero of N' and

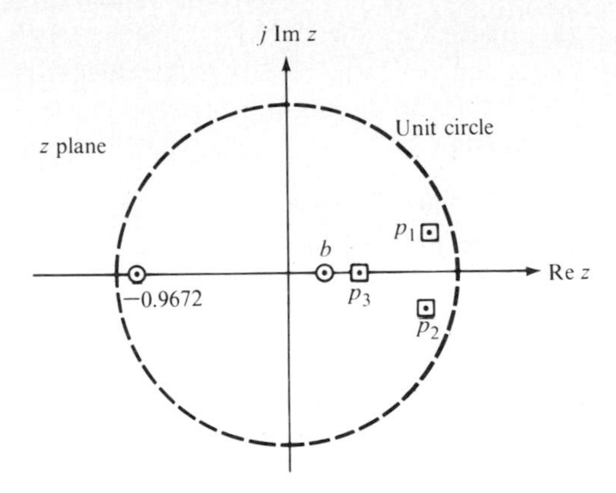

Figure 12-3 The plot of the poles and zeros of Eq. (12-22).

the other zero is a pole of $H_c(z)$. Thus for the desired dominant poles of Eq. (12-16) the model control ratio is

$$\left[\frac{C(z)}{R(z)}\right]_M = \frac{\{K_x[1-(1-\epsilon^{-T})]z - K_x[T-(1-\epsilon^{-T})]\}(z-z_2)}{(z-p_1)(z-p_2)(z-p_3)}$$

$$= \frac{0.0048374K_x(z+0.967255)(z-z_2)}{(z-0.57210 \pm j0.34870)(z-p_3)} = \frac{N_M}{D_M} \qquad (12\text{-}22)$$

where the concept of placing a third real dominant pole with respect to a dominant zero of $C(z)/R(z)$ to achieve the desired specifications is used in locating the dominant zero z_2 and pole p_3. Thus a configuration of poles and zeros for Eq. (12-22), as shown in Fig. 12-3, is used.

As a first trial, let $p_3 = 0.5$ and $z_2 = 0.4$; thus Eq. (12-22) becomes

$$\left[\frac{C(z)}{R(z)}\right]_M = \frac{0.0048374K_x(z+0.967255)(z-0.4)}{(z^2-1.1442z+0.44889)(z-0.5)}$$

$$= \frac{0.0048374K_x(z^2+0.567255z-0.386902)}{z^3-1.6442z^2+1.02099z-0.224445} = \frac{N_M}{D_M} \qquad (12\text{-}23)$$

Applying the final-value theorem to $C(z)_M$, for $R(z) = z/(z-1)$, yields from Eq. (12-23)

$$c^*(t)_{ss} = \lim_{z\to 1}\left[\frac{z-1}{z}C(z)_M\right] = \lim_{z\to 1}\left[\frac{C(z)}{R(z)}\right]_M = 0.0374767K_x \qquad (12\text{-}24)$$

Setting Eq. (12-24) equal to 1 yields

$$K_x = 26.683 \qquad (12\text{-}25)$$

Thus $C(z)$ becomes

$$C(z) = R(z)\left[\frac{C(z)}{R(z)}\right]_M = R(z)\frac{0.129066(z+0.967255)(z-0.4)}{(z-0.5721 \pm j0.3487)(z-0.5)} \qquad (12\text{-}26)$$

which yields the following values for the figures of merit: $M_p \approx 1.07 < 1.1$, $t_p \approx 0.6$ s, and $t_s \approx 0.9 < 1$ s. Since this model control ratio results in the specifications being met, then substituting the numerator and denominator polynomials of Eqs. (12-18) and (12-23) into Eq. (12-21) yields

$$H_c(z) = \frac{5.1183(z - 0.66415)(z - 0.31334)(z + 0.967255)}{(z - 0.4)(z + 0.967255 \pm j0.6 \times 10^{-7})} \tag{12-27}$$

which is approximated as

$$H_c(z) = \frac{5.1183(z - 0.66415)(z - 0.31334)}{(z - 0.4)(z + 0.967255)} \tag{12-28}$$

It should be noted that by substituting from Eqs. (12-18) and (12-22) (in the form where the values of K_x and T are unspecified) into Eq. (12-21) reveals, for this example, the following:

$$N'D_M = K_x[T - (1 - \epsilon^{-T})]z^4 + (\alpha)z^3 \cdots$$
$$D'N_M = K_x[T - (1 - \epsilon^{-T})]z^4 + (\beta)z^3 \cdots$$

Thus $N'D_M - D'N_M = (\alpha - \beta)z^3 + \cdots$ [*a third-order numerator for Eq. (12-18)*]. *A term in* z^4 *may occur in the numerator of Eq. (12-27) due to roundoff error. Thus it is neglected. Since the value of* K_x *is determined by the synthesis of the model* [*see Eqs. (12-23) to (12-26)*], *provision must be made for a gain adjustment in the final* $H_c(z)$ *which is actually used in order to satisfy* $c^*(t)_{ss} = 1$ *for a unit-step input.*

The system time-response characteristics using the feedback compensator of Eq. (12-28) are essentially the same as those for the model control ratio.

12-3B Root-Locus Approach

The open-loop transfer function of the system of Fig. 12-1, where

$$H_c(z) = K_{zc}\frac{z - c}{z - b} \tag{12-29}$$

and where $G_y(z)$ is given by Eq. (12-18), is

$$G_y(z)H_c(z) = \frac{K_x K_{zc}[0.0048374z + 0.00467884](z - c)}{(z - 1)(z - 0.9048)(z - b)}$$

$$= \frac{0.0048374K_x K_{zc}(z + 0.967255)(z - c)}{(z - 1)(z - 0.9048)(z - b)} = -1 \tag{12-30}$$

The known poles and zeros of Eq. (12-30) are plotted in Fig. 12-4. From an analysis of this figure, based upon Secs. 4-3 and 5-3D, it is noted that the zero of $H(z)$ is restricted to lie on the real axis at or to the left of $|z| = \epsilon^{-\sigma T} = 0.67$ [see Eq. (12-16)]. This constraint ensures that the resulting real root is not more dominant than the desired dominant roots $p_{1,2}$. Thus let $c = 0.67$ and by applying the angle condition at p_1, by use of a calculator, the pole of $H_c(z)$ is found to be at $b =$

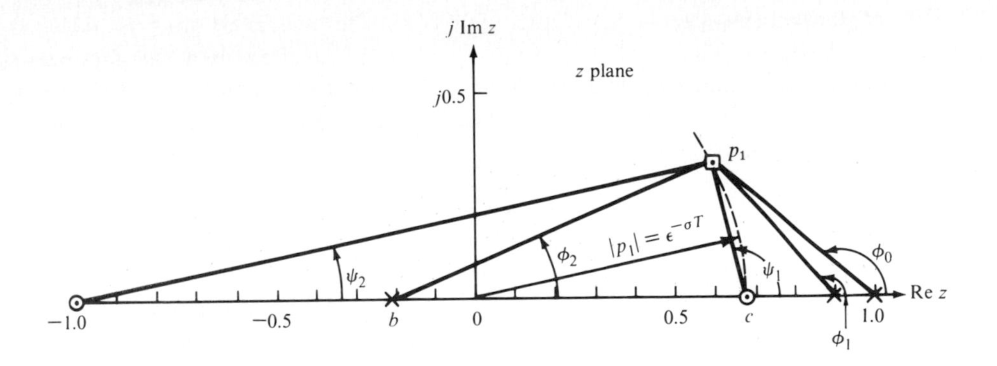

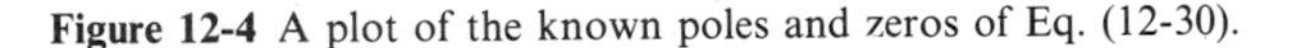

Figure 12-4 A plot of the known poles and zeros of Eq. (12-30).

-0.21478. These values of c and b yield a value of static loop sensitivity, at p_1, of $0.0048374K_xK_{zc} = 0.4003970$ and the control ratio

$$\frac{C(z)}{R(z)} = \frac{0.0048374K_x(z + 0.967255)(z - 0.21478)}{(z - 0.57218 \pm j0.34797)(z - 0.14527)} \qquad (12\text{-}31)$$

The required value of K_x, to achieve $e(\infty) = 0$ for a step input, is determined by applying the final-value theorem to $C(z)$. This results in $K_x = 34.78890$ and $K_{zc} = 2.37924$. The control ratio of Eq. (12-31) yields the following values for the figures of merit: $M_p \approx 1.1$, $t_p \approx 0.6$ s, and $t_s \approx 0.4$ s which meet the desired system performance specifications.

12-4 DIG TECHNIQUE

The PCT control-system approximation of the sampled-data system of Fig. 12-1 is shown in Fig. 12-5, where

$$G_x(s) = \frac{K_x}{s(s + 1)} \qquad (12\text{-}32)$$

and

$$H_x(s) = H_A(s)H_c(s) = \frac{20K_{sc}N(s)}{(s + 20)D(s)} \qquad (12\text{-}33)$$

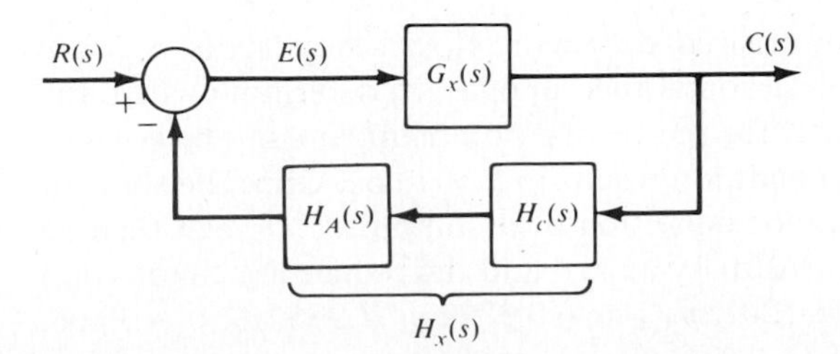

Fig. 12-5 PCT system approximation of the sampled-data control system of Fig. 12-1.

where $H_{pa}(s) = 2/(s + 20)$ is the Padé approximation of $H_{zo}(s)$ and $H_M(s) = H_{pa}(s)/T = 10H_{pa}(s)$. The Guillemin-Truxal technique resulted in a satisfactory $H_c(s)$ with two zeros and two poles, whereas the root-locus technique resulted in a satisfactory $H_c(s)$ with one zero and one pole. For the approach of this subsection the $H_c(s)$ is assumed to be of the form

$$H_c(s) = \frac{K_{sc}(s + a)^2}{(s + b)(s + c)} \tag{12-34}$$

Although the desired dominant roots $p_{1,2_s} = 4 \pm j5.4738$ are not within the shaded area of Fig. 7-7 for a very good Tustin approximation, they still lie within the overall region where the condition of Eq. (7-26), $|\sigma_{1,2}| \ll 2/T = 20$, is satisfied and the Tustin approximation may yield a satisfactory system design. In general, the poles and zeros of $H_c(s)$ are selected to be real. For $H_c(s)$ to effectively act as two "simple lead networks" in cascade the poles of Eq. (12-34) are located as far to the left of the zeros as practical. They are also selected in such a manner that when they are mapped into the z domain, via the Tustin transformation, they lie between 0 and +1 on the positive real axis of the z domain. Substituting from Eq. (7-17) into Eq. (12-34) yields

$$H_c(z) = \frac{(2 + aT)^2 K_{sc}}{(2 + bT)(2 + cT)} \frac{\left(z - \dfrac{2 - aT}{2 + aT}\right)^2}{\left(z - \dfrac{2 - bT}{2 + bT}\right)\left(z - \dfrac{2 - cT}{2 + cT}\right)} \tag{12-35}$$

For positive real poles and zeros to exist between 0 and +1 for Eq. (12-35), then $aT < 2$, $bT < 2$, and $cT < 2$. Assume that the poles in Eq. (12-35) are at the origin, that is,

$$2 - bT = 2 - cT = 0 \tag{12-36}$$

which results in $c = b = 2/T$. Note that poles at $z = 0$ in the z domain, by means of the transformation $z = \epsilon^{sT}$, correspond to poles at $-\infty$ in the s plane.

The open-loop transfer function, with $T = 0.1$ s and $c = b = 2/T$, of Fig. 12-5 is

$$G_x(s)H_x(s) = \frac{20K_{sc}K_x(s + a)^2}{s(s + 1)(s + 20)^3} \tag{12-37}$$

The poles of Eq. (12-37) and the desired dominant closed-loop roots $p_{1,2}$ are plotted in Fig. 12-6. The angle condition is then applied to determine where the zeros at $-a$ must be located in order for the desired dominant roots to be poles of $C(s)/R(s)$. The application of this condition results in $a \approx 7.062$. Once the value of a has been determined, the magnitude condition or a computer[14,15] can then be used to determine the static loop sensitivity at $p_{1,2}$ and the remaining closed-loop poles; for this example, $20K_{sc}K_x \approx 5201$, $p_{3,4} \approx -8.9772 \pm j8.976$, $p_5 \approx -35.06$,

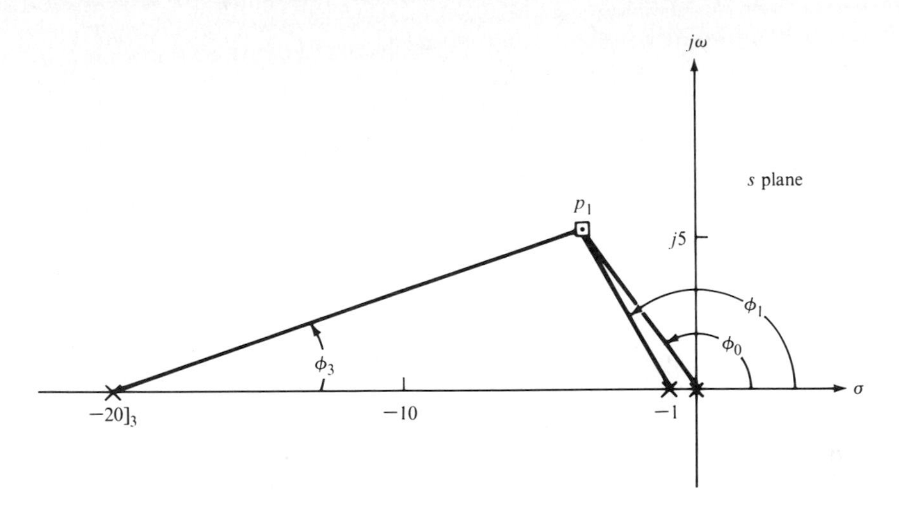

Figure 12-6 The plot of the poles of Eq. (12-37).

and

$$\frac{C(s)}{R(s)} = \frac{K_x(s+20)^3}{(s-p_1)(s-p_2)(s-p_3)(s-p_4)(s-p_5)} \tag{12-38}$$

To satisfy the requirement of $c(t)_{ss} = 1$, for $r(t) = u_{-1}(t)$, the final-value theorem is applied to the expression for $C(s)$ to determine the required value of $K_x \approx 32.425$ so that this condition can be met. This value of K_x results in $K_{sc} \approx 8.02$. The figures of merit for the control ratio of Eq. (12-38) are given in Table 12-1. To determine the ramp error coefficient it is necessary to solve for $G_{eq}(s)$ from

$$\frac{C(s)}{R(s)} = \frac{32.425(s+20)^3}{s^5 + 61s^4 + 1260s^3 + 14{,}400s^2 + 81{,}460s + 259{,}400}$$

$$= \frac{G_{eq}(s)}{1 + G_{eq}(s)}$$

Thus

$$G_{eq}(s) = \frac{32.425(s+20)^3}{s^5 + 61s^4 + 1227.6s^3 + 12{,}455s^2 + 42{,}550s} \tag{12-39}$$

Table 12-1 Figures of merit of DIG technique: Control system of Figs. 12-1 and 12-5

Domain	M_p	t_p, s	t_s, s	K_1, s^{-1}
s	1.134	0.544	0.86+	6.096
z	1.088	0.6	0.9+	4.505

and

$$K_1 = \lim_{s \to 0} sG_{eq}(s) = 6.096 \text{ s}^{-1}$$

Although M_p is larger than desired for the PCT system, the Tustin transformation of $H_c(s)$ is obtained next to determine if this first trial design results in a sampled-data system whose system performance is close to the desired specifications. Substituting the values of T, a, b, c, and K_{sc} into Eq. (12-35) yields

$$H_c(z) = \frac{K_{zc}(z - 0.48710)^2}{z^2} = \frac{N(z)}{D(z)} \tag{12-40}$$

where $K_{zc} = 3.6709$. Substituting Eq. (12-40) into Eq. (12-13), with $T = 0.1$ s, yields

$$\frac{0.0048374K_xK_{zc}(z + 0.967255)(z - 0.48710)^2}{z^2(z - 1)(z - 0.90480)} = -1 \tag{12-41}$$

where $K = 0.0048374K_xK_{zc} = 0.0048374(32.425)(3.6709) = 0.575791$. With this value of K the roots of the characteristic equation can be determined. Thus substituting from Eqs. (12-18) and (12-40) into Eq. (12-19) with K_x and K_{zc} unspecified and inserting the poles as determined by Eq. (12-20) yields

$$\frac{C(z)}{R(z)} = \frac{0.0048374K_xz^2(z + 0.967255)}{z^2(z - 1)(z - 0.9048) + K(z + 0.967255)(z - 0.4871)^2}$$

$$= \frac{0.0048374K_xz^2(z + 0.967255)}{(z - 0.60086 \pm j0.31144)(z - 0.063643 \pm j0.53202)}$$

$$= \frac{G_{eq}(z)}{1 + G_{eq}(z)} \tag{12-42}$$

where $K = 0.0048374K_xK_{zc} = 0.575791$. The reason for not using the value of K_x determined in the PCT system design is because of the warping effect. This effect is noted by the comparison of the desired z-domain roots $p_{1,2_z} = \epsilon^{p_{1,2_s}T} = 0.5721 \pm j0.3487$ with the roots $p_{1,2_z} = 0.6001 \pm j0.3112$ obtained as a result of the DIG technique for which $p_{1,2_s}$ do not lie in the shaded area of Fig. 7-7. The value of K_x is determined by satisfying the condition

$$c(\infty) = \lim_{z \to 1} [(1 - z^{-1})C(z)] = 1$$

where $R(z) = z/(z - 1)$. Thus Eq. (12-42) requires that $K_x = 31.09728$. To satisfy the required value of $K = 0.57579$ with this value of K_x, then $K_{zc} = 3.82763$.

The expression for $G_{eq}(z)$ (see Example 6-5) is determined from Eq. (12-42) to be

$$G_{eq}(z) = \frac{0.15115z^2(z + 0.967255)}{(z - 1)(z - 0.482043)(z + 0.001003 \pm j0.522266)}$$

Thus

$$K_1 = \frac{1}{T} \lim_{z \to 1} \left[\frac{z - 1}{z} G_{eq}(z) \right] = 4.50998 \text{ s}^{-1}$$

As noted from Table 12-1, the z-domain figures of merit are within the required specifications: thus the design is satisfactory. The required digital controller is

$$H_c(z) = \frac{I(z)}{C(z)} = \frac{3.82763(z - 0.4871)^2}{z^2} = 3.82763(1 - 0.4871z^{-1})^2 \quad (12\text{-}43)$$

which results in the difference equation

$$i(kT) = 3.82763c(kT) - 3.7288877c[(k-1)T] + 0.90816803c[(k-2)T] \quad (12\text{-}44)$$

This discrete control law (algorithm) is translated into a software program to be implemented by the digital computer (controller). Note that the third coefficient requires at least a 30-bit computer word.

12-5 UNWANTED DISTURBANCES

The design approach for minimizing unwanted disturbances for continuous-time systems is discussed in Sec. 5-7. This section discusses how the design approach for continuous-time systems can be used to minimize unwanted disturbances for sampled-data systems by both the DIG and DIR techniques. The design procedure of this section is based upon the system output *not* following the disturbance input. This is just the opposite situation for the design procedures given in Chaps. 5 and 11 and the previous sections of this chapter.

12-5A DIG Technique

Figure 12-7 (same as Fig. 5-21) is the PCT control-system representation of Fig. 12-1. In Example 5-3 the feedback compensator $H_c(s) = K_c/s$ resulted in an unstable system. An analysis of Fig. 5-23 indicates that $H_c(s)$ must contain at least one zero to pull the root-locus branches into the left-half s plane.

The results of Prob. 5-2 indicate that to minimize the magnitude of all the coefficients of

$$c(t) = A_1\epsilon^{p_1 t} + \cdots + A_n\epsilon^{p_n t} \quad (12\text{-}45)$$

which is desired for minimizing the effect of an unwanted disturbance on $c(t)$, it is necessary to locate as many of the poles of $C(s)/R(s)$ as far to the left of the s plane

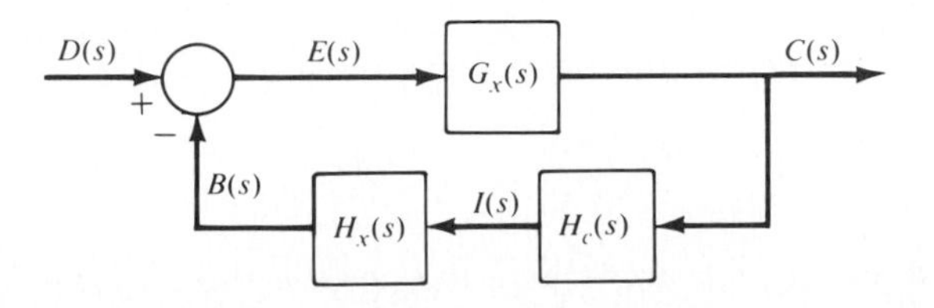

Figure 12-7 A control system with a disturbance input.

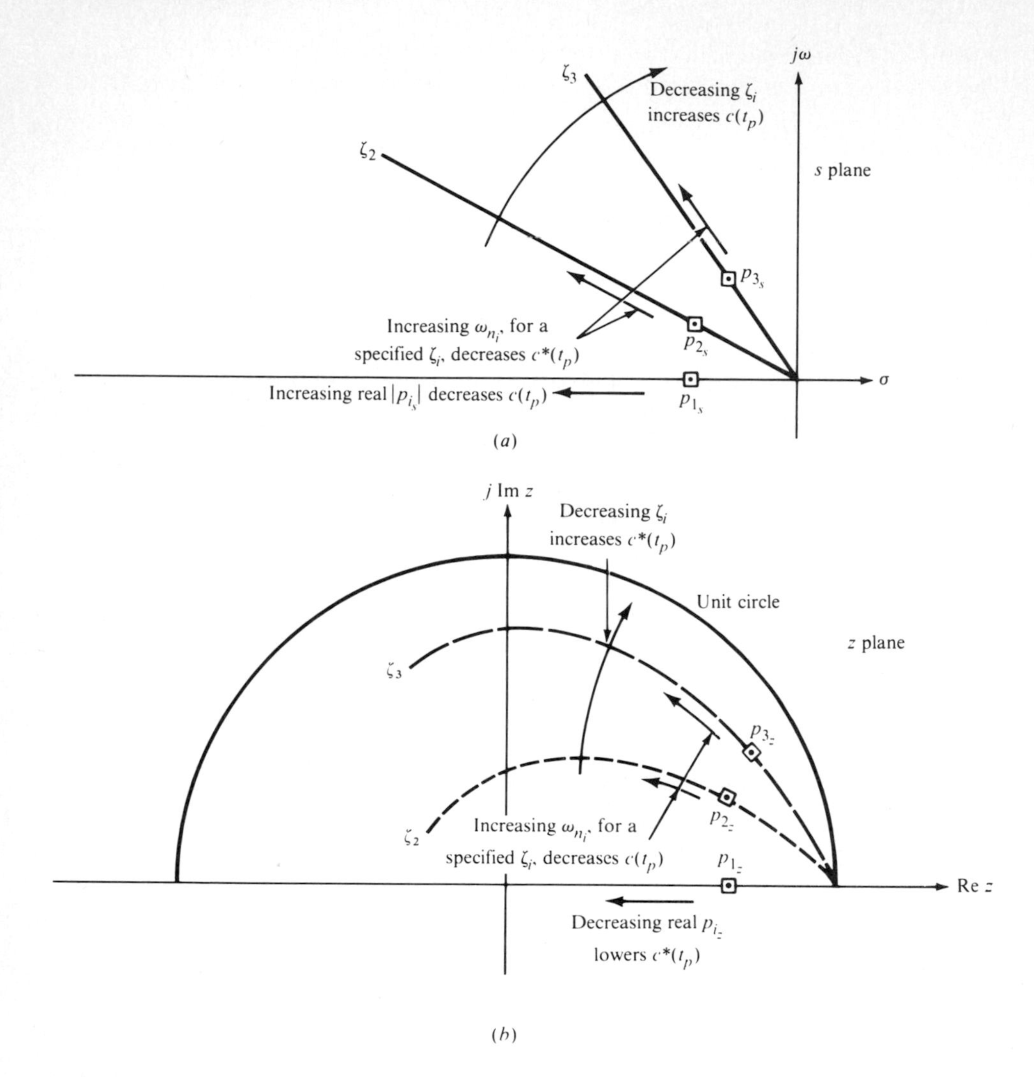

Figure 12-8 General directional changes in the parameters ζ_i and ω_{n_i} of complex poles, and the values of the real poles of the control ratio for decreasing the peak overshoot to a disturbance input. (*a*) *s* plane; (*b*) *z* plane.

as possible. Figure 12-8 graphically describes the result of the analysis made in Prob. 5-2 for the desired placement of the control ratio poles. To achieve the desired location of the poles and yet pull the root-locus branches as far as possible, to the left, assume the following structure for the feedback compensator:

$$H_c(s) = \frac{K_{sc}(s + a_1)(s + a_2)}{s(s + b_1)(s + b_2)} \tag{12-46}$$

Thus, as a second trial let $a_1 = 4$ and $a_2 = 8$ and where the poles b_1 and b_2 are chosen as far left as possible and yet lie in the Tustin region 0 to $-2/T$ (for this example, $T = 0.01$ s; therefore, $-2/T = -200$) of Fig. 7-7. The values of b_1 and b_2 are both set equal to 200; therefore the open-loop transfer function of Fig. 12-7, with $b_1 = b_2$, is

$$G_x(s)H_x(s)H_c(s) = \frac{200K_{sc}(s+4)(s+8)}{s^2(s+1)(s+200)^3} \tag{12-47}$$

where $G_x(s) = 1/[s(s+1)]$ and $H_x(s) = 200/(s+200)$. The poles and zeros of Eq. (12-47) are plotted in Fig. 12-9. For the plot of Lm $1/\mathbf{H}_x(j\omega)\mathbf{H}_c(j\omega) \leq -54$ dB, where $|\mathbf{C}(j\omega)/\mathbf{R}(j\omega)| \approx |1/\mathbf{H}_x(j\omega)\mathbf{H}_c(j\omega)|$ over the range $0 \leq \omega \leq \omega_b = 10$ rad/s, requires that $200K_{sc} \geq 50 \times 10^7$. This value of static-loop sensitivity yields the set of roots

$$p_1 = -3.8469 \qquad p_2 = -10.079$$

$$p_{3,4} = -23.219 \pm j66.914 \qquad p_{5,6} = -270.32 \pm j95.873$$

and are shown in Fig. 12-9. This choice of gain resulted in a satisfactory design; that is, $M_p = 0.001076 \leq 0.002(-54 \text{ dB})$, $t_p = 0.155$ s (< 0.2 s), $c(t)_{ss} = 0$, and $0 \leq \omega \leq \omega_b = 10$ rad/s.

Applying the Tustin transformation to Eq. (12-46), with $a_1 = 4$, $a_2 = 8$, and the two poles at -200, yields

$$H_c(z) = \frac{0.3315 \times 10^{+4}(z-0.9231)(z-0.9608)(z+1)}{z^2(z-1)} \tag{12-48}$$

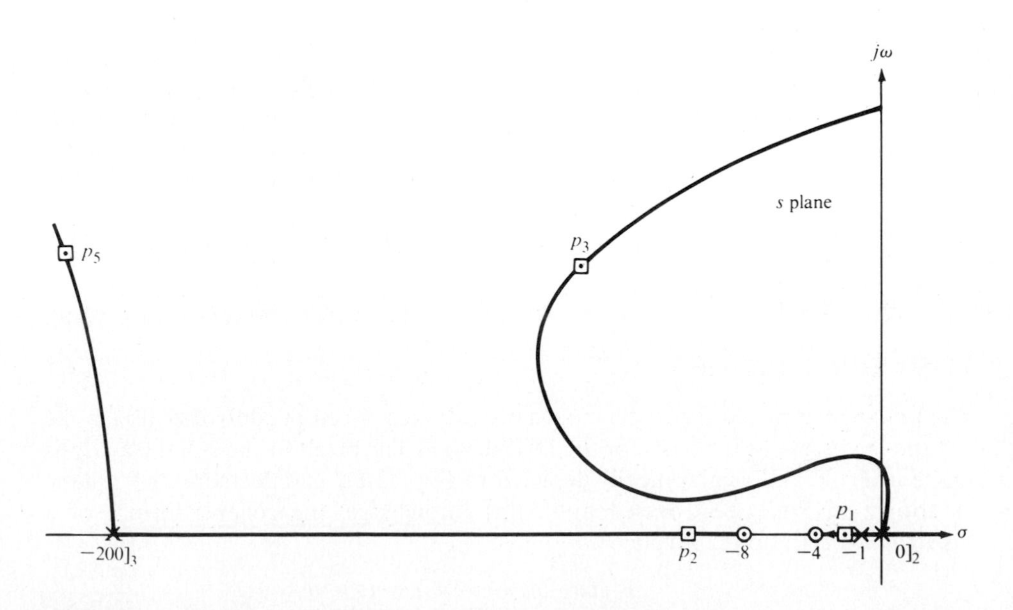

Figure 12-9 A possible root locus for Eq. (12-47). Not to scale.

where $K_{zc} = 0.3315 \times 10^{+4}$. For the system of Fig. 12-1, with $r(t) = u_{-1}(t)$ and $T = 0.01$ s,

$$G_xR(z) = \mathscr{Z}[G_x(s)R(s)] = \frac{0.4983 \times 10^{-4} z(z + 0.9967)}{(z - 1)^2(z - 0.9900)} \tag{12-49}$$

$$H_{zo}G_x(z) = \mathscr{Z}[H_{zo}(s)G_x(s)] = \frac{0.4983 \times 10^{-4}(z + 0.9967)}{(z - 1)(z - 0.9900)} \tag{12-50}$$

The characteristic equation, from Eq. (12-7), yields

$$H_{zo}G_xH_c(z) = -1 \tag{12-51}$$

Substituting from Eqs. (12-48) and (12-50) into Eq. (12-51) yields

$$\frac{0.16519(z - 0.9231)(z - 0.9608)(z + 0.9967)(z + 1)}{z^2(z - 0.99)(z - 1)^2} = -1 \tag{12-52}$$

from which the root locus may be obtained. The roots for the specified value of static-loop sensitivity yields the following expression for $C(z)$:

$$C(z) = \frac{4.983 \times 10^{-5}(z + 0.9967)z^2}{(z - 0.90403)(z - 0.96227)(z - 0.60787 \pm j0.53203)(z + 0.25723)} \tag{12-53}$$

Thus the performance characteristics for the sampled-data control system of Fig. 12-1 are

$$M_p = 1.08 \times 10^{-3}\ (\approx -59.3\ \text{dB}) \qquad \text{and} \qquad t_p \approx 0.16\ \text{s}$$

$$M_m = 1.304 \times 10^{-3}\ (\approx -57.7\ \text{dB}) \qquad \text{for} \qquad 0 \le \omega \le 10\ \text{rad/s}$$

The gain $K_{zc} = 3315$ may be considered to be "high" and thus present a problem with respect to its implementation on the digital controller. If this is the case, then a portion of this gain may be allocated to other components in the feedback path. Also, the gain can be reduced until either $M_p = 0.002$ with $t_p \le 0.2$ s or $t_p = 0.2$ s with $M_p \le 0.002$ is achieved. Although not specified in this problem the loop transmission frequency ω_ϕ must also be taken into account in the design of a practical system.

12-5B DIR Technique

The knowledge gained from designing a disturbance rejection control system in the s plane may serve as the basis for the DIR design. The result of the s-plane analysis made in Prob. 5-2, as graphically depicted in Fig. 12-8a, is reflected in the z plane as shown in Fig. 12-8b. Based upon this knowledge, the general format of a disturbance rejection controller is

$$H_c(s) = \frac{K_{zc}(z - d_1)(z - d_2)\cdots(z - d_j)}{z^i(z - 1)} \tag{12-54}$$

where $i = 1, 2, 3, \ldots \geq j - 1$. For the example of Sec. 12-5A, the controller is

$$H_c(z) = \frac{K_{zc}(z - d_1)(z - d_2)(z - d_3)}{z^2(z - 1)} \tag{12-55}$$

Based upon the frequency bandwidth requirement (see Sec. 5-7), some or all of the zeros should be placed "near" the $1 + j0$ point. By trial and error, the zeros of Eq. (12-48) may be selected or another set may be determined that may yield equally satisfactory results. Another controller that results in a satisfactory design is

$$H_c(z) = \frac{0.37036(z - 0.9231)(z - 0.9608)}{z^2(z - 1)} \tag{12-56}$$

Note: The term $z + 1$ in Eq. (12-48) is replaced by its dc gain to yield Eq. (12-56). For this controller $M_p \approx 0.00013$ and $t_p \approx 0.04$ s.

In summary, the main emphasis of this section is the presentation of a logical approach to the design of a control system that minimizes the system's response to an unwanted disturbance input. Design methods for such systems, in general, involve some trial and error in the design of an appropriate $H_c(\cdot)$.

12-6 SUMMARY

The conventional procedures (see Chap. 5) for the design of cascade compensators for sampled-data control systems are used in the examples of Chap. 11. They involve a certain amount of trial and error and the exercise of judgment on past experience. The improvements produced by each compensator are shown by application to a specific system. In cases where the conventional design procedures for cascade digital compensators (controllers, filters) do not produce the desired results, then the method of Sec. 11-6 may be used. The feedback-compensation design procedures of this chapter may be used when a cascade compensator cannot produce the desired system performance, or where a "simpler" compensator is preferred. These design procedures for cascade and feedback compensators, given in Chaps. 5 and 11 and in Secs. 12-2 and 12-3, respectively, are based on the system output *tracking* or *following* of the system input.

As mentioned in the previous chapter, the designer, when designing a digital controller, is not restricted to the use of the nominal upper bound of $\alpha = 10$ for a lag compensator and the nominal lower bound of $\alpha = 0.1$ for a lead compensator in the s or w domain. Larger upper and lower bounds, respectively, can be used based upon the flexibility of the computer hardware in terms of word-length accuracy (see Chap. 10). The factors stated in Sec. 12-1 should be taken into account when deciding between cascade and feedback compensation. The use of several compensators in cascade and/or in feedback may be required to achieve the desired pole-zero locations for the control ratio.

The general dual second-order compensator program of Sec. 11-8 can also be used in a feedback mode. For example, a rate-and-position feedback digital controller can easily be implemented by using the given program. Again, accuracy requirements as related to computer word length must be analyzed as discussed in Chap. 10.

The design procedures of a feedback compensator for minimizing the effects of an unwanted disturbance on the output are more involved and laborious. These procedures are based upon the system output not following the disturbance input. Generally, since high gain is required in the feedback path, a more complicated compensator (a higher-order numerator and denominator) is required to maintain the desired degree of system stability. The minimization of the mean squared error between the frequency response of the designed feedback compensator and the desired frequency response may result in a better design as presented in Sec. 11-6. Another form of compensation is the design of a "noise" filter that attenuates or rejects high-frequency system modes as discussed in Sec. 11-7.

The methods of compensation in this chapter can become laborious because there is some trial and error involved. Thus, as stated previously, the use of a CAD control system program[14,15] to solve design problems should be used to minimize the work involved. Using a computer means that an acceptable design of the compensator can be determined in a shorter time. It is still useful to obtain by an analytical or graphical method a few values of the solution to ensure that the problem has been properly defined and that the computer program is operating correctly.

As it may have been observed throughout this chapter, the number of significant digits used were not the same. The number of nonzero digits shown depended upon the CAD package used and the selected sampling time. As a simple illustration consider $T = 0.1$ s. Thus the CAD package prints time to only one significant digit to the right of the decimal point.

CHAPTER

THIRTEEN

SOFTWARE ENGINEERING IN DIGITAL CONTROL SYSTEMS

13-1 INTRODUCTION

Software engineering is a phrase that encompasses the proper development and maintenance of software. Specifically it means the *analysis, design, coding, testing, and maintenance of efficient and effective software*. Such an all-encompassing term implies that software engineering is a discipline within itself, which it is. However, the applicational aspect means that software engineering is also an interdisciplinary subject. The intent of this chapter is to introduce the more pertinent tools and techniques of software engineering as applied to digital control systems. Further reading and study is definitely suggested in order that one may be conversant with the varied aspects of software engineering.

The high cost of software development and maintenance is well-documented.[50] One of the major reasons for this phenomenon is the belief that software is easy to design, generate, and test. This conception is totally and completely wrong! *Software programs in most instances are more complex and sophisticated than the hardware on which they are executed.* Engineering tools and techniques must be applied to software development to lower costs.

Another major aspect associated with the high cost of software is the belief that software can easily be updated with little cost. Yes, software is more flexible from the point of view of updating hardware. However, the configuration control associated with software must also be extensive to ensure reliable systems. Moreover, proposed changes should be thoroughly tested before incorporation into an existing system. Even though software is more flexible, updating or modification costs are still very high.

In digital control systems, considerable effort must be expended in the initial development of specifications, design schematics, and software implementation of digital control laws.[107] Of course, if the controller involves only a simple second-order compensator with one control variable, the amount of configuration control documentation is somewhat limited but is still required. As digital control systems get more complex—i.e., three-axis control with stochastic estimation—more extensive documentation is required. Another complex example requiring intensive configuration management is process control software systems that automatically supervise and sequence many variables and parameters. Other examples of automated control sequences are found in chemical plants, space vehicles, robot control, etc.

The intent of this chapter is to enlarge the context of digital control software to involve many control variables and parameters. The complexity and interaction of such systems require proper and extensive use of software engineering tools and techniques. A major component of these tools and techniques is documentation. It is important to remember that *software only exists in its documentation*!

The emphasis in generating computer programs is placed upon effective means of communicating the program requirements, design, code, test results, and system integration procedures. The associated documentation is evaluated in terms of *quality*.

13-2 SOFTWARE QUALITY

Good software quality is the objective of software engineering. However, for any objective of this kind, quality judgments must be based on measurements. The measurement values validate the achievement of the desired quality performance. Thus, software measures must be defined. In achieving software quality, the primary measurement focus is on the program and its associated documentation. In general, good program characteristics (measurement criteria) are:

The program is *easy to understand.*
The program *meets its technical specification.*
The program is *easy to test.*
The program is *easy to modify.*

Note that the understandability (clarity) of a program is listed first. This is not done inadvertently. If a program is understandable even if it is not correct, testing can be easier and logical errors can be found in a reasonable and straightforward manner. Poor program documentation results in even a correct program being very difficult to test and validate in terms of the specifications.

In continuing the quest for software quality measurements, the concept of understandability leads to a human value judgment. To be specific, *at various stages of software development, documentation should be reviewed (audited) in accordance with established guidelines.* The method for these reviews is a

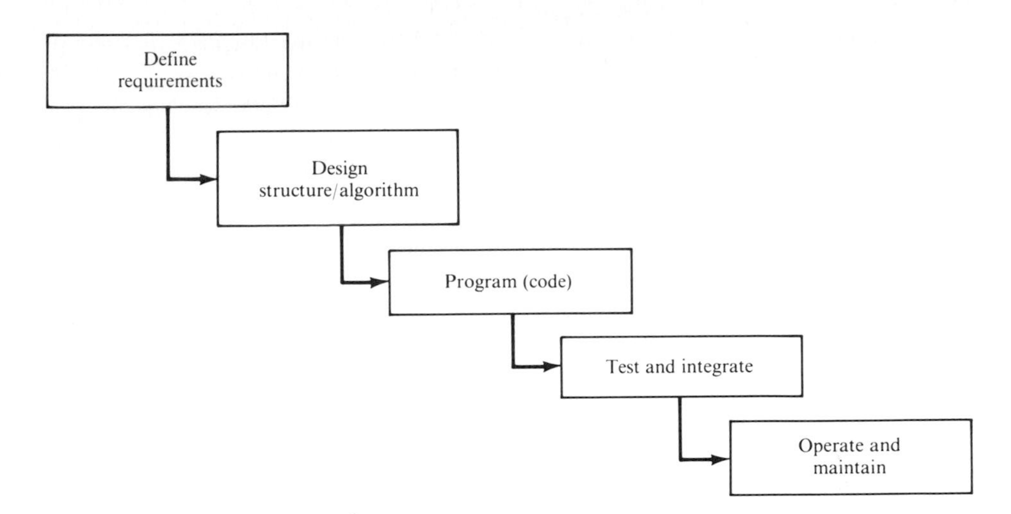

Figure 13-1 Software life cycle.

management process, but the software engineers must discipline themselves in documenting the details of the specific software development process and evaluating its quality. Thus, evaluation phases or milestones are usually defined.[51] Based on considerable experience, the general software development process has five major phases:

Define requirements.
Design structure and algorithms (general and detailed designs).
Program (code).
Test and integrate.
Operate and maintain.

These general phases constitute the *software life cycle* (see Fig. 13-1). Appropriate documentation should be available for review at the end of each phase. Engineering and management evaluate the documentation for completeness, nonambiguous statements, understandability, testability, etc. Checklists are suggested as a basis for the evaluation.[51] If approved, the documents are configuration-controlled and the next phase is initiated. In any real-world environment, of course, questions or interpretations arise in various phases. Based upon the answers, appropriate revisions of the applicable documents are made so that a completely nonambiguous documented record is continually maintained.

It should be noted that some of the most costly problems in the software life cycle are attributed to design and requirement errors found in the operational environment. Thus, extensive documentation has to be changed, and considerable manual time is usually spent correcting errors of this type. If a good analysis of requirements and design structure is accomplished, a much smaller number of logical errors are found during operation, thus reducing cost. Again, it must be remembered that humans are not perfect! *Logical errors must be assumed to exist in software documentation.* Procedures for finding the majority of these errors should be an integral part of any software development project. Note that it is nearly impossible to guarantee that a program is perfect.

Another important aspect of developing quality software lies in the concept of a *software engineering environment.* Such an environment must contain software development tools that provide for ease of documentation generation in the software life-cycle phases. Although the documentation can generally be done manually, the use of interactive tools provides an efficient capability for generation, analysis, and storage.

An excellent way for providing effective documentation at the end of each phase is to use appropriate (i.e., standard) tools and techniques. The following sections attempt to present some of these contemporary methods for generating good-quality digital-control-system software. Many of the techniques are graphical (two-dimensional) because humans are very good at analyzing two-dimensional representations of processes but have some difficulty with linear (one-dimensional) processes, i.e., many lines of words (prose).

In general, the concept of abstraction is an important quality in conveying the requirements and the design. The language employed should easily represent (1) the user's point of view in requirements and (2) the designer's point of view in requirements and design. In general, a computer system (hardware and software) consists of levels of abstractions as presented in previous chapters. The more levels of abstractions there are, theoretically, the easier it is to implement a complex function.

13-3 REQUIREMENTS ANALYSIS

This phase of software development is concerned with the development of a set of:

Consistent requirements
Practical requirements
Complete requirements
Unambiguous representations

This set of laudable objectives is easy to define, but usually difficult to measure. The difficulties lie in the general use of free-form English, which is extremely laborious to use and analyze with respect to the above objectives. A more constrained language is desired. Before presenting some examples, it is necessary to continue the presentation of objectives for such a language. The language should be able to describe *what* the proposed system should do (requirements), not how to do it (i.e., design and implementation). The requirements language should permit not only the ability to describe the process requirements but should define the *context* in which it must function. Note that the requirements development not only contains the *context analysis*, but also the *system constraints*, and finally the *functional requirements specifications*. Specific subprocesses within these three general aspects that are sometimes forgotten or left out of requirements documents are:

Initialization/terminization of a task
Error checking and processing
Configuration/reconfiguration
Flexibility/extensibility
Testability

These particular operations are usually integral elements of digital control systems and as such must be emphasized at the beginning to help ensure a reliable system implementation. The requirements language, then, has to be able to describe a wide spectrum of activities.

The justification for the use of a requirements language is that design and implementation can be validated, the software development effort can be easily monitored for problem areas, and cost can be minimized. Fewer errors should then be found in each of the following software life-cycle phases.

Various requirements languages have been proposed and employed.[52] A constrained English approach is used in the *Information System Design and Optimization System* (ISDOS)[51] and in the *Software Requirements Engineering Program* (SREP).[51] Another general technique is to display the requirements in the form of a two-dimensional graphics presentation. Examples include the *Structured Analysis and Design Technique* (SADT),[53] *Integrated Computer Aided Manufacturing Definition Language* (IDEF),[54] *data flow diagrams* (DFDs),[55] and the *Hierarchical-Input-Process-Output* (*HIPO*) method.[56]

It is difficult with limited space to discuss all the tools and techniques available for a software requirements definition. The cited references provide more details and familiarity with the languages. However, one of the tools, i.e., data flow diagrams, is presented owing to its general application to digital control systems and its similarity with the block diagram concept of control theory. The others, to a lesser or greater extent, depict the same information in different documentation structures.

13-3A Data Flow Diagrams (DFDs)

The data flow concept is an integrated component of DeMacro's structured analysis[55] approach to software requirements definition. Data flow diagrams (see Fig. 13-2) are used to portray processes and their interrelationships. Various levels of diagrams are used to build a hierarchy of processes. This hierarchy permits digital-control-system overview at the high levels and process details at the low levels.

Other elements of DeMacro's technique include the development of a data dictionary. *A data dictionary* (*DD*) *indexes the meaning of all variables and*

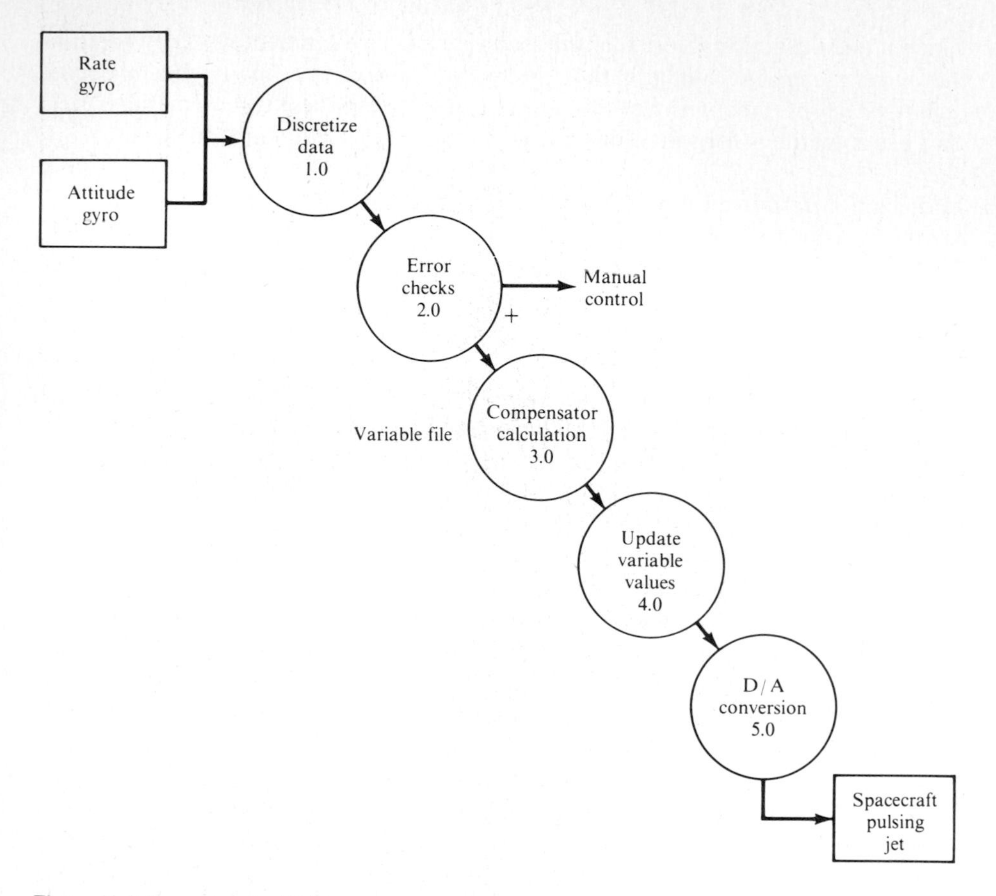

Figure 13-2 Example of a DFD.

constants and their location in the diagram levels. Using the DD, a digital control engineer can easily find the location of a digital control algorithm variable, for example.

As the DFD levels develop, the question "when does a DFD no longer define detailed requirements, but design?" must be answered. The answer to this question is discussed later in this chapter. But the interface between detailed requirements (functional specifications) and initial design is usually a "fuzzy" area in terms of semantics.

The activities or processes at the lowest-level DFD are defined by a structured *pseudo-English representation* or a *decision table* (*or decision tree*)[51] or flowchart which should relate directly to the operations in the selected computer implementation language. This lowest level interfaces design and defines program algorithms.

A DFD is a graphical representation of a system emphasizing the data flow of the overall process. The important aspect of the DFD is its ability (if it is used properly) to represent the component processes of the system and also their

Table 13-1 DFD syntax/semantics

Data flow: A pipeline through which data packets of known composition flow—*arrows*

Process: A transformation of incoming data flow into outgoing data flow(s)—a *circle*

File: A temporary repository of data—a *bar*

Data source/sink: A net originator or receiver of system data outside the requirements context—a *rectangle*

respective interfaces. *Many logical errors can arise in digital control software design if the information being transferred from process to process is not specified properly!* The two-dimensional DFD representation combined with the data dictionary should permit easier interface analysis. The basic representation of a process is through the use of circles or bubbles as shown in Fig. 13-2. The complete set of syntax and semantics is presented in Table 13-1. The other symbols used in Fig. 13-2 should now be comprehensible.

13-3B Generation of DFDs

In generating a data flow diagram, it is best to focus first on the input and output data flows in order to establish the DFD context. Next, the additional circles (processes) are added to connect the input and output, resulting in an overall data flow. As the circles are added, it is necessary to try to *define processes as activities at the same level* (i.e., manipulating the same level of data structures, or the same level of actions). In the case of digital control systems, the input activities are usually concerned with sensors in a data acquisition subsystem. General examples include analog signals such as position, speed, humidity, temperature, volume, and discrete switch settings. Output activity usually relates to controlling these variables through the use of D/A converters and transducers.

The high-level digital-control-system DFD should include the following basic elements:

Data sources (input)
Data input processes
Input error check process
Compensator processes
Variable update process
Output error check process
Data output processes
Data sinks (output)

In defining the processes as well as the data items, *labels are used that represent the physical reality of the overall system.* A check is made to ensure that the interface data flows are correct. Each circle should have a strong single-action verb and data object. The objective of understandability dictates these constraints.

To provide the capability of representing complex and sophisticated systems, a diagram hierarchy is developed as shown in the control-system example of Fig. 13-2. Each of the circles (processes) can be decomposed into lower-level diagrams that describe in more detail the actual process. For example, the compensator calculation could be decomposed as shown in Fig. 13-3. Each lower-level diagram is then contained within a context itself, i.e., *a parent-child relationship*. The functionality of each diagram at any level becomes self-evident. The understandability objective continues to be the primary focus as supported by the level hierarchy. It should be noted that the other requirements-definition techniques also have hierarchical organization.[51]

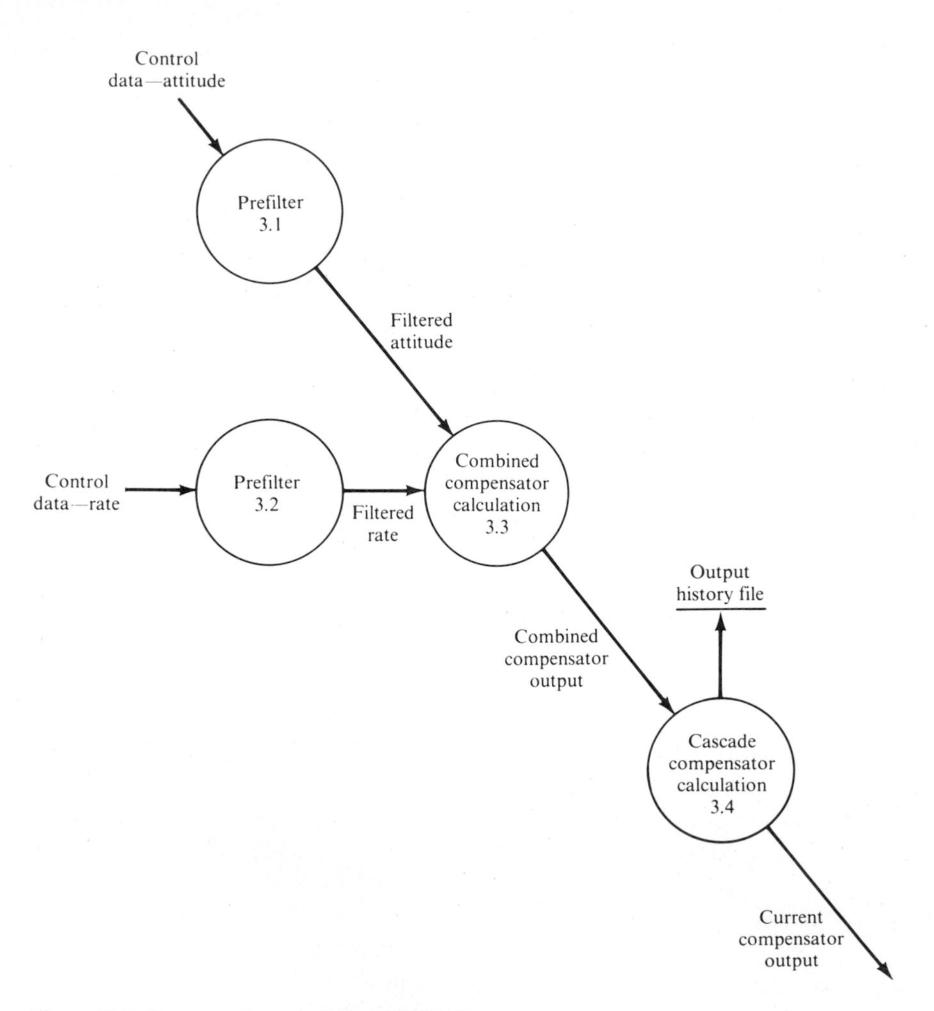

Figure 13-3 Compensator calculation (DFD3.0).

DFD-0	Context diagram for single-axis spacecraft controller
DFD0	Single-axis spacecraft controller
DFD1	Data input conversion
DFD2	Input data error checking
DFD3	Compensator calculation
DFD3.1	Attitude prefilter
DFD3.2	Rate prefilter
DFD3.3	Combined compensator calculation
DFD3.4	Cascade compensator calculation
DFD4	Variable updata
DFD5	Output data error checking
DFD6	Data output conversion

Figure 13-4 Hierarchy of diagram for digital control example.

The same circle-level concept can be applied to the arrows; i.e., at the high levels data are grouped (pipelined) and at the lower levels they are decomposed into specific data items. The data dictionary (DD) should reflect the decomposition of the arrows and circles. Note that a *maximum of seven circles per diagram* (one page) is suggested for understandability.

This top-down decomposition approach is a graphical technique that is organized as a tree structure with a parent-child relationship. Again, the control example of Figs. 13-2 and 13-3 can be expanded and described by a table hierarchy of diagrams as shown in Fig. 13-4. The top-level diagram contains only one circle as shown in Fig. 13-5, and its purpose is to represent the context and highest level of operation. The associated DD entries can contain constraint information such as timing requirements, weight, power limits, and spatial dimensions.

Conceptual files can also be annotated on the DFD. The file or abstract storage location (remember these are requirements, not implementations) is shown only at one level and *all* references to it are made at that diagram level. The DD indicates the variables that are stored in the file and the variable composition. The file should not represent a realization (implementation) structure. It indicates only the definition of the data, not its implementation structure.

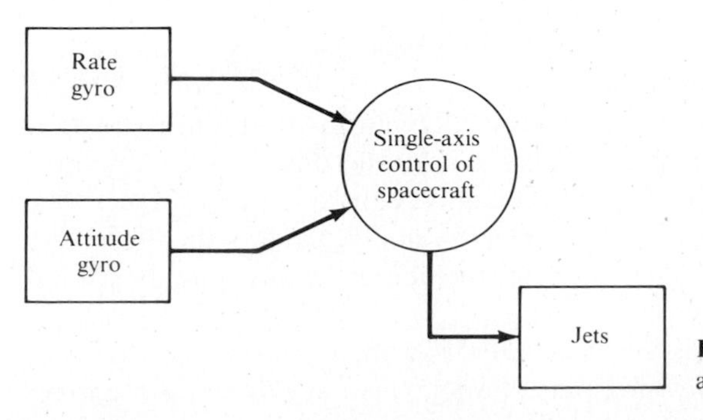

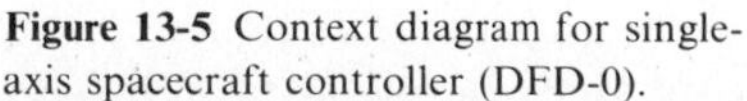
Figure 13-5 Context diagram for single-axis spacecraft controller (DFD-0).

Table 13-2 Data dictionary entries (per data flow diagram)

Data flow names (activity):
Name and number
Aliases
Description
Composition
Notes

Variable names—inputs/outputs (data elements or data structures):
Aliases
Variable description/type
Variable values (ranges)
Sources/Destination
Notes

File names:
Aliases
Composition
Notes

Process description:
Process number
Process description (I/O, libraries, decision tables/trees, structured English, flowchart)
Notes (constraints, data base file accesses)

In reference to terminating the level of DFD expansion, consideration should be given to the selected HOL implementation language. For example, if a HOL such as FORTRAN, PASCAL, or ADA is the selected implementation language, the pseudo-English dialogue defining the lowest level should use the same statement structure as the HOL. Further decomposition is irrelevant because of the implementation level. If assembly language is selected, then a more extensive hierarchy is required.

As defined earlier, associated with the DFD is a data dictionary (DD). A DD contains an entry for each data flow name (every level), all data element names, file names, and process names. Later it is augmented with design and code information. *The data dictionary must be organized in some logical fashion* such that data, activity names, module identifiers, and variables are easily found and understood. A general composition is shown in Table 13-2. Table 13-3 presents some aspects of the example digital controller DD. The use of notes in the DD permits the specification of timing, cost, and space constraints and other required performance criteria as mentioned earlier. The DD can also establish standard terminology (symbols, names, cross references, aliases, acronyms) which permits universal understanding in regard to a specific digital-control-system project. The DD is, of course, a central-managed document for changes and is therefore very useful in evolutionary system development including testing, training, and modification.

Continuing refinement is required as the DFDs are being developed. Analysis should focus on the search for missing data flows, extra data flows, and correct

Table 13-3 Partial example of simple spacecraft controller data dictionary

---

Name and number: compensator calculation—DFD3.0
  Aliases: Filter
  Composition: Prefilter (DFD3.1)
    Prefilter
    Combined compensator calculation (DFD3.3)
    Cascade compensator calculation (DFD3.4)
  Notes: This calculation has to be executed in less than 0.001 s.

Variable names:
  Control data attitude: digitized value of attitude measurement, range $\pm 10°$
  Control data rate: digitized value of rate measurement, range $\pm 20°/s$
  Filtered attitude: control data attitude filtered by first-order transfer function
  Filtered rate: control data rate filtered by second-order transfer function
  Combined compensator output: filter rate and filtered attitude combined and filtered by a sixth-order compensator
  Current compensator output: combined compensator output filtered by a second-order compensator

File names:
  Output history file: collection of all current compensator outputs for the last 10,000 samples
    Composition: time of sample, value of sample
    Notes: It is intended that data be statistically analyzed

Process description:
  Prefilter—{A second-order filter routine (DFD3.2 process) is described in pseudo-English}

---

decomposition. Peer review can be very helpful in this regard prior to the initiation of the design phase. The DFD and DD should eventually be configuration-controlled; i.e., modifications of requirements have to be reviewed and approved before incorporation. Also note that the DD is used for program and module descriptions generated during the design and implementation phases.

Whether DFDs and DDs or other appropriate techniques are used, disciplined development and documentation is required. *The requirements-definition process takes considerable time and effort!*

13-4 SOFTWARE DESIGN

In the past, the implementation of computer programs was done in a haphazard manner by attempting to immediately generate code from unstructured requirements. This approach led to quick results, but the programs probably had a considerable number of logical errors and were extremely difficult to understand. Software maintenance and modification costs were very high. Using the structured requirements of Sec. 13-3, the next phase of the software life cycle, i.e., design, is presented as a serious attempt to structure and document the overall program organization. This approach should provide for better software understanding, testability, maintainability, and lower costs.

Various techniques have evolved over the past three decades to provide a better structure to software programs. They include the concept of *subprograms* (subroutines, functions, procedures, macros, etc.). An extension of this concept led to general *modularity*, whereby separate program modules are separately translated, are limited in size (one page?), are limited in the number of input and output parameter/variables, and are modules that may be usable in other applications. *Structured programming* followed with emphasis on structured program flow. *The basic flow elements of structured programming are sequence, selection, and iteration which define program flow paths.*

Taking the structure concept a step up from implementation (structured programming), a structural software system design approach must be formulated and developed first. The structured-design approach relies on a number of important concepts that are presented in this section. They include *functional decomposition*[51] *of design* (top-down), *stepwise refinement*,[56] and *module coupling and cohesion*.[57] Other detailed approaches also exist and are briefly mentioned and compared at the end of this section.

The result of the software design phase should be an overall software systems architecture. However, the resulting organization should reflect the nature of the problem being solved, not just any structure. In other words, an understandable and correct structure is to be generated that is tied to the problem, in this case, specific digital control algorithms within an applications context.

A design approach must translate the requirements specifications into a system design. The tools and techniques used in this process should provide continuous validation of the design by documenting the current configuration. This objective provides audit paths or traceability between the requirements and the detailed design. This characteristic also provides for ease of analysis and modification.

The general design procedure is as follows:

Step 1. To identify and clearly state the basic functions as documented from the requirements specifications

Step 2. To structure (divide and conquer) the overall program organization around these functions by defining functional modules and their general interaction

Step 3. To iterate (redivide and reconnect—hierarchical modular structure) the design for better overall structure based upon design heuristics and module coupling

Step 4. To detail the internal control flow structure of each module

Step 5. To design the detail data structures for each module

Step 6. To iterate design for each module based upon module cohesion heuristics

Discipline is required to accomplish these design objectives. The continual evaluation of the current design configuration is an integral aspect of this approach as can be seen from the above steps. Iteration of the design is an extremely important step in achieving quality software. The early one-track approach to a quick design can result in a poor design that is difficult to maintain. The design approach suggested above attempts to solve this problem.

The *top-down approach* focuses initially on the overall modular design of the software structure. Module interfaces are then analyzed for simplicity, with structure changes made as required. The details of each module are generated in a top-down fashion starting from the top of the overall structure. Each module in turn is implemented and tested in this top-down fashion.

Lower-level module details are ignored initially in this top-down approach. In some cases, however, specific lower-level modules are defined and examined if they might constitute a high risk of performance failure (examples include numerical precision routines and high-speed operations) as found in some digital control applications.

In general, each software module should perform a clearly defined function such that it can be coded, translated, tested, and executed as an independent subprogram (cohesion).

Module relationships (*coupling*) are defined in terms of information types that are transferred between modules. The information transfers are references from one module to another. Analysis of different information types for implementation provides insight into the principles of quality design. *Effective modularity results in minimal connections (low coupling) and maximum module independence (high cohesion).* Before developing an approach for an overall structured organization, the detailed definitions for module coupling and cohesion are presented in the following sections. *The design objective is to develop the modular hierarchical structure and the overall structure so as to minimize software system complexity.*

13-4A Module Coupling

Coupling is a measure of the relationship between modules or the strength of interconnection between modules. The higher the coupling, the more knowledge required by one module to understand another. The objective is to minimize coupling.

A quantative measure of coupling must be defined in order to compare designs and improve them. A hierarchical classification of information coupling has been developed[58] as follows:

Content coupling (worst)
Common coupling
External coupling
Control coupling
Data structure coupling
Data element coupling
No direct coupling (best)

The various classes relate to the information structure involved with module coupling. The following paragraphs describe each level except the coupling associated with the absence of any coupling, that is, no direct coupling. Modules reflecting this characteristic, of course, are defined as independent and normally have little use in an overall hierarchical structure for a digital control system.

Content coupling means that one module makes a direct reference to the contents of another module. This coupling is considered the worst since the modification of one module affects the other (side effects). Therefore it is difficult to isolate logical errors and correct code. Examples include branching from one module directly into the middle of another, referencing data in another module, and modifying program statements in another module. A specific example is a direct branch into an error routine which may be required in a real-time system. This type of coupling should be minimized.

Common coupling refers to the referencing of a shared global data structure (central repository). Examples include a shared communication region, a file, and use of blank common (FORTRAN). It may include heterogenous data items within a data structure. Again, modifying one module with this type of coupling may affect all others using the common data. Users of a file, for example, assume no other users. Also, modules tend to be special-purpose and therefore are not very useful in other applications. Usually recompiling or reassembly is required. An example in digital control is a repository for compensator coefficients stored in a common array.

External coupling refers to a module referencing an externally declared symbol (data item) in another module. It is similar to common coupling except that references are homogenous data items, not data structures (arrays, files, etc.). Difficulties with this type of coupling include module modifications that affect other modules with possibly no knowledge of other module users of the data item. The above example with compensator coefficients stored as individual data items in an initiation module represents external coupling. In regard to the digital-controller example of Fig. 11-34, there is external data coupling where several modules have access to (reference) a globally declared data variable. The modules assume the correct data is stored when they access the variables such as ADSR which is used by both MAIN and INPUT.

Control coupling is evidenced by the passing of control parameters or arguments between modules. Here a control parameter refers to a boolean variable which is used in a program decision (that is, IF statement, etc.). The value of an argument directly then influences the flow of control in a specific module execution sequence. Examples include function codes, flags, switches. An example is the selection of a first- or second-order filter in a single module by the use of an argument that is passed. The problem with this level of coupling is that the calling module must have some knowledge of the internal processing of the called module.

Data structure coupling or *stamp coupling* means that one module references the same data structures as another except that the data structure identifier is passed to the module as a parameter. Thus, the data structure is not "common." A weakness of data structure coupling is that it is easy to use the data structures in a different context within a module, thus software modifications can be difficult. Note that this approach does not solve the problems of common coupling but only reduces their effect. Using the compensator example in a module of this type, the coefficients are passed to the module as an array.

Data element coupling requires, by definition, that all input and output to and from a module be passed as homogenous arguments. All arguments are data elements or items (no control elements or data structures). In general, data element coupling reflects highly independent modules that are readable and understandable. However, the argument lists can be excessively long. For example, there is the passing of numerous filter outputs in a process control environment or the passing of the individual compensator coefficients in an nth-order filter.

In implementing a second-order compensator module, one of the various forms ($1D$, $2D$, $3D$, $4D$ of Sec. 10-3) can be employed to generate the associated difference equation. The selected form depends upon the accuracy requirements, timing requirements, and other performance objectives. However, to minimize the computation time, only the current input, $x(kT)$, should be processed to the extent necessary to generate the output, $y(kT)$. The other manipulations required for the complete second-order filter are executed in the remaining time left in the previous sampling interval, $(k-1)T$, during which the CPU is waiting for the next sample to be taken at kT. The flowchart on page 438 indicates the control flow of the general process. The specific structure of each associated module and the overall system software structure for this process depends upon performance requirements. Note the conversion times and computation time associated with specific modules.

For specific processing of the second-order filter realization, the following equations as derived from Figs. 10-9, 10-10, and 10-11 are used where $v_i(\ )$'s are the immediate variables as shown in the figures. Note that the output calculation for $y(kT)$ in each case requires one multiply and one addition.

$1D$:

preprocessing (previous sampling interval)—

$$\text{temp1} = b_1 v_0[(k-1)T] + b_2 v_0[(k-2)T]$$

$$\text{temp2} = a_1 v_0[(k-1)T] + a_2 v_0[(k-2)T]$$

output calculation (current sampling interval)—

$$v_0(kT) = x(kT) - \text{temp1}$$

$$y(kT) = a_0 v_0(kT) + \text{temp2}$$

$2D$:

preprocessing—

$$v_1[(k-1)T] = a_2 x[(k-1)T] - b_2 y[(k-1)T]$$

$$v_2[(k-1)T] = a_1 x[(k-1)T] - b_1 y[(k-1)T] + v_1[(k-2)T]$$

output calculation—

$$y(kT) = a_0 x(kT) + v_2[(k-1)T]$$

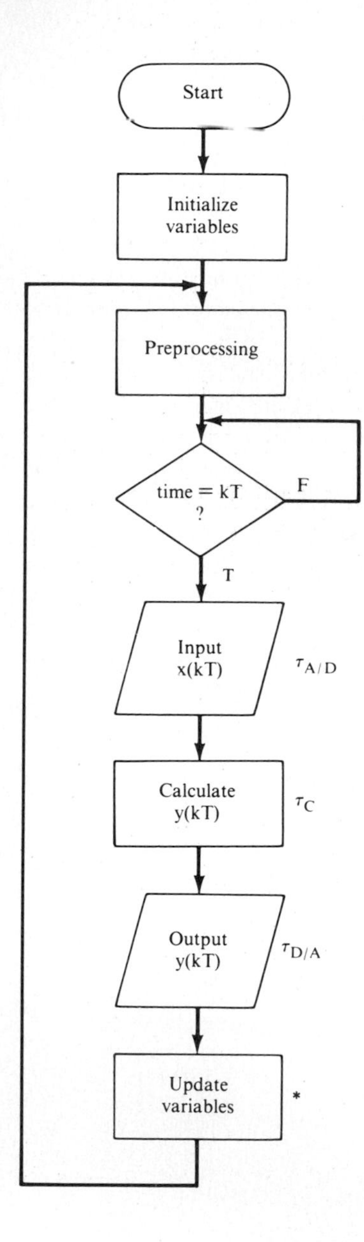

* History files can be generated at this processing point in the program.

$3D$:

preprocessing—

$$\text{temp1} = a_1x[(k-1)T] + a_2x[(k-2)T] - b_1y[(k-1)T] - b_2y[(k-2)T]$$

output calculation—

$$y(kT) = a_0x(kT) + \text{temp1}$$

Note that the $3D$ representation requires the least number of overall arithmetic operations. Scaling, however can still have an impact on the selection of a filter representation as it relates to finite word-length constraints as discussed in Chap. 10.

Assuming that only the input, $x(kT)$, is passed to the second-order compensator module and that all coefficients are local to the module, the module coupling would be of a data element type, a quality choice based on software engineering objectives. Other configurations can provide for easier modification and more general-purpose modules through the use of filter coefficient passing (data elements, array, calculations, ...) between modules. Of course, the coupling type becomes less desirable.

Due to extensive experience,[58] the previous coupling definitions have been ordered based upon the following criteria:

Reusability
Error proneness
Testing difficulty
Debugging difficulty
Maintainability
Extensibility

Of course, these criteria are relative to the application. In many cases, environmental considerations such as memory space and execution speed override the above criteria. Timing may play also a critical role in meeting performance objectives as in a digital control system. However, only after detailed analysis should this criteria be slighted. The specific impact of such a decision should be documented. In this case, as in other cases, experience is the best teacher.

13-4B Module Cohesion

To achieve software quality, a measure of internal module characteristics is also required. The term associated with these meaures is *module cohesion*. It is a measure of the relationship among the internal module operations, data items, and data structures. These measures relate to the concepts of module strength, internal module binding, and module functionality. The design objective is to have high module cohesion for ease of testing and understanding.

Module cohesion has the following hierarchical order of cohesion levels:

Coincidental (lowest)
Logical
Temporal
Procedural
Communication
Sequential
Information
Functional (highest)

Again this hierarchy is based upon the same experience as module coupling, i.e., module reusability, error proneness, independence, maintainability, and extensibility. The following paragraphs introduce each cohesion type starting with the lowest level.

Coincidental cohesion refers to the condition that no meaningful relationships exist among module elements. Historical rationale for this situation includes memory-size limitations, modularization of an existing program with lack of objectives, and consolidation of existing code. In a module with coincidental cohesion, many statements have no apparent relationships to others, a complete function (or functions) may not be executed, and the module is very specialized, i.e., not general in nature. Usually difficult to test modules in this form.

In *logical cohesion* or logical binding, there is some logical relationship between module elements. In addition, the module performs by definition one and only one of a class of related functions when invoked with a specific set of data. Problems with logical cohesion include "fuzzy" boundaries between specific function statements within the module and the lack of explicit understanding. An example includes various error routines that are integrated together and embedded in one module. Another example is a module consisting of various combined filters, only one of which is to be invoked. In many cases, this approach leads to tricky code that is difficult to modify. Moreover, unnecessary parameter values are probably being passed.

Temporal cohesion or classical cohesion refers to a series of operations that are executed in sequence (time-based) due to functional relationships. Usually, no argument values are transferred (coupled) to the next module. Examples of this type of cohesion include the placement of initialization and termination functions within the same module.

Procedural cohesion also involves multisequential functions, but they are also explicitly related to the problem procedure, i.e., a direct mapping of the data flow diagram. The form in module organization is then data control flow. The realization of procedural cohesion results from a flowchart structure (process-oriented). Observations of this approach indicate that modules often contain several functions or various parts. Control flow or procedure-oriented modules often cut across functional lines. An example of this type is a module that does the entire digital control computation for a given spacecraft axis (say, pitch, yaw, or roll).

Communication cohesion is procedural binding along with data communication between module elements. All the module elements operate upon the same input data set. This cohesion structure emphasizes the activities that can be done with the input data set. Examples include a position feedback input that is filtered and also used to create a value of rate which is filtered, the two filter outputs being recombined for the module output.

Sequential cohesion modules process data in a serial operation from one statement to another, i.e., a constricted communication module. An example is a module that checks the input data for being in range, filters them, and checks the output data for saturation.

Information cohesion refers to a module with multiple functions dealing with a single data structure. All the functions are related by concept. Usually one entry point exists for each function. Modules of this type have the same input information regardless. On the other hand, logical modules have different input data depending upon what function is to be executed. An error routine based upon an error status word can be an example of information cohesion modules.

Functional cohesion requires that all module elements be essential to the performance of a single function. It usually has only one input and one output. To identify a functional cohesion, the activity defining the module should be a single transitive verb with a nonplural object. Examples include a module to filter rate, error check, a software multiplication of two numbers, or the second-order compensator module in the last section. This type of cohesion is regarded as the best because of understandability.

A specific cohesion type can be determined by analyzing the specific module activity. For sequential, communicational, or logical binding, look for compound sentence instruction sequences with multiple verbs. Time-oriented actions (first, next, last, etc.) usually define a temporal or procedural module. Multiple objects within a module description should be analyzed for defining logical cohesion. Finally, temporal binding can be involved with initialization routines and termination routines.

For the digital-controller example of Fig. 11-34, the cohesion is sequential in that each module processes data in sequence from the previous module. Time (temporal) cohesion occurs in that an A/D conversion must be done before the input is checked for over-under flow. This activity must be done before the input data can be filtered. The filter operation must occur before the digital data is converted to analog data.

In summary, a module can be viewed as a mixture of several cohesion types; i.e., cohesion levels are not mutually exclusive. Also, cohesion is a continuous measure rather than a discrete measure like coupling. The previous definitions and associated levels provide only guidelines and insight. In practice, however, *do not lower the cohesion level for the sake of efficiency without considering the consequences for software quality*.

A somewhat different view of module cohesion[101] is defined as *object-oriented structuring*. In this particular case, the focus is on hiding the exact module control and data structure from other calling modules. This technique is known as *information hiding*. The object module is called by requesting that certain operations (functions) execute on a specific data structure whose specific organization is hidden within the object module. Note that the data structure values are not passed to the calling module by using an object-oriented design approach. An example of an object module would be a module that executes filtering (compensator) routines in a matrix format. The calling module would specify the type of filter but would have no knowledge of the internal data structure. Note in this example that a different object-oriented module would transform a scalar input into the matrix format. Thus a pointer or location address is passed to the object-oriented modules, indicating the location of the data to be processed (filtered). Thus the filtering module in this case is of the general informational cohesion type.

The data structure hiding is the unique aspect of this type of object-oriented modular design and can be used with various cohesion types. The information-hiding concept leads to a hierarchical organization of the data. At a specific level, only the *abstract data types* are provided that are needed to perform the indicated functions. This general technique permits easier testing and software modification due to the limited data and control module interfaces between each level. Considerable design discipline is required to develop object-oriented modules because of the levels of abstraction.

There is a strong relationship between module strength and module size. In general, a smaller module is stronger than a larger one. By making the modules smaller, the number of connections between them tends to increase. A trade-off exists based upon a given performance objective. In structuring software, it must be remembered that the danger of module complexity lies mainly in clouding the understanding of the way that the program works. It is easier to understand how a small module operates! Before analyzing each module internally and externally, an overall structure must be developed. This is the objective of the next subsection.

13-4C Design Method

The general software design method emphasizes a hierarchical-structured modularity. The ease of understanding (software quality) is provided by the *top-to-bottom structured approach*. The specific structural approaches along with the module coupling and cohesion definitions form a consistent and strong basis for software design.

The structured-design method[58] is based upon the data flow of the software requirements approach and the transformations (activities) performed. Each function or transformation contained in a DFD must be analyzed. The transformations are classified or grouped together depending upon the general type of activity (i.e., input, output, or computational). The type of activity helps to define an overall system software structure consisting of activity modules, functional modules, etc.

Structural types are usually *centered* around a specific aspect (view) of the overall requirements functions. The general types include:

Transaction-centered
Transform-centered
Procedure-centered
Data-centered

The language symbols used in a software structure chart based upon one of these structural types are portrayed in Table 13-4.

The structured-design method presented results in a structure chart (a block diagram portraying software modules and the interfaces between them) using the symbols of Table 13-4. The generation of a structure chart is an integral component of the software development process shown in the following diagram:

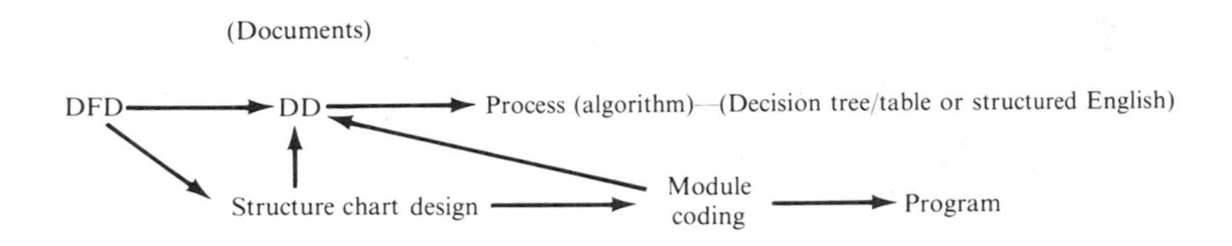

The general structured types are introduced next and detailed discussion follows.

Transaction-centered transaction analysis (see Fig. 13-6) structures are developed around a sequence of modules which can perform processing tasks. The transaction center activity defines which one of the many processes is to be executed. The transaction center determines what type of process is to be performed. The hierarchy of the software structure as shown in the figure requires a TYPE module to sequence the required process ACTIONs. Each action is performed by selective DETAIL modules. As appropriate, information is passed

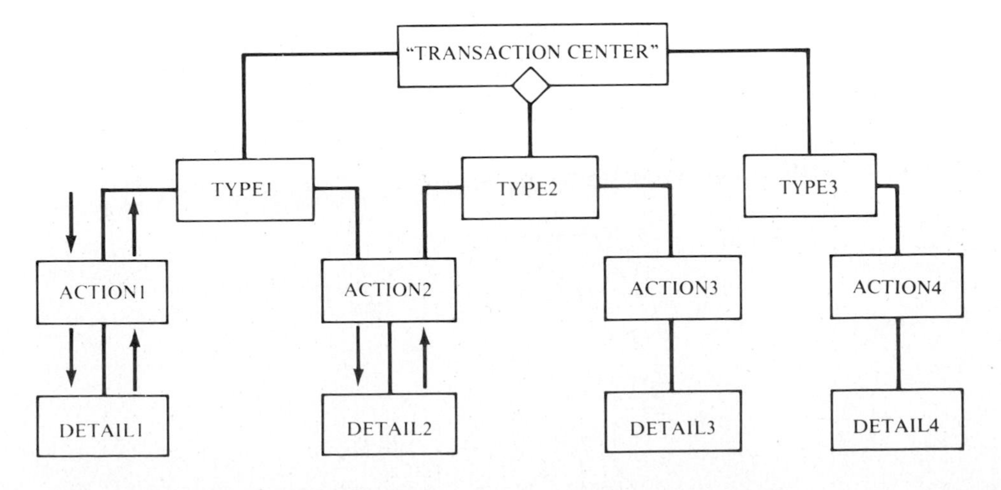

Figure 13-6 Transaction-centered structure.

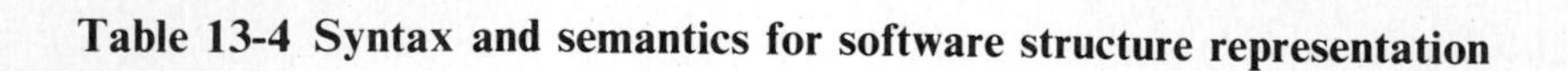

Table 13-4 Syntax and semantics for software structure representation

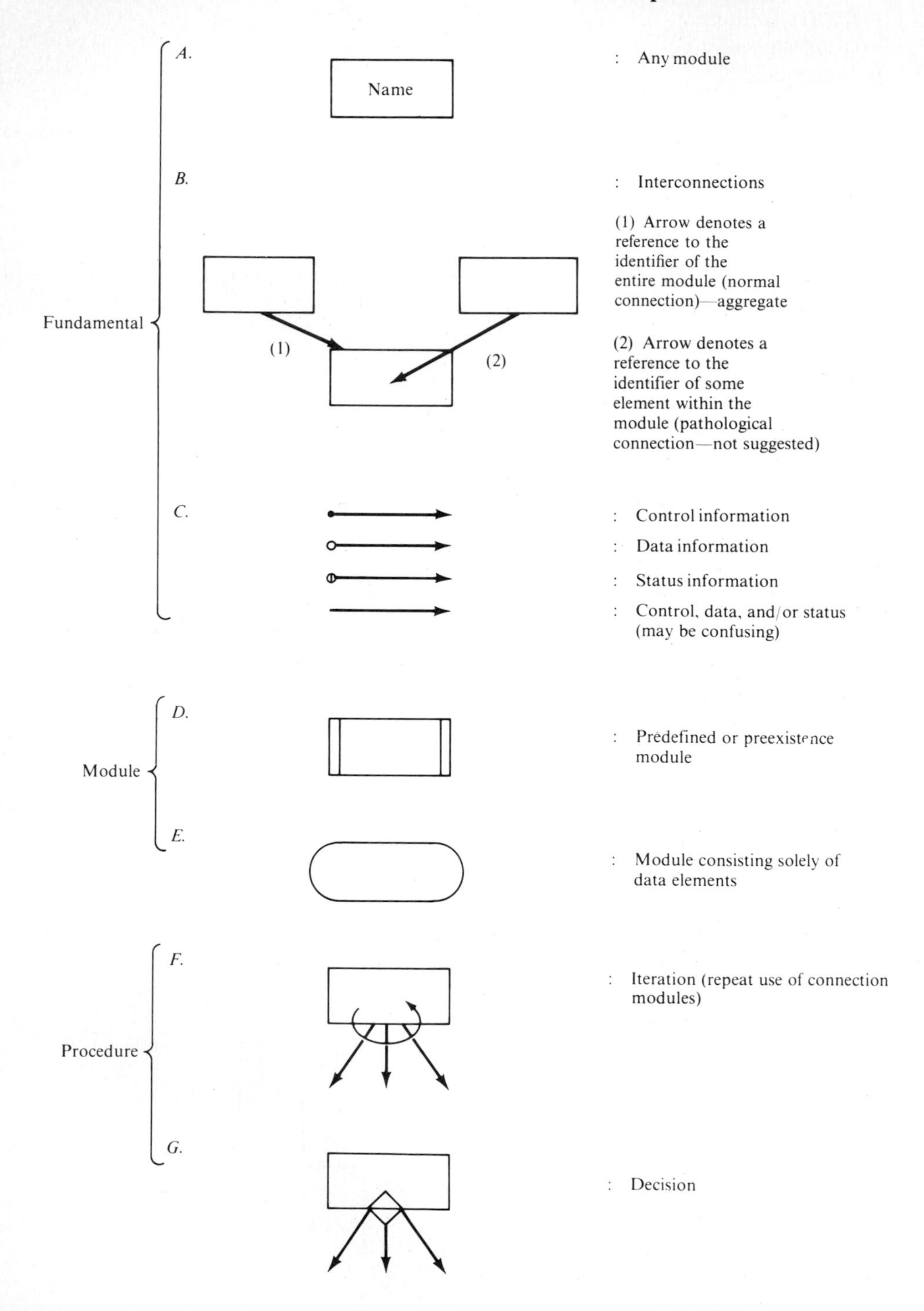

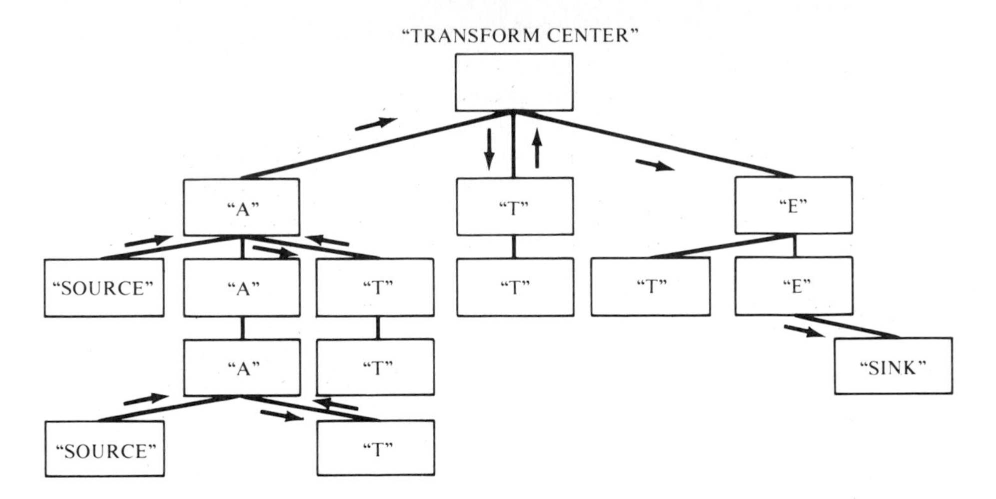

Figure 13-7 Transform-centered structure. "A" = afferent operation (input); "T" = transform operation; "E" = efferent operation (output).

between each module up and down the structure. However, all detailed data are usually obtained from sources at the lower levels and distributed to sinks also at the lower levels. This type of software structure would be used in a conditional process control environment.

Transform-centered (transform analysis) structures are based upon major system functions that are viewed as central transforms which manipulate major inputs and generate major outputs as shown in Fig. 13-7. The various operational terminology is discussed in more detail after presenting the next two software design structures. This structure would be used in a sequential process control environment.

Procedure-centered (procedural analysis) structures are derived from procedural representation (i.e., flowcharts) of a system's operation. A top-level module calls submodules (subordinates) as shown in Fig. 13-8. This approach

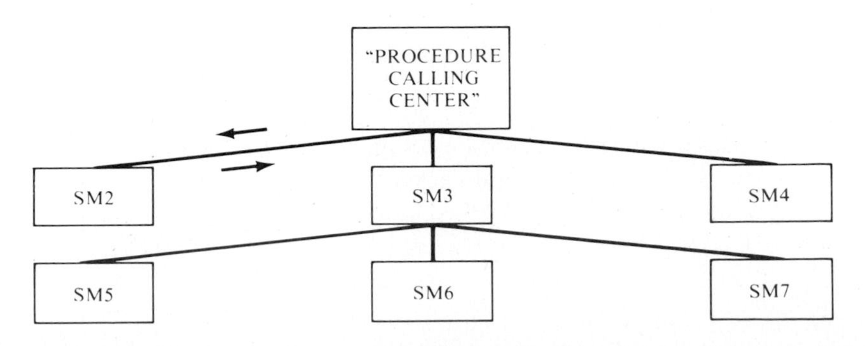

Figure 13-8 Procedure-centered structure. SM = submodules (subroutines).

usually results in modules with temporal or procedural cohesion and is useful in "simple" systems.

Data-centered organizations are derived from the data structure of the problem[57] with the premise that the structure of the program should mirror the correspondences between the input and output data structures. This type of structure is found in large-scale data processing applications emphasizing data manipulation, but not generally in process control applications.

To develop some details of the structuring approach, specific terminology must be understood. An *executive module* by definition does not perform any of the system tasks but controls and coordinates their performance by lower-level modules. A particular structure is *completely factored* if all but the bottom-level modules are executive modules. To discuss the initial development of a structural design technique, four general classifications of modules are presented. They are afferent, transform, efferent, and coordinate modules.

An *afferent module* obtains information from subordinates, and then passes it upward to its superordinate (input process or source flow). The following diagram defines an afferent module with an "A" and upward flow with a data variable:

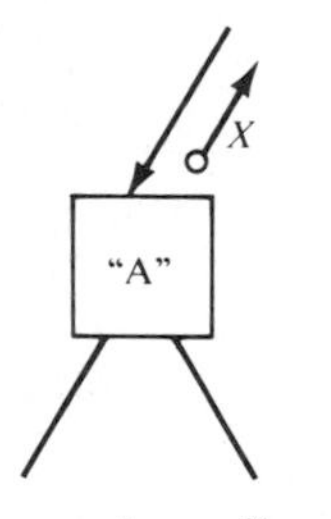

An *efferent module* takes information from its superordinate module and passes it on to its subordinate (output process or sink flow). The following diagram shows a data variable X flowing through an efferent module:

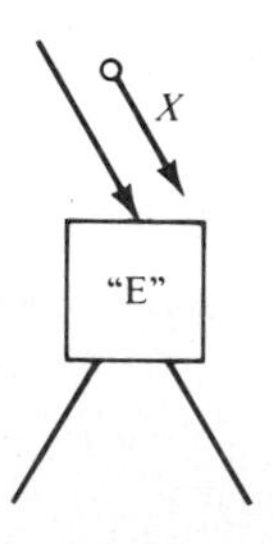

A *transform module* changes by definition its input data into some other data form or value as shown in the following diagram:

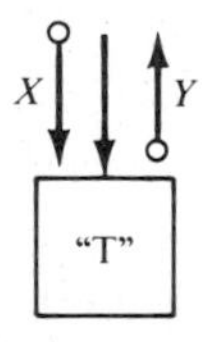

Finally, a *coordinate module*, an *executive module*, manages the affairs of other modules. The other three types can take on this function. For example, an afferent coordination module is

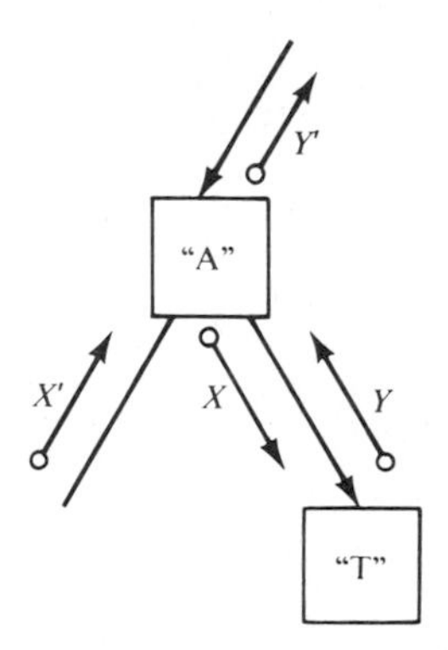

where "A" is coordinating the data transfer to and from "T."

In studying the overall software structure, the *scope of effect* is used to denote the collection of all modules containing any processing that is conditional upon a particular module's decision. Similarly, the *scope of control* is a module itself and all its subordinates. These two scopes are useful in structuring and analyzing a software system. In general, minimization of the scope of effect and scope of control is desired for understandability and testability.

An *input-driven* system is one that obtains all its inputs in elementary (raw) form at or near the top of the hierarchy. Examples include systems where critical response time or a major change in processing is required such as found in digital control systems. An *output-driven* system is one which is oriented toward the production of a complex output data item, i.e., data processing.

13-4D Design Heuristics

Processing distribution (scope of effect) is an important design heuristic. In a well-designed system, decision elements (i.e., conditional statements) tend to decrease toward the bottom of the hierarchy. The final structure approaches a completely factored design (actual data manipulation accomplished at bottom-level modules). Top-level modules deal with global matters. Lower-level modules deal with smaller and smaller aspects of global data matters.

Well-designed systems tend toward a "mosque" shape, i.e., higher fan-out at top, higher fan-in at bottom. Usually, the structure is neither input- nor output-driven. It is *balanced.*

Module size is another important design heuristic associated with quality software. Generally, a module should be small such that the program statements fit on a single sheet of paper. Another view says that a module should contain 40 to 60 statements in any programming language used. One must remember, however, that some statements are more complex than others. Whatever "standard" size is

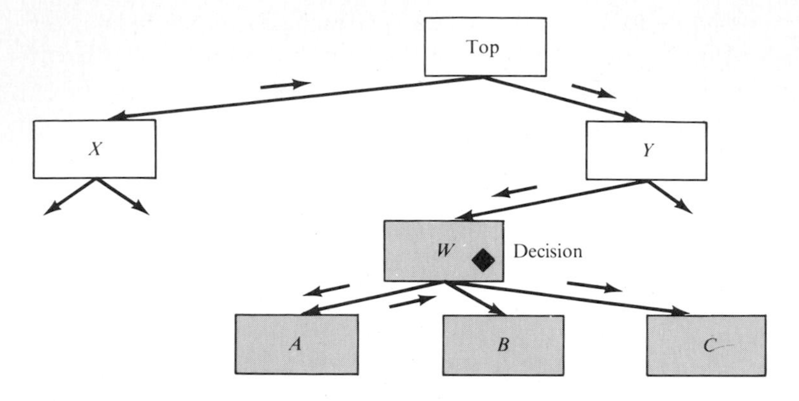

Figure 13-9 Proper placement of decision within software structure.

selected, adherence to the associated format should continually be reviewed for understandability.

Three factors affecting statement complexity are the amount of information that must be understood, the accessibility of information within the module, and the structure of the information. The scope of effect should be a subset of the scope of control. That is, *all modules affected by a decision should be subordinate to the module making the decision.* Decisions should be no higher in the hierarchy than is necessary to place the scope of effect within the scope of control! The block diagram of Fig. 13-9 illustrates an ideal solution.

The consequences of scope violation are duplicated decision making, control coupling (flags being passed to other scope-of-effect modules), and pathological control flow or data flow. The design approach to a scope violation situation (in the early design phase) is to compress a low-level module into its superordinate. This compression either moves the affected module down into the scope of control or moves the decision module to a higher-level module. Note that correction of scope violation may not be possible without weakening the cohesion of some modules. In general, the approach employed is correct only if it does not decrease module cohesion. In practice, the number of exceptions to this heuristic should be limited; other constraints are probably more important.

Another important design heuristic is the objective of minimizing software complexity. One way of depicting program complexity is to determine the number of paths through a module (or the entire hierarchical structure). A graph is a good way of representing these paths. The more paths there are in general the more complex the program. Decomposing a complex design[101] into a lower degree of complexity is, however, a difficult heuristic task and is beyond the scope of this text.

13-4E Transform Analysis

The major structural approach is the transform analysis technique. It generally develops a good design that is highly factored, balanced, and has high-cohesion modules. Only minor restructuring is required to arrive at a final design. The transform-centered approach identifies immediately the primary processing functions (activities), high-level inputs, and high-level outputs. The concept is based upon a data flow model vs. a procedural model.

Transform analysis involves the following steps:

Step 1. Restate the requirements as a data flow diagram.
Step 2. Identify the afferent and the efferent data elements.
Step 3. Generate the first-level structure based upon these elements.
Step 4. Factor afferent, efferent, and transform branches to form a hierarchical structure.
Step 5. Refine structure by continuing the same process at each level.

Step 1 restates the problem as a DFD by using the structure from the requirements definition phase and ignoring minor inputs and outputs until the next step. Data elements should be labeled carefully, but do not develop a flowchart. If this flowchart technique is utilized, modular structures tend to be input-driven.

Example 13-1 Generate a software structure for an assumed DFD which would result from Step 1, that is:

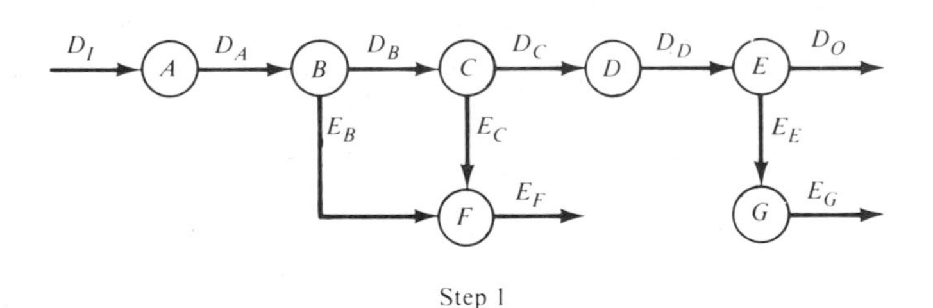

Step 1

Step 2 encompasses the identifiers of the afferent and the efferent data elements. Start at physical inputs and move inward until the data stream can no longer be considered as incoming data; a value judgment for defining the afferent domain is thus required. Using the same judgments, start with the output data and move backward in the DFD. When the data can no longer be considered output data, the efferent domain is defined. The number of DFD data lines (arrows) cut should be limited to minimize coupling.

The afferent, efferent, and central transform domains are thus defined within the software structure (note that it is possible to have no central transforms). Note in this example also that the error indication E_O is not considered as an efferent element possibly owing to a required "immediate" error message requirement as defined in the DD (data dictionary).

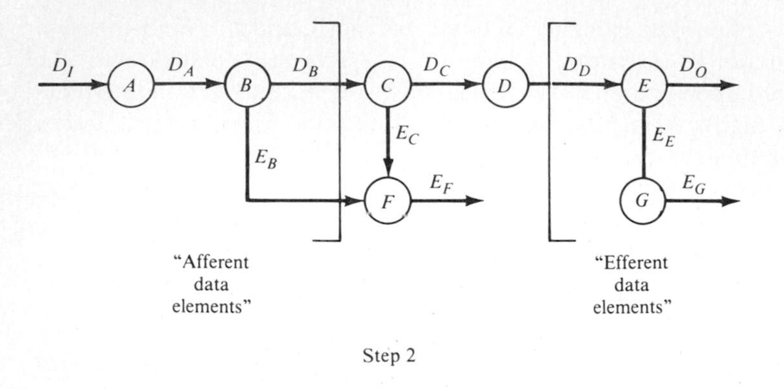

Step 2

Step 3 revolves around first-level factoring. A main module (EXEC) for the entire system is defined. An immediate subordinate afferent module is defined for each afferent data element. A subordinate efferent module for each efferent data element is included in the initial structure. A subordinate transform module for each central transform is also included as required. The first-level factoring is shown in the following figure:

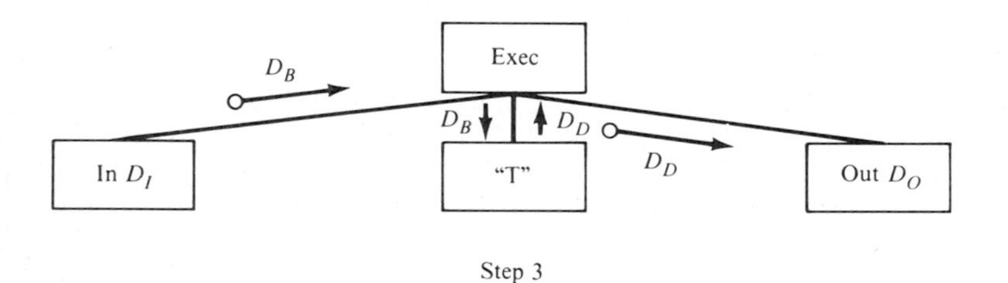

Step 3

Step 4 factors the afferent, efferent, and transform branches. Each afferent module is factored into at least one afferent module and one transform module which are closer to the input (data source). This afferent module receives the more abstract (manipulated) input after the transform module transforms the data into a more abstract form. The reverse process should occur for the efferent module. Note that the factoring of central transform modules is based primarily on the design heuristics including coupling and cohesion. Also note that the different methods are used for data transfers to sinks as shown in the example. (See Step 4 on opposite page.)

Step 5 refines the software structure in an iterative fashion until the software engineer is satisfied based upon heuristic criteria.

Note that afferent modules within central transforms and efferent domains are possible. The objective of this design approach is to make the software structure relate directly to the problem requirements as closely as possible. Design trade-offs (coupling, cohesion, design heuristics) are weighted based upon experience, performance requirements, and system constraints.

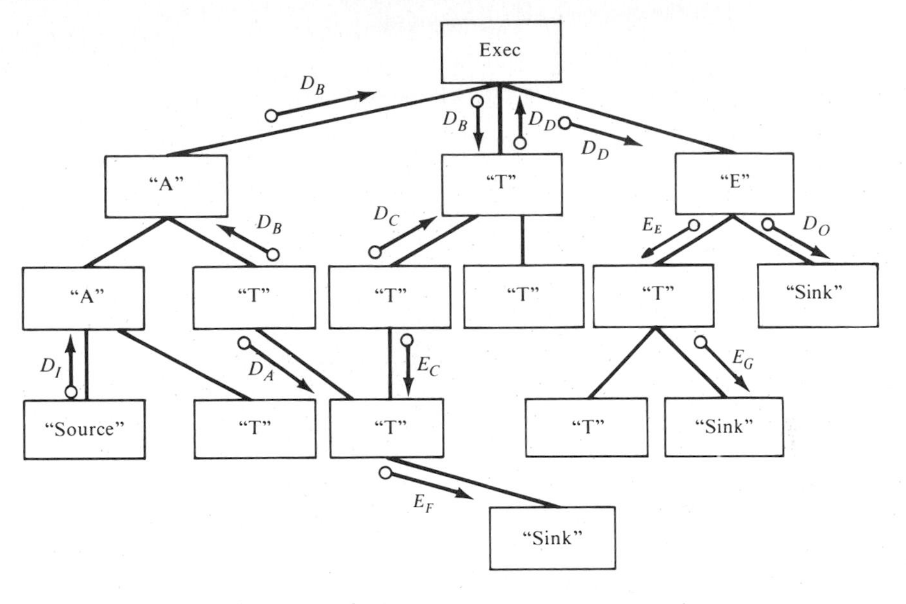

Step 4

The identification of a module name is also very important in understanding the software structure. Some of the module names come from the DFDs, while others must be selected. Some suggested names for the variables within various domains as used in a digital control system are:

Afferent: initialize, prefilter, develop, form, find, load, get, obtain, input, accept, generate, create, convert, etc.
Efferent: put, produce, save, output, store, write, deliver, control, convert, etc.
Transform: analyze, transform, convert, compute, filter, calculate, do, perform, process, etc.

Since the DFDs are hierarchical (consist of levels of diagrams), these levels are transversed as the software modular design structure is generated. The determination of structure diagram depth depends upon these situations:

A transform with no clear discernible subtasks has been reached based upon the language selected for implementation.
A library subroutine is available.
A module has been defined that interfaces with hardware.

Example 13-2 The transform structure for the digital controller of Sec. 11-8 is shown in the following diagram. Note the notation for global variables and the possible testing problems with this type of module coupling. Also, the initialization operations are not indicated.

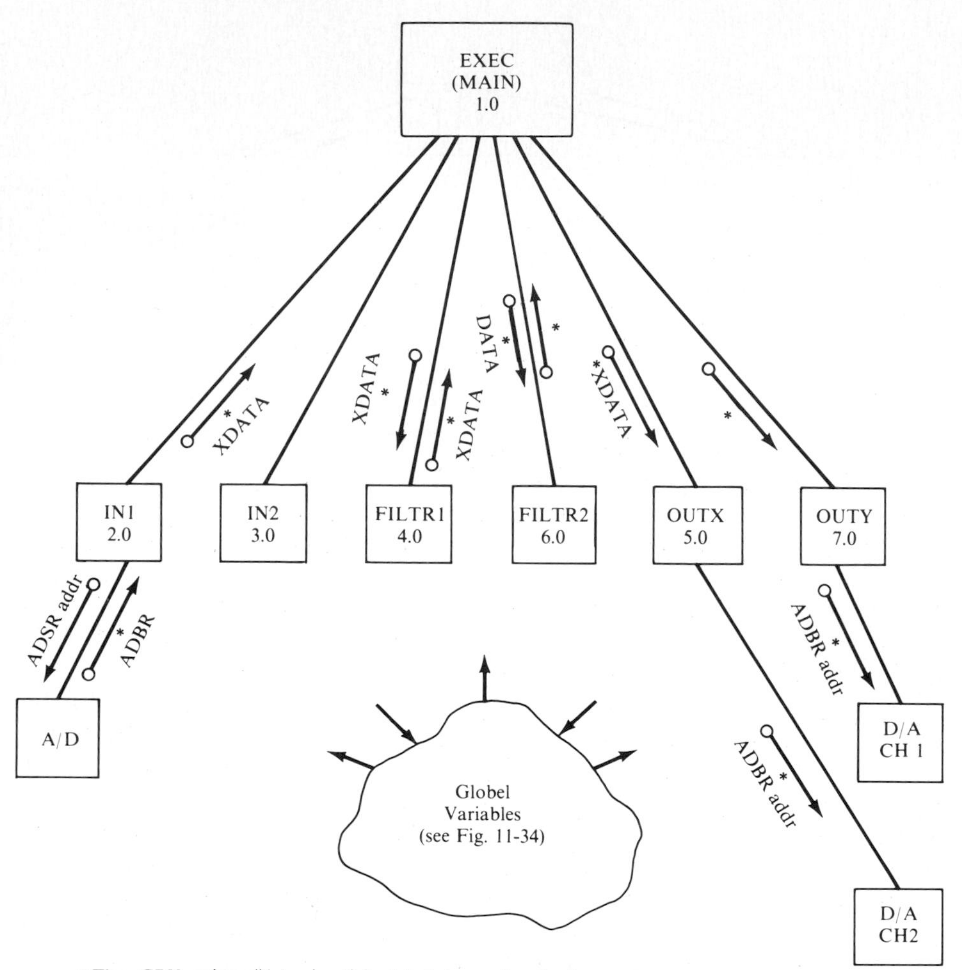

* Thru CPU register #1 (an implicit global storage location).

In conclusion, the transform analysis technique *generates* a one-dimensional hierarchical structure based upon DFDs and is not initially concerned with a single-processor implementation (many processors can be used to implement the structure). Its use requires judgment and appreciation of afferent, transform, and efferent data domains. In general, it is the major design suggested for complex requirements, such as those found in a major digital-control-system development.

13-4F Transaction Analysis

Transaction analysis or the transaction-centered approach focuses on requirements that essentially have equally weighted subprograms. The selected subprogram depends upon a decision or selection process at some level in the overall

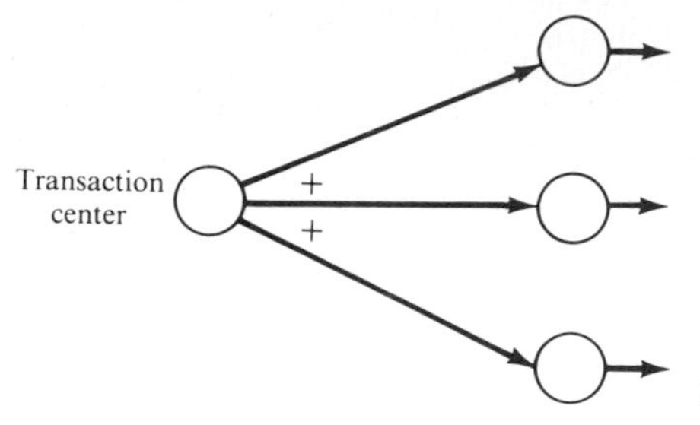

Figure 13-10 Example of transaction-centered DFD.

software structure. The specific level depends upon the importance of the decision (scope of effect). A transaction-centered module determines the specific (usually one) subordinate module to execute in the set of subordinates. Note that a transaction center can be part of a general transform structure.

A transaction element is any information (data, control, signal, event) or change of state that initiates some action or sequence of actions (real-time systems, switch, escape character, interrupt). From an appropriate DFD, an input data stream splits into several discrete output substreams. This is a data-driven selection process. Note that a transaction selection may occur within a central transform, afferent, or efferent branch in any software structure. An example of a typical transaction center in a DFD is shown in Fig. 13-10.

The functions of a transaction center are to respond to transaction information. The software must analyze each input transaction request to determine its type and the nature of the specific transaction before dispatching to a lower-level module. The lower-level modules complete the processing of each transaction. Note that *this approach could generate modules that are highly coupled with low cohesion. Be careful!* A model of a transaction-centered structure (orthodox form) is shown in Fig. 13-11.

Note that the scope of control can be high when using transactional approach. Cohesion here usually ranges from logical to communicational with an independent horizontal structure.

The steps in transaction analysis are:

Step 1. Identify the source of the transaction.
Step 2. Specify the appropriate transaction-centered organization.
Step 3. Identify the transactions and their defining actions (define the process).
Step 4. Note potential situations in which modules can be combined (trade-offs in coupling/cohesion).
Step 5. A vertical evaluation: For each transaction, or cohesive collection of transactions, define a module to completely process the transactions. Evaluate resulting structure based upon design heuristics.
Step 6. Horizontal evaluation: Specify for each specific transaction activity a module subordinate to the appropriate transaction module(s), combine functions into modules by looking for common activities, define required

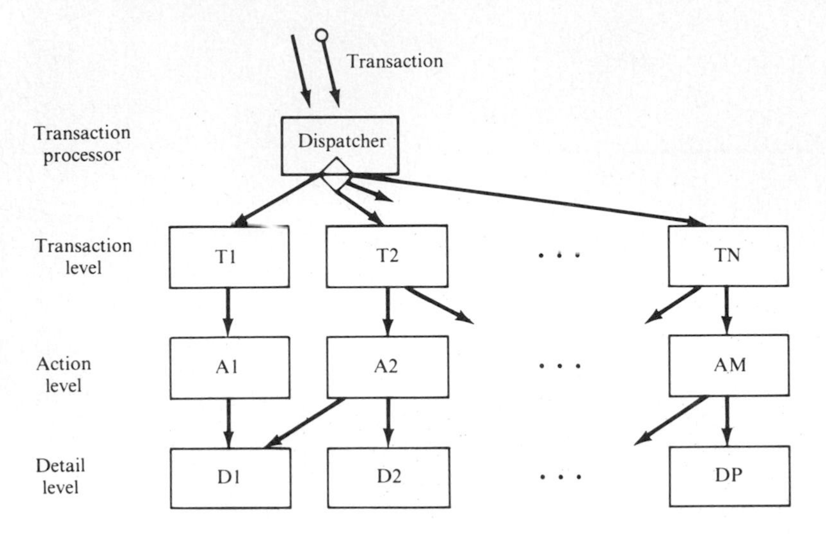

Figure 13-11 Example of transaction-centered software modular structure.

module detail; factoring continued to lower levels, again looking for commonality. Evaluate resulting structure based upon design heuristics.

The outputs of a transaction center are usually a converted, formatted version of the desired input transaction. A digital control example would be the selection of a scaling process based upon the input value being within a specific range.

In conclusion, the designer should not be too rigid (orthodox) in applying the transaction analysis approach. The approach generally injects independence into modules at a high level and thus is guided by coupling, cohesion, and design heuristics for placement of transaction center. A high-level transaction center indicates that the overall system is controlled by environment (response time). A low-level transaction center indicates that the system controls itself internally (high throughput being an objective).

In decomposing a set of requirements and generating a structure chart, the previous techniques in themselves have a great impact on the type of module cohesion. The primary transform approach indicated uses the DFD which can result in communicational and sequential cohesion. As the design structure chart is developed top to bottom, decomposition may focus on time order, control flow, or other criteria. Time ordering usually generates modules with temporal cohesion. Control flow structures usually represent flowcharts, resulting in logical cohesion. The type of resulting module coupling depends not only on the decomposition approach but on the specific information (data, control, and status) being implemented. Again various decompositions should be attempted and analyzed from design heuristics and other contextual points of view in order to obtain a "better" solution.

13-4G Other Design Techniques

Data-centered approaches emphasize the direct transformation of data structures into program structures. In other words, a specific data structure is transferred on a one-to-one basis into a program module. A specific data structure or module can, for example, be a data element, file, or a record. Two very similar data-center approaches are Jackson's technique[59] and the Warnier-Orr technique.[60, 61]

In most requirements, there are normally two separate data structures, the input and the output. The above two techniques initially focus on the output data structure. The output data are defined in a top-down hierarchical fashion in the form of a vertical tree in Jackson's method. In the Warnier-Orr technique, the structure is depicted in a horizontal fashion with brackets (a horizontal tree). Special symbols are used to depict data element repetition and unique data element selection. Note the use of repetition and selection concepts in the data domain. The input data structure is treated in the same manner. The objective is to generate a program structure that matches the data structure by using the sequence, iteration, and selection control flows. The desired output data structure incorporates the required elements of the input data structure. The program diagram is essentially a flowchart which can easily be translated into a programming language, i.e., coded.

A principal advantage of these data-centered approaches is the ability to easily understand the structure of the data and program. Both program and data decompositions are usually simpler than the previous approaches. The disadvantage of the two data-centered approaches is the general requirement for defining the one-to-one mapping from the output data structure to the program structure. This objective is difficult to satisfy where the relationship of the input data structure to output data structure is not obvious. Intermediate data structures have to be created, which complicates the design. Thus, the data-centered approach is useful for "simple" data structures. For example, it is useful to define a simple activity explicitly at the module level such as a filter.

Another well-known design technique is called HIPO (Hierarchical-Input-Process-Output).[51] This method is primarily a documentation procedure for taking the overall software system module organization at the highest level and decomposing the high-level module into detailed functional design, that is, top-down design. Again, the heuristics previously mentioned apply.

In conclusion, the tools and techniques presented for digital systems software design can lead to a well-structured organization and thus quality software. *Discipline is required to properly document and analyze a number of possible design alternatives before racing into coding.* The central-managed data dictionary normally includes all the terminology involved in the software life cycle including the design phase. The other techniques briefly discussed are also adequate for structuring a design within their own constraints. Other techniques[62, 63] also exist. The main transform approach presented in this chapter, however, should be adequate for digital system software design.

13-5 STRUCTURED PROGRAMMING[51]

The understanding of a software program is affected by the flow of control through the modules and through the statements in the modules themselves (see Sec. 11-8). Within a module, *complexity must be limited by keeping the control paths simple.* If there are many jumps to other parts of the same module, both forward and backward in the code, the module becomes difficult to understand. In other words, it is difficult to follow logical flow and, in addition, hard to define the state (value) of the variables at any point during module execution. Although this is generally referred to as the *GOTO problem*, it is really a *COME FROM problem.* Structured programming if used properly develops and maintains simple control paths in the module code. Structured programming provides a discipline that uses only a few standard control structures (sequence, selection, and iteration) as discussed in this section.

The primary goals of structured programming are to develop quality software programs, to minimize the effort required to correct errors, to minimize the life-cycle software cost, and to reduce software complexity. Structured programming is thus a major element in the software engineering discipline.

Starting with the module structure chart of the last section, the basic procedure of structured-program construction consists of a sequence of refinement steps. Each step of the construction consists of breaking a given task (module) into a number of subtasks. This refinement in the program description must be accompanied by a parallel development and refinement in the information structures as well, which constitutes the communication process between subtasks within each design module.

13-5A Program Control Structures

Considering the infinite variety of programming applications, the set of three basic program control structures (program flow structures) plus a fourth call exit (abnormal) defines four structure classes[51] as follows:

1. *Sequence:* START
   BEGIN
2. *Selection:* IF-THEN-ELSE
   IF-OR IF-ELSE
   CASE
   POSIT
3. *Iteration:* DO-WHILE
   DO-UNTIL
4. *Exit:* ESCAPE
   CYCLE

The sequence structure consists of a serial ordering of statements or instructions

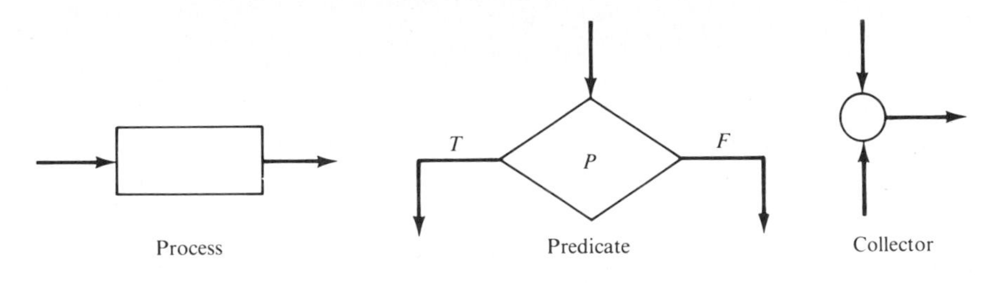

Figure 13-12 Program control structure primitive elements in graphical form.

in regard to their execution. The sequence is normally identified by START or BEGIN. Note that *all the program control structures can be nested.*

The selection structure refers to a binary decision process usually represented in a HOL by "IF logical statement is true, THEN execute the following statements, ELSE (logical statement false) execute another set of sequence of statements" (that is, IF-THEN-ELSE). In lower-level languages such as assembly language, a conditional branch statement (jump) is used instead of this selection primitive. Other forms of the selection structure as indicated in class 2 are discussed later.

Iteration structures reflect conditional "looping" through program statements. The condition (logical statement predicate) is tested each time through the loop. If it is false, the loop is exited. The generally used HOL forms for this structure are DO-WHILE or DO-UNTIL. An example is a wait-loop for A/D conversion.

The *exit* class is considered a new category of program control structures and is used to abnormally exit an iteration loop. Examples of the various classes are shown in the following paragraphs.

Note that all these control structures are composed of three primitive graphical elements (or nodes): a process node, a predicate (T or F) node, and a collector node as shown in Fig. 13-12. The graphical elements are used in an abstract flowchart formulation to depict and analyze program control flow as presented in Chap. 2.

The sequence approach by definition is the concatenation of multiprocess nodes into a linear sequence. The other structure classes have various representations. The following paragraphs present some of the other representations:

One-of-two alternatives (selection structure) are executed for the IF-THEN-ELSE (see Fig. 13-13) as follows:

```
IF        (predicate true) THEN
           process A
ELSE      (predicate false)
           process B
END IF
```

(Note the value of indentation in understanding a specific structure!)

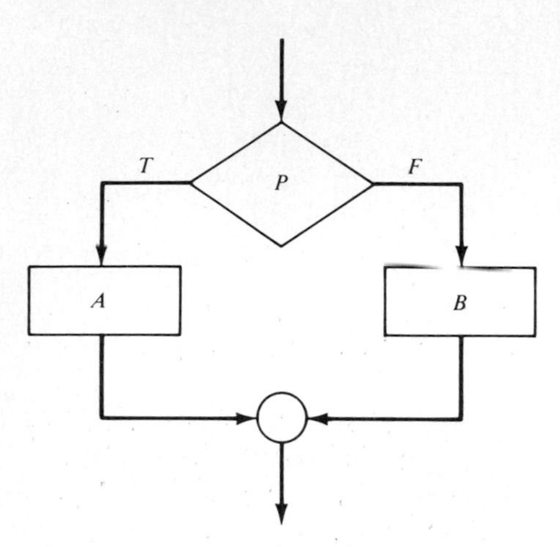

Figure 13-13 IF-THEN-ELSE flowchart.

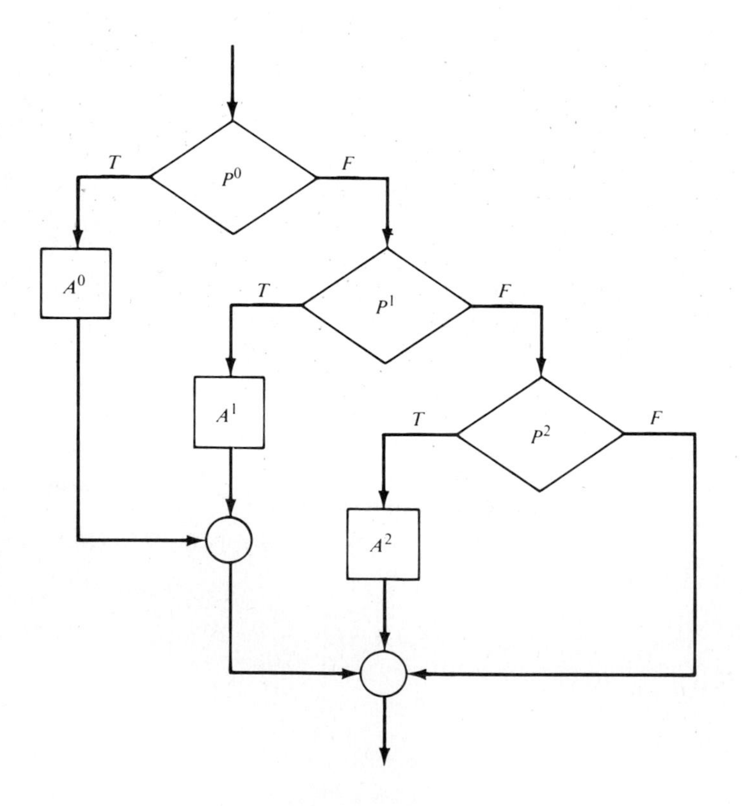

Figure 13-14 IF or IF-ELSE flowchart.

One-of-many alternatives (selection structure) are executed for the IF or IF-ELSE (see Fig. 13-14). Here a series of mutually exclusive predicates are associated with a single process, that is, $(P^0, P^1, P^2, \ldots)$. If the predicates, for example, are associated directly with a real-time clock, then a particular process would be executed at a specific time. Of course, in this example the computation time of the selection structure is considered to be much much less than the individual process times.

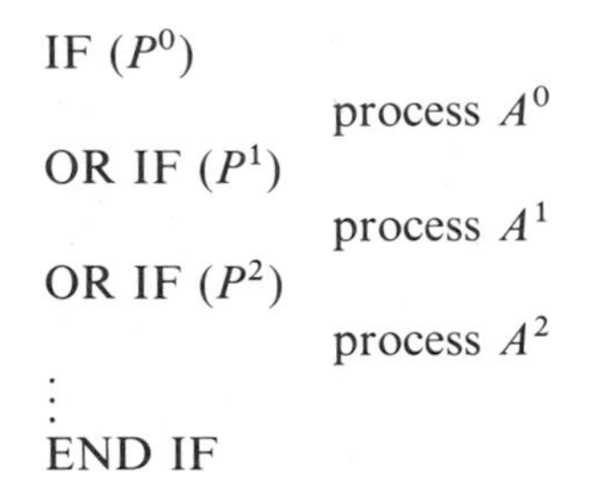

```
IF (P⁰)
                    process A⁰
OR IF (P¹)
                    process A¹
OR IF (P²)
                    process A²
:
END IF
```

Note that it is difficult to understand nested control structures beyond three levels deep. Thus, a *good software engineering principle is to limit the nested control structures to three.*

One-of-many alternatives (selection structure) are executed in CASE form as based on a single-variable value defined by a numerical index. In case of variable = 1, do process A; in case of variable = 2, do process B; and so forth. For example,

```
CASE OF (variable)
CASE 1
                    process A
CASE 2
                    process B
CASE 3
                    process C
:
END CASE
```

In the POSIT† selection structure, all alternatives are executed in sequence unless a specific predicate is false, in which case an exception process is executed and the posit structure terminated. In a sense, this control structure is based upon the logical inverse of selection, for example,

```
POSIT
    Processes with predicate checks (IF)
ELSE
    Exception process
END POSIT
```

† Assume as a fact, affirm, postulate.

Specific example of a POSIT structure In a process control environment, consider logical processes A, B, and C and an exception process X. By construction:

Process A generates a value for predicate P^A.
Process B generates a value for predicate P^B.
Process C generates a value for predicate P^C.

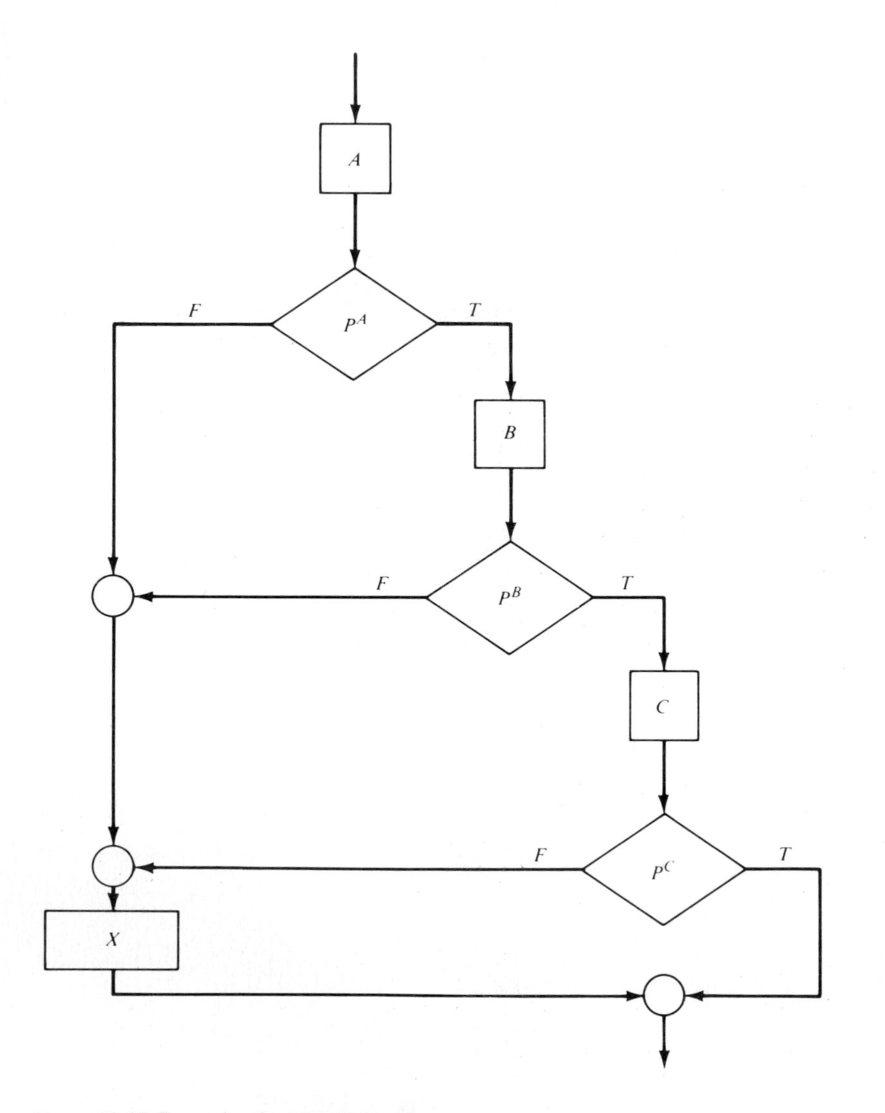

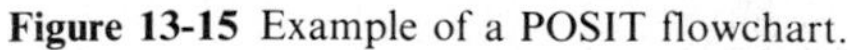
Figure 13-15 Example of a POSIT flowchart.

If any predicate is false, then the exception process is executed. Possible implementation is as follows and is shown in Fig. 13-15 as a flowchart:

```
POSIT
     process A
     IF P^A    THEN process B
               (else process X)
     IF P^B    THEN process C
               (else process X)
     IF P^C    THEN end POSIT
               (else process X)
ELSE
     Process X
END POSIT
```

The iteration (looping, repetition) structure usually consists of one of the following forms (see Fig. 13-16):

```
WHILE (P)
     process A
END WHILE
```

or

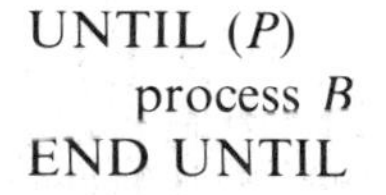

```
UNTIL (P)
     process B
END UNTIL
```

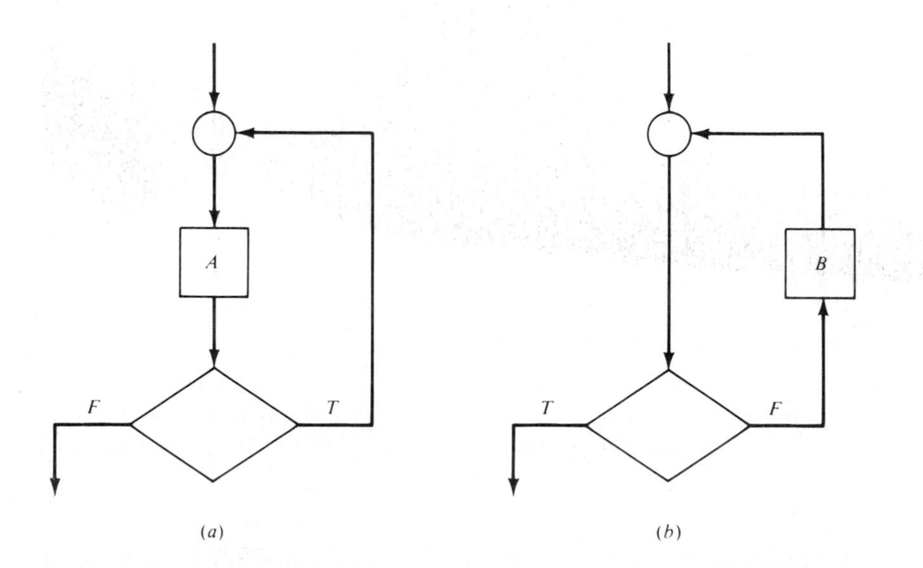

Figure 13-16 Iteration structure flowchart. (*a*) DO-WHILE; (*b*) DO-UNTIL.

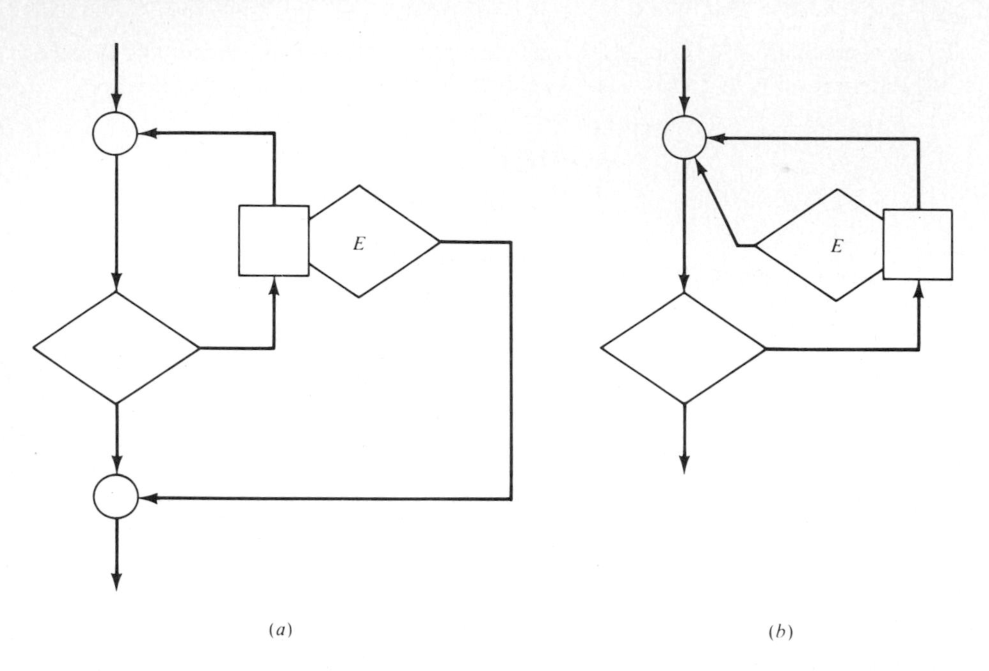

Figure 13-17 Loop exit program control structures. (*a*) Escape structure; (*b*) cycle structure.

In many process control applications, the associated program must exit the current process if an error occurs. The POSIT structure represents a simple way of organization for this situation. However, in loop structures, there are two other approaches: ESCAPE exit (unconditional branch to end of the current executing structure) and the CYCLE exit (unconditional branch to the predicate controlling the next loop iteration). The CYCLE exit continues looping, but could use new data, start over, or just continue next iteration. Figure 13-17 shows their flowchart representation. The implementation of either exit structure depends upon the type (severity) of error encountered and the desired corrective action. Proper use of this and other error procedures can support a *fault-tolerant digital control system* by providing levels of operational performance based upon current executing resources.

13-5B Structured Programming Utilization

The theoretical framework for structured programming rests on a constructive theorem of computer science which demonstrates that it is possible to write any program by using the basic three simple forms of control (sequence, selection, iteration). If program flow is restricted to these constructs, it is possible to write programs such that:

They can be read from top to bottom without ever jumping back to something executed earlier.

Each section of code (module) has only one entry point and one exit point (design-heuristic).
It is possible to see from the top of a loop precisely the conditions for loop termination.

Programs using this structure are accordingly much easier to understand. Because of the increased clarity, it is easier to validate the correctness of the program than it may be if there is a more complicated flow of control. To write programs with this restricted set of control structures requires a strong discipline but gives rewards in increased program quality. The other control structures should not be slighted, however. Where they can be employed to add clarity and utility to a module, do not hesitate to use them.

The discipline required to write structured programs also means making a choice of the particular parts of a programming language that are to be used in the program code. It is wise to select those language constructs that are suitable for the application and to omit those parts that add complication. The enhancement of the simplicity and thus the quality of the program are improved by using this approach. The choice of language statements should be based on criteria that are not limited to questions of efficiency, but based upon program control flow structures and data structures that represent the problem. *Clarity of expression and simplicity of program control flow are paramount.*

There are some commonly used languages that do not provide the three basic control structures in a convenient manner. FORTRAN, BASIC, and assembly language are examples. Nevertheless, structured programming can be practiced in these languages by adopting a standard implementation of the control structures. Also, some implementations of these languages offer the basic control structures through the use of subroutines, macros, and even language extensions.[64,105] The selection of a language or several languages is an important decision with considerable impact. Languages can have a major impact on coding time, testing time, memory requirements, execution speed, and maintainability.

In conclusion, *programs written with structured programming techniques are better-documented, produced more quickly with experience, are more reliable (systems point of view), and are easier to modify.*

13-5C Programming Style

Since the primary objective of structured programming is understandability, the programming style associated with the use of the basic program structures is very relative to their objective. In the implementation of a real-time program such as those in digital control systems, program understanding may be time-critical to proper modification. Programming style involves proper use of the implementation language as mentioned in the previous section. It also includes specific aspects of program clarity, the concept of efficiency, use of program comments, and proper selection of data structures.

The concept of *clarity* refers in part to the selection of appropriate program variable names that directly reflect the essence of the term. Programming approaches to be avoided for the sake of clarity include the use of temporary variables, use of too many statement labels, use of constants as variable arguments, and shorthand notation permitted by the implementation language. Peer review can be very helpful in improving program clarity.

In implementing digital control algorithms, the codes may attempt to "optimize" (time/space) immediately. It is suggested, however, that efficiency concepts be ignored until the program is generating proper variable values. Note that the simplest algorithms can be the fastest and use less space. Let the compiler or assembler provide the initial optimization. Then, approach efficiency objectives in a micromanner with testing if the real-time operational constraints are not being met. Continually generate detailed measurement data through testing to determine if real-time performance has been achieved (software testing is discussed in the next section).

Program Title:
Filename:
Version:
Date:
Use:
Contents:
Function:
Computer System:
Operating System:
Language:
Owner:

Author:
History:

(*a*)

Module Name:
Module Number:
Version:
Date:
Function:
Inputs:
Outputs:
Global Variables:
Modules Called:
Calling Modules:

Author:
History:

(*b*)

Figure 13-18 Examples of Software Documentation Standards for Headers: (*a*) program header, (*b*) module header.

Program Title: Two Axis Digital Controller
Filename: DIGCONT
Version: 1.2
Date:
Use: Example of digital controller implementation
Contents: 7 modules (EXEC, IN1, IN2, FILTR1, FILTR2, OUTX, OUTY)
Function: Provides a general purpose two input—two output digital controller with the provision for second order filtering in each axis
Computer System: LSI-11/2
Operating System: RT-11
Language: MACRO-11
Owner: Houpis and Lamont
Author: Gary Lamont

Figure 13-19 Program Header for Sec. 11-8 Digital Controller Software.

In the coding process using the proper control structures, program comments are still required due to computer language restrictions. Use of comments can, however, be overemphasized, resulting in overdocumentation. In general, develop a standard for employing comments including headers, level of comments per line or lines of code, and comment offsets. *Note that program comments should relate to the algorithm being implemented, not the language primitives.* Reference to the data dictionary and appropriate libraries should also be included. An example of standard documentation formats for program headers and module headers is shown in Fig. 13-18. Specific entries can be deleted or others added as appropriate for a given application. Figures 13-19 through 13-23 utilize these standard headers for some of the modules in the digital control software example of Sec. 11-8. Note that this documentation enhances the understanding of this software program beyond the comments in Fig. 11-34 and thus provides for easier modification as well. In

Module Name: EXECutive
Module Number: 1.0
Version: 1.2
Date:
Function: This executive program is the main driver for the digital controller. The executive calls other programs to input (A/D), filter (second order compensator) and output (D/A) appropriate signals for each axis. The executive also initializes the clock counter at −10 and the real-time clock base frequency at 1 kHz resulting in an A/D conversion frequency of 100 Hz since the clock mode is set for repeated interval in this module. The A/D mode is set to bipolar with offset binary encoding. The input channel is initially defined as channel 0 with the change to channel 1 accomplished in the IN2 routine.
Inputs: none
Outputs: none
Global Variables: DATA (ch 1), ADSR (A/D status reg), CKB (clock preset buffer), CKSB (clock status buffer)
Modules Called: IN1, IN2, FILTR1, FILTR2, OUTX, OUTY
Calling Modules: none

Figure 13-20 Module header for EXEC module in Sec. 11-8 example

Module Name: IN1
Module Number: 2.0
Version: 1.0
Date:
Function: This program module initially checks the status of the A/D converter (channel 0) until the conversion is complete. The binary input data is then moved to the CPU where it is checked for both positive and negative overflow. If an input overflow is detected, the system user is signaled by ringing a bell (terminal). This routine also converts from offset binary to sign plus 2s complement number representation and sign extends to 16 bits from the 10 bit A/D input.
Inputs: CPU register 1
Outputs: CPU register 1
Global Variables: ADSR
Modules Called: none
Calling Modules: EXEC

Figure 13-21 Module header for IN1 in Sec. 11-8 example.

Module Name: FILTR1
Module number: 4.0
Version: 1.0
Date:
Function: The second order difference equation is implemented by this routine. The five coefficients require five multiplies, two additions and two subtractions to complete the filter (compensator) mathematical operations. The routine can be used to generate a simple gain, a first order filter or a second order filter by proper selection of the coefficients. In this routine, a fixed point multiply routine (MPY) is used permitting the coefficients to assume values in the range +32.99902344 to −32.99902344 in steps of .00097. The binary format for the coefficients is XXXXXX.XXXXXXXXXX. The multiply routine is an implementation of Booth's Algorithm.
Inputs: CPU register 1
Outputs: CPU register 1
Global Variables: filter coefficients—N0, N1, N2, M1, M2
past input—X1, X2
past output—Y1, Y2
current input—X
current output—TOT
Modules Called: MPY
Calling Modules: EXEC

Figure 13-22 Module header for FILTR1 in Sec. 11-8 example.

Module Name: OUTX
Module Number: 5.0
Version: 1.0
Date:
Function: This routine converts the sign plus 2s complement representation to offset binary. Since the D/A has 10 bit accuracy, the least significant 10 bits of the 16 bit number are used.
Inputs: CPU register 1
Outputs: CPU register 1
Global Variables: none
Modules Called: none
Calling Modules: EXEC

Figure 13-23 Module header for IN1 in Sec. 11-8 example.

many cases, the program documentation is longer than the actual programming language code! Note also that the suggested documentation formats can be used in the development of a data dictionary (DD).

The selection of data structures for a digital control algorithm implementation has partially been discussed in Chaps. 2 and 10 and Sec. 13-4. The use of specific filtering structures is shown in Chap. 10 to affect performance depending on coefficient values. Also, the form of these coefficient values such as an array, vector, or discrete data element, and the associated argument passing affect module coupling and cohesion is presented in Sec. 13-4. Some additional aspects of data structure selection and utilization include the use of unique variables names for unique purposes, definition of all variable attributes in a data dictionary, and an explicit declaration in the program of all variables.

Programming style especially in the development of real-time programs must be disciplined. Standards for real-time programming should be generated and adhered to if software quality is to be achieved.

13-6 SOFTWARE TESTING

To determine if a program is correct based upon the requirements specifications, exhaustive testing is required. If certain specifications are not met, code testing may not determine the associated logical errors. *Testing can demonstrate the presence of errors ("bugs"), not their absence.* Testing or evaluation at all phases of the software life cycle must be an integral aspect of any software engineering approach. Debugging or software error connection must also proceed in an orderly and structured fashion.

Since the requirements and design phases result in extensive documentation, the evaluation of such documentation is paramount to achieving quality software. *The correctness, completeness, consistency, and clarity of requirements, the design structure, and the coded program are tested by detailed analysis and review.* Continuity between the customer objectives and the detailed design organization must be achieved. Each software module and function must be necessary and realistically feasible. Consistency of the test plan should also be addressed. This analytical evaluation approach for documentation is called *static testing.*

Testing of software programs follows from the structured requirements that define the performance. Performance requirements in a real-time environment such as digital control are usually very extensive. Therefore, program-testing procedures must be generated throughout the design and implementation phases so that a relatively efficient and effective testing procedure is achieved relating defined input to expected output. Documentation in all the software life-cycle phases supports the evaluation of a consistent and detailed program *test plan.* This plan is in itself documentation also and should be continually evaluated.

Static testing of coded programs also involves the evaluation of structured programming principles as stated in the previous section. Peer groups including the software engineer should study the module routines for adherence to specified standards, correct symbolic execution, correct expression evaluation, and correct data types.

Top-down structured implementation involves the coding and testing of high-level modules after the software organization is developed. Module testing should include both dynamic and static analysis. The *dynamic testing* or execution of high-level modules is frequently accomplished before the low-level modules are defined and implemented. By the use of *program stubs* in place of low-level modules, it is possible to have a preliminary executing version of the program. The primary function of stubs is to record the fact that they are invoked. Top-down implementation allows for top-down testing where the high-level modules are validated before the low-level modules are implemented.

The use of these top-down development techniques allows for management control of the software development by providing continuous product visibility. Through the use of techniques that record the up-to-date state of the programs and test data (for example, through the use of a *program librarian*), the test data are no longer hidden in the software engineers' desks and minds but documented. Management can have a reliable measure of what part of the system is defined, designed, tested, and operational.

The defining of program stubs is difficult and important and relates directly to module-coupling design. If done properly, coupling can be tested and evaluated efficiently. Some types of stubs are exit, return a constant output (saturation), turn on an alarm, terminate, execute a timing loop (estimate), and provide a primitive implementation of an actual function (accuracy). Benefits of the top-down approach include the realization that major interfaces are tested early, users see a working demonstration, milestones are easily established, testing is easier, machine test time is better distributed, and software engineer morale is improved. The approach can also eliminate the need for expensive hardware test harnesses. Disadvantages of the top-down approach are the difficult identification of high-risk components and reusable components in the early stages. The resulting program may also be dependent on the particular machine architecture.

In a bottom-up approach, low-level modules are generated first and tested by using "drivers." Modules are then interpreted upward, resulting in the overall software structure implementation. *Combinations of top-down and bottom-up are normally used in testing complex system programs and should be defined in a test plan.*

13-6A Testing Approaches

Selecting test case data is an important aspect of digital control system software testing. Common sense suggests that test cases be divided into the following three categories: (1) the normal cases, (2) the extremes, and (3) the exceptions. Extreme and exception measurements can consist of maximum and minimum values, special values (0, 1, blank, the empty string, the null pointer, and the dummy file), and the wrong discrete value or character. Although input data for digital control systems are normally numerical values, process control input can also involve character strings (commands).

Also, test data can be derived from the internal program control structure. Initialization testing errors, for example, mean that coded statements are wrong or missing. Another error is the failure to reinitialize. Construction errors also can occur due to improper condition (predicate) logic in program flow. Particular statements can perform the wrong action when the predicate is true. Missing path analysis, for example, should be based on requirements specifications.

An overall strategy for test case design can be chosen from one or all of the following:

Logic coverage testing (path logic)
Equivalence partitioning
Boundary-value analysis
Cause-effect graphing
Error guessing

The first goal of *path logic testing* is to provide test cases that, in aggregate, exercise each path at least once between decision statements. In other words, every statement is executed at least once and every decision is brought to each possible outcome at least once. Another goal is to provide test cases that exercise all pairs of these decision-to-decision program control path sequences that are executable in a simple test case. Additional goals include test cases that exercise various combinations of decision paths within groups of overall program control flows. Of course, the cost of these tests can become prohibitive as the number of combinations becomes large. Defining all possible paths then gives insight to program complexity. Specific structures to look for are missing paths, wrong path selection, and incorrect actions. Note that path testing is effective in finding logical errors such as in a manned spacecraft control system.

Another strategy is *equivalence partitioning*. The technique is to select a reasonable small subset of test cases out of the large set of possible test cases. The selected subset should have a high probability of identifying errors within an equivalence class, i.e., those tests that test the same performance requirement. The technique consists of identifying equivalence classes, defining a large set of associated tests, and finally selecting one appropriate test case for a class based upon some criteria (execution time, memory requirements, etc.).

Boundary-value analysis is a testing strategy consisting of test cases focusing on the maximum and minimum of each variable. An example includes test cases involving input values that are just at or beyond the ends of the acceptable ranges. A boundary-value testing technique defines the minimum and maximum values of the input values, and then one value below and one value above each of the two. Similarly, for the output variables, test cases can be generated by using the end values of the ranges. Internal program variables can be tested in a like manner. Note that the values may be numerical or alphabetic or even special symbols. This approach can be used for real-time error testing for D/A saturation or input transducer ranging.

Cause-effect graphing is another strategy. The approach requires the definition (list) of all system outcomes (effects) caused by possible inputs (causes). These relationships define a graph which includes an arrow for each cause and each effect. Arrows are used between nodes to relate all possible causes and effects. These nodes could be put in a decision table, if desired, for a different viewpoint. Test cases should attempt to reflect each of the arrows. Again, the cost of testing has to be considered.

Finally, intuition should not be disregarded as a partial strategy. Some *error guessing* or intuitive tests should be incorporated into the final test plan. This approach does have some merit, since in practice not all tests can be conducted.

When errors are documented as they should be, they can be classified for evaluation and analysis. Specific error classifications are the type (arithmetic-logic-I/O), severity, cause, or source, detection method, location time or stage of detection, stage of error generation, and correction technique. On analyzing errors, prevention techniques should be addressed as well as earlier detection methods for future software development.

Techniques do exist[51] that permit a mathematical proof of program correctness. These techniques based upon predicate calculus are extensive, costly, manual, and generally require much time and experience.

Note that it is impossible to run tests that execute all paths in a complex program, use all possible input values, and try every combination of internal program events. *Judicious selection, however, of specific tests using the previous techniques can result in a high level of confidence in software quality.*

13-6B Debugging

The determination of the cause of program errors and the proper modification to eliminate the errors are the objectives of *debugging*. Debugging involves discipline, detailed analysis of test results and supporting design and requirements documentation, and creative investigative approaches using the testing approaches of the last section. Of course, there is never any guarantee that all program errors have been found and corrected.

Approaches to debugging include the "brute force" technique, induction, deduction, backtracking, and general testing as discussed in the previous section. The *brute force approach* involves "dumping" or printing of digital-control-system data continually as the system is in execution and then studying the massive amount of data in detail. Little thought is used in this approach, which is probably the most costly and yet the slowest. In general, pertinent selection of input and output data is less costly and generally more helpful in finding errors.

The *induction approach* to debugging involves generating an error hypothesis based upon general performance inconsistencies and then developing a procedure to test the hypothesis. The test is performed and revised until the hypothesis is shown to be correct or false. This approach requires careful analysis of performance data and the continued process of data elimination and hypothesis modification. A specific digital control situation requiring this approach is the problem of "poor" numerical accuracy around the operating point (position, velocity, etc.).

Deduction requires the enumeration of all possible error causes (hypothesis). Then, testing procedures are selected to eliminate the incorrect hypothesis. Of course, the deduction process is continuously defined and detailed as individual language statements are selected. As in any debugging approach, individual statements or only groups of statements can be executed through the use of address "breakpoints." This deductive approach might be useful in determining the reason for a digital control system going unstable after a certain period of time or with a specific input.

Backtracking, of course, requires the reverse of program execution. From the program point of improper performance, backtracking logically flows through the program in reverse, attempting to find the statement where something just went wrong. Generally, the variable or variables in error are backtracked to the point of incorrect generation. This debugging process is good for small, complex modules such as those found in a digital control system.

In correcting an error, document the reason, correct the source, and retest to make sure other errors are not present or the correction has not introduced new errors. Also, retesting may require not only the monitoring of variable values but the studying of waveforms (timing diagrams) and their interrelationships. As accuracy performance becomes very restrictive, this waveform level of analysis is sometimes required to ascertain the accomplishment of performance objectives.

Extensive testing is critical in those environments where loss of life or major physical damage may occur due to subtle errors in digital process control software. Fault tolerant software must be properly developed to meet this objective[102] using life-cycle procedures similar to those in this chapter.

13-7 SUMMARY

The maintenance of software, although sometimes demoted to a secondary role, requires the same discipline discussed in this chapter. The high cost of software is in most cases caused by high maintenance costs. The high maintenance costs can generally be attributed to lack of structuring techniques in the earlier software life-cycle phases.

The requirements, design, and program organization resulting from a structure approach should mirror the problem context for ease of understanding.

The use of software engineering tools and methods requires discipline by the software engineers and managers. Digital systems software is not unique and not immune from all the software problems mentioned throughout the chapter. The data flow diagrams, structured design charts, structured programs, and effective testing schemes presented in this chapter should be very helpful in developing consistent and high-quality digital control systems. The *stepwise refinement* from requirements through design and implementation is a powerful and long-range approach to achieving quality software. The associated management process[103,106] including technical reviews should attempt to find errors as early as possible in the software life cycle.

As implied, many different software programs can probably perform, from a black box point of view, the same digital control law. Although not unique, only a few may meet the criteria presented in this chapter in terms of structured programming objectives, design heuristics, and software life-cycle documentation. Discipline then permits acceptable and quality software to be developed.

The different software life-cycle tools and techniques presented are continually being improved and enhanced. Interactive and self-documentation tools should be developed and tuned to the particular digital systems software environment. The use of computer-aided design (CAD) packages[14,15] is an example of self-documenting software utilities.

CHAPTER

FOURTEEN

REAL-TIME OPERATING SYSTEMS FOR DIGITAL CONTROL

14-1 INTRODUCTION

By definition, a digital-computer operating system in general provides a user with conveniences for developing software and scheduling program execution. Thus, an operating system processes user commands (edit, file transfer, data transfer, program execute, etc.) and manages computer system resources (CPU, memory, and peripherals). The extent to which an operating system can perform all these functions depends upon its design and operational environment. For example, in a real-time environment (see the DFD of Fig. 14-1) specific tasks (program elements) must be executed according to a predefined periodic schedule. Generally, the scheduling of these tasks is critical. Therefore, the minimization of processing time (overhead) required by the operating system scheduler may be essential in meeting

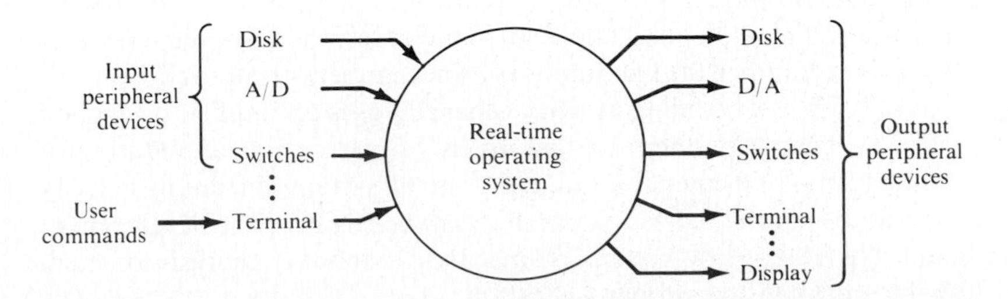

Figure 14-1 Real-time operating system data flow diagram.

performance requirements. Considerable effort in designing or using an existing real-time operating system necessarily involves a detailed understanding of system performance requirements and the operating system overhead. The objective then of this chapter is to provide insight into the structure of a real-time operating system as related to digital control systems.

A real-time operating system by definition has to process certain tasks within specified time intervals. Examples include data acquisition systems, flight controllers, and process control applications. In contrast, a non-real-time system's main objective is to process tasks in an efficient manner (time and space) without regard to initiating task activity at a precise time instant.

In observing and analyzing the various computer system resources in a given system environment, many peripheral devices including memory can operate (do useful work) in parallel (concurrent operation). The operating system attempts to optimize the concurrent utilization of these resources. In many situations, however, the various real-time programs and their associated tasks can not operate concurrently due to the task relationships and the internal organization of the computer hardware. Analysis of task sequencing and proposed computer hardware configurations usually can provide better performance in these situations.

One of the main problems in real-time digital-control-system design lies in task interaction. Some real-time tasks must be performed in a sequential manner; others can be executed concurrently but must be completed before the overall process continues. This situation leads to the requirement for communication or *synchronization* between tasks through the use of status flags (bit information) or equivalent data structures. The requirement is critical in a multicomputer distributed control system such as found in industrial plants and large contemporary vehicles.

A general operating system consists of schedulers, resource managers, editors, program language translators (compilers, assemblers, interpreters), and utilities (diagnostic programs, disk file managers, general input-output programs, etc.). To discuss all the components of a general operating system is beyond the scope or intent of this text. This chapter focuses on the nucleus or so-called operating system executive or monitor in a real-time digital control environment. The executive entity of the operating system usually consists of the scheduler and the individual resource managers, which are presented in the following sections.

The scheduling and resource management activities usually have extensive interaction. This interaction relates to scheduling the use of a resource as required by an application task and managing its use. In some cases, the resource scheduler in reality consists of a number of scheduling routines embedded in each resource management activity. The scheduling activity generally uses an implicit or explicit queuing structure to determine the next task to use a resource. The updating of these queues can be a major element of a digital-control-system scheduling activity where each task can be considered as a separate plant controller. The development of a distributed digital control system[104] involves extensive analysis of each computer's operation to ensure proper operation.

14-2 REAL-TIME OPERATING SYSTEM REQUIREMENTS

A real-time operating system must provide coordination activity between the real-time programs or tasks, the computer system hardware, and the associated data files (usually on disks). The general requirements are as follows:

Manage computer system resources: CPU, memory, peripherals (A/D, D/A, disk, switches, etc.)
Schedule periodic tasks
Coordinate communication between tasks
Control system integrity (error checking, protection, etc.)

The extent to which these functions are realized in software and hardware depends upon the environment and the given applications. Note that the real-time executive has a variety of tasks that are different from those of the real-time application programs in the sense of managing (controlling) internal computer system resources. The computer system must be available to execute all tasks effectively and efficiently whether internal or external.

The complexity of a real-time operating system depends on the amount and level of computer resources that must be controlled and the number of real-time programs and their interrelationships. For singular programs or for a small number of real-time programs, a rather simple sequential executive system such as that presented in Chap. 11 is adequate. However, different operating structures even at this simple level can influence performance as shown in Chap. 13. The example of Sec. 11-8 in this case has embedded into the digital control program the scheduling and input-output functions.

Complex and sophisticated commercial real-time operating systems are available that can perform a wide variety of scheduling and resource management activities. As always, a particular selection depends upon the environment and the nature of the application. Of course, the software engineer can generate a unique real-time operating system if performance requirements require absolute minimization of overhead time. In this chapter the scheduling and resource management elements of some general real-time operating systems are presented as they relate to real-time digital control systems.

Figure 14-2 depicts the higher-level activities involved in a real-time operating system executive using the data flow diagram organization of Chap. 13. In the figure, the executive schedules periodic real-time application tasks. As these application tasks execute, they use system resources (memory and I/O) by calling executive tasks that manage those resources. Although not explicit, *error checking is an integral element of each real-time executive activity.* Historical data on the digital-control-system operation can be stored on a data file. As mentioned earlier, depending upon computer system hardware capabilities, resource management can be extensive or relatively simple. With limited memory, memory management may be relatively easy. On the other hand, large disk space requires a sophisticated control mechanism of which a direct memory access subsystem is an integral part.

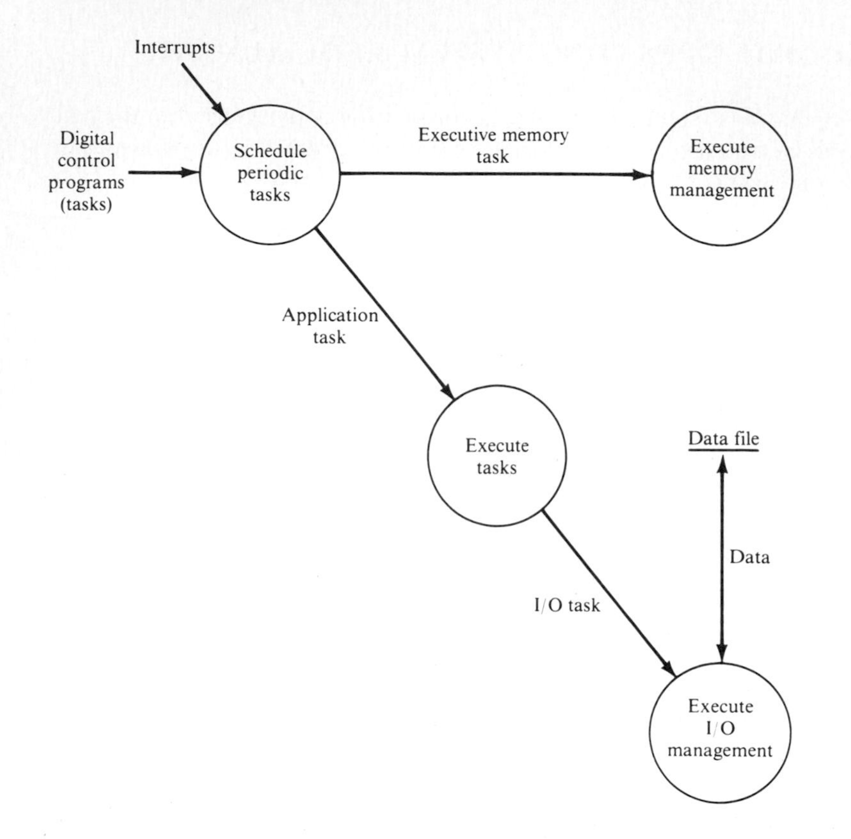

Figure 14-2 Real-time executive functions (DFD-1). *Note*: Error routines should be embedded in all activities.

The following paragraphs introduce the various activities involved with scheduling and resource management activities and then examples of real-time operating systems are presented.

14-3 INPUT-OUTPUT MANAGEMENT

Input and output data management includes routines that generate hardware addresses and perform data buffering, data transfer, error checking (correct data type, check for within range, etc.), data scaling, and status checking as discussed in Chap. 8. Since one of the major time-consuming activities in a real-time digital-control-system environment is the input and output data transfers (A/D, D/A, etc.), the computer hardware-software structure should attempt to minimize the transfer time and permit more concurrent operations. A computer bus-structured organization permits direct transfers from memory to the input-output devices, and vice versa, with proper hardware interfaces. Such an interface is used for direct memory access (DMA) which can considerably improve large-volume data transfer speeds. A DMA interface provides more direct data transfers between I/O devices and main memory by addition of intermediate buffer storage and data paths. The CPU then transfers data from the high-speed main memory to its registers for processing compensator calculations.

As in any resource management operation, the I/O management process must allocate the resource, let the requesting task use the resource, and deallocate the resource. Queuing mechanisms and interrupt capabilities are usually employed in this process to attempt to "optimize" I/O performance.

14-4 TASK SCHEDULER

The executive task scheduler provides a hierarchical set of operations defined by the following:

Create a task schedule
Maintain status and priority of tasks
Initiate task execution
Suspend task execution due to I/O or higher-priority task
Reactivate a task from suspension
Complete task execution (usually done by task itself)
Cancel scheduled task execution
Handle interrupts
Maintain the state of each task

To schedule a real-time task, the scheduler must first have knowledge of the current task status (state). The various task state definitions are defined as:

Inactive (not scheduled)
Inactive (scheduled)
Active (task execution)
Waiting (suspended due to task waiting for I/O or higher-priority task completion)

Figure 14-3 depicts the *state diagram* (not a DFD) for a specific task t_i. The transfer from state to state is associated with the scheduler operations as defined previously. The task state information is stored in an associated task status word similar to the status words for the hardware components. The state information for all tasks is used by the scheduler to select the next task to be executed in some type of sequential order.

A task is created as an inactive scheduled task by some external input at initialization. Then, based upon the occurrence of a specific time or event, the task is initiated (active task). When an active task is suspended, certain computer information must be retained so that when the task is reactivated the values in specific computer registers are correct for the continuation of the task. Thus,

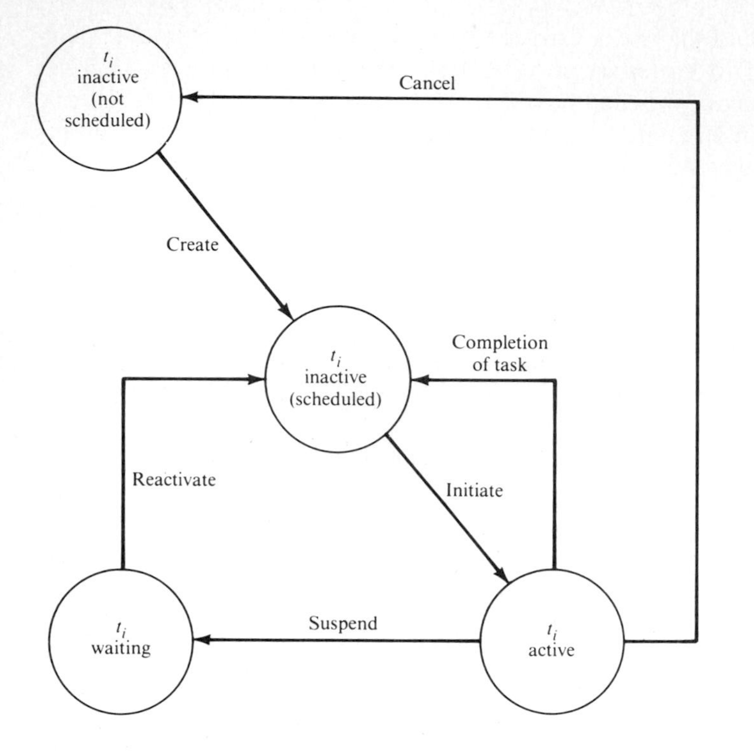

Figure 14-3 Individual task state (status) diagram.

the current state of the computer (i.e., selected register values) must be saved when a task is suspended. This state information is called the *task-suspended computer state* and is stored along with each task's state word, priority, and task address in a link element. A general data structure is presented in Fig. 14-4, which is a linked list structure.

The task location in storage (main memory, disk) must be known also if the scheduler is to initially start executing the task. If the task is suspended, then this area in the link element can be used to save the address of the next task instruction to be executed when the task is reactivated. Sometimes this area is considered as part of the suspended computer state. Thus, a dashed line is used in Fig. 14-4. This linked-list data structure stores task parameters that are used by the scheduler. In the structure of Fig. 14-4, the scheduler can start at the beginning of the linked list by checking task parameters to determine if a specific task should be initiated or reactivated. The precise method depends upon specific task relationships and the scheduling algorithm. Various examples are discussed later in this chapter. Note, however, that if the link elements are arranged in order of priorities, then the scheduler can just proceed sequentially from the beginning to the end of the list. The link element priority information may not be needed for this case. If new tasks are to be inserted in the list, then priority information must be retained. Through

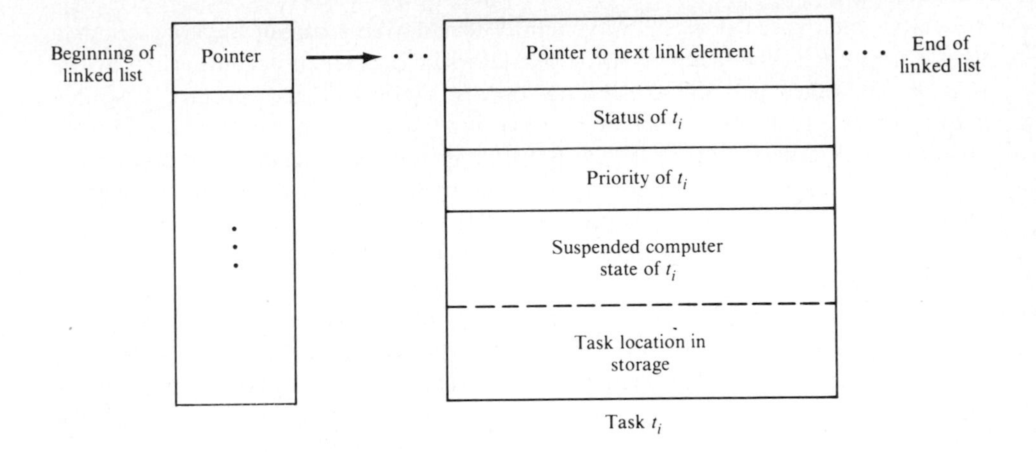

Figure 14-4 Task link element and associated list.

the link element for each task, the scheduler maintains individual task state and task priority. The synchronized interrupt handler under the scheduler performs the storage of the suspended computer state information in the link element when a task is suspended.

The interrupt handler within the scheduler performs the following executive tasks:

Acknowledge interrupts (usually hardware)
Identify the specific time or event interrupt source (usually hardware)
Determine interrupt priority (usually hardware)
Save state of current computer system before executing handling task (software)
Execute interrupt-handling task (software)
Delay handling-task execution depending upon priority (software-hardware)
Complete interrupt-handling task and return (termination)

The details of interrupt handling are presented in Chap. 8.

14-5 STORAGE MANAGEMENT

Storage management in a real-time executive consists of managing the use of interface buffers, main memory, and mass storage.[65] The interrelationships of these storage capabilities has a large impact on system performance within a given environment, because the vast majority of the operations in any real-time computation involve strictly data transfers as contrasted with data manipulation. An I/O handler (service routine) can usually store data in an interface buffer for later transfer to main memory under a separate operation. In other structures, a CPU address register permits data transfer through the CPU. However, data from relatively high speed devices are generally transferred to main memory under a DMA approach. The task allocation of buffers or registers is accomplished by the storage manager per the requesting task in terms of its specified resource requirements. Tasks must indicate to this executive function that they need or are through with a particular resource. Before any data-transfer activity occurs, the memory manager must have allocated or assigned enough memory to the specific task. I/O storage is usually managed by the I/O resource manager, not the memory manager. In both cases, processing overhead is required and may be too extensive for a real-time environment.

In a digital control environment, certain buffers, sections of main memory, and mass storage may have to be permanently allocated to specific tasks in order to achieve performance requirements. Various real-time operating systems achieve their intended goals by incorporating unique storage allocation approaches, for example, large I/O buffers for array storage or a burst of serial I/O data.

If a considerable number of tasks are to be executed within a relatively short time interval, it might be desirable to keep all of them in main memory at one time if space permits. Another design objective is to maximize the possible concurrent task executions between the CPU, memory, and I/O. One way of achieving both these goals is to dynamically control task execution within main memory. When a particular task is suspended (blocked) due to an I/O operation being requested by this task, the memory manager can store the state of this task and initiate another. If more than one task is simultaneously stored in memory, the new task can quickly and efficiently be put in execution. This technique of having more than one task in memory at the same time with the ability through the scheduler to switch from one to the other is called *multiprogramming*. In essence, the computer system resources are divided in time and space between many tasks based upon priorities.

The executive can be completely stored in main memory (primary storage), or certain executive routines can be in secondary storage (disk, etc.) and called into main memory only when needed. In a real-time executive, most routines are always resident in main memory because of speed requirements. The overhead of executive tasks can be minimized if the entire executive is always located in the primary memory; i.e., no disk I/O is required. If some executive routines are located in secondary storage, application program processing can be delayed until the desired operating system task is transferred into main memory and executed. This situation occurs with most commercial minicomputer and microcomputer operating systems.

Another name for multiprogramming is *time-division multiplexing*. Time-division multiplexing can be controlled directly or indirectly. For example, if all tasks in main memory have the same priority, then each task is allocated the same amount of finite execution time with this approach. The system programmer must ensure that each task can complete execution in the specified time interval. This scheduling technique is sometimes called *time sharing*.

Another scheduling technique is based upon an interrupt scheme (indirect scheduling). In this case, *I/O task-driven* scheduling is defined. A task in execution either runs to completion or is suspended due to its own I/O activity request. The task must then wait until it is rescheduled in the linked-list priority chain. Initially, the scheduler starts the scheduling process with the highest-priority task. When this task is completed, the scheduler selects the current highest-priority task. When any other task is suspended (completion, I/O activity, etc.), the scheduler selects the next highest-priority task, and so forth. Note that a low-priority active task with no I/O activity can control resources for a long time (relative) in this scheduler configuration. In this scheduler all the software engineer must ensure is that the I/O data associated with a suspended task are not lost due to the execution of a higher-priority task. In general, however, this approach is not used since low-priority tasks can prohibit high-priority tasks from executing.

A third type of scheduling is built upon the previous techniques. In addition to rescheduling when a task is completed or requires I/O activity, this scheduler also reschedules when an I/O activity is completed (*interrupt-event-driven*). In all cases, the current highest-priority task becomes active. Other events can initiate a new schedule based upon interrupts from an alarm, a timer, a terminal request for information, or other external activities. In many cases, these interrupts can be masked based upon specific priorities. For example, when a task is put in the waiting state due to an I/O operation, the interrupt mask is modified to permit the specific I/O device to interrupt the CPU on completion of the I/O subtask. Again, the operating system must protect any suspended-task I/O data while other tasks are active.

Since the general type of scheduling described above (see Fig. 14-5 for three tasks) permits a wide and flexible spectrum of determining task execution, it is reasonable to use it in a real-time digital control application. Note that the timing diagram of Fig. 14-5 assumes very little overhead for executing the priority scheduling algorithm! In reality, it may be rather large.

Other schedulers also use the techniques of interrupt-event-driven activities, but they transfer application programs from secondary storage to main memory. For fast data movement, a DMA interface is usually employed. This external storage occurs because large storage requirements for certain real-time application programs are beyond the capacity of main memory. The operating system must

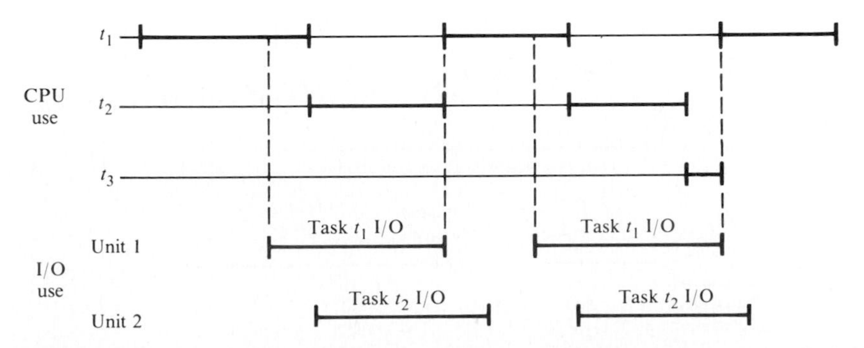

Figure 14-5 Timing diagram for real-time schedule with three tasks—t_1, t_2, t_3 (note concurrent activity).

keep track of where everything is located for proper operation, i.e., memory management.

14-6 SOME EXAMPLES OF REAL-TIME OPERATING SYSTEMS

The intent of this section is to describe the organization of various real-time operating systems. A simplified example is presented first, followed by more extensive examples that are progressively more complex.

14-6A Simple Real-Time Operating System

The diagram of Fig. 14-6 depicts a scheduler that retains *all* needed task information (status, task main memory address) in main memory as a link element. The basic link element here is shown in Fig. 14-7. Since all tasks are active by

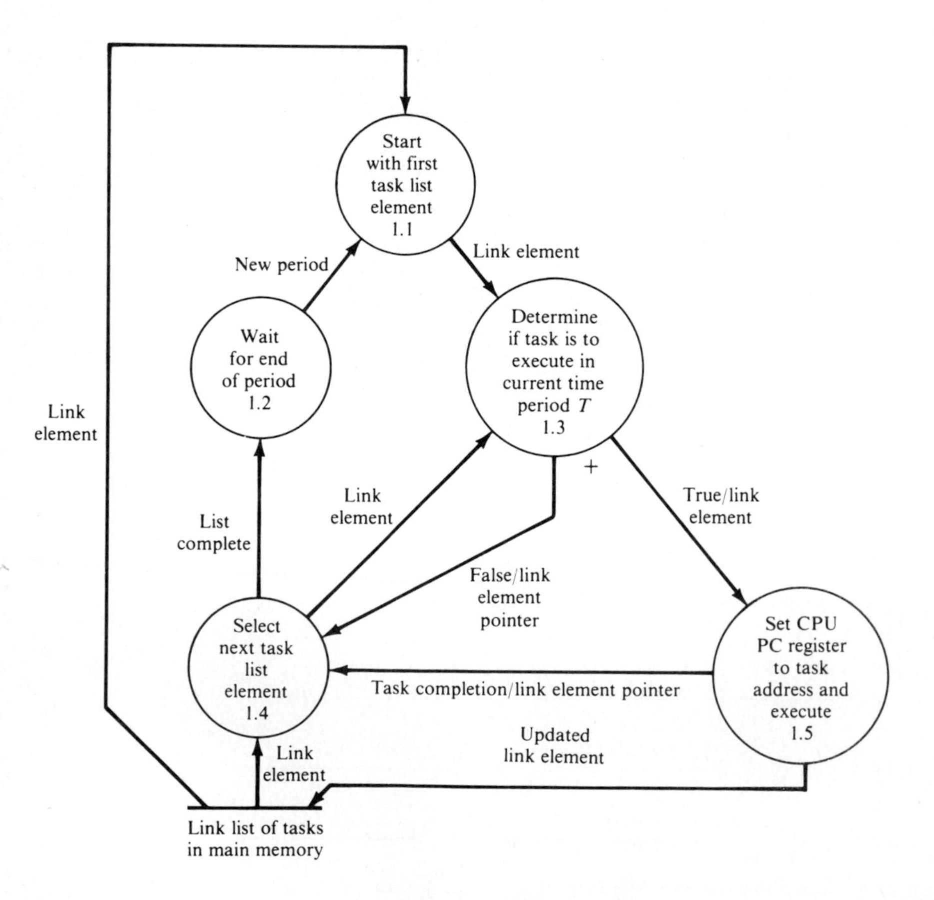

Figure 14-6 The diagram for a simple real-time operating system.

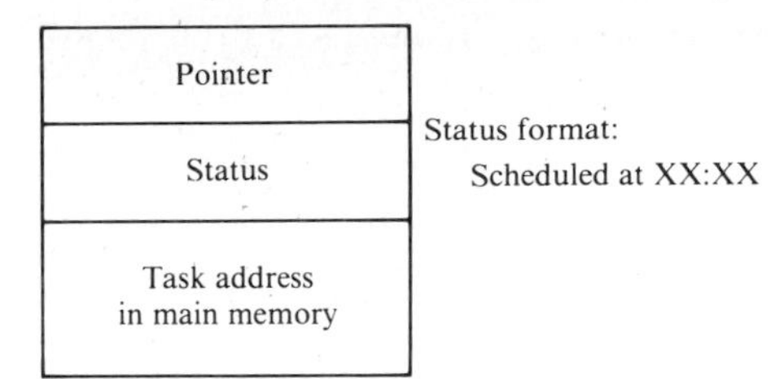

Figure 14-7 Link list element for simple real-time system.

definition, the only status information required is the next clock period in which the task should execute. Thus, a real-time clock must be an integral hardware component. The current value (current period kT) of the real-time clock is compared by the scheduler with each task status word to determine if the task should execute during the current time period. T, for example, may be 10 s. The value of the link element status word may be set in 10-s increments over a 24-h interval. The task when executing can update its own status word with the next time of execution.

The priority information in the general status word is not required here since the priority is contained in the sequential list of link elements through the pointers (implicit priority). Also, the suspended state information is not needed since each task executes until completion with no interrupts—no suspension. However, the scheduler must know the main memory address of the task so that it can be put into the CPU program counter (PC) and task execution initiated. Thus, it is included in the link element.

It is assumed that each task can complete within the time period T. Consider a selected subset of tasks (five and two in this case), all completing within the same time interval T (see Fig. 14-8). Note that no scheduling overhead is shown in this figure.

It is the software engineer's responsibility then to make sure that all required tasks t_i can complete execution within a period T. As shown, some T intervals may contain a large number of tasks, others only a few. The software engineer must require each task to reschedule itself, but again the totality of task times within a period must not be greater than T as constrained by scheduler overhead. Sequencing of nonindependent tasks is done by the proper use of pointers and time-period scheduling. A method for generating an "optimal" schedule of this type requires considerable mathematical background and therefore is beyond the scope of this

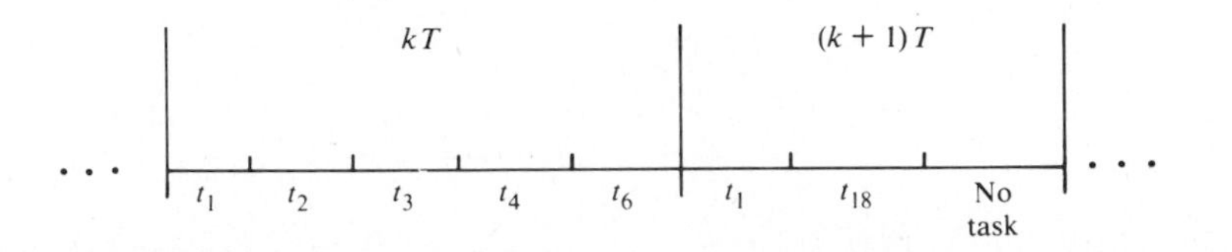

Figure 14-8 Schedule for simple real-time operating system.

text.[69] *One must be very careful especially when empirically setting up a sequence of task executions to ensure that each task is scheduled properly.*

To initially start the scheduling process, each individual task must have its associated link list element initialized. An interactive software program is usually employed to perform this operation. This initialization program can also let the user request that each task be executed independently for testing purposes. The program can also request that the current time be loaded into the real-time clock.

Each task in this simple operating system when active executes until completion. No interrupt or suspension technique is used to put the task in a wait state due to an I/O request. Considerable CPU time can be lost because of this situation. However, scheduler overhead is kept relatively low since only a little supervision is involved. Even so, it may be that overhead is still, say, 5 percent of the period for each task.

This simple scheduler requires little in sophisticated hardware. Tasks are not dynamically reloaded, which negates the requirement for multiple-addressing registers in the CPU. Simple microprocessors can be used to implement an executive of this type[67] for sequential tasks in a digital control system.

14-6B Extended Real-Time Operating System

Consider the simple real-time operating system of the last section with the additional element of an interrupt timer. The timer interrupts the current executing task when $(k + 1)T$ is reached and returns the scheduler to the first element in the task link list. In reality the timer sets an interrupt flag. The hardware acknowledges the interrupt, and the interrupt handler is invoked. The interrupt handler saves the current computer state of the associated task in the associated link element (Fig. 14-9). Also, the status word must now contain the fact that the task has been interrupted, i.e., is in a wait state.

Again, task priority information is not required in the link element since the priority is embedded in the pointers.

The scheduling process in this environment initially starts with the first task element in the list. It determines if the task should be scheduled based upon the current time and the value stored in the status word. If the task should be scheduled, it is. If not, the next task linked-list element is evaluated. After a certain number of tasks have executed, the timer interrupts (Fig. 14-10). To develop

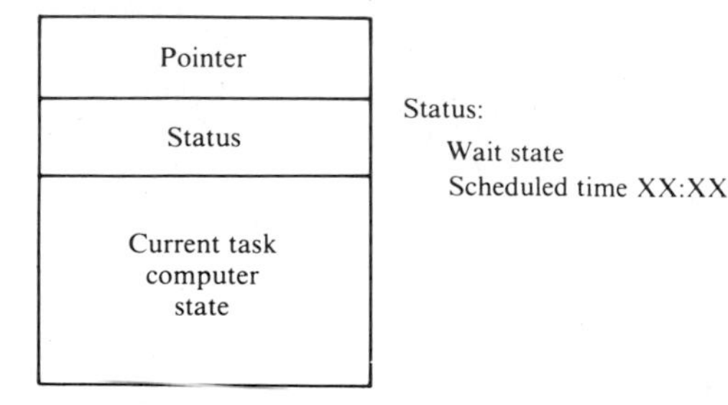

Fig. 14-9 Extended link element.

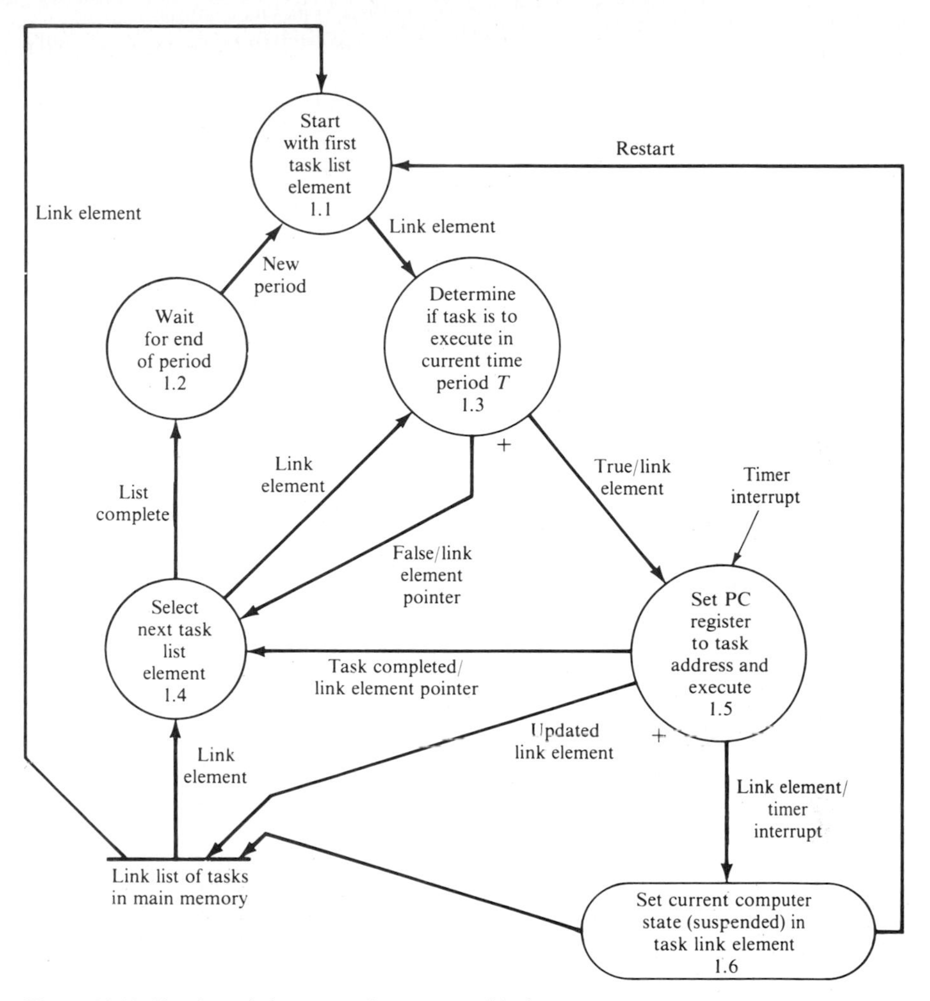

Figure 14-10 Simple real-time operating system with timer.

confidence that all tasks in the link list execute, the software engineer must do considerable testing. The setting of the specific timer interval value T depends primarily upon the high-priority tasks that must be executed periodically. For example, certain digital controllers have coefficients which are based upon T. The other tasks (lower priority) are executed in increments as "background" programs. Examples include terminal interaction, statistical updating, and data reduction within a real-time environment. Without proper analysis, some tasks may never complete execution or even start!

14-6C Real-Time Operating System with I/O Interrupts

The general multiprogramming organization as defined in Sec. 14-5 is now employed. When a task requires I/O, it is put in a wait state. When the I/O has

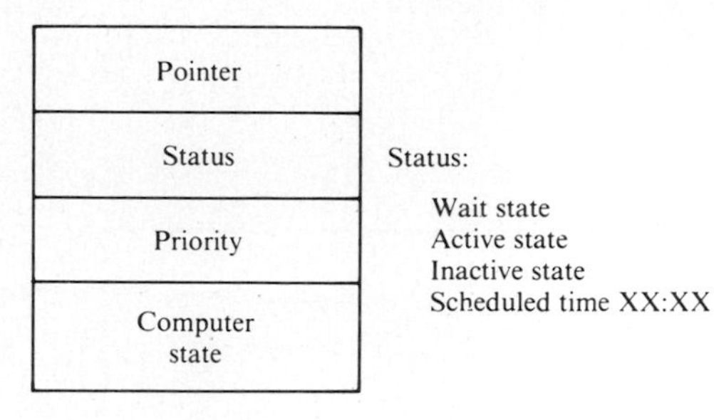

Figure 14-11 Task link element for event-driven-interrupt multiprogramming real-time operating system.

completed, the associated device can interrupt the current executing task such that the waiting task can complete if its priority is higher than the current executing task. In this situation, more information must be available to the scheduler. First of all, the priority of the task must be explicitly known. The task link element for this system contains the explicit priority (see Fig. 14-11). The status word now contains all state information as well as the scheduled internal task execution time (kT).

The scheduler (Fig. 14-12), by means of the interrupt handler, has to explicitly unmask the specific I/O device interrupt register when a task requires that resource. This permits the I/O device to interrupt the processor after completing the I/O task. By definition, when an I/O device is allocated to a specific I/O task, it runs to completion, i.e., no internal I/O device interrupts. Therefore, if the current executing task requests an I/O operation using a current executing I/O device, the I/O request is stored in a table or matrix. Of course, the current executing task is still put in a wait state. The table must reflect a queuing mechanism for each device as shown in Fig. 14-13. In the table, the priority queue is based upon a task index; i.e., the task with the lowest assigned number is given the desired I/O device when the I/O device becomes available. From the table t_2 can be allocated I/O device 12 before task t_3 and so forth.

Instead of a table, a queue data structure[65] may be programmed for each I/O device. A first-come–first-serve mechanism can suffice. However, if the higher-priority task should use the device first, then the tabular approach or a special-priority queue for each task may be appropriate.

A further extension of this scheduler can permit changes in task priority. Such changes may be based upon the time of day, current process, external management changes, or current reliability-operational state of overall computer system. This priority flexibility may also be monitored by another scheduler subprogram.

14-6D Real-Time Operating System with Secondary Storage

The previous real-time operating system examples assume *all* software elements are resident in main memory continually. As mentioned previously when introducing memory management concepts, some environments require a secondary storage for application programs. In these situations, all programs cannot be resident in

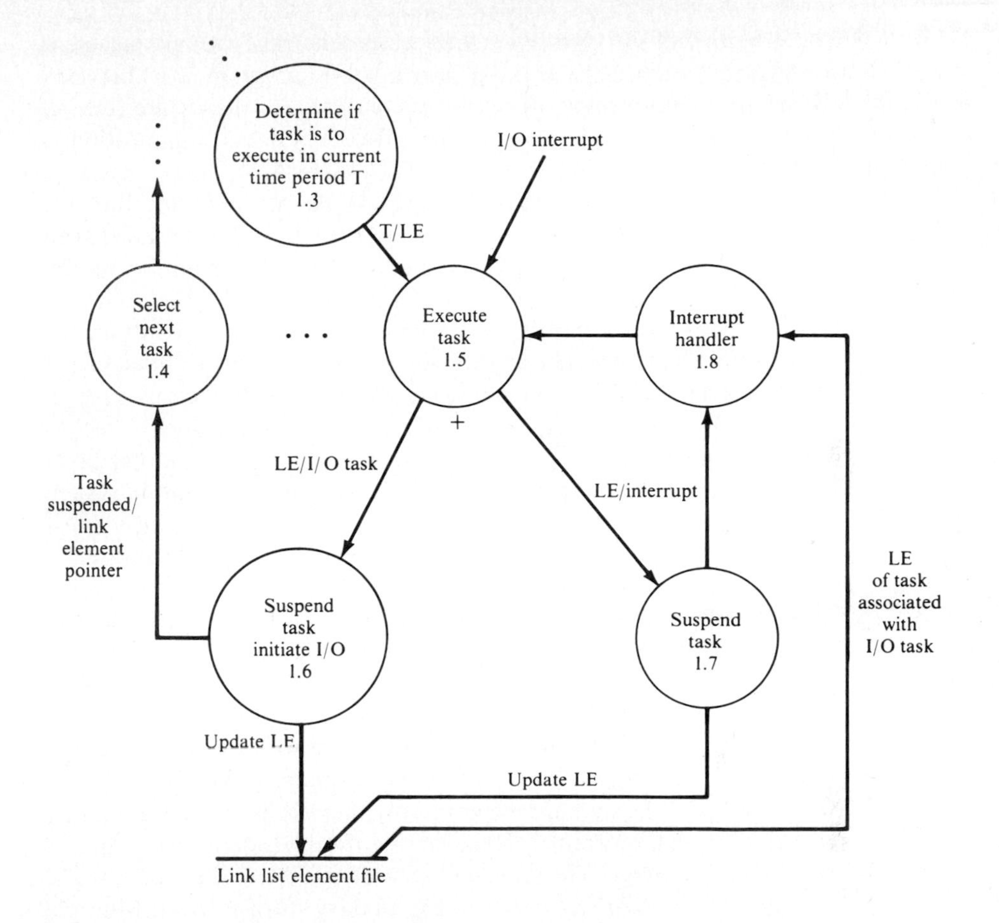

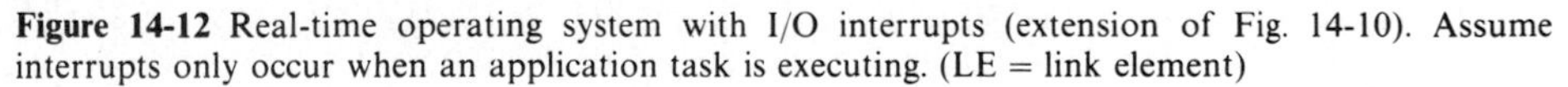
Figure 14-12 Real-time operating system with I/O interrupts (extension of Fig. 14-10). Assume interrupts only occur when an application task is executing. (LE = link element)

I/O device	Task requesting use				
	t_1	t_2	t_3	...	t_i
11	—	—	x		
12	—	x	x		
13	x	—	—		
⋮					

Figure 14-13 Table for "Request I/O Device" as used in multiprogramming environment. (x = request)

main memory simultaneously owing to their size. The storage management structure must attempt to efficiently transfer information (program, etc.) between secondary storage and main memory. The process requires fast hardware (DMA, etc.) and efficient software. In this section, the software process organization is described.

An appropriate scheduler can provide, using an extension of the link list concept, an answer to the secondary storage problem. Each link element must keep track of the location of the task (program) in secondary storage. Other information must also be maintained.

The transfer or *swapping* of application programs is the responsibility of the software engineer in this structure. The application program developer must incorporate into each associated task the swapping activity. This approach uses, then, the swapping concept of *overlays*. An overlay is a task (routine) stored in secondary storage that is configured to replace (overlay) other tasks (or another task) stored in main memory. The swapping is accomplished by use of operating system I/O disk routine calls in the application programs. The software engineer must explicitly request that certain routines be overlaid at specific times by using these I/O calls. Note that a single application program in main memory can itself call various overlays. These tasks on secondary storage may be transferred one at a time as called by the application program. Of course, that part of the original application program that calls the overlay task must remain in main memory. Note that overlays can call other overlays. The timing of the overlaid task transfers is again the responsibility of the software engineer. Considerable analysis is usually required to meet digital control system real-time performance objectives.

Imagine that the scheduler in main memory only knows the existence of tasks resident continuously in main memory. Then each of these resident tasks can call overlays from secondary storage. The scheduler still retains the link list of resident task for scheduling and has no direct control of secondary storage tasks. Multiple level organizations of this type of scheduling may be easier to design, but hard to implement.

Since the swapping process requires considerable time, the previous concept of a task blocking or suspension due to an I/O activity is used. When a particular task calls an overlay, the scheduler suspends that task and selects the next highest-priority task. The overlay is transferred to main memory concurrently with CPU operation, and an I/O interrupt occurs when the transfer is complete. If the calling task has higher priority than the current execution task, then overlay is executed.

The software engineer can control directly the scheduling of application task overlays without relying on the commercial operating system scheduler. The resident task scheduler may be part of a general real-time operating system and will probably not "optimize" the specific application schedule. Overlays, then, can play an important role in meeting performance objectives since the scheduling is inherently done by the applications program.

Another approach to the scheduling problem with secondary storage is to have each link list element contain either the address of the task in main memory or the encoded address (name, etc.) in secondary storage. The scheduler then schedules based upon the link list pointers. When a task to be executed is in secondary storage, the scheduler issues an I/O request and finds the next highest-priority task to execute. When the secondary storage (I/O) interrupts, signifying that the task has been DMA transferred to main memory, the current executing task is put in a wait state and the higher-priority task executed. Task priority can be dynamic and static as before.

The software engineer using this second approach must attempt to have all tasks (main task and subroutines) involved with a particular process such as aircraft control or navigation resident in main memory at the same time. This organization obviously executes faster than having to go to secondary storage for associated tasks. The initial loading of main memory and subsequent task executions must be thoroughly analyzed to ensure the desired real-time system performance.

For example, each time a new independent process is to be scheduled, a new set of tasks are loaded (transferred) into main memory from secondary storage. This set includes the main task (executive), all subroutines, and all pertinent I/O and interrupt handlers involved with the specific process. Each link list element can be associated with a particular process and in reality contains another pointer field to the first subtask, that subtask pointing to another subtask, and so forth (see Fig. 14-14). A two-dimensional link list structure results.

The main scheduler focuses on loading process tasks. A secondary scheduler schedules individual tasks. Note, however, that an individual task can require a reload as well. That is, a hierarchy of link list structures may exist. In reality a large load (say, 32K) can take a half a second, which means that scheduling with leveled task transfers should be kept to a minimum.

With a timer interrupt (Sec. 14-6B) in this type of executive, the current resident multitask process must be returned to secondary storage and the new high-priority process loaded. The information required to store and analyze programs is similar to that described in Sec. 14-6C. Process information is retained just like data associated with a single task. It is then advantageous to limit the number of interrupts permitted. Masking interrupts is one way of doing this. Another is to analyze the scheduling objective and generate an "optimal" schedule.[66,68] Because of the length of time to perform a process swap between main memory and secondary storage, only critical time processes are permitted to interrupt. In many sophisticated and complex real-time applications, the more main memory available, the better the performance; i.e., the more main memory, the less complex the memory management task.

One problem may arise with a multiprogramming executive of this type. Certain processes must communicate with others such as an overall process control system or an aircraft electronic navigation and guidance system. In these cases, the communication area can be in main memory or secondary storage. If it is in secondary storage due to main memory space limitation, execution delays predominate. Also, adherence to software engineering principles (Chap. 13) must be an integral aspect for complex process control applications.

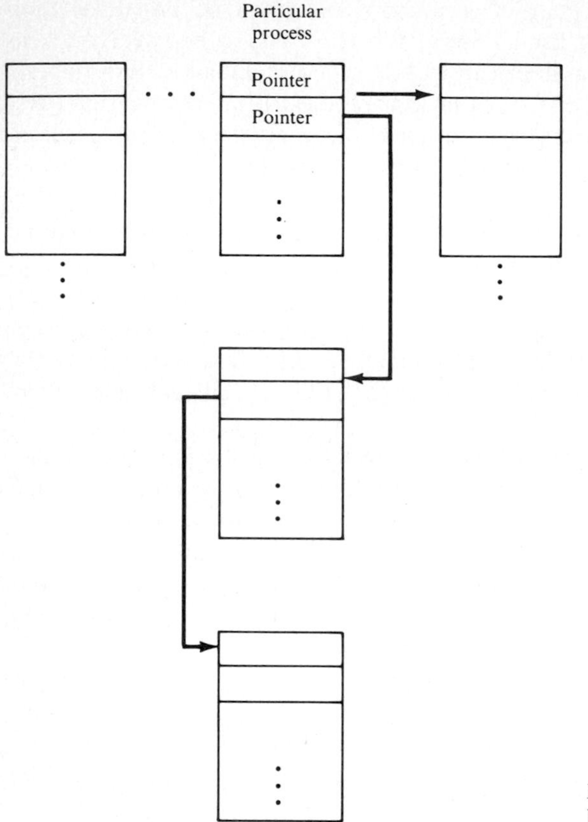

Figure 14-14 A two-dimensional link list data structure.

14-6E Real-Time Operating System with Memory Partitions

In the previous link list elements, the location of each task in main memory is at a fixed location (address). The same thing is true for a process (collection of interrelated tasks) residing in main memory. If main memory is large enough, more than one process can reside in memory simultaneously. Then the timing advantages of multiprogramming may exist even for these large real-time environments. To allocate more than one process to main memory at the same time requires a more elaborate memory management system. To develop the concept of multiprogramming with process loads from secondary storage, *memory partitions* or areas must be organized.

Each memory partition contains by definition one set of process tasks. Note that these partitions can be fixed in size (length) or variable. The *fixed partition* requires less memory allocation activities since the process address within the partition is fixed. Fixed-partition execution is discussed first.

Any number of fixed partitions can exist, depending upon the size of main memory. Since each partition is fixed, all the processes that are to be put into this partition must fit—the responsibility of the software engineer. The software engineer selects those processes that are to be assigned to the nth partition based upon some priority. Generally speaking, each partition is assigned a priority in the sense that all processes using that partition have the same priority. Thus, once a set of process tasks is loaded into the assigned partition, the process completes before another transfer from secondary storage is transferred into this partition. Owing to this type of scheduling activity, excessive numbers of main memory loads do not occur. When higher-order tasks are to be executed via a timer, the current partition (low-priority tasks) is interrupted. A transfer from secondary storage is not required. Again, when a current process requires an I/O operation, the current process is put in a wait state and execution control transferred to the next lower-priority process (partition).

In attempting to use the CPU effectively, the software engineer must attempt to schedule only one process per partition during the current time period. If more than one process is scheduled, then delays occur due to high process-program transfer from secondary storage. In a digital control system with various processes, one or two partitions can have a permanent process assigned. Only lower-priority processes may be swapped from secondary storage into a partition.

In some cases, the totality (size) of the various process tasks that are to be put into the same partition ranges over a wide set of values. The size of the fixed partition then must be matched to the maximum process size. Thus, since main memory locations are not used in all process executions, in order to be more efficient in terms of main memory storage, the concept of *variable partitions* is employed.

In essence, fixed partitions are still used in a variable-partition execution. However, the fixed partitions are much smaller and do not have to be large enough to fit the largest process. To generate a variable partition, a variable number of small fixed partitions are combined to minimize the memory management overhead. The size of the fixed partition is selected to be adequate to store some small processes. Processes are still allocated to a partition. When a specific process requires more than one fixed partition, the required fixed partitions are allocated to the specific task. When memory management allocates the other partitions, the processes normally loaded into these areas are blocked. This information is kept in a table for each small fixed partition. If a higher-priority task is using the partition, then a lower-priority task can not execute. The scheduler then checks this table before executing a lower-priority task.

Even in this variable-partition executive, critical high-priority processes must continually be stored in main memory in order to meet timing performance objectives. However, variable partitioning can lead to more effective use of main memory, assuming low-priority processes can be somewhat delayed. Note that careful process scheduling must be accomplished since all processes within their partition run to completion. That is, a low-priority process can use a partition for a considerable length of time before completion and in so doing prohibit a higher-priority task from using the same partition. Remember also that the variable partition can be required to be continuous, i.e., adjacent fixed partitions.

14-6F Other Real-Time Operating Systems

As indicated over and over again in the previous examples, the major element influencing real-time operating system performance is the capacity and capability of the hardware. Specifically, the storage management or allocation is complex if memory resources are very limited in a multitasking environment. Also, since considerable relocation of memory occurs in such an environment, the hardware addressing modes should easily implement such a process. Implementation may require base registers and index registers for more complex addressing modes.

Other and more sophisticated real-time operating systems require extensive hardware resources, such as various storage technologies (disk, tape, floppy disk, cache memory, large I/O buffers, etc.), special displays (graphics, lights, needle indicaters, etc.), and possibly multiple CPUs. Such organizations permit extensive concurrent processing. As shown in the previous sections, the scheduling and control of one CPU and a limited number of I/O devices is complex, especially in an interrupt environment. The extended environment is much more sophisticated. To retain information concerning status, priority, and queues not only of tasks but also of the I/O devices and CPUs requires a tremendous amount of bookkeeping. In addition, error checkings must be done to help ensure overall system reliability. Many medium-size computer systems (16 bit and 32 bit) reflect this type of system.

Another possible improvement is the use of *virtual memory addressing*[7] where hardware and software provide the task instruction addressing, thus freeing the software engineer from the addressing responsibility. Again, overhead may be too great to meet specific performance requirements in a digital control system.

14-7 COMMON RESOURCE ACCESS

As mentioned earlier, the current executing task may require the use of an I/O device currently performing an I/O activity. The current I/O activity was initiated by a previous task in execution which is now in a wait state. The associated sequencing of the I/O device utilization seems to be manageable if enough time is available for each I/O device to be properly assigned to a given task and the I/O subtask (activity) executed to completion. However, if an executing task is interrupted at the moment (very small time interval) it is assigned an I/O device and the new task requests the same I/O device, the new I/O request is blocked. But, the I/O device is not busy since the old I/O subtask is never initiated due to the task interrupt.

The solution to this arbitration problem involves the control of reusable resources, I/O devices. The objective is to grant I/O device access to at most one task at a time and that the I/O task must execute to completion. The general term associated with this problem is *mutual exclusion*. All other I/O tasks requests are excluded from using an I/O device if it has been assigned and a current I/O task is executing. To retain the information concerning tasks which are currently waiting for the use of the I/O device, a queue or similar data structure is required. The queueing mechanism must not only provide for storing the current list of requesting tasks but must also assign the I/O device to one and only one I/O task that runs to completion. Certain operations within this process must not be interruptible as mentioned earlier if the system is to execute properly.

A set of noninterruptible primitives to solve the mutual exclusion problem require routines for enqueueing and dequeueing for each requesting task. The enqueueing routine E must keep track of an I/O task requesting the specific I/O device. This routine must also permit only one I/O task to be assigned at a time and initiate the execution of the specific I/O task. The dequeueing routine D updates the queue so that another I/O task can use the I/O device. To define the queueing mechanism, consider an indexed queue $Q[n]$ that has $n + 1$ elements and a pointer p which are both global variables, that is, storage accessible from all tasks. The queue Q is used to store the order of n requesting tasks in a first-in–first-out sequence through the use of the pointer p. This is accomplished by defining the task identifier index as an integer number i. Then in location $Q[i]$ is stored the index of the task following the ith task. p points to the last task requesting the I/O device. The value of $Q[0]$ defines the current task (index) using the I/O device. Note that a simple linked list has been constructed. The routines E_i and D_i for each task i, $1 \leq i \leq n$, with initial values $p = 0$ and $Q[0] = 0$ (an empty queue) are:

E_i: 1. If I/O device request, $Q[p] \leftarrow i$, $p \leftarrow i$;request from ith ;task for specific ;I/O device

2. Check status of I/O device; if busy return to task scheduler, if not continue
3. Select I/O task from position $Q[0]$, assign to I/O device, and initiate I/O task execution and return to task scheduler

D_i: ;current I/O task ;completed (by interrupt)

4. If $p \neq i$, then $Q[0] \leftarrow Q[i]$ and return ;update queue ;and process interrupt

else $Q[0] = 0$; $p = 0$ and return

To ensure that all information in this queueing mechanism is always current and that the I/O tasks are executed in a first-in–first-out basis, Steps 1, 3, and 4 must be noninterruptible (primitive). Of course interrupts can be masked by including masking explicitly in the various task implementations (code). Other primitives such as noninterruptible memory reads and writes can be used to implement the first-in–first-out mechanism.[66]

Example 14-1 Consider a mutual exclusion implementation using a queue with dimension $n = 8$. Assume that tasks t_2, t_4, t_1, t_7, t_5 request I/O device service in the order indicated by the sequence. Also assume that all I/O device requesting is done before the service begins. Show the generation of the queue values and the allocation of the I/O device for each I/O subtask.

SOLUTION

setup –

$$
\begin{array}{ll}
Q[0 - - - - - - - -], & p = 0 \\
Q[2 - - - - - - - -], & p = 2 \\
Q[2 - 4 - - - - - -], & p = 4 \\
Q[2 - 4 - 1 - - -], & p = 1 \\
Q[2\ 7\ 4 - 1 - - -], & p = 7 \\
Q[2\ 7\ 4 - 1 - - -], & p = 5
\end{array}
$$

service –

$$
\begin{array}{ll}
Q[4\ 7\ 4 - 1 - - 5], & p = 5 \\
Q[1\ 7\ 4 - 1 - - 5], & p = 5 \\
Q[7\ 7\ 4 - 1 - - 5], & p = 5 \\
Q[5\ 7\ 4 - 1 - - 5], & p = 5 \\
Q[0\ 7\ 4 - 1 - - 5], & p = 0
\end{array}
$$

Note that requesting tasks use the I/O device resource in the order in which they enter the queue (step 1). The values stored in an empty queue are not important except for $Q[0] = 0$. The other locations in the queue are written over as appropriate for requesting tasks. Due to the primitive operations in the mutual exclusion routine, once the I/O device is allocated to a task, the I/O subtask runs to completion.

Another problem associated with the concurrent I/O task execution is the proper sequencing or ordering of all the tasks. Some I/O tasks must be synchronized with non-I/O tasks such that the overall program flow is correct. The associated primitives must provide for controlling precedence relations that are inherent in the overall task system. Since such relationships are unique to the given environment, synchronization primitives are usually embedded in each task requiring synchronization with another. As required, each task communicates with another through integer variables known as *semaphores* or *status flags*.

A semaphore is a flag that contains a positive, negative, or zero value. The semaphore is manipulated by two noninterruptible primitives. These operations are embedded in a task execution sequence requiring task synchronization. One operation is defined by *Wait* (s_i), the other by *Send* (s_i), where s_i stands for the semaphore being manipulated and associated with the ith resource, I/O device. These operations are sometimes called P and V operations, respectively based upon the original work by Dijkstra.[66] Wait (s_i) is the operation that is used by the ith task to indicate whether or not all its predecessors have completed. By definition, then,

Wait (s_i):

Step 1. $s_i = s_i - 1$.

Step 2. If $s_i = 0$ BEGIN "next task" ELSE "Check again later." If all predecessors checked by the ith task are not ready, check suspended task later as scheduled.

Wait (s_i) decrements the semaphore, and the ith task sits in a wait loop waiting for all predecessors to complete. If the ith task is suspended, then the scheduler reexecutes step 2 each time the ith task appears with the highest priority. When $s_i = 0$, the ith task is permitted to use the I/O device.

Send (s_j) is the operation that is used by the completed tasks for updating the successor semaphore s_j. By definition,

Send(s_j): $s_j = s_j + 1$

which increments the semaphore s_j. The Send operation indicates through the semaphore incrementation that the associated predecessor tasks have completed. The initial value of the semaphore is the negative of 1 less than the total number of predecessors.

Semaphores can also be used to control the use of resources. In this situation, the value of the semaphore for the ith device is associated with a number of tasks. If s_i is negative, the magnitude of s_i is the number of tasks suspended because the device is busy. Assuming only one task can use the device at a time, s_i may be zero, but not positive. Note that a queue is needed to retain information on the next task to use the device.

Example 14-2 The precedence graph (Fig. 14-15) results in the following augmented tasks T_i with the indicated synchronization primitives:

T_1: T_1, send (S_2), send (S_3)

T_2: wait (S_2), T_2, send (S_4)

T_3: wait (S_3), T_3, send (S_5), send (S_6)

T_4: wait (S_4), T_4

T_5: wait (S_5), T_5

T_6: wait (S_6), T_6

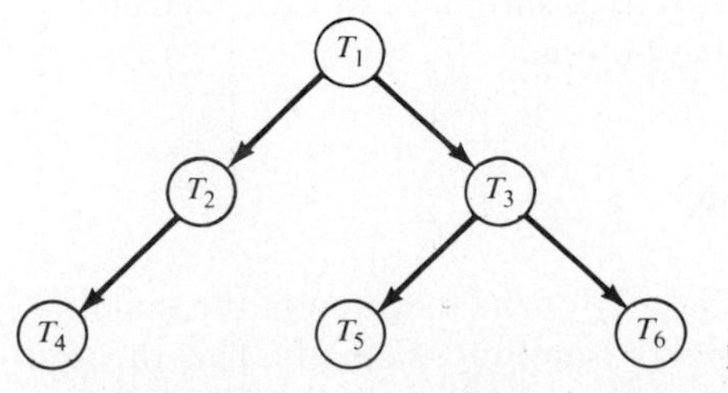

Figure 14-15 Precedence graph.

Note that the updating of the semaphores in storage must be indivisible (non-interruptible). This can be accomplished by unique hardware, instruction sets, or proper masking of interrupts.

Finally, when using mutual exclusion and synchronization primitives, a specific problem can occur called *deadlock*. In this situation two or more tasks are holding resources (I/O devices) that are needed by other tasks, and vice versa. The task system is stuck in a "circular wait." In the general computer system organization (one CPU, one memory, many I/O devices), the I/O tasks since they execute to completion can have subtasks that use other I/O devices. Thus, a deadlock situation may occur if proper analysis and task sequencing is not accomplished. Special algorithms[65] can be used to prohibit deadlock but considerable overhead results.

14-8 WATCHDOG TIMERS

A watchdog timer is an interval timer (counter) that generates a high-priority interrupt unless periodically reinitialized by the scheduler or a periodic task. It is normally utilized to detect a software or hardware failure. For example, individual watchdog counters can be associated with specific tasks whose counters are decremented at some defined clock cycle. When a task has not completed within a certain interval of time, a counter value of zero generates an interrupt. The task is suspended indefinitely until an error routine is processed. Error routines may be executed for system reinitialization, switching panic lights, shutdown, or other reevaluation of the condition in accordance with the programming structure of Chap. 13.

A general watchdog timer capability may provide for appropriate tasks to call for watchdog monitoring. The calling statement may include three parameters. The first may be the value of the counter (interval time). The second is the clock cycle (1 ms, 10 ms, etc.) for counter decrementing. The third may be the address (or name) of the specific error procedure. When the counter value becomes zero, the priority interrupt is activated and the system schedules the error routine. Use of the POSIT coding structure from Chap. 13 can be used in this application.

A watchdog timer, then, is used in critical situations where a task must be accomplished within a certain interval or the system will not function. Of course, with proper design and implementation this watchdog timer should never be activated. However, in reality, certain situations outside the context of the system may occur causing the delay. Thus, a watchdog timer is just another technique for checking proper interval values. Also, error routines should be integral elements of any real-time system to prohibit catastrophic failure.

14-9 SAMPLING-TIME SELECTION

As presented in previous chapters, the foundation for the selection of the sampling frequency or the sampling time T is Shannon's sampling theorem. This theorem specifies that a sampling rate be greater than twice the highest signal frequency of interest. The theorem assumes two important conditions:

1. The sampled signal is to be reconstructed.
2. Infinite accuracy of computations.

In the design and implementation of digital control systems, these two conditions are not true or valid. Also in a real-time digital control system, there is an inherent time delay in the conversion and digital computation processes even though the signal is usually not reconstructed (Chap. 8). It can be assumed that a slower sampling frequency may be implemented if the signal is not to be reconstructed. The computer, therefore, could be slower (cheaper) or, on the other hand, the computer could execute more independent process control functions in a given time interval. The selection of a sampling frequency, however, depends on various system characteristics, and overall accuracy requirements really determine the specific sampling time. The purpose of this section is to provide some additional insight into the selection of T.

The finite word length of the computer affects accuracy performance through controller coefficients, A/D and D/A conversion, and computer arithmetic operations (Chaps. 10 and 13). Other accuracy considerations include compensator structure (Chap. 10), disturbance signals, measurement noise, uncontrolled body-bending mode signals detected by system sensors, and the inherent time delay. In essence, *the primary influence on sampling time is the bandwidth characteristics of these signals.*

Starting with the system's highest frequency of interest, Shannon's theorem is applied to determine a sampling frequency. In most digital-controller implementations, this frequency is increased by a factor of 5 to 10, maybe 20, to define the sampling frequency. This heuristic approach bypasses a difficult, if not impossible, nonlinear analysis of the discrete nature of computer operations and the finite-word-length constraint. Simulation can be used to attempt to lower the sampling frequency for a specific application. *The determination of an "optimal" sampling frequency, however, is a compromise between many parameters*, i.e., cost, flexibility, accuracy. Specific areas that dictate the sampling rate are:

The desired system response to a given control input[40]
The desired response to external disturbance (deterministic or stochastic)[24]
The sensitivity to the plant model (body-bending, structural changes, etc.)[72]

Using $\mathscr{Z}$-domain root-locus techniques, the effects of sampling time on system performance can be analyzed since the $\mathscr{Z}$-domain poles and zeros are explicit functions of T. The previous chapters have integrated the analysis of T to provide a tool for sampling-frequency selection.

14-9A Single-Rate Sampling

To limit the influence of measurement noise and body-bending mode signals on accuracy and sampling-time selection, the use of an analog prefilter as developed in Chap. 11 is suggested. Body-bending modes are those open-loop plant dynamics that are not being controlled. These dynamics are usually quite different and at higher frequencies than those of the closed-loop dynamics or control bandwidth. Using a frequency band-pass prefilter (analog) for the control bandwidth (see Fig. 14-16), the sampling frequency can be lower than that required by a digital compensator which also contains digital filters for unknown inputs with higher-frequency components.

In most applications, the control bandwidth is less than 100 Hz, but higher frequencies can fold into the control bandwidth if not properly filtered (analog). Note also that a source of noise sometimes overlooked in determining sampling frequency is the power supply ripple (60 cycle, 400 cycle, etc.). For these noise sources as well as other disturbances, analog notch filters can be employed. The effect of these filters on phase must be analyzed since they can influence the performance within the control bandwidth.

The complete computational delay $\tau_{A/D} + \tau_c + \tau_{D/A}$ is assumed to be much less than the sampling time T so that the theoretical development can proceed (Chap. 8). This assumption may not be true in a given implementation due to hardware constraints or considerable multiprocessing. To ascertain if computational delay or another nonlinear characteristic is causing an accuracy problem, hybrid simulation is probably the best approach. Another approach is to include the delay model ϵ^{τ} in the general analytical model. This results, however, in a very difficult mathematical model to manipulate and analyze.

The D/A output can also cause accuracy problems owing to its staircase

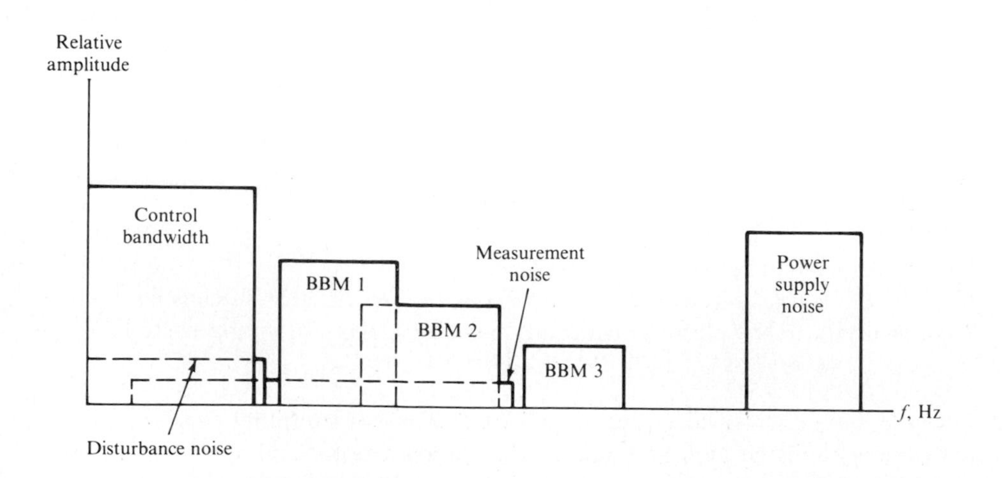

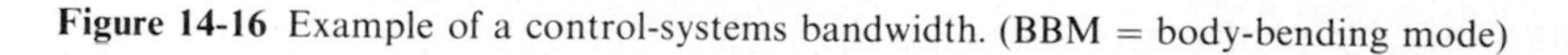
Figure 14-16 Example of a control-systems bandwidth. (BBM = body-bending mode)

characteristic. This staircase function is "smoothed" by the output transducer (stepper motor, heater, cooler, etc.) due to the transducer's finite bandwidth. This smoothing may permit less wear on hardware components but can cause accuracy problems. Thus, the transducer transfer function should be included in the original plant model for complete analysis and selection of T. One technique for studying the effect of the staircase function is to generate its amplitude rate of change relative to time and relate these values to other performance criteria. Note that good transducer smoothers generate additional phase delay in the system model but may not affect accuracy.

In general, the larger the order of the system with steeper control-bandwidth roll-off, the less effect higher frequencies have on performance. This occurs since frequencies above the bandwidth limit are being decreased more in amplitude. Although these frequencies are folded into the primary strip due to sampling (Chap. 4), their amplitude can be so low such that their effect is negligible within the control band. The steeper the control-bandwidth roll-off, then, the less restriction on T. Ten to fifty percent decreases in the value of T may occur in a specific application as the closed-loop order increases. In particular, a graphical relationship can be portrayed between the amplitude I/O ratio, system order, and sampling frequency for a given system.[89]

14-9B Multirate Sampling

The sampling-rate design procedures presented in previous chapters and based on Shannon's theorem assume single-rate sampling. *No explicit and complete methodology exists, however, which can exploit the unique characteristics of multirate sampling*, as shown in the examples of Fig. 3-9. Multirate sampling is used sometimes for non-real-time computational purposes such as simulation. Some real-time multirate systems, however, actually exist, for example, in flight-control systems, where the control computer samples at a rate different from that of the radar subsystem. Also, examples occur in real-time multiprogramming systems. In these cases, the bandwidths of the various subsystem components are quite different. In general, it is more appropriate to sample a slow-varying signal at a low sampling rate, while a signal with a high-frequency content should be sampled at a high rate. Another example involves the cruise condition of an aircraft which requires low sample rates, while during maneuvering, landing, or tracking, the sample rate should be increased because of accuracy requirements.

Multirate systems are, in reality, time-varying systems although they are seldom referred to as such. Systems that are time-varying are quite complex to analyze. Some time-variant problems can be solved from the point of view of a time-invariant or fixed system through modeling and the use of the $\mathscr{Z}$ transform. To solve some multirate problems, the modified $\mathscr{Z}$ transform can be employed. In addition, the use of the w' domain can be used for the simplification of the mathematical analysis. The practical application of multirate systems arises out of problems connected with the analysis and synthesis of standard single-rate sampled-data systems. Several investigators have used such systems as convenient mathematical models.

Stability analysis for multirate systems consists mainly of the contributions made by Kranc.[73,74] His method is based on replacing samplers operating at various rates by equivalent samplers. Delay configurations are implemented in the model which contain samplers at only one rate. The analysis proceeds by using modified $\mathscr{L}$ transforms to handle the effects of the induced time delays.

A second procedure for stability analysis has been developed[75] based upon the use of identities expressing the Laplace transforms of sampled signals in terms of shifted transforms. The method is better known as *switch decomposition*[76] in the frequency domain.

Both the previous procedures require the solution of a set of simultaneous equations with rational polynomial coefficients.

A third method that has been developed is based on vector operators and signal-flow graph analysis. The approach used is similar to the switch decomposition method. The advantage of using this method is that a transfer function of a feedback system may be obtained without solving a system of simultaneous equations.[77,78] Other references on multirate sampling are Refs. 78–81.

Although all these studies are analytically detailed, a combined theory has not been developed.[82] Currently, the main technique employed for evaluating multirate digital control systems is simulation. Digital simulation of a complete system requires that the plant continuous model be discretized. This can be accomplished using the numerical analysis techniques proposed in Chaps. 1 and 3 relating to generating difference equations from differential equations. As a rule of thumb, the continuous plant simulation should run at least ten times faster than the digital controller simulation in order to have a high degree of confidence in the results. This suggestion is based upon the reality of finite word length and the desire that the simulated plant measurements be accurate. Also, hybrid (analog and digital) simulation can also be effective (Chap. 10) in determining sampling rate or rates.

In the selection of the sampling frequency for single-rate or multirate sampling, the control-system cost is probably paramount. Yet, the system must be able to track specific inputs, reject others, and be nonsensitive to changes in plant parameters. These objectives along with other engineering considerations and constraints usually result in a *sampling frequency between 5 and 10 times the control bandwidth* ω_b *of the system.*

14-10 USER INTERFACES TO REAL-TIME OPERATING SYSTEMS

Many real-time environments do not permit the system to be stopped or shut down while a controller (task) is modified. Examples include chemical plants, spacecraft, and general process control environments. In these environments, parameter changes are usually initiated on-line. To make changes, five objectives should be met in designing the overall system:

1. The user should be able to modify (add, delete, change) control systems by putting the data (coefficients, variable declarations, etc.) into dummy or nonfunctioning tables.
2. Error checks should be made on all new information entering these tables for conformance to value intervals and other appropriate bounds.
3. The new information should also be checked by running a simulation. The simulation may not be part of the real-time software.
4. To bring the new system on-line, the operating system must provide for movement of the new table data into the current operating tables in a scheduled manner. The scheduling process for the new data may be manually controlled or automatic. Normal integration permits a step-by-step incorporation.
5. An easily understood interactive terminal dialogue should be provided. Note that it can be used for documenting the update process as well.

The general user interface capabilities mentioned above require extensive and complex software routines. Such routines are usually unique to the application because with real-time constraints to develop control software of this type, a disciplined software engineering approach (Chap. 13) is employed with considerable emphasis on testing due to the critical nature of the application.

14-11 SUMMARY

The purpose of this chapter is to introduce the major aspects in analyzing and designing a real-time operating system for use in a digital control environment. In many digital control applications with a limited number of fixed tasks, some of the concepts presented may not have to be implemented. In fact, the overhead (time and space) may be too extensive for the system to function properly in a real-time environment. A single-variable control implementation may be such an example.

However, other real-time control systems require monitoring capabilities for multitasking (multiprogramming). Digital control systems with numerous I/O devices and multiple tasks generally require a complex scheduler and storage management capability. Various scheduling algorithms are shown to depend upon the storage management scheme employed, that is, the storage of tasks and their link list element information. System performance depends upon the proper integration of these two functions.

The selection of an acceptable sampling time is a complex engineering function depending upon many factors and compromises, leading to a cost-effective implementation. The major elements of the selection are related to the application and include the bandwidth of external signals, plant parameters, variations, and closed-loop bandwidth. Normally, a selection between 5 to 10 times the highest frequency (bandwidth) is chosen for single-variable control. Multirate sampling requires considerable analysis of intersystem relationships through simulation.

The interweaving of many real-time tasks seems to be a relatively simple problem at first. However, the many dimensions (number of tasks, task times, memory size, sampling time, computer speed, etc.) of the problem quickly make the problem quite difficult. Very logical and thorough analysis, design, implementation, and testing procedures must be employed, possibly including simulation.

CHAPTER

FIFTEEN

INTRODUCTION TO DISCRETE STATE-VARIABLE MODEL

15-1 INTRODUCTION

The analytical methods of sampled-data control systems presented in the earlier chapters are classified under conventional control theory. These methods rely on the use of the $\mathscr{L}$ and $\mathscr{Z}$ transforms, associated transfer functions, and block diagrams to model scalar systems. Another classification of system analysis is multivariable modern control theory which uses a state-variable (SV) model. This modeling technique which uses vectors and matrices permits a simple notation for large-dimensional systems. The popularity of this analytical method was made possible by the advent of large digital computers for which the higher-level programming languages easily accept and process vector and matrix notation. Thus, the purpose of this chapter is to develop specific techniques for generating and simulating SV models. Another purpose of this chapter is to present one application of state-variable analysis in the minimization of plant variation upon the system output parameter, i.e., *robustness* of SISO systems.

The important advantages of the SV method as compared to the conventional approach are:

The speed and accuracy of the analysis and design of MIMO systems.
It is ideally suited for digital computer MIMO computations.
It is ideally suited for designing optimal control systems.†

† An optimal control system is one whose performance is optimal in terms of some defined performance criterion. For example, the energy control quantity $\sum_{k=0}^{\infty} u^2(kT)$ shall be a minimum as constrained by the plant model equations and desired final plant conditions. The objective is to derive a control law $u(kT)$ that minimizes the performance index subject to the constraints.

It is ideally suited for obtaining the MIMO plant time response in between sampling instants.

To develop SV models for sampled-data systems, a specific terminology is required which is similar to that used for SV models for continuous-data systems.

A continuous-data system can be modeled by a set of first-order differential (SV) equations which are called *state equations*. Also, a discrete-time control system composed entirely of discrete-data components can be described by a set of state equations that are only first-order difference (SV) equations. Thus, for a *hybrid* system (which contains both discrete- and continuous-data components) the state equations consist of both first-order differential and first-order difference equations.[2,18] It should be noted that in conventional control theory the letters e, E and c, C are used to denote the actuating and output signals, respectively, for time and transform domains. In modern multivariable control theory the letters u, U and y, Y are used to denote the actuating and output signals, respectively.

15-2 STATE-VARIABLE REPRESENTATION OF CONTINUOUS-DATA CONTROL SYSTEMS[1]

The state of a system (henceforth referred to only as state) is defined by Kalman[69] as follows: The state is defined as a mathematical structure containing a set of n variables $x_1(t), x_2(t), \ldots, x_i(t), \ldots, x_n(t)$, called the state variables, such that the initial values $x_i(t_0)$, where $i = 1, 2, \ldots, n$, of this set and the system input variables $u_j(t)$, where $j = 1, 2, \ldots, r$, are sufficient to uniquely describe the system's future response $y_q(t)$, where $q = 1, 2, \ldots, p$, for $t \geq t_0$. There is a minimum set of independent state variables which is required to represent the system accurately. The r inputs $u_1(t), u_2(t), \ldots, u_j(t), \ldots, u_r(t)$ are assumed to be deterministic; i.e., they have known specific values for all values of time $t \geq t_0$. The MIMO continuous-data system of Fig. 15-1, which is LTI, has r input and p output variables and may be described by a set of n first-order linear differential equations, called *state equations*, of the form

$$\dot{x}_i(t) \equiv f_i[x_1(t), x_2(t), \ldots, x_n(t), u_1(t), u_2(t), \ldots, u_r(t)] \tag{15-1}$$

f_i represents the linear functional relationships between $\dot{x}_i(t)$ and all the state variables and the input variables. To completely describe the dynamic behavior of the total system, the following *system output equation* is needed to relate the output variables to the state variables and the input variables:

$$y_q(t) = f_q[x_1(t), x_2(t), \ldots, x_n(t), u_1(t), u_2(t), \ldots, u_r(t)] \tag{15-2}$$

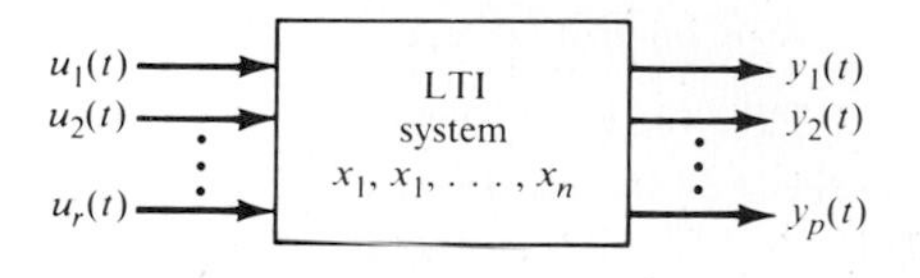

Figure 15-1 An LTI MIMO continuous-data system with n state variables, r inputs, and p outputs.

where f_q is a linear functional and $q = 1, 2, \ldots, p$. The variables $x_i(t)$, $u_j(t)$, and $y_q(t)$ can be represented in vector notation as

$$\mathbf{x}(t) \equiv \begin{bmatrix} x_1(t) \\ x_2(t) \\ \vdots \\ x_n(t) \end{bmatrix} \qquad \mathbf{u}(t) \equiv \begin{bmatrix} u_1(t) \\ u_2(t) \\ \vdots \\ u_r(t) \end{bmatrix} \qquad \mathbf{y}(t) \equiv \begin{bmatrix} y_1(t) \\ y_2(t) \\ \vdots \\ y_p(t) \end{bmatrix} \tag{15-3}$$

where the dimensions of these column vectors are $n \times 1$, $r \times 1$, and $p \times 1$, respectively. The vector-matrix formulation of Eqs. (15-1) and (15-2) which represent the functions $f_i(\cdots)$ and $f_q(\cdots)$ are

$$\dot{\mathbf{x}}(t) = \mathbf{A}\mathbf{x}(t) + \mathbf{B}\mathbf{u}(t) \qquad \textit{State equation} \tag{15-4}$$

$$\mathbf{y}(t) = \mathbf{C}\mathbf{x}(t) + \mathbf{D}\mathbf{u}(t) \qquad \textit{Output equation} \tag{15-5}$$

where $\mathbf{A} = n \times n$ *plant matrix*
$\mathbf{B} = n \times r$ *control matrix*
$\mathbf{C} = p \times n$ *output matrix*
$\mathbf{D} = p \times r$ *feedforward matrix*

Note that

$$\dot{\mathbf{x}}(t) = \begin{bmatrix} \dot{x}_1(t) \\ \dot{x}_2(t) \\ \vdots \\ \dot{x}_n(t) \end{bmatrix}$$

Equations (15-1) and (15-5) are linear in nature and are referred to as the system *dynamic equations*. The methods for determining the elements of the **A**, **B**, **C**, and **D** matrices can be found in the literature.[3] The phase state-variable method,[1] for example, is presented in Example 15-1.

The solution for $\mathbf{x}(t)$ may be obtained by taking the Laplace transform of Eq. (15-4); thus,

$$s\mathbf{X}(s) - \mathbf{x}(0) = \mathbf{A}\mathbf{X}(s) + \mathbf{B}\mathbf{U}(s)$$

and rearranging and solving for $\mathbf{X}(s)$ yields

$$[s\mathbf{I} - \mathbf{A}]\mathbf{X}(s) = \mathbf{x}(0) + \mathbf{B}\mathbf{U}(s)$$

$$\mathbf{X}(s) = [s\mathbf{I} - \mathbf{A}]^{-1}\mathbf{x}(0) + [s\mathbf{I} - \mathbf{A}]^{-1}\mathbf{B}\mathbf{U}(s) \tag{15-6}$$

where $[s\mathbf{I} - \mathbf{A}]$ is assumed to be nonsingular† so that its matrix inverse, $[s\mathbf{I} - \mathbf{A}]^{-1}$, may be found. The resolvent matrix designated by $\mathbf{\Phi}(s)$ is defined by the above inverse; that is,

$$\mathbf{\Phi}(s) \equiv [s\mathbf{I} - \mathbf{A}]^{-1} = \frac{\operatorname{adj}[s\mathbf{I} - \mathbf{A}]}{|s\mathbf{I} - \mathbf{A}|} \tag{15-7}$$

† It can be shown that $[s\mathbf{I} - \mathbf{A}]$ is nonsingular;[1] that is, $|s\mathbf{I} - \mathbf{A}| \neq 0$.

where the adjoint, adj$[s\mathbf{I} - \mathbf{A}]$, is the transpose of the cofactor matrix and $|s\mathbf{I} - \mathbf{A}|$ is the determinant of $[s\mathbf{I} - \mathbf{A}]$. The cofactor matrix is formed by replacing each element of $[s\mathbf{I} - \mathbf{A}]$ by its cofactor. The inverse Laplace transform of Eq. (15-6) gives

$$\mathbf{x}(t) = \boldsymbol{\phi}(t)\mathbf{x}(0) + \int_0^t \boldsymbol{\phi}(t - \tau)\mathbf{B}\mathbf{u}(\tau)\,d\tau = \boldsymbol{\phi}(t)\mathbf{x}(0) + \boldsymbol{\gamma}(t) \tag{15-8}$$

where $t \geq 0$, the *state transition matrix* (STM) $\boldsymbol{\phi}(t)$ is defined by

$$\boldsymbol{\phi}(t) \equiv \mathscr{L}^{-1}[\boldsymbol{\Phi}(s)] = \epsilon^{\mathbf{A}t} \tag{15-9}$$

and where $\boldsymbol{\gamma}(t)$ is defined as

$$\begin{aligned}\boldsymbol{\gamma}(t) &\equiv \int_0^t \boldsymbol{\phi}(t - \tau)\mathbf{B}\mathbf{u}(\tau)\,d\tau \\ &\equiv \mathscr{L}^{-1}[\boldsymbol{\Gamma}(s)] = \mathscr{L}^{-1}\{[s\mathbf{I} - \mathbf{A}]^{-1}\mathbf{B}\mathbf{U}(s)\}\end{aligned} \tag{15-10}$$

where

$$\boldsymbol{\Gamma}(s) = [s\mathbf{I} - \mathbf{A}]^{-1}\mathbf{B}\mathbf{U}(s)$$

The above development is based upon the initial time $t_0 = 0$. Before obtaining a general solution of $\mathbf{x}(t)$ for $t_0 \neq 0$, it is necessary to list some properties of the STM for LTI systems. They are:

1. $\boldsymbol{\phi}(t)$ is nonsingular.
2. Nonsingular for all finite values of t.
3. $\boldsymbol{\phi}(0) = \mathbf{I}$ (15-11)
4. $\boldsymbol{\phi}(t_2 - t_1)\boldsymbol{\phi}(t_1 - t_0) = \boldsymbol{\phi}(t_2 - t_0)$ (15-12)
5. $\boldsymbol{\phi}^{-1}(t) = \boldsymbol{\phi}(-t)$ (15-13)
6. $\boldsymbol{\phi}^k(t) = \boldsymbol{\phi}(kT) \qquad k = 0, 1, 2, 3, \ldots$ (15-14)

Assuming $t = t_0 > 0$, substitute $t = t_0$ into Eq. (15-8) and solve for $\mathbf{x}(0)$. The resulting expression for $\mathbf{x}(0)$ is then substituted into Eq. (15-8) to yield

$$\mathbf{x}(t) = \boldsymbol{\phi}(t - t_0)\mathbf{x}(t_0) + \int_{t_0}^t \boldsymbol{\phi}(t - \tau)\mathbf{B}\mathbf{u}(\tau)\,d\tau \tag{15-15}$$

for which $t \geq t_0$. This equation, which is referred to as the *state transition equation* of the system described by Eq. (15-3), lends itself to modeling and to the analysis of sampled-data control systems as illustrated in both the following example and the next section.

Example 15-1 Given

$$G_x(s) = \frac{Y(s)}{U(s)} = \frac{K_G}{s^2 + a_1 s + a_0} \tag{15-16}$$

and

$$u(t) = R_0 u_{-1}(t)$$

(*a*) The state transition equation for this plant and forcing function, where $a_0 = 0$, $a_1 = 2$, and $K_G = 2$, is obtained for $t_0 = 0$ and $t_0 \neq 0$ as follows:

The **A**, **b**′, and **c**′ matrix and vectors, respectively, for an *n*th-order system, using the phase state-variable representation, are

$$\mathbf{A} = \begin{bmatrix} 0 & 1 & 0 & \cdots & 0 \\ 0 & 0 & 1 & \cdots & 0 \\ \cdot & \cdot & \cdot & \cdot & \cdot \\ 0 & 0 & 0 & \cdots & 1 \\ -a_0 & -a_1 & -a_2 & \cdots & -a_{n-1} \end{bmatrix} \tag{15-17a}$$

$$\mathbf{b}' = [0 \quad 0 \quad 0 \quad \cdots \quad K_G] \tag{15-17b}$$

$$\mathbf{c}' = [1 \quad 0 \quad 0 \quad \cdots \quad 0] \tag{15-17c}$$

where **A** is referred to as the *companion* matrix and is $n \times n$, **b** is $n \times 1$, **c** is $n \times 1$, and the prime notation is used to denote the transpose of a column vector or a matrix. Thus, for this example, the system state equation is

$$\dot{\mathbf{x}} = \mathbf{A}\mathbf{x} + \mathbf{b}\mathbf{u} = \begin{bmatrix} 0 & 1 \\ -a_0 & -a_1 \end{bmatrix}\mathbf{x} + \begin{bmatrix} 0 \\ K_G \end{bmatrix}\mathbf{u} = \begin{bmatrix} 0 & 1 \\ 0 & -2 \end{bmatrix}\mathbf{x} + \begin{bmatrix} 0 \\ 2 \end{bmatrix}\mathbf{u} \tag{15-18}$$

which yields, for $t_0 = 0$,

$$\mathbf{\Phi}(s) = [s\mathbf{I} - \mathbf{A}]^{-1} = \begin{bmatrix} s & -1 \\ 0 & s+2 \end{bmatrix}^{-1} = \frac{\begin{bmatrix} s+2 & 1 \\ 0 & s \end{bmatrix}}{s(s+2)} = \begin{bmatrix} \dfrac{1}{s} & \dfrac{1}{s(s+2)} \\ 0 & \dfrac{1}{s+2} \end{bmatrix}$$

$$\mathbf{\phi}(t) = \mathscr{L}[\mathbf{\Phi}(s)] = \begin{bmatrix} 1 & 0.5(1 - \epsilon^{-2t}) \\ 0 & \epsilon^{-2t} \end{bmatrix} \tag{15-19a}$$

and

$$\gamma(t) = \mathscr{L}^{-1}[\mathbf{\Gamma}(s)] = \mathscr{L}^{-1}\{[s\mathbf{I} - \mathbf{A}]^{-1}\mathbf{B}U(s)\}$$

$$= \left[\mathscr{L}^{-1}\left\{[s\mathbf{I} - \mathbf{A}]^{-1}\mathbf{B}\frac{1}{s}\right\}\right]R_0 = \{\mathscr{L}^{-1}[\mathbf{\Theta}(s)]\}R_0$$

where

$$\mathbf{\Theta}(s) \equiv \frac{[s\mathbf{I} - \mathbf{A}]^{-1}\mathbf{B}}{s}$$

Thus

$$\gamma(t) = \mathscr{L}^{-1}\begin{bmatrix} \dfrac{2}{s^2(s+2)} \\ \dfrac{2}{s(s+2)} \end{bmatrix} R_0 = \begin{bmatrix} 0.5(2t - 1 + \epsilon^{-2t}) \\ 1 - \epsilon^{-2t} \end{bmatrix} R_0 = \boldsymbol{\theta}(t) R_0 \quad (15\text{-}19b)$$

where

$$\boldsymbol{\theta}(t) = \begin{bmatrix} 0.5(2t - 1 + \epsilon^{-2t}) \\ 1 - \epsilon^{-2t} \end{bmatrix} \quad (15\text{-}19c)$$

Thus, substituting Eqs. (15-19*a*) and (15-19*b*) into (15-8) yields, for $t_0 = 0$,

$$\mathbf{x}(t) = \boldsymbol{\phi}(t)\mathbf{x}(0) + \gamma(t) = \boldsymbol{\phi}(t)\mathbf{x}(0) + \boldsymbol{\theta}(t)R_0$$

$$= \begin{bmatrix} 1 & 0.5(1 - \epsilon^{-2t}) \\ 0 & \epsilon^{-2t} \end{bmatrix}\mathbf{x}(0) + \begin{bmatrix} 0.5(2t - 1 + \epsilon^{-2t}) \\ 1 - \epsilon^{-2t} \end{bmatrix} R_0 \quad (15\text{-}20)$$

For $t_0 > 0$, let $t = t_0$ in Eq. (15-20), which results in

$$\mathbf{x}(t_0) = \boldsymbol{\phi}(t_0)\,\mathbf{x}(0) + \boldsymbol{\theta}(t_0)R_0$$

Solving for $\mathbf{x}(0)$ generates

$$\mathbf{x}(0) = \boldsymbol{\phi}(t_0)^{-1}\mathbf{x}(\mathbf{t}_0) - \boldsymbol{\phi}(t_0)^{-1}\boldsymbol{\theta}(t_0)R_0 \quad (15\text{-}21)$$

Substituting from Eq. (15-21) into (15-20) yields

$$\mathbf{x}(t) = \boldsymbol{\phi}(t)\boldsymbol{\phi}(t_0)^{-1}\mathbf{x}(t_0) - \boldsymbol{\phi}(t)\boldsymbol{\phi}(t_0)^{-1}\gamma(t_0)R_0 + \gamma(t)R_0 \quad (15\text{-}22)$$

Utilizing Eqs. (15-12) and (15-13) by letting $t = t_2$ and $t_1 = 0$ results in the identity

$$\boldsymbol{\phi}(t)\boldsymbol{\phi}(t_0)^{-1} = \boldsymbol{\phi}(t)\boldsymbol{\phi}(-t_0) = \boldsymbol{\phi}(t_2 - t_1)\boldsymbol{\phi}(t_1 - t_0)$$
$$= \boldsymbol{\phi}(t_2 - t_0) = \boldsymbol{\phi}(t - t_0)$$

To obtain $\boldsymbol{\phi}(-t_0)$, substitute $t = -t_0$ into Eq. (15-19*a*). Thus

$$\boldsymbol{\phi}(t - t_0) = \boldsymbol{\phi}(t)\boldsymbol{\phi}(-t_0) = \begin{bmatrix} 1 & 0.5(1 - \epsilon^{-2t}) \\ 0 & \epsilon^{-2t} \end{bmatrix}\begin{bmatrix} 1 & 0.5(1 - \epsilon^{2t_0}) \\ 0 & \epsilon^{2t_0} \end{bmatrix}$$

$$= \begin{bmatrix} 1 & 0.5(1 - \epsilon^{-2(t-t_0)}) \\ 0 & \epsilon^{-2}(t - t_0) \end{bmatrix} \quad (15\text{-}23)$$

In a similar manner, from Eq. (15-19*c*) obtain

$$\boldsymbol{\theta}(t - t_0) = -[\boldsymbol{\phi}(t)\boldsymbol{\phi}(-t_0)\boldsymbol{\theta}(t_0) - \boldsymbol{\theta}(t)]$$

$$= \begin{bmatrix} 0.5[2(t - t_0) - 1 + \epsilon^{-2(t-t_0)}] \\ 1 - \epsilon^{-2(t-t_0)} \end{bmatrix} \quad (15\text{-}24)$$

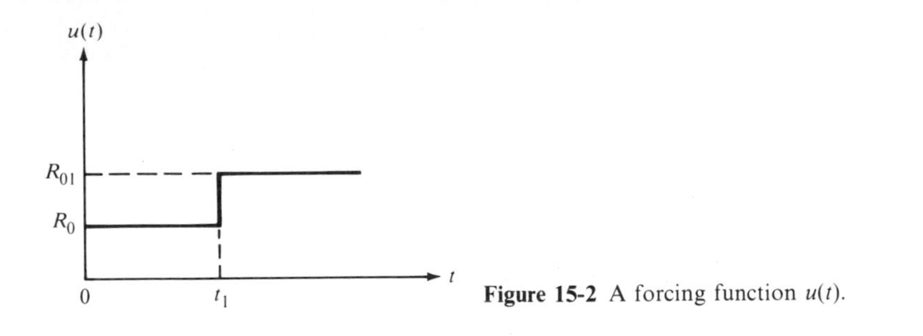

Figure 15-2 A forcing function $u(t)$.

Note that Eqs. (15-23) and (15-24) can be obtained by replacing t in Eqs. (15-19a) and (15-19c) by $t - t_0$. Thus, substituting from Eqs. (15-23) and (15-24) into Eq. (15-22) yields the state transition equation for $t_0 > 0$ as follows:

$$\mathbf{x}(t) = \boldsymbol{\phi}(t - t_0)\mathbf{x}(t_0) + \boldsymbol{\theta}(t - t_0)R_0 \tag{15-25}$$

(*b*) For the system of part *a*, the forcing function $u(t)$ changes in magnitude instantaneously at some time $t = t_1$ as shown in Fig. 15-2. It is desired to determine $\mathbf{x}(t)$ *for* $t \geq t_1$.

It is first necessary to evaluate $\mathbf{x}(t)|_{t=t_1}$ for $u(t) = R_0 u_{-1}(t)$, where $0 \leq t < t_1$. Once this is done Eq. (15-25) is reinitialized; i.e., let $t_0 = t_1$ and set $\mathbf{x}(t_0) = \mathbf{x}(t_1)$ in Eq. (15-25) to yield

$$\mathbf{x}(t) = \boldsymbol{\phi}(t - t_1)\mathbf{x}(t_1) + \boldsymbol{\theta}(t - t_1)R_{01} \tag{15-26}$$

where R_{01} replaces R_0 and where $\boldsymbol{\phi}(t - t_1)$ and $\boldsymbol{\theta}(t - t_1)$ are obtained, respectively, from Eqs. (15-23) and (15-24) by substituting $t_1 = t_0$ into these equations.

(*c*) For the system of part *a* the step forcing function changes in value every *T* s; that is

$$u(t) = R_0(kT)\{u_{-1}(t - kT) - u_1[t - (k + 1)T]\} \qquad \text{for } kT \leq t < (k + 1)T$$

or

$$\begin{aligned} u(kT + \hat{t}) &= m(kT + \hat{t}) \\ &= u(kT) = m(kT) \qquad \text{for } 0 \leq \hat{t} < T,\ kT \leq t < (k + 1)T \end{aligned}$$

where $R_0(kT)$ is given by $m(kT)$ (see Fig. 15-3). It is now necessary to derive the expression for $\mathbf{x}(t)$ and $y(t)$ for the time span of $0 \leq t < NT$, where $N = k + 1$.

Note that Eq. (15-25) determines $\mathbf{x}(t)$ for $0 \leq t < T$ due to $\mathbf{x}(0)$ and $m(0) = R_0(0) = R_0$, where $k = 0$ and $t_0 = 0$. Thus Eq. (15-25) is rewritten as

$$\mathbf{x}(t) = \boldsymbol{\phi}(t)\mathbf{x}(0) + \boldsymbol{\theta}(t)m(0) \qquad \text{for } 0 \leq t < T \tag{15-27}$$

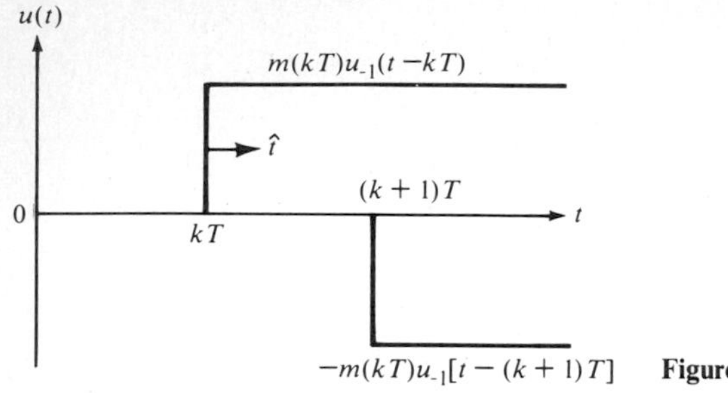

Figure 15-3 Plot of $u(kT + \hat{t})$.

Next, note that Eq. (15-26) determines $\mathbf{x}(t)$ for $T \le t < 2T$ due to $\mathbf{x}(t_1) = \mathbf{x}(T)$ and $m(t) = R_0(t) = R_{01}$. To determine $\mathbf{x}(t)$ in this time period, let $t_1 = T$ in Eq. (15-26). Thus

$$\mathbf{x}(t) = \boldsymbol{\phi}(t - T)\mathbf{x}(T) + \boldsymbol{\theta}(t - T)m(T) \qquad \text{for } T \le t < 2T \qquad (15\text{-}28)$$

In a similar manner, $\mathbf{x}(t)$ for the time period $2T \le t < 3T$ due to $\mathbf{x}(2T)$ and $m(2T)$ is given by

$$\mathbf{x}(t) = \boldsymbol{\phi}(t - 2T)\mathbf{x}(2T) + \boldsymbol{\theta}(t - 2T)m(2T) \qquad \text{for } 2T \le t < 3T \qquad (15\text{-}29)$$

Thus from Eqs. (15-27) to (15-29) it is possible to deduce the expression for $\mathbf{x}(t)$ in the time period $0 \le t < NT$ due to a change in value in the step forcing function every T s; that is,

$$\mathbf{x}(t) = \boldsymbol{\phi}(t - kT)\mathbf{x}(kT) + \boldsymbol{\theta}(t - kT)m(kT)$$

$$\text{for } kT \le t < (k + 1)T = NT \qquad (15\text{-}30)$$

where $k = 0, 1, 2, \ldots$. For this example, $\boldsymbol{\phi}(t - kT)$ and $\boldsymbol{\theta}(t - kT)$ can be evaluated from Eqs. (15-23) and (15-24), respectively, by letting $t_0 = kT$. For an all-pole plant, the phase-variable representation of the output $y(t)$ is

$$y(t) = x_1(t) = [1 \quad 0]\mathbf{x}(t) = \mathbf{c}'\mathbf{x} \qquad \text{for } kT \le t < (k + 1)T \qquad (15\text{-}31)$$

(*d*) For parts *a* and *b*, plot $y(t)$ vs. t for $\mathbf{x}(0) = 0$, $R_0 = 1$, $R_{01} = 2$, and $t_1 = 0.1$ s. From Eqs. (15-20) and (15-31), for $0 \le t < t_1$,

$$y(t)|_{t=0.1} = x_1(t)|_{t=0.1} = (t - 0.5 + 0.5\epsilon^{-2t})_{t=0.1} = 0.009365$$

$$x_2(t)|_{t=0.1} = (1 - \epsilon^{-2t})|_{t=0.1} = 0.18127$$

From Eqs. (15-26) and (15-31), for $t \ge t_1$ where

$$\mathbf{x}(t_0)|_{t_0=t_1} = \begin{bmatrix} 0.009365 \\ 0.181269 \end{bmatrix}$$

$$y(t) = x_1(t)$$

$$= 0.009365 + 0.090635[1 - \epsilon^{-2(t-0.1)}] + 2(t - 0.1) - 1 + \epsilon^{-2(t-0.1)}$$

Table 15-1 Data for $y(t)$ of Example 15-1*d*.

t	$y(t) = x_1(t)$	t	$y(t) = x_1(t)$
$0 \le t \le t_1$		$t_1 < t$	
0	0	0.12	0.013709
0.02	0.000395	0.14	0.019453
0.04	0.001558	0.16	0.026534
0.06	0.003460	0.18	0.034906
0.08	0.006072	0.2	0.044524
0.1	0.009365	↓	↓

The data in Table 15-1 may be used to plot $y(t)$ vs. t.

15-3 TIME-DOMAIN STATE AND OUTPUT EQUATIONS FOR SAMPLED-DATA CONTROL SYSTEMS[4,24]

The sampled-data system of Fig. 15-4, as stated in Chap. 1, is referred to as a *hybrid* system since it contains both discrete and continuous signals, whereas a system in which all signals are discrete is referred to as a *discrete-time* system. For the hybrid system of Fig. 15-4, the output signals of the ZOH units are described by

$$\begin{aligned} m_j(kT + \hat{t}) &= m_j(kT) \\ &= u_j(kT) \qquad \text{for } 0 \le \hat{t} < T,\ kT \le t < (k+1)T \end{aligned} \tag{15-32}$$

where $k = 0, 1, 2, 3, \ldots$ and $j = 1, 2, \ldots, r$. That is, the signal $m_j(kT + \hat{t})$ is constant between the sampling instants and is equal to the value of $u_j(kT)$. The value of $\mathbf{x}(t)$ in the time interval $kT \le t < (k+1)T$ can be determined from Eq.

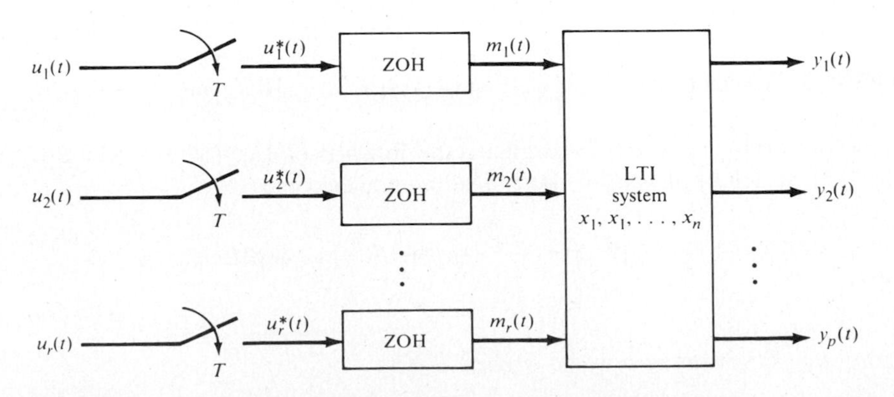

Figure 15-4 An LTI MIMO sampled-data control system with ZOH.

(15-15) by letting $t_0 = kT$ and $u(\tau) = m(\tau) = m(kT) = u(kT)$. Thus

$$\mathbf{x}(t) = \boldsymbol{\phi}(t - kT)\mathbf{x}(kT) + \left[\int_{kT}^{t} \boldsymbol{\phi}(t - \tau)\mathbf{B}\, d\tau\right]\mathbf{u}(kT) \qquad \text{for } kT \le t < (k+1)T \tag{15-33}$$

This equation may be evaluated by the Laplace transform technique of the previous section. The resulting equation is identical to Eq. (15-6) except that $\mathbf{x}(0)$ and $\mathbf{U}(s)$ are replaced, respectively, by $\mathbf{x}(kT)$ and $\mathbf{u}(kT)/s$. That is,

$$\mathbf{X}(s) = [s\mathbf{I} - \mathbf{A}]^{-1}\mathbf{x}(kT) + [s\mathbf{I} - \mathbf{A}]^{-1}\,\mathbf{B}\frac{\mathbf{u}(kT)}{s} \tag{15-34}$$

where now

$$\{\boldsymbol{\phi}(\hat{t}) = \mathscr{L}^{-1}[s\mathbf{I} - \mathbf{A}]^{-1}\}_{\hat{t}=t-kT}$$

and

$$\left\{\boldsymbol{\gamma}(\hat{t}) = \mathscr{L}^{-1}\left[[s\mathbf{I} - \mathbf{A}]^{-1}\mathbf{B}\frac{\mathbf{u}(kT)}{s}\right]\right\}_{\hat{t}=t-kT}$$

Note that Eq. (15-15) for the continuous-time system can be used for a sampled-data system by merely reinitializing the initial conditions and the forcing functions at the beginning of each sampled period. Thus the values of $\mathbf{x}(t)$ *at successive sampling instants can be derived by setting* $t = (k+1)T$ *and* $t_0 = kT$ *in Eq. (15-15). Therefore*

$$\mathbf{x}[(k+1)\mathrm{T}] = \boldsymbol{\phi}(T)\mathbf{x}(kT) + \int_{\tau=kT}^{(k+1)T} \boldsymbol{\phi}[(k+1)T - \tau]\mathbf{B}\mathbf{m}(\tau)\, d\tau \tag{15-35}$$

where $t - t_0 = (k+1)T - kT = T$. *Since* $\mathbf{m}(\tau) = \mathbf{u}(kT)$ *for* $kT \le \tau < (k+1)T$ *then* $\mathbf{m}(\tau) = \mathbf{m}(kT)$ *is constant within the time limits of the integration. Thus, Eq. (15-35) may be put into the following format:*

$$\mathbf{x}[(k+1)T] = \boldsymbol{\phi}(T)\mathbf{x}(kT) + \left\{\int_{kT}^{(k+1)T} \boldsymbol{\phi}[(k+1)T - \tau]\mathbf{B}\, d\tau\right\}\mathbf{m}(kT) \tag{15-36}$$

A change of variable is made to evaluate the integral of Eq. (15-36). Let $\beta = (k+1)T - \tau$, which implies $d\beta = -d\tau$. The integral becomes

$$\int_{kT}^{(k+t)T} \boldsymbol{\phi}[(k+1)T - \tau]\mathbf{B}\, d\tau = -\int_{T}^{0} \boldsymbol{\phi}(\beta)\mathbf{B}\, d\beta = \int_{0}^{T} \boldsymbol{\phi}(\beta)\mathbf{B}\, d\beta \equiv \boldsymbol{\theta}(T) \tag{15-37}$$

Therefore, Eq. (15-36) may now be written as

$$\mathbf{x}[(k+1)T] = \boldsymbol{\phi}(T)\mathbf{x}(kT) + \boldsymbol{\theta}(T)\mathbf{m}(kT) \tag{15-38}$$

where

$$\boldsymbol{\phi}(T) = \{\mathscr{L}^{-1}[s\mathbf{I} - \mathbf{A}]^{-1}\}_{t=T} \tag{15-39}$$

and

$$\boldsymbol{\theta}(T) = \mathscr{L}^{-1}\left\{[s\mathbf{I} - \mathbf{A}]^{-1}\mathbf{b}\frac{1}{s}\right\}_{t=T} \tag{15-40}$$

Equation (15-38) is referred to as the *discrete state transition equation* since it represents a set of first-order difference equations which describe the state variables at discrete instants of time. Note, as shown in Sec. 15-4, that this equation is of the form

$$\mathbf{x}[(k+1)T] = \mathbf{F}\mathbf{x}(kT) + \mathbf{H}\mathbf{u}(kT) \tag{15-41}$$

where the matrices **F** and **H** have the order of $n \times n$ and $n \times r$, respectively. Also, the dimensions of the vectors $\mathbf{x}[(k+1)T]$, $\mathbf{x}(kT)$, and $\mathbf{u}(kT)$ are $n \times 1$, $n \times 1$, and $r \times 1$, respectively. The system output equation is given by

$$\mathbf{y}(kT) = \mathbf{C}\mathbf{x}(kT) + \mathbf{D}\mathbf{u}(kT) \tag{15-42}$$

Equations (15-41) and (15-42) are the dynamic equations of the sampled-data system which is represented by the block diagram of Fig. 15-5.

From Eq. (15-38) the homogeneous equation of the system is

$$\mathbf{x}[(k+1)T] = \boldsymbol{\phi}(T)\mathbf{x}(kT) \tag{15-43}$$

which may be evaluated as follows:

$0 \le t < T, k = 0$: $\quad \mathbf{x}(T) = \boldsymbol{\phi}(T)\mathbf{x}(0)$

$T \le t < 2T, k = 1$:

$$\begin{aligned}\mathbf{x}(2T) &= \boldsymbol{\phi}(T)\mathbf{x}(T) \\ &= \boldsymbol{\phi}(T)\boldsymbol{\phi}(T)\mathbf{x}(0) \\ &= \boldsymbol{\phi}(2T)\mathbf{x}(0)\end{aligned}$$

$2T \le t < 3T, k = 2$:

$$\begin{aligned}\mathbf{x}(3T) &= \boldsymbol{\phi}(T)\mathbf{x}(2T) \\ &= \boldsymbol{\phi}(T)\boldsymbol{\phi}(2T)\mathbf{x}(0) \\ &= \boldsymbol{\phi}(3T)\mathbf{x}(0)\end{aligned}$$

$\vdots \qquad \vdots$

$(N-1)T \le t < NT, k = N-1$:

$$\begin{aligned}\mathbf{x}(NT) &= \boldsymbol{\phi}(T)\mathbf{x}[(N-1)T] \\ &= \boldsymbol{\phi}(NT)\mathbf{x}(0)\end{aligned} \tag{15-44}$$

where $N = k + 1 = 1, 2, 3, \ldots$. Equation (15-44) is another way of expressing Eq. (15-43) and represents a *recursive procedure* for evaluating $\mathbf{x}(NT)$.

The complete solution of Eq. (15-38) is obtained by manipulating this equation in a similar manner as for the homogeneous case to obtain

$$\mathbf{x}(NT) = \boldsymbol{\phi}(NT)\mathbf{x}(0) + \sum_{k=0}^{N-1} \boldsymbol{\phi}[(n-k-1)T]\boldsymbol{\theta}(T)\mathbf{m}(kT) \tag{15-45}$$

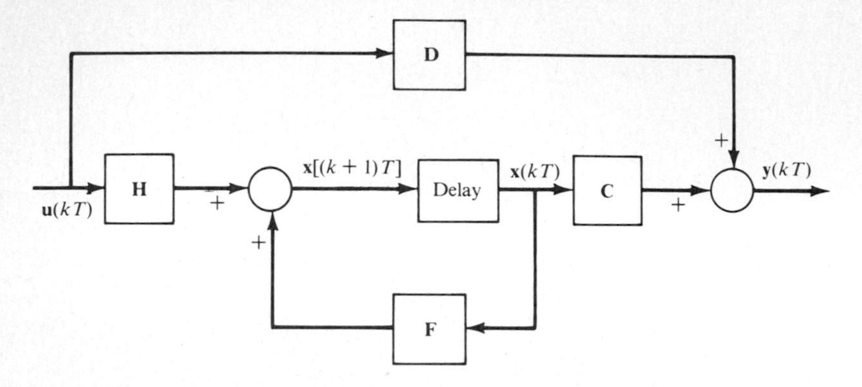

Figure 15-5 Block diagram representation of Eqs. (15-41) and (15-42).

where, from Eq. (15-14), $\boldsymbol{\phi}(NT) = \boldsymbol{\phi}^N(T)$. Thus Eq. (15-38) may be readily solved for $\mathbf{x}(NT)$ by the simple recursive procedure represented by Eq. (15-45).

Example 15-2 The technique of this section is now used to obtain a numerical solution for a specified forcing function. Consider the SISO sampled-data open-loop system shown in Fig. 15-6, where the LTI system is the one given in Example 15-1. For a time-varying input function $u(t)$, the output $m(t)$ of the ZOH unit is a staircase function as shown in Fig. 3-21 [note that $u(t)$ corresponds to $e(t)$ in this figure]. Thus $m(t)$ is described by Eq. (15-32). Therefore, $\mathbf{x}(t)$ and $y(t)$ for this system are given by Eqs. (15-30) and (15-31), respectively. The output of the fictitious sampler yields values of $y(t)$ at the sampling instants kT; that is,

$$y^*(t) = \sum_{k=0}^{\infty} [y(t)]_{t=kT}\, \delta(t - kT) = \sum_{k=0}^{\infty} y(kT)\, \delta(t - kT) \tag{15-46}$$

The values of $y(kT) = x_1(kT)$ are obtained from Eq. (15-38), where $\boldsymbol{\phi}(T)$ and $\boldsymbol{\theta}(T)$ are obtained from Eqs. (15-23) and (15-24), respectively, by substituting $t = (k + 1)T$ and $t_0 = kT$, or from Eq. (15-20) by replacing t by T. Thus

$$\boldsymbol{\phi}(T) = \begin{bmatrix} 1 & 0.5(1 - \epsilon^{-2T}) \\ 0 & \epsilon^{-2T} \end{bmatrix} \tag{15-47}$$

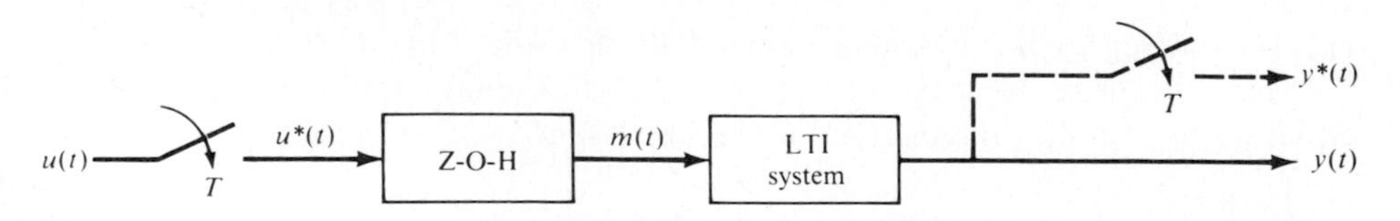

Figure 15-6 A SISO open-loop sampled-data system.

$$\boldsymbol{\theta}(T) = \begin{bmatrix} \mathrm{T} - 0.5 + 0.5\epsilon^{-2T} \\ 1 - \epsilon^{-2T} \end{bmatrix} \tag{15-48}$$

For $T = 0.1$ s and for the values of $m(kT)$ and $\mathbf{x}(0)$ given in part d of Example 15-1, the values of $y(kT) = x_1(kT)$ for $k = 1$ and 2 (that is, $t = kT = 0.1, 0.2$) are, respectively, from Table 15-1, 0.009365 and 0.044524. These values may also be obtained from Eq. (15-38). Thus for $T = 0.1$ s, Eqs. (15-47) and (15-48) become

$$\boldsymbol{\phi}(T) = \boldsymbol{\phi}(0.1) = \begin{bmatrix} 1 & 0.090635 \\ 0 & 0.818731 \end{bmatrix} \tag{15-49}$$

$$\boldsymbol{\theta}(T) = \boldsymbol{\theta}(0.1) = \begin{bmatrix} 0.009365 \\ 0.181269 \end{bmatrix} \tag{15-50}$$

From Eq. (15-38),

$$\mathbf{x}[(k+1)T] = \begin{bmatrix} 1 & 0.090635 \\ 0 & 0.818731 \end{bmatrix} \mathbf{x}(kT) + \begin{bmatrix} 0.009365 \\ 0.181269 \end{bmatrix} m(kT) \tag{15-51}$$

For $k = 0$, from Eq. (15-51),

$$\mathbf{x}(0.1) = \begin{bmatrix} 0.009365 \\ 0.181269 \end{bmatrix}$$

For $k = 1$, from Eq. (15-51),

$$\mathbf{x}(0.2) = \begin{bmatrix} 1 & 0.090635 \\ 0 & 0.818731 \end{bmatrix} \begin{bmatrix} 0.009365 \\ 0.181269 \end{bmatrix} + \begin{bmatrix} 0.009365 \\ 0.181269 \end{bmatrix} (2)$$

$$= \begin{bmatrix} 0.044524 \\ 0.529511 \end{bmatrix}$$

Thus, the values for $y(0.1)$ and $y(0.2)$ obtained by using Eqs. (15-38) and (15-42) are identical to those obtained by the method of Example 15-1 within the accuracy of the selected calculator.

15-4 STATE-VARIABLE REPRESENTATION IN THE TIME DOMAIN OF A DISCRETE-TIME SISO SYSTEM[18]

Consider the z-transform function of an all-discrete-data system given by the transfer function

$$G(z) = \frac{Y(z)}{U(z)} = K_D \frac{z^w + f_{w-1}z^{w-1} + \cdots + f_1 z + f_0}{z^n + d_{n-1}z^{n-1} + \cdots + d_1 z + d_0} \tag{15-52}$$

where $w \le n$, the coefficients of the z^w and z^n terms are unity, and zero initial conditions exist. If the coefficient of z^n is not unity, then the denominator polynomial should be divided by this coefficient. This enables the solution of the associated discrete-time equation on the computer to be solved without a division operation. In real-time control applications the solution of Eq. (15-52) should be relatively fast without division operations. This *n*th-order equation may be manipulated in a number of different ways to obtain an SV representation of the discrete-time system in terms of first-order difference equations. A variety of methods exist for defining the state variables of a discrete-time system. The specific method employed depends upon the analysis and design techniques that are used. For example, it may be necessary to identify as many physical variables of the system as possible with corresponding state variables. Other representation techniques are employed to facilitate algebraic manipulations and analysis, but the three most common SV representations are presented in this text; they are the phase-variable, canonical-variable, and physical-variable representations. Equation (15-2) may represent a discrete-time system or the discretization of a hybrid system.

As is the case for the continuous-time SV representation, a *state transition* matrix can be defined for discrete-time systems in a similar manner.

15-4A Phase-Variable Method

In the phase-variable technique all the state variables differ from each other only by a discrete-time delay or advance of T s. To generate the phase-variable representation, Eq. (15-52), where $w = n - 1$, is manipulated as follows. Let

$$G(z) = \frac{Y(z)}{X(z)} \cdot \frac{X(z)}{U(z)}$$

$$\frac{X(z)}{U(z)} = \frac{K_D}{z^n + d_{n-1}z^{n-1} + \cdots + d_1 z + d_0} \tag{15-53}$$

and

$$\frac{Y(z)}{X(z)} = z^w + f_{w-1}z^{w-1} + \cdots + f_1 z + f_0 \tag{15-54}$$

Then, from Eqs. (15-53) and (15-54) obtain, respectively,

$$z^n X(z) + d_{n-1}z^{n-1}X(z) + \cdots + d_1 zX(z) + d_0 X(z) = K_D U(z) \tag{15-55}$$

and

$$Y(z) = f_0 X(z) + f_1 zX(z) + \cdots + f_{w-1}z^{w-1}X(z) + z^w X(z) \tag{15-56}$$

since

$$\mathscr{Z}^{-1}[zE(z)] = e[(k+1)T]$$

$$\mathscr{Z}^{-1}[z^2 E(z)] = e[(k+2)T]$$

$$\vdots$$

Then, the $\mathscr{Z}^{-1}$ of Eqs. (15-55) and (15-56) yield, respectively, the difference equations

$$x[(k+n)T] + d_{n-1}x[(k+n-1)T] + \cdots + d_1 x[(k+1)T] + d_0 x(kT) = K_D u(kT) \quad (15\text{-}57)$$

and

$$y(kT) = f_0 x(kT) + f_1 x[(k+1)T] + \cdots + f_{w-1}x[(k+w-1)T] + x[(k+w)T] \quad (15\text{-}58)$$

These manipulations reveal that the z-transfer function of a system represents a set of difference equations.

Defining the first phase state variable as

$$x_1(kT) \equiv x(kT) \qquad \text{where } X_1(z) = X(z)$$

then Eqs. (15-57) and (15-58) can be written, respectively, as

$$x_1[(k+n)T] = -d_0 x_1(kT) - d_1 x_1[(k+1)T] - d_2 x_1[(k+2)T] - \cdots - d_{n-1}x_1[(k+n-1)T] + K_D u(kT) \quad (15\text{-}59)$$

$$y(kT) = f_0 x_1(kT) + f_1 x_1[(k+1)T] + f_2 x_1[(k+2)T] + \cdots + f_{w-1}x_1[(k+w-1)T] + x_1[(k+w)T] \quad (15\text{-}60)$$

Thus the remaining phase variables are defined as follows:

$$\begin{aligned}
x_2(kT) &\equiv x_1[(k+1)T] & &\rightarrow & X_2(z) &= zX_1(z) = zX(z) \\
x_3(kT) &\equiv x_2[(k+1)T] = x_1[(k+2)T] & &\rightarrow & X_3(z) &= zX_2(z) \\
& & & & &= z^2X_1(z) = z^2X(z) \\
&\vdots & & & &\vdots \qquad (15\text{-}61) \\
x_n(kT) &\equiv x_{n-1}[(k+1)T] = x_1[(ek+n-1)T] & &\rightarrow & X_n(z) &= zX_{n-1}(z) \\
x_{n+1}(kT) &\equiv x_n[(k+1)T] = x_1[(k+n)T] & &\rightarrow & X_{n+1}(z) &= zX_n(z)
\end{aligned}$$

Making the substitution of Eq. (15-61) into Eqs. (15-59) and (15-60) yields, respectively, the following for $w = n - 1$:

$$x_n[(k+1)T] = -d_0 x_1(kT) - d_1 x_2(kT) - d_2 x_3(kT) - \cdots - d_{n-1}x_n(kT) + K_D u(kT) \quad (15\text{-}62)$$

and

$$y(kT) = f_0 x_1(kT) + f_1 x_2(kT) + f_2 x_3(kT) + \cdots + f_{w-1}x_{n-1}(kT) + x_n(kT) \quad (15\text{-}63)$$

Equations (15-61) to (15-63) yield the following discrete state and system output equations:

$$\mathbf{x}[(k+1)T] = \mathbf{F}\mathbf{x}(kT) + \mathbf{h}u(kT) \tag{15-64}$$

$$y(kT) = \mathbf{c}'\mathbf{x}(kT) \tag{15-65}$$

where $\mathbf{x}$ is $n \times 1$, $\mathbf{F}$ is $n \times n$, $\mathbf{h}$ is $n \times 1$, $\mathbf{c}'$ is $1 \times n$, and

$$\mathbf{x}[(k+1)T] \equiv \begin{bmatrix} x_1[(k+1)T] \\ x_2[(k+1)T] \\ \vdots \\ x_n[(k+1)T] \end{bmatrix} \tag{15-66}$$

$$\mathbf{x}(kT) \equiv \begin{bmatrix} x_1(kT) \\ x_2(kT) \\ \vdots \\ x_n(kT) \end{bmatrix} \tag{15-67}$$

$$\mathbf{F} = \begin{bmatrix} 0 & 1 & 0 & \cdots & 0 \\ 0 & 0 & 1 & \cdots & 0 \\ \cdot & \cdot & \cdot & \cdot & \cdot \\ \cdots & \cdots & \cdots & \cdots & 1 \\ -d_0 & -d_1 & -d_2 & \cdots & -d_{n-1} \end{bmatrix} \tag{15-68}$$

$$\mathbf{h} \equiv \begin{bmatrix} 0 \\ 0 \\ \vdots \\ K_D \end{bmatrix} \tag{15-69}$$

and

$$\mathbf{c}' \equiv [f_0 \quad f_1 \quad \cdots \quad f_{w-1} \quad 1] \qquad \text{for } w = n - 1 \tag{15-70}$$

For the case $w < n - 1$, the output vector is given by

$$\mathbf{c}' \equiv [f_0 \quad f_1 \quad f_2 \quad \cdots \quad f_{w-1} \quad 1 \quad 0 \quad \cdots \quad 0] \tag{15-71}$$

Note that Eq. (15-64) represents a system of n first-order difference equations and that the matrix $\mathbf{F}$ is defined as being in *companion form*. It is referred to as being the companion matrix since the elements of the nth row are the identical coefficients of the associated nth-order denominator of Eq. (15-52). Equations (15-64) and (15-65) yield the SV diagram of Fig. 15-7 for the case $w = n - 1$. Some of the advantages of this representation are that Eqs. (15-68) to (15-70) can be readily obtained from Eq. (15-52) and the zero elements appearing in $\mathbf{F}$ facilitate easier matrix manipulations.

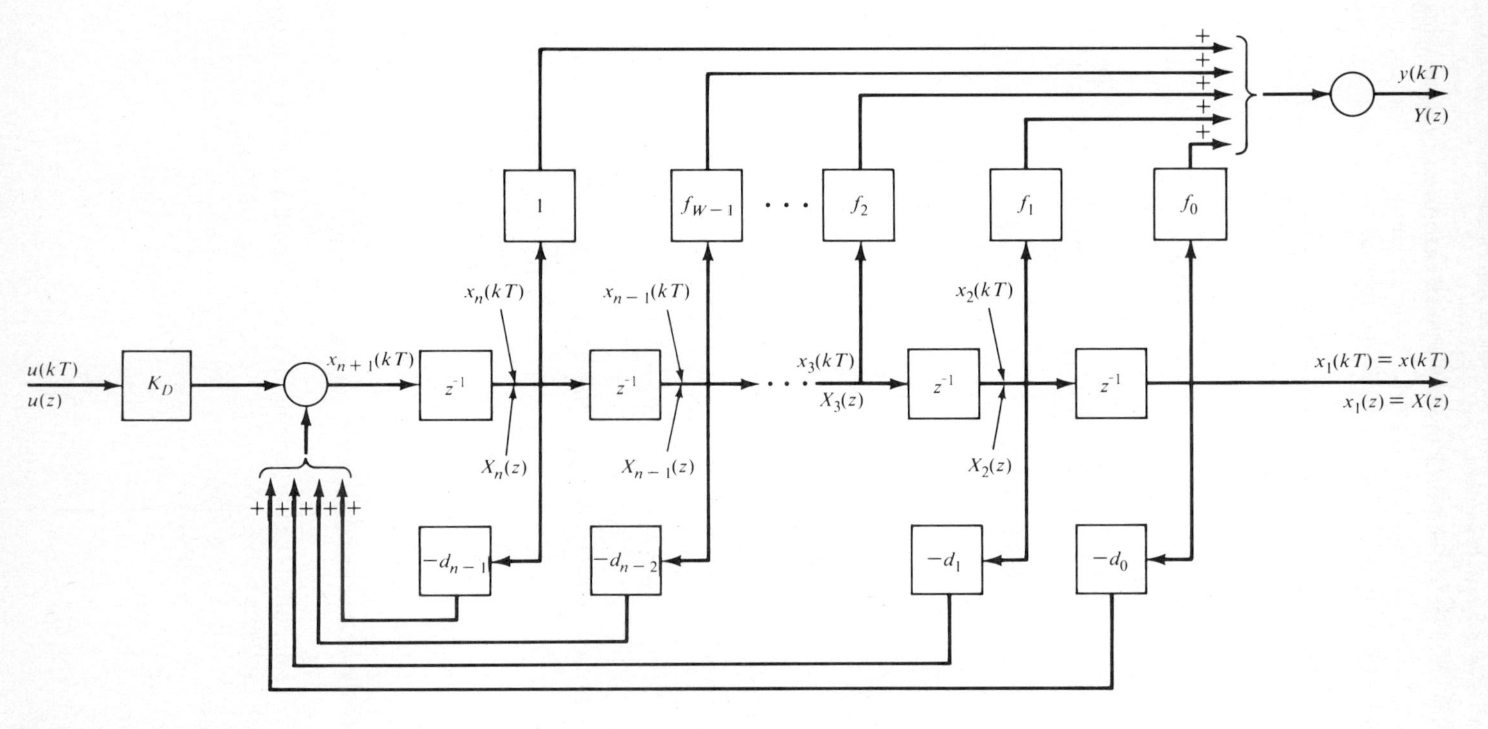

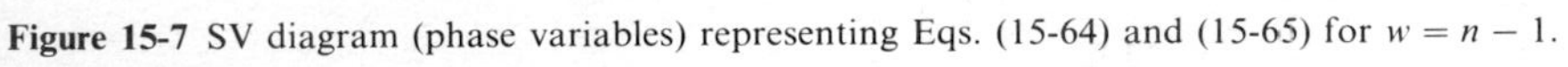

Figure 15-7 SV diagram (phase variables) representing Eqs. (15-64) and (15-65) for $w = n - 1$.

If $n = w$, then $G(z)$ must be put into *proper form* before proceeding with the determination of the output equation. That is, divide the numerator polynomial of $G(z)$ by its denominator polynomial to obtain the format

$$G(z) = \frac{Y(z)}{U(z)} = K_D\left[1 + \frac{f'_{n-1}z^{n-1} + f'_{n-2}z^{n-2} + \cdots + f'_0}{z^n + d_{n-1}z^{n-1} + \cdots + d_1 z + d_0}\right]$$

$$= K_D[1 + G'(z)] = K_D + K_D G'(z) \tag{15-72}$$

The $K_D G'(z)$ portion of Eq. (15-72) is rearranged to the format of Eq. (15-52) by factoring out f'_{n-1} from $G'(z)$; that is,

$$K_D G'(z) = K_D f'_{n-1}\left[\frac{z^{n-1} + (f'_{n-2}/f'_{n-1})z^{n-2} + \cdots + (f'_0/f'_{n-1})}{z^n + d_{n-1}z^{n-1} + \cdots + d_1 z + d_0}\right] \tag{15-73}$$

Note that the bracketed term of Eq. (15-73) is of the format of Eq. (15-52), where $(f'_{n-2}/f'_{n-1}) \approx f_{w-1}$ and $(f'_0/f'_{n-1}) \approx f_0$, etc. Thus Fig. 15-7 can represent the case of $n = w$, Eq. (15-72), by adding a feedforward path from $u(kT)$ to $y(kT)$ with a feedforward transfer function equal to K_D and replacing K_D by $K_D f'_{n-1}$. The discrete state equation for this case is determined from Eq. (15-73), where now the output equation is

$$y(kT) = c'x(kT) + K_D u(kT) \tag{15-74}$$

Example 15-3 The forward-loop transfer function of the open-loop sampled-data system of Example 15-2 is

$$G(s) = G_{z_0}(s)G_x(s) = \frac{K_G(1 - \epsilon^{-sT})}{s^2(s + 2)}$$

Thus, by use of entry 8 in Table 4-1, the $\mathscr{Z}[G(s)]$ is

$$G(z) = \frac{K_G[(T + 0.5\epsilon^{-2T} - 0.5)z + (0.5 - 0.5\epsilon^{-2T} - T\epsilon^{-2T})]}{2(z - 1)(z - \epsilon^{-2T})} \tag{15-75}$$

For $T = 0.1$ s and $K_G = 2$, Eq. (15-75) becomes

$$G(z) = \frac{0.009365z + 0.008762}{(z - 1)(z - 0.818731)} = \frac{0.009365(z + 0.935611)}{z^2 - 1.818731z + 0.818731} \tag{15-76}$$

Utilizing Eqs. (15-68) to (15-70), the state and output equations (15-64) and (15-65) are obtained for the system represented by Eq. (15-76) in terms of phase variables. Thus

$$\mathbf{F} = \begin{bmatrix} 0 & 1 \\ -0.81873 & 1.81873 \end{bmatrix} \tag{15-77}$$

$$\mathbf{h} = \begin{bmatrix} 0 \\ 0.009365 \end{bmatrix} \tag{15-78}$$

and

$$\mathbf{c}' = [0.935611 \quad 1] \tag{15-79}$$

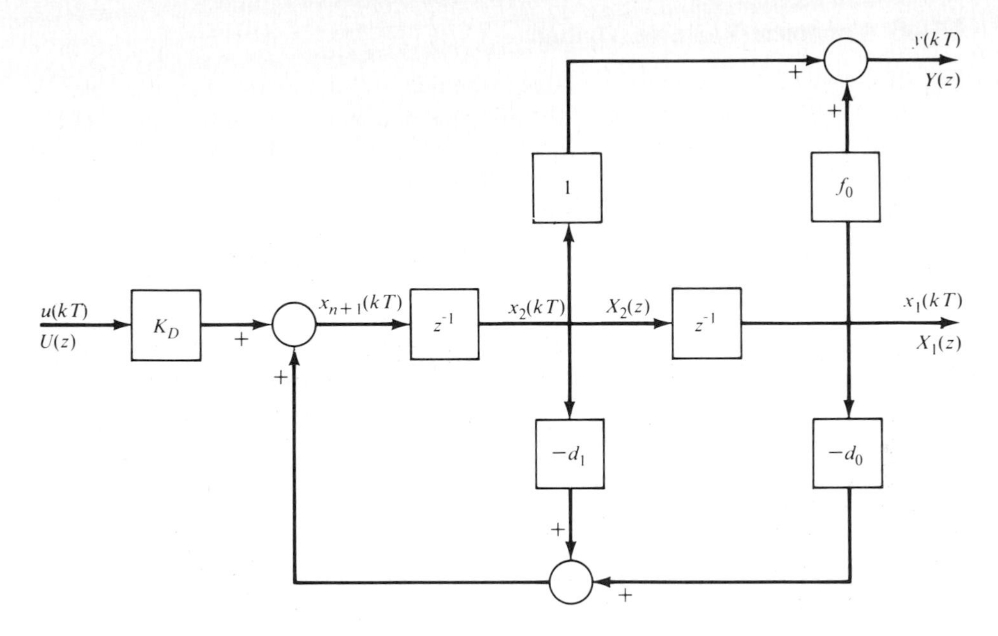

Figure 15-8 The phase SV diagram for Eq. (15-76).

The SV diagram representing Eq. (15-76), where $n = 2$ and $w = 1$, is shown in Fig. 15-8.

For $\mathbf{x}(0) = \mathbf{0}$, $u(0) = 1$, and $u(0.1) = 2$, $y(0.1)$ and $y(0.2)$ are determined by using Eqs. (15-64), (15-65), and (15-77) to (15-79). Thus, *for* $k = 0$, Eqs. (15-64) and (15-65) yield

$$\mathbf{x}(0.1) = \begin{bmatrix} 0 \\ 0.009365 \end{bmatrix}$$

$$y(0.1) = [0.935611 \quad 1]\begin{bmatrix} 0 \\ 0.009365 \end{bmatrix} = 0.009365$$

and *for* $k = 1$,

$$\mathbf{x}(0.2) = \begin{bmatrix} 0 & 1 \\ -0.81873 & 1.81873 \end{bmatrix}\begin{bmatrix} 0 \\ 0.009365 \end{bmatrix} + \begin{bmatrix} 0 \\ 0.009365 \end{bmatrix}(2) = \begin{bmatrix} 0.009365 \\ 0.035762 \end{bmatrix}$$

$$y(0.2) = [0.935611 \quad 1]\begin{bmatrix} 0.009365 \\ 0.035762 \end{bmatrix} = 0.0445244$$

These values of $y(0.1)$ and $y(0.2)$ are the same as those obtained in Examples 15-1 and 15-2.

15-4B Canonical-Variable Method[1]

The objective of the canonical-variable technique is to define the state variables in such a manner that the elements of the principal diagonal of **F** contain the system's *eigenvalues* (roots of the system's characteristic equation).† The canonical-variable method requires that the denominator of Eq. (15-52) be in factored form in order that the partial-fraction expansion technique can be used. Assuming that $w = n$, then $G(z)$ must be put into the *proper form* before proceeding with the expansion technique; thus

$$G(z) = \frac{Y(z)}{U(z)} = K_D\left[1 + \frac{f'_{n-1}z^{n-1} + f'_{n-2}z^{n-2} + \cdots + f'_0}{(z-p_1)(z-p_2)\cdots(z-p_n)}\right]$$

$$= K_D[1 + G'(z)] = K_D + K_D G'(z) \tag{15-80}$$

Case 1: $G(z)$ has distinct poles For this case the partial-fraction expansion of Eq. (15-80) is

$$\frac{Y(z)}{U(z)} = K_D + \frac{A_1}{z-p_1} + \cdots + \frac{A_n}{z-p_n} \tag{15-81}$$

where the coefficients $A_1, A_2, \ldots, A_n$ are determined in the manner described in Chap. 4. This equation is rearranged to

$$Y(z) = \left(K_D + \sum_{i=1}^{n} \frac{A_i}{z-p_i}\right)U(z) \tag{15-82}$$

The SV diagram representing Eq. (15-81) is shown in Fig. 15-9, where the outputs of each of the $1/(z-p_i)$ blocks are identified as an SV in the manner shown. Note that each first-order block is part of an overall parallel structure. Thus

$$X_i(z) \equiv \frac{1}{z-p_i}U(z) \qquad \text{for } i = 1, 2, \ldots, n \tag{15-83}$$

which after manipulation yields

$$zX_i(z) - p_iX(z) = U(z)$$

and

$$x_i[(k+1)T] = p_ix_i(kT) + u(kT) \tag{15-84}$$

Therefore, from Eq. (15-84) and Fig. 15-9, the discrete state and system output equations are obtained as follows:

$$\mathbf{x}[(k+1)T] = \begin{bmatrix} p_1 & 0 & \cdots & 0 & 0 \\ 0 & p_2 & \cdots & 0 & 0 \\ & & \ddots & & \\ 0 & 0 & \cdots & 0 & p_n \end{bmatrix}\mathbf{x}(kT) + \begin{bmatrix} 1 \\ 1 \\ \vdots \\ 1 \end{bmatrix}\mathbf{u}(kT) \tag{15-85}$$

† This is true for a completely controllable and observable system.[1]

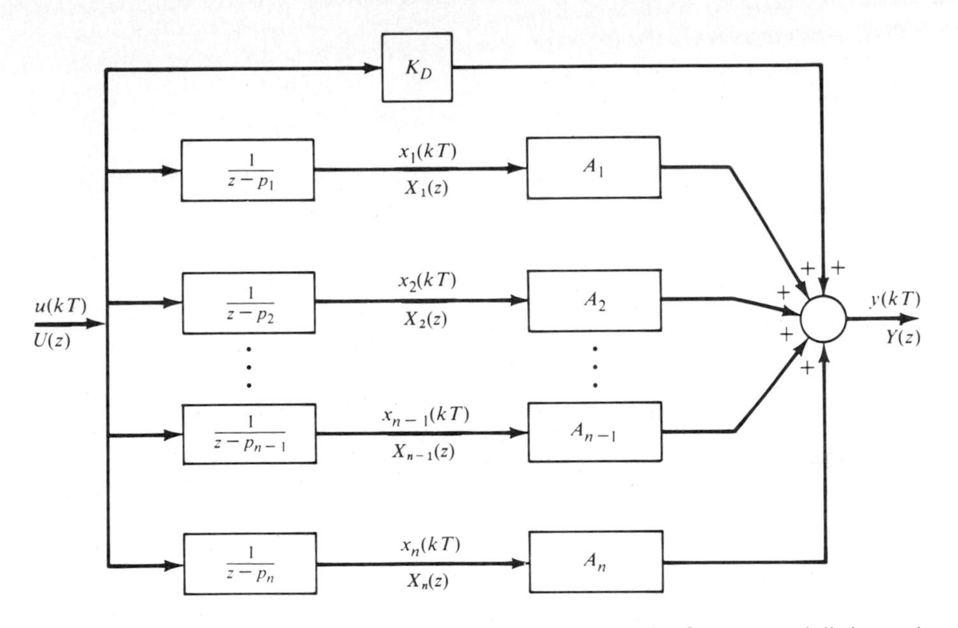

Figure 15-9 SV diagram (canonical variable) representing Eq. (15-81) for $n = w$ and distinct poles.

$$\mathbf{y}(kT) = [A_1 \quad A_2 \quad \cdots \quad A_n]\mathbf{x}(kT) + K_D\mathbf{u}(kT) \tag{15-86}$$

where $\mathbf{x}[(k+1)T]$ and $\mathbf{x}(kT)$ are defined by Eqs. (15-66) and (15-67), respectively, and where the matrix $\mathbf{F}$ and the vectors $\mathbf{h}$ and $\mathbf{c}'$ are given by

$$\mathbf{F} = \begin{bmatrix} p_1 & 0 & \cdots & 0 \\ 0 & p_2 & \cdots & 0 \\ & & \ddots & \\ 0 & & & p_n \end{bmatrix} \tag{15-87}$$

$$\mathbf{h} = \begin{bmatrix} 1 \\ 1 \\ \vdots \\ 1 \end{bmatrix} \tag{15-88}$$

and

$$\mathbf{c}' = [A_1 \quad A_2 \quad \cdots \quad A_n] \tag{15-89}$$

Note that the matrix $\mathbf{F}$ is now in Jordan canonical (or diagonal) form.

Case 2: $G(z)$ has multiple poles For this case it is assumed that $w < n$, that the pole $z = -p_1$ of $G(z)$ has multiplicity q, and that all other poles of $G(z)$ are distinct; that is,

$$G(z) = G'(z) = \frac{f'_w z^w + f'_{w-1} z^{w-1} + \cdots + f'_0}{(z - p_1)^q (z - p_2) \cdots (z - p_n)} \tag{15-90}$$

In the event $n = w$, first put $G(z)$ into its proper form. For either situation, expanding $K_D G'(z)$ by the partial-fraction expansion yields

$$K_D G'(z) = \frac{A_{1q}}{(z - p_1)^q} + \frac{A_{1(q-1)}}{(z - p_1)^{q-1}} + \cdots + \frac{A_{12}}{(z - p_1)^2} + \frac{A_{11}}{z - p_1} + \frac{A_2}{z - p_2} + \frac{A_3}{z - p_3} + \frac{A_{n-q+1}}{z - p_{n-q+1}} \tag{15-91}$$

where the coefficients are obtained by the rules of this expansion technique. Equation (15-91), assuming $n > w$, is rearranged to

$$Y(z) = \frac{U(z)}{z - p_1} A_{11} + \frac{U(z)}{(z - p_1)^2} A_{12} + \cdots + \frac{U(z)}{(z - p_1)^q} A_{1q} + \frac{U(z)}{z - p_2} A_2 + \frac{U(z)}{z - p_3} A_3 + \cdots + \frac{U(z)}{z - p_{n-q+1}} A_{n-q+1} \tag{15-92}$$

From Eq. (15-92) the state variables $X_1(z), X_2(z), \ldots, X_q(z)$ are defined as follows:

$$X_1(z) = \frac{U(z)}{z - p_1}, \quad X_2(z) = \frac{U(z)}{(z - p_1)^2} = \frac{U(z)}{z - p_1} \frac{1}{z - p_1} = \frac{X_1(z)}{z - p_1}, \quad \ldots,$$

$$X_q(z) = \frac{U(z)}{(z - p_1)^q} = \frac{U(z)}{(z - p_1)^{q-1}} \frac{1}{z - p_1} = \frac{X_{q-1}(z)}{z - p_1}$$

which yield the following transfer functions:

$$\frac{X_1(z)}{U(z)} = \frac{1}{z - p_1}, \quad \frac{X_2(z)}{X_1(z)} = \frac{1}{z - p_1}, \quad \ldots, \quad \frac{X_q(z)}{X_{q-1}(z)} = \frac{1}{z - p_1} \tag{15-93}$$

Also, from Eq. (15-92) the remaining $n - q$ state variables are defined as follows:

$$X_{q+1}(z) = \frac{U(z)}{z - p_2}, \quad X_{q+2}(z) = \frac{U(z)}{z - p_3}, \quad \ldots, \quad X_n(z) = \frac{U(z)}{z - p_{n-q+1}} \tag{15-94}$$

which yield the following transfer functions:

$$\frac{X_{q+1}(z)}{U(z)} = \frac{1}{z - p_2}, \quad \frac{X_{q+2}(z)}{U(z)} = \frac{1}{z - p_3}, \quad \ldots, \quad \frac{X_n(z)}{U(z)} = \frac{1}{z - p_{n-q+1}}$$

Utilizing Eqs. (15-92) to (15-94) the SV diagram representing Eq. (15-91) is obtained as shown in Fig. 15-10. Equations (15-93) and (15-94) are

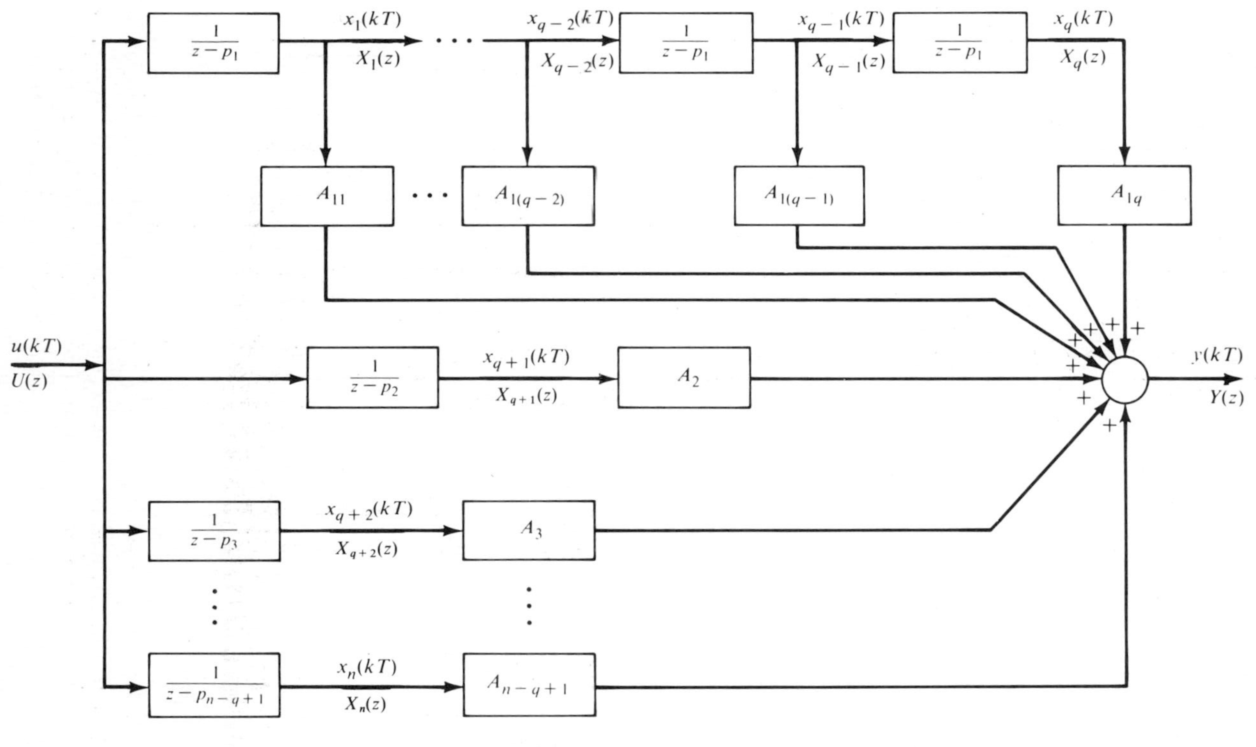

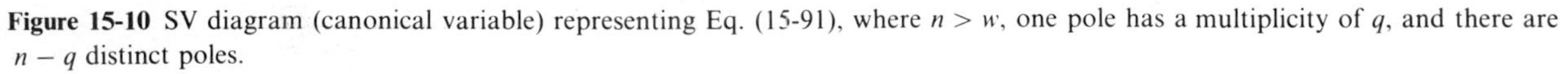

Figure 15-10 SV diagram (canonical variable) representing Eq. (15-91), where $n > w$, one pole has a multiplicity of q, and there are $n - q$ distinct poles.

manipulated in a similar manner as for the distinct-pole case to obtain the following discrete state and system output equations:

$$\underbrace{\begin{bmatrix} x_1[(k+1)T] \\ x_2[(k+1)T] \\ \vdots \\ x_{q-1}[(k+1)T] \\ x_q[(k+1)T] \\ x_{q+1}[(k+1)T] \\ \vdots \\ x_n[(k+1)T] \end{bmatrix}}_{\mathbf{x}[(k+1)T]} = \underbrace{\left[\begin{array}{cccccc|cccc} p_1 & 0 & 0 & \cdots & 0 & 0 & \cdots & \cdots & \cdots & 0 \\ 1 & p_1 & 0 & & 0 & 0 & \cdots & \cdots & \cdots & 0 \\ 0 & 1 & p_1 & & 0 & 0 & \cdots & \cdots & \cdots & 0 \\ & & & \ddots & & & & & & \\ 0 & 0 & 0 & \cdots & & p_1 & 0 & \cdots & \cdots & 0 \\ \hline 0 & 0 & 0 & \cdots & & 0 & p_2 & 0 & \cdots & 0 \\ 0 & & & & & 0 & 0 & p_3 & \cdots & 0 \\ \vdots & & & & & \vdots & \mathbf{0} & & \ddots & \mathbf{0} \\ 0 & \cdots & \cdots & \cdots & \cdots & 0 & \cdots & \cdots & \cdots & p_{n-q} \end{array}\right]}_{\mathbf{F}}$$

$$\times \underbrace{\begin{bmatrix} x_1(kT) \\ x_2(kT) \\ \vdots \\ x_{q-1}(kT) \\ x_q(kT) \\ x_{q+1}(kT) \\ \vdots \\ x_n(kT) \end{bmatrix}}_{\mathbf{x}(kT)} + \underbrace{\begin{bmatrix} 1 \\ 0 \\ \vdots \\ 0 \\ 0 \\ 1 \\ \vdots \\ 1 \end{bmatrix}}_{\mathbf{h}} u(kT) \qquad (15\text{-}95)$$

$$y(kT) = \underbrace{[A_{11} \quad \cdots \quad A_{1(q-2)} \quad A_{1(q-1)} \quad A_{1q} A_2 \quad \cdots \quad A_{n-q+1}]}_{\mathbf{c}'} \mathbf{x}(kT) \qquad (15\text{-}96)$$

The matrix **F** and the vectors $\mathbf{x}[(k+1)T]$, $\mathbf{x}(kT)$, **h**, and $\mathbf{c}'$ are denoted in Eqs. (15-95) and (15-96). Note that when $G(z)$ has multiple poles the system matrix **F** is also in the *Jordan* canonical form. The advantages of the canonical form of **F** is the ease of mathematical manipulation due to the diagonal form. Also, the modes (Ref. 1) $\epsilon^{p_1 kT}$ of the system are readily available.

Example 15-4 Example 15-3 is now redone by using canonical variables. Since Eq. (15-76) has distinct poles, then the Case 1 method is used. The partial-fraction expansion is performed on Eq. (15-76) since $n > w$ [that is, $G(z) = K_D G'(z)$], which yields

$$G(z) = \frac{0.1}{z-1} + \frac{0.0906349}{z - 0.818731} \tag{15-97}$$

Utilizing Eqs. (15-85) to (15-89) yields

$$\mathbf{x}[(k+1)T] = \begin{bmatrix} 1 & 0 \\ 0 & 0.818731 \end{bmatrix} \mathbf{x}(kT) + \begin{bmatrix} 1 \\ 1 \end{bmatrix} u(kT) \tag{15-98}$$

$$y(kT) = [0.1 \quad -0.090635]\mathbf{x}(kT) \tag{15-99}$$

These equations are now used, as follows, to determine $y(0.1)$ and $y(0.2)$. Thus, *for* $k = 0$,

$$\mathbf{x}(0.1) = \begin{bmatrix} 1 \\ 1 \end{bmatrix}$$

$$y(0.1) = [0.1 \quad -0.090635] \begin{bmatrix} 1 \\ 1 \end{bmatrix} = 0.009365$$

and *for* $k = 1$,

$$\mathbf{x}(0.2) = \begin{bmatrix} 1 & 0 \\ 0 & 0.81873 \end{bmatrix} \begin{bmatrix} 1 \\ 1 \end{bmatrix} + \begin{bmatrix} 1 \\ 1 \end{bmatrix} (2) = \begin{bmatrix} 3 \\ 2.81873 \end{bmatrix}$$

$$y(0.2) = [0.1 \quad -0.090635] \begin{bmatrix} 3 \\ 2.81873 \end{bmatrix} = 0.044525$$

These values agree, as expected, with those obtained previously. The canonical SV diagram for this open-loop system is shown in Fig. 15-11.

It should be noted that if canonical state variables are defined in the s domain for $G_x(s)$, then they carry over into the z domain when a ZOH device is used in conjunction with $G_x(s)$. Thus, for distinct eigenvalues λ_i, the normal-form s-

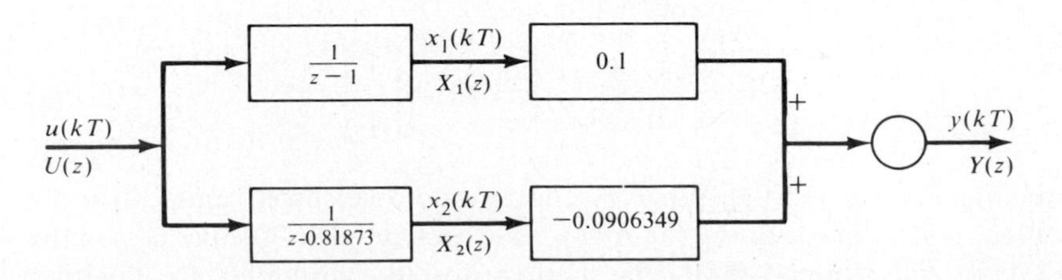

Figure 15-11 The canonical SV diagram for Eq. (15-97).

domain system matrix and state equation are, respectively, given by

$$\mathbf{A}_s = \begin{bmatrix} \lambda_1 & & & \mathbf{0} \\ & \lambda_2 & & \\ & & \ddots & \\ \mathbf{0} & & & \lambda_n \end{bmatrix} \tag{15-100}$$

$$\dot{\mathbf{x}}(t) = \mathbf{A}_s\mathbf{x}(t) + \mathbf{b}_s\mathbf{u}(t) \tag{15-101}$$

The corresponding normal-form z-domain system matrix and state equation are given by

$$\mathbf{F}_z = \begin{bmatrix} \epsilon^{\lambda_1 T} & & & \mathbf{0} \\ & \epsilon^{\lambda_2 T} & & \\ & & \ddots & \\ \mathbf{0} & & & \epsilon^{\lambda_n T} \end{bmatrix} = \begin{bmatrix} p_1 & & & \mathbf{0} \\ & p_2 & & \\ & & \ddots & \\ \mathbf{0} & & & p_n \end{bmatrix} \tag{15-102}$$

and

$$\mathbf{x}[(k+1)T] = \mathbf{F}_z\mathbf{x}(kT) + \mathbf{h}_z\mathbf{u}(kT) \tag{15-103}$$

where $\mathbf{h}_z = \mathbf{b}_s$. To illustrate this characteristic of canonical state-variable representation, consider the partial-fraction expansion of the plant transfer function of Example 15-1,

$$G_x(s) = \frac{1}{s(s+2)} = \frac{1}{2}\left(\frac{1}{s} - \frac{1}{s+2}\right)$$

which yields

$$\dot{\mathbf{x}}(t) = \mathbf{A}_s\mathbf{x}(t) - \mathbf{b}_s\mathbf{u}(t) = \begin{bmatrix} 0 & 0 \\ 0 & -2 \end{bmatrix}\mathbf{x}(t) + \begin{bmatrix} 1 \\ 1 \end{bmatrix}u(t)$$

and $\qquad y(t) = [0.5 \quad -0.5]\mathbf{x}(t)$

Using Eqs. (15-102) and (15-103), where $\lambda_1 = 0$ and $\lambda_2 = -2$, yields

$$\begin{aligned} \mathbf{x}[(k+1)T] &= \mathbf{F}_z\mathbf{x}(kT) + \mathbf{h}_z\mathbf{u}(kT) \\ &= \begin{bmatrix} 1 & 0 \\ 0 & 0.81873 \end{bmatrix}\mathbf{x}(kT) + \begin{bmatrix} 1 \\ 1 \end{bmatrix}u(kT) \end{aligned} \tag{15-104}$$

which is identical to Eq. (15-98). When $\mathbf{F}_z$ contains complex eigenvalues, then the method of sec. 5-11 (Transforming the **A** Matrix with Complex Eigenvalues) in the text by D'Azzo and Houpis[1] should be used to obtain a discrete state equation containing real numbers.

15-4C Physical-Variable Method

The objective of the physical-variable technique is to identify as many of the physical variables of the discrete system as state variables. The mathematical manipulations of the physical-variable method used for continuous-data system can also be utilized to obtain an SV representation for a discrete-time system. This method requires that both the numerator and denominator of $G'(z)$ be in factored form, i.e., for $n = w$,

$$G(z) = \frac{Y(z)}{U(z)} = K_D[1 + G'(z)]$$

$$= K_D\left[1 + \frac{(z - z_1)(z - z_2)\cdots(z - z_{w-1})}{(z - p_1)(z - p_2)\cdots(z - p_n)}\right] \tag{15-105}$$

For $n > w$ and where the poles and zeros are all real, the transfer function $G(z)$ may be represented by a block diagram, Fig. 15-12. This diagram model contains n blocks in *cascade* where:

1. The transfer function of each block, due to the factoring, is one of the following forms: either

$$A_i \frac{z - z_i}{z - p_i} \tag{15-106}$$

or

$$\frac{A_j}{z - p_j} \tag{15-107}$$

2. The input signal $U(z)$ feeds into a block having a transfer function of the form given by Eq. (15-107).
3. The pairing of the w zeros with w poles in order to obtain transfer functions of the form of Eq. (15-106) may be immaterial. It is immaterial if a given z_i is not physically associated in the system with a given p_j.
4. $K_D = A_1 A_2 \cdots A_n$

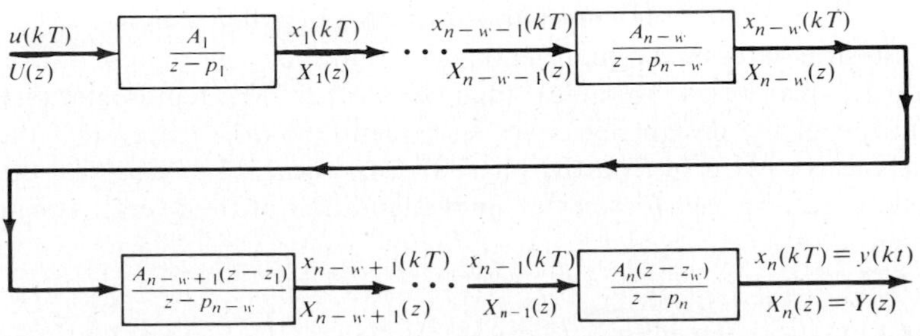

Figure 15-12 SV diagram representing Eq. (15-105), where $n > w$.

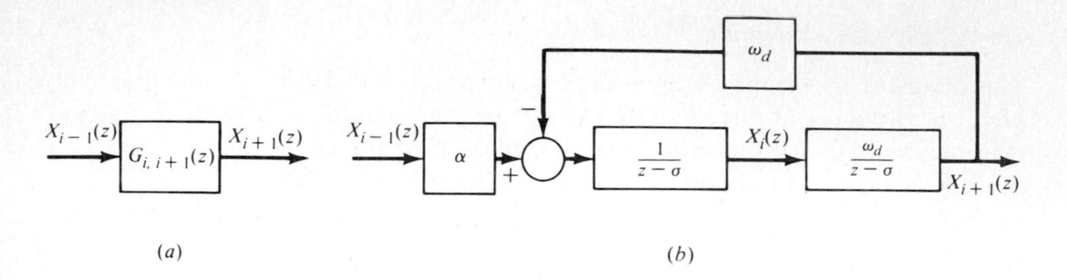

Figure 15-13 Complex-pole plant representation.

The state variables are identified in the manner shown in Fig. 15-12. From this diagram, n equations may be written that relate the output and the input for each of the n blocks; that is,

$$X_i(z) = A_i \frac{z - z_j}{z - p_i} X_{i-1}(z)$$

or (15-108)

$$X_j(z) = \frac{A_j}{z - p_j} X_{j-1}(z)$$

These equations may then be manipulated, in the same manner as used for canonical variables, to obtain the discrete state equation for the physical-variable representations. The system output equation is

$$y(kT) = [0 \quad 0 \quad \cdots \quad 0 \quad 1]\mathbf{x}(kT) = \mathbf{c}'\mathbf{x}(kT) \tag{15-109}$$

For Fig. 15-12, if $A_i/(z - p_i)$ and $A_{i+1}/(z - p_{i+1})$ correspond to a pair of complex-conjugate poles of $G'(z)$, recombine these terms to form

$$G_{i,i+1}(z) = \frac{X_{i+1}(z)}{X_{i-1}(z)} = \frac{A_i A_{i+1}}{z^2 - 2\sigma z + \omega_n^2}$$

This equation is represented by the block diagram of Fig. 15-13*a*. Figure 15-13*b* is the equivalent block diagram representation of Fig. 15-13*a*, where $\alpha\omega_d = A_i A_{i+1}$.

Since the physical-variable SV representation is more closely representative, in general, of the real-world variables than the other SV representations, the implementation of the design objectives is straightforward. However, if the analysis and design is carried out via the phase or canonical state variables, the design parameters must be transformed for implementation of the overall design objectives.

Example 15-5 Example 15-3 is repeated by using physical variables. Utilizing Eq. (15-76) the physical SV diagram for this system is shown in Fig. 15-14.

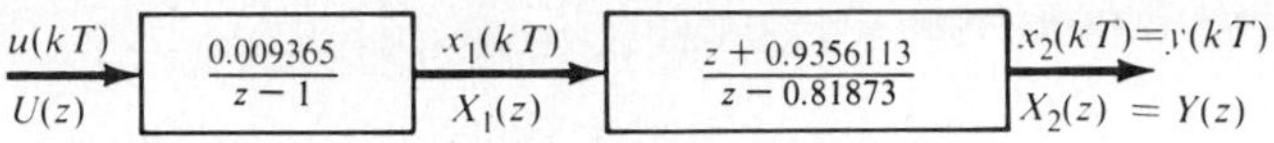

$$u(kT),\ U(z) \rightarrow \boxed{\frac{0.009365}{z-1}} \xrightarrow{x_1(kT),\ X_1(z)} \boxed{\frac{z+0.9356113}{z-0.81873}} \rightarrow x_2(kT)=y(kT),\ X_2(z) = Y(z)$$

Figure 15-14 The physical SV diagram for Eq. (15-76).

From this diagram the following equations are obtained:

$$\frac{X_1(z)}{U(z)} = \frac{0.009365}{z-1} \rightarrow zX_1(z) - X_1(z) = 0.009365U(z)$$

$$\rightarrow x_1[(k+1)T] = x_1(kT) + 0.009365u(kT) \qquad (15\text{-}110)$$

$$\frac{X_2(z)}{X_1(z)} = \frac{z+0.9356113}{z-0.818731} \rightarrow$$

$$zX_2(z) - 0.818731X_2(z) = zX_1(z) + 0.9356113X_1(z)$$

$$= X_1(z) + 0.009365U(z) + 0.935611X_1(z)$$

$$\rightarrow x_2[(k+1)T] = 1.935611x_1(kT) + 0.818731x_2(kT)$$

$$+ 0.009365u(kT) \qquad (15\text{-}111)$$

Thus from Eqs. (15-110) and (15-111),

$$\mathbf{x}[(k+1)T] = \begin{bmatrix} 1 & 0 \\ 1.935611 & 0.818731 \end{bmatrix} x(kT) + \begin{bmatrix} 0.009365 \\ 0.009365 \end{bmatrix} u(kT) \qquad (15\text{-}112)$$

$$y(kT) = [0 \quad 1]x(kT) \qquad (15\text{-}113)$$

Equations (15-112) and (15-113) yield the following values of $y(0.1)$ and $y(0.2)$ *for* $k = 0$ *and* 1, *respectively*:

$$\mathbf{x}(0.1)\Big|_{k=0} = \begin{bmatrix} 0.009365 \\ 0.009365 \end{bmatrix}$$

$$y(0.1)\Big|_{k=0} = x_2(0.1) = 0.009365$$

$$\mathbf{x}(0.2)\Big|_{k=1} = \begin{bmatrix} 0 & 0 \\ 1.935611 & 0.818731 \end{bmatrix} \begin{bmatrix} 0.009365 \\ 0.009365 \end{bmatrix} + \begin{bmatrix} 0.009365 \\ 0.009365 \end{bmatrix}(2)$$

$$= \begin{bmatrix} 0.018730 \\ 0.044524 \end{bmatrix}$$

$$y(0.2)\Big|_{k=2} = 0.044533$$

15-4D State Transition Equation

For all three SV methods of representing a discrete-time system, the discrete state and system equations for SISO systems have the same mathematical format given by

$$\mathbf{x}[(k+1)T] = \mathbf{F}\mathbf{x}(kT) + \mathbf{h}u(kT) \tag{15-114}$$

$$y(kT) = \mathbf{c}'\mathbf{x}(kT) \tag{15-115}$$

For a MIMO system these equations, which are also the same for each method, become

$$\mathbf{x}[(k+1)T] = \mathbf{F}\mathbf{x}(kT) + \mathbf{H}\mathbf{u}(\mathrm{kT}) \tag{15-116}$$

$$\mathbf{y}(kT) = \mathbf{C}\mathbf{x}(kT) \tag{15-117}$$

The discrete state equation (15-116) may be solved as follows. For

$k = 0$: $\quad \mathbf{x}(T) = \mathbf{F}\mathbf{x}(0) + \mathbf{H}\mathbf{u}(0)$

$k = 1$:
$$\begin{aligned} \mathbf{x}(2T) &= \mathbf{F}\mathbf{x}(T) + \mathbf{H}\mathbf{u}(T) \\ &= \mathbf{F}[\mathbf{F}\mathbf{x}(0) + \mathbf{H}\mathbf{u}(0)] + \mathbf{H}\mathbf{u}(T) \\ &= \mathbf{F}^2\mathbf{x}(0) + \mathbf{F}\mathbf{H}\mathbf{u}(0) + \mathbf{H}\mathbf{u}(T) \end{aligned}$$

$k = 2$: $\quad \mathbf{x}(3T) = \mathbf{F}^3\mathbf{x}(0) + \mathbf{F}^2\mathbf{H}\mathbf{u}(0) + \mathbf{F}\mathbf{H}\mathbf{u}(T) + \mathbf{H}\mathbf{u}(2T)$

$\vdots \qquad\qquad \vdots$

By repeating this recursive procedure up to and including $k = N - 1$, the general solution of $\mathbf{x}(NT)$ is obtained as

$$\mathbf{x}(NT) = \mathbf{F}^N\mathbf{x}(0) + \sum_{k=0}^{N-1} [\mathbf{F}^{N-k-1}\mathbf{H}\mathbf{u}(kT)] \tag{15-118}$$

which is defined as the *state transition equation* of the discrete-time system described by Eq. (15-116). For a SISO system, Eq. (15-118) becomes

$$\mathbf{x}(NT) = \mathbf{F}^N\mathbf{x}(0) + \sum_{k-0}^{N-1} [\mathbf{F}^{N-k-1}\mathbf{h}u(kT)] \tag{15-119}$$

where the output equation is now given by

$$y(NT) = \mathbf{c}'\mathbf{x}(NT) \tag{15-120}$$

Note that Eq. (15-119) usually does not yield as accurate results as the format of Eq. (15-45) due to the inherent derivation.

Example 15-6 The state transition equation (15-119) is used for the canonical-variable representation of Example 15-4. Since in this example $\mathbf{x}(0) = \mathbf{0}$, Eq.

(15-119) reduces to

$$\mathbf{x}(NT) = \sum_{k=0}^{N-1} [\mathbf{F}^{N-k-1}\mathbf{h}u(kT)]$$

which yields, for $k = 0$ ($N = 1$),

$$\mathbf{x}(T) = \mathbf{F}^0\mathbf{h}u(0) = \mathbf{h}u(0)$$

Thus

$$\mathbf{x}(0.1) = \begin{bmatrix} 1 \\ 1 \end{bmatrix}$$

and from Eq. (15-120),

$$y(0.1) = \mathbf{c}'\mathbf{x}(0.1) = 0.009365$$

and for $k = 1$ ($N = 2$),

$$\mathbf{x}(2T) = \sum_{k=0}^{1} [\mathbf{F}^{2-k-1}\mathbf{h}u(kT)] = \mathbf{F}\mathbf{h}u(0) + \mathbf{F}^0\mathbf{h}u(T)$$

$$= \begin{bmatrix} 1 & 0 \\ 0 & 0.81873 \end{bmatrix}\begin{bmatrix} 1 \\ 1 \end{bmatrix}(1) + \begin{bmatrix} 1 \\ 1 \end{bmatrix}(2) = \begin{bmatrix} 3 \\ 2.81873 \end{bmatrix}$$

$$y(0.2) = 0.044525$$

15-4E SV Representation Summary

The SV representation methods of Secs. 15-3 and 15-4 can be summarized as follows:

1. The SV representation methods of Sec. 15-4 are developed for an all-discrete-data system.
2. For a hybrid sampled-data system, the SV representation may be obtained by either of two general approaches; i.e.,
   a. By discretizing the continuous state variables resulting in Eqs. (15-4) and (15-5), as done in Secs. 15-2 and 15-3
   b. or by using the $\mathscr{Z}$-transform representation of the hybrid system, as given by Eq. (15-2)

   The former requires the selection of a specific SV representation for the continuous-time system, and the latter requires the selection of a specific SV representation for the hybrid system utilizing Eq. (15-2).
3. Equation (15-5) or (15-118) may be used to evaluate the discrete state equation that represents a hybrid system.

In general, it is not necessary to specify initially the value of T to use in Eq. (15-45), whereas for the purpose of mathematical simplification, the value of T should be initially specified to obtain $G(z)$ from which Eq. (15-118) is derived. Thus, in general, T does not appear explicitly as a variable in the **F** and **H** matrices. (*Note*: This does not imply that $T = 1$ s.)

15-5 STATE-VARIABLE REPRESENTATION OF A SAMPLED-DATA CONTROL SYSTEM IN THE z DOMAIN

The discrete state equation (15-38) or (15-116) may also be solved by means of the $\mathscr{Z}$-transform method. Note that these equations are solved previously in the time domain to derive the solutions Eqs. (15-45) and (15-118). *Thus, the objective of this section is to present the z-domain method for generating these same solutions.* Also, a z-domain method for obtaining the system transfer functions is developed. The solution to Eq. (15-38) in the z domain is considered first. The $\mathscr{Z}$ transform of

$$\mathbf{x}[(k+1)T] = \boldsymbol{\phi}(T)\mathbf{x}(kT) + \boldsymbol{\theta}(T)\mathbf{m}(kT)$$

is taken on both sides and the result manipulated to yield $\mathbf{X}(z)$ as shown below:

$$z\mathbf{X}(z) - z\mathbf{x}(0) = \boldsymbol{\phi}(T)\mathbf{X}(z) + \boldsymbol{\theta}(T)\mathbf{M}(z) \qquad (15\text{-}121)$$

$$[z\mathbf{I} - \boldsymbol{\phi}(T)]\mathbf{X}(z) = z\mathbf{x}(0) + \boldsymbol{\theta}(T)\mathbf{M}(z) \qquad (15\text{-}122)$$

$$\mathbf{X}(z) = [z\mathbf{I} - \boldsymbol{\phi}(T)]^{-1}z\mathbf{x}(0) + [z\mathbf{I} - \boldsymbol{\phi}(T)]^{-1}\boldsymbol{\theta}(T)\mathbf{M}(z) \qquad (15\text{-}123)$$

The time solution of the discrete state equation as a function of $\mathbf{x}(0)$ is obtained by taking the inverse $\mathscr{Z}$ transform of Eq. (15-123), which is

$$\mathbf{x}(kT) = \mathscr{Z}^{-1}\{[z\mathbf{I} - \boldsymbol{\phi}(T)]^{-1}z\mathbf{x}(0)\} + \mathscr{Z}^{-1}\{[z\mathbf{I} - \boldsymbol{\phi}(T)]^{-1}\boldsymbol{\theta}(T)\mathbf{M}(z)\} \qquad (15\text{-}124)$$

To correlate Eq. (15-45) with Eq. (15-124), first apply Eq. (4-1) to the STM, $\boldsymbol{\phi}(kT)$, which yields

$$\mathscr{Z}[\boldsymbol{\phi}(kT)] = \sum_{k=0}^{\infty} \boldsymbol{\phi}(kT)z^{-k} = \mathbf{I} + \boldsymbol{\phi}(T)z^{-1} + \boldsymbol{\phi}(2T)z^{-2} + \cdots \qquad (15\text{-}125)$$

Premultiply both sides of this equation by $\boldsymbol{\phi}(T)z^{-1}$ and subtract the resulting equation from Eq. (15-125) to yield

$$[\mathbf{I} - \boldsymbol{\phi}(T)z^{-1}]\mathscr{Z}[\boldsymbol{\phi}(kT)] = \mathbf{I} \qquad (15\text{-}126)$$

Manipulating this equation yields

$$\mathscr{Z}[\boldsymbol{\phi}(kT)] \equiv [z\mathbf{I} - \boldsymbol{\phi}(T)]^{-1}z \qquad (15\text{-}127)$$

or

$$\boldsymbol{\phi}(kT) = \mathscr{Z}^{-1}\{[z\mathbf{I} - \boldsymbol{\phi}(T)]^{-1}z\} \equiv \mathscr{Z}^{-1}[z\mathbf{P}(z)] \qquad (15\text{-}128)$$

where

$$\mathbf{P}(z) \equiv [z\mathbf{I} - \boldsymbol{\phi}(T)]^{-1} \qquad \text{and} \qquad \mathscr{Z}^{-1}[\mathbf{P}(z)] \equiv \mathbf{W}(k'T)$$

Multiplying both sides of Eq. (15-127) by z^{-1} yields

$$z^{-1}\mathscr{Z}[\boldsymbol{\phi}(kT)] = \mathbf{P}(z) \tag{15-129}$$

Taking the inverse $\mathscr{Z}$ transform of both sides yields

$$\boldsymbol{\phi}[(k-1)T] \equiv \mathbf{W}(k'T) \tag{15-130}$$

where $k' = k - 1$. Next, the real convolution theorem, Eq. (4-32), must be used to obtain the inverse $\mathscr{Z}$ transform of

$$[z\mathbf{I} - \boldsymbol{\phi}(T)]^{-1}\boldsymbol{\theta}(T)\mathbf{M}(z) = \mathbf{P}(z)\mathbf{S}(z) \tag{15-131}$$

where $\mathbf{S}(z) \equiv \boldsymbol{\theta}(T)\mathbf{M}(z)$. This theorem, when applied to a MIMO system, in terms of matrix functions, yields

$$\mathbf{E}_1(z)\mathbf{E}_2(z) = \mathscr{Z}\left[\sum_{i=0}^{k'} \mathbf{E}_1(k'T - iT)\mathbf{E}_2(iT)\right] \tag{15-132}$$

Equation (15-132) is rewritten in terms of the variables of Eq. (15-131) by letting $\mathbf{E}_1(z) \equiv \mathbf{P}(z)$ and $\mathbf{E}_2(z) \equiv \mathbf{S}(z)$. Thus Eq. (15-132) is rewritten as

$$\mathbf{P}(z)\mathbf{S}(z) = \mathscr{Z}\left\{\sum_{i=0}^{k'} \mathbf{W}[(k-i)T]\boldsymbol{\theta}(T)\mathbf{m}(iT)\right\} \tag{15-133}$$

or, by using Eq. (15-130),

$$\mathbf{P}(z)\mathbf{S}(z) = \mathscr{Z}\left\{\sum_{i=0}^{k-1} \boldsymbol{\phi}[(k-i-1)T]\boldsymbol{\theta}(T)\mathbf{m}(iT)\right\} \tag{15-134}$$

Therefore, utilizing Eqs. (15-128) and (15-134), Eq. (15-124) can now be rewritten as

$$\mathbf{x}(kT) = \boldsymbol{\phi}(kT)\mathbf{x}(0) + \sum_{i=0}^{k-1} \boldsymbol{\phi}[(k-i-1)T]\boldsymbol{\theta}(T)\mathbf{m}(iT) \tag{15-135}$$

for $k = 0$, this equation becomes

$$\mathbf{x}(0) = \boldsymbol{\phi}(0)\mathbf{x}(0) + \sum_{i=0}^{0} \boldsymbol{\phi}[-(i+1)T]\boldsymbol{\theta}(T)\mathbf{m}(iT) = \mathbf{x}(0)$$

since $\boldsymbol{\phi}(0) = \mathbf{I}$ and $\boldsymbol{\phi}[-(i+1)T] = 0$ for $t < 0$. By replacing k by N in Eq. (15-135), it becomes identical to Eq. (15-45).

Using the same procedure as above for the discrete state equation of Eq. (15-41), which is repeated below,

$$\mathbf{x}[(k+1)T] = \mathbf{F}\mathbf{x}(kT) + \mathbf{H}\mathbf{u}(kT)$$

yields

$$\mathbf{x}(kT) = \mathscr{Z}^{-1}\{[z\mathbf{I} - \mathbf{F}]^{-1}z\}\mathbf{x}(0) + \mathscr{Z}^{-1}\{[z\mathbf{I} - \mathbf{F}]^{-1}\mathbf{H}\mathbf{U}(z)\} \tag{15-136}$$

The correlation between Eqs. (15-118) and (15-136) may be made in a similar manner as is done for the correlation between Eqs. (15-45) and (15-124).

For a MIMO system that is described by the discrete state equation of Eq. (15-136) and the system output equation

$$\mathbf{y}(kT) = \mathbf{C}\mathbf{x}(kT) + \mathbf{D}\mathbf{u}(kT) \tag{15-137}$$

it is desired to obtain transfer functions and weighting functions $g_{ij}(kT)$ that relate an output $y_i(kT)$ to an input $u_j(kT)$. This is accomplished by first taking the $\mathscr{Z}$ transform of both sides of Eq. (15-136) and then manipulating this result to yield

$$\mathbf{X}(z) = [z\mathbf{I} - \mathbf{F}]^{-1}z\mathbf{x}(0) + [z\mathbf{I} - \mathbf{F}]^{-1}\mathbf{H}\mathbf{u}(z) \tag{15-138}$$

Next, substituting Eq. (15-138) into the $\mathscr{Z}$ transform of Eq. (15-137), where $\mathbf{x}(0)$ must be assumed to be zero in order to obtain the transfer functions, results in

$$\mathbf{Y}(z) = \mathbf{C}[z\mathbf{I} - \mathbf{F}]^{-1}\mathbf{H}\mathbf{U}(z) + \mathbf{D}\mathbf{U}(z) = [\mathbf{C}[z\mathbf{I} - \mathbf{F}]^{-1}\mathbf{H} + \mathbf{D}]\mathbf{U}(z) \tag{15-139}$$

This equation is of the form

$$\mathbf{Y}(z) = \mathbf{G}(z)\mathbf{U}(z) \tag{15-140}$$

where

$$\mathbf{G}(z) = [\mathbf{C}[z\mathbf{I} - \mathbf{F}]^{-1}\mathbf{H} + \mathbf{D}] \tag{15-141}$$

The $m \times 1$ vector $\mathbf{Y}(z)$, the $m \times n$ matrix $\mathbf{G}(z)$, and the $r \times 1$ vector $\mathbf{U}(z)$ are given, respectively, by

$$\mathbf{Y}(z) \equiv \begin{bmatrix} Y_1(z) \\ Y_2(z) \\ \vdots \\ Y_i(z) \\ \vdots \\ Y_m(z) \end{bmatrix} \tag{15-142}$$

$$\mathbf{G}(z) \equiv \begin{bmatrix} G_{11}(z) & G_{12}(z) & \cdots & G_{1r}(z) \\ G_{21}(z) & G_{22}(z) & \cdots & G_{2r}(z) \\ \cdots & \cdots & \cdots & \cdots \\ G_{m1}(z) & G_{m2}(z) & \cdots & G_{mn}(z) \end{bmatrix} = [G_{ij}(z)] \tag{15-143}$$

$$\mathbf{U}(z) \equiv \begin{bmatrix} U_1(z) \\ U_2(z) \\ \vdots \\ U_r(z) \end{bmatrix} \tag{15-144}$$

and where

$$G_{ij}(z) = \frac{Y_i(z)}{U_j(z)} = K_D \frac{z^w + f_{w-1}z^{w-1} + \cdots + f_1 z + f_0}{z^n + d_{n-1}z^{n-1} + \cdots + d_1 z + d_0} \tag{15-145}$$

Remember that Eqs. (15-45) and (15-118) are obtained for systems whose actuating signal $\mathbf{u}(t)$ is sampled. For the open-loop SISO system of Fig. 4-12*a* of Sec. 4-7A the transfer function $G(z)$ exists, whereas for the open-loop system of Fig. 4-12*c* it does not exist. Note in Chap. 4 that the actuating signal is denoted by $e(t)$, and for the open-loop systems of Fig. 4-12, the actuating signal is also the system input signal. If the system is described by discrete state equation (15-38), the matrices $\mathbf{F}$ and $\mathbf{H}$ are replaced by $\boldsymbol{\phi}(T)$ and $\boldsymbol{\theta}(T)$, respectively. Taking the inverse $\mathscr{Z}$ transform of both sides of Eq. (15-141), utilizing Eqs. (15-128) to (15-130), yields the impulse weighting function

$$\mathbf{g}(kT) = \mathbf{C}\boldsymbol{\phi}[(k-1)T]\mathbf{H} + \mathbf{D} \qquad \text{for } k \geq 1 \tag{15-146}$$

where *for* $k = 0$,

$$\boldsymbol{\phi}[(k-1)T] = 0 \tag{15-147}$$

Then

$$\mathbf{g}(kT) = \mathbf{D} \qquad \text{for } k = 0 \tag{15-148}$$

The discrete state and system output equations (15-138) and (15-139), respectively, and the open-loop transfer function, Eq. (15-141), when applied to a SISO system become

$$\mathbf{X}(z) = [z\mathbf{I} - \mathbf{F}]^{-1}z\mathbf{x}(0) + [z\mathbf{I} - \mathbf{F}]^{-1}\mathbf{h}\mathbf{U}(z) \tag{15-149}$$

$$\mathbf{Y}(z) = \{\mathbf{c}'[z\mathbf{I} - \mathbf{F}]^{-1}\mathbf{h} + \mathbf{d}\}\mathbf{U}(z) \tag{15-150}$$

$$G(z) = \mathbf{c}'[z\mathbf{I} - \mathbf{F}]^{-1}\mathbf{h} + \mathbf{d} = \frac{Y(z)}{U(z)} \tag{15-151}$$

The control ratio $Y(z)/R(z)$ for the unity-feedback SISO system of Fig. 4-15*a*, in which the actuating signal $u(t) = e(t)$ is sampled and $y(t) = c(t)$, is obtained in the following manner. First, taking the inverse $\mathscr{Z}$ transform of $u^*(t)$ yields

$$U(z) = R(z) - Y(z) \tag{15-152}$$

Next, Eq. (15-152) is substituted into Eq. (15-150) to give

$$Y(z) = \{\mathbf{c}'[z\mathbf{I} - \mathbf{F}]^{-1}\mathbf{h} + d\}[R(z) - Y(z)]$$

which is manipulated to yield

$$\frac{Y(z)}{R(z)} = \frac{\{\mathbf{c}'[z\mathbf{I} - \mathbf{F}]^{-1}\mathbf{h} + d\}}{1 + \{\mathbf{c}'[z\mathbf{I} - \mathbf{F}]^{-1}\mathbf{h} + d\}} \tag{15-153}$$

Example 15-7 The discrete state and output equations describing a sampled-data open-loop system are given by Eqs. (15-98) and (15-99). The $\mathscr{Z}$-transform method of Sec. 15-5 is used to solve for $y(kT)$ of Example 15-2. To use

Eq. (15-136), the following functions are derived:

$$[z\mathbf{I} - \mathbf{F}]^{-1} = \begin{bmatrix} z-1 & 0 \\ 0 & z-0.818731 \end{bmatrix}^{-1} = \begin{bmatrix} \dfrac{1}{z-1} & 0 \\ 0 & \dfrac{1}{z-0.818731} \end{bmatrix} \tag{15-154}$$

$$U(z) = \mathscr{Z}[u^*(t)] = \frac{R_0 z}{z-1}$$

Note: The expression for $u(t)$ is not given for $0 \le t < \infty$. Only the values of $u(kT)$ for $k = 0$ and 1 are specified along with the values of $\mathbf{x}(t_0)$ for $(t_0) = 0$. Thus, the following procedure is required to determine the values of $y(0.1)$ and $y(0.2)$. First, solve for $\mathbf{x}(0.1)$ and $y(0.1)$ with $k = 1$, $\mathbf{x}(0) = \mathbf{0}$, and $R_0 = 1$ by using Eq. (15-136). Then, reinitialize the initial conditions; i.e., let zero time be established at $t = kT|_{k=1} = 0.1$ s. Therefore the new initial conditions are

$$\mathbf{x}(t')\Big|_{\substack{t'=0 \\ k'=k-1}} = \mathbf{x}(t)\Big|_{t=0.1} \qquad \text{for } k = 1 \text{ where now } R_0 = 2$$

Equation (15-136) is again applied by using $\mathbf{x}(t')|_{t'=0}$ as the initial conditions, $R_0 = 2$, and $k' = k - 1$ to determine $\mathbf{x}(k'T)$ and $y(k'T)$ for $k' = k - 1 = 1$; that is, $k = 2$. It should be realized that in reality $\mathbf{x}(k'T)$ and $y(k'T)$ for $k' = 1$ are the values of $\mathbf{x}(kT)$ and $y(kT)$ for $k = 2$. Thus, for the system of Example 15-4,

$$[z\mathbf{I} - \mathbf{F}]^{-1}\mathbf{h}U(z) = \begin{bmatrix} \dfrac{z}{(z-1)^2} \\ \dfrac{z}{(z-1)(z-0.818731)} \end{bmatrix} R_0 \tag{15-155}$$

$$\mathscr{Z}^{-1}\{[z\mathbf{I} - \mathbf{F}]^{-1}z\} = \mathscr{Z}^{-1}\begin{bmatrix} \dfrac{z}{z-1} & 0 \\ 0 & \dfrac{z}{z-0.818731} \end{bmatrix}$$

$$= \begin{bmatrix} \delta_T(t) & 0 \\ 0 & \epsilon^{-2kT} \end{bmatrix} \tag{15-156}$$

$$\mathscr{Z}^{-1}\{[z\mathbf{I} - \mathbf{F}]^{-1}\mathbf{h}U(z) = \begin{bmatrix} k \\ \dfrac{1}{0.18127}[\delta_T(t) - \epsilon^{-2kT}] \end{bmatrix} R_0 \tag{15-157}$$

Before proceeding, it should be noted from Table 4-1 that the $\mathscr{Z}$

transform of the function $e(t) = t$ is

$$E(z) = \frac{Tz}{(z-1)^2} \tag{15-158}$$

Thus, the top element in the vector of Eq. (15-155) differs from Eq. (15-158) by the factor T. This element is manipulated in the following manner to arrive at its inverse $\mathscr{Z}$ transform of k in Eq. (15-157):

$$e(kT) = \frac{1}{T}\mathscr{Z}^{-1}\left[\frac{Tz}{(z-1)^2}\right] = \frac{1}{T}(kT) = k$$

Therefore, from Eqs. (15-136) and (15-99)

$$\mathbf{x}(kT) = \begin{bmatrix} \delta_T(t) & 0 \\ 0 & \epsilon^{-2kT} \end{bmatrix}\mathbf{x}(0) + \begin{bmatrix} k \\ \dfrac{1}{0.18127}[\delta_T(t) - \epsilon^{-2kT}] \end{bmatrix} R_0 \tag{15-159}$$

$$y(kT) = [0.1 \quad -0.0906349]\mathbf{x}(kT) \tag{15-160}$$

For $k = 0$, $\mathbf{x}(0) = \mathbf{0}$, and $R_0 = 1$ Eq. (15-158) yields $\mathbf{x}(0) = 0$, which verifies Eq. (15-159).

For $k = 1$, Eqs. (15-159) and (15-160), with $R_0 = 1$, yield, respectively,

$$\mathbf{x}(0.1) = \begin{bmatrix} 1 \\ 1 \end{bmatrix} \qquad \text{and} \qquad y(0.1) = 0.009365$$

For $k = 2$ and $R_0 = 2$, it is necessary to reinitialize the initial conditions of Eq. (15-159). This is done by letting $k' = k - 1$ and $t' = (k-1)T = k'T$ to yield

$$x(t')\bigg|_{t'=0} = \begin{bmatrix} 1 \\ 1 \end{bmatrix}$$

where

$$t' = t - 0.1\bigg|_{t=0.1} = 0$$

Thus, the reinitialized Eq. (15-159) is

$$\mathbf{x}(k'T)\bigg|_{k'T=0.1} = \begin{bmatrix} \delta_T(t) & 0 \\ 0 & \epsilon^{-2k'T} \end{bmatrix}\begin{bmatrix} 1 \\ 1 \end{bmatrix} + \begin{bmatrix} k' \\ \dfrac{1}{0.18127}[\delta_T(t') - \epsilon^{-2k'T}] \end{bmatrix}(2)$$

$$= \begin{bmatrix} 1 \\ 0.818731 \end{bmatrix} + \begin{bmatrix} 2 \\ 2 \end{bmatrix}$$

$$= \begin{bmatrix} 3 \\ 2.81873 \end{bmatrix} = \mathbf{x}(kT)\bigg|_{kT=0.2} \tag{15-161}$$

Substituting from Eq. (15-161) into Eq. (15-160) yields

$$y(0.2) = [0.1 \quad -0.090635]\mathbf{x}(t')\Big|_{t'=0.1} = 0.044533$$

Applying Eq. (15-151) yields

$$G(z) = \mathbf{c}'[z\mathbf{I} - \mathbf{F}]^{-1}\mathbf{h} = [0.1 \quad -0.0906349]\begin{bmatrix} \dfrac{1}{z-1} \\ \dfrac{1}{z-0.81873} \end{bmatrix}$$

$$= \frac{0.009365z + 0.008786}{(z-1)(z-0.81873)}$$

which, of course, must agree with Eq. (15-76).

Example 15-8 A unit-step input is applied to the sampled-data system of Example 15-2. For $\mathbf{x}(0) = \mathbf{0}$ and $T = 0.1$, the output response is now determined in the z domain; i.e., use Eq. (15-124). To apply this equation, $[z\mathbf{I} - \boldsymbol{\phi}(T)]^{-1}$ is first determined. From Eq. (15-49),

$$[z\mathbf{I} - \boldsymbol{\phi}(0.1)]^{-1} = \begin{bmatrix} z-1 & -0.090635 \\ 0 & z-0.81873 \end{bmatrix}^{-1}$$

$$= \frac{\begin{bmatrix} z-0.81873 & 0.090635 \\ 0 & z-1 \end{bmatrix}}{(z-1)(z-0.81873)}$$

$$= \begin{bmatrix} \dfrac{1}{z-1} & \dfrac{0.090635}{(z-1)(z-0.81873)} \\ 0 & \dfrac{1}{z-0.81873} \end{bmatrix} \tag{15-162}$$

Using Eqs. (15-50) and (15-162), Eq. (15-124) yields

$$\mathbf{x}(kT) = \mathscr{Z}^{-1}\left\{\begin{bmatrix} \dfrac{1}{z-1} & \dfrac{0.090635}{(z-1)(z-0.81873)} \\ 0 & \dfrac{1}{z-0.81873} \end{bmatrix}\begin{bmatrix} 0.009365 \\ 0.18127 \end{bmatrix}\frac{z}{z-1}\right\}$$

$$= \mathscr{Z}^{-1}\left\{\begin{bmatrix} \dfrac{0.009365z}{(z-1)^2} + \dfrac{0.016429z}{(z-1)^2(z-0.81873)} \\ \dfrac{0.18127z}{(z-1)(z-0.81873)} \end{bmatrix}\right\}$$

$$\mathbf{x}(kT) = \begin{bmatrix} 0.1k + 0.5\epsilon^{-0.2k} - 0.5 \\ 1 - \epsilon^{-0.2k} \end{bmatrix}$$

$$y(kT) = [1 \quad 0]\mathbf{x}(kT) = x_1(kT) = 0.1k + 0.5\epsilon^{-0.2k} - 0.5 \qquad \text{for } k \geq 0$$

15-6 SYSTEM STABILITY

As stated previously, the stability of a linear-feedback control system depends upon the location of the poles of the transfer function that relates its output to its input (see Sec. 6-2). This section presents the method for determining the sampled-data system's control ratio and characteristic equation from its state and output equations.

Consider first the open-loop linear system whose transfer function is

$$G(z) = \frac{Y(z)}{U(z)} = \frac{f_w z^w + f_{w-1} z^{w-1} + \cdots + f_1 z + f_0}{z^n + d_{n-1} z^{n-1} + \cdots + d_1 z + d_0} \tag{15-163}$$

From this equation is obtained the system's difference equation, which is

$$y[(k+n)T] + d_{n-1} y[(k+n-1)T] + \cdots + d_1 y[(k+1)T] + d_0 y(kT)$$
$$= f_w u[(k+w)T] + f_{w-1} u[k+w-1)T] + \cdots + f_1 u[(k+1)T] + f_0 u(kT) \tag{15-164}$$

The denominator of the open-loop system transfer function (15-163), when set equal to zero, is defined as the *characteristic equation* of the open-loop system; that is,

$$z^n + d_{n-1} z^{n-1} + \cdots + d_1 z + d_0 = 0 \tag{15-165}$$

Equation (15-165) may also be obtained from Eq. (15-150), which is derived from the system's dynamic equations. First, note the definition

$$[z\mathbf{I} - \mathbf{F}]^{-1} = \frac{\text{adj}[z\mathbf{I} - \mathbf{F}]}{|z\mathbf{I} - \mathbf{F}|} \tag{15-166}$$

where $\text{adj}[z\mathbf{I} - \mathbf{F}]$ is the transpose of the cofactor matrix and $|z\mathbf{I} - \mathbf{F}|$ is the determinant of $[z\mathbf{I} - \mathbf{F}]$. The cofactor matrix is formed by replacing each element of $[z\mathbf{I} - \mathbf{F}]$ by its cofactor. Now substitute Eq. (15-166) into Eq. (15-150) to obtain

$$G(z) = \frac{Y(z)}{U(z)} = \frac{\mathbf{c}'\text{adj}[z\mathbf{I} - \mathbf{F}]\mathbf{h} + d|z\mathbf{I} - \mathbf{F}|}{|z\mathbf{I} - \mathbf{F}|} \tag{15-167}$$

The denominator of $G(z)$ is set equal to zero to yield the characteristic equation

$$|z\mathbf{I} - \mathbf{F}| = 0 \tag{15-168}$$

This is purely an algebraic equation which has *n roots* or *eigenvalues*. These eigenvalues are referred to as the eigenvalues of the matrix **F** and are the poles of

$G(z)$. Note that $G(z)$ and the characteristic equation can be expressed in terms of $\phi(T)$. For example, replacing **F** by $\phi(T)$ in Eq. (15-168) yields

$$|z\mathbf{I} - \phi(T)| = 0 \tag{15-169}$$

Thus, applying this equation to Example 15-2 yields the eigenvalues of the open-loop system [poles of $G(z)$]—see Eq. (15-76).

Next, consider a closed-loop discrete-data control system that is described by the control ratio of Eq. (15-153). The denominator of the closed-loop system transfer function (15-153), when set equal to zero, is defined as the characteristic equation of the closed-loop system; that is,

$$1 + \mathbf{c}'|z\mathbf{I} - \mathbf{F}|^{-1}\mathbf{h} + d = 0 \tag{15-170}$$

Therefore the roots of this equation are the poles of $Y(z)/R(z)$.

The roots of Eq. (15-170) may also be derived as follows. Utilizing

$$\mathbf{x}[(k+1)T] = \mathbf{F}\mathbf{x}(kT) + \mathbf{h}u(kT)$$

$$y = \mathbf{c}'\mathbf{x}(kT) + du(kT)$$

and

$$u(kT) = r(kT) - y(kT) = r(kT) - \mathbf{c}'\mathbf{x}(kT) - du(kT)$$

thus,

$$u(kT) = \frac{r(kT)}{1+d} - \frac{\mathbf{c}'\mathbf{x}(kT)}{1+d}$$

The following closed-loop state equation is obtained

$$\begin{aligned}\mathbf{x}[(k+1)T] &= \left[\mathbf{F} - \frac{\mathbf{h}\mathbf{c}'\mathbf{I}}{1+d}\right]\mathbf{x}(kT) + \frac{1}{1+d}\mathbf{h}r(kT)\\ &= \mathbf{F}_{c1}\mathbf{x}(kT) + \mathbf{h}_{c1}r(kT)\end{aligned} \tag{15-171}$$

The *closed-loop eigenvalues* are given by

$$|z\mathbf{I} - \mathbf{F}_{c1}| = 0 \tag{15-172}$$

Schur's formulas,[70]

$$\begin{vmatrix}\mathbf{P} & \mathbf{Q}\\ \mathbf{R} & \mathbf{S}\end{vmatrix} = \left\{\begin{matrix}|\mathbf{P}|\,|\mathbf{S} - \mathbf{R}\mathbf{P}^{-1}\mathbf{Q}|\\ \text{or}\\ |\mathbf{S}|\,|\mathbf{P} - \mathbf{Q}\mathbf{S}^{-1}\mathbf{R}|\end{matrix}\right\} = 0 \tag{15-173}$$

provide the equivalent expressions for a determinant whose elements are matrices. They are now used to show that Eq. (15-172) yields Eq. (15-170). Note that **P** is a $p \times p$ matrix, **Q** is a $p \times v$ matrix, **R** is a $v \times p$ matrix, and **S** is a $v \times v$ matrix. The top relationship of Eq. (15-173) is used by letting $\mathbf{P} = z\mathbf{I} - \mathbf{F}$, $\mathbf{R} = -\mathbf{c}'$, $\mathbf{Q} = \mathbf{h}$, and $\mathbf{S} = 1 + d$, where $p = n$ and $v - 1$. Making these substitutions into

Eq. (15-173) yields

$$|z\mathbf{I} - \mathbf{F}|\,|1 + d + \mathbf{c}'[z\mathbf{I} - \mathbf{F}]^{-1}\mathbf{h}| = 0 \tag{15-174}$$

Noting Eq. (15-166) and comparing Eq. (15-174) to Eq. (15-170), it is seen that the roots of Eq. (15-170) and Eq. (15-174) are identical.

Having determined the poles of the system transfer function as the eigenvalues of the characteristic equation, the conditions for a stable response for an LTI discrete-data system are now presented.

To determine the stability of a linear system it is necessary to consider only the homogeneous solution of Eq. (15-124), i.e., for the open-loop system,

$$\mathbf{x}(kT) = \mathscr{Z}^{-1}\{[z\mathbf{I} - \boldsymbol{\phi}(T)]^{-1}z\mathbf{x}(0)\} \tag{15-175}$$

$$\boldsymbol{\phi}(kT) = \mathscr{Z}^{-1}\{|z\mathbf{I} - \boldsymbol{\phi}(T)|^{-1}z\} = \mathscr{Z}^{-1}\left\{z\frac{\operatorname{adj}[z\mathbf{I} - \boldsymbol{\phi}(T)]}{|z\mathbf{I} - \boldsymbol{\phi}(T)|}\right\} \tag{15-176}$$

The requirement for a stable system is that all state variables in Eq. (15-175) must approach zero as $k \to \infty$, that is,

$$\lim_{k\to\infty} \mathbf{x}(kT) = \mathbf{0} \tag{15-177}$$

For an open-loop system this requires that

$$\lim_{k\to\infty} \boldsymbol{\phi}(kT) = \lim_{k\to\infty}\left\{\mathscr{Z}^{-1}\left[z\frac{\operatorname{adj}[z\mathbf{I} - \boldsymbol{\phi}(T)]}{|z\mathbf{I} - \boldsymbol{\phi}(T)|}\right]\right\} = \mathbf{0} \tag{15-178}$$

which is satisfied when the roots of the characteristic equation (15-169) all lie within the unit circle in the z plane (see Fig. 4-8).

In a similar manner, it may be shown that for a stable closed-loop sampled-data control system, the roots of the characteristic equation (15-170) must all lie within the unit circle in the z plane.

Example 15-9 The plant of Example 15-2 is used for a unity-feedback sampled-data system as shown in Fig. 15-15. The objectives of this example are (*a*) to derive the characteristic equation and (*b*) to determine $Y(z)/R(z)$ in terms of K_G and T (note that the values of K_G and T are unspecified by means of the SV approach).

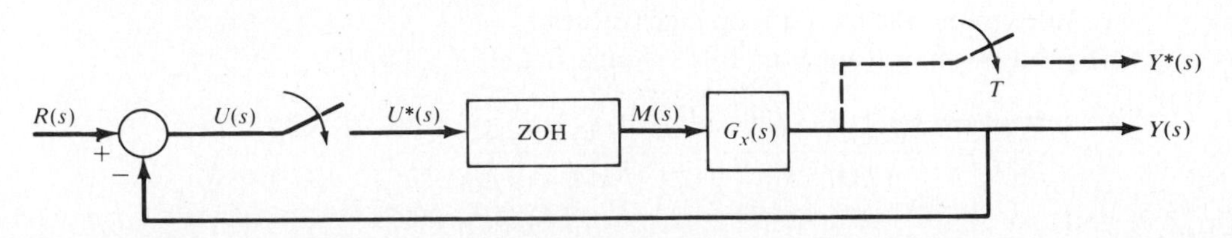

Figure 15-15 A unity-feedback sampled-data system.

(*a*) Equation (15-170) is utilized by replacing **F** by $\boldsymbol{\phi}(T)$ and **h** by $\boldsymbol{\theta}(T)$ to determine the system's characteristic equation. Using Eq. (15-47) yields

$$[z\mathbf{I} - \boldsymbol{\phi}(T)]^{-1} = \begin{bmatrix} \dfrac{1}{z-1} & \dfrac{0.5(1-\epsilon^{-2T})}{(z-1)(z-\epsilon^{-2T})} \\ 0 & \dfrac{1}{z-\epsilon^{-2T}} \end{bmatrix} \tag{15-179}$$

Since the value of K_G is unspecified, then using Eqs. (15-40) and (15-179), where

$$\mathbf{A} = \begin{bmatrix} 0 & 1 \\ 0 & -2 \end{bmatrix} \quad \text{and} \quad \mathbf{b} = \begin{bmatrix} 0 \\ K_G \end{bmatrix}$$

yields

$$[z\mathbf{I} - \boldsymbol{\phi}(T)]^{-1}\boldsymbol{\theta}(T) = \begin{bmatrix} \dfrac{T - 0.5 + 0.5\epsilon^{-2T}}{2(z-1)} + \dfrac{0.25(1-\epsilon^{-2T})^2}{(z-1)(z-\epsilon^{-2T})} \\ 0.5\dfrac{1-\epsilon^{-2T}}{z-\epsilon^{-2T}} \end{bmatrix} K_G \tag{15-180}$$

Since $\mathbf{c}' = [1 \quad 0]$, then

$$\mathbf{c}'[z\mathbf{I} - \boldsymbol{\phi}(T)]^{-1}\boldsymbol{\theta}(T) = \left[\frac{T - 0.5 + 0.5\epsilon^{-2T}}{2(z-1)} + \frac{0.25(1-\epsilon^{-2T})^2}{(z-1)(z-\epsilon^{-2T})}\right] K_G \tag{15-181}$$

Substituting from Eq. (15-180) into Eq. (15-170) yields the characteristic equation

$$z^2 - [(1+\epsilon^{-2T}) - 0.5K_G(T - 0.5 + 0.5\epsilon^{-2T})]z + \epsilon^{-2T} + 0.5K_G(0.5 - 0.5\epsilon^{-2T} - T\epsilon^{-2T}) = 0 \tag{15-182}$$

This equation can be analyzed by the techniques of Chap. 6 to determine the effect of K_G and T on system stability or for specified values of K_G and T to determine the closed-loop eigenvalues.

(*b*) Substituting from Eq. (15-181) into Eq. (15-153) yields

$$\frac{Y(z)}{R(z)} = \frac{0.5K_G[(T - 0.5 + 0.5\epsilon^{-2T})z + (0.5 - 0.5\epsilon^{-2T} - T\epsilon^{-2T})]}{z^2 - [(1-\epsilon^{-2T}) - 0.5K_G(T - 0.5 + 0.5\epsilon^{-2T})]z + \epsilon^{-2T} + 0.5K_G(0.5 - 0.5\epsilon^{-2T} - T\epsilon^{-2T})} \tag{15-183}$$

Equation (15-183) can be readily verified by substituting Eq. (15-76) into

$$\frac{Y(z)}{R(z)} = \frac{G(z)}{1 + G(z)} = \frac{N(z)}{D(z)} \tag{15-184}$$

Therefore, a sampled-data unity-feedback stable control system whose plant $G_x(s)$ is Type 1 has the same steady-state performance characteristics as a stable unity-feedback continuous-time control system (see Sec. 6-4). In a similar manner, an analysis for parabolic, etc., inputs can be made for other Type m plants. [For a parabolic input, the stable system of Example 15-9 yields $u^*(\infty) = \infty$.]

15-7 TIME RESPONSE IN BETWEEN SAMPLING INSTANTS

The modified $\mathscr{Z}$-transform method of Sec. 4-9 is a conventional control-theory technique for evaluating the system response between sampling instants. A modern control-theory technique, utilizing a similar approach, is now presented for evaluating the system response between sampling instants.

The equation for the state vector $\mathbf{x}(t)$, Eq. (15-33), when $\mathbf{x}(kT)$ and $\mathbf{u}(kT)$ are specified, is

$$\mathbf{x}(t) = \boldsymbol{\phi}(t - kT)\mathbf{x}(kT) + \left[\int_{kT}^{t} \boldsymbol{\phi}(t - \tau)\mathbf{B}\, d\tau\right]\mathbf{u}(kT) \tag{15-185}$$

where $k = 0, 1, 2, \ldots, t_0 = kT, kT \le t < (k + 1)T$, and $\mathbf{u}(kT) = \text{constant}$. For the specified time interval, let

$$\boldsymbol{\theta}(t - kT) = \int_{kT}^{t} \boldsymbol{\phi}(t - \tau)\mathbf{B}\, d\tau \tag{15-186}$$

Then Eq. (15-185) may be written as

$$\mathbf{x}(t) = \boldsymbol{\phi}(t - kT)\mathbf{x}(kT) + \boldsymbol{\theta}(t - kT)\mathbf{u}(kT) \tag{15-187}$$

To determine the response in between the sampling instants, let

$$t = (k + \rho)T \tag{15-188}$$

where $0 \le \rho < 1$. Substituting Eq. (15-188) into Eq. (15-187) yields

$$\mathbf{x}[(k + \rho)T] = \boldsymbol{\phi}(\rho T)\mathbf{x}(kT) + \boldsymbol{\theta}(\rho T)\mathbf{u}(kT) \tag{15-189}$$

which by varying ρ between 0 and 1 permits the evaluation of $\mathbf{x}(t)$ for the period $kT \le t < (k + \rho)T$.

Example 15-10 For the sampled-data system of Example 15-2 the system dynamic equations are

$$\mathbf{x}[(k + 1)T] = \boldsymbol{\phi}(T)\mathbf{x}(kT) + \boldsymbol{\theta}(T)m(kT)$$

$$y(kT) = [1 \quad 0]\mathbf{x}(kT)$$

for $kT \le t < (k+1)T$, where

$$\boldsymbol{\phi}(T) = \begin{bmatrix} 1 & 0.5(1 - \epsilon^{-2T}) \\ 0 & \epsilon^{-2T} \end{bmatrix} \tag{15-190}$$

$$\boldsymbol{\theta}(T) = \begin{bmatrix} T - 0.5 + 0.5\epsilon^{-2T} \\ 1 - \epsilon^{-2T} \end{bmatrix} \tag{15-191}$$

and

$$\mathbf{x}[(k+\rho)T] = \boldsymbol{\phi}(\rho T)\mathbf{x}(kT) + \boldsymbol{\theta}(\rho T)u(kT) \tag{15-192}$$

For $u(t) = u_{-1}(t)$, $\mathbf{x}(0) = \mathbf{0}$, and $T = 0.1$ s, determine $y(0.05)$ and $y(0.15)$. To utilize Eq. (15-189), replace T by ρT in Eqs. (15-190) and (15-191), where $0 \le \rho < 1$. Thus,

$$\boldsymbol{\phi}(\rho T) = \begin{bmatrix} 1 & 0.5(1 - \epsilon^{-2\rho T}) \\ 0 & \epsilon^{-2\rho T} \end{bmatrix} \tag{15-193}$$

$$\boldsymbol{\theta}(\rho T) = \begin{bmatrix} \rho T - 0.5 + 0.5\epsilon^{-2\rho T} \\ 1 - \epsilon^{-2\rho T} \end{bmatrix} \tag{15-194}$$

which are independent of the variable k. For $\rho = 0.5$ and $T = 0.1$ s, Eqs. (15-193) and (15-194) become

$$\boldsymbol{\phi}(0.5T)\Big|_{T=0.1} = \begin{bmatrix} 1 & 0.047658 \\ 0 & 0.904837 \end{bmatrix}$$

$$\boldsymbol{\theta}(0.5T)\Big|_{T=0.1} = \begin{bmatrix} 0.00242 \\ 0.09516 \end{bmatrix}$$

For $(k+\rho)T = (k+\rho)0.1$, Eq. (15-189) yields, for $k = 0$,

$$\mathbf{x}(0.05) = \boldsymbol{\theta}(0.05) = \begin{bmatrix} 0.00242 \\ 0.09516 \end{bmatrix}$$

Thus

$$y(0.05) = [1 \quad 0]\mathbf{x}(0.05) = x_1(0.05) = 0.00242$$

From Eq. (15-51), for $k = 1$,

$$\mathbf{x}(T) = \mathbf{x}(0.1) = \begin{bmatrix} 0.009365 \\ 0.181269 \end{bmatrix}$$

Thus Eq. (15-189), for $k = 1$ and $\rho = 0.5$, yields

$$\mathbf{x}(0.15) = \begin{bmatrix} 1 & 0.047658 \\ 0 & 0.904837 \end{bmatrix}\begin{bmatrix} 0.009365 \\ 0.181269 \end{bmatrix} + \begin{bmatrix} 0.00242 \\ 0.09516 \end{bmatrix} 1 = \begin{bmatrix} 0.020395 \\ 0.259179 \end{bmatrix}$$

and $y(0.15) = x_1(0.15) = 0.259179$

15-8 STATE-VARIABLE FEEDBACK: PARAMETER INSENSITIVITY

The output of a state-feedback SISO control system, under high-forward-gain (hfg) operation, can be designed to have a high degree of insensitivity to parameter variation.[1] This design, via a state-feedback approach, results in a feedback unit, $H_{eq}(s)$, that may not be physically realizable when it is used in a continuous-time control system. When this situation arises, it may be possible to achieve the desired degree of insensitivity (robustness), with an acceptable loop transmission frequency (0-dB crossover frequency), of the output to parameter variation. This is accomplished by transforming the designed state-feedback continuous-time system to a nonunity-feedback digital control system. It should be stressed that this section does not consider unmodeled plant dynamics and the associated stability considerations of the system. The main emphasis is to present a method for converting an unrealizable s-domain compensator into a realizable one in the z domain, and, also, to illustrate that the Tustin transformation is a "good design tool" for "low" frequency (with respect to f_s) plants.

15-8A State-Feedback Continuous-Time Control System

The output $Y(s) = \{G_x(s)/[1 + G_x(s)]\}R(s)$ of the control system of Fig. 15-16*a* is affected by the variation of the plant parameters. To make the output as insensitive as possible to plant parameter variations, one approach is to design a hfg complete state-feedback system (see Fig. 15-16*b*). This design is based upon satisfying the system performance specifications by using the nominal values of the plant parameters. By means of block diagram manipulations, as shown in Fig. 15-16*c*, Fig. 15-16*b* is transformed to its equivalent output-feedback representation of Fig. 15-16*d*. In this manipulative process the feedback units k_2 and k_3 are shifted to the output as shown in Fig. 15-16*c*. Thus, $H_{eq}(s) = H_c(s) + k_1$ represents a feedback compensator, and Fig. 15-16*d* is referred to as the state-feedback *H equivalent* of Fig. 15-16*b*. The configuration of Fig. 15-16*d* is now shown to be optimal for minimizing the sensitivity of the output response with a parameter variation between x_n and y.[1] Physical state variables are indicated in Fig. 15-16 which require that the transfer functions $G_1(s), G_2(s), \ldots, G_{n-1}(s)$ have the form

$$\frac{A_i}{s} \qquad \frac{A_i}{s+a} \qquad \text{or} \qquad \frac{A_i(s+b)}{s+a}$$

and $G_n(s)$, whose input is $U(s)$, must have the form

$$\frac{A_i}{s} \qquad \text{or} \qquad \frac{A_i}{s+a}$$

For the system of Fig. 15-16, let

$$G_x(s) = G_1(s)G_2(s)G_3(s) = \frac{10A}{s(s-p_2)(p-p_3)} \tag{15-195}$$

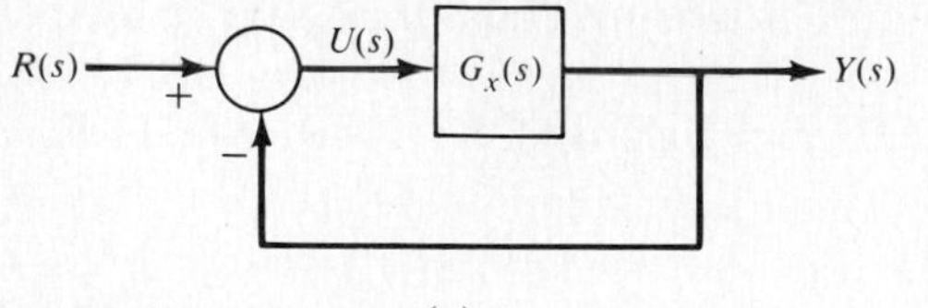

(a)

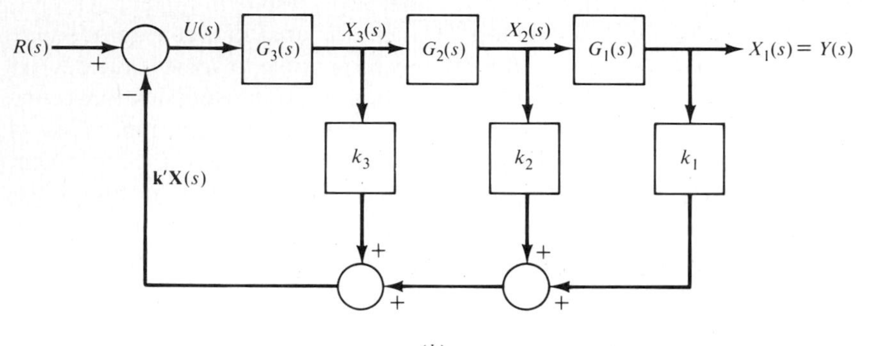

(b)

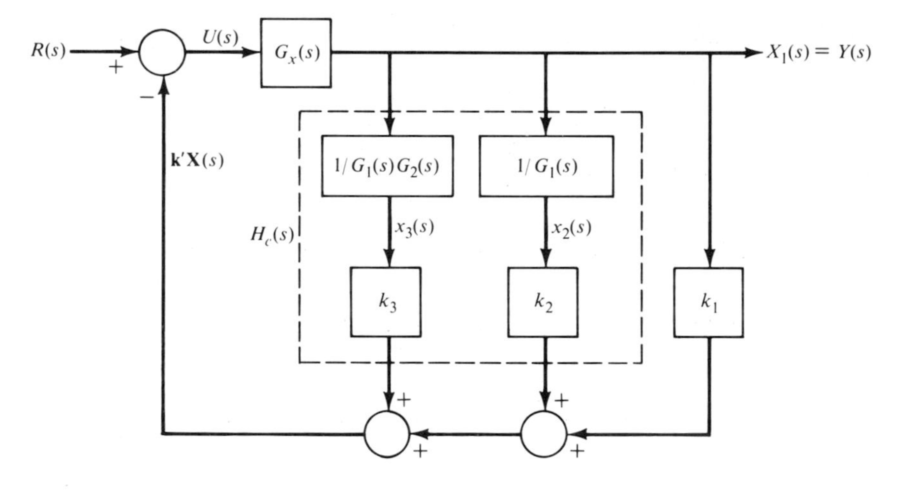

(c)

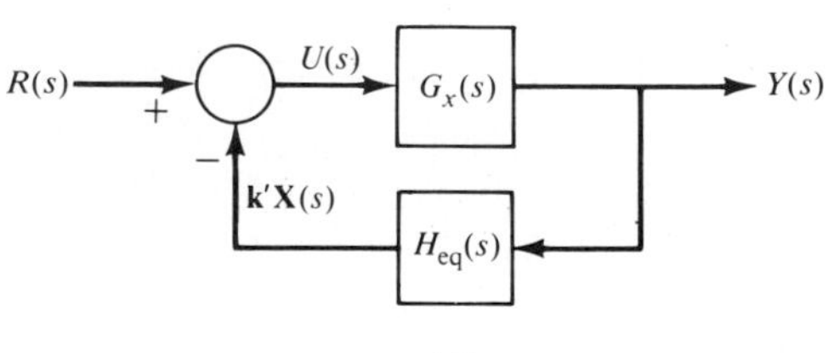

(d)

Figure 15-16 Control systems. (*a*) Unity feedback; (*b*) state feedback; (*c*), (*d*) state feedback, H equivalent.

where

$$G_1(s) = \frac{1}{s} \qquad G_2(s) = \frac{5}{s - p_2} \qquad G_3(s) = \frac{2A}{s - p_3}$$

and

$$H_{eq}(s) = \frac{\mathbf{k}'\mathbf{X}(s)}{Y(s)} = \frac{\mathbf{k}'\mathbf{X}(s)}{\mathbf{c}^T\mathbf{X}(s)} \tag{15-196}$$

where

$$\mathbf{k}' = [k_1 \quad k_2 \quad k_3] \tag{15-197}$$

$$Y(s) = \mathbf{c}'\mathbf{X}(s) = [1 \quad 0 \quad 0]\mathbf{X}(s) \tag{15-198}$$

$$\mathbf{c}' = [1 \quad 0 \quad 0] \tag{15-199}$$

$$\frac{X_3(s)}{X_1(s)} = \frac{1}{G_1(s)G_2(s)} \tag{15-200}$$

$$\frac{X_2(s)}{X_1(s)} = \frac{1}{G_1(s)} \tag{15-201}$$

Substituting from Eqs. (15-197) to (15-201) into Eq. (15-196) yields

$$H_{eq}(s) = \frac{k_3}{G_1(s)G_2(s)} + \frac{k_2}{G_1(s)} + k_1$$

$$= \frac{k_3 s^2 + (5k_2 - k_3 p_2)s + 5k_1}{5} \tag{15-202}$$

and

$$\left[\frac{Y(s)}{R(s)}\right]_A = \frac{G_x(s)}{1 + G_x(s)H_{eq}(s)}$$

$$= \frac{10A}{s^3 + (2Ak_3 - p_2 - p_3)s^2 + [p_2 p_3 + 2A(5k_2 - k_3 p_2)]s + 10Ak_1} \tag{15-203}$$

Assume the nominal values for this control system to have the values $p_2 = -5$ and $p_3 = -1$ and that the desired control ratio, based upon a set of desired conventional control figures of merit, is

$$[M(s)]_D = \frac{Y(s)}{R(s)} = \frac{100}{s^3 + 101s^2 + 142.7s + 100}$$

$$= \frac{100}{(s + 0.70659 + j0.70771)(s + 99.987)} \tag{15-204}$$

Achieving a satisfactory degree of output insensitivity to parameter variation requires that $[M(s)]_D$ must contain at least one nondominant pole, as illustrated by Eq. (15-204). Making this pole more nondominant increases the value of gain A required to maintain zero steady-state error for a step input. Thus, the requirement of hfg operation is satisfied and the system becomes more insensitive to gain variation. The dominant poles of $[M(s)]_D$ are determined from the system specifications as illustrated in Chap. 5. Equating the corresponding coefficients in Eqs. (15-203) and (15-204) yields the nominal values $A = 10$, $k_1 = 1$, $k_2 = -3.393$, and $k_3 = 4.77$. These values are used to plot the root locus of

$$G(s)H_{eq}(s) = \frac{95.4(s + 0.721698 \pm j0.726202)}{s(s + 1)(s + 5)} = -1 \qquad (15\text{-}205)$$

in Fig. 15-17. Note that the zeros of Eq. (15-205) are the zeros of $H_{eq}(s)$ which are determined by the nominal values of the plant and $[M(s)]_D$. As illustrated by the root locus, the state-variable feedback system is stable for all $A > 0$. Since the roots are very close to the zeros of $G(s)H_{eq}(s)$, any increase in gain from the value $A = 10$ produces negligible changes in the location of the closed-loop roots. In other words, for hfg operation, $A > 10$, the system's response is essentially unaffected by variations in A, provided that the feedback coefficients remain fixed. Thus, the system is more insensitive to parameter variation when $H_{eq}(s)$ is used than when actual state feedback $\mathbf{k}'\mathbf{x}(s)$ is utilized. The same effect is noted, with hfg operation, when the parameters p_2 or p_3 vary, provided that $H_{eq}(s)$ is invariant, and the control system configuration of Fig. 15-16*d* is used to implement the system design. Thus, to achieve a high degree of insensitivity to a plant parameter variation, the $H_{eq}(s)$ unit of Fig. 15-16*d* is built by using the design obtained based on the

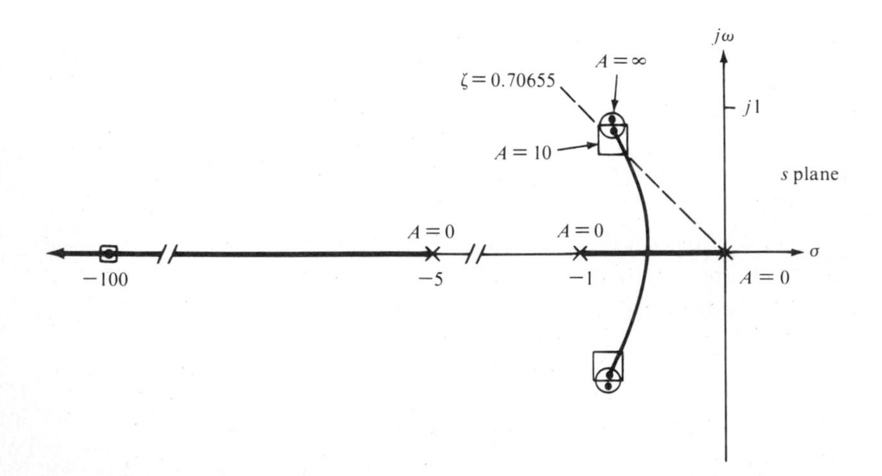

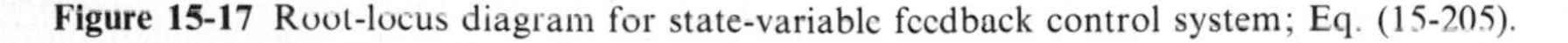
Figure 15-17 Root-locus diagram for state-variable feedback control system; Eq. (15-205).

Table 15-2 Time-response data: continuous-time control system

Case	$M_p(t)$	t_p, s	t_s, s	Parameter change
1 (nominal)	1.043−	4.46	5.968	
2	1.043+	4.39+	5.9−	$A = 20$
3	1.04−	4.62	6.09	$p_2 = -10$
4	1.035−	4.63−	5.92−	$p_3 = -2$

nominal plant parameters $[M(s)]_D$ and Fig. 15-16*b*. Thus

$$H_{eq}(s) = \frac{4.77s^2 + 6.885s + 5}{5} \tag{15-206}$$

To illustrate the insensitivity of the control system to plant parameter variations, the values of k_1, k_2, and k_3 determined on the basis of the nominal parameter values are inserted into Eq. (15-203) to yield

$$\left[\frac{Y(s)}{R(s)}\right]_A = \frac{10A}{s^3 + (9.54A - p_2 - p_3)s^2 + [p_2p_3 + 2A(6.885)]s + 10A} \tag{15-207}$$

Table 15-2 presents the figures of merit for the control system represented by Fig. 15-16*d* and Eq. (15-207) for the cases (1) nominal values; (2) $A = 20, p_2 = -5$, and $p_3 = -1$; (3) $A = 10$, $p_2 = -10$, and $p_3 = -1$; and (4) $A = 10$, $p_2 = -5$, and $p_3 = -2$. Note that the response for the state-feedback *H*-equivalent system is essentially unaffected by plant parameter variation. The 0-dB crossing frequency $\mathbf{G}(j\omega)\mathbf{H}_{eq}(j\omega)$ is 95.2 rad/s. When the components of the feedback unit are selected so that the transfer function $H_{eq}(s)$ is invariant, the implementation of the control system of Fig. 15-16*c* uses the minor-loop compensator $H_c(s) = (k_3/5)s^2 + (k_3 + k_2)s$. This compensator requires both first- and second-order derivative action in the minor loop. In many applications, this type of continuous-time active compensator is not feasible because of noise and implementation problems. It may be possible to add poles to $H_{eq}(s)$ that are "far out" in the left-half *s* plane in order to implement $H_{eq}(s)$ with *R* and *C* components along with Op amps.[21] When this is not feasible then the next subsection indicates how this active compensator requirement may be satisfied by transforming the continuous controller into a digital controller. Another approach to achieving the desired robustness is by the use of an observer.[3,4] Other definitions of robustness, such as minimizing the effect of unmodeled dynamics, results in different design techniques and is not discussed here.

15-8B State-Feedback *H*-Equivalent Digital Control System

When the dominant poles and zeros of $[M(s)]_D$ and $G_x(s)$ lie in the shaded area of Fig. 7-7, then the continuous-time control system of Fig. 15-16*d* may be converted

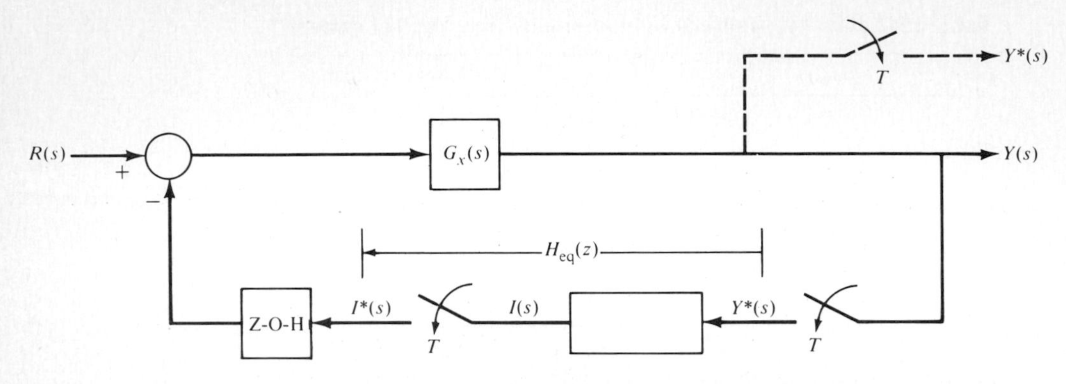

Figure 15-18 Conversion of the continuous-time control system of Fig. 15-16*d* into a hybrid digital control system.

into a digital control system as shown in Fig. 15-18. For a unit-step input,

$$Y(z) = \frac{G_x R(z)}{1 + H_{z0}G_x(z)H_{\text{eq}}(z)} \tag{15-208}$$

where

$$G_x R(z) = \mathscr{Z}\left[\frac{10A}{s^2(s - p_2)(s - p_3)}\right] = \frac{K_x z(z + a)(z + b)}{(z - 1)^2(z - c)(z - d)} \tag{15-209}$$

$$H_{z0}G_x(z) = (1 - z^{-1})\mathscr{Z}\left[\frac{10A}{s^2(s - p_2)(s - p_3)}\right]$$

$$= \frac{K_x(z + a)(z + b)}{(z - 1)(z - c)(z - d)} \tag{15-210}$$

$$H_{\text{eq}}(z) = [H_{\text{eq}}(z)]_{\text{TU}} = \text{Tustin transformation of } H_{\text{eq}}(s) \tag{15-211}$$

For the control system of Sec. 15-8A, a value of $T = 10$ ms is chosen to ensure that the dominant poles lie in the shaded region of Fig. 7-7. It is assumed that a ZOH unit is not utilized in the input of the $H_{\text{eq}}(z)$ block. Thus, Eq. (15-211) yields

$$[H_{\text{eq}}(z)]_{\text{TU}} = \frac{K_{\text{TU}}(z - p \pm jq)}{(z + 1)^2} \tag{15-212}$$

The root-locus plot of $H_{z0}G_x(z)[H_{\text{eq}}(z)]_{\text{TU}} = -1$ reveals that the digital control system is completely unstable due to the two poles at -1 contributed by $[H_{\text{eq}}(z)]_{\text{TU}}$; that is, two branches are outside the unit circle. Therefore, as a good first "engineering" modification to Eq. (15-212), to ensure a conditionally stable system, the factor $(z + 1)^2$ is replaced by $\alpha(z + \xi)^2$ (see Secs. 7-3 and 11-3), where α is the dc gain factor; that is,

$$[(z + 1)^2]_{z=1} = [\alpha(z + \xi)^2]_{z=1} \tag{15-213}$$

$\epsilon > \xi > -1$, and $\epsilon > 0$ (a very small positive number to ensure that the poles ξ are nondominant). Note that the converted control system is a fifth-order system.

Cases 1 and 4 of Sec. 15-8A are used to illustrate how the conversion of the continuous-time control system into a digital control system, by use of $[H_{eq}(s)]_{TU}$, maintains the desired parameter insensitivity with a physically realizable digital feedback compensator. The arbitrary choice of $\xi = 0.5$ is made in order that the $H_{eq}(z)$ poles are nondominant and yield a conditionally stable system. For this value of ξ results in $\alpha = 16/9$, so that

$$H_{eq}(z) = \frac{K_T(z - p \pm jq)}{(z + 0.5)^2} \tag{15-214}$$

and $K_T = 9K_{TU}/16$. Thus, for $T = 0.01$ s, the parameters of Eqs. (15-210) and (15-214) are:

	Case 1	Case 4
a	3.676668598	3.667485636
b	0.2639470221	0.2632881595
c	0.9900498337	0.9801986733
d	0.9512294245	0.9512294245
p	0.9927828829	
q	0.007209801628	
K_x	$1.641922848733 \times 10^{-5}$	1.637822×10^{-5}
K_T	21,620.475	
K_xK_T	$3.549915189 \times 10^{-1}$	$3.541049393 \times 10^{-1}$

Table 15-3 presents the figures of merit of the continuous-time and digital control systems for Cases 1 and 4 with the configurations of Figs. 15-16*d* and 15-18, respectively. The results in Table 15-3 indicate that the converted continuous-time control system of Fig. 15-18 maintains the desired degree of parameter insensitivity, i.e., the desired performance specifications with a realizable digital feedback controller $H_{eq}(z)$. A root locus of

$$H_{z0}G_x(z)H_{eq}(z) = -1 \tag{15-215}$$

Table 15-3 Time-response data

	M_p	t_p, s	t_s, s
Case 1:			
Fig. 15-16*d*	1.043 −	4.46	5.968
Fig. 15-18	1.043	4.45 +	5.94
Case 4:			
Fig. 15-16*d*	1.035 −	4.63 −	5.92
Fig. 15-18	1.035	4.63	5.88

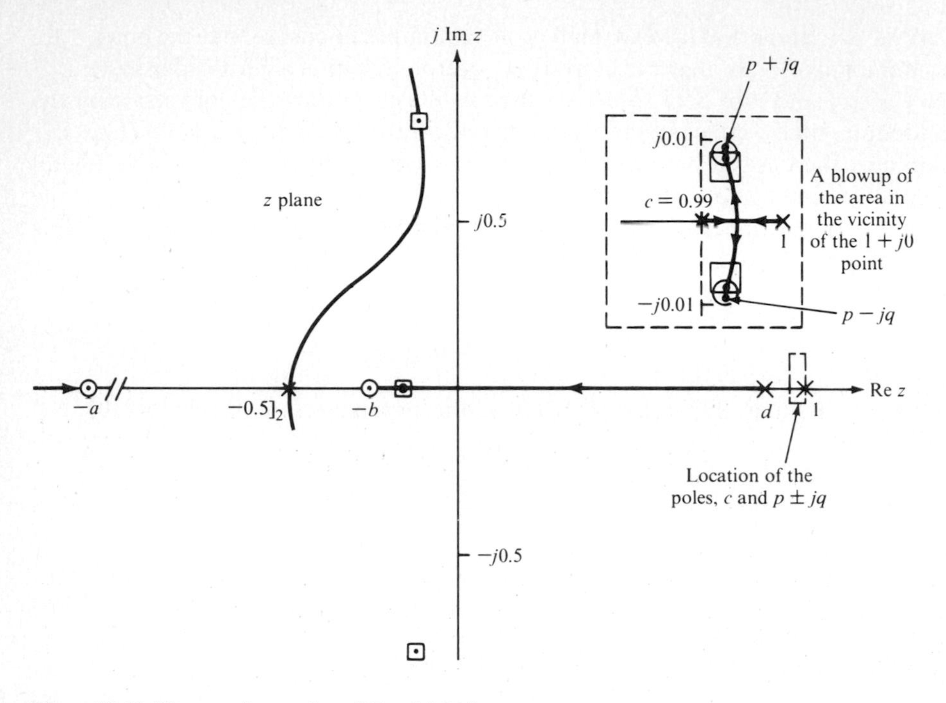

Figure 15-19 The root-locus plot of Eq. (15-215).

is shown in Fig. 15-19 for Case 1. It is seen from this figure that this root locus has equivalent characteristics as the root locus for the continuous-time system in Fig. 15-17; that is, the dominant roots are very close to the "fixed zeros" of $H_{eq}(z)$ and $H_{eq}(s)$, respectively.[1] It is this characteristic which achieves the degree of parameter insensitivity that is desired for the system design. The additional two poles, of the digital control system, due to $[H_{eq}(z)]_{TU}$, are nondominant and thus do not appreciably affect the time-response characteristics. The loop transmission frequency, ω_ϕ, of Im $\mathbf{H}_{z0}\mathbf{G}_x(j\omega)\mathbf{H}_{eq}(j\omega)$ is 107.7 rad/s.

A continuous-time controller $H(s)$ equivalent to Eq. (15-206) can yield results similar to those in Table 15-3. The feasibility of using $H(s)$ in a continuous-time control system lies in its implementation as mentioned previously.

15-8C Design Procedure

The implementation of the state-feedback H equivalent, Fig. 15-16d, does not require that the design of $H_{eq}(s)$ be obtained from the physical-variable representation. The design, whether it is modeled by a physical, phase, or canonical state-variable representation, results in the same $H_{eq}(s)$. This can be shown as follows: Since Eq. (15-196) is expressed in terms of physical variables, then to express it in terms of phase variables, consider the transformation

$$\mathbf{X}(s) = \mathbf{T}\mathbf{X}_c(s) \tag{15-216}$$

where $\mathbf{T}$ is the transformation matrix that transforms the physical variables $\mathbf{X}(s)$ into phase variables $\mathbf{X}_c(s)$.[1] Thus, substituting from Eq. (15-216) into Eq. (15-196) yields

$$\mathbf{H}_{eq}(s) = \frac{\mathbf{k}'\mathbf{X}(s)}{\mathbf{c}'\mathbf{X}(s)} = \frac{\mathbf{k}'\mathbf{T}\mathbf{X}_c(s)}{\mathbf{c}'\mathbf{T}\mathbf{X}_c(s)} = \frac{\mathbf{k}_c'\mathbf{X}_c(s)}{\mathbf{c}_c'\mathbf{X}_c(s)} \tag{15-217}$$

where $\mathbf{k}_c' = \mathbf{k}'\mathbf{T}$ and $\mathbf{c}_c' = \mathbf{c}'\mathbf{T}$. Using the nominal values for the system of Fig. 15-16 yields

$$\mathbf{T} = \begin{bmatrix} 1 & 0 & 0 \\ 0 & 1 & 0 \\ 0 & 1 & 0.2 \end{bmatrix} \tag{15-218}$$

$$\mathbf{k}' = \mathbf{k}'\mathbf{T} = [1 \quad 1.377 \quad 0.954] \tag{15-219}$$

$$\mathbf{c}' = \mathbf{c}'\mathbf{T} = [1 \quad 0 \quad 0] \tag{15-220}$$

Since $X_i(s) = sX_{i=1}(s)$ for the phase-variable representation, then from Eq. (15-217),

$$H_{eq}(s) = [1 \quad 1.377 \quad 0.954] \begin{bmatrix} X_1(s) \\ sX_1(s) \\ s^2X_1(s) \end{bmatrix} \frac{1}{X_1(s)}$$

$$= \frac{4.77s^2 + 6.885s + 5}{5} \tag{15-221}$$

which is identical to Eq. (15-206).

The specific design procedure for $H_{eq}(s)$ is as follows:

Step 1 The plant transfer function is expressed in polynomial format:

$$G_x(s) = \frac{Y(s)}{U(s)} = \frac{K_G(s^w + c_{w-1}s^{w-1} + \cdots + c_1s + c_0)}{s^n + a_{n-1}s^{n-1} + \cdots + a_1s + a_0} \tag{15-222}$$

This yields the state and output equations in terms of phase variables as follows:

$$\dot{\mathbf{x}} = \mathbf{A}_c\mathbf{x} + \mathbf{b}_c u \tag{15-223}$$

$$y = \mathbf{c}_c'\mathbf{x} \tag{15-224}$$

where

$$\mathbf{A}_c = \begin{bmatrix} 0 & 1 & 0 & \cdots & 0 \\ 0 & 0 & 1 & \cdots & 0 \\ \cdots & \cdots & \cdots & \cdots & \cdots \\ \cdots & \cdots & \cdots & \cdots & 1 \\ -a_0 & -a_1 & -a_2 & \cdots & -a_{n-1} \end{bmatrix} \tag{15-225}$$

$$\mathbf{b}_c = \begin{bmatrix} 0 \\ 0 \\ \vdots \\ K_G \end{bmatrix} \tag{15-226}$$

$$\mathbf{c}_c' = [c_0 \quad c_1 \quad \cdots \quad c_{w-1} \quad 1 \quad 0 \quad \cdots \quad 0] \tag{15-227}$$

Thus Eqs. (15-217) and (15-227) yield

$$[H_{eq}(s)]_c = \frac{k_n s^{n-1} + k_{n-1} s^{n-2} + \cdots + k_2 s + k_1}{s^w + c_{w-1} s^{w-1} + \cdots + c_1 s + c_0} \tag{15-228}$$

From Fig. 15-16d and substituting from Eqs. (15-222) and (15-228) into

$$\left[\frac{Y(s)}{R(s)}\right]_A = \frac{G_x(s)}{1 + G_x(s) H_{eq}(s)} \tag{15-229}$$

where $H_{eq}(s) = [H_{eq}(s)]_c$, yields

$$\left[\frac{Y(s)}{R(s)}\right]_A = \frac{K_G(s^w + c_{w-1} s^{w-1} + \cdots + c_0)}{s^n + (a_{n-1} + K_G k_n) s^{n-1} + \cdots + (a_0 + K_G k_1)} \tag{15-230}$$

which is the closed-loop system control ratio.

Step 2 Based upon the system's performance specifications, as discussed in Chap. 5, obtain $[M(s)]_D$ which contains at least one nondominant pole in order to achieve hfg operation. The minimum value of T, T_{min}, allowable should be considered when synthesizing $[M(s)]_D$. That is, the dominant poles and zeros of $G_x(s)$ and $[M(s)]_D$ should be in the shaded area of Fig. 7-7, and path A of Fig. 11-2 (with D_c' replaced by H_{eq}') applies. If this condition is not satisfied, a satisfactory design may still be achievable. For this situation, path B, $[M(s)]_D$ is synthesized, using T_{min}, based only upon the performance specifications. It may be necessary to refine the design technique for $H_{eq}(s)$ based upon the PCT system of Fig. 15-18. (See Prob. 15-19.)

Step 3 Set Eq. (15-230) equal to $[M(s)]_D$ and equate the corresponding coefficients to obtain K_G and k_i.

Step 4 By use of the Tustin transformation, transform $H_{eq}(s)$ to

$$[H_{eq}(z)]_{TU} = \frac{K_{TU}(z^{n-1} + \gamma_{n-2} z^{n-2} + \cdots + \gamma_1 z + \gamma_0)}{(z+1)^{n-1-w}(z^w + \beta_{w-1} + \cdots + \beta_1 z + \beta_0)} \tag{15-231}$$

Step 5 Replace $(z+1)^{n-1-w}$ in Eq. (15-231) by the dc equivalent, i.e., by

$$[(z+1)^{n-1-w}]_{z=1} = [\alpha(z+\xi)]_{z=1}^{n-1-w} \tag{15-232}$$

to obtain

$$H_{eq}(z) = \frac{K_T(z^{n-1} + \gamma_{n-2}z^{n-2} + \cdots + \gamma_1 z + \gamma_0)}{(z+\xi)^{n-1-w}(z^w + \beta_{w-1}z^{w-1} + \cdots + \beta_1 z + \beta_0)} \tag{15-233}$$

where $K_T = K_{TU}/\alpha$, $\epsilon > \xi > -1$, and $\epsilon = 0+$.

Step 6 Obtain $G_xR(z)$, $H_{z0}G_x(z)$, and

$$Y(z) = \frac{G_xR(z)}{1 + H_{z0}G_x(z)H_{eq}(z)} \tag{15-234}$$

For path B, determine the dominant roots for the desired ζ by use of the root-locus method; that is,

$$H_{z0}G_x(z)H_{eq}(z) = -1 \tag{15-235}$$

From the root-locus analysis, the remaining roots and static loop sensitivity are determined. For this case the value of A and K_T must be adjusted since, in general, they are not the same as for path A. If the loop transmission frequency is considered to be too large, for example, due to noise, then repeat step 2 with a new $[M(s)]_D$ that may yield a lower value for ω_ϕ or use another design technique.

This section has illustrated how a physically unrealizable $H_{eq}(s)$ controller may be transformed to an equivalent physically realizable digital controller $H_{eq}(z)$, which may be easily implemented, that maintains the desired insensitivity of the system output to plant parameter variations. The approach of this section of implementing an improper H_{eq} may not be feasible when unmodeled system dynamics are present.

15-9 SUMMARY

This chapter discussed various methods of representing discrete systems in state-variable forms. Using these forms, a discrete system can be depicted in the time or z domain. In addition, time response and stability phenomena can be analyzed. The evaluation of MIMO time-response values in between sampling instants is made easier by use of the state-variable representation. The last section presents a method for designing a nonunity-feedback digital control system whose output may have a high degree of insensitivity (robustness) to plant parameter variations. The SV formalization can be extended in conjunction with desired performance objectives to generate appropriate "optimal" controllers.[2-4,18] CAD packages (see Appendix E) are available[14,15] to aid in the extensive mathematical manipulations required for MIMO system analysis and synthesis.

BIBLIOGRAPHY

1. D'Azzo, J. J., and C. H. Houpis: "Linear Control System Analysis and Design: Conventional and Modern," 2d ed., McGraw-Hill, New York, 1981.
2. Kuo, B. C: "Discrete-Data Control Systems," Prentice-Hall, Englewood Cliffs, N.J., 1970.
3. Maybeck, P.: "Stochastic Models, Estimation, and Control," Academic, New York, vol. 1, 1979, vols. 2 and 3, 1982.
4. Jacquot, R.: "Modern Digital Control Systems," Dekker, New York, 1981.
5. Boyce, C.: "Microprocessor and Microcomputer Basics," Prentice-Hall, Englewood Cliffs, N.J., 1979.
6. Klingman, E.: "Microprocessor Systems Design," Prentice-Hall, Englewood Cliffs, N.J., 1977.
7. Mano, M.: "Computer Systems Architecture," Prentice-Hall, Englewood Cliffs, N. J., 1982.
8. Taub, H.: "Logic Design and Microprocessors," McGraw-Hill, New York, 1981.
9. Sloan, M. E.: "Introduction to Minicomputers and Microcomputers," Addison-Wesley, Reading, Mass., 1980.
10. Wakerley, F.: "Micro-Computer Architecture and Programming," Wiley, New York, 1981.
11. Oppenheim, A. V., and R. W. Schafer: "Digital Signal Processing," Prentice-Hall, Englewood Cliffs, N.J., 1975.
12. Shannon, C. E., B. M. Oliver, and J. R. Pierce: The Philosophy of Pulse Code Modulation, *Proc. IRE*, vol. 36, no. 11, November 1948, pp. 1324–1331.
13. Carnaham, B., H. A. Luther, and J. O. Wilkes: "Applied Numerical Methods," Wiley, New York, 1969.
14. Gembrowski, C. J.: "Continued Development of an Interactive Computer-Aided Design (CAD) Package for Discrete and Continuous Control System Analysis and Synthesis," M.S. thesis, GE/EE/82D, School of Engineering, Air Force Institute of Technology, Wright-Patterson AFB, Ohio, 1982.
15. Larimer, S. J.: "An Interative Computer-Aided Design Program for Digital and Continuous System Analysis and Synthesis (TOTAL)," M.S. thesis, GE/GGC/EE/78–2, School of Engineering, Air Force Institute of Technology, Wright-Patterson AFB, Ohio, 1978. Available from Defense Documentation Center (DDC), Cameron Station, Alexandria, Va. 22314.
16. Taylor, A. E.: "Advanced Calculus," Ginn, Boston, 1955.
17. Churchill, R. V.: "Complex-Variables and Applications," 2d ed., McGraw-Hill, New York, 1960.
18. Cadzow, J. A., and H. Martens: "Discrete-Time and Computer Control Systems," Prentice-Hall, Englewood Cliffs, N.J., 1974.

19. Tou, J. C.: "Modern Control Systems," McGraw-Hill, New York, 1976.
20. Kuo, B. C.: "Analysis and Synthesis of Sampled-Data Control Systems," Prentice-Hall, Englewood Cliffs, N.J., 1963.
21. D'Azzo, J. J., and C. H. Houpis: "Feedback Control System Analysis and Synthesis," 2d ed., McGraw-Hill, New York, 1966.
22. Jury, E. I., and J. Blanchard: A Stationary Test for Linear Discrete Systems in Table Form, *Proc. IRE*, vol. 49, no. 12, December 1961, pp. 1947–1948.
23. "Master Library, II Programwide 58/59," Texas Instruments, Inc., Dallas, Tex., 1977.
24. Franklin, G. F., and J. David Powell: "Digital Control of Dynamic Systems," Addison-Wesley, Reading, Mass., 1980.
25. Tustin, A.: A Method of Analyzing the Behavior of Linear Systems in Terms of Time Series, *JIEE* (London), vol. 94, pt. IIA, 1947.
26. Thaler, G. J., and M. P. Pastel: "Analysis and Design of Nonlinear Feedback Control Systems," McGraw-Hill, New York, 1962.
27. Garrett, H.: "Analog Systems for Microprocessors and Minicomputers," Reston Publishing Co., Reston, Virginia, 1978.
28. VanDoren, H.: "Data Acquisition Systems," Reston, Reston, Va., 1982.
29. Bibberro, J.: "Microprocessors in Instruments and Control," Wiley, New York, 1977.
30. de Sa, A.: "Principles of Electronic Instrumentation," Wiley, New York, 1981.
31. Singh, G., J. Elloy, R. Mezencev, and N. Munro: "Applied Industrial Control," Pergamon, Oxford, England, 1980.
32. Wrigley W., W. M. Hollister, and W. G. Denhard: "Gyroscopic Theory, Design, and Instrumentation," M.I.T., Cambridge, Mass., 1969.
33. Davenport, W. B.: "Probability and Random Processes," McGraw-Hill, New York, 1970.
34. Papoulis, A.: "Probability, Random Variables, and Stochastic Processes," McGraw-Hill, New York, 1965.
35. Myers, R. H., and R. E. Walpole: "Probability and Statistics for Engineers and Scientists," 2d ed., Macmillan, New York, 1978.
36. Feller, W.: "An Introduction to Probability Theory and Its Applications," 3d ed., Wiley, New York, vol. 1, 1966, vol. 2, 1968.
37. Hodges, J. L., Jr., and E. L. Lehmann: "Basic Concepts of Probability and Statistics," 2d ed., Holden-Day, San Francisco, 1970.
38. Miller, I., and J. E. Freund: "Probability and Statistics for Engineers," 2d ed., Prentice-Hall, Englewood Cliffs, N.J., 1977.
39. Dahlin, E. G.: Designing and Tuning Digital Controllers, *Instrum. Control Syst.*, vol. 41, no. 6, June 1968.
40. Rabiner, L. R., and B. Gold: "Theory and Application of Digital Signal Processing," Prentice-Hall, Englewood Cliffs, N.J., 1975.
41. Fletcher, R., and M. J. D. Powell: A Rapidly Convergent Descent Method for Minimization, *Comput. J.*, vol. 6, no. 2, 1963, pp. 163–168.
42. Gugeler, D. R., and B. M. Hall: "Delta Inertial Guidance System Digital Filter Analysis," McDonnel Douglas Astronautics Co. – West, Huntington Beach, Calif., paper no. WD 2111, June 1973.
43. Rabiner, L. R., and K. Steiglitz: The Design of Wide-Band Recursive and Nonrecursive Digital Differentiators, *IEEE Trans. Audio Electroacoust.*, vol. 18, no. 2, June 1970, pp. 204–209.
44. Ratten, K. S.: "Digitization of Existing Continuous-Data Control Systems," Air Force Flight Dynamics Laboratory, Air Force Wright Aeronautical Laboratories, Wright-Patterson AFB, Ohio, report AFWAL-TM-80-105-FIGC, September 1980.
45. Auslander, D. M., and P. Sagues: "Microprocessors for Measurement and Control," McGraw-Hill, New York, 1981.
46. Gault, J. W., and R. L. Pimmel: "Introduction to Micro-Computer-Based Digital Systems," McGraw-Hill, New York, 1982.
47. Kernighan, W., and Dennis M. Ritchie: "The C Programming Language," Prentice-Hall, Englewood Cliffs, N.J., 1978.

48. McCracken, D.: "A Guide to P1/M Programming for Microcomputer Applications," Addison-Wesley, Reading, Mass., 1978.
49. Eckhouse, R. H.: "Minicomputer Systems: Organization and Programming (PDP-11)," Prentice-Hall, Englewood Cliffs, N.J., 1975.
50. "Classics in Software Engineering," Yourdon, New York, 1978.
51. Jensen, R. W., and C. C. Tonies: "Software Engineering," Prentice-Hall, Englewood Cliffs, N.J., 1979.
52. Bergland, G. D.: A Guided Tour of Program Development Methodologies, *Computer*, vol. 14 no. 10, October 1981.
53. Ross, D., and K. Schoman: Structured Analysis for Requirements Definition, *IEEE Trans. Software Eng.*, vol. 3, no. 1, January 1977, pp. 6–15.
54. "Integrated Computer Aided Manufacturing, Dynamic Modeling Manual," Materials Laboratory, Air Force Wright Aeronautical Laboratories, Air Force Systems Command, Wright-Patterson AFB, Ohio, June 1981.
55. DeMarco, T.: "Structured Analysis and System Specification," Yourdon, New York, 1978.
56. Wirth, N.: "Systematic Programming: An Introduction," Prentice-Hall, Englewood Cliffs, N.J., 1976.
57. Yourdon, E., and L. L. Constantine: "Structured Design," Prentice-Hall, Englewood Cliffs, N.J., 1979.
58. Meyers, J.: "Software Reliability: Principles and Practices," Wiley, New York, 1976.
59. Jackson, A.: "Principles of Program Design," Academic, New York, 1975.
60. Orr, K. T.: "Structured Systems Development," Yourdon, New York, 1977.
61. Warnier, J. D.: "Logical Construction of Programs," Stenfert Kroese B.V., Leiden, Holland, 1974.
62. Gilb, T.: "Software Metrics," Winthrop, Cambridge, Mass., 1977.
63. Higgins, D.: "Program Design and Construction," Prentice-Hall, Englewood Cliffs, N.J., 1979.
64. Kernighan, B. W., and W. Plauger: "The Elements of Programming Style," McGraw-Hill, New York, 1974.
65. Blakelock, J. H.: "System Analysis and Hybrid Simulation of an Integrated Flight Fire Control System Including Movable Gun and Gimbaled Line of Sight Tracker," Ph.D. dissertation, University of Dayton, Dayton, Ohio, December 1979. Appendix J references the first printing of the manuscript of this text.
66. Coffman, Edward G., and Peter J. Denning: "Operating Systems Theory," Prentice-Hall, Englewood Cliffs, N.J., 1973.
67. Dahmke, Mark: "Microcomputer Operating Systems," Byte Books, Peterborough, N.H., 1982.
68. Conzalez, M. J.: Deterministic Processor Scheduling, *Comput. Surv.*, vol. 9, no. 3, 1977.
69. Seward, W. D.: "Optimal Multiprocessor Scheduling of Periodic Tasks in a Real-Time Environment," Ph. D. dissertation, DS/EE/79-2, School of Engineering, Air Force Institute of Technology, Wright-Patterson AFB, Ohio, 1979.
70. Kalman, R. E.: When Is a Linear Control System Optimal?, *J. Basic Eng.*, pp. 51–60, March 1964.
71. Barnett, S.: "Matrices in Control Theory," Van Nostrand Reinhold, New York, 1971.
72. Kutz, P., and J. D. Powell: Sample Rate Selection for Aircraft Digital Control, *AIAA J.*, vol. 13, no. 8, 1973.
73. Kranc, G. M.: Input-Output Analysis of Multirate Feedback Systems, *IRE Trans. Auto. Control*, November 1957, pp. 21–28.
74. Kranc, G. M.: Compensation of an Error-Sampled System by a Multirate Controller, *AIEE Trans.*, November 1957, pp. 149–158.
75. Coffey, T. C., and I. J. Williams: Stability Analysis of Multiloop, Multirate Sampled Systems, *AIAA J.*, vol. 4, December 1966, pp. 2178–2190.
76. Kuo, B. C.: "Digital Control Systems," Holt, New York, 1980.
77. Boykin, W. H., and B. D. Frazier: Analysis of Multiloop Multirate Sampled-Data System, *AIAA J.*, vol. 13, no. 4, April 1975, pp. 453–456.

78. Boykin, W. H., and B. D. Frazier: Multirate Sampled-Data Systems Analysis via Vector Operators, *IEEE Trans. Autom. Control*, AC-20, August 1975, pp. 548–551.
79. Jury, E. I.: Synthesis and Critical Study of Sampled-Data Control Systems, *AIEE Trans.*, vol. 75, pt. II, July 1956, pp. 141–151.
80. Jury, E., and W. Schroeder: Discrete Compensation of Sampled-Data and Continuous Control Systems, *AIEE Trans.*, vol. 75, pt. II, January 1957, pp. 315–317.
81. Sklansky, J., and J. R., Ragazzine: Analysis of Error in Sample Data Feedback Systems, *AIEE Trans.*, vol. 74, pt. II, May 1955, pp. 65–71.
82. Stanford, D. P.: Stability for Multirate Sample Data Systems, *SIAM J. Optimum Control*, vol. 17, no. 3, May 1979, pp. 390–399.
83. Whitbeck, R. F., and L. G. Hofmann: "Analysis of Digital Flight Control Systems with Flying Qualities Applications, Vol II," Technical Report, AFFDL-TR-78-115, 1978.
84. Ragazzini, J. R., and G. F. Franklin: "Sampled-Data Control Systems," McGraw-Hill, New York, 1958.
85. Porter, B.: Deadbeat State Reconstruction of Linear Multivariable Discrete-Time Systems, *Electron. Lett.*, vol. 9, 1973, pp. 176–177.
86. Iserman, Rolf: "Digital Control Systems," Springer-Verlag, New York, 1981.
87. Calingnert, P.: "Operating System Elements, A User Perspective," Prentice-Hall, Englewood Cliffs, N. J., 1982.
88. Roberts, S. K.: "Industrial Design with Microcomputer," Prentice-Hall, Englewood Cliffs, N.J., 1982.
89. Katz, P.: "Digital Control Using Microprocessors," Prentice-Hall, Englewood Cliffs, N.J., 1981.
90. Powell, J. D., E. Parsons, and M. G. Tashker: "A Comparison of Flight Control Design Methods," Guidance and Control Conference, San Diego, Calif., August 1976.
91. Bode, H.: "Network Analysis and Feedback Amplifier Design," Van Nostrand, New York, 1945.
92. Bracewell, R. N.: "The Fourier Transform and Its Applications," 2d ed., McGraw-Hill, New York, 1978.
93. Daniels, R. W.: "Approximation Methods for Electronic Filter Design," McGraw-Hill, New York, 1974.
94. Evans, W. R.: Control System Synthesis by Root Locus Methods, *AIEE Trans.*, vol. 69, pt. 2, 1950, pp. 66–69.
95. Hamming, R.: "Numerical Methods for Scientists and Engineers," McGraw-Hill, New York, 1962.
96. Hnatek, E. R.: "A User's Handbook of D/A and A/D Converters," Wiley, New York, 1976.
97. Jury, E. I.: "Theory and Application of the z-Transform Method," Wiley, New York, 1964.
98. Mantey, P. E., and G. F. Franklin: Comment on Digital Filter Design Techniques in the Frequency Domain, *Proc. IEEE*, vol. 55, no. 12, 1967, pp. 2196–2197.
99. Mayr, O.: "The Origins of Feedback Control," M.I.T., Cambridge, Mass., 1970.
100. Truxal, J. G.: "Automatic Feedback Control System Synthesis," McGraw-Hill, New York, 1955.
101. Parnes, D. L.: On the Criteria to Be Used in Decomposing Systems into Modules, *Commun. ACM*, vol. 5, no. 12, December, 1972.
102. Leveson, Nancy G.: Software Safety in Computer-Controlled Systems, *IEEE Computer*, vol. 17, no. 2, February 1984.
103. Matsumoto, Yoshihiro: Management of Industrial Software Production, *IEEE Computer*, vol. 17, no. 2, February 1984.
104. Schoeffler, James D.: Distributed Computer Systems for Industrial Process Control, *IEEE Computer*, vol. 17, no. 2, February 1984.
105. Steusloff, Hartwig U.: Advanced Real-time Languages for Distributed Industrial Process Control, *IEEE Computer*, vol. 17, no. 2, February 1984.
106. Walter, Claudio: Control Software Specification and Design: An Overview, *IEEE Computer*, vol. 17, no. 2, February 1984.

107. Weide, Bruce W., Mark Brown, Jayashree Ramanathan, and Karsten Schwan: Process Control: Integration and Design Methodology Support, *IEEE Computer*, vol. 17, no. 2, February 1984.
108. Nagle, H. T., and V. P. Nelson: Digital Filter Implementation on 16-Bit Microcomputers, *IEEE Micro*, vol. 1, no. 1, February 1981.
109. Andrews, M.: "Programming Microprocessor Interfaces for Control and Instrumentation," Prentice-Hall, Englewood Cliffs, N.J., 1982.

PROBLEMS

Chapter 2

2-1 Convert the following positive numbers to base 10 numbers:

(a) 101110_2 (*f*) 1111_8
(*b*) 11111110_2 (*g*) 7777_8
(*c*) 1011.011_2 (*h*) $ABCD_{16}$
(*d*) 10101.010101_2 (*i*) 1111_{16}
(*e*) 6751.7_8 (*j*) $FOFO_{16}$

2-2 Convert the (*a*) to (*j*) numbers in Prob. 2-1 to base 16 numbers.

2-3 Add the following pairs of binary numbers and show carries:

(*a*) 100111, 11101 (*c*) 101110.001, 1101.11
(*b*) 00111, 00010 (*d*) 10.11101, 0.11110

2-4 Subtract the binary pairs in Prob. 2-3 (the second number is the subtrahend).

2-5 Add the following pairs of octal numbers and show carries:

(*a*) 7777, 1111 (*c*) 010110, 010001
(*b*) 6541, 1456 (*d*) 4152.64, 312.75

2-6 Subtract the octal pairs in Prob. 2-5 (the second number is the subtrahend).

2-7 Add the following pairs of hexadecimal numbers and show carries:

(*a*) 7777, 1111 (*c*) 7A2B, 5F4C
(*b*) ABCD, 8DCB (*d*) ADEF.AB, 1725.16

2-8 Subtract the hexadecimal pairs in Prob. 2-7 (the second number is the subtrahend).

2-9 Find the 2s complement of the following numbers assuming 7 bits:

(*a*) 011101.1 (*c*) 1111111 (*e*) 000000011.1101
(*b*) 101.10 (*d*) 1010101

2-10 For the following decimal numbers, give the three possible number representations assuming 16-bit representation:

(*a*) $+31$ (*b*) -804 (*c*) -61.9 (*d*) -0.12 (*e*) -9.61 (*f*) -1.11

2-11 Which of the following 8-bit summations generate an overflow?

(*a*) 10110111 + 00101111 (*c*) 011011.11 + 011111.01
(*b*) 01101.101 + 1010.1010 (*d*) 1101110.0 + 001001.01

2-12 Multiply the following binary pairs using the three binary number representations:

(*a*) +1101, +1011 (*c*) +1101, −1011
(*b*) −1101, +1011 (*d*) −1101, −1011

2-13 How many address bits are required to address 64,000 and 513,000 words of memory, respectively?

2-14 What is the advantage of a synchronous counter over a ripple counter?

2-15 Explain how to modify a ripple counter so that it divides by N, where N is not a power of 2.

2-16 How do you measure frequency with a counter?

2-17 Draw the block diagram of a four-stage synchronous counter with ripple carry. (*a*) Explain its operation. (*b*) What is the maximum frequency of operation? Please define symbols used in the associated boolean equations. Repeat for a parallel carry.

2-18 Consider a 5 × 7 matrix to represent an alphanumeric character. (*a*) Draw in block diagram form a system for generating such characters and for writing them on a CRT. (*b*) Indicate the horizontal (X), the vertical (Y), and the intensity (Z) waveforms for a single character. (*c*) Explain the operation of the system. (*d*) How is this system modified so as to write a line of N characters on the CRT? (*e*) Calculate the clock frequency if each line of 40 characters is to be refreshed at a 60-Hz rate.

2-19 Determine if

(*a*) $\overline{(A + \bar{B})\bar{C}D} = \bar{A}B + C + \bar{D}$

(*b*) $\overline{A\bar{B}C} = ABC + \overline{\bar{A}B}C + \bar{A}BC$

(*c*) $\bar{A}\bar{B} + AB = \overline{\bar{A}B + A\bar{B}}$

(*d*) $ABC = \overline{A\bar{B}\bar{C} + A}$

2-20 Generate a boolean expression and an AND/OR logic diagram for the following statement:

"The garage door opens if the override switch is off, the house alarm signal from Sec. 2-3 is off, and the following code is entered on a keyboard: I,N,2."

Use 8-bit ASCII codes for the coded input.

2-21 Develop a boolean expression and a NAND/NOR logic diagram for the following chemical process:

1. Load into cooker W lb barley, X lb wheat, Y gal water, and Z cups yeast.
2. Cook at 197° for 173 h.
3. Drain cooker contents into cooling containers.
4. Cool to 38° and leave for 80 h.
5. Drain into 8-oz cans and seal.
6. When cooling container is empty, return to step 1.

Hint: Define switches for each sequential activity. Assume when a measurement value indicates that a subprocess should commence, a switch is set true. The continuous dynamics of the system are not controlled by this sequence, i.e., lower-level control systems provide this capability. Note that this is a hierarchical process control system.

2-22 Study the block diagram hardware architecture of various microprocessor integrated circuit systems in terms of the five basic computer units (see Refs. 7 to 9). Draw simplified bus diagrams of each.

2-23 Describe, by means of a block diagram, a home digital heating and air-conditioning control system. Use the basic five units of a computer system and define explicitly each functional block. Again, represent the system at a process switching level with a boolean equation.

2-24 Using Refs. 7 to 9 or a visit to a local process control facility, model a discrete switching network as a set of boolean logic equations and generate a logic diagram equivalent.

2-25 A process control system controls the converter speed and direction of a hot iron ingot as well as the rollers that compress its thickness. Measurements of roller interval boundaries, speed, speed direction, ingot thickness, and ingot width are employed. An alarm sounds if the speed is too high, the boundaries are exceeded, the ingot thickness is too high or low, or the width is too large. Develop a boolean equation and a NAND/NOR logic diagram.

2-26 Expand the process control loop of Fig. 1-5 into a block diagram using the five basic computer system units. Describe the switching process in terms of a boolean equation.

Chapter 3

3-1 For the SISO sampled-data systems characterized by the following equations, determine which coefficients must be zero for the system to be

(*a*) Linear (*b*) Time-invariant

(1) $b_1\ddot{c}^2(t) + b_2\ddot{c}(t) + (b_3 + b_4 \cos t)\dot{c}(t) + b_5 c(t) = r(t)$

(2) $b_1 c^2[(k + 2)T] + [b_2 + b_3 c(kT) + b_4 \sin kT]c(kT) = r(kT)$

(3) $c(kT) = \dfrac{b_1}{T}\{r(kT) - r[(k - 1)T]\} + 2b_2$

3-2 The transfer function of a time function $e(t)$ is

$$E(s) = \frac{10(s + 2)}{s(s^2 + 2s + 2)}$$

Based upon the Shannon sampling theorem, determine the maximum value of T that can be used to reconstruct $e(t)$.

3-3 Given $e(t) = 10 \sin 6.28t$

$e(t)$ —sampler (T)→ $e^*(t)$

Based upon Shannon's sampling theorem, determine the value of T.

3-4 Given $e(t) = \sin \omega_c t = \sin 2\pi t$. A value of $T = 1$ s is chosen as the sampling time. Is this value T a good choice? Explain (briefly) with the aid of a sketch.

3-5 Utilize the rth-backward difference equations to simulate the following continuous-time system on a digital computer. Solve for $c(kT)$.

$$G(s) = \frac{C(s)}{E(s)} = \frac{10}{s + 1}$$

3-6 Utilize the rth-backward difference equations to simulate the following systems on a digital computer.

(1) $G(s) = \dfrac{C(s)}{E(s)} = \dfrac{2(s + 2)}{s(s + 1)}$ (2) $G(s) = \dfrac{C(s)}{E(s)} = \dfrac{10s}{s^2 + s + 1}$

(*a*) Solve for $c(kT)$.

(*b*) If $e(t) = 2u_{-1}(t)$, determine the appropriate value of T for accurately evaluating values of $c(t)$ at $t = kT$.

(*c*) Determine $c(kT)$ for $0 \le k \le 5$.

(*d*) Compare your values of $c(kT)$ with those obtained from $c(t) = \mathscr{L}^{-1}[C(s)]$.

3-7 Given $G(s) = C(s)/E(s) = 1/(s + 1)$ and $E(s) = 1/s^2$. Determine $c(t)$ at $t = 0.5$ by use of rectangular integration as represented by Eq. (3-79). (Utilize Table 3-2.)

3-8 Figure *a* depicts a continuous-time system where $G_x(s) = 1/(T_x s + 1)$, along with its output response $y_c(t)$ and its input $u(t) = u_{-1}(t)$. Figure *b* is a sampled-data system for the same $G_x(s)$. By the method of Sec. 3-7, derive the expressions for $y_s(kT)$ and $y_c(t)$ when:

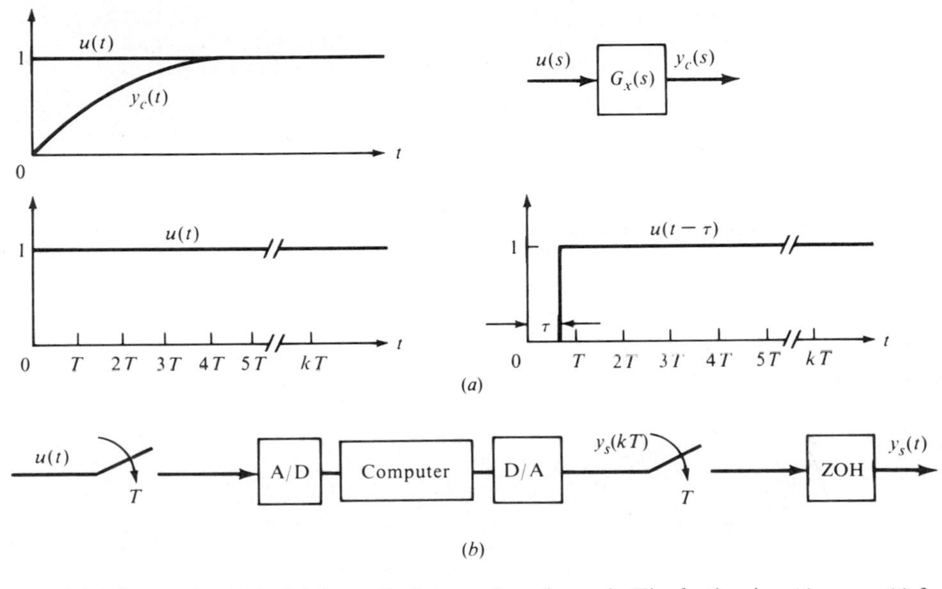

(*a*) The forcing function is applied at $t = 0$ as shown in Fig. *b*; that is, $u(t) = u_{-1}(t)$ for $t > 0$, or

$$u(t) = \begin{cases} 0 & \text{for } t \le 0 \\ 1 & \text{for } t > 0 \end{cases}$$

(*b*) The forcing function is applied at $t = \tau$ (delayed τ s) as shown in Fig. *b*; that is,

$$u(t - \tau) = \begin{cases} 0 & \text{for } t \le \tau \\ 1 & \text{for } t > \tau \end{cases}$$

(*c*) For each case where $T_x = T = 1$ s and $\tau = 0.6$ s, sketch $y_s(t)$. *Note*: For most practical systems the delay of τ s occurs with respect to the forcing function. However, in this text, unless otherwise specified, it is assumed to be zero.

3-9 Repeat Prob. 3-8 where the forcing function is $u(t) = \sin 0.2t$.

Chapter 4

4-1 For Prob. 3-4 determine $E(z)$ with T unspecified. Correlate the resulting $E(z)$ function, for $T = 1$ s, with the answer to Prob. 3-4.

4-2 Utilize the definition of the one-sided $\mathscr{Z}$ transform given by Eq. (4-3) to determine the $\mathscr{Z}$ transform in closed form of the following discrete-time functions:

$$(a)\ e(kT) = \begin{cases} 0 & \text{for } k < 2 \\ a^{k-1} & \text{for } k \geq 2 \end{cases}$$

$$(b)\ e(kT) = \begin{cases} 0 & \text{for } k < 0 \\ \epsilon^{-akT} - \epsilon^{akT} & \text{for } k \geq 0 \end{cases}$$

where a is a constant. *Note*:

$$F(x) = 1 + x^{-1} + x^{-2} + x^{-3} + \cdots = \frac{1}{1 - x^{-1}} \qquad \text{for } |x| > |1|$$

4-3 Given $e(t) = \cos 2\pi t$.

(*a*) A value of $T = 0.5$ s is chosen as the sampling time. Determine $E(z)$.

(*b*) Is $T = 0.5$ a good choice? Explain (briefly) with the aid of a sketch.

4-4 Derive $E(z)$ by use of Eq. (4-12) or (4-14) for

$$(a)\ E(s) = \frac{1}{s(s+1)^2} \qquad (c)\ E(s) = \frac{s+c}{(s+a)(s+b)}$$

$$(b)\ E(s) = \frac{10(s+2)}{s(s+4)^2} \qquad (d)\ E(s) = \frac{K(1-\epsilon^{-sT})}{s^2(s+1)(s+2)}$$

4-5 From the graphs of $e(kT)$ as shown in the following figures, determine $E(z)$.

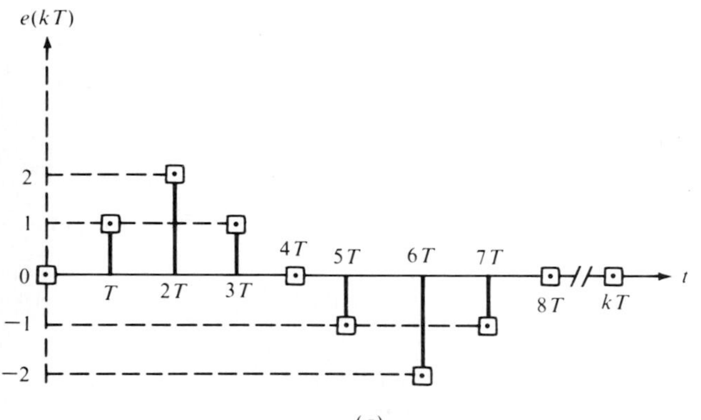

(*a*)

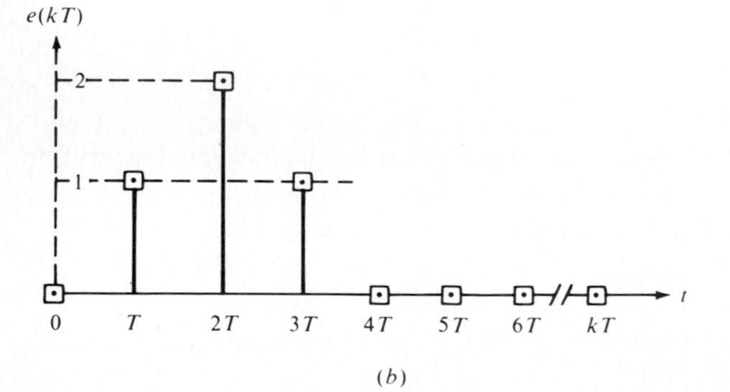

(*b*)

4-6 (*a*) Two sets of complex-conjugate poles, $p_{1,2}$ and $p_{3,4}$, are shown, respectively, in the figure.

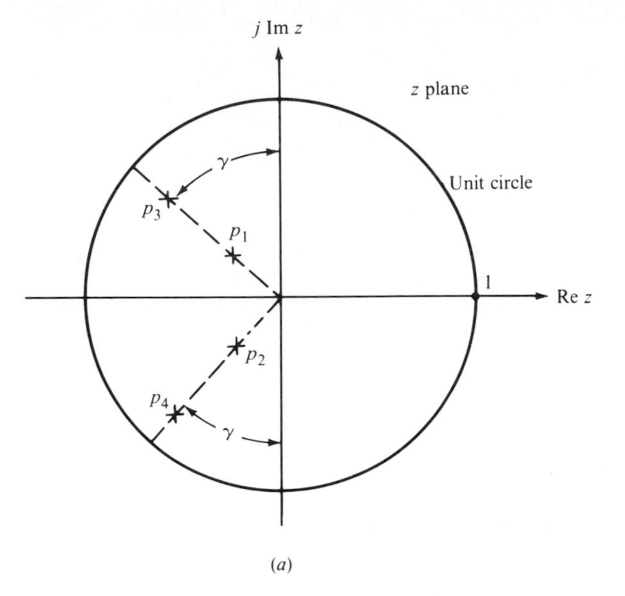

(*a*)

Describe the time-response characteristics corresponding to each set of poles and compare the time responses. Your answer is to be based upon a mathematical analysis (see Sec. 4-3).

(*b*) By means of a sketch, indicate the time-response characteristic of $e_i(kT)$ due to each of the real poles and $e_{i,j}(kT)$ for the complex-conjugate pairs shown, respectively, in the figure.

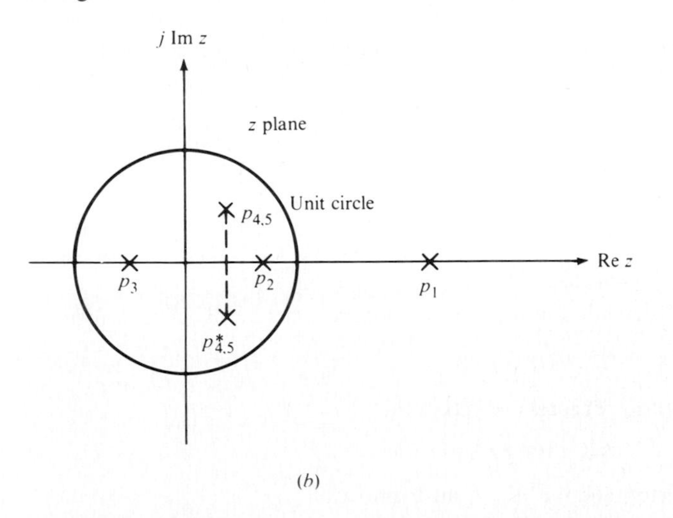

(*b*)

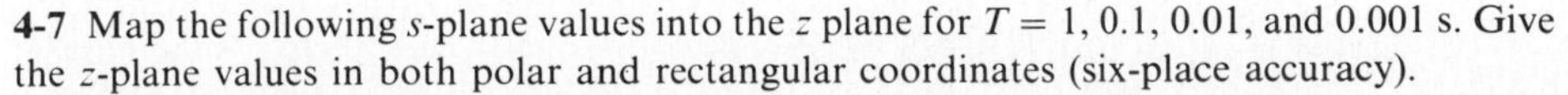

4-7 Map the following s-plane values into the z plane for $T = 1, 0.1, 0.01$, and 0.001 s. Give the z-plane values in both polar and rectangular coordinates (six-place accuracy).

(*a*) $s_1 = -0.5 + j0.5$ (*c*) $s_3 = -0.2$

(*b*) $s_2 = -0.1 + j0.1$ (*d*) $s_4 = -2$

4-8 (*a*) Locate the *s*-plane values of Probs. 4-7*a* and *b* on the appropriate ζ curve of Fig. 4-9 for each value of T.

(*b*) Comment on the ability to distinguish between z_1 and z_2, for $T = 0.001$ s, on Fig. 4-9.

4-9 Determine $e(0)$ and $e(\infty)$ for the following $E(z)$ functions:

$$(a)\ E(z) = \frac{z(z+2)}{(z-1)(z^2-3z+1)} \qquad (b)\ E(z) = \frac{Kz(z+1)}{(z-1)(z^2-0.5z+0.5)}$$

$$(c)\ E(z) = \frac{z(z-2)}{(z-1)(z+0.5)(z+2)(z^2+0.5z+0.4)}$$

$$(d)\ E(z) = \frac{K[z^3 - 2az^2 + (a^3 - a^2 + a)z]}{(z-1)(z-a)(z-a^2)}$$

$$(e)\ E(z) = \frac{z(z^2+z-0.25)}{(z-1)(z-0.5)^2}$$

$$(f)\ E(z) = \frac{Kz(z+b)}{(z-1)(z+a \pm jc)}$$ where a, b and c are real numbers.

4-10 For Probs. 4-9*d* and *f*, assuming a stable response, determine what must be the value of K for $e(\infty) = 1$.

4-11 By use of Theorem 3, determine $C(z)/R(z)$ for the system characterized by the following linear difference equations:

$$(a)\ c(kT) = -2c[(k-1)T] + c[(k-2)T] + r(kT) - 0.4r[(k-2)T]$$

$$(b)\ c(kT) = -0.1c[(k-1)T] + 0.16c[(k-2)T] - 0.2c[(k-3)T] + K\{r[(k-1)T] + 1.5r[(k-2)T] + 0.5r[(k-3)T]\}$$

4-12 Obtain the linear difference equation for the sampled-data systems described by the following control ratios by solving for $c(kT)$:

$$(a)\ \frac{C(z)}{R(z)} = \frac{z^{-1} + 3z^{-2} + 2z^{-3}}{1 + 1.5z^{-1} + 1.5z^{-2} + 0.5z^{-3}}$$

$$(b)\ \frac{C(z)}{R(z)} = \frac{1 - 0.5z^{-2}}{1 + 2z^{-1} - z^{-2}}$$

Hint: Use Theorem 3.

4-13 By use of the partial differentiation theorem, Theorem 8, derive the $\mathscr{Z}$ transform of

$$f(t) = t \sin \omega t$$

4-14 By use of the $\mathscr{Z}$ transform, Theorem 4, determine

$$E_A(z) \text{ for } e_A(t) = t^2 \epsilon^{+At}$$

4-15 By use of Table 4-1, determine the $\mathscr{Z}^{-1}$ transform for

$$(a)\ E(z) = \frac{z}{z+d} \qquad (b)\ F(z) = z^{-1}E(z) = \frac{1}{z+d}$$

Express your answer in terms of d where $d > 0$. *Hint*: Let

$$-d = \exp(-\mathbf{a}T) \qquad \text{where } \mathbf{a} = a_1 + ja_2$$

4-16 Given

$$\frac{C(z)}{R(z)} = \frac{z+2}{z^2+4z+8} \qquad \text{and} \qquad R(z) = \frac{z}{z-1}$$

(*a*) By use of the $\mathscr{Z}$-transform theorems, determine $c(0)$ and $c(\infty)$.
(*b*) By use of the power-series method, determine $c(3T)$.

4-17 Determine $e(kT)$ by the use of the partial-fraction expansion method for the following functions:

(*a*) $E(z) = \dfrac{z(z+0.1)}{(z-1)(z-0.2)}$

(*b*) $E(z) = \dfrac{0.5z}{(z-1)(z-0.5)}$

(*c*) $E(z) = \dfrac{z^2+z-0.25}{(z-0.5)^2}$

(*d*) $E(z) = \dfrac{z^3+2z^2+z+1}{(z-1)(z-0.5)^2} = \dfrac{z^3+2z^2+z+1}{z^3-2z^2+1.25z-0.25}$

(*e*) $E(z) = \dfrac{z(z-0.25)}{(z-1)(z-0.4 \pm j0.3)}$

(*f*) $E(z) = \dfrac{z(1.3z+0.55)}{(z-1)(z^2+0.6z+0.25)}$

(*g*) $E(z) = \dfrac{2}{(z-1)(z-0.5)}$

4-18 Repeat Prob. 4-17 by the power-series method. Obtain values of $e(k') = e(kT)$ for $k' = 0, 1, \ldots, 4$ where $k' = kT$. Verify the value of $e(3)$ with the value obtained in Prob. 4-17 for $T = 1$ s for each function.

4-19 If $I(s) = 1$ (unit impulse), determine $O_a(z)$ and $O_b(z)$, respectively, where

$$G_1(s) = \frac{1}{s(s+1)} \qquad \text{and} \qquad G_2(s) = \frac{1}{s(s+2)}$$

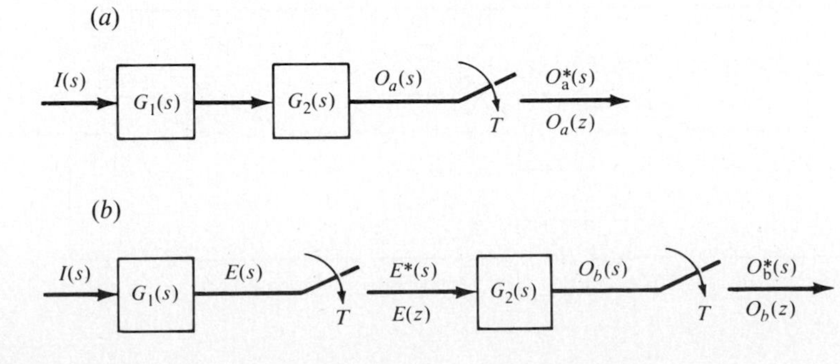

4-20 For the sampled-data systems shown, determine $C(z)$ and $C(z)/R(z)$, if possible.

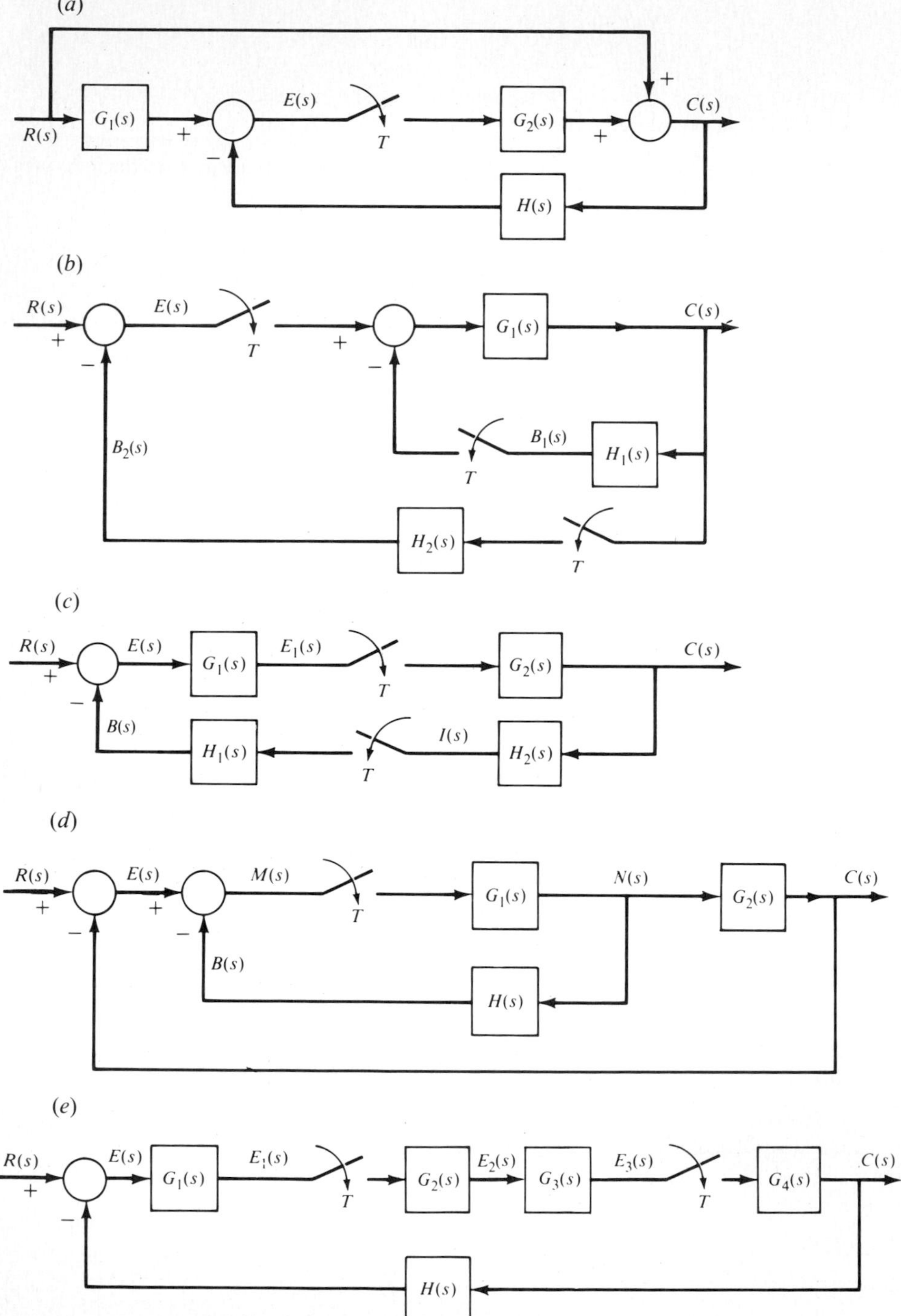

4-21 A discrete data system has the following transfer functions. By use of the translation theorem, determine $c(kT)$.

$$(a)\ \frac{C(z)}{R(z)} = \frac{2(1-0.8z^{-1})}{1+z^{-1}+2z^{-2}-0.5z^{-4}}$$

$$(b)\ \frac{C(z)}{R(z)} = \frac{15z^{-2}(1-0.89z^{-1})(1+0.9z^{-1})}{(1+0.5z^{-1})(1+0.1z^{-1})(1+0.5z^{-1}+0.4z^{-2})}$$

$$(c)\ \frac{C(z)}{E(z)} = \frac{4+3z^{-1}+z^{-2}}{1+5z^{-1}+6z^{-2}}$$

4-22 A digital computer implements the following linear difference equations.

$$e_2(kT) = 2e_1(kT) + 0.2e_3(kT)$$

$$e_3(kT) = e_3[(k-1)T] + Te_1(kT)$$

By use of the translation theorem, determine the computer transfer function

$$D(z) = \frac{E_2(z)}{E_1(z)}$$

4-23 By use of the translation theorem, determine the transfer function $C(z)/R(z)$ from the difference equation

$$c(kT) = d_1c[(k-1)T] - d_2c[(k-2)T] + d_3c[(k-3)T] + a_1r[(k-1)T] + a_2r[(k-2)T]$$

4-24 Apply the initial- and final-value theorems to the following functions to determine $c(0)$ and $c(\infty)$:

$$(a)\ C(z,m) = \frac{(\epsilon^{-m}-\epsilon^{-2m})z + (0.368\epsilon^{-2m}-0.135\epsilon^{-m})}{z^2-0.5z+0.05}$$

$$(b)\ C(z,m) = \frac{0.25[2(1-\epsilon^{-m})z + (2\epsilon^{-m}-0.736)]}{z^2-1.368z+0.368}$$

4-25 Find the modified $\mathscr{Z}$ transform of the following functions:

$$(a)\ \frac{1}{s^2} \qquad (b)\ \frac{1}{s(s+a)(s+b)} \qquad (c)\ \frac{2}{s(s^2+2s+2)}$$

Do the modified $\mathscr{Z}$ transforms of these functions check with their $\mathscr{Z}$ transforms when $m = 1$? Determine $e[(k-\delta)T]$ for $k = 1$ and $\delta = 1/2$. *Hint*: For (*b*) obtain $E(z, 1)$ in the expanded form and as a single fraction. On the single-fraction expression of $E(z, 1)$ perform a partial-fraction expansion on $E(z, 1)/z$ and then compare $E(z, 1)$ to the expanded form of $\mathscr{Z}[E(s)]$.

4-26 For the modified $\mathscr{Z}$ transforms of Prob. 4-25, where $T = 0.1$ s, determine $e(kT, m)$ (*a*) by use of Eq. (4-114) and (*b*) by the power-series method for $k = 1, 2, 3$. (*c*) For parts *a* and *b*, determine the values of $e(0.025)$, $e(0.125)$, and $e(0.225)$.

4-27 Derive Eq. (4-104) from Eq. (4-103). *Hint*: Let $k' = k - 1$.

Chapter 5

5-1 Given

$$C(s) = \frac{K(s+a_1)\ldots(s+a_w)}{s(s+\zeta_1\omega_{n_1} \pm j\omega_{d_1})(s+\zeta_3\omega_{n_3} \pm j\omega_{d_3})(s+b_5)\ldots(s+b_n)}$$

All the a_i's and b_k's are positive and real and $\zeta_1\omega_{n_1} = \zeta_3\omega_{n_3}$ with $\omega_{n_3} > \omega_{n_1}$. Determine the effect on the partial-fraction coefficients as the value of ω_{n_3} is allowed to approach infinity with all other parameters held constant. *Hint*: Analyze

$$\lim_{\omega \to \infty} A_3 = \lim_{\omega \to \infty} [(s - p_k)Y(s)]_{s = p_3 = -\zeta_3\omega_{n,3} + j\omega_{d,3}}$$

K has the value required for $c(t)_{ss} = 1$ for a unit-step input. Consider the cases (1) $n = w$, (2) $n > w$.

5-2 Repeat Prob. 5-1 with $a_1 = 0$, $w = 2$, $n = 6$, and K is a fixed value.

5-3 It is desired that a control system have a damping ratio of 0.5 for the dominant complex roots. Using the root-locus method, (*a*) add a lag compensator, with $\alpha = 10$, so that this value of ζ can be obtained; (*b*) add a lead compensator with $\alpha = 0.1$; (*c*) add a lag-lead compensator with $\alpha = 10$. Indicate the time constants of the compensator in each case. Compare the results obtained by the use of each type of compensator with respect to the error coefficient, ω_d, T_s, and M_p.

$$(1)\quad G(s) = \frac{K}{(s+1)(s+2)(s+5)} \qquad (2)\quad G(s) = \frac{K}{s(s+2)(s+5)(s+7)}$$

5-4 A control system has the forward transfer function

$$G_x(s) = \frac{K_x}{s^2(1 + 0.1s)}$$

The closed-loop system is to be made stable by adding a compensator $G_c(s)$ and an amplifier A in cascade with $G_x(s)$. A ζ of 0.5 is desired with a value of $\omega_n \approx 1.6$ rad/s. By use of the root-locus method, determine the following: (*a*) What kind of compensator is needed? (*b*) Select an appropriate α and T for the compensator. (*c*) Determine the value of the error coefficient K_2. (*d*) Plot the compensated locus. (*e*) Plot M vs. ω for the compensated system. (*f*) From the plot of part *e* determine the M_m and ω_m. With this value of M_m determine the effective ζ of the system by the use of $M_m = (2\zeta\sqrt{1-\zeta^2})^{-1}$. Compare the effective ζ and ω_m with the values obtained from the dominant pair of complex roots. *Note*: The effective $\omega_m = \omega_n\sqrt{1 - 2\zeta^2}$. (*g*) Obtain $c(t)$ for a unit-step input.

5-5 With

$$(1)\quad G_x(s) = \frac{K}{s(s+10)(s^2 + 30s + 625)} \qquad H(s) = 1$$

$$(2)\quad G_x(s) = \frac{K(s+25)}{s(s+10)(s^2 + 30s + 625)} \qquad H(s) = 1$$

find K_1, ω_n, M_p, T_p, T_s, N for each of the following cases: (*a*) original system; (*b*) lag compensator added, $\alpha = 10$; (*c*) lead compensator added, $\alpha = 0.1$; (*d*) lag-lead compensator added, $\alpha = 10$. Use $\zeta = 0.5$ for the dominant roots. *Note*: Except for the original system it is not necessary to obtain the complete root locus for each type of compensation. When the lead and lag-lead compensators are added to the original system, there may be other dominant roots in addition to the complex pair. When a real root is also dominant, a $\zeta = 0.3$ may produce the desired improvement (see Sec. 5-3).

5-6 A unity-feedback control system contains the forward transfer function shown below. The system specifications are $\zeta = 0.6$ and $T_s \le 1.2$ s. Without compensation the value of K_x

is 57.33, and the control ratio has an undesirable dominant real pole. The proposed cascade compensator has the form indicated.

$$G_x(s) = \frac{K_x(s+2)}{s(s+5)(s+7)} \qquad G_c(s) = A\frac{s+a}{s+b}$$

(*a*) Determine the values of *a* and *b* such that the desired complex-conjugate poles of $C(s)/R(s)$ are truly dominant. (*b*) Determine the values of *A* and α for this compensator. (*c*) Is this $G_c(s)$ a lag or lead compensator?

5-7 For the control system of Example 5-2 the desired specifications are $1 < M_p \leq 1.1$, $t_s \leq 2$ s, $t_p \leq 1.5$ s, and $K_1 \geq 3\text{ s}^{-1}$. If these specifications cannot be achieved by the basic system, then (*a*) synthesize a desired control ratio that satisfies these specifications. (*b*) Using the Guillemin-Truxal method (see Sec. 5-5), determine the required compensator $G_c(s)$.

5-8 (*a*) Using the Guillemin-Truxal method, determine a passive cascade compensator $G_c(s)$ for a unity-feedback system. The desired closed-loop time response for a step input is achieved by the system having the control ratio

$$M(s) = \frac{20(s+3)}{(s^2+2s+5)(s+2)(s+6)}$$

where

$$G_x(s) = \frac{10}{s(s^2+4s+8)}$$

(*b*) The ideal compensator $G_c(s)$ is to be approximated by one having the mathematical form of

$$G_c(s) = \frac{A(s+a)}{s+b}$$

Determine *A*, *a*, and *b*. Compare M_p, t_p, and t_s for the desired and actual $M(s)$.

5-9 The following block diagram is proposed to minimize the effect of the unwanted disturbance T_L on the output response $c(t)$:

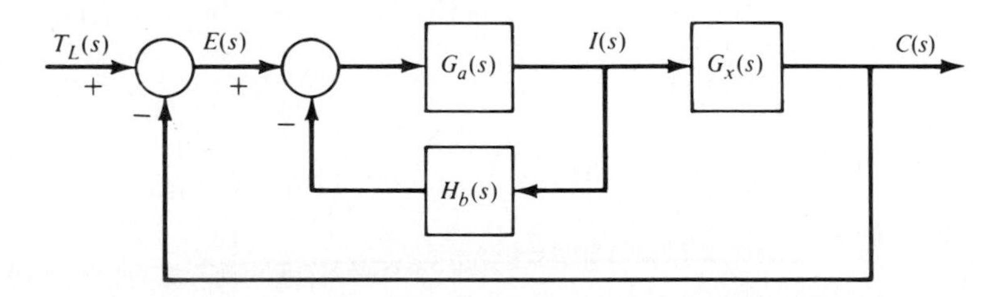

(*a*) Determine $C(s)/T_L(s)$ only in terms of the symbols $G_a(s)$, $G_x(s)$, and $H_b(s)$. That is, do not use the actual transfer functions.

(*b*) *Determine the condition*, in the frequency domain, that must be satisfied so that H_b essentially determines whether the following specifications can be met:

$$\text{Lm}\,\frac{\mathbf{C}(j\omega)}{\mathbf{T}_L(j\omega)} \leq -60\text{ dB} \qquad \text{for frequency range } 0 \leq \omega \leq 4 \text{ rad/s}$$

Use *only* the straight-line approximation technique. *Hint*: Determine $C(s)/T_L(s)$ so that it is directly a function of $1/H_b(s)$, and again do not use the actual transfer functions.

(*c*) Determine the value of K_b that can satisfy the frequency-domain specifications where

$$G_x(s) = \frac{1}{s+1} \qquad G_a(s) = \frac{1}{s} \qquad H_b(s) = \frac{K_b(s+2)}{s(s+100)}$$

(*d*) Does this value of K_b yield a stable system? Illustrate your answer by a root-locus sketch.

(*e*) If the value of K_b results in a stable system, determine $c(t_p)$, t_p, and t_s [where $c(t)_{ss} = 2 \times 10^{-5}$] for $T_L(t) = 1$.

5-10 The figure below represents a disturbance-rejection control system where

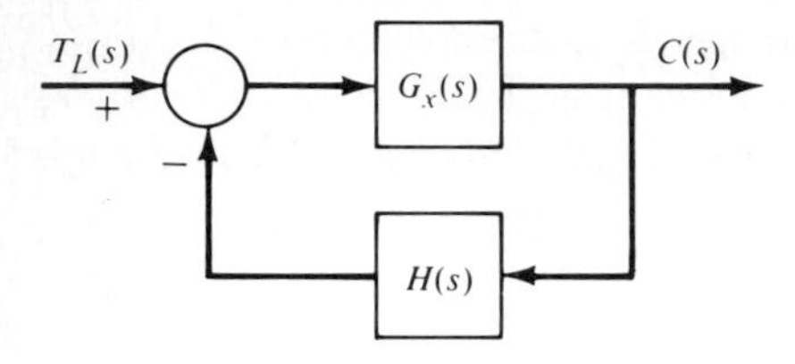

$$G_x(s) = \frac{50(s+5)}{s(s+0.5)(s+10)}$$

The feedback unit

$$H(s) = \frac{K_h(s+1)}{s(s+5)} \qquad K_h > 0$$

is proposed to meet the specifications of $c(t)_{ss} = 0$ for a unit-step disturbance and

$$\text{Lm}\,\frac{\mathbf{C}(j\omega)}{\mathbf{T}_L(j\omega)} \le -40 \text{ db} \qquad \text{for } 0 \le \omega \le 5 \text{ rad/s}$$

(*a*) *Graphically*, by use of straight-line asymptotes, determine a value of K_h that results in these specifications being met with the specified $H(s)$ unit.

(*b*) With this value of K_h plot Lm $\mathbf{C}(j\omega)/\mathbf{T}_L(j\omega)$ vs. ω.

(*c*) Does this value of K_h yield a stable system?

5-11 Given the nonunity-feedback system:

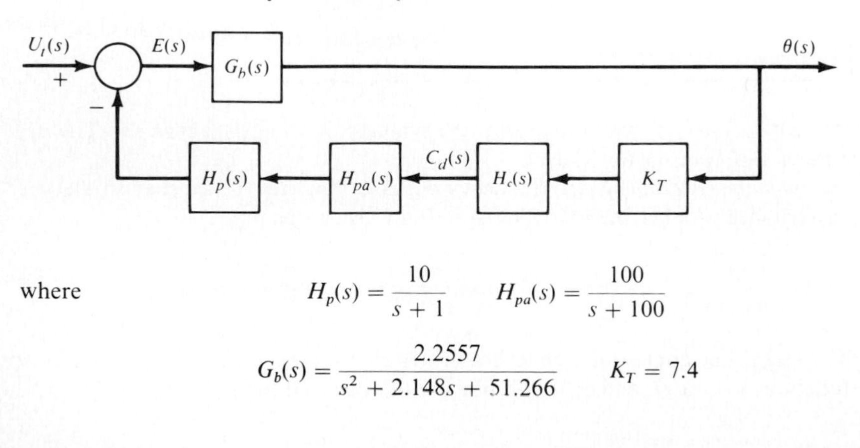

where

$$H_p(s) = \frac{10}{s+1} \qquad H_{pa}(s) = \frac{100}{s+100}$$

$$G_b(s) = \frac{2.2557}{s^2 + 2.148s + 51.266} \qquad K_T = 7.4$$

The system specifications are $\theta(t_p) \leq 0.01$ for $u_t(t) = 1.25u_{-1}(t)$, $\theta(t)_{ss} = 0$, $\omega_b \approx 125$ rad/s (bandwidth), and $t_s \leq 4$ s [t_s is based upon $\theta(t_s) = 0.0002$, that is, 2 percent of maximum allowable $\theta(t_p)$]. $H_c(s)$ is a third-order over a third-order feedback compensator.

(*a*) Determine $H_c(s)$ that results in the system achieving the desired specifications.

(*b*) For your final design obtain the following plots: $\theta(t)$ vs. t, Lm $\mathbf{\theta}(j\omega)/\mathbf{U}_t(j\omega)$ vs. ω, and the root locus. [*Note*: This problem is part of the DIG (digitization) technique method of designing the sampled-data control system of Probs. 6-16, 7-11, and 12-7.]

5-12 By means of a root-locus analysis, find the range of gain that yields a stable response for the control system of Prob. 5-11, where:

(*a*) $H_c(s) = K_D$ (*b*) $H_c(s) = K_D/s$

Chapter 6

6-1 The output of the system represented by the block diagram is given by

$$C(z) = \frac{GR(z)}{1 + GH(z)}$$

where

$$GR(z) = \frac{0.393Kz}{(z-1)(z-0.607)} \qquad GH(z) = \frac{0.393K}{z - 0.607}$$

$$r(t) = u_{-1}(t)$$

$$T = 0.5 \text{ s}$$

For what values of K is the system shown stable? *Hint*: Use the approach of Example 6-1.

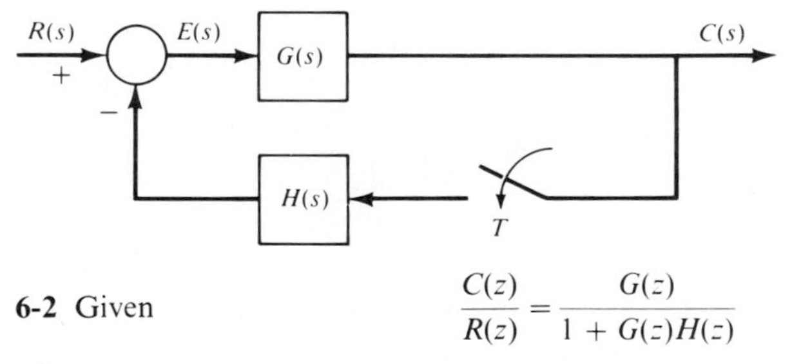

6-2 Given

$$\frac{C(z)}{R(z)} = \frac{G(z)}{1 + G(z)H(z)}$$

where

$$G(z) = \frac{Az}{z-1} \qquad H(z) = \frac{2(z-0.8)}{z} \qquad A > 0$$

Using the approach of Example 6-1, determine the range of A that maintains a stable system.

6-3 Given

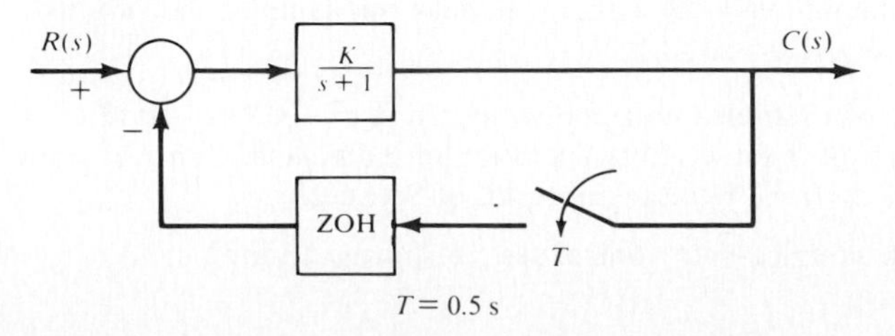

(*a*) Determine $C(z)$.

(*b*) For what range of K is the above system stable? *Hint*: Use the approach of Example 6-1.

(*c*) Determine the value of K so that when $R(s) = 1/s$, $c(\infty) = 1$, and

$$c(kT) = \begin{cases} 1 & \text{for } k \geq 1 \\ 0 & \text{for } k < 1 \end{cases}$$

results in a *deadbeat response*; that is, $r(kT) - c(kT) = 0$ for a minimum value of $k(=k_{\min})$ and $k \geq k_{\min}$.

6-4 $G(z)$ for the system shown is

$$G(z) = \frac{0.0214K(z + 0.857)}{(z - 1)(z - 0.606)}$$

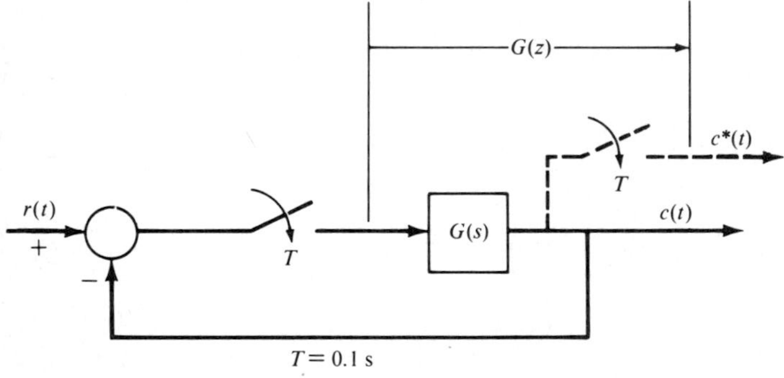

Using the approach of Example 6-1 and where $K > 0$, determine the maximum value of K for a stable system.

6-5 Given

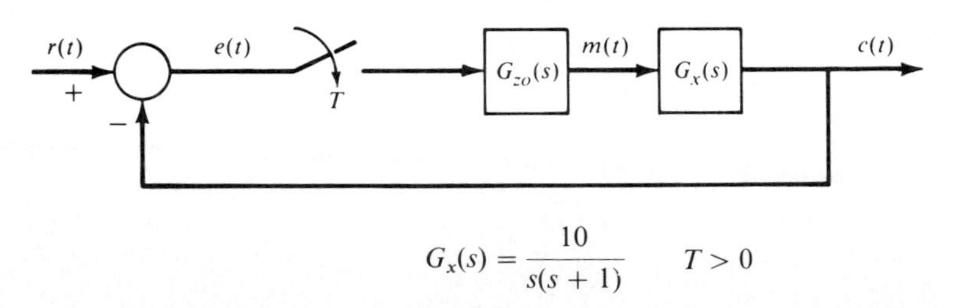

$$G_x(s) = \frac{10}{s(s + 1)} \qquad T > 0$$

Determine the approximate maximum value of T that will make this sampled-data control system unstable.

6-6 The characteristic equation of a sampled-data control system is $z^2 - 0.7Az + 0.12A^2 = 0$. Determine the range of values of A for a stable operation of the system. Consider both positive and negative values of A. *Hint*: Use the approach of Sec. 6-2.

6-7 For the nonunity-feedback sampled-data control system shown, determine:

(*a*) The expression for $C(z)$.

(*b*) The system's characteristic equation, where $r(t) = u_{-1}(t)$ and

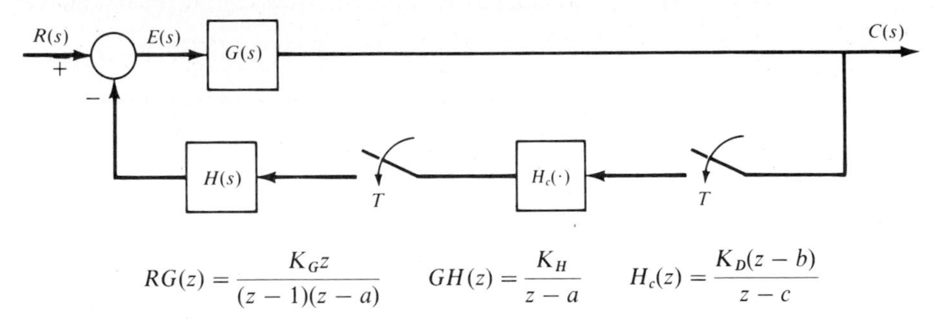

$$RG(z) = \frac{K_G z}{(z-1)(z-a)} \qquad GH(z) = \frac{K_H}{z-a} \qquad H_c(z) = \frac{K_D(z-b)}{z-c}$$

(*c*) $G_{eq}(z)$ and the expression for $K_H K_D$ in terms of a, b, c, and K_G in order to achieve a Type 1 system.

6-8 By means of Jury's stability test, determine the stability of the sampled-data systems whose characteristic equations are:

(*a*) $z^3 + 4z^2 + 3z + 2 = 0$
(*b*) $z^3 - 1.5z^2 - 2z + 3 = 0$
(*c*) $z^3 - 0.1z^2 + z - 0.5 = 0$
(*d*) $z^4 - 0.1z^3 + 0.2z^2 + 0.05z - 0.1 = 0$
(*e*) $2z^4 + z^3 + z^2 + z + 1 = 0$

6-9 A sampled-data control system's characteristic equation is

$$z^3 - 0.1z^2 + 0.2Az - 0.1A = 0$$

where $A > 0$. Determine by Jury's stability test if there is a value of A that ensures a stable system performance.

6-10 Repeat Prob. 6-9 for the characteristic equation

$$z^3 - 0.1z^2 + 0.3Az - A = 0$$

6-11 For the system shown below, where

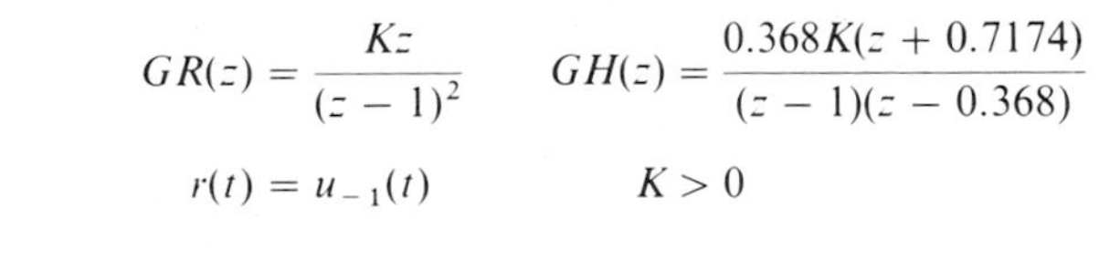

$$GR(z) = \frac{Kz}{(z-1)^2} \qquad GH(z) = \frac{0.368K(z+0.7174)}{(z-1)(z-0.368)}$$

$$r(t) = u_{-1}(t) \qquad K > 0$$

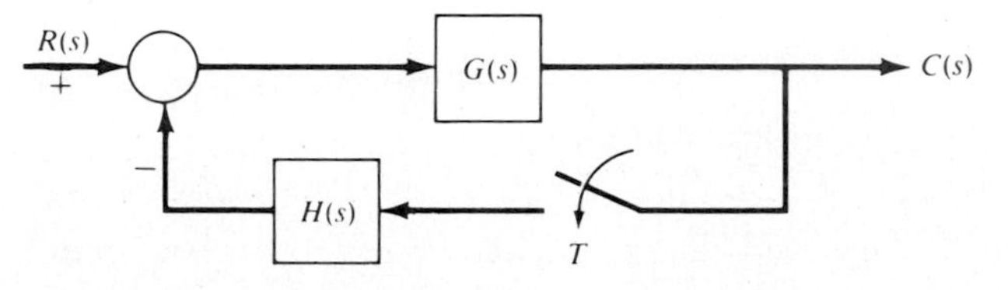

determine the possible maximum system type. From a root-locus analysis, determine the maximum value of K for a stable system. Plot the root locus.

6-12 (*a*) Plot a root locus for the sampled-data control system shown on page 580 for $T = 1$ s and $T = 0.1$ s. (*b*) What can you say about the stability of the system for $T > 0$ and $K_x > 0$? (*c*) Determine the system type. *Note*: $D_c(z)$ corresponds to inserting an ideal integrator in cascade with $G_x(s)$ in a continuous-time system.

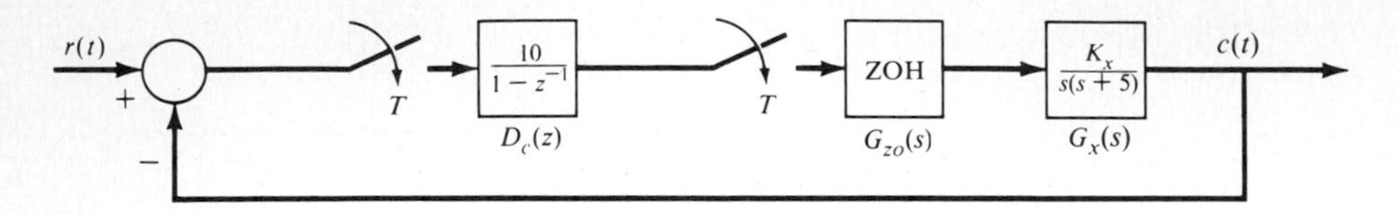

6-13 For Prob. 4-20*c* let

$$G_1(s) = \frac{K_x}{s} \qquad G_2(s) = \frac{K_y(1 - \epsilon^{-sT})}{s(s+2)}$$

$$H_1(s) = \frac{K_H(1 - \epsilon^{-sT})}{s(s+1)} \qquad H_2(s) = 1 \qquad r(t) = u_{-1}(t)$$

(*a*) Derive $C(z)$ in terms of T where $K_x = K_y = K_H = 1$. *Hint*: In substituting the various transfer functions into the expression for $C(z)$ the $z - 1$ factors must remain distinct. Also do not simplify the numerator and denominator into their respective polynomial formats.

(*b*) With K_H unspecified and $T = 1$ s, plot the root locus.

(*c*) Determine if this system is stable for $K_H = 1$.

(*d*) If stable, determine $c(kT)$.

(*e*) Determine the maximum value $K = K_x K_y K_H$ can have for a stable system performance.

(*f*) Using the expression for $C(z)$ from part *a*, with T unspecified, determine $G_{eq}(z)$ and derive the expression for K_H that yields the value it must have in order for the system to be Type 1.

(*g*) For $T = 1$ s, and the value of K_H from part *f*, determine K_p, K_v, and K_a for this Type 1 system.

(*h*) Analyze error $= r(\infty) - c(\infty)$ for step, ramp, and parabolic inputs, respectively (i.e., for the value of K_H of part *f*).

6-14 Repeat Prob. 6-13 (parts *b* to *e*, *g*, and *h*) for $T = 0.1$ s.

6-15 (*a*) For positive values of gain, sketch the root locus for unity-feedback sampled-data systems having the following open-loop transfer functions:

(1) $$G(z) = \frac{K(z + 0.1)^2}{(z-1)(z-0.9)(z-0.1)}$$

(2) $$G(z) = \frac{K(z - 0.2 \pm j0.1)}{(z-1)(z-0.9)(z-0.5)}$$

(3) $$G(z) = \frac{K}{(z-1)(z^2 - 1.6z + 0.65)}$$

(4) $$G(z) = \frac{Kz}{(z-1)(z^2 - 1.6z + 0.65)}$$

(5) $$G(z) = \frac{K(z - 0.2 \pm j0.1)}{(z-1)^2(z-0.5)}$$

(6) $$G(z) = \frac{K(z - 0.2 \pm j0.2)}{(z-1)^2(z-0.5)}$$

(*b*) For what value or values of gain does the system become unstable in each case?

6-16 The minor loop of a seismic isolation platform digital control system is shown below, where

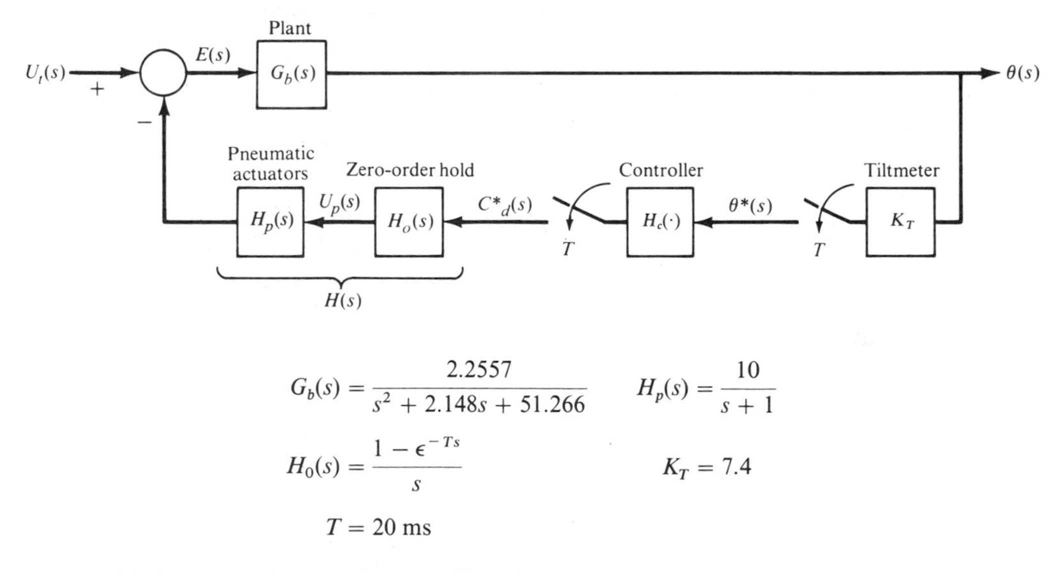

$$G_b(s) = \frac{2.2557}{s^2 + 2.148s + 51.266} \qquad H_p(s) = \frac{10}{s+1}$$

$$H_0(s) = \frac{1 - \epsilon^{-Ts}}{s} \qquad K_T = 7.4$$

$$T = 20 \text{ ms}$$

(*a*) Assume $H_c(z) = K_D$ is a variable-gain controller. By means of a root-locus analysis, find the range of gain for a stable response.

(*b*) Repeat part *a*, where

$$H_c(z) = \frac{TK_D}{2}\frac{z+1}{z-1}$$

6-17 (*a*) Obtain a root-locus plot in the *w* and *z* planes for the systems shown below.

(*b*) For $\zeta = 0.7$, determine the time-response characteristics, using both the *z*- and *w*-domain methods. *Note*: $c(kT) = \mathscr{Z}^{-1}[C(z)]$ and $c(t) \approx \mathscr{L}^{-1}[C(w)]$.

(*c*) Examine the steady-state error of each system for a step and ramp input when $T = 0.1$ and 1.0 s.

(*d*) Obtain plots of $|\mathbf{C}(j\omega)/\mathbf{R}(j\omega)|$ vs. ω for $0 \le \omega < \omega_s/2$ and $0 \le \omega < 2\omega_s$ for both values of T.

System 1:

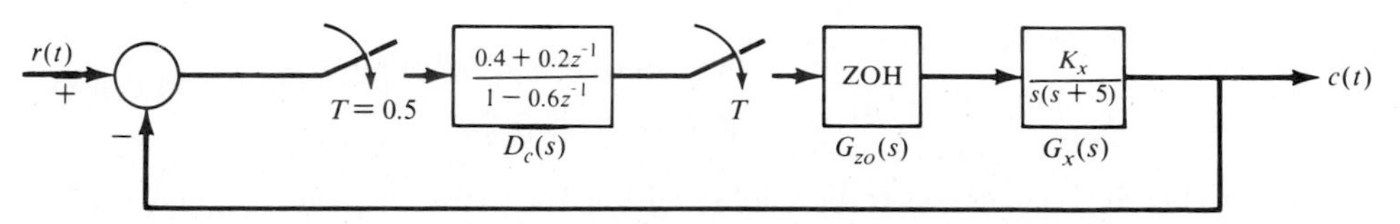

System 2:

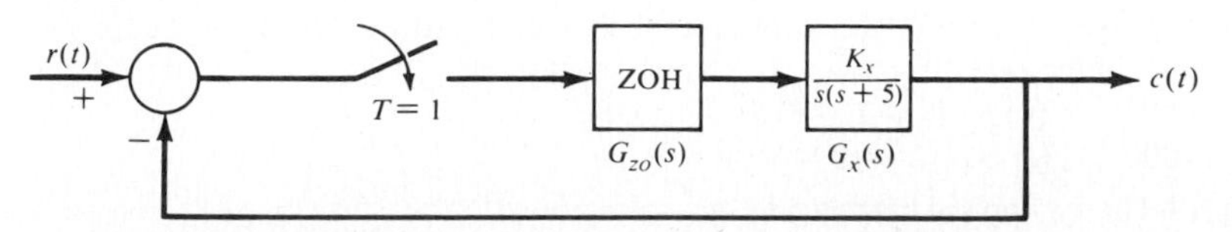

6-18 Repeat Probs. 6-8 to 6-10, for $T = 2$ s, by means of the bilinear transformation (z- to w-domain transformation).

6-19 A unity-feedback continuous-time control system whose open-loop transfer function is

$$G_x(s) = \frac{31{,}650(s + 25)}{s(s + 100)(s^2 + 30s + 625)}$$

has the following figures of merit: $M_p = 1.0776$, $t_p = 0.167$ s, $t_s = 0.352$ s, and $K_1 = 12.66\ \text{s}^{-1}$. This plant is to be used in the sampled-data control system of Fig. 6-5. Create tables similar to those of Tables 6-3 and 6-4 for $T = 0.01$, 0.05, 0.1, and 1.0 s. Include the value of K_1 for each value of T.

6-20 The unity-feedback continuous-time control system whose forward-loop transfer function is

$$G_x(s) = \frac{2172(s + 25)}{s(s + 10)(s^2 + 30s + 625)}$$

has the following figures of merit: $M_p = 1.197$, $t_p = 0.363$ s, $t_s = 0.819$ s, and $K_1 = 8.688\ \text{s}^{-1}$. Repeat Prob. 6-19 for this control system.

6-21 Repeat Prob. 6-15 in the w plane assuming $T = 0.1$ s.

Chapter 7

7-1 Determine by means of the Tustin transformation, $[D_c(z)]_{\text{TU}}$ that has the following s-plane transfer functions. Also, map each pole and zero of $D_c(s)$ into the z plane via $z = \epsilon^{sT}$. Compare the values of the z domain poles and zeros obtained by both approaches.

(*a*) $D_c(s) = K_{sc}\dfrac{s + 0.1}{s + 0.01}$ for $T = 0.01$, 0.1, and 1 s

(*b*) $D_c(s) = K_{sc}\dfrac{s + 1}{s + 10}$ for $T = 0.01$, 0.1, and 1 s

(*c*) $D_c(s) = K_{sc}\dfrac{s + 1}{s(s + 10)}$ for $T = 1$ s

(*d*) $D_c(s) = \dfrac{K_{sc}}{s(s + 10)}$ for $T = 0.1$ and 1 s

(*e*) $D_c(s) = K_{sc}\dfrac{(s + 1)(s + 2)(s + 3)}{s(s + 10)(s + 100)}$ for $T = 0.01$ and 1 s

7-2 (*a*) For Prob. 7-1*d*, obtain $D_c(z) = \mathscr{Z}[D_c(s)]$ for $T = 0.01$ s.
(*b*) Compare $D_c(z)$ with $[D_c(z)]_{\text{TU}}/T$ with respect to their initial and final values.

7-3 Determine if prewarping is required for the following s-plane poles (or zeros) when Eq. (7-7) is used to map these poles into the z domain for $T = 0.01$ and 1 s. Do these poles lie in the shaded region of Fig. 7-7?
(*a*) $s_1 = -1$ (*c*) $s_3 = -2 + j0.5$
(*b*) $s_2 = -1000$ (*d*) $s_4 = -0.1 + j0.1$

7-4 Derive Eq. (7-12) by approximating the series for $z = \epsilon^{sT}$.

7-5 Derive Eq. (7-17) by use of (*a*) $s = (\ln z)/T$ and (*b*) the trapezoidal integration (s^{-1}) method.

7-6 Given

$$G(s) = \frac{K(s+2)(s+10)}{s(s+1)(s+2 \pm j2)(s+10 \pm j20)}$$

and $T = 0.04$ s. Map each pole and zero of $G(s)$ by use of Eq. (7-7) and $z = \epsilon^{sT}$. Compare and analyze the results with respect to Fig. 7-7 and the warping effect.

7-7 Repeat Prob. 7-6, where $T = 0.01$ s, for:

(*a*) $s + 1000$

(*b*) $s^2 + 2000s + 1{,}002{,}500 = s + 1000 \pm j50$

7-8 Equation (7-24) is rearranged to

$$b = \frac{a-1}{a+1}$$

where $b = \hat{\sigma}_{sp}T/2$ and $a = \epsilon^{\sigma_{sp}T}$.

(*a*) Obtain a plot of $\sigma_{sp}T/2$ vs. b (horizontal axis is b).

(*b*) From the plot determine the range where $\hat{\sigma}_{sp}T \approx \sigma_{sp}T$.

7-9 The poles and zeros of the following desired control ratio of the PCT system have been specified:

$$\left[\frac{C(s)}{R(s)}\right]_M = \frac{K(s+7)}{(s+5 \pm j10)(s+6)}$$

(*a*) What must be the value of K in order for $y(t)_{ss} = r(t)$ for a step input?

(*b*) A value of $T = 0.1$ s has been proposed for the sampled-data system whose design is to be based upon the above control ratio. Do all the poles and zeros lie in the good Tustin approximation region of Fig. 7-7? If the answer is no, what should be the value of T so that all the poles and zeros do lie in this region?

7-10

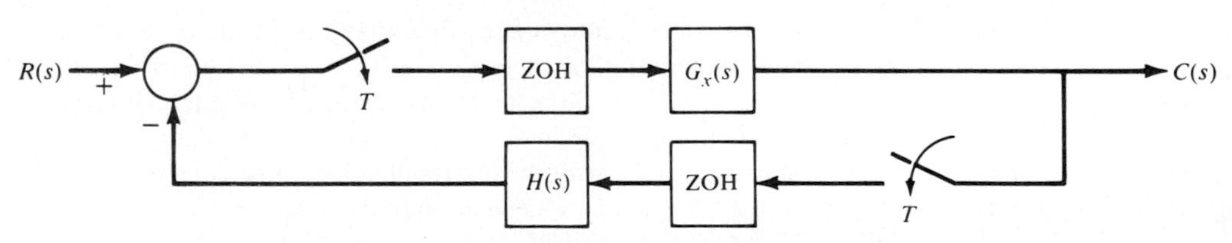

(*a*) By use of the PCT domain analysis determine the range of values for T with $K_x = 1$, for a stable system. Given

$$G_x(s) = \frac{K_x}{s+1} \qquad H(s) = \frac{1}{s}$$

(*b*) For $T = 0.1$ s, $K_x = 0.5$, and a unit-step input, obtain $c(t)$ for the PCT control system and $c^*(t)$ for the sampled-data system by the DIR technique. Compare the peak values and peak times of $c(t)$ and $c^*(t)$.

(*c*) For part *b* compare the settling times of $c(t)$ and $c^*(t)$. The settling time is to be based upon the point where the output reaches 0.02 of its maximum value and remains within this value.

7-11 Obtain the PCT control system of the sampled-data control system of Prob. 6-16. (*a*) Compare the individual transfer functions of the PCT system with those of Prob. 5-11 with $H_c(\cdot)$ unspecified. (*b*) Compare the results of Probs. 5-12*a* and 6-16*a*. (*c*) Compare the results of Probs. 5-12*b* and 6-16*b*. [*Note*: $H_c(z)$ is the Tustin transformation of $H_c(s) = K_D/s$. This problem is the forerunner to Prob. 12-7.]

7-12 Obtain the PCT control system of:

(*a*) Prob. 4-20*a*, where $G_2(s)$ includes a ZOH device.
(*b*) Prob. 4-20*b*, where $G_1(s)$ and $H_2(s)$ include a ZOH device.
(*c*) Prob. 4-20*c*, where $G_2(s)$ and $H_1(s)$ include a ZOH device.
(*d*) Prob. 4-20*d*, where $G_1(s)$ includes a ZOH device.

7-13 The corresponding continuous-time control system of the following sampled-data control system has the control ratio

$$\left[\frac{C(s)}{R(s)}\right]_M = \frac{AG_x(s)}{1 + [A + H_c(s)]G_x(s)} = \frac{88.4(s + 20)}{(s + 2.558 \pm j3.227)(s + 104.3)}$$

with the figures of merit $M_p = 1.08493$, $t_p = 0.92675$ s, and $t_s = 1.40876$ s, and $K_1 = 4.0\ \mathrm{s}^{-1}$ where

$$G_x(s) = \frac{44.2}{s(s + 1)} \qquad H_c(s) = \frac{2s(s + 5)}{s + 20} \qquad A = 2$$

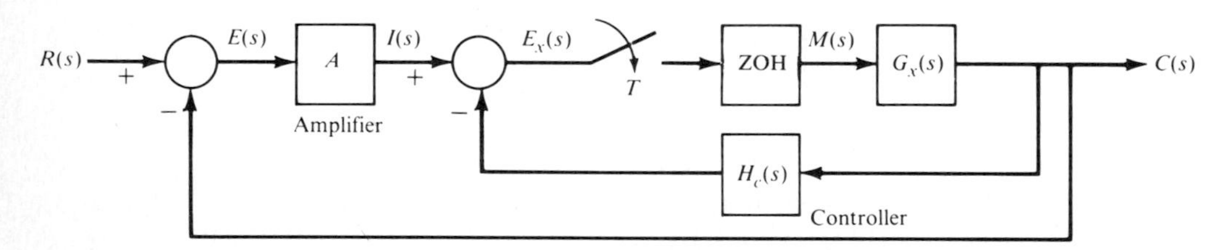

(*a*) Obtain the detailed block diagram for the PCT control system of this sampled-data control system.

(*b*) For the PCT system, determine $C(s)/R(s)$ and its corresponding figures of merit for $T = 0.01$ and 0.005 s. Maintain 4-decimal-digit accuracy.

(*c*) Determine $C(z)/R(z)$ by the DIR technique, and its corresponding figures of merit for each value of T for $r(t) = u_1(t)$. Maintain 4-decimal-digit accuracy. Repeat using 8 digits.

(*d*) Compare the results of parts *b* and *c* with the figures of merit of the corresponding continuous-time control system for both 4- and 8-decimal-digit accuracy.

7-14 The continuous-time control system of Prob. 7-13 is converted to the following sampled-data control system:

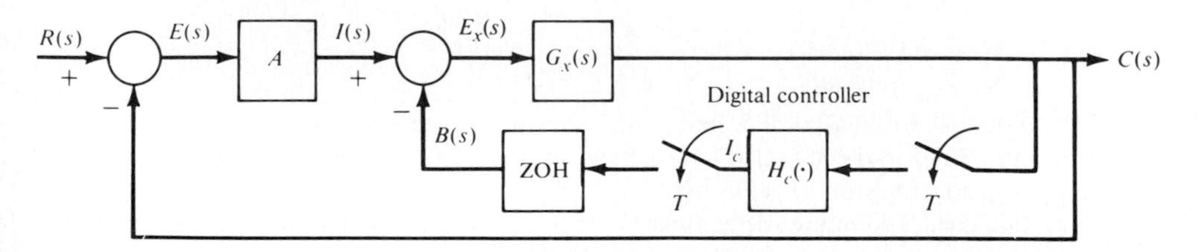

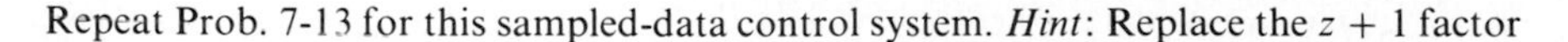

Repeat Prob. 7-13 for this sampled-data control system. *Hint*: Replace the $z + 1$ factor of $[H_c(z)]_{TU}$ by $1.25(z + 0.6)$ in order to remove the pole at $-1 + j0$ from the UC and to ensure a stable system.

7-15 Analyze the sampled-data control system of Prob. 6-12 in the s plane by means of the PCT system approximation for $T = 0.1$ and 1 s. For each value of T determine $K_{x_{max}}$ for a stable system. Compare these values with those determined in Prob. 6-12.

Chapter 8

8-1 Draw a schematic diagram of a 12-bit WCS ladder D/A converter. (*a*) Use resistance values whose ratios are multiples of 2. (*b*) Explain the operation of the converter. (*c*) Discuss accuracy as compared to a 4-bit converter.

8-2 Determine the volts per bit for a 10-bit A/D with a ± 5-V encoding.

8-3 A temperature transducer with an accuracy of ± 0.1 percent FS has a range of 100 to 10,000°C. What is the uncertainty in a measurement of 3216°C? How would this uncertainty relate to the percentage error of any temperature?

8-4 With a sampling frequency of 200 kHz, what is the D/A output, assuming the following analog frequencies in kilohertz at the input (controller transfer function has a gain of 1):

(*a*) 157.6 (*b*) 261 (*c*) 184.2 (*d*) 54.9 (*e*) 500.7

8-5 A 14-bit sequential-approximation A/D converter has an aperture time of 50 ns. What is the highest frequency that can be sampled to within $\pm(1/2)q$?

8-6 A 10-bit weighted-ladder network D/A converter has an MSB resistor value of 2000 ohms.

(*a*) What is the value of the LSB resistor?

(*b*) A digital value of 0011101110 generates a voltage value of what amount?

8-7 A ± 5-V A/D converter with 12 bits of resolution has inputs of 2.71 and -3.15 V. What are the digital outputs for both these inputs assuming sign plus 2s complement encoding?

8-8 What analog outputs are generated by a 10-bit sign plus 2s complement D/A converter with inputs 1011011011 and 00101110101 and a $+10$-V reference?

8-9 An 8-bit D/A converter should have a -9-V output when all the input bits are 1. What is the required D/A reference voltage?

8-10 A 12-bit A/D converter (sign plus 2s complement encoding) with a reference of ± 5 V has output of 111011011110. What is the input analog voltage and associated range of unactivity (error)?

8-11 A 10-bit parallel converter contains how many comparators?

8-12 Given an encoder disk, at what minimum speed (w) must it rotate to handle the sampling of a 100-Hz signal? From this, develop a transfer function to simulate the encoder disk. Use T_m for the time constant of the drive motor (second-order).

8-13 A closed-loop transfer function $118.53/[s(s + 165)]$ is transferred into a discrete function by using a sampling rate of $T = 0.01$ s and a zero-order hold. Determine the frequency response and quantization error for this sampled-data system. In addition, what is the minimum ZOH capacitor value (use Fig. 8-18) which results in almost ideal rectangular pulses? (Draw your results for an input unit step.)

8-14 In how many seconds does a pressure transducer with the transfer function

$$\frac{K_p}{s + 0.2} \qquad K_p = 6\ \text{mV}/(\text{lb}/\text{ft}^2)$$

reach 91 percent of a step input (assuming no saturation)?

8-15 A digital control system must sequentially sample 50 continuous parameters in order to control 15 other continuous parameters. Assume that 20 assembly-language instructions are required to execute each A/D operation including error checks and that 12 instructions are required for each D/A output. Assume each channel requires 70 instructions for compensation. What is the maximum sampling time for each input parameter and control parameter? How do these values relate to the overall digital controller sampling time T? Assume 2.3 μs per instruction, an A/D conversion time (sequential approximation) of 21.4 μs, and a D/A conversion time of 11.3 μs.

8-16 A rate gyro is used to measure a vehicle rate over a distance of 2×10^6 km. The rate gyro has a gain of 0.4 mV/(ft/s) with an error of ± 0.1 percent. If it takes 8 h ± 2 percent to travel the above distance, what is the output and associated uncertainty of the rate gyro assuming constant speed?

8-17 A continuous-temperature transducer has the transfer function $K_T/(s + 5)$, with $K_T = 0.2$ mV/°F. If the temperature changes from -5 to $+62$°F, what is the transducer output after 0.1, 0.5, 1.0, 2.0, and 10.0 s?

8-18 Assuming a first-order transfer function, find the pressure transducer time constant based upon a step input change from 10 to 76 psi. The pressure transducer output measures 41 psi at 3 s.

8-19 If an A/D resolution of 0.01°C is required, what are the number of bits needed, assuming a temperature transducer gain of 10 mV/°C and a range of -100 to $+100$°C? Generate at least a 10-row table of A/D input temperatures vs. digital values.

8-20 Consider the errors associated with a simple cascade digital control system with a two-signal multiplexer, a 10-bit A/D converter, a simple gain of 1 (control law), and a 6-bit D/A converter. The error of each of the three devices can be defined in terms of the average error, the random error, and the quantization error. If the total random error is the rms of the individual random errors and all the others are additive, what is the overall error equation for this system? Also, if the average errors are ± 0.01 percent and the random errors are ± 0.005 percent, what is the overall system error?

8-21 Why do Planck's constant and the uncertainty principle (Heisenberg) *not* play a critical role in control systems from a data-acquisition point of view?

For generating and testing the following programs, assume unit-step and unit-ramp inputs. Use good programming techniques, error routines, organized testing, and extensive documentation methods. For HOL implementations assume I/O routines exist with appropriate parameter passing.

8-22 Write a program in a selected assembly language to test the A/D and the D/A converter in an 8-bit system. Focus on linearity and range. Document with flowcharts.

8-23 Write a general program in assembly and in an HOL to control the temperature of an oven, assuming an analog temperature gauge and a first-order filter. Define range assumptions. Show a block diagram.

8-24 Write an assembly and an HOL program to calibrate and control a pressure device. Assume a second-order compensator. Show a block diagram.

8-25 Theoretically and empirically determine single-loop execution speed for the programs in Probs. 8-22 to 8-24. Compare and analyze the various coded implementation structures.

8-26 Write a program to tell time in terms of 1.0, 0.1, 0.01, and 0.001 s. Use the CPU clock as associated with instruction execution time. Display results continuously if possible.

8-27 Expand Prob. 8-26 to stop when reaching 10 s under a wait-and-test loop and an interrupt structure.

8-28 Write a program to display interactive terminal information for the oven-temperature control problem of Prob. 8-23. Show temperature, change in temperature, and display out-of-temperature readings, current temperature average, etc.

8-29 Using a timing algorithm, develop an HOL program to control house temperature in seven rooms during different periods of the day. Include maximum and minimum error boundaries with appropriate error messages.

8-30 Determine the transfer function of the first-order hold (FOH) of Eq. (8-6). Generate a frequency-domain plot and comment on its form as compared to a ZOH. Is the first-order-hold hardware realizable? If not, how would you make it hardware realizable?

8-31 Consider the following transfer function as a fractional-order hold:

$$G_{\mathrm{FOH}}(s) = (1 - \alpha\epsilon^{-Ts})\frac{(1 - \epsilon^{-Ts})}{s} + \frac{(\alpha - \alpha\epsilon^{-Ts})^2}{Ts^2} \qquad 0 \leq \alpha \leq 1$$

Plot the frequency and time responses of this function as parameterized by α. Relate directly to the ZOH and FOH in terms of specific values for α. Is this polynomial extrapolator realizable?

8-32 Study contemporary integrated circuits that provide for on-chip A/D and D/A conversion as well as provision for filtering/compensation. Evaluate in terms of word length, maximum filter order, instruction speed, number of CPU registers, memory, etc.

Chapter 9

9-1 Let A, B, C, and D be four events associated with one experiment. In the notation of this chapter, express the following probability statements:

Exactly one event occurs.
Exactly three events occur.
At least one event occurs.
Not more than three events occur simultaneously.

9-2 A quantity of 100 digital-controller integrated circuits (ICs) has 37 with minor defects and 21 with critical defects. From this quantity 10 ICs are selected. What is the probability that this group of 10 has (*a*) no defects, (*b*) 2 minor defects, (*c*) 4 major defects, or (*d*) any defect?

9-3 To analyze the effects of roundoff or truncation in a digital controller, the quantizer random process $x_q(n)$ can be modeled as the sum of the correct value (infinite accuracy) $x_c(n)$ and the quantization error $e(n)$, that is,

$$x_q(n) = x_c(n) + e(n) \qquad n = 0, 1, \ldots$$

If the quantization error autocorrelation has the form

$$E[e(n)e(n + m)] = e^2\delta(m)$$

what statistical assumptions have been made? Consider $E[e_n e_{n+m}]$.

9-4 If the probability density function for the rounding error and the truncation error are both defined as uniform, what are the associated parameters or bounds? Also, calculate the means, variances, and standard deviations for both rounding and truncation errors by using the uniform distribution.

9-5 Consider a linear discrete system with transfer function $h(n)$, input $x(n)$, and output $y(n)$. If $\mathbf{\Phi}_{xx}(n, n + m) = \sigma_x^2(m)$, $m_x = 0$, find the autocorrelation of y, the cross correlation of

x and y, and the cross covariance of x and y. Discuss the meanings of each in terms of linear systems.

9-6 If $\{x_n\}$ is a real stationary random process with mean m_x and variance σ_x^2, show that:

(*a*) $\Psi_{xx}(m) = \Phi_{xx}(m) - m_x^2$
(*b*) $\Psi_{xx}(o) = \sigma_x^2$
(*c*) $\Phi_{xx}(m) = \Phi_{xx}(-m)$
(*d*) $\Psi_{xx}(m) = \Phi_{xx}(-m)$
(*e*) $|\Phi_{xx}(m)| \leq \Phi_{xx}(o)$
(*f*) $|\Psi_{xx}(m)| \leq \Psi_{xx}(o)$

9-7 Consider the following linear system:

$$y_1(n) = x(n)*h_1(n)$$

$$y_2(n) = y_1(n)*h_2(n)$$

If $x(n)$ is a real and stationary random process with zero mean and variance σ_x^2, find $\sigma_{y_1}^2$ and $\sigma_{y_2}^2$ in terms of h_1 and h_2.

9-8 Consider a cascade controller with three separate transfer functions $H_i(z)$; $i = 1, 2, 3$. If the input to this controller is a white-noise sequence with zero mean and variance σ^2, what is the mean and variance of the output? If $H_i(z)$ is a first-order transfer function, what is the form of the output variance equation?

9-9 Relate the definition of the probability mass function to the probability density function for random processes in terms of operators.

9-10 Process control temperature measurements are measured every 2 h for a 24-h period with the resulting °C values: 29.5, 31.6, 35.7, 30.7, 35.2, 41.0, 32.6, 28.7, 30.7, 33.1, 40.7, and 39.7. Determine the mean and standard deviation. Discuss the associated PDF.

9-11 One way to filter a constant noisy signal is to average over n measurements. In this case, the new signal-to-noise ratio is multiplied by $\sqrt{n}$, where a zero mean noise source is assumed. If the original signal-to-noise ratio is 4 to 1, what is the ratio after 4096 samples?

9-12 Show that the following relationships are true:

$$P(E \cup E') = P(E) + P(E') - P(E \cap E')$$

and

$$P(\bigcup_i E_i) \leq \sum_i P(E_i) \quad \text{for a sequence of events.}$$

9-13 Using the definition of *conditional probability*

$$P(E/E') = P(E \cap E')/P(E')$$

show that if $P(E') = 0$ and the events E and E' are disjoint, then

$$P(E/E') = P(E)$$

9.14 For joint probability distributions, develop all the associated probability equations using event set notation for discrete random variables.

9-15 Develop the *mixed probability distribution* equations for two random variables, one discrete and the other continuous. Assume that a *joint probability mass/density function* exists.

Chapter 10

10-1 Consider a first-order system where

$$H(z) = \frac{1}{1 - az^{-1}} \qquad a = 0.125$$

Assume the digital controller is to be implemented by using fixed-point arithmetic with an 8-bit word length and $T = 1$ s. Determine the system's unit-sampled and step responses for a small and large number of samples. Assume the coefficients are represented in:

(*a*) Sign plus magnitude with two integer bits.

(*b*) Sign plus 2s complement with one integer bit.

(*c*) Sign plus 1s complement with four integer bits.

Note the different response due to the assumed binary point position and multiplication roundoff or truncation.

10-2 Do Prob. 10-1 with a 16-bit word length and compare the result with an 8-bit word.

10-3 Consider a second-order system, where output $y(kT)$ is

$$y[kT] = ay[(k-1)T] + by[(k-2)T] + cx[kT]$$

with zero initial conditions and $a = -0.1097$, $b = 0.20614$, and $c = 9671.7$. Repeat Probs. 10-1 and 10-2, assuming an appropriate T.

10-4 Determine the form of Eq. (10-28) for multiple system transfer-function roots.

10-5 Using the system transfer function of Prob. 10-3, add a 10-bit A/D and a 10-bit D/A. Determine the system response to a square-wave input (frequency = 1 kHz, amplitude = 10 units), assuming sign plus 2s complement representation.

10-6 Consider a first-order controller; that is,

$$y(k) = ay(k-1) + bu(k-1) \qquad \text{where } a = 0.5867 \text{ and } b = 0.8266$$

Assuming a rounding operation to two significant digits ($+0.01$), generate a table of values for $k = 1, 20$. Note the small output oscillations or limit cycle due to the quantization. What values of controller gain reduce the limit-cycle amplitude? Repeat this problem if the input $u(k-1)$ changes to $u(k)$.

10-7 Using the control system transfer function

$$H(z) = \frac{K(1 + az^{-1})}{(1 + b_0 z^{-1} + b_1 z^{-2})(1 + c_0 z^{-1} + c_1 z^{-2})}$$

where $a = 0.967$, $b_0 = 1.8$, $b_1 = 0.8189$, $c_0 = 1.229$, $c_1 = 0.2937$, and $K = 5.188 \times 10^{-3}$

(*a*) Determine the various general 4th-order configurations for implementing this digital controller and an appropriate value for T.

(*b*) Determine the maximum output value possible for each configuration using a unit-step function input. Select an appropriate fixed-point word length for each configuration.

(*c*) For each configuration then, along with a 10-bit A/D and an 8-bit D/A, determine the best error performance to a square-wave input. State assumptions.

10-8 How can arithmetic overflow be analyzed for a given system transfer function? How does this analysis depend upon the input? How does overflow analysis relate to stability?

10-9 How does scaling play a role in designing for no overflow or underflow?

10-10 Discuss the impact of the solutions to Prob. 10-4 in terms of quantization sensitivity.

10-11 Develop the bounds of the quantization error by assuming a floating-point representation of numbers (coefficients, sums, products).

10-12 Assuming a 24-bit (mantissa = 16) floating-point representation of numbers, do Prob. 10-1.

10-13 Find the sensitivity for the $H(z)$ poles and zeros of Prob. 6-11 due to controller coefficient values in a 16- and 8-bit processor.

10-14 Find the sensitivity for the $H(z)$ poles and zeros of Prob. 7-13 duc to controller coefficient values in a 16- and 8-bit processor.

10-15 Find the sensitivity for the $H(z)$ poles and zeros of Prob. 7-14 due to controller coefficient values in a 16- and 8-bit processor.

10-16 Realize the $H(z)$ of Prob. 10-7 in a cascade structure with second-order subfilters, and answer parts *b* and *c*.

10-17 Realize the $H(z)$ of Prob. 10-7 in a cascade structure with first-order subfilters, and answer parts *b* and *c*.

10-18 Develop the signal scaling bound per subfilter section by using the L_p norm; that is,

$$F = \frac{1}{\omega} \int_{+\omega}^{-\omega} [F(\omega')^p \, d\omega']^{1/p} \qquad p > 0$$

10-19 Generate the scaling bounds for Probs. 10-16 and 10-17.

10-20 Develop the general limit-cycle bound equations for an nth-order controller transfer function. *Hint*: Extend the formulas developed for the second-order model.

10-21 Consider a second-order controller

$$H(z) = \frac{1 + a_1 z^{-1} + a_2 z^{-2}}{1 + b_1 z^{-1} + b_2 z^{-2}} \qquad T = .001 \text{ s}$$

where $a_1 = 0.36$, $a_2 = 0.69$, $b_1 = -0.8$, and $b_2 = +0.1788$. Determine if a limit cycle can exist with sign plus 2s complement arithmetic and rounding quantization using 16 bits.

10-22 If possible, simulate the appropriate problems in this chapter and compare computer simulated results to the analysis. Example problems can be 10-7, 10-16, 10-17, and 10-21.

10-23 Generate an absolute limit-cycle bound for a second-order compensator with repeated roots.

10-24 Develop the general transfer functions and difference equations for the 1*D*, 2*D*, 3*D*, and 4*D* filter structures using the symbols indicated in the figures.

10-25 Generate an assembly language program and an HOL program for each of the filter structures of Prob. 10-24.

10-26 Consider an implementation of a second-order compensator where the real and imaginary parts of the poles and zeros are explicitly in the formulation. This structure can be of two forms, 1*X* and 2*X* (see Ref. 108). Develop these two forms each of which considers cross-coupling of two first-order filters with complex poles and zeros.

10-27 Do Prob. 10-25 for the 1*X* and 2*X* structures of Prob. 10-26.

10-28 For the six structures (1*D*, 2*D*, 3*D*, 4*D*, 1*X*, 2*X*) generate a table comparing the number of delay elements, multiplies and additions and comment on differences. Consider both theory and implementation.

10-29 Another technique for solving the scaling problem, is the pole-zero pairing approach. In this method, poles and zeros are combined that are nearest to each other. Poles are matched with other poles that are farthest away. The process starts with poles nearest the $z = +1$ point (why?). The approach attempts to balance the dc gain of each subfilter. Using this approach, analyze Prob. 10-7.

Chapter 11

11-1 The transfer function of the basic plant and ZOH unit of Fig. 11-1*b* is

$$G_z(z) = \frac{0.002306(z + 0.9672)}{(z - 1)(z - 0.9048)} \qquad \text{where } T = 0.1 \text{ s}$$

and the system's desired control ratio, for $\zeta = 0.7071$, is defined as

$$\left[\frac{C(z)}{R(z)}\right]_M = \frac{0.2087(z + 0.9672)}{z - 0.4409 \pm j0.312}$$

(*a*) By use of the Guillemin-Truxal method, determine the cascade controller $D_c(z)$ that yields the desired control ratio.

(*b*) Approximate your $D_c(z)$ as

$$[D_c(z)]_A = \frac{K_{zc}(z + a)}{z + b}$$

and with this controller (1) adjust K_{zc} to determine the closed-loop dominant roots for a $\zeta = 0.7071$, (2) determine $C(z)/R(z)$, and (3) determine the time-response characteristics for a unit-step function input.

(*c*) Compare the time-response characteristics of part *b* with those of Sec. 11-4C. Also, compare $C(z)/R(z)$ of part *b* with $[C(z)/R(z)]_M$.

11-2 For the sampled-data control system of Fig. 11-1*b*, $G_x(s) = 2/[s(s + 1)]$. The desired control ratio of the PCT control system is

$$\left[\frac{C(s)}{R(s)}\right]_M = \frac{140(s + 2)}{(s^2 + 2s + 2)(s + 1.4)(s + 100)}$$

(*a*) Draw the PCT block diagram for this control system.

(*b*) Determine an appropriate (maximum) value of T that yields a satisfactory $D_c(z)$ by use of the DIG technique.

(*c*) Determine $D_c(s)$ by applying the Guillemin-Truxal method in the *s* domain. Note that a zero, z_{1_s}, of $D_c(s)$ cancels a pole, p_{1_s}, of $G_x(s)$.

(*d*) Determine $[D_c(z)]_{TU}$ and obtain the corresponding

$$\frac{C(z)}{R(z)} = \frac{[D_c(z)]_{TU} G_z(z)}{1 + [D_c(z)]_{TU} G_z(z)}$$

Caution: The Tustin transformation z_{1_z} of the *s*-plane zero z_{1_s} *does not exactly equal* the value of the pole p_{1_z} of $G_z(z) = \mathscr{Z}[G_{zo}(s)G_x(s)]$ that corresponds to p_{1_s} of $G_x(s)$. This is taken into account by setting z_{1_z} of the controller equal to p_{1_z} in order that this pole-zero combination cancels in $G(z) = [D_c(z)]_{TU} G_z(z)$.

(*e*) Apply the Guillemin-Truxal method in the *z* domain to $[C(z)/R(z)]_{TU}$ {the Tustin approximation of $[C(s)/R(s)]_M$} to derive $[D_c(z)]_{TU}$. Determine a physically realizable digital control from $[D_c(z)]_{TU}$. With this controller, determine $C(z)/R(z)$ and the corresponding figures of merit for a unit-step input.

(*f*) For part *e*, using the physically realizable $D_c(z)$, adjust the controller gain to achieve $\zeta = 0.7071$ for the dominant complex-conjugate poles of $C(z)/R(z)$. Obtain the corresponding figures of merit and compare with those of part *e*.

(*g*) Compare and evaluate the figures of merit of parts *d* to *f* with those for $[C(s)/R(s)]_M$.

11-3 Apply the w-plane DIG technique to the example of Sec. 11-3; that is:

(*a*) Obtain $[C(w)/R(w)]_M$ from $[\text{Eq. (11-9)}]_{w \approx s}$.

(*b*) Obtain $G_z(w)$ by substituting Eq. (11-17) into $G_z(z) = \mathscr{Z}[G_{zo}(s)G_x(s)]$.

(*c*) Apply the Guillemin-Truxal method to obtain $[D_c(w)]_{\text{TU}}$.

(*d*) Obtain $D_c(z)$ by substituting Eq. (11-19) into $[D_c(w)]_{\text{TU}}$.

(*e*) Generate

$$\frac{C(z)}{R(z)} = \frac{D_c(z)G_z(z)}{1 + D_c(z)G_z(z)}$$

and obtain the corresponding figures of merit for $r(t) = u_{-1}(t)$. Compare these values with those in Sec. 11-3.

11-4 For System 2 of Prob. 6-17†:

(*a*) Determine M_p, t_p, t_s, and K_1 for $\zeta = 0.7$ and $T = 0.01$ s by the DIR method.

(*b*) Obtain the PCT control system of part *a* and the corresponding figures of merit.

(*c*) Generate a digital compensator $D_c(z)$ (with no ZOH) that reduces the value of t_s of part *a* by approximately one-half for $\zeta = 0.7$ and $T = 0.01$ s. Do first by the DIG method (s plane) and then by the DIR method. Obtain M_p, t_p, and t_s for $c(t)$ of the PCT and $c^*(t)$ (for both the DIG and the DIR designs) and K_m. Draw the plots of $c^*(t)$ vs. kT for both designs.

(*d*) Find a digital compensator $D_c(z)$ (with no ZOH) that increases the ramp error coefficient of part *a*, with $\zeta = 0.7$ and $T = 0.01$ s with a minimal degradation of the transient-response characteristics of part *a*. Do first by the DIG method (s plane) and then by the DIR method. Obtain M_p, t_p, and t_s for $c(t)$ of the PCT and $c^*(t)$ for both the DIG and the DIR designs and K_m. Draw a plot of $c^*(t)$ vs. kT for both designs.

(*e*) Summarize the results of this problem in tabular form and analyze.

(*f*) Obtain the frequency-response plots of parts *a* to *d* and evaluate.

11-5 Repeat parts *a* and *c* to *f* of Prob. 11-4 by use of the DIG method (w plane). For parts *c* and *d*, determine $D_c(w)$ and its corresponding $[D_c(z)]_{\text{TU}}$. *Note*: In the w plane, $c(t) = \mathscr{L}^{-1}[C(w)]$.

11-6 An inexperienced engineer has designed a digital control system for a drawbridge as shown below:

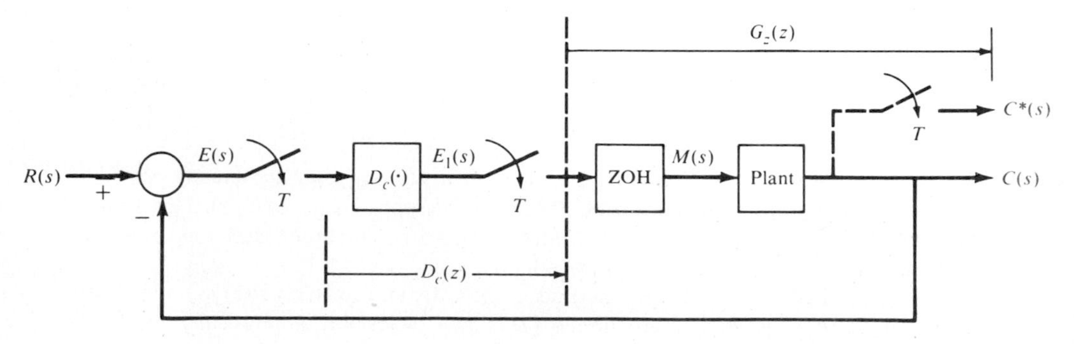

Given that

$$G_z(z) = \frac{C(z)}{E_1(z)} = \frac{3(z + 1)}{(z - 1)^2}$$

† For part *b* use 4-decimal-digit accuracy; for parts *c* and *d* use 4-decimal-digit accuracy then repeat using 8-digit accuracy. Compare the results.

the designer maintains that a pure gain will be sufficient for the system, that is, $D_c(z) = K_D$, such that the output follows a step input with zero steady-state error.

(*a*) Show why this does not work.

(*b*) Suggest and explain a *specific* algorithm for $D_c(z)$ that does work.

(*c*) For your $D_c(z)$ of part *b* verify that $e(\infty) = 0$.

11-7 The plant $G_x(s)$ of Prob. 6-20 is used in the sampled-data control system of Fig. 11-1*b*. It is desired to design a digital controller, without a ZOH device, that reduces t_s to 0.6 s while maintaining $M_p = 1.2$. By use of the DIG technique, $T = 0.01$ s:

(*a*) Draw the PCT block diagram for this control system.

(*b*) Select an appropriate controller $D_c(s)$ that satisfies the desired specifications.

(*c*) Obtain $C(s)/R(s)$ for the compensated PCT system and the corresponding figures of merit for $r(t) = u_{-1}(t)$.

(*d*) Obtain the corresponding $[D_c(z)]_{\mathrm{TU}}$ and

$$\frac{C(z)}{R(z)} = \frac{[D_c(z)]_{\mathrm{TU}} G_z(z)}{1 + [D_c(z)]_{\mathrm{TU}} G_z(z)}$$

(*e*) For $r(t) = u_{-1}(t)$ determine the figures of merit of the compensated sampled-data control system and compare these values to those of part *b* and the uncompensated system.

11-8 Repeat Prob. 11-7 (parts *b* to *e*) by applying the DIG technique in the *w* plane. Compare with the results of Prob. 11-7.

11-9 Repeat Prob. 11-7 by use of the DIR technique; that is:

(*a*) Select an appropriate $D_c(z)$ that satisfies the specifications.

(*b*) Determine the control ratio $C(z)/R(z)$ for the compensated system and the figure of merit for $r(t) = u_{-1}(t)$.

(*c*) Compare with the results of Probs. 11-7 and 11-8.

11-10 Repeat Probs. 11-7 to 11-9 for $T = 0.1$ s.

11-11 For the example in Sec. 11-5A, where $T = 0.1$ s, replace the lag compensator by

$$D_c(z) = \frac{K_{zc}(z - 0.999)}{z - 1}$$

(*a*) Obtain the root locus for $D_c(z)G_z(z) = -1$.

(*b*) For $\zeta = 0.7071$, determine K_z, K_{zc}, K_m, $C(z)/R(z)$, and the figures of merit for $r(t) = u_{-1}(t)$.

(*c*) Compare the time-response characteristics of part *b* with those in Sec. 11-5A.

(*d*) Compare the compensated system of this problem with that of Sec. 11-5A with respect to system type and $e(t)_{ss}$ for both ramp and parabolic inputs.

11-12 The lag and lead compensators of Prob. 11-4, designed by the DIR method, are to be used in cascade as a lag-lead compensator.

(*a*) Determine the compensator gain required to achieve $\zeta = 0.7$ for the dominant roots.

(*b*) Obtain $c^*(t)$ for $r(t) = u_{-1}(t)$ and the figures of merit.

(*c*) Compare these figures of merit with those of Prob. 11-4.

11-13 Repeat Prob. 11-12 by using a PID controller. Base your design on achieving $\zeta = 0.7$ for the dominant poles of $C(z)/R(z)$. Compare the results with those of Prob. 11-12.

11-14 Obtain and compare the frequency-response plots of $D_c(s)$ and $D_c(z)$ of Prob. 11-4 for the lag and lead compensators.

11-15 For the Guillemin-Truxal example in Sec. 11-3, determine the effect of replacing $(z+1)^3$ by (*a*) $4(z+1)$ and (*b*) 8 in Eq. (11-11) on achieving the desired performance.

11-16 By use of the approximations of differentiation and integration of Chap. 3, derive Eq. (11-58) from Eq. (11-57).

11-17 Using Example 11-5 from Sec. 11-7, generate the discrete Butterworth filter frequency response by using the impulse-invariant approach. Compare the result with the filter resulting from the Tustin approach of Fig. 11-25.

11-18 For the filter specifications in Example 11-5 of Sec. 11-7, generate (*a*) the continuous Chebyshev filter and (*b*) the discrete Chebyshev equivalent filter by using (1) the Tustin, (2) the impulse-invariant, and (3) the first-difference methods. Compare the results and evaluate the frequency responses (magnitude and phase).

11-19 Analyze the impact of the Butterworth phase response in Example 11-5 in terms of overall system stability.

11-20 Compare the discrete Chebyshev filter responses of Prob. 11-17 for word-length implementations of 32, 24, 16, and 8 bits. Coefficients as well as computations must reflect the specified word length.

11-21 Considering Example 10-7 and the example of Sec. 11-5A, what are the stability margins associated with the two implementations

$$D_c(z) = 1.0088\,\frac{Z - 0.9999}{Z - 0.9998996}$$

and

$$D_c(Z) = 1.0088\,\frac{Z - 0.999}{Z - 0.9999}$$

11-22 Find the sensitivity for the $D_c(z)$ poles and zeros of Prob. 11-1 due to controller coefficient values in a 16- and 8-bit processor.

11-23 Find the sensitivity for the $D_c(z)$ poles and zeros of Prob. 11-2 due to controller coefficient values in a 16- and 8-bit processor.

11-24 Find the sensitivity for the $D_c(z)$ poles and zeros of Prob. 11-11 due to controller coefficient values in a 16- and 8-bit processor.

Chapter 12

12-1 Given:

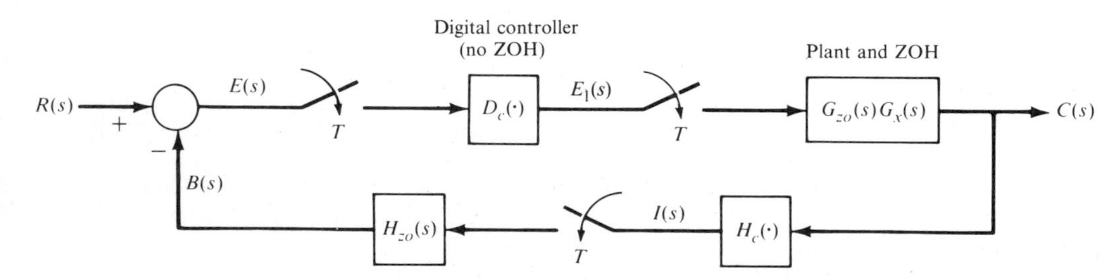

where $G_x(s) = 1/[s(s+1)]$ and $H_c(s) = K_H(s+4)$. $H_c(s)$ is a proportional plus dc tachometer device. It is desired that the corresponding PCT control system control ratio have a dominant complex-conjugate pair of poles $p_{1,2}$ at $-4 \pm j4$.

(*a*) Draw the PCT block diagram for this sampled-data control system.

(*b*) Determine the value (maximum) of T necessary for an accurate Tustin transformation of the required $D_c(s)$ to achieve the desired values of $p_{1,2}$. Suggested procedure: (1) Select your initial value of T based upon $p_{1,2}$ ignoring the required locations of the pole p_c and the zero z_c of $D_c(s)$. (2) Do a root-locus analysis to determine the required values of p_c and z_c to achieve the desired values of $p_{1,2}$. (3) With these initial values of p_c and z_c, determine a new value of T. (4) With this value of T repeat Step 2.

(*c*) Is $D_c(s)$ a lag or a lead compensator?

(*d*) Determine K_{sc}, K_H, and $C(s)/R(s)$ and the corresponding figures of merit for $r(t) = u_{-1}(t)$ and $c(\infty) = 1$.

(*e*) Determine $[D_c(z)]_{TU}$ and $C(z)/R(z)$. Note that the Tustin transformation is to be used only to determine $[D_c(z)]_{TU}$. For $r(t) = u_{-1}(t)$, determine the figures of merit and compare them to part *d*.

12-2 Repeat Prob. 12-1 by the DIR technique, where $T = 5$ ms and the required $p_{1,2}$ poles are mapped onto the z plane via $z = \epsilon^{sT}$. Compare results.

12-3 Repeat Prob. 12-2 for $T = 25$ ms.

12-4 Repeat Probs. 12-1 to 12-3 by the DIG technique (w plane).

12-5 Given:

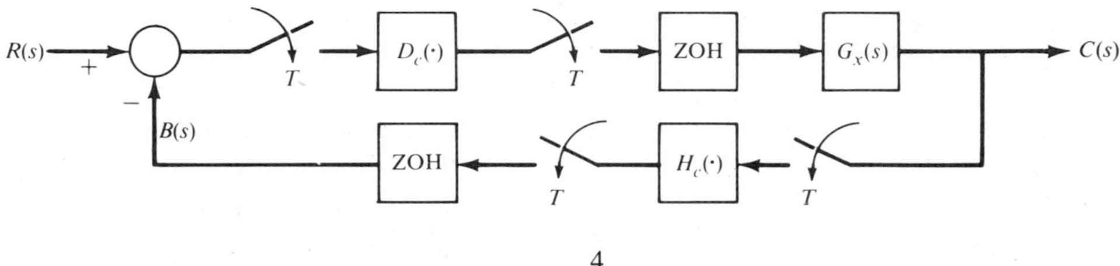

where
$$G_x(s) = \frac{4}{s(s+1)}$$

and D_c and H_c are digital controllers that do not utilize a ZOH device.

(*a*) Draw the PCT control-system block diagram representing the sampled-data system shown above. Specify the transfer functions for the blocks of your diagram.

(*b*) It is desired to determine a $D_c(s)$ that yields the desired control ratio

$$\left[\frac{C(s)}{R(s)}\right]_M = \frac{K}{(s^2 + 4s + 8)(s + 200)^2}$$

where $K = 320{,}000$ and $H_c = K_D$, a gain. Does the value of $T = 0.1$ s yield a satisfactory $C(z)/R(z)$, by use of the Tustin approximation for D_c and H_c, such that the sampled-data system has essentially the same time-response characteristics as $[C(s)/R(s)]_M$? If not, specify a value of T that permits the use of the Tustin approximation in the design of this sampled-data system. *Note*: $T \geq 0.01$ s. Give your reasons for your choice of T by aid of an s-plane diagram.

(*c*) Design $D_c(s)$ so that it yields the desired dominant complex poles of the control ratio. Is it a lag, lead, or lag-lead compensator?

(*d*) For your $D_c(s)$, determine K_D and the resulting control ratio $C(s)/R(s)$.

(*e*) Compare the figures of merit, for $r(t) = u_{-1}(t)$, of $C(s)/R(s)$ with those of $[C(s)/R(s)]_M$.

(*f*) Obtain $[D_c(z)]_{TU}$ and $C(z)/R(z)$.

(*g*) Compare the figures of merit of your sampled-data system with those of part *e*.

12-6 For the feedback-compensated sampled-data control system of Fig. 12-1, $G_x(s) = K_x/[s(s+1)(s+5)]$. The system specifications for a unit-step input are $\zeta = 0.65$, $t_s \leq 2$ s, $e(t)_{ss} = 0$, and $T = 0.2$ s. (*a*) Design H_c and determine the value of K_x required so that these specifications are met. (*b*) Obtain $[H_c(z)]_{TU}$, $C(z)$, $[C(z)/R(z)]_{Pseudo}$ and the figures of merit. (*c*) Compare the results of part *a* with the results of part *b*.

12-7 For Prob. 6-16, design $H_c(z)$ to satisfy the requirement of $\theta(t)_{ss} = 0$, $\theta(t_p) \leq 0.01$ arc s, and $t_s \leq 4$ s [± 2 percent $\theta(t_p) = 0.01$] for $u_t(t) = 1.25u_{-1}(t)$.

(*a*) From Prob. 5-11, obtain $[H_c(z)]_{TU}$ and the z-plane root locus.

(*b*) If $[H_c(z)]_{TU}$ does not meet the specifications via path *A* of Fig. 11-2, then adjust K_{zc} (path *B*) until they are met.

(*c*) Design $H_c(z)$ by use of the DIR technique where $H_c(z)$ is a third-order over a third-order controller.

Chapter 13

Some of the following problems require that appropriate references be consulted for detailed physical models. The intent of the problems, however, is to focus on the structure of the requirements, design, implementation, and test phases. For testing the following programs, assume unit-step and unit-ramp inputs. Use good structured programming, error routines, organized testing, and documentation methods.

13-1 What are the phases of the software life cycle? What are other possible views of the software life cycle? What type of documentation should be available at the end of each phase?

13-2 Continue the development of the DFDs for the single-axis spacecraft control system of this chapter to four levels.

13-3 Considering the spacecraft digital-control software example, extend the requirements to three axes (pitch, yaw, and roll). Generate DFDs and a DD for the 0.0, 1.0, and 1.X levels. Define a structure chart and analyze module coupling for each module and relate to design heuristics.

13-4 Consider the design of a single-channel automated manufacturing process such as packaging or soda pop bottle filling. Develop a requirements DFD and design a structure chart. Discuss proper documentation.

13-5 Consider an elaborate robot process control system such as found in a car assembly line or a chemical plant and develop the DFD, DD, and the structure charts. Use object-oriented modules.

13-6 Develop DFDs, a structure chart, and a coded program for digitally controlling a stepper motor (second-order subsystem) with position and velocity feedback.

13-7 Generate DFDs, structure chart, and a DD for controlling a house environment (temperature, humidity, and airflow).

13-8 Find a digital-controller software implementation. Analyze and evaluate the overall software organization, using the design heuristics, module coupling, and module cohesion objectives.

13-9 Are requirements clearly separable from design? Give examples of both situations. Compare the two situations in terms of intrinsic differences and similarities.

13-10 Generate a documented test plan to the level of individual variables for the digital-control software example of Sec. 11-8.

13-11 Develop a test plan for the spacecraft control Prob. 13-3.

13-12 Apply structured-programming concepts to the example of Sec. 11-8. Discuss program flow. Modify the program for better coupling and cohesion and D/A output saturation control.

13-13 Write a program in a selected assembly language to test the A/D or the D/A converter in an 8-bit system. Document with DFDs and a DD.

13-14 Develop the DFDs, DD, and the structure chart and write a program to control the temperatures of 32 ovens, assuming analog temperature gauges.

13-15 Write a documented general program† to calibrate various digital control system transducers. Focus on linearity and range.

13-16 Theoretically and empirically determine single-loop execution speed for the programs in Probs. 13-13 to 13-15.

13-17 Write a program† to tell time in terms of 1.0, 0.1, 0.01 and 0.001 s, using a real-time clock. Display results continuously if possible.

13-18 Expand the program in Prob. 13-17 to stop when reaching 7.92 s. Discuss the use of this program in a wait-and-test loop and an interrupt structure for critical process control (vehicle landing, staging, mode changes, ...). Reconsider the implementation using a watch-dog timer and POSIT structure.)

13-19 Write a program† to display interactive terminal information for the oven temperature control problem of Prob. 13-14. Show temperature, change in temperature, out-of-temperature readings, current temperature average, etc. Revise DFDs and structure chart of Prob. 13-14.

13-20 For the example in Sec. 11-8, reconstruct the program so that only two arithmetic operations (1 ADD, 1 MUL) are required on the current input before outputting C(K). Compare execution speed of the old and new structure.

13-21 Using a periodic program,† develop and document a system to control building (10 levels) temperature during different periods of the day. Include maximum and minimum error boundaries with appropriate error messages.

13-22 Instead of using two separate filter programs for the example of Sec. 11-8, implement an array structure of coefficients with proper calls. Analyze timing and space performance for the two implementations.

13-23 Add an additional interactive module to the example of Sec. 11-8 such that the coefficients of the filter can be changed through the use of indirect addressing (pointers)

13-24 If the coefficients of Prob. 13-23 are to be changed on-line, i.e., while the program is executing, what are the implementation problems? How can you structure and analyze such a system?

13-25 Is there a relationship between the length of digital-control-system data (operands) and the length of an instruction in a given computer? Give examples.

13-26 Write a program to perform the switching logic of Probs. 2-8 and 2-19.

13-27 Write a program to perform the switching logic and measurements of Prob. 2-13.

13-28 Generate a coded program† from the results of Prob. 13-3 by using structured-programming concepts.

13-29 Generate a coded program† from the structure chart and requirements of Prob. 13-4 by using structured-program concepts and the POSIT operation.

† The language chosen can be an HOL or assembly language (LSI-11 or others).

13-30 Develop a test plan for the software of Probs. 13-28 and 13-29.

13-31 A seismic isolation platform is used as a platform for the testing and evaluation of inertial instruments. The effect of external disturbances upon the stability of this platform must be eliminated or minimized.

The requirements are to develop a control system to control the movement of this platform in three axes (pitch, roll, and yaw) over the frequency band of 0 to 20 Hz. The motion of the platform must be controlled to within $+0.001$ arc s of tilt and 10^{-5} arc s/s of angular velocity for all three axes. The external disturbance consists of a step input of 1.25 ft.·lb. Gyros are used for both rate and position information. Second-order actuators are used for position control and velocity control of the platform. Develop the DFDs to level 1.0. Assuming the use of a procedural-centered software design structure, comment on the coupling and cohesion that would result from this structural method.

13-32 Write a structured assembly-language program for Prob. 13-31.

13-33 Write documented module routines† for the six second-order compensator structures that minimize computational delay. Comment on coupling and cohesion variations for each general structure implementation.

Chapter 14

14-1 Define the memory map (structure) for each of the following examples of real-time operating systems:

(*a*) Sec. 14-6A (*c*) Sec. 14-6C (*e*) Sec. 14-6E
(*b*) Sec. 14-6B (*d*) Sec. 14-6D

14-2 Generate the data flow diagram for the scheduler of the following examples:

(*a*) Sec. 14-6D (*b*) Sec. 14-6E

14-3 Generate an explicit scheduler design for the example of Sec. 11-8.

14-4 Generate an explicit scheduler design for the three-axis control Prob. 13-3.

14-5 Explicitly define how semaphores can be used in

(*a*) Prob. 13-3
(*b*) Prob. 13-5

14-6 Incorporate watchdog timers into the three-axis control systems of Prob. 13-3.

14-7 to 14-11 Implement multitasking executives by using the five real-time operating DFDs listed in Prob. 14-1. Integrate each executive into a commercial microprocessor operating system. Document by using good software engineering principles (structure charts, data dictionaries, etc.). This is an extensive project per executive.

14-12 Study and analyze commercial real-time operating system structures in terms of the concepts presented in this chapter.

14-13 Recalculate the minimum sampling rate for the following problems:

(*a*) Prob. 11-4 (*e*) Prob. 11-9
(*b*) Prob. 11-5 (*f*) Prob. 12-1
(*c*) Prob. 11-7 (*g*) Prob. 12-5
(*d*) Prob. 11-8

14-14 Plot the input-output amplitude ratio for the control problem of Sec. 11-3 as a function of f_s/f_p. f_s is the sampling frequency and f_p is the control-bandwidth breakpoint frequency.

† The language chosen can be an HOL or assembly language (LSI-11 or others).

The amplitude ratio is specifically defined as the ratio between the amplitude of the highest frequency of interest (noise, disturbance, body-bending modes) and the amplitude of the associated output due to this input. Now, delete the poles ($p = -10$) from the continuous model [Eq. (11-9)], design the controller, and again plot the amplitude ratio. The plots can provide a visual method for selection of T based upon control-bandwidth accuracy requirements. Relate this analysis to Prob. 11-20.

Chapter 15

15-1 For the sampled-data control system of Prob. 6-5:

(*a*) Determine the state and output equations by using the phase-variable representation (see Sec. 15-3).

(*b*) For $T = 0.1$ s determine $\boldsymbol{\phi}(0.1)$ and $\boldsymbol{\theta}(0.1)$.

(*c*) If $r(t) = 2u_{-1}(t)$ and $\mathbf{x}(0) = \mathbf{0}$, determine $y(kT)$ for $k = 0$, 1, and 2.

(*d*) Repeat part *c* if $\mathbf{x}'(0) = [0.1 \quad 0]$.

(*e*) Determine $Y(z)/R(z)$ and system stability by use of Eqs. (15-153) and (15-169), respectively.

15-2 Repeat Prob. 15-1, where

$$G_x = \frac{2(s+2)}{s(s+1)(s+5)} \qquad T = 1 \text{ s}$$

and for part *d* if $\mathbf{x}'(0) = [-1 \quad 0 \quad 0.1]$.

15-3 Repeat Prob. 15-1 for system 2 of Prob. 6-17 with $K_x = 3$ and $T = 1$ s.

15-4

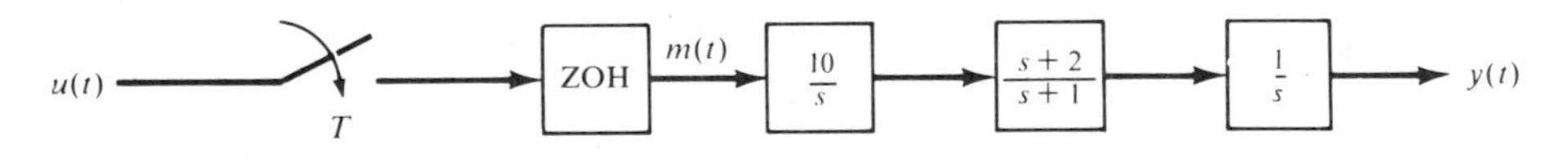

Starting from $\dot{\mathbf{x}}(t) = \mathbf{A}\mathbf{x}(t) + \mathbf{b}u(t)$ and $y(t) = \mathbf{c}'\mathbf{x}(t)$, determine the state and output difference equations for this system by using

(*a*) Phase variables

(*b*) Physical variables

(*c*) Determine the eigenvalues for $T = 0.1$ s and $T = 1$ s utilizing Eq. (15-169).

(*d*) Determine $G(z)$ by use of Eq. (15-151).

15-5 Starting with the $G(z)$ of Prob. 15-4*d*, determine the state and output difference equations for this system (see Sec. 15-4) by using

(*a*) Phase variables

(*b*) Physical variables

(*c*) Determine the eigenvalues utilizing Eq. (15-168).

15-6 For system 2 of Prob. 6-17, where $K_x = 3$, determine

(*a*) The state difference equations for both the open-loop, using the methods of Sec. 15-4, and closed-loop systems, that is,

$$\mathbf{x}[(k+1)T] = \mathbf{F}\mathbf{x}(kT) + \mathbf{h}\mathbf{u}(kT)$$

$$\mathbf{x}[(k+1)T] = \mathbf{F}_{\text{CL}}\mathbf{x}(kT) + \mathbf{h}\mathbf{r}(kT)$$

using phase and canonical state variables in obtaining the open-loop state equations

(*b*) The eigenvalues for the open-loop and closed-loop systems
(*c*) The output difference equation
(*d*) $G(z)$ and $Y(z)/R(z)$ by use of Eqs. (15-151) and (15-153), respectively

15-7 For Prob. 15-6, determine $y(kT)$ for $k = 0, 1, 2$, where $r(t) = u_{-1}(t)$ and $\mathbf{x}(0) = \mathbf{0}$.

15-8 Given an open-loop sampled-data system whose transfer function is

$$G(z) = \frac{Y(z)}{U(z)} = \frac{6z(z - 0.5)}{(z - 1)(z - 0.8)(z - 0.4)}$$

determine the canonical state-variable representation of the open-loop system.

15-9 Given

$$(1)\ G(z) = \frac{2(z^3 - 1.5z^2 + 1.24z - 0.64)}{z^3 - 2.5z^2 + 2.04z - 0.54}$$

$$(2)\ G(z) = \frac{10}{z^2 + 2z + 2}$$

Determine the state and output difference equations for these systems by using (*a*) phase variables, (*b*) physical variables, and (*c*) canonical variables.

15-10 Given $\mathbf{x}[(k + 1)T] = \mathbf{F}\mathbf{x}(kT)$, where

$$\mathbf{F} = \begin{bmatrix} 0.5 & 1 & 0 \\ 0 & 1 & 0 \\ 2 & 0.5 & 1 \end{bmatrix}$$

and $\mathbf{x}'(0) = [1 \quad 0 \quad 1]$. Determine $\mathbf{x}(kT)$ for $k > 0$ from $\mathbf{x}(kT) = \mathscr{Z}^{-1}[\mathbf{L}(z)]$, where $\mathbf{L}(z)$ is just a matrix which is a function of z.

15-11 For the sampled-data control system of Fig. 15-15, the state and output equations for $G_x(s)$ are

$$\dot{\mathbf{x}}(t) = \begin{bmatrix} -1 & 0 \\ 1 & 0 \end{bmatrix} \mathbf{x}(t) + \begin{bmatrix} 1 \\ 0 \end{bmatrix} m(t)$$

$$y(t) = [0 \quad 1]\mathbf{x}(t)$$

Determine (*a*) $\boldsymbol{\phi}(T)$ and $\boldsymbol{\theta}(T)$, (*b*) the closed-loop system discrete state equation, (*c*) $G(z)$, and (*d*) whether the output can follow the input, $r(t) = tu_{-1}(t)$, assuming that the system is stable.

15-12 Given the following state and output equations for a unity sampled-data system:

$$\mathbf{x}[(k + 1)T] = \begin{bmatrix} -0.7 & 1 \\ -0.12 & 0 \end{bmatrix} \mathbf{x}(kT) + \begin{bmatrix} 1 \\ 0 \end{bmatrix} r(kT)$$

$$y(kT) = [1 \quad -1]\mathbf{x}(kT)$$

(*a*) Determine $Y(z)/R(z)$.
(*b*) Determine if this system yields a stable response.
(*c*) If $\mathbf{x}(0) = \mathbf{0}$ and $r(kT) = 1$ for $k \geq 0$, determine $y(T)$ and $y(2T)$.

15-13 A sampled-data control system is described by the state equation

$$\mathbf{x}[(k + 1)T] = \mathbf{F}\mathbf{x}(kT) + \mathbf{h}r(kT)$$

where

$$\mathbf{F} = \begin{bmatrix} 0.5 & -1 \\ -K & -1 \end{bmatrix} \qquad \mathbf{h} = \begin{bmatrix} 1 \\ 1 \end{bmatrix}$$

Determine the value of K for the system to be asymptotically stable.

15-14 Given

$$\mathbf{x}[(k+1)T] = \begin{bmatrix} T & -0.1 \\ -0.1 & T \end{bmatrix} \mathbf{x}(kT) + \begin{bmatrix} 0 \\ 1 \end{bmatrix} r(kT)$$

$$c(kT) = [1 \quad -1]\mathbf{x}(k) + 0.5r(kT)$$

Determine the range of values that the sampling time $T\,(>0)$ can have for the system to be stable.

15-15 Given

$$\mathbf{x}[(k+1)T] = \begin{bmatrix} 0.2 & -0.1 \\ -0.1 & 0.2 \end{bmatrix} \mathbf{x}(kT) + \begin{bmatrix} 0 \\ 1 \end{bmatrix} r(kT)$$

$$c(kT) = [1 \quad -1]\mathbf{x}(kT) + 0.5r(kT)$$

$$r(kT)\begin{cases} = kT & \text{for } k \geq 0 \\ = 0 & \text{for } k < 0 \end{cases} \qquad \mathbf{x}(0) = \begin{bmatrix} 0.1 \\ 0 \end{bmatrix}$$

(*a*) Determine $c(0)$, $c(1)$, $c(2)$ where $T = 1$ s.
(*b*) Determine if this system is stable.

15-16 A deadbeat response, that is, $u(kT)_{ss} = 0$ for a step input and for the minimum value $k = 1$, can be achieved for a unity-feedback sampled-data system if its control ratio $Y(z)/R(z)$ is equal to $1/z$. Determine $D(z)$ for the following sampled-data control system by the Guillemin-Truxal method that will ensure a deadbeat response.

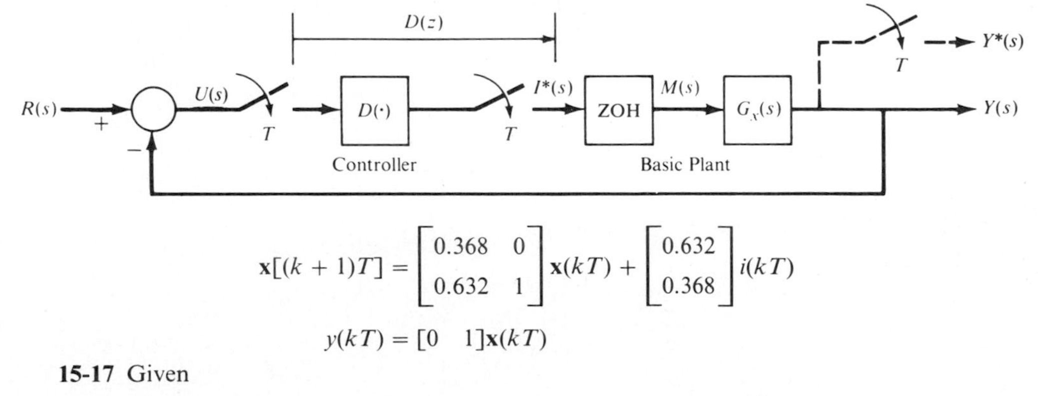

$$\mathbf{x}[(k+1)T] = \begin{bmatrix} 0.368 & 0 \\ 0.632 & 1 \end{bmatrix} \mathbf{x}(kT) + \begin{bmatrix} 0.632 \\ 0.368 \end{bmatrix} i(kT)$$

$$y(kT) = [0 \quad 1]\mathbf{x}(kT)$$

15-17 Given

$$\mathbf{x}[(k+1)T] = \begin{bmatrix} 2\epsilon^{-T} - \epsilon^{-2T} & \epsilon^{-T} - \epsilon^{-2T} \\ -2\epsilon^{-T} + 2\epsilon^{-2T} & -\epsilon^{-T} + 2\epsilon^{-2T} \end{bmatrix} \mathbf{x}(kT)$$

$$+ \begin{bmatrix} 0.5 - \epsilon^{-T} + 0.5\epsilon^{-2T} \\ \epsilon^{-T} - \epsilon^{-2T} \end{bmatrix} u(kT)$$

$$y(kT) = [1 \quad -1]\mathbf{x}(kT)$$

$\mathbf{x}(0) = \mathbf{0}$, $u(0) = 1$, $u(T) = 0.2$, $u(2T) = 0.05$, $u(kT) = 0$ for $k \geq 3$, $T = 1$ s. Determine $y(1.2)$.

15-18 A unity-feedback sampled-data control system has the state and output difference equations of Prob. 15-17. Given that $\mathbf{x}(0) = \mathbf{0}$ and $r(t) = u_{-1}(t)$ for $t \geq 0$, determine $y(1.2)$ and $y(2.2)$.

15-19 Discuss the application of the PCT technique to the example of Sec. 15-8A. *Hint*: First convert Fig. 15-16*d* to a sampled-data control system then obtain the PCT system.

APPENDIX

A

FOURIER TRANSFORM

A-1 INTRODUCTION

The Fourier transform is basically a mapping from the time domain to the frequency domain. This transform by definition explicitly defines the frequency content (magnitude and phase) of a signal which can be periodic or nonperiodic. If it is periodic, the frequency is known in the time domain. In digital control systems, as in most applications, the input and output signals are not periodic. The Fourier transform then provides a method of representing the frequency components of the time-domain signal. Note, however, that no new information is being obtained by transferring to the frequency domain. The frequency domain explicitly indicates frequency information (magnitude and phase), whereas the time domain (time response) displays frequency implicitly.

A-2 DEFINITION

The Fourier transform is defined mathematically as

$$\mathbf{E}(j\omega) = \int_{-\infty}^{+\infty} e(t)\epsilon^{-j\omega t}\,dt \qquad \textbf{Case 1} \qquad \text{(A-1)}$$

The function $\mathbf{E}$ is only a function of frequency (radians per second) since the integral is integrated from minus infinity to plus infinity in the time domain. The function $\mathbf{E}(j\omega)$ exists if $e(t)$ is of exponential order.[A1] For the Fourier transform to be useful, it must generate a unique $\mathbf{E}(j\omega)$ such that $e(t)$ can uniquely be recovered from $\mathbf{E}(j\omega)$. Thus, the other part of the Fourier transform is the equation for $e(t)$, the inverse transform:

$$e(t) = \frac{1}{2\pi}\int_{-\infty}^{+\infty} \mathbf{E}(j\omega)\, \epsilon^{j\omega t}\, d\omega \qquad \textbf{Case 1} \tag{A-2}$$

These two equations then constitute the Fourier transform pair. The plot of the magnitude $\mathbf{E}(j\omega)$ of the Fourier transform is also called the *amplitude frequency response*, and the associated phase is sometimes defined as the *phase angle response*. An example of a magnitude plot is shown in Fig. A-1 for a continuous signal.

The definition of the Fourier transform is very similar to the Laplace transform. The main difference is the integration in Eq. (A-1). The Fourier transform is found by integration only along the imaginary axis (frequency axis) in the s domain; that is, $s = j\omega$.

A-3 EXTENSION

One excellent property of the Fourier transform is its utility in performing equivalent time-domain convolution in the frequency domain by only having to multiply the two associated Fourier transforms. The same property is also true of Laplace and $\mathscr{Z}$ transforms. Its use in linear system theory is self-evident when the linear system output response $y(t)$ is desired based on a known input $e(t)$. The output frequency response $\mathbf{Y}(j\omega)$ is the product of the input frequency-domain description $\mathbf{E}(j\omega)$ and the system transfer function defined in the frequency domain $\mathbf{G}(j\omega)$; that is, $\mathbf{Y}(j\omega) = \mathbf{G}(j\omega)\mathbf{E}(j\omega)$. The output frequency response completely defines the steady-state response resulting from the sinusoidal components of any input. Remember that in a linear system the sinusoidal input components are only changed in magnitude and phase as determined by the linear system transfer function characteristics.

In the previous definition, the function $e(t)$ is assumed to be continuous. In digital control systems, analysis of discrete signals and compensators still requires that a frequency-domain method be available. From Chap. 3, $e(t)$ in the discrete form becomes a function of impulses; that is,

$$e(t) = \sum_{k=0}^{\infty} e(t_k)\, \delta(t - t_k) \qquad e(t_k) = 0, \quad t_k < 0 \tag{A-3}$$

or

$$e(t) = \sum_{k=0}^{\infty} e(kT)\, \delta(t - kT) \qquad k = 0, 1, \ldots$$

which yields for $\mathbf{E}(j\omega)$ from Eq. (A-1),

$$\mathbf{E}(j\omega) = \sum_{k=0}^{\infty} e(kT)\, \epsilon^{-j\omega kT} \tag{A-4}$$

Equation (A-4) is in reality the $\mathscr{Z}$ transform of $e(kT)$ evaluated at $z = \epsilon^{sT}|_{s=j\omega} = \epsilon^{j\omega T}$.

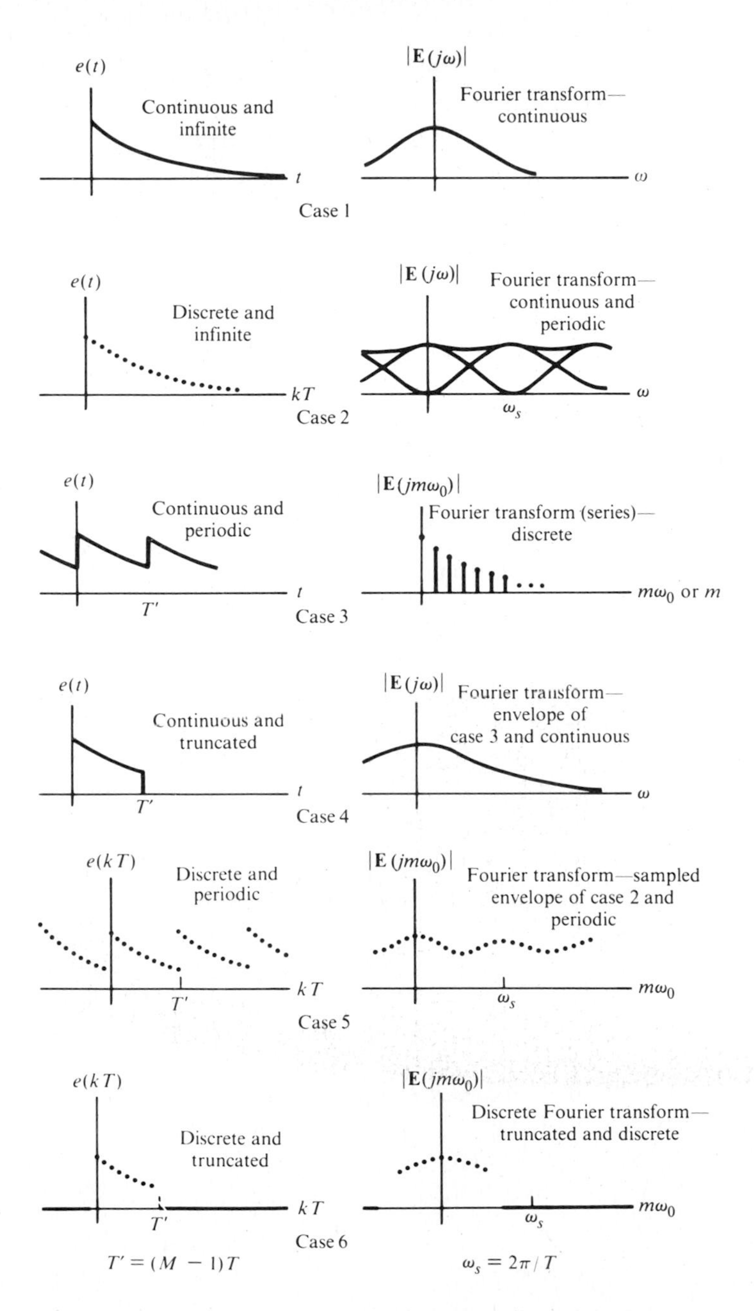

Figure A-1 Fourier transform for different forms of $e(t)$ and $e(kT)$.

Thus,

$$E(z)\Big|_{z=\epsilon^{j\omega T}} = \sum_{k=0}^{\infty} e(kT)z^{-k}\Big|_{z=\epsilon^{j\omega T}} \tag{A-5}$$

Equation (A-5) may not be proper if the infinite series does not converge; i.e., the $\mathscr{Z}$ transform does not exist. The following discussion presents the various forms of the Fourier transform as associated with signals modeled in digital control systems.[A2] All the transform pairs can be derived from the basic definition, Eqs. (A-1) and (A-2).

Consider a sampled continuous signal, where $\omega_s = 2\pi/T$; then the transform pair is

$$\mathbf{E}^*(j\omega) = \frac{1}{T}\sum_{r=-\infty}^{+\infty} \mathbf{E}\left(j\omega + \frac{j2\pi r}{T}\right) \tag{A-6}$$

or

$$\mathbf{E}^*(j\omega) = \sum_{k=0}^{+\infty} e(kT)\,\epsilon^{-j\omega kT} \qquad \textbf{Case 2}$$

$$e(kT) = \frac{1}{2\pi}\int_{-\infty}^{+\infty} E^*(j\omega)\,\epsilon^{j\omega kT}\,d\omega \qquad e(kT)=0,\ k<0 \tag{A-7}$$

Equation (A-6) reflects the folding phenomena due to sampling as discussed and developed in Chap. 3. An example is shown in Fig. A-1 for Case 2.

If the signal is a periodic signal with period $2\pi/\omega_0$, then the transform pair is

$$\mathbf{E}(j\omega) = \sum_{m=-\infty}^{\infty} \mathbf{C}(jm\omega_0)\delta(\omega - m\omega_0) \qquad \textbf{Case 3} \tag{A-8}$$

where the former coefficients $\mathbf{C}(jm\omega_0)$ are determined from

$$\mathbf{C}(jm\omega_0) = \frac{\omega_0}{2\pi}\int_{-\pi/\omega_0}^{\pi/\omega_0} e(t)\,\epsilon^{-j\omega_0 mt}\,dt \qquad \textbf{Case 3} \tag{A-9}$$

and $\delta(\omega - m\omega_0)$ is an impulse in the frequency domain. The inverse form is

$$e(t) = \sum_{m=-\infty}^{+\infty} \mathbf{C}(jm\omega_0)\,\epsilon^{j\omega_0 mt} \qquad \textbf{Case 3} \tag{A-10}$$

which is, of course, the standard Fourier series representation of $e(t)$, a sine or cosine series. An example pair is shown in Fig. A-1 for Case 3.

If a continuous truncated signal with length T' is to be Fourier-transformed, the transform pair to use is

$$\mathbf{E}(j\omega) = \int_{-T'}^{T'} e(t)\,\epsilon^{-j\omega t}\,dt \tag{A-11}$$

Case 4

$$e(t) = \begin{cases} \dfrac{1}{2\pi}\displaystyle\int_{-\infty}^{+\infty} \mathbf{E}(j\omega)\,\epsilon^{j\omega t}\,d\omega & \text{(A-12)} \\ 0 \qquad \text{for } t<0,\ t\geq T' & \text{(A-13)} \end{cases}$$

An example of this representation is shown in Fig. A-1 as Case 4.

If the signal is a sampled periodic sequence with the period $MT = T'$, the transform pair is

$$\mathbf{E}(j\omega) = \sum_{m=-\infty}^{+\infty} \mathbf{C}(jm\omega_0)\delta(\omega - m\omega_0) \tag{A-14}$$

$$\mathbf{C}(jm\omega_0) = \frac{1}{MT}\sum_{k=0}^{\infty} e(kT)\,\epsilon^{-j\omega_0 mkT} \qquad \textbf{Case 5} \tag{A-15}$$

$$e(kT) = \sum_{m=0}^{M-1} \mathbf{C}(jm\omega_0)\,\epsilon^{-j\omega_0 mkT} \tag{A-16}$$

where M is the number of samples in one period.

An example for Case 5 is depicted in Fig. A-1. The $1/MT$ term in Eq. (A-15) is a normalization factor similar to $1/2\pi$ in Eq. (A-2). From Eq. (A-14), the Fourier representation consists of discrete points in the frequency domain for a sampled periodic time-domain sequence.

Now, consider the truncation of the sampled periodic sequence in Eqs. (A-14) to (A-16), i.e., a sampled truncated sequence. The transform pair is defined as

$$\mathbf{E}(jm\omega_0) = \begin{cases} \dfrac{1}{MT}\displaystyle\sum_{k=0}^{M-1} e(kT)\,\epsilon^{-j\omega_0 mkT} & 0 \le m \le M-1 \\ 0 & \text{otherwise} \end{cases} \tag{A-17}$$

Case 6

$$e(kT) = \begin{cases} \displaystyle\sum_{m=0}^{M-1} \mathbf{E}(jm\omega_0)\,\epsilon^{-j\omega_0 mkT} & 0 \le k \le M-1 \\ 0 & \text{otherwise} \end{cases} \tag{A-18}$$

Figure A-1 presents an example of Case 6. This transform pair is called the *discrete Fourier transform* (DFT).

To determine the frequency-domain representation of a z-domain transfer function $G(z)$, z is set equal to $\epsilon^{j\omega T}$ and ω varied over the frequency bands of interest. Note the role that the sampling time T plays in the frequency-axis weighting for z-domain transfer functions. The frequency axis is shifted by a factor of T. Thus, frequency plots can be normalized for T if desired. The generation of frequency plots for $G(z)$ is similar to setting s equal to $j\omega$ for generating Bode plots (magnitude and phase) in continuous control-theory studies.

A-4 SUMMARY

In digital-control-theory applications, sampled measurements are used to determine the frequency content of signals. Since measurements can not be taken forever, the sequence of measurements is truncated. The real-world truncation operation is based upon theoretically assuming that the signal is periodic. Thus, Case 6 reflects the correct model, a sampled truncated signal. Equations (A-17) and (A-18) can be used to transform the time-domain sampled values into a frequency-domain spectrum.

It should be mentioned that Eqs. (A-17) and (A-18) can be manipulated so that the number of terms and arithmetic operations can be minimized under certain assumptions. This manipulation results in the so-called fast Fourier transform (FFT).[A3, A4] Many digital frequency analyzers (hardware and software) use this implementation concept.

REFERENCES

A1. Jong, M. T.: "Methods of Discrete Signal and System Analysis," McGraw-Hill, New York, 1982.

A2. Oppenheim, A. V., and R. W. Schafer: "Digital Signal Processing," Prentice-Hall, Englewood Cliffs, N.J., 1975.

A3. Papoulis, Athanasios: "Probability, Random Variables, and Stochastic Processes," McGraw-Hill, New York, 1965.

A4. Rabiner, L. R., and B. Gold: "Theory and Application of Digital Signal Processing," Prentice-Hall, Englewood Cliffs, N.J., 1975.

APPENDIX

B

CONVOLUTION

B-1 INTRODUCTION

Of primary interest in the analysis of linear systems is the determination of the system response (output) to an input signal. If the system is modeled or described by a linear differential or difference equation, the solution to the nonhomogeneous equation represents the output response. Another approach is based upon the linear system superposition principle leading to the convolution operation. The system output then is represented by an integral operation, namely, the *convolution or superposition integral.* This integral approach permits another viewpoint offering new insight and intuition.[B1]

B-2 DEFINITION

For a linear system with system function $g(t)$, the output $c(t)$ in response to an input $e(t)$ is defined by the convolution interval;[B2] that is,

$$\begin{aligned} c(t) &= \int_{-\infty}^{+\infty} g(\tau)e(t-\tau)\,d\tau \\ &= g(t) * e(t) \end{aligned} \tag{B-1}$$

In other words, $g(t)$ is convolved, $*$, with $e(t)$.

If the Laplace or Fourier transform is taken of both sides of Eq. (B-1), the resulting output transformations are

$$C(s) = G(s)E(s) \tag{B-2}$$

or

$$\mathbf{C}(j\omega) = \mathbf{G}(j\omega)\mathbf{E}(j\omega) \tag{B-3}$$

respectively. $C(\ )$, $G(\ )$, and $E(\ )$ are the respective transforms of the output, system function, and the input.

For discrete systems, the system output $c(kT)$ is defined by a convolution sum[B1] with input $e(kT)$ and system function $g(kT)$; that is,

$$\begin{aligned} c(kT) &= \sum_{m=-\infty}^{+\infty} g(mT)e(kT - mT) \\ &= g(kT)*e(kT) \end{aligned} \tag{B-4}$$

Of course, T being a constant, it can be explicitly deleted from this equation without loss of generality. A frequent way of considering the operation defined by Eq. (B-4) is to picture the operation as a shifting and multiplying of two discrete functions.[B3]

Taking the $\mathscr{Z}$ transform of both sides of Eq. (B-4) yields

$$C(z) = G(z)E(z) \tag{B-5}$$

the transform of the discrete system output.

B-3 SUMMARY

In general, the time-domain convolution operation as represented in the Fourier, s, and z domains provides relatively easy ways of representing system input-output relationships and, thus, easier methods of analysis. Arithmetic operations are used instead of integral (summation) or differential (difference) operations. Examples using these approaches involve the response of a series of cascaded linear systems, the use of Bode plot analysis, and manipulation of system equations or models. Various chapters used convolution structure in a transformed domain to develop system models.

REFERENCES

B1. Antoniou, Andreas: "Digital Filters Analysis and Design," McGraw-Hill, New York, 1979.

B2. Papoulis, Athanasios: "Probability, Random Variables, and Stochastic Processes," McGraw-Hill, New York, 1965.

B3. Roden, Martin S.: "Analog and Digital Communication Systems," Prentice-Hall, Englewood Cliffs, N.J., 1979.

APPENDIX

C

PADÉ APPROXIMATIONS

The Padé method produces rational-fractions approximations of the form

$$F(x) = \frac{A_m(x)}{B_m(x)} \qquad m = 0, 1, 2, \ldots \tag{C-1}$$

for the differentiable power series

$$d(x) = c_0 + c_1 x + c_2 x^2 + c_3 x^3 + \cdots \tag{C-2}$$

where $c_0 \neq 0$ and $F(x) \approx g(x)$. $A_m(x)$ and $B_m(x)$ are polynomials of the form

$$A_m(x) = a_0 + a_1 x + a_2 x^2 + \cdots + a_m x^m \tag{C-3}$$

$$B_m(x) = b_0 + b_1 x + b_2 x^2 + \cdots + b_m x^m \tag{C-4}$$

where, for this discussion, the coefficients a_i and b_j are finite values other than zero. Let

$$\mathbf{a} = \begin{bmatrix} a_0 \\ a_1 \\ a_2 \\ \vdots \\ a_m \end{bmatrix} \tag{C-5}$$

$$\mathbf{b} = \begin{bmatrix} b_0 \\ b_1 \\ b_2 \\ \vdots \\ b_m \end{bmatrix} \tag{C-6}$$

$$\mathbf{C}_{p,q} = \begin{bmatrix} c_q & c_{q-1} & \cdots & c_{q-p} \\ c_{q+1} & c_q & \cdots & c_{q-p+1} \\ \cdots & \cdots & \cdots & \cdots \\ c_{p+q-1} & c_{q+p-2} & \cdots & c_{q-1} \end{bmatrix} \tag{C-7}$$

where the elements of Eq. (C-7) whose subscripts are negative, that is, $i - j < 0$, are all zero.

The coefficients of Eqs. (C-3) and (C-4) may be solved from the following matrix equations:

$$\mathbf{C}_{m,m+1}\mathbf{b} = \mathbf{0} \tag{C-8}$$

$$\mathbf{C}_{m+1,0}\mathbf{b} = \mathbf{a} \tag{C-9}$$

The matrices $\mathbf{C}_{m,m+1}$ and $\mathbf{C}_{m+1,0}$ are obtained from Eq. (C-7), where for the former $p = m$ and $q = m + 1$ and for the latter $p = m + 1$ and $q = 0$. Note that $\mathbf{a}$ and $\mathbf{b}$ are $(m + 1) \times 1$ vectors, $\mathbf{C}_{m,m+1}$ is an $m \times (m + 1)$ matrix, and $\mathbf{C}_{m+1,0}$ is an $m \times m$ matrix.

The following two examples are given to illustrate the Padé method.

Example C-1 The power series of ϵ^x is expanded about zero (MacLaurin series), resulting in

$$d(x) = \epsilon^x = 1 + x + \frac{x^2}{2!} + \frac{x^3}{3!} + \cdots \tag{C-10}$$

$d(x)$ is to be approximated by the rational fraction

$$F(x) = \frac{a_0 + a_1 x}{b_0 + b_1 x} \tag{C-11}$$

where $m = 1$, $c_0 = 1$, $c_1 = 1$, $c_2 = 1/2$, $c_3 = 1/6$, etc. From Eqs. (C-5) to (C-7) the following are obtained:

$$\mathbf{a} = \begin{bmatrix} a_0 \\ a_1 \end{bmatrix} \qquad \mathbf{b} = \begin{bmatrix} b_0 \\ b_1 \end{bmatrix}$$

$$\mathbf{C}_{1,2} = [c_2 \quad c_1] = [1/2 \quad 1]$$

$$\mathbf{C}_{2,0} = \begin{bmatrix} c_0 & 0 \\ c_1 & c_0 \end{bmatrix} = \begin{bmatrix} 1 & 0 \\ 1 & 1 \end{bmatrix}$$

Inserting these vectors and matrix into Eqs. (C-8) and (C-9) results in

$$\mathbf{C}_{1,2}\mathbf{b} = [1/2 \quad 1]\begin{bmatrix} b_0 \\ b_1 \end{bmatrix} = 0 \rightarrow b_1 = \frac{-b_0}{2}$$

$$\mathbf{C}_{2,0}\mathbf{b} = \begin{bmatrix} 1 & 0 \\ 1 & 1 \end{bmatrix}\begin{bmatrix} b_0 \\ b_1 \end{bmatrix} = \begin{bmatrix} a_0 \\ a_1 \end{bmatrix} \rightarrow \begin{cases} a_0 = b_0 \\ b_0 + b_1 = a_1 \end{cases}$$

By letting $a_0 = b_0 = 1$ yields $b_1 = -1/2$ and $a_1 = 1/2$. Inserting these values into Eq. (C-11) yields

$$F(x) = \frac{1 + x/2}{1 - x/2} \approx d(x) \tag{C-12}$$

Example C-2 The results of Example C-1 are used to obtain the rational-fraction approximation for the ZOH device (see Chap. 4).

$$G_{zo}(s) = \frac{1 - \epsilon^{-sT}}{s} \approx \frac{1 - d(x)}{s} = G_{pa}(s) \tag{C-13}$$

where $x = -sT$. Substituting Eq. (C-12) into Eq. (C-13) yields

$$G_{zo}(s) \approx \frac{2T}{Ts + 2} = G_{pa}(s) \tag{C-14}$$

[see Eq. (7-27)].

Example C-3 The power series of Eq. (C-10) is to be approximated by the rational fraction

$$F(x) = \frac{a_0 + a_1 x + a_2 x^2}{b_0 + b_1 x + b_2 x^2} \tag{C-15}$$

where $m = 2$. From Eqs. (C-5) to (C-7) the following are obtained:

$$\mathbf{a} = \begin{bmatrix} a_0 \\ a_1 \\ a_2 \end{bmatrix} \qquad \mathbf{b} = \begin{bmatrix} b_0 \\ b_1 \\ b_2 \end{bmatrix}$$

$$\mathbf{C}_{2,3} = \begin{bmatrix} c_3 & c_2 & c_1 \\ c_4 & c_3 & c_2 \end{bmatrix} = \begin{bmatrix} 1/6 & 1/2 & 1 \\ 1/24 & 1/6 & 1/2 \end{bmatrix}$$

$$\mathbf{C}_{3,0} = \begin{bmatrix} c_0 & 0 & 0 \\ c_1 & c_0 & 0 \\ c_2 & c_1 & c_0 \end{bmatrix} = \begin{bmatrix} 1 & 0 & 0 \\ 1 & 1 & 0 \\ 1/2 & 1 & 1 \end{bmatrix}$$

Inserting these vectors and matrixes into Eqs. (C-8) and (C-9) results in

$$\mathbf{C}_{2,3}\mathbf{b} = \begin{bmatrix} 1/6 & 1/2 & 1 \\ 1/24 & 1/6 & 1/2 \end{bmatrix} \begin{bmatrix} b_0 \\ b_1 \\ b_2 \end{bmatrix} \rightarrow \begin{cases} b_0/6 + b_1/2 + b_2 = 0 \\ b_0/24 + b_1/6 + b_2/2 = 0 \end{cases}$$

$$\mathbf{C}_{3,0}\mathbf{b} = \begin{bmatrix} 1 & 0 & 0 \\ 1 & 1 & 0 \\ 1/2 & 1 & 1 \end{bmatrix} \begin{bmatrix} b_0 \\ b_1 \\ b_2 \end{bmatrix} = \begin{bmatrix} a_0 \\ a_1 \\ a_2 \end{bmatrix} \rightarrow \begin{cases} a_0 = b_0 \\ b_0 + b_1 = a_1 \\ b_0/2 + b_1 + b_2 = a_2 \end{cases}$$

By letting $a_0 = b_0 = 1$ yields $a_1 = 1/2$, $a_2 = b_2 = 1/12$, and $b_1 = -1/2$. Inserting these values into Eq. (C-15) yields

$$F(x) = \frac{1 + x/2 + x^2/12}{1 - x/2 + x^2/12} = \frac{12 + 6x + x^2}{12 - 6x + x^2} \tag{C-16}$$

The errors in the Padé approximation increase with distance away from the power-series expansion point. Note in Example C-2 that $|x| = |-sT|$ is relatively small for "small" T. For a more complete discussion of the Padé method, the reader is referred to H. S. Wall, "Analytic Theory of Continued Fractions," Van Nostrand, New York, 1948.

APPENDIX

D

POWER SERIES

D-1 INTRODUCTION

The $\mathscr{Z}$ transform can be represented by a power series; that is,

$$E(z) = \sum_{k=0}^{\infty} e(kT)z^{-k}$$

where $e(kT)$ defines a sequence of values for $k = 0, 1, \ldots$. This one-sided $\mathscr{Z}$ transform is a power series in the variable z^{-1}, where z is a complex variable. $E(z)$ is a power series with convergence properties depending upon $e(kT)$. The constant T may be left out of the formulation since power series are only functions of an index parameter, in this case k. However, T is left in the functional representation since digital control systems are being discussed in this text. The same representation is also used in the Fourier transform discussion of Appendix A.

D-2 CONVERGENCE PROPERTIES

The intent of this appendix is to focus on the convergence properties of the $\mathscr{Z}$ transform. From complex-variable theory, the negative power series (Laurent series)

$$\sum_{k=0}^{\infty} e_k(z - z_0)^{-k} \tag{D-1}$$

where $e_k = e(kT)$, converges absolutely outside a circle of radius r centered at z_0. By absolute convergence is meant that

$$\sum_{k=0}^{\infty} |e_k(z - z_0)^{-k}| \tag{D-2}$$

also converges. For the $\mathscr{L}$ transform, the value of z_0 is zero. The convergence circle then is centered at the origin. The radius r is found by using the ratio test or root test, respectively; that is,

$$r = \lim_{k \to \infty} \left| \frac{e_{k+1}}{e_k} \right| \qquad \text{if the limit exists} \tag{D-3}$$

or

$$r = \lim_{k \to \infty} \sqrt[k]{|e_k|} \qquad \text{if the limit exists} \tag{D-4}$$

The proof of this theorem may be found in any text on power-series convergence and their properties.[D1, D2] The power series in z^{-1} can also converge on the boundary of the convergence circle. Every case has to be analyzed separately in regard to the boundary convergence.

This convergence concept of the $\mathscr{Z}$-transform power series may be confusing in relationship to the region of convergence and the analysis of the $\mathscr{Z}$ transform within the unit circle. The theorem says that the function $E(z)$ only exists for $z > r$; that is, the power series converges only for $z > r$. Considering the manner in which $E(z)$ is generated from the infinite power series, this theorem is mathematically correct. However, the various design and analysis procedures in the z plane focus on the unit circle without regard to the convergence region. The reason that this unit circle analysis is possible and is also mathematically correct is based upon the concept of analytical continuation. By analytical is meant that $E(z)$ is defined (i.e., has finite values as a function of z) in a region that does not contain any of its poles. Thus, once $E(z)$ is formed in terms of a ratio of polynomials, it is analytical not only outside the region of convergence but also inside except for the poles of $E(z)$ inside this region. Therefore, $E(z)$ is analytically continued across the circle boundary from the $\mathscr{Z}$-transform convergence region to inside the nonconvergence circle. A similar situation occurs in the s plane when considering the Laplace transform in regard to existence of the integral transform. Note that the boundary for the Laplace transform is a vertical line in the s plane.

D-3 SUMMARY

When manipulating $E(z)$ by using the various text procedures, power-series convergence regions are not explicitly considered because of the analytical continuation phenomena. However, careful consideration should be given when using initial- and final-value theorems as to the limits as z goes to 0 or 1, respectively. Chapter 6 presents some of these difficulties.

REFERENCES

D1. Ahlfors, L. V.: "Complex Analysis," McGraw-Hill, New York, 1953.
D2. Sneldon, E. H.: "The Use of Integral Transforms," McGraw-Hill, New York, 1972.

APPENDIX

E

INTERACTIVE COMPUTER-AIDED-DESIGN (CAD) PROGRAMS FOR DIGITAL AND CONTINUOUS CONTROL-SYSTEM ANALYSIS AND SYNTHESIS

E-1 INTRODUCTION

Various CAD software packages[E1–E15] are available for the development of continuous- and discrete-time control systems. These packages can assist the control engineer in performing time-consuming calculations for both scalar and multivariable models. Some of the available CAD packages provide the user with an integrated set of design tools and perform a broad range of calculations applicable to control-system analysis and synthesis. The computer program called TOTAL[E9] (TOTAL-I) is presented in this appendix as an example of a specific CAD package. Note that this CAD package is the one employed in the text examples. Extension [E5,E10] to TOTAL is also summarized.

E-2 OVERVIEW OF TOTAL

TOTAL is designed as a tool to be used with as much speed, agility, and confidence as a familar hand calculator. To assist the user in obtaining an overview of TOTAL, the following information is provided. A complete description of this FORTRAN program, with examples of input and output data, is given in Ref. E9, which can be obtained from the Defense Documentation Center (DDC), Cameron Station, Alexandria, Va. 22314.

1. TOTAL is built around twelve general-purpose polynomials and seven general-purpose matrices.
2. Eight of the polynomials may be paired to form the numerators and denominators of four general-purpose transfer functions.
3. With just these polynomials, matrices, and transfer functions the user is able to use the entire spectrum of TOTAL's capabilities, which include:
   a. Adding, subtracting, multiplying, dividing, factoring, expanding, and copying polynomials
   b. Adding subtracting, multiplying, inverting, transposing, obtaining eigenvalues, presetting, and copying matrices
   c. Obtaining root locus, frequency response, and time response of open- and closed-loop transfer functions in both the continuous and discrete domains
   d. Performing block diagram reduction
   e. Computing transfer functions from the state-space equations
   f. Transforming between continuous and discrete domains using a variety of methods
4. A built-in 20-register scientific calculator with a four-register stack is available to the user at all times.
5. The user can quickly list, transfer, or modify any variable, at any time, anywhere in the program.
6. Complete error detection, diagnostics, and abnormal-termination protection are provided. Specific help is available at any time by simply typing a question mark.

The TOTAL package currently executes on the CDC Cyber, SEL 32, and the DEC VAX-11/780.

E-3 INTRODUCTION TO TOTAL

TOTAL contains over 100 options and a few special commands with which the user can manipulate a data base of 12 polynomials, 7 matrices, and roughly 60 scalar variables. Each option uses information stored in these variables to perform a particular function and saves the results for output or future use. During execution of TOTAL, the user has complete freedom to select options, modify variables, and give commands at will. The following sections summarize the information needed to make use of the full power of this program.

E-4 TOTAL's INPUT MODES

TOTAL has three modes in which it pauses for the user to input information: OPTION, DATA, and CALCULATOR. Each mode has its own vocabulary of allowable inputs and its own characteristic prompt to the user. The OPTION mode is the primary command mode of TOTAL. When TOTAL pauses in this mode, the user is free to type any command, select any option, modify any variable, or list any data. It is from this mode that the user controls the program. When an option has been selected and data are needed, TOTAL will ask for them. This is called the DATA mode. The CALCULATOR mode can be entered at any time by typing C. The calculator operates like an HP-45 calculator (using reverse Polish notation). The CALCULATOR mode is terminated by typing another C.

E-5 TOTAL'S OPTIONS

TOTAL presently contains over 100 options which are divided into groups of 10 according to general function. Table E-1 lists some of the options currently available.

Table E-1 Some TOTAL options

No.	Option
	Transfer-function input options
0	LIST OPTIONS
1	RECOVER ALL DATA FROM FILE MEMORY
2	POLYNOMIAL FORM—GTF (forward transfer function)
3	POLYNOMIAL FORM—HTF (feedback transfer function)
4	POLYNOMIAL FORM—OLTF (open-loop transfer function)
5	POLYNOMIAL FORM—CLTF (closed-loop transfer function)
6	FACTORED FORM—GTF
7	FACTORED FORM—HTF
8	FACTORED FORM—OLTF
9	FACTORED FORM—CLTF
	Matrix input options for state equations
10	LIST OPTIONS
11	AMAT—CONTINUOUS PLANT MATRIX
12	BMAT—CONTINUOUS INPUT MATRIX
13	CMAT—OUTPUT MATRIX
14	DMAT—DIRECT TRANSMISSION MATRIX
15	KMAT—STATE VARIABLE FEEDBACK MATRIX
16	FMAT—DISCRETE PLANT MATRIX
17	GMAT—DISCRETE INPUT MATRIX
18	SET UP STATE SPACE MODEL OF SYSTEM
19	EXPLAIN USE OF ABOVE MATRICES
	Block diagram manipulation and state-space options
20	LIST OPTIONS
21	FORM OLTF=GTF*HTF (IN CASCADE)
22	FORM CLTF=(GAIN*GTF)/(1+GAIN*GTF*HTF)
23	FORM CLTF=(GAIN*OLTF)/(1+GAIN*OLTF)
24	FORM CLTF=GTF+HTF (IN PARALLEL)
25	GTF(*s*) AND HTF(*s*) FROM CONTINUOUS STATE SPACE MODEL
26	GTF(*z*) AND HTF(*z*) FROM DISCRETE STATE SPACE MODEL
27	WRITE ADJOINT (*s***I**−AMAT) TO FILE ANSWER
28	FIND HTF FROM CLTF & GTF FOR CLTF=GTF*HTF/(1+GTF*HTF)
29	FIND HTF FROM CLTF & GTF FOR CLTF=GTF/(1+GTF*HTF)

Table E-1 Some TOTAL options—*continued*

No.	Option
	Time-response options, continuous $F(t)$ and discrete $F(kT)$†
30	LIST OPTIONS
31	TABULAR LISTING OF $F(t)$ OR $F(kT)$
32	PLOT $F(t)$ OR $F(kT)$ AT USER'S TERMINAL
33	PRINTER PLOT (WRITTEN TO FILE ANSWER)
34	CALCOMP PLOT (WRITTEN TO FILE PLOT)
35	PRINT TIME OR DIFFERENCE EQUATION [$F(t)$ OR $F(kT)$]
36	PARTIAL FRACTION EXPANSION OF CLTF (OR OLTF)
37	LIST T-PEAK, T-RISE, T-SETTLING, T-DUP, M-PEAK, FINAL VALUE
38	QUICK SKETCH AT USER'S TERMINAL
39	SELECT INPUT: STEP, RAMP, PULSE, IMPULSE, SIN ωT
	Root-locus options
40	LIST OPTIONS
41	GENERAL ROOT LOCUS
42	ROOT LOCUS WITH A GAIN OF INTEREST
43	ROOT LOCUS WITH A ZETA (DAMPING RATIO) OF INTEREST
44	LIST N POINTS ON A BRANCH OF INTEREST
45	LIST ALL POINTS ON A BRANCH OF INTEREST
46	LIST LOCUS ROOTS AT A GAIN OF INTEREST
47	LIST LOCUS ROOTS AT A ZETA OF INTEREST
48	PLOT ROOT LOCUS AT USER'S TERMINAL
49	LIST CURRENT VALUES OF ALL ROOT LOCUS VARIABLES
	Frequency-response options
50	LIST OPTIONS
51	TABULAR LISTING
52	TWO CYCLE SCAN OF MAGNITUDE (OR DB)
53	TWO CYCLE SCAN OF PHASE (DEGREES OR RADIANS)
54	PLOT $F(j\omega)$ AT USER'S TERMINAL
55	CALCOMP PLOT—LINEAR FREQUENCY AXIS
56	CALCOMP PLOT—LOG FREQUENCY AXIS
57	TABULATE POINTS OF INTEREST: PEAKS, BREAKS, ETC.
58	CALCOMP PLOT—NICHOLS LOG-MAG/ANGLE PLOT
	Polynomial operations
60	LIST OPTIONS
61	FACTOR POLYNOMIAL (POLYA)
62	ADD POLYNOMIALS (POLYC = POLYA + POLYB)
63	SUBTRACT POLYNOMIALS (POLYC = POLYA − POLYB)
64	MULTIPLY POLYS (POLYC = POLYA*POLYB)
65	DIVIDE POLYS (POLYC + REM = POLYA/POLYB)
66	STORE (POLY_) INTO POLYD
67	EXPAND ROOTS INTO A POLYNOMIAL
68	$(s+a)^n$ EXPANSION INTO A POLYNOMIAL
69	ACTIVATE POLYNOMIAL CALCULATOR

† Option 37 is not available for the discrete domain.

Table E-1 Some TOTAL options—*continued*

No.	Option
	Matrix operations
70	LIST OPTIONS
71	ROOTA=EIGENVALUES OF AMAT
72	CMAT=AMAT+BMAT
73	CMAT=AMAT−BMAT
74	CMAT=AMAT*BMAT
75	CMAT=AMAT INVERSE
76	CMAT=AMAT TRANSPOSED
77	CMAT=IDENTITY MATRIX I
78	DMAT=ZERO MATRIX 0
79	COPY ONE MATRIX TO ANOTHER
	Digitization options
80	LIST OPTIONS
81	CLTF(s) TO CLTF(z) BY IMPULSE INVARIANCE
82	CLTF(s) TO CLTF(z) BY FIRST DIFFERENCE APPROXIMATION
83	CLTF(s) TO CLTF(z) BY TUSTIN TRANSFORMATION
84	CLTF(z) TO CLTF(s) BY IMPULSE INVARIANCE
85	CLTF(z) TO CLTF(s) BY INVERSE FIRST DIFFERENCE
86	CLTF(z) TO CLTF(s) BY INVERSE TUSTIN
87	FIND FMAT AND GMAT FROM AMAT AND BMAT
89	CLTF(X) TO CLTF(Y) BY X = ALPHA*(Y+A)/(Y+B)
	Miscellaneous options
90	LIST OPTIONS
91	REWIND AND UPDATE MEMORY FILE WITH CURRENT DATA
92	ERASE SCREEN (TEKTRONIX GRAPHICS TERMINALS ONLY)
93	LIST CURRENT SWITCH SETTINGS (ECHO, ANSWER, ETC.)
94	HARD COPY (TEK TERMINAL WITH HARD COPY UNIT ONLY)
95	STAR TREK GAME
96	LIST SPECIAL COMMANDS ALLOWED IN OPTION MODE
97	LIST VARIABLE NAME DIRECTORY
98	LIST MAIN OPTIONS OF TOTAL
99	PRINT NEW FEATURES BULLETIN
199	AUGMENT AMAT:[AMAT]=[AMAT]−GAIN*[BMAT]*[KMAT]
131	(INTEGRAL OF (CLTF SQUARED))/2PI=
	Double-precision discrete transform options
*140	LIST OPTIONS
*141	CLTF(s) TO CLTF(z) (IMPULSE VARIANCE)
*142	CLTF(s) TO CLTF(w) W=(Z−1)/(Z+1)
*143	CLTF(s) TO CLTF(w') W′=(2/T)(Z−1)/(Z+1)
*144	HI-RATE CLTF(z) TO LO-RATE CLTF(z)
*145	HI-RATE CLTF(w) TO LO-RATE CLTF(w)
*146	HI-RATE CLTF(w') TO LO-RATE CLTF(w')
*147	OPTION 144 (AVOIDING INTERNAL FACTORING)
*148	OPTION 144 (ALL CALCULATIONS IN Z PLANE)

E-6 HELP

Help is available to the user at all times and can be requested in two ways. In option mode, typing the command "HELP, option number" gives the user a short explanation of the option number specified. In all modes, the user may type "?" for assistance.

E-7 TOTAL ENHANCEMENTS

The TOTAL CAD package was developed as an interactive tool based upon a line terminal. Because of the increasing use of CRT terminals, TOTAL has been rehosted on a DEC VAX-11/780 with a screen-oriented interface permitting easier and faster utilization.[E5, E10] This new package is called ICECAP and is written in the FORTRAN and PASCAL languages. The referenced documents are also available from the Defense Documentation Center (DDC), Cameron Station, Alexandria, Va. 22314. Currently TOTAL is being modified and expanded and is designated as TOTAL-II.

REFERENCES

E1. A Computer Graphics Program for the Analysis and Design of Digital Filters, *Proc. Int. Conf. Cybernetics Soc.*, October 1980, pp. 1053–1057.
E2. Brubaker, A.: "Development of Improved Design Methods for Digital Filtering Systems," AFAL-TR-77-207, Colorado State University, Fort Collins, 1977.
E3. Colgate, A.: "INTERAC—An Interactive Software Package for Direct Digital Control Design," M.S. thesis, GGC/EE/77-1, School of Engineering, Air Force Institute of Technology, Wright-Patterson AFB, Ohio, 1977.
E4. Emami-Naeini, A., and G. F. Franklin: Interactive Computer-Aided Design of Control Systems, *IEEE Control Syst. Mag.*, vol. 10, December 1981, pp. 31–35.
E5. Gembrowski, C. J.: "Continued Development of an Interactive Computer-Aided Design (CAD) Package for Discrete and Continuous Control System Analysis and Synthesis," M.S. thesis, GE/EE/82D, School of Engineering, Air Force Institute of Technology, Wright-Patterson AFB, Ohio, 1982.
E6. Herget, C. J., and P. Weis: "Linear Systems Analysis Program User's Manual," UCID-30184, Lawrence Livermore Laboratory, Livermore, Calif., 1980.
E7. Hernandez, G.: "An Interactive Computational Aerodynamics Analysis Program," M.S. thesis, GAE/AA/80D-9, School of Engineering, Air Force Institute of Technology, Wright-Patterson AFB, Ohio, 1980.
E8. Kennedy, A.: "The Design of Digital Controllers for the C-141 Aircraft Using Entire Eigenstructure Assignment and the Development of an Inter-Active Computer Design Program," M.S. thesis, GGC/EE/79-1, School of Engineering, Air Force Institute of Technology, Wright-Patterson AFB, Ohio, 1979.
E9. Larimer, S. J.: "An Interactive Computer-Aided Design Program for Digital and Continuous System Analysis and Synthesis (TOTAL)," M.S. thesis, GE/GGC/EE/78-2, School of Engineering, Air Force Institute of Technology, Wright-Patterson AFB, Ohio, 1978.
E10. Logan, G. T.: "Development of an Interactive Computer-Aided Design Program for Digital and Continuous Control System Analysis and Synthesis," M.S. thesis, GE/EE/82M-2, School of Engineering, Air Force Institute of Technology, Wright-Patterson AFB, Ohio, 1982.

E11. Manchini, J.: "Computer Aided Control System Design Using Frequency Domain Specifications," Naval Postgraduate School, Monterey, Calif., 1976.
E12. O'Brien, L.: "A Consolidated Computer Program for Control System Design," M.S. thesis, GE/EE/76-38, School of Engineering, Air Force Institute of Technology, Wright-Patterson AFB, Ohio, 1976.
E13. Spielman, C.: "Frequency Methods in Computer Aided Design of Control Systems," M.S. thesis, University of Illinois, Urbana, 1976.
E14. Vines, P.: "Computer Aided Design of Systems," Naval Postgraduate School, Monterey, Calif., 1975.
E15. "Multivariable Analysis Package (MAP)," Lockheed-Georgia Company, Marietta, Ga., 1983.

APPENDIX
F

PDP-11/LSI-11 ASSEMBLY LANGUAGE

F-1 INTRODUCTION

The Digital Equipment Corporation (DEC) PDP-11 minicomputer family consists of a hierarchy of software and hardware architectures. The various 16-bit minicomputers within the family offer a wide range of technologies, cost, and performance. Examples of the family include PDP-11/70, PDP-11/60, PDP-11/44, PDP-11/24, and the single-board computers, the LSI-11, LSI-11/23 and MICRO/PDP-11.

The following sections introduce the PDP-11/LSI-11 general hardware structure and assembly-language instructions. It should be noted that some additional specialized instructions are available on emerging CPU hardware but are of no concern in the text examples.

F-2 HARDWARE STRUCTURE

The basic organizational structure of a DEC minicomputer/microcomputer is bus-structured (address bus–16 bits, data bus–16 bits, control bus). The control bus consists of various dedicated signals for sequencing instruction execution. The three buses comprise the UNIBUS (PDP-11s) or the QBUS (LSI-11). The UNIBUS and QBUS are different structurally, but since the instruction sets are basically the same, the programmer need not be concerned with this implementation level.

It should be noted that neither the PDP-11 nor the LSI-11 use dedicated I/O instructions, but instead use memory addresses for I/O devices, namely, the upper 8K bytes of storage. Total storage is 2^{16} (64K) without extended memory

Table F-1 Processor status word (PSW) bit associations

PSW bit	Designator	Meaning
0	C	Carry
1	V	Overflow
2	Z	Zero
3	N	Nonzero
4	T	Flag
5–7	PRI	Priority interrupt mask or special purpose depending upon model. Not used in LSI-11.

capability. This memory-mapped I/O provides for ease of programming and hardware architecture expansion. The 64K addresses are byte-oriented with the odd address assigned to the lower-order byte and the even to the higher-order byte of a 16-bit word. It should be noted that some members of the family have $2^{18} = 256K$ memory locations, i.e., extended memory addressing. Selected software supports this additional memory space.

The PDP-11/LSI-11 processor has eight general-purpose 16-bit registers, R0–R7. Register R7 is used as the program counter (PC), and register R6 is the stack pointer (SP) which is used for subroutine calls and interrupt operators. Also, a 16-bit processor status word (PSW) is used for encoding four condition codes (CC) plus priority. These condition codes reflect the result of an arithmetic operation and are used as indicated in Table F-1.

The machine instruction formats are shown in Fig. F-1. Various addressing modes can be used to define a source (src) or destination (dst) field within an instruction. Note that an address can specify a word (even address) or a byte (odd or even address). Because of the variety of applications for a general-purpose

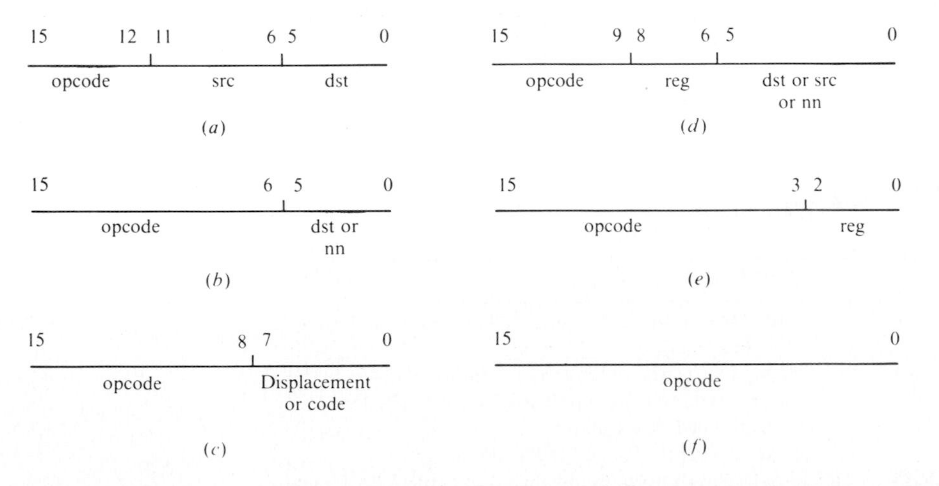

Figure F-1 Instruction formats. (*a*) Double operand; (*b*) single operand; (*c*) program control; (*d*) program control and fixed-point arithmetic; (*e*) return from subroutine; and (*f*) no operand (miscellaneous instructions).

minicomputer/microcomputer, instruction lengths vary from one to three words. The addressing mode used dictates the number of words employed in the specific instruction.

F-3 PDP-11/LSI-11 ASSEMBLY LANGUAGE

Instead of having to use the fixed-format (Fig. F-1) restrictive machine-language formats with binary programming, an easier implementation structure is available, assembly language. The assembly-language formats do not require specific spaces for delimiters between fields, i.e., free format and permit use of labels.

Important features of the PDP-11/LSI-11 assembly language are summarized as follows:

1. There is no origin statement such as ORG or START or BEGIN directive; instead, a statement like ".=1000" sets the starting value of the program counter (PC). Note that the PC is denoted by a period.
2. Register numbers are distinguished from memory addresses by preceding them with a percent sign (SP=%6, PC=%7, R1=%1). Newer versions of the assembler (3.0 and later) also accept R0–R7.
3. An equal sign is used instead of an EQU or SET instruction.
4. Integer constants are in octal; for decimal, the number is followed by a period.
5. One-byte ASCII constants are defined by preceding an ASCII character with a single quote. Two-byte ASCII constants are defined by a preceding double quote.
6. Assembly-language pseudoinstruction names begin with a period (Table F-2).

F-3A Addressing Modes

Addresses of source (src) or destination (dst) operands are defined by a 6-bit field (Fig. F-1). The first 3 bits define one register (R0–R7), and the last 3 bits denote the addressing mode to be used. The mode defines how the value in the specified

Table F-2 PDP-11 assembly-language pseudoinstruction names

Name	Definition
.ASCII	Store ASCII values of characters into successive memory words.
.BLKW	Reserve a number of words of memory.
.BYTE	Store constant values into successive memory bytes.
.END	Indicates end of assembly
=	Equate symbol with 16-bit expression.
.EVEN	Forces the PC to an even value (useful after .BYTE and .ASCII).
.WORD	Store constant values into successive memory words.

register is to be used to determine the operand address. Thus, by definition, mode 0 uses the register value as the operand. The other modes (1–7) generate the effective address from the specified register value. Also, by using register R7 (PC) in conjunction with various modes, immediate addressing, absolute addressing, and PC relative addressing are implemented. The various modes are summarized as follows:

Mode	Name	Description
0	Register	The operand is the register value
1	Register indirect	The effective address of the operand is the register value.
2	Autoincrement	The effective address of the operand is taken from register value and then the register value is incremented by the length of the operand (1 or 2).
3	Autoincrement indirect	The indirect address is defined by the register value. The effective address is taken from the contents of memory at the indirect address. The register value is then incremented by 2.
4	Autodecrement	The register value is decremented by the length of the operand (1 or 2) and then the effective address of the operand is this register value.
5	Autodecrement indirect	The register value is decremented by 2. The indirect address is this decremented register value. The effective address is taken from the contents of memory at the indirect address.
6	Indexed	The register value and the next word in the instruction stream are added to form the effective address of the operand set. The PC is incremented by 2. Unless the register is R7 = PC, the value of the register is not changed.
7	Indexed indirect	The register value and the next word in the instruction stream are added to form an indirect address. The effective address is taken from the contents of memory at the indirect address. The PC is then incremented by 2.
2-PC†	Autoincrement PC (immediate)	The effective address is the value of PC, and the PC is thus incremented by 2.
3-PC†	Autoincrement PC indirect (absolute)	The effective address is taken from the next word in the instruction sequence and the PC is thus incremented by 2.
6-PC†	PC-indexed (relative)	The next word in the instruction sequence is read and saved; PC is incremented by 2. Then the saved value and the new PC value are added to form the effective address. The PC is then again incremented by 2.
7-PC†	PC-indexed indirect (relative indirect)	The next word in the instruction sequence and the new PC value are added to form an indirect address. The effective address is taken from the contents of memory at the indirect address. The PC is then incremented by 2.

† PC addressing: when the PC is used with the modes above, special addressing operations occur as indicated.

Table F-3 Addressing mode summary

Mode	Register	Notation	Name
0	0–7	R	Register
1	0–7	(R) or @R	Register indirect
2	0–6	(R)+	Autoincrement
3	0–6	@(R)+	Autoincrement indirect
4	0–7	−(R)	Autodecrement
5	0–7	@−(R)	Autodecrement indirect
6	0–6	X(R)	Indexed
7	0–6	@X(R)	Indexed indirect
2	7	#n	Immediate (PC autoincrement)
3	7	@#A	Absolute (PC autoindirect)
6	7	A	Relative (PC-indexed)
7	7	@A	Relative indirect (PC-indexed)

R = register, X, n, A stored as next word in instruction stream.

Addressing modes and assembly-language notation are summarized in Table F-3.

A memory management unit is available on some PDP-11s to map a 16-bit logical address into an 18- or 22-bit physical address, depending on the model. The PDP-11/24, PDP-11/44, and LSI-11/23 use the 18-bit mapping technique. PDP-11 models with memory management also have user and supervisor operating modes to support operating systems. In addition, there are separate stack pointers and memory maps for user programs and the operating system. Fields in the PSW indicate the current operating mode and which memory map (64K unit) is being used.

F-3B Instructions

Instructions can be classified into five types: double operand, single operand, program control, miscellaneous, and extended (see Fig. F-1). The instructions are introduced in the following subsections. All instructions require an operation (opcode) and information for locating operands if they are required.

Double operand Double-operand instructions, listed in Table F-4, require an opcode and both a source (src) and a destination (dst) for the operation. The operation order in standard assembly language is left to right. Most operations may be performed on either words or bytes (B). All instructions set, clear, or hold condition-code bits NZVC[F1], according to the result of the operation.

Single operand Single-operand instructions are listed in Table F-5 and require only an opcode and destination (dst). All instructions manipulate the condition code bits depending upon the specific operation.

Table F-4 Double-operand instructions

Mnemonic	Operands	Format	Description
MOV(B)	src, dst	a	Copy src to dst
ADD	src, dst	a	Add src word to dst
SUB	src, dst	a	Subtract src word from dst
CMP(B)	src, dst	a	Set NZVC according to src minus dst
BIS(B)	src, dst	a	Set corresponding "1" bits of src in dst
BIC(B)	src, dst	a	Clear corresponding "1" bits of src in dst
BIT(B)	src, dst	a	Set NZVC according to src AND dst
XOR†	reg, dst	b	EXCLUSIVE OR reg into dst

† Not on 11/54, 11/65, or 11/10.

Program control Program-control instructions are listed in Tables F-6a and b. They include branch, jump, subroutine and trap instructions types. A branch machine instruction has an 8-bit opcode and an 8-bit signed displacement. If the branch is taken, the displacement is multiplied by 2 and added to the PC to obtain the address of the next instruction. When this addition takes place, the processor increments the PC to point to the next instruction in the sequence. Range of branches is -128 to $+127$ words from PC. Thus, a displacement of -1, not 0, causes a machine instruction to branch to itself. In assembly language branch instructions and the other types use labels for operands permitting easier programming.

Extended instructions There are two groups of extended instructions, addressing and arithmetic. The extended addressing instructions are used to move data between different address spaces in models[F1] with memory management units. Extended arithmetic instruction provides fixed-point arithmetic as shown in Table F-7. Fixed-point operations are of the double operand type. Floating-point

Table F-5 Single-operand instructions (format b)

Mnemonic	Operand	Description
CLR(B)	dst	Clear dst
COM(B)	dst	Complement dst (1s complement)
INC(B)	dst	Increment dst by 1
DEC(B)	dst	Decrement dst by 1
NEG(B)	dst	Negate dst (2s complement)
TST(B)	dst	Set NZVC according to value of dst
ASR(B)	dst	Arithmetic shift right dst
ASL(B)	dst	Arithmetic shift left dst
ROR(B)	dst	Rotate dst right with C
ROL(B)	dst	Rotate dst left with C
ADC(B)	dst	Add C to dst
SBC(B)	dst	Subtract C from dst
SWAB	dst	Swap bytes of dst word
SXT†	dst	Extend sign (N) into dst word

† Not on 11/64. 11/65, 11/10.

Table F-6a Program-control instructions

Mnemonic	Operands	Format	Description
JMP	dst	d	Jump to effective address of dst
JSR	reg, dst	d	Jump to subroutine
RTS	reg	e	Return from subroutine
MARK	nn	b†	Stack cleanup and subroutine return
SOB	reg, nn	d†	Subtract one and branch if not zero
EMT	code	c	Emulator trap (not recommended)
TRAP	code	c	User trap
BPT	No operand	f	Breakpoint trap
IOT	No operand	f	Input-output trap
RTI	No operand	f	Return from interrupt, allow trace
RTT	No operand	f	Return from interrupt, inhibit trace

† Not on 11/64, 11/65, 11/10.

Table F-6b Program-control instructions for branching (format c)

Mnemonic	Description
BR	branch (unconditional)
BNE	branch if not equal (to zero)
BEQ	branch if equal (to zero)
BPL	branch if plus
BMI	branch if minus
BVC	branch if overflow is clear
BVS	branch if overflow is set
BCC	branch if carry is clear
BCS	branch if carry is set
BGE	branch if greater than or equal (to zero)
BLT	branch if less than (zero)
BGT	branch if greater than (zero)
BLE	branch if less than or equal (to zero)
SOB	subtract one and branch (if not =0)
BHI	branch if higher
BLOS	branch if lower or same
BHIS	branch if higher or same
BLO	branch if lower

Table F-7 Fixed-point arithmetic instructions (format d)

Mnemonic	Operands	Description
MUL	src, reg	Multiply reg by src, product in reg† and next reg in sequence
DIV	src, reg	Divide reg by src, quotient in reg and next reg in sequence
ASH	src, reg	Arithmetic shift reg according to src (LS 6 bits)
ASHC	src, reg	Shift double reg and next reg according to src (LS 6 bits)

† If reg odd numbered, low-order product stored only.

Table F-8 Miscellaneous instructions (no operands—format f)

Mnemonic	Description
NOP	No operation
CL(N, Z, V, C)	Clear CC bit(s)
SE(N, Z, V, C)	Set CC bit(s)
RESET	Reset UNIBUS
HALT	Halt the processor
WAIT	Wait for interrupt

hardware should be a reality in future microprocessor systems permitting less concern with scaling but timing considerations continue.

Miscellaneous instructions Miscellaneous instructions are given in Table F-8. They include condition code set or clearing and special CPU operations.

F-4 SUMMARY

The presentation of PDP-11/LSI-11 assembly-language instructions (opcodes) is a brief introduction for purposes of understanding text example programs. Detailed discussions of instruction execution, execution times, details of stack structures, and interrupt processing are found in the references. Manuals of this type must be consulted in determining compensator performance and selection of possible improvements as well as contemporary DEC hardware improvements.

REFERENCES

F1. "Programming Text Books on Various PDP-11 Models," Digital Equipment Corporation, Maynard, Mass.

F2. Eckhouse, R. H., and L. R. Morris: "Minicomputer Systems," Prentice-Hall, Englewood Cliffs, N.J., 1979.

F3. Stone, H. S., and D. P. Siewiored: "Introduction to Computer Organization and Data Structures: PDP-11 Edition," McGraw-Hill, New York, 1975.

ANSWERS TO SELECTED PROBLEMS†

Chapter 2

2-1 (*a*) 46_{10} (*e*) 3561.8625_{10} (*j*) 61920_{10}

2-3 (*b*) 01001_2, carries $= 00100_2$

2-6 (*d*) 3637.67_8

2-9 (*c*) 0000001_2 (*e*) 0000000_2

2-12 (*a*) 10001111_2

2-14 A synchronous counter has *reduced propagation delay time*. The carry propagation delay is the time required for a counter to complete its response to an input pulse. The carry time of a ripple counter is longest when each stage is in the 1 state. In this situation, the next pulse must cause all previous flip-flops to change state. Any particular stage will not respond until the preceding stage has nominally completed its transition. The clock pulse effectively "ripples" through the chain. Therefore the carry time for a ripple counter will be of the order of magnitude of the sum of each propagation delay time.

However, if the asynchronous operation of a ripple counter is changed so that all flip-flops are clocked simultaneously (synchronously) by the input pulses, the propagation delay time may be reduced considerably. Repetition rate is limited by the delay of any one flip-flop plus the propagation times of any control gates required.

Chapter 3

3-1 (3) System is not linear but is time-invariant.

3-4 No; the sine wave is being sampled at its zero-valued points; $\omega_c = 2\pi = \omega_s$, which violates Shannon's sampling theorem.

† The accuracy of numerical results in the various examples in the text and in the problems depends upon the CAD software package used and whether the problem is solved in the continuous or discrete time domain (on the specific value of T). As appropriate, enough significant digits are presented to compare results of various design and implementation approaches.

3-6 (1) (*a*) $c(kT) = a\{(2 + T)c[(k - 1)T] - c[(k - 2)T] + T(2 + 4T)e(kT) - 2Te[(k - 1)T]\}$

where $a = 1/(1 + T)$.

(*b*) Since $c(t) = \mathscr{L}^{-1}[C(s)]$ represents an overdamped ramp response, then select $T \ll T_s = 4/|\sigma| = 4$. Try $T = T_s/10 \approx 0.4$ s.

Chapter 4

4-2 (*a*) $E(z) = az^{-2}/(1 - az^{-1})$.

4-4 (*a*) $E(z) = \dfrac{z}{z-1} - \dfrac{z^2 - (1 - T)\epsilon^{-T}z}{(z - \epsilon^{-T})^2}$

4-7 (*b*) $0.905\underline{/5.73^\circ}$, $0.99005\underline{/0.573^\circ}$, $0.99900\underline{/0.0573^\circ}$, $0.999900\underline{/0.00573^\circ}$.

4-9 (*a*) $e(0) = 0$; response is unstable (one positive real root), $e(\infty) = \infty$.

(*e*) $e(0) = 1$; response is stable, $e(\infty) = 7$.

4-12 (*b*) $c(kT) = -2c[(k - 1)T] + c[(k - 2)T] + r(kT) - 0.5r[(k - 2)T]$.

4-17 (*b*) $e^*(t) = \sum_{k=0}^{\infty} [1 - (0.5)^k]\delta(t - kT)$ for $k \geq 0$.

4-18 (*b*) $e(3) = 7/8$.

4-20 (*d*) $\dfrac{C(z)}{R(z)} = \dfrac{G_1G_2(z)}{1 + G_1G_2(z) + G_1H(z)}$

Chapter 5

5-3

System (1)	Dominant roots	Other roots	K_1, s^{-1}	T_p, s	T_s, s	M_o
Basic	$3.57 \pm j6.19$	$-16.4 \pm j19.2$	5.23	0.558	1.19	0.167
Lag-compensated: $\alpha = 10$, $T = 10$	$-3.6 \pm j6.4$	$-16.4 \pm j19.2$ -0.10175	53.2	0.54	1.55	0.195
Lead-compensated: $\alpha = 0.1$, $T = 0.1$	$-9.6 \pm j16.6$	-11.5 -99.4	6.7	0.324	0.273	0.0025
Lag-lead-compensated: $\alpha = 10$, $T_1 = 10$, $T_2 = 0.1$	$-9.6 \pm j16.8$	-0.101 -11.3 -99.4	67.8	0.324	0.268	0.0071
Tachometer-compensated: $A = 15$, $K_t = 1$	$-8.8 \pm j15.17$	$-11.2 \pm j9.85$	6.34	0.36	0.26	0.1105

5-8 (*b*) $G_c(s) = \dfrac{2(s + 3)}{s + 6}$

With this cascade compensator $M_p = 1.081$, $t_p = 1.9$ s, $t_s = 3.65$ s, and $K_1 = 1.25\ s^{-1}$. The specified $M(s)$ yields $M_p = 1.127$, $t_p = 1.95$ s, $t_s = 3.9$ s, and $K_1 = 1.36\ s^{-1}$.

5-10 (*a*) 500 (*c*) No.

Chapter 6

6-1 $-1 < K < 4.08$

6-8 (*c*) Unstable (*e*) Stable

6-11 With $K = 1$, Type 2. For a stable system, $K < 3.7878$

6-17 (*b*) System 2 ($T = 1$ s)

Domain	M_p	t_p, s	t_s, s	K_1, s^{-1}
w	1.066	2.6	3.788	0.9415
z	1.048	3	4-5	0.75074

(*c*) For $T = 1$, $e^*(\infty) = 0$ for a step input and $e^*(\infty) = R_1/K_1 = R_1/0.75074$.

Chapter 7

7-1 (*c*) $[D_c(z)]_{TU} = \dfrac{3K_{sc}(z - 1/3)(z + 1)}{(z - 1)(z + 4)}$

7-6

	z-domain value	
s-plane value	Eq. (7-7)	$z = \epsilon^{sT}$
-1	0.96078	0.96079
-2	0.92308	0.92312
-10	0.66667	0.67032
$-2 \pm j2$	$0.92024 \pm j0.07385$	$0.92016 \pm j0.07377$
$-10 \pm j20$	$0.5 \pm j0.5$	$0.46702 \pm j0.48086$

7-10 (*a*) $0 < T < 1.678+$

7-13 (*b*) $$\frac{C(s)}{E(s)} = \left[\frac{17680(s + 20)}{s(s^3 + 221s^2 + 21{,}900s + 92{,}400)}\right]_{T=0.01}$$

$$= \left[\frac{17680(s + 20)}{s(s^3 + 421s^2 + 43{,}780s + 184{,}800)}\right]_{T=0.005}$$

(*d*)

	M_p	t_p, s	t_s, s	K_1, s^{-1}
		s domain		
Model	1.08493	0.92675	1.40876	4.0000
$T = 0.005$	1.08509	0.9258	1.40764	↓
$T = 0.01$	1.08523	0.925	1.40655	↓
		z domain		
$T = 0.005$	1.085+	0.925	1.405	3.8291
$T = 0.01$	1.085+	0.92	1.40	3.82684

Note: 10-digit accuracy required for these values of T.

Chapter 8

8-1 The schematic V_R diagram is

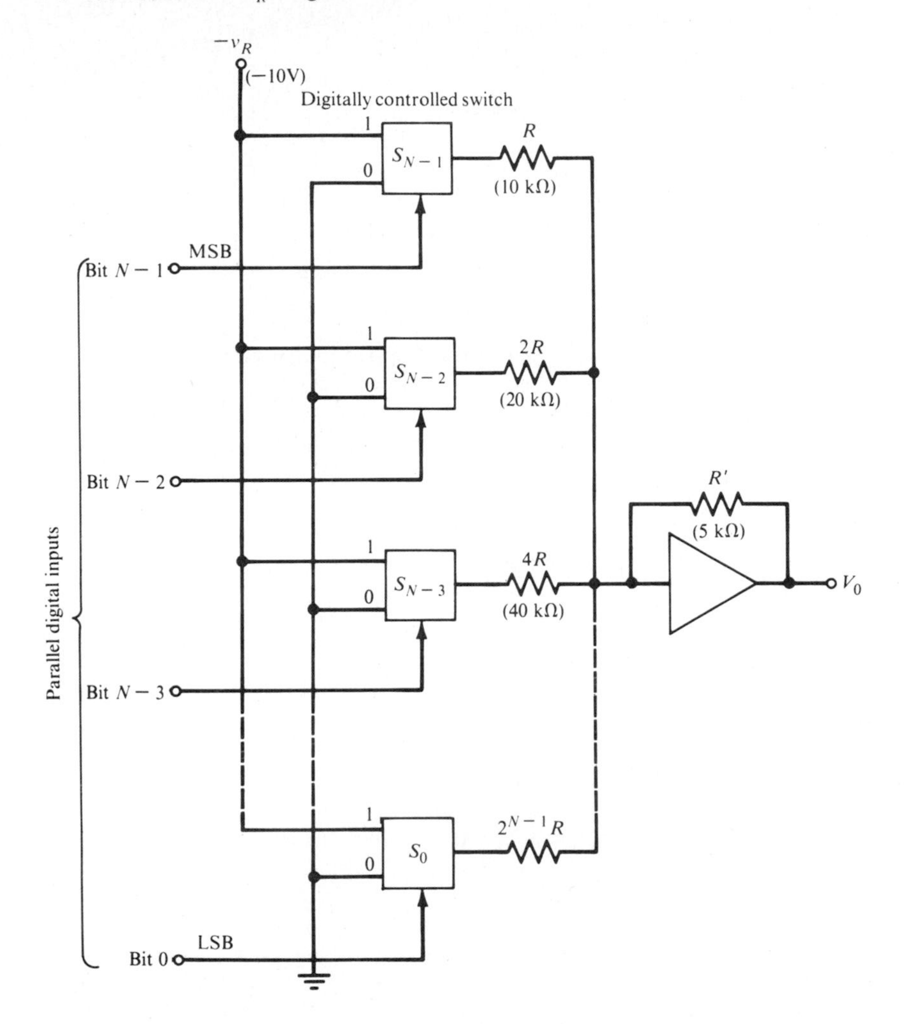

The output V_0 of an N-bit D/A converter is given by the equation

$$V_0 = (a_{N-1}2^{-1} + a_{N-2}2^{-2} + a_{N-3}2^{-3} + \cdots + a_0 2^{-N})V_R$$

where the concatination of the coefficients a_n represents the binary word. The voltage V_R is the stable reference voltage used in the circuit.

The MSB is that corresponding to a_{N-1}, and its weight is $V_R/2$, while the LSB corresponds to a_0, and its weight is $V_R/2^N$.

From the figure, the blocks $S_0, S_1, S_2, \ldots, S_{N-1}$ are digitally controlled electronic switches. For example, when a 1 is present on the MSB line, switch S_{N-1} connects the 10-kΩ resistor to the reference voltage $V_R(-10\text{ V})$; conversely, when a 0 is present on the MSB line, the switch connects the resistor to the ground line. Thus, the switch is a single-pole, double-throw (SPDT) electronic switch. The op amp acts as a current-to-voltage converter. The analog output is proportional to the digital input.

8-12 $\omega = 2\pi f = 2\pi f_s$, where $f_s > 2f$ and $f = 100$ Hz.

$$\begin{aligned} \omega &> 2\pi(200) \\ &> 400\pi \text{ rad/s (minimum)} \end{aligned}$$

Use spherical coordinates for the disk model; thus

$$r \sin \theta \approx r\theta \qquad \text{if } \theta \text{ is small}$$

Then

$$\text{Velocity} = r\, d\theta$$

(velocity relates to binary coding). The transfer function for the encoder is $K/(1 + T_m)$.

8-21 Planck's constant $= h = 6.626 \times 10^{-34}$ J · s (joule-seconds) and $h < \Delta x\, \Delta p$, where p is momentum, x is displacement, and the Δ's are the uncertainty in x and p. Since

$$\Delta p \geq \frac{h}{\Delta x} \qquad \text{as } x \to \infty$$

then $\Delta p \geq 0$. Therefore, for large displacements $\Delta p > 0$ is always true and data measurements can be made.

Chapter 9

9-2 *Assumptions:*

1. m independent, identical trials.
2. Each trial can have one of two possible outcomes (Bernoulli trials).
3. Probability of success on any trial is p and remains the same throughout the experiment. Assume m is large; in this case $m = 100$. Probability of failure on each trial is $q = 1 - p$.
4. Binomial probability distribution

$$P(X = x) = \binom{m}{x} p^x q^{m-x}$$

where X = total number of successes in m trials
m = number of trials

(*a*) *Probability of no defects:*

$$P(\text{good IC}) = 0.42$$

$$X \equiv \text{good IC}$$

$$P(X = 10) \approx \binom{10}{10} 0.42^{10}\ (0.58)^0$$

$$P(X = 100) \approx 0.0001708019814$$

Actually, to be more precise,

P(no defects for the 10 ICs)

$$= \frac{42}{100} \cdot \frac{41}{99} \cdot \frac{40}{98} \cdot \frac{39}{97} \cdot \frac{38}{96} \cdot \frac{37}{95} \cdot \frac{36}{94} \cdot \frac{35}{93} \cdot \frac{34}{92} \cdot \frac{33}{91}$$

$$= \frac{5.33957226 \times 10^{15}}{6.281565096 \times 10^{19}} = 8.50038514 \times 10^{-5}$$

(assume without replacement.)

(*b*) *Probability of exactly 2 minor defects:*

$$P(\text{minor defect}) = 0.37$$

$$X \equiv \text{IC with a minor defect}$$

$$P(X = 2) \approx \binom{10}{10} (0.37)^2 (0.63)^8$$

$$\approx \frac{10 \times 9 \times 8}{2 \times 8!} (0.37)^2 (0.63)^8$$

$$\approx 45(0.1369)(0.024815578)$$

$$\approx 0.1528763685$$

(*c*) *Probability of exactly 4 major defects:*

$$P(\text{bad defect}) = 0.21$$

$$X \equiv \text{IC with a major defect}$$

$$P(X = 4) \approx \binom{10}{4} (0.21)^4 (0.79)^6$$

$$\approx 210(0.21)^4 (0.79)^6$$

$$\approx 210(1.944810002 \times 10^{-3})(0.2430874558)$$

$$\approx 0.0992793722$$

(*d*) *Probability of any defect:*

$$P(\text{good IC}) = 0.42$$

$$X \equiv \text{IC with a major or minor defect}$$

$$P(\cdot) = 0.58$$

$$\text{Probability of any defect} \approx 1 - \text{probability of no defect}$$

$$\approx 1 - 0.0001708019814 \qquad \text{(from part } a\text{)}$$

$$\approx 0.9998291981$$

9-3 If $E[e_k e_{k+m}] = \sigma e^2 \delta(m)$, then the error random variables are all linearly independent in time. Then

$$E[e_k e_{k+m}] = \Phi_{ee}(m)$$

where e_k is a stationary random process and is assumed ergotic and the error sequence is linearly independent or uncorrelated with $X_c(k)$.

9-5 Given a linear discrete system,

$$h(k) = \text{transfer function}$$
$$x(k) = \text{input function}$$
$$y(k) = \text{output function}$$

If

$$\Phi_{xx}(k, k+m) = \sigma_x^2\delta(m)$$

then the input random variables are all *linearly independent in time* but they are correlated. Then

$$\Phi_{xx}(m) = \sigma_x^2\delta(m)$$

and $x(k)$ is a stationary random process and also assumed ergotic since

$$E[X_k, X_{k+m}] = \Phi_{xx}(m)$$

The expected value of $x_k = E[X_k] = m_x = 0$ (constant). Therefore, the expected value of $Y_k = E[Y_k] = m_y = 0$ (constant). Now the autocorrelation of Y is

$$\Phi_{yy}(k, k+m) = E(Y_k Y_{k+m})$$
$$= \sum_{y_k} \sum_{y_{k+m}} y_k y_{k+m} f(y_k, k, y_{k+m}, k+m)$$

Since this is a stationary process where the mean and variance are independent of time, then

$$\Phi_{yy}(k, k+m) = \Phi_{yy}(m) \qquad \text{which is one-dimensional}$$

The cross covariance of X and Y is

$$\Psi_{xy}(k, k+m) = E[(X_k - m_k)(Y_{k+m} - m_{k+m})]$$

Since both X and Y are zero mean processes,

$$\Psi_{xy}(k, k+m) = \Phi_{xy}(k, k+m)$$

Chapter 10

10-2 $H(z) = \dfrac{z}{z-a} = \dfrac{z}{z-0.125}$

Thus

$$c(kT) = 0.125^{kT}$$

or

$$c(k) = 0.125^k \qquad \text{with } N = 0, 1, 2, \ldots$$

The difference equation can also be generated in a different way (same result):

$$\frac{C(z)}{R(z)} = \frac{1}{1 - az^{-1}}$$
$$C(z) - aC(z)z^{-1} = R(z)$$

or
$$c(k) = ac(k-1) + r(k)$$
$$= 0.125c(k-1) + r(k)$$

For
$$r(k) = \begin{cases} 1 & k = 0 \\ 0 & k > 0 \end{cases}$$

and assuming $c(k) = 0$ for $k < 0$, taking a few samples with a 7-bit accuracy on a digital calculator:

$$c(0) = 0.125c(-1) + r(0)$$
$$c(0) = 1$$
$$c(1) = 0.125c(0) + r(1)$$
$$c(1) = 0.125$$
$$c(2) = 0.125c(1)$$
$$c(2) = 0.015625$$
$$c(3) = 1.953125 \times 10^{-3}$$
$$\vdots$$
$$c(\infty) = 0$$

With sign plus magnitude, the coefficient for 0.125 using 16-bit format is

```
   XXXXXX   XXXXXXXXXX
MSB └──┬──┘ └────┬────┘
sign 5 bits      10 bits
bit  for         of
     non-        fractional part
     fractional
     part
```

$$0.125 \rightarrow 0/000/000/010/000/000$$
$$= 000200_8$$
$$1 \rightarrow 0/000/010/000/000/000$$
$$= 002000_8$$
$$c(0) \rightarrow 002000_8 \qquad \text{for } k = 0$$
$$c(1) \rightarrow 000200_8 \qquad \text{for } k = 1$$
$$c(2) \rightarrow 000020_8 \qquad \text{for } k = 2$$
$$0.015625 \times 8 = 0.125$$
$$0.125 \times 8 = 1.000 \rightarrow 0/000/000/010/000$$
$$c(3) \rightarrow 000002_8 \qquad \text{for } k = 3$$
$$1.953125 \times 10^{-3} \times 8 = 0.0156235$$
$$0.015625 \times 8 = 0.125$$
$$0.125 \times 8 = 1.000 \rightarrow 0/000/000/010/000$$

$$c(4) \to 000000_8 \qquad \text{for } k = 4$$

$$2.44140625 \times 10^{-4} \times 8 = 1.953125 \times 10^{-3}$$

$$1.953125 \times 10^{-3} \times 8 = 0.015625$$

$$0.015625 \times 8 = 0.125$$

$$0.125 = 1.000 \to 0/000/000/000/000 \times 8$$

With more fractional bit accuracy, the result will be accurate for perhaps one or two more samples before the result $c(k)$ reaches 0. With 8 bits, only 7 bits can be used for the fractional part, assuming the result is less than 1 (nonfractional bits).

No changes occur in the analysis for sign plus 2s complement forms because all the coefficients are positive.

10-3 Let

$$A = 0.5 \qquad C = -0.125$$

$$B = -0.25 \qquad T = 0.1 \text{ s}$$

Consider $y[kT] = 0.5y[(k-1)T] - 0.25y[(k-2)T]y - 0.125x[(kT)] \to$

$$y(k) = 0.5y(k-1) - 0.25y(k-2) - 0.125x(k)$$

Consider a unit sample response:

k	$x(k)$	$y(k)$
0	1	−0.125
1	0	−0.0625
2	0	0
3	0	0.015625
4	0	0.0078125
5	0	0

Consider a unit step $x(k) = 1\ k > 0$:

k	$x(k)$	$y(k)$
0	1	−0.125
1	1	−0.1878
2	1	−0.1875
3	1	−0.265625
4	1	−0.2109375
5	1	−0.1640625

For an 8-bit machine—since all values are less than 1 (magnitude)—suggest a word structure.

Sign plus magnitude:

$$A = 100_8$$

$$B = 240_8$$

$$C = 220_8$$

Sign plus 2s complement:

$$A = 100_8$$
$$B = 340_8$$
$$C = 360_8$$

For a unit impulse:

Sign plus magnitude:

$k = 0$: $\quad y(0) = -0.125 \times 1 = -0.125 = 220_8 = -0.125$

$k = 1$: $\quad y(1) = (0.5)(-0.125) = 01000000 \times \underline{10010000}$

$= 10001000$

$= -0.0625$ (correct)

$k = 2$: $\quad y(2) = (0.5)(-0.0625) + (-0.25)(-0.125)$

$= (01000000)(10001000) + (10100000)(10010000)$

$= 10000100 + 00000100$

$= 0$ (assumed computer handles sign bit correctly)

$k = 3$: $\quad y(3) = (-0.25)(-0.0625)$

$= 0.015625$ (this is tracking correctly)

Sign plus 2s complement:

$k = 0$: $\quad y(0) = -0.125$

$k = 1$: $\quad y(1) = (0.5)(-0.125)$

$= (01000000)(11110000)$

$= 10001000 = -0.0625$

$k = 2$: $\quad y(2) = (0.5)(-0.0625) + (-0.25)(-0.125)$

$= (01000000)(10001000) + (11100000)(11110000)$

$= 10111100 + 01011010$

$= 00.010110$

$= 0.171875$ vs. 0 (note difference due to truncation)

For a unit step, all terms add:

$(-0.125)(1) = 220_8$ (sign plus magnitude)

$= 360_8$ (sign plus 2s comp)

Sign plus magnitude:

$k = 0$: $x(0) = -0.125$

$k = 1$: $y(1) = (0.5)(-0.125) - (0.125)$

$= 10001000 + 10100000$

$= 10101000 = -0.1875$

$k = 2$: $y(2) = (0.5)(0.1875) + (-0.25)(-0.125) + (-0.125)$

$= 10010100 + 00000100 + 10100000$

$= 1011000 = -0.375$ vs. -0.1875 exact error! (due to multiplication truncation)

Sign plus 2s complement:

$k = 0$: $y(0) = -0.125$

$k = 1$: $y(1) = (0.5)(-0.125) + (-0.125)$

$= 10001000 + 11110000$

$= 01111000 \rightarrow$ overflow

For 16-bit machines, more flexibility in placing the radix point exists, reducing the overflow errors.

Chapter 11

11-2 (*b*) $T = 0.05$ s

(*c*) $$D_c(s) = \frac{1.75(s+1)(s+2)(s+40)}{(s+1.707 \pm j0.7171)(s+99.99)}$$

(*d*) $$[D_c(z)]_{TU} = \frac{0.9897z(z-0.9048)(z-0.9512)}{(z-0.9176 \pm j0.03297)(z+0.4285)}$$

$$G(z) = [D_c(z)]_{TU}G_z(z) = \frac{2.43367 \times 10^{-3}z(z+0.9835)(z-0.9048)}{(z-0.9176 \pm j0.03297)(z-1)(z+0.4285)}$$

$$= \frac{KN(z)}{D(z)}$$

$$\frac{C(z)}{R(z)} = \frac{KN(z)}{(z-0.9500 \pm j0.04758)(z-0.9324)(z+0.4282)}$$

(*e*) $$[D_c(z)]_{TU} \approx \frac{0.2454(z-0.9048)(z-0.9512)(z+1)^2}{(z-0.9176 \pm j0.03297)(z+0.4285)(z+0.9835)}$$

$$\frac{C(z)}{R(z)} = \frac{1.2069 \times 10^{-3}(z-0.9048)(z+1)^2}{(z-0.9515 \pm j0.04821)(z-0.9312)(z+0.4287}$$

(*f*) $$\frac{C(z)}{R(z)} = \frac{1.1650 \times 10^{-3}(z-0.9048)(z+1)^2}{(z-0.9510 \pm j0.04679)(z-0.9323)(z+0.4287)}$$

(g)

Part	M_p	t_p, s	t_s, s	K_1, s^{-1}
(b)	1.0231	3.6483 –	4.0576 –	0.8168
(d)	1.023 +	3.60–3.65	4.00–4.05	0.81683
(e)	1.029	3.55	4.25	0.94767
(f)	1.023	3.70	4.05 +	0.91502

10 digits were used for calculations. The above values are as read from the computer print-out without requesting the 10 digit values.

11-4 (a) $G_y(z) = \dfrac{6.13 \times 10^{-4}(z + 0.9835)}{(z - 1)(z - 0.9512)}$

(b) $G_p(s) = \dfrac{2492.8}{s(s + 5)(s + 200)}$

(c) DIG: $D_c(s) = \dfrac{80.83(s + 5)}{s + 50}$

$[D_c(z)]_{TU} = \dfrac{66.28(z - 0.9512)}{z - 0.6000}$

DIR: $D_c(z) = \dfrac{66.745(z - 0.9512)}{z - 0.6000}$

(d) DIG: $D_c(s) = \dfrac{0.99807(s + 0.01)}{s + 0.001}$

$[D_c(z)]_{TU} = \dfrac{0.9981(z - 0.9999)}{z - 0.99999}$

DIR: $D_c(z) = \dfrac{1.01339(z - 0.9999)}{z - 0.99999}$

(e)

Part		Domain	M_p	t_p, s	t_s, s	K_1, s^{-1}
(a) DIR		z	1.046	1.245	1.70	2.4916 –
(b) DIG		s	1.0461 –	1.2519	1.700	2.4928
(c)	DIR	z	1.046	0.14	0.19 –	20.289
(c)	DIG	s	1.04538	0.1462 –	0.19678	20.142
(c)	DIG	z	1.045	0.14	0.19 –	20.147
(d)	DIR	z	1.05	1.26	1.755	25.251 –
(d)	DIG	s	1.05 –	1.2545	1.7595	24.88
(d)	DIG	z	1.05	1.24	1.75	24.867

See footnote to table in Prob. 11-2g.

11-5 (*a*) $G_y(w) = \dfrac{-2.083 \times 10^{-7}(w - 200)(w + 24000)}{w(w + 4.999)}$

(*c*) $D_c(w) = \dfrac{0.9996(w + 0.01)}{w + 0.001}$

(*d*) $D_c(w) = \dfrac{82.6774(w + 5)}{w + 50}$

(*e*)

Part	Domain	M_p	t_p, s	t_s, s	K_1, s^{-1}
(*a*)	w	1.04604	1.25184	1.7000	2.4892
(*c*)	w	1.04977	1.25440	1.75934	24.879
	z	1.050	1.25	1.755	24.903
(*d*)	w	1.04654	0.141563	0.191198	20.5769
	z	1.048	0.135	0.19–	20.6095

See footnote to table in Prob. 11-2*g*.

Chapter 12

12-1 (*b*) (1) Initial value: $0.1/T \le 4 \rightarrow T \le 0.25$ s.

(2) $G(s)H(s) = \dfrac{(2/T)^2(s + 4)D_c(s)}{s\left(s + \dfrac{2}{T}\right)^2 (s + 1)} = -1$

$$(\phi_0 + \phi_1 + \phi_2 + \phi_3) - \psi_0 + (\phi_c - \psi_c) = 180°$$

$$(\phi_c - \psi_c) = 2.13°$$

$$D_c(s) = \frac{K_{sc}(s + 7.272)}{s + 7}$$

(*Note*: Assume pole p_c of D_c is at -7.)

(3) Let $-z_c = 0.1/T = 7.272 \rightarrow T \le 0.0136$ s; thus let $T = 0.01$ s which results in $z_c = -7.691$.

(*c*) A lag compensator.

(*d*) $K_{sc} = 24.99$, $K_H = 0.25$

$$\frac{C(s)}{R(s)} = \frac{4999(s + 7.691)(s + 200)}{(s + 4.004 \pm j4.0038)(s + 6.4036)(s + 161.07)(s + 232.52)}$$

$M_p = 1.03283$ $\quad t_p = 0.848088$ s $\quad t_s = 1.05729$ s $\quad K_1 = 3.5535\ s^{-1}$

(*e*) $[D_c(z)]_{TU} = \dfrac{25.07(z - 0.9259)}{z - 0.9324}$

$$\frac{C(z)}{R(z)} = \frac{0.001249(z - 0.9259)(z + 0.9967)}{(z - 0.9606 \pm j0.03754)(z - 0.9376)}$$

$$M_p = 1.033 \qquad t_p = 0.88 \text{ s} \qquad t_s = 1.07 \text{ s} \qquad K_1 = 3.6182 \text{ s}^{-1}$$

Chapter 13

13-31 Requirements analysis A seismic isolation platform is used as a platform for the testing and evaluation of inertial instruments. The effect of external disturbances upon the stability of this platform must be eliminated or minimized.

The requirements are to develop a control system to control the movement of this platform in three axes (pitch, roll, and yaw) over the frequency band of 0 to 20 Hz. The motion of the platform must be controlled to within +0.001 arc s of tilt and 10^{-5} arc s/s of angular velocity for all three axes. The external disturbance consists of a step input of 1.25 ft·lb.

Data flow diagram. Level 0

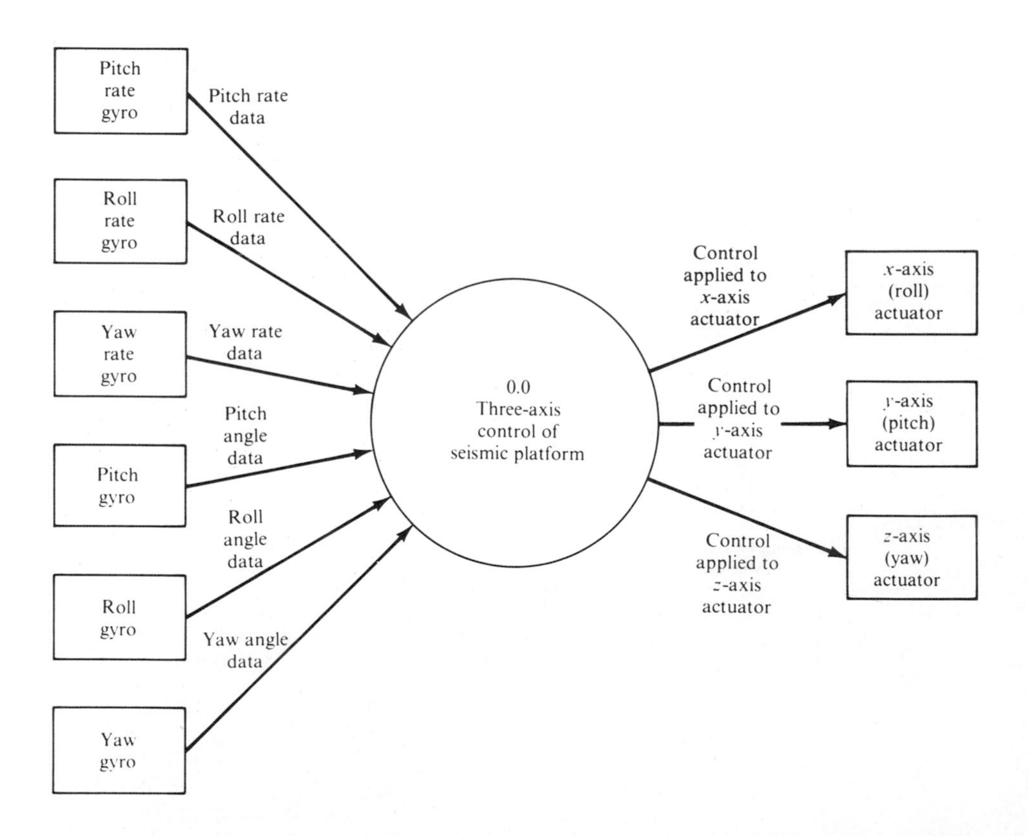

Each axis is assumed to act independently such that the DFD can be decomposed into three separate DFDs, one each for the x, y, and z axis.

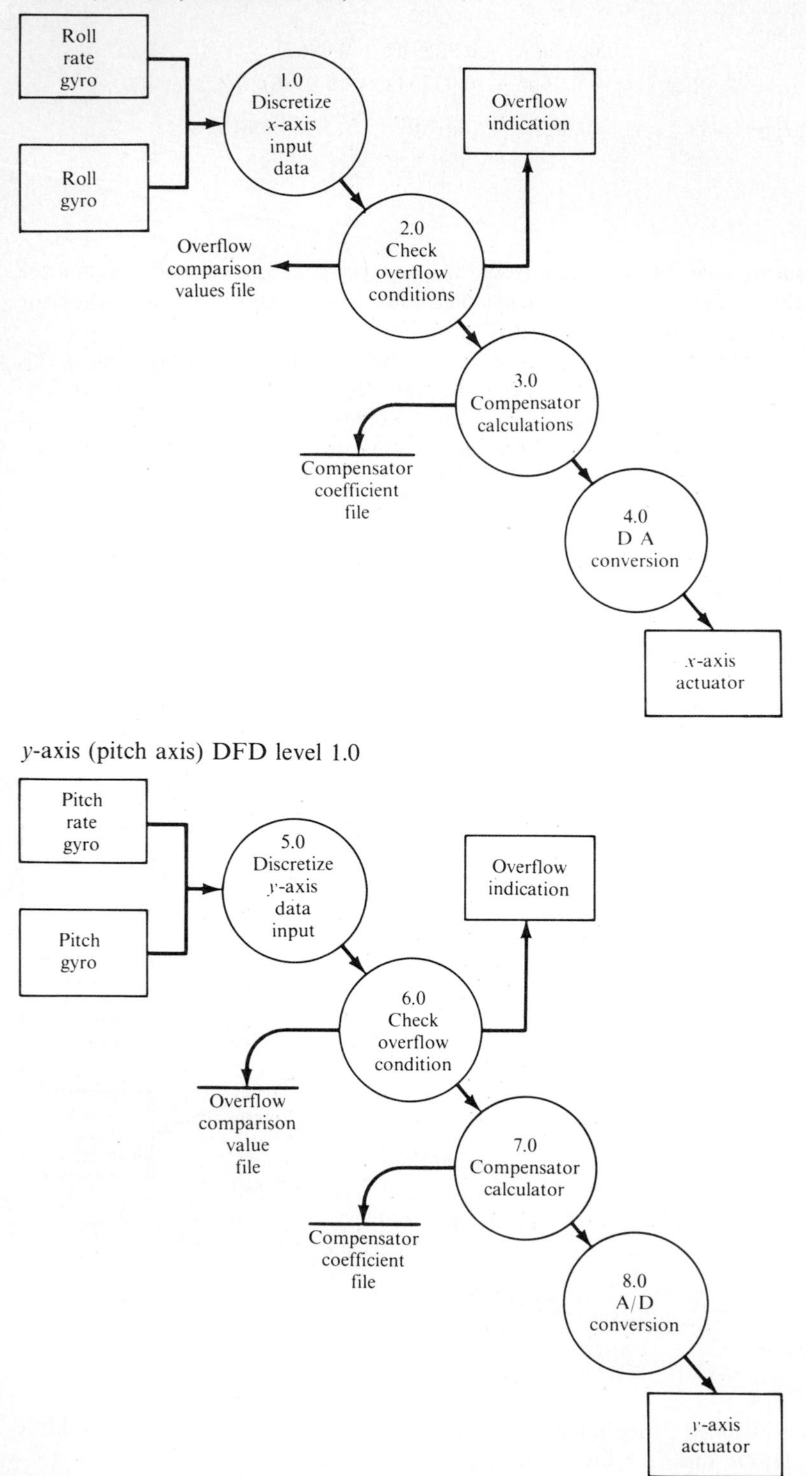
x-axis (roll axis) DFD level 1.0)
Roll rate gyro
Roll gyro
1.0 Discretize x-axis input data
Overflow indication
2.0 Check overflow conditions
Overflow comparison values file
3.0 Compensator calculations
Compensator coefficient file
4.0 D A conversion
x-axis actuator
y-axis (pitch axis) DFD level 1.0
Pitch rate gyro
Pitch gyro
5.0 Discretize y-axis data input
Overflow indication
6.0 Check overflow condition
Overflow comparison value file
7.0 Compensator calculator
Compensator coefficient file
8.0 A/D conversion
y-axis actuator

z-axis (yaw) DFD level 1.0

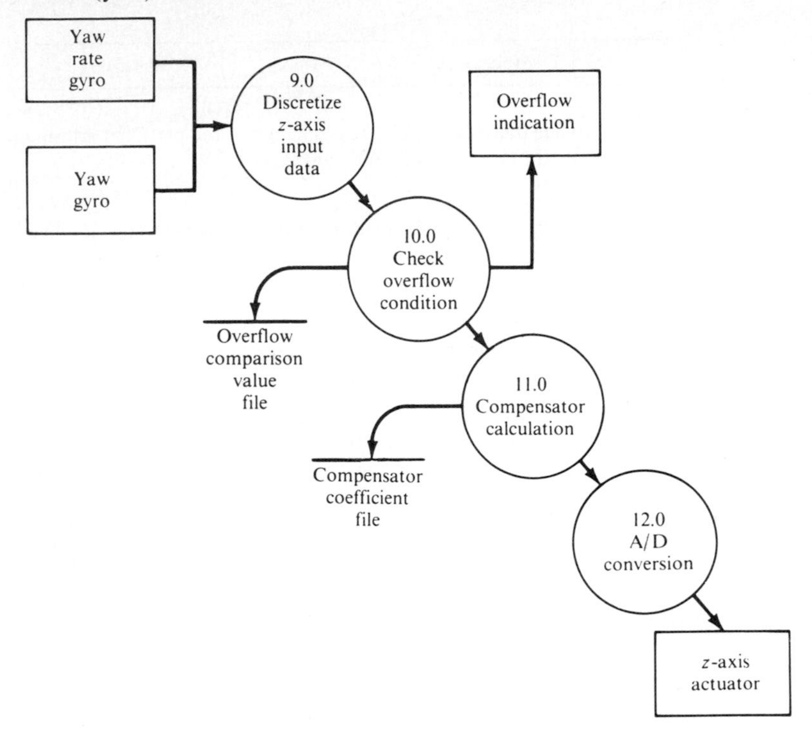

Now select one of the 1.0 levels to decompose into lower-level diagrams. Since the overflow process is the same for each of the three axes, DFDs 2.0, 6.0, and 10.0 can be represented in the same manner.

Using the 2.0 process:

x-axis DFD level 1.X

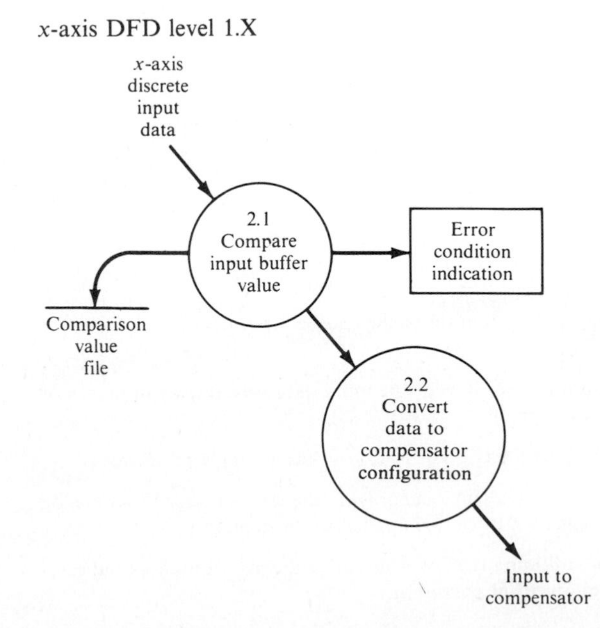

	Data Dictionary Level 0.0
Input description:	
Pitch rate gyro Roll rate gyro Yaw rate gyro	Provide data about the rate of change of the platform attitude about the pitch, roll, and yaw axes
Pitch gyro Roll gyro Yaw gyro	Provide data about the platform attitude relative to the nominal position with respect to the pitch, roll, and yaw axes
Process description:	
0.0 Three-axis control of seismic platform	Determines the amount of control to be applied to the x-, y-, and z-axis actuators to counter external disturbances.
Data flow names:	
Pitch rate data Roll rate data Yaw rate data	Input data containing information from rate gyros concerning rate of change of platform attitude
Pitch data Roll data Yaw data	Input data containing information from gyros concerning current attitude of platform
Control applied to pitch actuator Control applied to roll actuator Control applied to yaw actuator	Output control signal to x-, y-, and z-axis actuators to counter disturbance input
Output description:	
X-axis actuator Y-axis actuator Z-axis actuator	Actuators or servos which apply force about the x, y, and z axes to level platform

	Data Dictionary Level 1.0
Input description:	
Pitch rate gyro Roll rate gyro Yaw rate gyro	As per DD for level 0.0
Pitch gyro Roll gyro Yaw gyro	As per DD for level 0.0
Process description:	
1.0 Discretize x axis (y or z) input data	The A/D conversion of the analog input data
2.0 Overflow check	Check for overflow of the input data with respect to processor word length
3.0 Compensator calculations	Calculate filter-compensation values to apply to actuators
4.0 D/A conversions	Convert the digital signal from the compensator to an analog signal which can be applied to the actuators
Overflow indication	An indication (bell or printout) if the overflow check indicates an overflow condition

Data Dictionary Level 1.0—(*cont.*)

Output description:	
x-axis actuator	As per DD for level 0.0
y-axis actuator	
z-axis actuator	
File names:	
Compensator coefficient file	This file contains the values of the compensator coefficients

Data Dictionary Level 1.X

Data flow names:	
x-axis discrete input data	This is the rate and angle data from the x-axis gyros in discrete form
Input to compensator	x-axis input data converted to the particular representation required by the compensator implementation
Process description:	
Compare input value	The magnitude of the value of the input data is compared with a prestored value to determine if the word-length limits are exceeded
Convert input data to compensator configuration	The discretized input data is converted to the representation required by the compensator (that is, sign plus 2s complement)
File names	File contains min and max values used by comparison process to check for overflow
Error condition indication	As per DD for level 1.0

Coupling and cohesion analysis

Coupling (*minimize*). There is a degree of common coupling between some elements at the DFD 1.0 level. The three overflow-condition check processes for the *x*, *y*, and *z* axes all share or could conceivably share the same overflow-value comparison file. Assuming all three axes are implemented on the same computer, then the word-length restrictions would be the same and logically all three overflow check processes would use the same comparison values stored in a common file. This could be overcome by creating three separate files. Another case of common coupling may occur if the three compensator calculation processes use the same coefficients or have them stored on a common file block. However, it is likely that each compensator would have unique coefficients, and this type of coupling could be avoided by assigning each its own file. The remainder of the coupling between the elements is data element coupling where input and output are passed as homogenous arguments.

Module cohesion (*maximize*). The majority of the DFDs show functional cohesion modules. All the elements are essential to the single function performed along the control line of each particular axis.

There is a degree of information cohesion in the overflow check modules which performs two functions related by concept. Also the compensator calculation, although not decomposed to lower levels, does have a number of functions it performs. These are all related by concept to the same task and are therefore information cohesion modules.

From this analysis, the simple DFD shown here compares favorably to the coupling and cohesion criteria, and those areas not maximized could be changed by the software engineer to further reduce coupling.

Chapter 14

14-1 (*a*) Since all tasks and related task information are resident in main memory and all tasks run to completion, there is no need of interrupts.

The task status word does not require any priority information since the priority is determined by the order of the link elements through the pointers.

There is no suspended-task information needed since each task executes to completion.

Scheduler
Task link element 1 Task address Task status—pointer
Task link element 2 Task address Task status—pointer
Task link element 3 Task address Task status—pointer
Task link element 4 Task address Task status—pointer
Task link element 5 Task address Task status—pointer
Task 1 Task 2 Task 3 Task 4 Task 5

14-3 For this example, the scheduler is an integral part of the executive or main program.

Flowchart of Process.

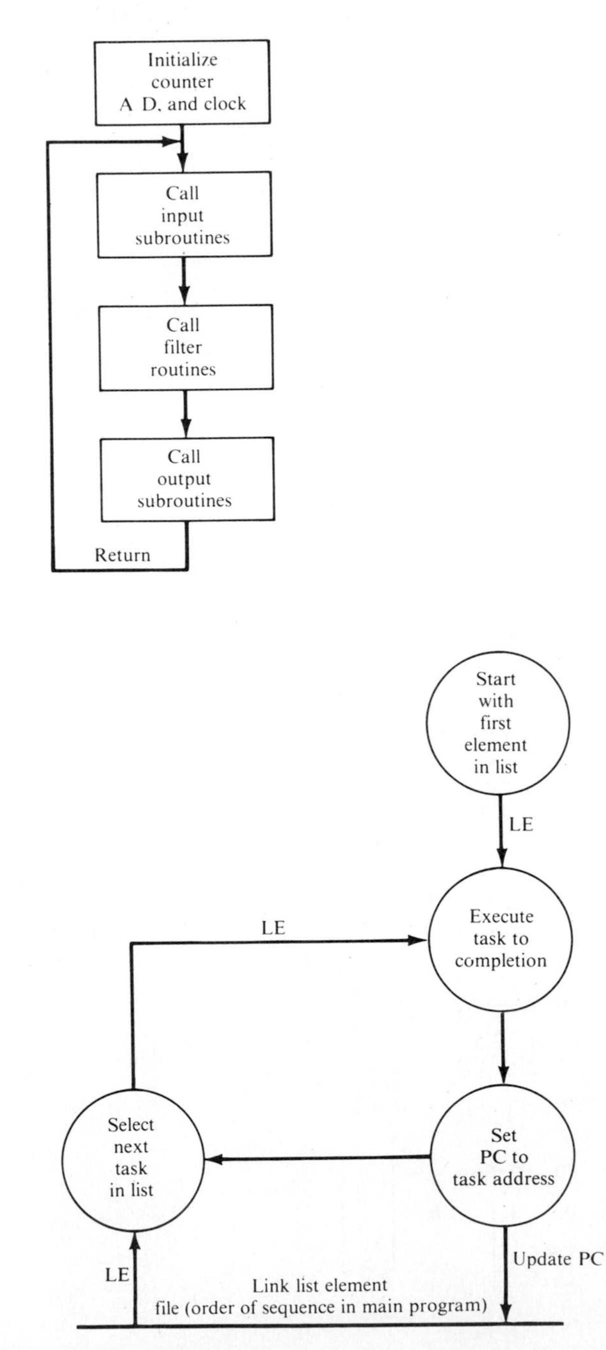

The order of the subroutine calls in the main program determines the priority of the tasks in the link list.

Chapter 15

15-4 (*a*)

$$\boldsymbol{\phi}(T) = \begin{bmatrix} 1 & T & T - 1 + \epsilon^{-T} \\ 0 & 1 & 1 - \epsilon^{-T} \\ 0 & 0 & \epsilon^{-T} \end{bmatrix}$$

$$\boldsymbol{\theta}(T) = 10 \begin{bmatrix} T^2 - T + 1 - \epsilon^{-T} \\ T - 1 + \epsilon^{-T} \\ 1 - \epsilon^{-T} \end{bmatrix}$$

$$\mathbf{c}' = [2 \quad 1 \quad 0]$$

(*b*)

$$\boldsymbol{\phi}(T) = \begin{bmatrix} 1 & 1 - \epsilon^{-T} & 2(T - 1 - \epsilon^{-T}) \\ 0 & \epsilon^{-T} & 2(1 - \epsilon^{-T}) \\ 0 & 0 & 1 \end{bmatrix}$$

$$\boldsymbol{\theta}(T) = \begin{bmatrix} 2T^2 - T + 1 - \epsilon^{-T} \\ 2T - 1 + \epsilon^{-T} \\ T \end{bmatrix}$$

$$\mathbf{c}' = [1 \quad 0 \quad 0]$$

(*c*) $T = 0.1$ s: $\lambda_1 = 1$ $\quad \lambda_2 = 1$ $\quad \lambda_3 = 0.90484-$
$T = 1$ s: $\lambda_1 = 1$ $\quad \lambda_2 = 1$ $\quad \lambda_3 = 0.36788-$

15-9 System (1):

(*a*)

$$\mathbf{F} = \begin{bmatrix} 0 & 1 & 0 \\ 0 & 0 & 1 \\ 0.54 & -2.04 & 2.5 \end{bmatrix} \qquad \mathbf{h} = \begin{bmatrix} 0 \\ 0 \\ 2 \end{bmatrix}$$

$$\mathbf{c}' = [-0.1 \quad -0.3 \quad 1] \qquad d = 2$$

(*b*)

$$\mathbf{F} = \begin{bmatrix} 0.6 & 0 & 0 \\ 0.4 & 0.9 & 0 \\ 0.4 & 1.4 & 1 \end{bmatrix} \qquad \mathbf{h} = \begin{bmatrix} 2 \\ 2 \\ 2 \end{bmatrix}$$

$$\mathbf{c}' = [0 \quad 0 \quad 1] \qquad d = 2$$

(*c*)

$$\mathbf{F} = \begin{bmatrix} 1 & & \mathbf{0} \\ & 0.9 & \\ \mathbf{0} & & 0.6 \end{bmatrix} \qquad \mathbf{h} = \begin{bmatrix} 1 \\ 1 \\ 1 \end{bmatrix}$$

$$\mathbf{c}' = [60 \quad -98/3 \quad 11/3]$$

$$p_1 = 1 \qquad p_2 = 0.9 \qquad p_3 = 0.6$$

15-12 (*a*) $\dfrac{Y(z)}{R(z)} = \dfrac{z + 0.12}{z^2 + 1.7z + 0.24}$

(*b*) $\lambda_1 = -0.155377$, $\lambda_2 = -1.5446$. System is unstable.

(*c*) $y(T) = 1$, $y(2T) = -0.58$.

15-17 $y(1.2) = 0.13936$

INDEX